D1266267

SCIENCE AND ULTIMATE REALITY

Quantum Theory, Cosmology, and Complexity

This volume provides a fascinating preview of the future of physics, covering fundamental physics at the frontiers of research. It comprises a wide variety of contributions from leading thinkers in the field, inspired by the pioneering work of John A. Wheeler. Quantum theory represents a unifying theme within the book, along with topics such as the nature of physical reality, the arrow of time, models of the universe, superstrings, gravitational radiation, quantum gravity, and cosmic inflation. Attempts to formulate a final unified theory of physics are discussed, along with the existence of hidden dimensions of space, spacetime singularities, hidden cosmic matter, and the strange world of quantum technology.

JOHN ARCHIBALD WHEELER is one of the most influential scientists of the twentieth century. His extraordinary career has spanned momentous advances in physics, from the birth of the nuclear age to the conception of the quantum computer. Famous for coining the term "black hole," Professor Wheeler helped lay the foundations for the rebirth of gravitation as a mainstream branch of science, triggering the explosive growth in astrophysics and cosmology that followed. His early contributions to physics include the *S* matrix, the theory of nuclear rotation (with Edward Teller), the theory of nuclear fission (with Niels Bohr), action-at-a-distance electrodynamics (with Richard Feynman), positrons as backward-in-time electrons, the universal Fermi interaction (with Jayme Tiomno), muonic atoms, and the collective model of the nucleus. His inimitable style of thinking, quirky wit, and love of the bizarre have inspired generations of physicists.

John Archibald Wheeler, 1987. (Photograph by Robert Matthews, courtesy of Princeton University.)

SCIENCE AND ULTIMATE REALITY

Quantum Theory, Cosmology, and Complexity

Edited by

JOHN D. BARROW
University of Cambridge

PAUL C. W. DAVIES
Macquarie University

and

CHARLES L. HARPER, Jr.
John Templeton Foundation

PUBLISHED BY THE PRESS SYNDICATE OF THE UNIVERSITY OF CAMBRIDGE
The Pitt Building, Trumpington Street, Cambridge, United Kingdom

CAMBRIDGE UNIVERSITY PRESS
The Edinburgh Building, Cambridge, CB2 2RU, UK
40 West 20th Street, New York, NY 10011–4211, USA
477 Williamstown Road, Port Melbourne, VIC 3207, Australia
Ruiz de Alarcón 13, 28014 Madrid, Spain
Dock House, The Waterfront, Cape Town 8001, South Africa

http://www.cambridge.org

First published 2004

Printed in the United Kingdom at the University Press, Cambridge

Typeface Times 11/14 pt. *System* LATEX 2ε [TB]

A catalog record for this book is available from the British Library

Library of Congress Cataloging in Publication data
Science and ultimate reality : quantum theory, cosmology, and complexity / edited by
John D. Barrow, Paul C. W. Davies, and Charles L. Harper, Jr.
p. cm.
Includes bibliographical references and index.
ISBN 0 521 83113 X
1. Quantum theory. 2. Cosmology. 3. Wheeler, John Archibald, 1911–
I. Barrow, John D., 1952– II. Davies, P. C. W. III. Harper, Charles L., 1958–
QC174.12.S4 2004
530.12 – dc22 2003055903

ISBN 0 521 83113 X hardback

The publisher has used its best endeavors to ensure that the URLs for external websites referred to in this book are correct and active at the time of going to press. However, the publisher has no responsibility for the websites and can make no guarantee that a site will remain live or that the content is or will remain appropriate.

Contents

Contributors

Andreas Albrecht, University of California, Davis, USA
John D. Barrow, University of Cambridge, UK
Raymond Y. Chiao, University of California, Berkeley, USA
Philip D. Clayton, Claremont School of Theology and Claremont Graduate University, Claremont, California, USA
Paul C. W. Davies, Macquarie University, Sydney, Australia
David Deutsch, University of Oxford, UK
Bryce S. DeWitt, University of Texas, Austin, USA
Freeman J. Dyson, Institute for Advanced Study, Princeton, New Jersey, USA
George F. R. Ellis, University of Cape Town, South Africa
Berthold-Georg Englert, National University of Singapore, Malaysia
Marcelo Gleiser, Dartmouth College, Hanover, New Hampshire, USA
Lucien Hardy, Perimeter Institute for Theoretical Physics and the University of Waterloo, Ontario, Canada
Serge Haroche, College of France, Paris
Stuart Kauffman, Santa Fe Institute, Santa Fe, New Mexico, USA
Paul G. Kwiat, University of Illinois, Champaign–Urbana, USA
Andrei Linde, Stanford University, Stanford, California, USA
Hideo Mabuchi, California Institute of Technology, Pasadena, USA
João Magueijo, Imperial College of Science, Technology and Medicine, London, UK
Juan M. Maldacena, Institute for Advanced Study, Princeton, New Jersey, USA
Fotini Markopoulou, Perimeter Institute for Theoretical Physics and the University of Waterloo, Ontario, Canada
Christopher R. Monroe, University of Michigan, Ann Arbor, USA
Juan Pablo Paz, University of Buenos Aires, Argentina
Jaroslav Pelikan, Yale University, New Haven, Connecticut, USA

Lisa Randall, Harvard University, Cambridge, Massachusetts, USA
Lee Smolin, Perimeter Institute for Theoretical Physics and the University of Waterloo, Ontario, Canada
Aephraim M. Steinberg, University of Toronto, Ontario, Canada
Max Tegmark, University of Pennsylvania, Philadelphia, Pennsylvania, USA
H. Dieter Zeh, University of Heidelberg, Germany
Anton Zeilinger, University of Vienna, Austria
Shou-Cheng Zhang, Stanford University, Stanford, California, USA
Wojciech H. Zurek, Los Alamos National Laboratory, Los Alamos, New Mexico, USA

Foreword

I am immensely pleased with this wonderful volume, and humbled by it. It demonstrates the incredible vibrancy of fundamental physics, both theoretical and experimental, as a new century gets under way. Just as unimagined vistas of the physical world were revealed in the early years of the twentieth century, so too we are encountering unimagined wonders a hundred years later. If there is an end to physics, an end to understanding the reasons for existence, it lies far in the future.

Who would have guessed in 1925, or even in 1950, that quantum mechanics would remain for so many decades such a fertile field of research? Who would have guessed then that its reason for being would remain mysterious for so long? Like many of the authors in this book, I remain convinced that some deeper reason for quantum mechanics will one day emerge, that eventually we will have an answer to the question, "How come the quantum?" And to the companion question, "How come existence?"

And who could have guessed in 1975 – when the black hole was coming to be accepted, when an explanation of pulsars was at hand, when primordial black-body radiation had been identified – who could have guessed then that an incredible confluence of deep thinking and stunning experimental techniques would push our understanding of cosmology – of the beginnings, the history, and the fate of the universe – to its present astonishing state?

Niels Bohr liked to speak of "daring conservatism" in pursuing physics. That is what I see in this volume. Nearly every chapter reveals a scientist who is hanging on to what is known and what is valid while, with consummate daring (or should I say derring-do?), pushing beyond the limit of what current observation confirms onward to the outer limit of what current theory allows. Here are scientists daring to share their visions of where future knowledge may lie.

The organizers of the symposium on which this book is based are to be congratulated for pulling together and so beautifully integrating the threads of quantum physics, cosmology, and the emergence of complexity. In the 1920s and 1930s, as my own career got started, I was inspired by Niels Bohr, Werner Heisenberg, Albert Einstein, and others. I hope that the young people who read this volume will find similar inspiration in it.

Princeton University
Princeton, New Jersey

John A. Wheeler

Editors' preface

This book project began as part of a special program, *Science and Ultimate Reality*, developed in honor of the ninetieth birthday of renowned theoretical physicist John Archibald Wheeler.[1] Having long yearned for a comprehensive, integrated understanding of the nature of the universe, Wheeler has blended scientific rigor with an unusually adventurous approach to research in physics and cosmology over a career spanning almost 70 years. Known for investigating many of the most fundamental and challenging issues in physics, Wheeler has often worked at the frontiers of knowledge where science and philosophy meet, probing the deep nature of physical reality. His vision, shaped in part by his influential mentor Niels Bohr, still flourishes today amid ongoing research activities pursued by several generations of those he has influenced over the course of much of the twentieth century.

With Wheeler as its inspiration, the *Science and Ultimate Reality* program was developed with a focus on the future. It brought together a carefully selected group of outstanding contemporary research leaders in the physics community to explore the frontiers of knowledge in areas of interest to Wheeler and to map out major domains and possibilities for far-reaching future exploration. Its two principal components – (1) this book and (2) a previously held symposium[2] – were developed to take Wheeler's vision forward into a new century of expanding discovery.

In addition to his role as a research leader in physics, Wheeler has been an inspirational teacher of many of the twentieth century's most innovative physicists. In this context, the program developers, many of whom contributed chapters to this volume, were asked to offer recommendations for their best candidates – not only

[1] Born July 9, 1911 in Jacksonville, Florida.

[2] The symposium, *Science and Ultimate Reality: Celebrating the Vision of John Archibald Wheeler*, was held March 15–18, 2002 in Princeton, New Jersey, United States. See www.metanexus.net/ultimate_reality/ for more information and to order the symposium proceedings on DVD. We wish to acknowledge the support of Dr. William Grassie, Executive Director of the Metanexus Institute, and his expert staff for helping to organize the symposium and for hosting this website. Also see Appendix A for a listing of the program committee members.

distinguished, well-established research leaders, but also highly promising, up-and-coming young innovators – to address the great questions of physical science in the twenty-first century. It is well known that, in physics and mathematics at least, the most powerful insights often come from surprisingly young people. By including young researchers in the Science and Ultimate Reality program, its developers hoped to identify future research leaders for the coming decades.[3]

In formulating the program, the developers purposely solicited research topics in areas close to some of Wheeler's most passionate interests. Some of the questions the developers kept in mind were: What can Wheeler's vision imply for the century ahead? What surprises lie in store for physics? What are the best ways to obtain deep insights into the heart of reality? What are the great unsolved problems of cosmology? How might the next generation of researchers tackle some of Wheeler's "Really Big Questions," such as: "Why the quantum?" "How come existence?" "It from bit?" "A participatory universe?" "What makes 'meaning'?"

This book is intended to stimulate thinking and research among students, professional physicists, cosmologists, and philosophers, as well as all scholars and others concerned with the deep issues of existence. Authors were invited to be bold and creative by developing themes that are perhaps more speculative than is usual in a volume of this sort. Specifically, they were asked to reflect on the major problems and challenges that confront fundamental science at this time and to animate their discussions by addressing the "Really Big Questions" for which Wheeler is so famous. This book is therefore more than a retrospective celebration of Wheeler's ideas and inspirations, or a simple survey of contemporary research. Rather, the editors sought to develop a collection of chapters that also point to novel approaches in fundamental research.

The book's first two chapters provide an overview of John Wheeler's contributions (Part I) and an historian's look at scientific speculation through the ages (Part II). The remaining twenty-eight chapters are grouped according to four themes:

Part III: Quantum reality: theory
Part IV: Quantum reality: experiment
Part V: "Big questions" in cosmology
Part VI: Emergence, life, and related topics.

The *Science and Ultimate Reality* program has provided a high-level forum in which some of the most visionary and innovative research leaders in science today could present their ideas. It continues to be an important mechanism for funding serious

[3] In conjunction with the program, a special Young Researchers Competition was held in which 15 young scientists chosen competitively from among applicants under age 32 presented short talks at the symposium. One of the researchers who tied for first place – Fotini Markopoulou – contributed a chapter to this volume. See www.metanexus.net/ultimate_reality/competition.htm/ for more information. Also see Appendix B for a listing of the competition participants and overseers.

scientific research and to engage with issues of "ultimate reality" in fascinating ways. Its overarching goal is to provide a risk-taking stimulus for research leading to (at least a few) major advances in knowledge of the nature of physical reality. We hope that John A. Wheeler's example will continue to stimulate the imaginations of new generations of the world's best and brightest students and researchers in science and that this book will serve to carry that vision forward.

University of Cambridge — *John D. Barrow*
Macquarie University — *Paul C.W. Davies*
John Templeton Foundation — *Charles L. Harper, Jr.*

Preface

My first encounter with John Archibald Wheeler was in the fall of 1945 in the reading room of the Science Library in London, a warm and comfortable place where anyone could walk in off the street to escape from rain and fog or to browse at leisure in scientific books and journals. I had just been released from war service and was eager to get back into science. I found the classic paper of Bohr and Wheeler, "The mechanism of nuclear fission," in volume 56 of the *Physical Review*, pages 426–450. It was published on September 1, 1939, the day on which Hitler's armies marched into Poland and the Second World War began. Bohr and Wheeler wrote the paper in Princeton, where Bohr was visiting in the spring of 1939, a few months after the discovery of fission. The paper is a masterpiece of clear thinking and lucid writing. It reveals, at the center of the mystery of fission, a tiny world where everything can be calculated and everything understood. The tiny world is a nucleus of uranium 236, formed when a neutron is freshly captured by a nucleus of uranium 235.

The uranium 236 nucleus sits precisely on the border between classical and quantum physics. Seen from the classical point of view, it is a liquid drop composed of a positively charged fluid. The electrostatic force that is trying to split it apart is balanced by the nuclear surface tension that is holding it together. The energy supplied by the captured neutron causes the drop to oscillate in various normal modes that can be calculated classically. Seen from the quantum point of view, the nucleus is a superposition of a variety of quantum states leading to different final outcomes. The final outcome may be a uranium 235 nucleus with a re-emitted neutron, a uranium 236 nucleus with an emitted gamma-ray, or a pair of fission-fragment nuclei with one or more free neutrons. Bohr and Wheeler calculate the cross-section for fission of uranium 235 by a slow neutron and get the right answer within a factor of two. Their calculation is a marvelous demonstration of the power of classical mechanics and quantum mechanics working together. By studying this process in detail, they show how the complementary views provided by classical and quantum pictures are both essential to the understanding of nature. Without the combined power of classical and quantum concepts, the intricacies of the fission process could never have been understood. Bohr's notion of complementarity is triumphantly vindicated.

The Wheeler whose dreams inspired this book is another Wheeler, different from the one I encountered in London. Throughout his life, he has oscillated between two styles of writing and thinking that I like to call prosaic and poetic. In the fission paper, I met the prosaic Wheeler, a master craftsman using the tools of orthodox physical theory to calculate quantities that can be compared with experiment. The prosaic Wheeler always has his feet on the ground. He is temperamentally conservative, taking the existing theories for granted and using them with skill and imagination to solve practical problems. But from time to time, we see a different Wheeler, the poetic Wheeler, who asks outrageous questions and takes nothing for granted. The poetic Wheeler writes papers and books with titles such as "Beyond the black hole," "Beyond the end of time," and "Law without law." His message is a call for radical revolution. He asks, "Should we be prepared to see someday a new structure for the foundations of physics that does away with time?" He proclaims, "Proud unbending immutability is a mistaken ideal for physics; this science now shares, and must forever share, the more modest mutability of its sister sciences, biology and geology." He dreams of a future when "as surely as we now know how tangible water forms out of invisible vapor, so surely we shall someday know how the universe comes into being."

The poetic Wheeler is a prophet, standing like Moses on the top of Mount Pisgah, looking out over the Promised Land that his people will one day inherit. Moses did not live long enough to lead them into the Promised Land. We may hope that Wheeler will live like Moses to the age of 120. But it is the young people now starting their careers who will make his dreams come true. This book is a collection of writings by people who take Wheeler's dreams seriously and dare to think revolutionary thoughts. But in science, as in politics and economics, it is not enough to think revolutionary thoughts. If revolutionary thoughts are to be fruitful, they must be solidly grounded in practical experience and professional competence. What we need, as Wheeler says here in his Foreword, is "daring conservatism." Revolutionary daring must be balanced by conservative respect for the past, so that as few as possible of our past achievements are destroyed by the revolution when it comes.

In the world of politics as in the world of science, revolutionary leaders are of two kinds, conservative and destructive. Conservative revolutionaries are like George Washington, destroying as little as possible and building a structure that has endured for 200 years. Destructive revolutionaries are like Lenin, destroying as much as possible and building a structure that withered and collapsed after his death.

The people who will lead us into the new world of physics must be conservative revolutionaries like Wheeler, at home in the prosaic world of practical calculation as well as in the poetic world of speculative dreams. The prosaic Wheeler and the poetic Wheeler are equally essential. They are the two complementary characters that together make up the John Wheeler that we know and love.

Institute for Advanced Study
Princeton, New Jersey

Freeman J. Dyson

Acknowledgments

The editors wish to acknowledge the John Templeton Foundation (see www.templeton.org), and Sir John Templeton personally, for making this project possible. Sir John was enthusiastic about recognizing John A. Wheeler as one of the greatest living scientific leaders exemplifying a bold and far-sighted vision combined with tough-minded scientific rigor. We would also like to thank The Peter Gruber Foundation (see www.petergruberfoundation.org) for additional financial support.

Freeman Dyson and Max Tegmark, contributors to this volume, played key roles in organizing the program. Artur Ekert, Robert Laughlin, Charles Misner, William Phillips, and Charles Townes also provided valuable program advisory assistance, as did many of the contributors to this volume. (See Appendix A for a listing of the program committee members.)

Pamela Bond, working with the John Templeton Foundation, deserves special thanks for her dedicated work. Pam helped to organize the symposium and oversaw and coordinated the communications, manuscripts, artwork, and other practical aspects of producing this book.

We wish to express our special gratitude to Kenneth Ford, a long-time Wheeler colleague, who served as Senior Program Consultant for both the book and the symposium. Ken worked tirelessly to bring a bold idea into fruition. His commitment, skillful diplomacy, and hard work are deeply appreciated.

Finally, we wish to thank Cambridge University Press, and particularly Tamsin van Essen, for their support of this book project.

Part I

An overview of the contributions of John Archibald Wheeler

1

John Archibald Wheeler and the clash of ideas

Paul C. W. Davies
Macquarie University

History will judge John Archibald Wheeler as one of the towering intellects of the twentieth century. His career spanned the transition from the celebrated Golden Age of physics to the New Physics associated with the Space Age, the information revolution and the technological triumphs of quantum and particle physics. His contributions, ranging from trailblazing work in nuclear physics to general relativity and astrophysics, are too numerous to list here.[1] His influence on three generations of physicists is immense.

But Wheeler has been more than a brilliant and influential theoretical physicist. The decision to hold a symposium *Science and Ultimate Reality* in his honor reflects the fact that he is also an inspiring visionary who brought to physics and cosmology a unique style of thought and mode of reasoning, compared by Jaroslav Pelikan in this volume to that of the Greek philosopher Heraclitus.

"Progress in science," Wheeler once remarked to me, "owes more to the clash of ideas than the steady accumulation of facts." Wheeler has always loved contradiction. After all, the Golden Age of physics was founded on them. The theory of relativity sprang from the inconsistency between the principle of relativity of uniform motion, dating back to Galileo, and Maxwell's equations of electromagnetism, which predicted a fixed speed of light. Quantum mechanics emerged from the incompatibility of thermodynamics with the continuous nature of radiation energy.

Wheeler is perhaps best known for his work in the theory of gravitation, which receives its standard formulation in Einstein's general theory of relativity. Although hailed as a triumph of the human intellect and the most elegant scientific theory

[1] See Wheeler's autobiography *Geons, Black Holes, and Quantum Foam: A Life in Physics* (W. W. Norton, New York, 1998) for more background information.

Science and Ultimate Reality, eds. J. D. Barrow, P. C. W. Davies and C. L. Harper, Jr. Published by Cambridge University Press.

known, the general theory of relativity was for decades a scientific backwater. Its renaissance in the 1950s and 1960s was due in large measure to the work and influence of John Wheeler and his many talented students. It was Wheeler who coined the terms *black hole*, *wormhole*, *spacetime foam*, *no-hair theorems*, and many other ubiquitous expressions of gravitational physics. And it was in the theory of gravitation that Wheeler confronted the starkest contradiction, the most iconoclastic clash of ideas in science, and one that underscored so much of his imaginative later work. In the 1960s, astronomical evidence began to accumulate that compact massive bodies such as the cores of burnt-out heavy stars could not avoid imploding, suddenly and catastrophically, under their own immense weight, a phenomenon dubbed gravitational collapse. Wheeler was fascinated by the paradox implied by the final stages of collapse – the formation of a so-called spacetime singularity, in which matter is squeezed into a single point of infinite density, and the gravitational field rises without limit. The fate of a star in this respect resembles on a small scale and in time reverse the origin of the entire universe in a big bang, where spacetime is hypothesized to have exploded into existence from an initial singularity.

Gravitational collapse evidently signals the end of . . . what? The general theory of relativity? The concept of spacetime? Physical law itself? Here was a state of affairs in which a physical theory contained within itself a prediction of its own demise, or at least its own inherent limitation. "Wheeler's style," a colleague once told me, "is to take a perfectly acceptable physical theory, and extrapolate it to the ultimate extreme, to see where it must fail." With gravitational collapse, that ultimate extreme encompassed not just the obliteration of stars but the birth and perhaps death of the entire universe.

Two words that recur frequently in the Wheeler lexicon are *transcendence* and *mutability*. At what point in the extrapolation of a theory would the physical situation become so extreme that the very concepts on which the theory is built are overtaken by circumstances? No topic better epitomizes this philosophy than the black hole. When a massive star collapses the gravitational field rises higher and higher until even light itself is trapped. The material of the core retreats inside a so-called event horizon and effectively disappears as far as the outside universe is concerned. What, one may ask, happens to the imploding matter? What trace does it leave in the outside universe of its erstwhile existence? Theory suggests that only a handful of parameters survive the collapse – mass, electric charge, and angular momentum being the three principal conserved quantities. Otherwise, cherished conservation laws are not so much violated as transcended; they cease to be relevant. For example, a black hole made of matter cannot be distinguished from one made of antimatter, or neutrinos, or even green cheese, if the few conserved parameters are the same.

Although the concept of the black hole had been implicit in the general theory of relativity for some decades, it was only in the late 1960s, in large part through the work of John Wheeler, together with Roger Penrose, Stephen Hawking, Robert Geroch, Brandon Carter, and others, that the extraordinary physical properties of such objects became understood. Already in those early days it was clear that one very basic law of physics – the second law of thermodynamics – was threatened by the existence of black holes, since they could apparently swallow heat and thus reduce the entropy of the universe. Conversely, if a black hole is perfectly black, its own entropy (normally measured as energy divided by temperature) would seem to be infinite. Again, physical theory extrapolated to the limit led to a nonsensical result. Wheeler brilliantly spotted that quantum mechanics might provide a way around this and come to the rescue of the second law of thermodynamics, in some ways the most cherished of all physical laws. Together with his student Jacob Bekenstein, Wheeler surmised that the event horizon area of the black hole constitutes a completely new form of entropy, so that when a black hole swallows heat, it swells in size, and its entropy will rise by at least the loss of heat entropy.

These early ideas were placed on a sound theoretical basis in 1975 when Hawking showed by applying quantum mechanics to black holes that they are not black at all, but glow with heat radiation at a temperature directly related to their mass. So quantum mechanics rescues the second law of thermodynamics from an apparent absurdity when it is combined with general relativity. This episode provides a wonderful example of the internal consistency of theoretical physics, the fact that disparate parts of the discipline cunningly conspire to maintain the deepest laws.

But what of the fate of the imploding matter? The general theory of relativity makes a definite prediction. If the core of the star were a homogeneous spherical ball, continued shrinkage can result in one and only one end state: all matter concentrated in a single point of infinite density – the famed *singularity*. Since the general theory of relativity treats gravitation as a warping or curvature of spacetime, the singularity represents infinite curvature, which can be thought of as an edge or boundary to spacetime. So here is a physical process that runs away to infinity and rips open space and/or time itself. After that, who can say what happens?

As a general rule, infinity is a danger signal in theoretical physics. Few physicists believe that any genuinely infinite state of affairs should ever come about. Early attempts to solve the problem of spacetime singularities by appealing to departures from symmetry failed – a wonky star may not implode to a single point, but, as Penrose proved, a spacetime singularity of some sort is unavoidable once the star has shrunk inside an event horizon or something similar.

To Wheeler, the singularity prediction was invested with far-reaching significance that conveyed his core message of mutability. He likened the history of physics to a staircase of transcendence, at each step of which some assumed physical property dissolved away to be replaced by a new conceptual scheme. Thus Archimedes established the density of matter as an important quantity, but later high-pressure technology showed it was not conserved. Nuclear transmutations transcended the law of conservation of the elements. Black holes transcended the law of conservation of lepton number. And so on. Wheeler went on to conclude:

> At the head of the stairs there is a last footstep of law and a final riser of law transcended. There is no law of physics that does not require "space" and "time" for its statement. Obliterated in gravitational collapse, however, is not only matter, but the space and time that envelop that matter. With that collapse the very framework falls down for anything one ever called a law of physics.

The lesson to be learned from this? "Law cannot stand engraved on a tablet of stone for all eternity . . . All is mutable." Indeed, mutability was Wheeler's own choice when I asked him, in the mid 1980s, what he regarded as his most important contribution to physics. He summed up his position with a typical Wheelerism: "There is no law other than the law that there is no law."

It has always been Wheeler's belief that gravitational collapse is a pointer to a deeper level of reality and a more fundamental physical theory, and that this deeper level will turn out to be both conceptually simple and mathematically elegant: "So simple, we will wonder why we didn't think of it before," he once told me. But delving beyond the reach of known physical law is a daunting prospect, for what can be used as a guide in such unknown territory? Wheeler drew inspiration from his "law without law" emblem. Perhaps there are no ultimate laws of physics, he mused, only chaos. Maybe everything "comes out of higgledy-piggledy." In other words, the very concept of physical law might be an emergent property, in two senses. First, lawlike behavior might emerge stepwise from the ferment of the Big Bang at the cosmic origin, instead of being mysteriously and immutably imprinted on the universe at the instant of its birth: "from everlasting to everlasting," as he liked to put it. In this respect, Wheeler was breaking a 400-year-old scientific tradition of regarding nature as subject to eternal laws. Second, the very appearance of lawlike behavior in nature might be linked in some way to our observations of nature – subject and object, observer and observed, interwoven. These were radical ideas indeed.

One beacon that Wheeler has employed throughout his search for a deeper level of reality is quantum mechanics. The quantum is at once both an obstacle and an opportunity. Twentieth-century physics was built largely on the twin pillars of the general theory of relativity and quantum mechanics. The former is a description of space, time, and gravitation, the latter a theory of matter. The trouble is, these two very different sorts of theories seem to be incompatible. Early attempts at combining them by treating the gravitational field perturbatively like the electromagnetic field, with hypothetical gravitons playing a role analogous to photons, ran into severe mathematical problems. Although a few physical processes could be satisfactorily described by this procedure, most answers were infinite in a way that could not be circumvented by simple mathematical tricks. The predictive power of this approach to quantum gravity was fatally compromised.

In spite of these severe conceptual and mathematical problems, Wheeler realized that at some level quantum effects must have a meaningful impact on gravitational physics, and since the gravitational field receives a geometrical interpretation in Einstein's general theory of relativity, the net result must be some sort of quantum spacetime dynamics, which Wheeler called quantum geometrodynamics. Here, spacetime geometry becomes not only dynamical, but subject to quantum rules such as Heisenberg's uncertainty principle.

One expression of the uncertainty principle is that physical quantities are subject to spontaneous, unpredictable fluctuations. Thus energy may surge out of nowhere for a brief moment; the shorter the interval the bigger the energy excursion. Simple dimensional analysis reveals that for durations as short as 10^{-43} s, known after Max Planck as the Planck time, these energy fluctuations should be so intense that their own gravity seriously distorts the microscopic structure of spacetime. The size scale for these distortions is the Planck length, 10^{-33} cm, about 20 powers of ten smaller than an atomic nucleus, and hence in a realm far beyond the reach of current experimental techniques. On this minuscule scale of size and duration, quantum fluctuations might warp space so much that they change the topology, creating a labyrinth of wormholes, tunnels, and bridges within space itself. As ever, Wheeler was ready with an ingenious metaphor. He compared our view of space to that of an aviator flying above the ocean. From a great height the sea looks flat and featureless, just as spacetime seems flat and featureless on an everyday scale of size. But with a closer look the aviator can see the waves on the surface of the sea, analogous to ripples in spacetime caused by quantum fluctuations in the gravitational field. If the aviator comes down low enough, he can discern the foam of the breaking waves, a sign that the topology of the water is highly complex and shifting on a small scale. Using this reasoning, Wheeler predicted in 1957 the existence of spacetime foam at the Planck scale, an idea that persists to this day.

In his later years Wheeler drew more and more inspiration from the quantum concept. In 1977 he invited me to spend some time with him in Austin at the University of Texas, where he was deeply involved with the nature of quantum observation. He was fond of quoting the words of his mentor Niels Bohr concerning the thorny question of when a measurement of a quantum system is deemed to have taken place. "No measurement is a measurement until it is brought to a close by an irreversible act of amplification," was the dictum. Accordingly, I delivered a lecture on quantum mechanics and irreversibility that I hoped might have some bearing on this project, but it was clear that Wheeler saw something missing in mere irreversibility per se. A true observation of the physical world, he maintained, even something as simple as the decay of an atom, must not only produce an indelible record, it must somehow impart *meaningful information*. Measurement implies a transition from the realm of mindless material stuff to the realm of knowledge. So it was not enough for Wheeler that a measurement should record a bit of information; that lowly bit had to *mean* something. Applying his usual practice of extrapolating to the extreme, he envisaged a community of physicists for whom the click of the Geiger counter amounted to more than just a sound; it was connected via a long chain of reasoning to a body of physical theory that enabled them to declare "The atom has decayed!" Only then might the decay event be accorded objective status as having happened "out there" in the physical world.

Thus emerged Wheeler's idea of the *participatory universe*, one that makes full sense only when observers are implicated; one that is less than fully real until observed. He envisaged a *meaning circuit*, in which atomic events are amplified and recorded and delivered to the minds of humans – events transformed into meaningful knowledge – and then conjectured a return portion of that meaning circuit, in which the community of observers somehow loops back into the atomic realm.

To bolster the significance of the "return portion" of the circuit, Wheeler conceived of a concrete experiment. It was based on an adaptation of the famous Young's two-slit experiment of standard quantum mechanics (see Fig. 1). Here a pinhole light source illuminates a pair of parallel slits in an opaque screen, closely spaced. The image is observed on a second screen, and consists of a series of bright and dark bands known as interference fringes. They are created because light has a wave nature, and the waves passing through the two slits overlap and combine. Where they arrive at the image screen in phase, a bright band appears; where they are out of phase a dark band results.

According to quantum mechanics light is also made up of particles called photons. A photon may pass through either one slit or the other, not both. But since the interference pattern requires the overlap of light from *both* slits there seems to be a paradox. How can particles that pass through either one slit or the other make a wave interference pattern? One answer is that each photon arrives at a specific spot on the image screen, and many photons together build up a pattern in a speckled sort of way. As we don't know through which slit any given photon has passed, and as the photons are subject to Heisenberg's uncertainty principle, somehow each photon "knows" about both slits. However, suppose a mischievous experimenter stations a detector near the slit system and sneaks a look at which slit each photon traverses? There would then be a contradiction if the interference pattern persists, because each photon would be seen to encounter just a single slit. It turns out that if this intervention is done, the pattern gets washed out. One way to express this is to say that the experimenter can choose whether the wavelike or particlelike nature of light shall be manifested. If the detector is positioned near the slits, light takes on the properties of particles and there is no wave interference pattern. But if the experimenter relinquishes information about which slit each photon traverses, light behaves like a wave and the pattern is observed. In this manner the experimenter helps determine the nature of light, indeed, the nature of reality. The experimenter *participates* in deciding whether light is made up of waves or particles.

Wheeler's distinctive adaptation of this experiment came from spotting that the experimenter can delay the choice – wave or particle – until *after* the light has passed through the slit screen. It is possible to "look back" from the image screen and deduce through which slit any given photon came. The conclusion is that the experimenter not only can participate in the nature of physical reality that is, but also in the nature of physical reality that *was*. Before the experimenter decided, the light was neither wave not particle, its status in limbo. When the decision was made, the light achieved concrete status. But . . . this status is bestowed upon the light at a time *before* the decision is made! Although it seems like retrocausation, it isn't. It is not possible for the experimenter to send information back in time, or to cause a physical effect to occur in the past in a controlled way using this set-up.

In typical Wheeler fashion, John took this weird but secure result and extrapolated it in the most extreme fashion imaginable. I first learned about this when I ran into Wheeler at a conference in Baltimore in 1986. "How do you hold up half the ghost of a photon?" he asked me quizzically. What he had in mind was this. Imagine that the light source is not a pinpoint but a distant quasar, billions of light years from Earth. Suppose too that the two slits are replaced by a massive galaxy that can bend the light around it by gravitational lensing. A given photon now has a choice of routes for reaching us on Earth, by skirting the galaxy either this way or that. In principle, this system can act as a gigantic cosmic interferometer, with dimensions measured in billions of light years. It then follows that the experimenter on Earth today can perform a delayed choice experiment and thereby determine the nature of reality that was *billions of years ago*, when the light was sweeping by the distant galaxy. So an observer here and now participates in concretizing the physical universe at a time that life on Earth, let alone observers, did not yet exist! John's query about holding up half the ghost of a photon referred to the technical problem that the light transit times around the galaxy might differ by a month or so, and light coming around one way might have to be coherently stored until the light coming round the other way arrives in order to produce an interference pattern.

So Wheeler's participatory universe ballooned out from the physics lab to encompass the entire cosmos, a concept elegantly captured by his most famous emblem, shown in Fig. 26.1 on p. 578. The U stands for universe, which originates in a Big

Bang, evolves through a long and complicated sequence of states to the point where life emerges, then develops intelligence and "observership." We, the observers, can look back at photons coming from the early universe and play a part in determining the reality that was, not long after the Big Bang. In this symbol, Wheeler seeks to integrate mind and cosmos via quantum physics in a dramatic and provocative manner. He is not claiming that the physical universe doesn't exist unless it is observed, only that past states are less than real (if by real one means possessing a full set of physical attributes, such as all particles having a definite position, motion, etc.), and that present observers have a hand in determining the actuality of the past – even the remote past.

The universe according to Wheeler is thus a "strange loop" in which physics gives rise to observers and observers give rise to (at least part of) physics. But Wheeler seeks to go beyond this two-way interdependence and turn the conventional explanatory relationship

$$\text{matter} \rightarrow \text{information} \rightarrow \text{observers}$$

on its head, and place observership at the base of the explanatory chain

$$\text{observers} \rightarrow \text{information} \rightarrow \text{matter}$$

thus arriving at his famous "it from bit" dictum, the "it" referring to a physical object such as an atom, and the "bit" being the information that relates to it. In it from bit, the universe is fundamentally an information-processing system from which the appearance of matter emerges at a higher level of reality.

The symposium *Science and Ultimate Reality* held in Princeton, March 15–18, 2002, sought to honor John Wheeler in his ninetieth birthday year and to celebrate his sweeping vision of the physical universe and humankind's place within it. In keeping with Wheeler's far-sightedness, the symposium dwelt less on retrospection and more on carrying Wheeler's vision into a new century. To this end, papers were presented by leading scientists working at the forefront of research in fields that have drawn inspiration from Wheeler's ideas. In addition, 15 young researchers were invited to deliver short addresses on their groundbreaking work. The presentations were organized around three broad themes: "Quantum reality," "Big questions in cosmology," and "Emergence." The full texts of those papers appear in the later chapters of this volume.[2]

In seeking a deeper level of reality, Wheeler returned again and again to quantum mechanics as a guide. The vast majority of scientists take quantum mechanics for

[2] See http://www.metanexus.net/ultimate_reality/competition.htm for more information on the competition and http://www.metanexus.net/ultimate_reality/main.htm for more information on the symposium.

granted. Students often ask *why* there is indeterminism in the atomic domain, or where Schrödinger's equation comes from. The standard answer is simply, "The world just happens to be that way." But some scientists, following Wheeler's lead, are not content to simply accept quantum mechanics as a God-given package of rules: they want to know *why* the world is quantum mechanical. Wheeler's insistent question, "How come the quantum?" received a good deal of attention at the symposium.

One way to approach this problem is to ask if there is anything special about the logical and mathematical structure of quantum mechanics that singles it out as a "natural" way for the universe to be put together. Could we imagine making small changes to quantum mechanics without wrecking the form of the world as we know it? Is there a deeper principle at work that translates into quantum mechanics at the level of familiar physics?

Lucien Hardy of the Centre for Quantum Computation at the University of Oxford has attempted to construct quantum mechanics from a set of formal axioms. Boiled down to its essentials, quantum mechanics is one possible set of probabilistic rules with some added properties. The question then arises as to whether it has specially significant consequences. Could it be that quantum mechanics is the simplest mechanics consistent with the existence of conscious beings? Or might quantum mechanics be the structure that optimizes the information processing power of the universe? Or is it just an arbitrary set of properties that the world possesses reasonlessly?

The flip side of the challenge "How come the quantum?" is to understand how the familiar classical world of daily experience emerges from its quantum underpinnings, in other words, "How come the classical?" Few disputes in theoretical physics have been as long-running or as vexed as the problem of how the quantum and classical worlds join together. Since the everyday world of big things is made up of little things, quantum mechanics ought to apply consistently to the world as a whole. So why don't we see tables and chairs in superpositions of states, or footballs doing weird things like tunnelling through walls?

Melding the madhouse world of quantum uncertainty with the orderly operation of cause and effect characteristic of the classical domain is a challenge that has engaged many theorists. Among them is Wojciech Zurek from the Los Alamos National Laboratory. In a nutshell, Zurek's thesis is that quantum systems do not exist in isolation: they couple to a noisy environment. Since it is central to the quantum description of nature that matter has a wave aspect, then quantum weirdness requires the various parts of the waves to retain their relative phases. If the environment scrambles these phases up, then specifically quantum qualities of the system are suppressed. This so-called "decoherence" seems to be crucial in generating a quasi-classical world from its quantum components, and Zurek has played a leading role in establishing the theoretical credibility of this explanation. There is, however, a snag. A decohering environment well explains how, for example, Schrödinger's cat will always be seen either alive or dead, never in a ghostly amalgam of the two conditions. But what if there is no environment? Suppose the system of interest is the universe as a whole?

The application of quantum mechanics to the entire universe is perhaps the most audacious extrapolation of physical theory in the history of science, and once again John Wheeler played a key role in its inception. When I was a student in the 1960s the ultimate origin of the universe was widely regarded as lying beyond the scope of physical science. To be sure, cosmological theory could be applied to the early moments of the universe following the Big Bang, but the initiating event itself seemed to be decisively off-limits – an event without a cause. The singularity theorems suggested that the Big Bang was a boundary to spacetime at which the gravitational field and the density of matter were infinite, and physical theory broke down. This meant that there was very little that could be said about how the universe came to exist from nothing, or why it emerged with the properties it has.

Wheeler drew the analogy between the instability of the classical atom, resulting in the emission of an infinite quantity of radiation, and the infinite spacetime curvature of gravitational collapse. He conjectured that just as quantum mechanics had saved classical mechanics from diverging quantities, and predicted a stable, finite atom, so might quantum mechanics ameliorate the Big Bang singularity – smearing it with Heisenberg uncertainty. But could one take seriously the application of quantum mechanics, a theory of the subatomic realm, to cosmology, the largest system that exists? Wheeler, with typical boldness, believed so, and with Bryce DeWitt produced a sort of Schrödinger equation for the cosmos. Thus was the subject of quantum cosmology born.

Today, most cosmologists agree that the universe originated in a quantum process, although the application of quantum mechanics to the universe as a whole remains fraught with both conceptual and technical difficulties. In spite of this, the subject of quantum cosmology received a fillip in the early 1980s from Alan Guth's inflationary universe scenario, which was predicted by applying the quantum theory of fields to the state of matter in the very early universe, while largely neglecting the quantum aspects of the expanding spacetime itself. Some cosmologists went on to suggest that quantum fluctuations in the inflationary era created the small primordial irregularities in the early universe that served as the seeds for its large-scale structure. If so, then the slight temperature variations detected in the cosmic background heat radiation – the fading afterglow of the Big Bang – are none other than quantum fluctuations from the dawn of creation inflated and writ large in the sky.

Andrei Linde, now at Stanford University, has been involved in the quantum cosmology program since the early days of inflation. Indeed, he developed his own variant of the inflationary theory, termed chaotic inflation, which has been lucidly explained in his expository articles and books. One fascinating prediction of chaotic inflation is that what we call "the universe" might be merely a "Hubble bubble" within a multiverse of vast proportions. There could be other "bubbles" out there, way beyond the scope of even our most powerful instruments, distributed in their infinite profusion.

If "our universe" is indeed but a minute component of a far more extensive and complex system, the philosophical consequences are profound. Since Copernicus, scientists have clung to the notion that there is nothing special about our location in the universe. We inhabit a typical planet around a typical star in a typical galaxy. But the multiverse theory suggests that, on a super-Hubble scale, our "bubble" might be far from typical. It might represent a rare oasis of habitability in a desert that is generally hostile to life. If so, perhaps many of the felicitous features we observe, including the fact that the physics of "our universe" seems so bio-friendly, might actually be the product of a cosmic selection effect. We live in such a special bubble only because most of the multiverse is unfriendly to life. This idea generally goes under the name *anthropic principle*, a misnomer since no special status is accorded to *Homo sapiens* as such. Alternatively, it may be that bio-friendly regions of the universe are also those that grow very large and so occupy the lion's share of space. Again, it would be no surprise to find ourselves living in one such region.

Why might our region of the universe be exceptionally bio-friendly? One possibility is that certain properties of our world are not truly fundamental, but the result of frozen accidents. For example, the relative strengths of the forces of nature might not be the result of basic laws of physics, but merely reflect the manner in which the universe in our particular spatial region cooled from a hot Big Bang. In another region, these numbers may be different. Mathematical studies suggest that some key features of our universe are rather sensitive to the precise form of the laws of physics, so that if we could play God and tinker with, say, the masses of the elementary particles, or the relative strengths of the fundamental forces, those key features would be compromised. Probably life would not arise in a universe with even slightly different values.

A more radical idea is Wheeler's *mutability*, according to which there are actually no truly fixed fundamental laws of physics at all. Many years ago Wheeler suggested that if the universe were to eventually stop expanding and collapse to a Big Crunch, the extreme conditions associated with the approach of the spacetime singularity might have the effect of "reprocessing" the laws of physics, perhaps randomly.

Imagine, then, that the universe bounces from this Big Crunch into a new cycle of expansion; then the laws of physics in the new cycle might be somewhat different from those with which we are familiar.

Is it credible to imagine "different" laws of physics in this way? Einstein once remarked that the thing that most interested him was whether God had any choice in the nature of his creation. What he was referring to is whether the physical universe necessarily has the properties it does, or whether it might have been otherwise. Obviously some of the fine details could be different, such as the location of this or that atom. But what about the underlying laws? For example, could there be a universe in which gravity is a bit stronger, or the electron a bit heavier?

Physicists remain divided on the issue. Some have flirted with the idea that if we knew enough about the laws of physics, we would find that they are logically and mathematically unique. Then to use Einstein's terminology, God would have had no choice: there is only one possible universe. However, most scientists expect that there are many – probably an infinite number – of possible alternative universes in which the laws and conditions are self-consistent. For example, the universe could have been Newtonian, consisting of flat space populated by hard spheres that fly about and sometimes collide. It would probably be a boring world, but it's hard to say why it is impossible.

Perhaps the most extreme version of a multiverse has been suggested by Max Tegmark, a theoretical physicist at the University of Pennsylvania. Tegmark proposes that all mathematically self-consistent world descriptions enjoy real existence. There is thus a sliding scale of cosmic extravagance, ranging from multiplying worlds with the mathematical laws fixed, to multiplying laws within an overall mathematical scheme, to multiplying the mathematical possibilities too. Some people conclude that invoking an infinity of unseen worlds merely to explain some oddities about the one we do see is the antithesis of Occam's razor. Others believe the multiverse is a natural extension of modern theoretical cosmology.

One of the basic properties of the world that seems to be crucial to the emergence of life is the dimensionality of space. Must space have three dimensions? A hundred years ago, Edwin Abbott delighted readers with his book *Flatland*, an account of a two-dimensional world in which beings of various geometries lived weird lives confined to a surface. Strange it may be, but logically there is no reason why the universe could not be like this. However, the physics of a two-dimensional world would be odd. For example, waves wouldn't propagate cleanly as they do in three dimensions, raising all sorts of problems about signaling and information transfer. Whether these differences would preclude life and observers – which depend on accurate information processing – is unclear. Other problems afflict worlds with more than three dimensions; for example, stable orbits would be impossible, precluding planetary systems. So a three-dimensional world might be the only one in which physicists could exist to write about the subject.

Is it just a lucky fluke that space has three dimensions, or is there a deeper explanation? Some theorists speculate that space emerged from the hot Big Bang with three dimensions just by chance, and that maybe there are other regions of the universe with different dimensionality, going unseen because they cannot support life. Others suggest that space is actually not three-dimensional anyway, but only appears that way to us – an idea that dates back to the 1920s.

There are two ways to conceal extra space dimensions. One is to "roll them up" to a tiny size. Imagine viewing a hosepipe from afar: it looks like a wiggly line. On closer inspection, the line is revealed as a tube, and what was taken to be a point on the line is really a little circle going around the circumference of the tube. Similarly, what we take to be a point in three-dimensional space could be a tiny circle going around a fourth dimension, too small to detect. It is possible to conceal any number of extra dimensions by such a "compactification" process.

The other way to conceal a higher dimension is to suppose that light and matter are restricted by physical forces to a three-dimensional "sheet" or "membrane," while allowing some physical effects to penetrate into the fourth dimension. In this way we see just three dimensions in daily life. Only special experiments could reveal the fourth dimension. Our "three-brane" space need not be alone in four dimensions. There could be other branes out there too. Recently it was suggested that the collision of two such branes might explain the Big Bang. Lisa Randall explains her take on these brane worlds in Chapter 25.

No starker clash of ideas, no more enduring a cosmological conundrum, can be given than that of the *arrow of time*, a topic with which Wheeler has a long association. The problem is easily stated. Almost all physical processes display a unidirectionality in time, yet the underlying laws of physics (with a minor exception in particle physics) are symmetric in time. At the molecular level, all processes are perfectly reversible; only when a large collection of molecules is considered together does directionality emerge. To express it Wheeler's way: "Ask a molecule about the flow of time and it will laugh in your face."

How does asymmetry arise from symmetry? This puzzle has exercised the attention of many of the world's greatest physicists since the time of James Clerk Maxwell and Ludwig Boltzmann. It was discussed at the symposium by Andreas Albrecht. Most investigators conclude that the explanation for time's arrow can be traced to the initial conditions of the system concerned. Seeking the ultimate origin of time asymmetry in nature inevitably takes one back to the origin of the universe itself, and the question of the cosmic initial conditions. It seems reasonable to assume that the universe was born in a low-entropy state, like a wound clock, from which it is sliding irreversibly toward a high-entropy state, or heat death, like a clock running down.

Therein lies a puzzle. From what we know about the state of the universe at, say, 1 second after the Big Bang, it was very close to thermodynamic equilibrium. Evidence for this comes from the cosmic background heat radiation, which has a so-called Planck spectrum, resembling the state of a furnace that has settled down to a uniform temperature. So at first sight the universe seems to have been born in a high-, rather than a low-, entropy state. However, first impressions are misleading, because it is necessary to take into account the thermodynamic effects of gravitation. Self-gravitating systems tend to go from smooth to clumpy, so the fact that matter in the early universe was spread evenly through space (disturbed only by the tiny ripples that triggered the growth of large-scale structure) suggests that, gravitationally, the universe actually did begin in a low-entropy state. The arrow of time derives ultimately from the initial cosmological smoothness.

Naturally that conclusion begs the question: "Why did the universe start out smooth?" One explanation is given by the inflation theory: when the universe jumped in size by a huge factor, any primordial irregularities would have been stretched away, just as an inflating balloon decrinkles. However, the inflation process is a product of laws of physics that are symmetric in time, so, plausible though this explanation may be, it appears to have smuggled in asymmetry from symmetry. Clearly there is still much room for disagreement on this historic and vexatious topic.

John Wheeler has contributed to the discussion of the arrow of time in a number of ways. Perhaps best known is the work he did in the 1940s with Richard Feynman, formulating a theory of electrodynamics that is symmetric in time. The arrow of time manifests itself in electromagnetic theory through the prevalence of retarded over advanced waves: a radio transmitter, for example, sends waves outwards (into the future) from the antenna. In the Wheeler–Feynman theory, a radio transmitter sends half its signal into the future and half into the past. Causal chaos is avoided by appealing to cosmology. It turns out that given certain cosmological boundary conditions – namely the ability of cosmic matter to absorb all the outgoing radiation and turn it into heat – a frolic of interference between the primary and secondary sources of electromagnetic radiation occurs. This serves to wipe out the advanced waves and build up the retarded waves to full strength. Thus the electromagnetic arrow of time derives directly from the thermodynamic arrow in this theory.

Many of these big questions of cosmology run into trouble because they appeal to physics that occurs at extremely high energies, often approaching the Planck values at which quantum gravitational effects become important. Producing a consistent and workable theory of quantum gravity has become something of a Holy Grail in theoretical physics. About 20 years ago it became fashionable to assume that a successful theory of quantum gravity would have to form part of a bigger amalgamation, one that incorporated the other forces of nature too. The misleading term *theories of everything* was adopted. The idea is that there exists a mathematical scheme, maybe even a formula simple enough to wear on a T-shirt, from which space, time, gravitation, electromagnetism, the weak and strong nuclear forces plus descriptions of all the varieties of subatomic particles would emerge, in more or less their familiar form, at low energies.

An early candidate for such a unified theory went by the name of superstrings. In simple terms, the idea is that the world isn't made up of tiny particles, in the tradition of Greek atomism. Instead, the ultimate building-blocks of the physical universe are little loops of string that wiggle about, perhaps in a space with 10 or 26 dimensions, in such a way as to mimic particles in three space dimensions when viewed on a larger scale. Since then, string theory has been incorporated into a more ambitious scheme known cryptically as M theory. It is too soon to proclaim that M theory has produced a final unification of physics, though some proponents are extremely upbeat about this prospect.

Part of the problem is that M theory is characterized by processes that occur at ultra-high energies or ultra-small scales of size (the so-called Planck scale). Testing the consequences of the theory in the relatively low-energy, large-scale world of conventional particle physics isn't easy. Another problem is the plethora of abstract mathematical descriptions swirling around the program. Not only does this make the subject impenetrable for all but a select few workers, it also threatens to smother the entire enterprise itself. There seem to be so many ways of formulating unified theories, that, without any hope of experimental constraint, there is a danger that the subject will degenerate into a battle of obscure mathematical fashions, in which progress is judged more on grounds of philosophical and mathematical appeal than conformity with reality.

But perhaps that is too cynical a view. There are hints of a meta-unification, in which the welter of unified theories are themselves unified, and shown to be merely alternative languages for the same underlying structure. The ultimate hope is that this meta-unification will stem from a deep physical principle akin to Einstein's principle of general covariance, and we will find that all the fundamental properties of the physical universe flow from a single, simple reality statement.

Although M theory is the currently popular approach to quantum gravity, it is by no means the only show in town. An alternative scheme, known as loop quantum gravity, has been painstakingly developed by Lee Smolin and others over many years, and is described in Chapter 22 of this volume.

Let me turn now to another major theme of the symposium, Wheeler's "It from bit," and the nature of information. Many of the contributors have been working on quantum information processing and the quest for a quantum computer. These researchers often take a leaf out of Wheeler's book and describe nature in informational terms. Some even go so far as to claim that the universe is a gigantic computer. The sociology underlying this line of thinking is interesting. There is a popular conception that science drives technology rather than the other way about, but a study of history suggests a more subtle interplay. There has always been a temptation to use the pinnacle of contemporary technology as a metaphor for nature. In ancient Greece, the ruler and compass, and musical instruments, were the latest technological marvels. The Greeks built an entire world view based on geometry and harmony, from the music of the spheres to the mystical properties associated with certain geometrical shapes. Centuries later, in Renaissance Europe, the clock represented the finest in craftsmanship, and Newton constructed his theory of the clockwork universe by analogy. Then in the nineteenth century the steam engine impressed everybody, and lo and behold, physicists began talking about the universe as a thermodynamic system sliding toward a final heat death.

In recent times the digital computer has served as a seductive metaphor for nature. Rather than thinking of the universe as matter in motion, one could regard it as information being processed. After all, when the planets orbit the Sun they transform input date (the initial positions and motions) into output data (the final positions and motions) – just like a computer. Newton's laws play the role of the great cosmic program.

As technology moves on, we can already glimpse the next step in this conceptual development. The quantum computer, if it could be built, would represent a technological leap comparable to the introduction of the original electronic computer. If quantum computers achieve their promise, they will harness exponentially greater information processing power than the traditional computer. Unfortunately at this stage the technical obstacles to building a functional quantum computer remain daunting. The key difficulty concerns the savage decohering effect of the environment, which I explained above.

In spite of, or perhaps because of, these technical problems, the race to build a quantum computer has stimulated a new program of research aimed at clarifying the nature of information and its representation in states of matter, as elucidated in Chapter 8 by Juan Pablo Paz of the Ciudad Universitaria in Buenos Aires. This work has led to the introduction of the so-called qubit (or qbit) as the "atom" of quantum information, the counterpart of the bit in classical information theory. Inevitably the question arises of whether the universe is "really" a vast assemblage of frolicking qubits, i.e., a cosmic quantum information processing system. It could be a case of "it from qubit" as Wheeler might put it.

If quantum information processing is really so powerful, it makes one wonder whether nature has exploited it already. The obvious place to look is within living organisms. A hundred years ago the living cell was regarded as some sort of magic matter. Today we see it as a complex information processing and replicating system using a mathematical organization very similar to a digital computer. Could quantum mechanics play a nontrivial role in this, at least at certain crucial stages such as the origin of life from nonlife? Does life manipulate qubits as well as bits?

No topic better illustrates the interplay of information, quantum, and gravitation than the quantum black hole. Hawking showed that the temperature of a black hole depends inversely on its mass, so as the hole radiates energy it gets hotter. The process is therefore unstable, and the hole will evaporate away at an accelerating rate, eventually exploding out of existence.

Theorists remain sharply divided about the ultimate fate of the matter that fell into the hole in the first place. In the spirit of Wheeler's "it from bit," one would like to keep track of the information content of the collapsing body as the black hole forms. If the matter that imploded is forever beyond our ken, then it seems as if all information is irretrievably lost down the hole. That is indeed Hawking's own favored interpretation. In which case, the entropy of the black hole can be associated with the lost information. However, Hawking's position has been challenged by Gerard 't Hooft and others, who think that ultimately the evaporating black hole must give back to the universe all the information it swallowed. To be sure, it comes out again in a different form, as subtle correlations in the Hawking radiation. But according to this position information is never lost to the universe. By contrast, Hawking claims it is.

Theorists have attacked this problem by pointing out that when matter falls into a black hole, from the point of view of a distant observer it gets frozen near the event horizon. (Measured in the reference frame of the distant observer it seems to take matter an infinite amount of time to reach the event horizon – the surface of the hole.) So from the informational point of view, the black hole is effectively two-dimensional: everything just piles up on the horizon. In principle one ought to be able to retrieve all the black hole's information content by examining all the degrees of freedom on the two-dimensional surface. This has led to the analogy with a hologram, in which a three-dimensional image is created by shining a laser on a two-dimensional plate.

Despite heroic attempts to recast the theory of quantum black holes in holographic language, the issue of the fate of the swallowed, or stalled, information is far from resolved. There is a general feeling, however, that this apparently esoteric technical matter conceals deeper principles that relate to string theory, unification, and other aspects of quantum gravity. Juan Maldacena, a theoretical physicist from Princeton's Institute for Advanced Study, discusses a specific model that examines a black hole in a surrounding model universe. Whilst artificial, this so-called anti-de Sitter universe admits of certain transparent mathematical properties that help elucidate the status of the holographic analogy, and the nature of black hole information and entropy.

At the symposium, there was no better example of Wheeler's clash of ideas than the conflict between locality and nonlocality. This is an aspect of the tension that exists between gravitation and relativity on the one hand, and quantum mechanics on the other. The theory of relativity requires one to specify the state of matter, and describe the gravitational field, or the geometry, at each spacetime point. In other words, it is a "local" theory. By contrast, quantum mechanics is "nonlocal" because the state of a quantum system is spread throughout a region of space. To take a simple illustration, in general relativity one might specify the gravitational field of the Sun, and compute the orbit of a planet moving in that field. But in an atom, there is no well-defined orbit for an electron going around a nucleus. Quantum uncertainty "smears out" the motion, so that it is simply not possible to say from one moment to the next where the electron is located. So what happens when we want to develop the theory of the gravitational effects of a quantum particle? Even harder is to describe the gravitational effects of a collection of quantum particles that may exist in a so-called entangled state – a state in which the whole is not merely the sum of its parts, but an indivisible, inseparable unity containing subtle correlations that may extend over a wide region of space.

The most dramatic examples of nonlocal quantum states are the quantum fluids, such as superconductors or Bose–Einstein condensates, in which the weird quantum effects extend over macroscopic distances. Such systems have been successfully created in the laboratory in recent years, and Raymond Chiao, a physicist at the University of California, Berkeley, used them to illustrate novel gravitational experiments. In particular, Chiao claims that quantum systems will make superefficient gravitational wave antennas. The search for gravitational waves is a major research theme in general relativity and astrophysics, using high-cost laser interferometers. Chiao's ideas could revolutionize that search as well as provide new insights into the apparent clash between gravitation theory and quantum physics.

The final theme of the symposium focused on the philosophical concept of emergence. In the foregoing I have referred to the quest to identify the fundamental building blocks of the physical world that began with the Greek atomists. The philosophy that underpins that project is reductionism: the conviction that everything in nature may ultimately be understood by reducing it to its elementary components. But reductionism in general, and the philosophy of particle physics in particular, has been criticized for committing what Arthur Koestler called the fallacy of "nothing-buttery." The problem is that a complete theory of the interactions of particles and forces would tell us little, for example, about the origin of life or the nature of consciousness. It may not even be of value in describing phenomena as basic as fluid turbulence or the properties of bulk matter as mundane as metals.

Many scientists recognize that new phenomena emerge and new principles may be discerned at each level of complexity in physical systems, principles that simply cannot be reduced to the science of lower levels. To take a familiar example, a person may be said to be living even though no atoms of their body are living. A reductionist might claim that the property of "being alive" is not really a fundamental or ultimately meaningful one, but merely a convenient way of discussing a certain class of unusual and complicated physical systems. But there is an alternative point of view that goes by the name of emergence. An emergentist would counter the reductionist by saying that it is just as scientifically meaningful to talk about life processes and the laws that describe them, as it is to talk about subatomic particles.

A growing body of expository literature, along with discoveries such as the fractional quantum Hall effect – an electromagnetic phenomenon occurring at extremely low temperatures – has sharpened the focus of debate between reductionism and emergence. It may even be the case that aspects of physics we have previously taken to be fundamental in the reductionist's sense will turn out to be emergent. This was the position argued by Robert Laughlin of Stanford University, who made a case that our familiar understanding of gravitation might one day emerge from a deeper level of description analogous to the way that superfluidity emerges from a study of the quantum properties of helium atoms.

It is clear that emergence has relevance wider than physical science. Other disciplines, particularly psychology, sociology, philosophy, and theology, are also vulnerable to reductionism. If emergent phenomena are taken seriously, then it seems we must take seriously not only life, but also consciousness, social behavior, culture, purpose, ethics, and religion. For instance, philosophers and theologians debate the thorny issue of whether right and wrong are just human conventions or whether the universe has a moral dimension, perhaps itself an emergent property, but nevertheless real. The cosmologist George Ellis of the University of Cape Town addressed these broader issues at the symposium.

Another aspect of emergence concerns the origin of things. Without a miracle, how can something come to exist that did not exist before? The ancient Greek philosophers were sharply divided on the issue. One school, represented by Heraclitus, maintained that everything in the world was in a constant state of flux, so that the world presents us with perpetual novelty. It is a philosophy summarized by the dictum that "you never step in the same river twice." The other position, associated with Parmenides, is that true change is impossible, since nothing can become what it is not. This ancient tension between being and becoming did not remain confined to philosophy. It came to pervade science too, and still provokes heated debate on matters concerning the arrow of time, chaos theory, and the psychological perception of a temporal flux.

The Greek atomists thought they had the answer to the being–becoming dichotomy. The universe, they said, consists of nothing but atoms moving in the void. All change is simply the rearrangement of atoms. The atoms represented being, their motion becoming. Thus began the long tradition of physical reductionism, in which true novelty is defined away. In the reductionist's universe, there can never be anything genuinely new. Apparently new systems or phenomena – such as living organisms or consciousness – are regarded as simple repackaging of already-existing components.

The philosophy of emergence, by contrast, takes change and novelty seriously. Emergentists suppose that genuinely new things can emerge in the universe, and bring with them qualities that simply did not exist before. Such a transformation may seem mysterious, and it often is. That is why some nonscientists home in on the origin of things to seek a breakdown of science – a gap into which they might slip divine intervention. The list of enigmatic gaps begins with the Big Bang origin of the universe, and goes on to include the origin of life and the origin of consciousness. These are all tough transitions for scientists to explain. (Curiously, I think the origin of the universe, which might be considered the most challenging, is the easiest to explain.) In some cases it seems as if the new systems spring abruptly and unexpectedly from the precursor state. Cosmologists think (at least they used to) that the Big Bang was the sudden spontaneous appearance of spacetime from nothing, a transformation that took little more than a Planck time. The origin of life might have been an equally amazing and sudden phase transition, or there again it might have involved a long sequence of transitional states extended over millions of years. Nobody knows. And as for the emergence of consciousness, this remains deeply problematic.

Emergence suffuses Wheeler's work at all levels. Perhaps the most arresting example concerns his "participatory universe" in which "observership" assumes a central place in the nature of physical reality, and presumably at some level must enter into physical theory. But what exactly constitutes a participant/observer? Is a particle detector enough? A living organism? An information gathering and utilizing system (IGUS)? A human being? A community of physicists?

Melding the participatory universe with "it from bit" reveals the key concept of information lying at the core. On the one hand an observation involves the acquisition and recording of information. On the other hand an observer, at least of the living variety, is an information processing and replicating system. In both cases it is not information per se that is crucial, but *semantic* information. An interaction in quantum mechanics becomes a true measurement only if it means something to somebody (made explicit in Wheeler's meaning circuit which I have already discussed). Similarly, the information in a genome is a set of instructions (say, to build a protein) requiring a molecular milieu that can recognize, decode, and act upon it. The base-pair sequence on a strand of DNA is just so much gobbledygook without customized cellular machinery to read and interpret it.

Where is there room in physics for the notion of information, not merely in the blind thermodynamic sense, but in the active life/observation/meaning sense? How does a lofty, abstract notion like meaning or semantic information emerge from the blundering, purposeless antics of stupid atoms?

Part of the answer must involve the concept of *autonomy*. Living organisms are recognized because they really do have "a life of their own." A cell is subject to the laws of physics, but it is not a slave to them: cells harness energy and deploy forces to suit their own ends. How does this quality of autonomy arise? Clearly the system must be *open* to its environment: there must be a throughput of matter, energy, and – crucially – information. But more is needed. When my computer plays chess, the shapes move around on the screen in accordance with the rules of chess. But my computer is also subject to the laws of physics. So are the rules of chess contained in the laws of physics? Of course not! The chess-playing regularities are an *emergent* property in the computer, manifested at the higher level of software, absent in the bottom level of hardware (atoms and electrons). Trace back how the rules of chess work in the computer and you will discover that *constraints* are the answer. The physical circuitry is constrained to embody the higher-level rules.

Stuart Kauffman has coined the term *autonomous agents* to characterize a program of research aimed at explaining how a system can have "a life of its own." He is a biophysicist and complexity theorist with his own theory of the origin of life based on autocatalytic cycles of chemical reactions. For Kauffman, constraints play a key role in the theory of autonomous agents. Another important quality is a type of Gödelian incompleteness that permits the system to display freedom or spontaneity in its behavior. Kauffman's ideas provide a new definition of life. They may even help us understand how, with increasing complexity, a physical system can leap from being mere clodlike matter to being an information-rich participator in a meaningful universe.

I leave the final word on this subject to John Wheeler himself, whose vision for the future beckons not merely an advance in science, but an entirely new type of science:

> We have to move the imposing structure of science over onto the foundation of elementary acts of observer-participancy. No one who has lived through the revolutions made in our time by relativity and quantum mechanics . . . can doubt the power of theoretical physics to grapple with this still greater challenge.

Part II

An historian's tribute to John Archibald Wheeler and scientific speculation through the ages

2

The heritage of Heraclitus: John Archibald Wheeler and the itch to speculate

Jaroslav Pelikan
Yale University

It has to be a jolting culture shock, or at any rate a severe case of the bends, for someone who has spent the past 60 years since completing the Gymnasium in 1942 studying, reading, translating, and interpreting St Augustine and St Thomas Aquinas, Martin Luther and the other sixteenth-century reformers, and the fourth-century Greek church fathers together with the Greek and Russian Orthodox tradition coming out of them, suddenly to be plunged into the rarefied atmosphere of this volume. Why, back where I come from, *quantum* is still a Latin interrogatory adjective in the neuter singular! One thing that I did learn, however, from Thomas Aquinas and his fellow scholastics was the doctrine of the *analogia entis*, the analogy of Being, which enables even a finite mind to speak by analogy about the Infinite (as the old proverb says, "A cat may look on a king"), because in some sense, at any rate in an analogous sense, it may be said that both of them "are," even though only God "is" noncontingently; it has been brilliantly discussed in the Gifford Lectures of Professor Etienne Gilson at Aberdeen (Gilson 1944). But from this Thomistic doctrine of *analogia entis* I have always thought it permissible to extrapolate to what I like to call the *analogia mentis*, the analogy of mind, by which it is still possible, *pace* C. P. Snow's *The Two Cultures*, to discern a significant affinity even between, for example, philology and physics, in that both of them seek by a rigorous intellectual process to tease out evidence from a puzzling body of material and then to make sense of that evidence in the light of some emerging larger hypothesis: Grimm's law or Boyle's law.

And I find myself not totally unprepared for this present task because of at least two other experiences as well. In 1999 I accepted the invitation of the American Association for the Advancement of Science to provide some historical perspective for a wide-ranging discussion and debate at the National Museum of Natural

Science and Ultimate Reality, eds. J. D. Barrow, P. C. W. Davies and C. L. Harper, Jr. Published by Cambridge University Press.

History of the Smithsonian Institution over three "Cosmic Questions," by relating them to the indispensable but largely unknown two thousand years of theological Denkexperimente coming out of both their Greek and their Hebrew roots; it has now been published by the New York Academy of Sciences (Pelikan 2001). In addition, I come to this assignment from the many conversations I have had over many years with my late friend Victor Weisskopf – one of my distinguished predecessors as President of the American Academy of Arts and Sciences – who presented me with a copy of his autobiography, *The Joy of Insight: Passions of a Physicist* (Weisskopf 1991), which has a cast of characters overlapping the one in John Wheeler's (1998) *Geons, Black Holes, and Quantum Foam*, including especially, of course, Niels Bohr; he graciously inscribed it: "To my spiritual brother Jary Pelikan, from Viki Weisskopf, Jan[uary] 1994."

In a moment of weakness, because of my respect for Professor John Archibald Wheeler, as well as my connections with Professor Freeman J. Dyson, the Chair of the Program Oversight Committee (I was born exactly two days after he was, whatever that may signify astrologically, and I share with him membership in the elite company of Gifford Lecturers at Aberdeen, along with the likes of Etienne Gilson and Karl Barth), and the Chair of the Organizing Committee, Dr Charles Harper (his wife Susan wrote her undergraduate thesis on St Bernard of Clairvaux under my direction at Yale), together with the other colleagues who were arranging this symposium, I rashly accepted their invitation. Prefacing that invitation with the note that "in twentieth-century physics, Wheeler's approach represents a modern part of a great and ancient tradition in Western history of the quest for knowledge of the deep nature of things and for ultimate comprehensive understanding," they asked me "to deliver a 45-minute talk [*sic*!] . . . to encapsulate an overview of the history of Western thought in quest of deep understanding of the major issues in ontology (such as the nature of time, the domain of reality, issues of infinity or finitude, eternity versus noneternity, materialism versus idealism, etc.)," and then to make that talk a chapter in this volume (F. J. Dyson and C. L. Harper Jr., pers. comm.). They added, apparently on the premise that flattery will get you almost anywhere: "In thinking, 'Who in the world could possibly do this?' there really was only one answer: Jary Pelikan." For good measure, the invitation went on to prescribe: "In that you would be addressing physicists who are not acquainted with the relevant scholarly literature, your task would be somewhat 'lyrical' in nature, aimed to help frame the current quest in physics within a diverse and fascinating historical, cultural, intellectual, and religious/theological matrix about which most physicists are not well aware." To all of this I reacted by making a vow to myself, which I am earnestly striving to honor, that I would not quote any more Greek and Latin than physicists used to quote equations when they were delivering papers to the stated meetings of the American Academy of Arts and Sciences in Cambridge while I was its President. (And absolutely no Russian, I promise!)

It is almost unavoidable to look over our shoulders as we gather this weekend to honor and celebrate the work of our friend John Archibald Wheeler in this his ninetieth birthday year. If we do so, we must certainly be overtaken by the powerful awareness that we are only the most recent cohort – and, we may also hope, not yet the last cohort – to be doing so. For in addition to all the other awards that have been heaped upon him over all these decades from so many sources both public and private, including the Enrico Fermi Award in 1968, which meant so much to him (Wheeler 1998), there have been, after all, not one but at least three Festschriften in his honor, one in 1972 (for his sixtieth birthday), another in 1988, and then another in 1995. These three volumes add up to a massive total of 2049 pages (Klauder 1972; Zurek *et al.* 1988; Greenberger and Zeilinger 1995). (But to put this page count into some rational perspective, it must be pointed out, even by a scholar who has never been accused of being excessively quantitative in his methodology, that 2049 does break down to about 10 or so pages for each month of Professor Wheeler's life, which seems on balance to be a rather modest quantity!)

An occasion like the publication of this celebratory volume – which will of course end up adding still more to that page count – is an almost irresistible invitation to attempt various kinds of comparison. And so, I suppose I could begin with the familiar opening question of Shakespeare's Sonnet 18: "Shall I compare thee to a summer's day?" Well, at our ages, I am forced to conclude that that is probably not the best possible comparison! In his delightful essay of 1990, "Can We Ever Expect to Understand Existence?" in *Information, Physics, Quantum: The Search for Links*, Professor Wheeler himself, without quite resorting to outright comparison as such, does refer to several earlier thinkers, not only to Bishop George Berkeley, Friedrich von Schelling, and the enigmatic C. S. Peirce (subject of a recent Pulitzer Prize book, which I reviewed for the *Los Angeles Times*), but to the pre-Socratic philosopher Parmenides of Elea (Wheeler 1990: 23). Which I shall happily construe as providing me with all the poetic license I need to look at other pre-Socratics. For of all the thinkers to whom John Archibald Wheeler might be compared, in what the Organizing and Program Oversight Committees call "the quest for knowledge of the deep nature of things and for ultimate comprehensive understanding" – Aristotle and Sir Isaac Newton are, of course, in some ways the obvious candidates – none provides more fascinating parallels than another pre-Socratic scientist and philosopher, Heraclitus of Ephesus (*c*. 540 – 480 BCE) (Marcovich 1982). But from the very beginning I should make it clear that there are also significant differences between them. For one thing, the Greek style of Heraclitus, beautiful and sometimes even "lyrical" as it is, is notoriously difficult to parse and understand, and it has brought down many an amateur Hellenic philologist (as I have reason to know, because that is really the most I can honestly claim to be). Apparently Heraclitus wrote this way on purpose (Cornford 1965): as Another once said, "that seeing they may see and not perceive; and hearing they may hear, and not understand" (Mark 4, v. 12). By contrast, the colleagues, students, and readers of John Wheeler, even those outside his own field, are unanimous in testifying that he has always had an uncanny pedagogical ability to suit the clarity of his language, both spoken and written, not alone to the complexity of the subject matter but to that mysterious and powerful force that Aristotle in his *Rhetoric* (I.ii.3 1356[a]) calls *pathos*, "the frame of mind of the audience," without an informed awareness of which, Aristotle tells us, no communicator can succeed.

"Mankind can hardly be too often reminded," John Stuart Mill writes in *On Liberty* (1859), "that there was once a man named Socrates." The older I get, the more profoundly I agree. Very few maxims, whether biblical or classical, can match the Socratic epigram, which (we "can hardly be too often reminded") he dared to articulate out loud to his accusers in his self-defense when he was on trial for his life, that "the unexamined life is not worth living" (Plato *Apology* 38A) – although, as a friend of mine, the sometime Chairman of the National Endowment for the Humanities and Chancellor Emeritus of the University of Massachusetts, Dr Joseph D. Duffey, once amplified it for me, "Remember, the *examined* life is no bed of roses either!" The obverse side of this prominence of Socrates is that sometimes it has tended to upstage figures like the sophists who deserve major attention in their own right, and to treat them as though they were mere spear-carriers during the Triumphal March in Verdi's *Aida.* To none does this apply more than to the group of scientists, mathematicians, and philosophers, including both the Parmenides whom John Wheeler cites and the Heraclitus of my title, whom the history of Greek science and philosophy has usually called "the *pre*-Socratics." But in many respects it would seem to be historically more fair to refer to them as pre-*Aristotelians*. For one thing, it was Aristotle rather than Socrates who took up again, as Wheeler has in our time, their concern with the relations between physics and metaphysics. Socrates reported, according to a precious autobiographical fragment appearing in Plato's *Phaedo* (96–100), that as a young man he began with a strong scientific interest and "a prodigious desire to know that department of philosophy which is called the investigation of nature or physics *[peri physeōs historia].*" This prompted him to read the scientist and philosopher Anaxagoras, the great friend of the noble Pericles, "as fast as I could in my eagerness to know the best and the worst" about cosmology. But, as Socrates continued a bit wistfully, he "failed in the study of physical realities [*ta onta*]" and abandoned science for the study of the soul, which was great for the study of the soul but not so good for the study of physics. It is also in some significant measure to Aristotle that we are indebted for the preservation of the quotations that Hermann Diels (1956), in his superb collection first published nearly a hundred years ago, called "die Fragmente der Vorsokratiker." After going through several revised editions, Diels was last revised in 1956 by Walther Kranz, in three substantial volumes; this is the edition of the Greek original being cited here. There has also been an attempt at an altogether new edition by Franz Josef Weber (1976) which, mercifully, has at least kept the numbering of Diels, while rearranging the order of the fragments themselves and, alas, omitting many of them; and a little over a year ago Oxford University Press published a very helpful edition and translation into English of large selections from the pre-Socratics as well as from the sophists, by Robin Waterfield, under the title *The First Philosophers* (Waterfield 2000).

Time and first things: *archē – aiōn*

All the way from Albert Einstein to Martin Heidegger, which is a long distance philosophically although not chronologically, the twentieth-century "itch to speculate" has been inspired, directly or indirectly, by Heraclitus. To mention one example that I must admit came as a surprise to me when I undertook the preparation for this paper: although Oswald Spengler was to become best known during the first half of the twentieth century, also in the English-speaking world, for his prophetic *The Decline of the West [Der Untergang des Abendlandes]*, which was published in the aftermath of the First World War and which I as a historian of ancient and medieval apocalyptic have studied with some care as a modern example of the same genre, he had already established himself as a philosophical thinker deserving of serious attention by first writing a monograph when he was in his early twenties entitled *The Fundamental Metaphysical Idea of the Philosophy of Heraclitus [Der metaphysische Grundgedanke der heraklitischen Philosophie]* (Spengler 1904). And the same is true of Martin Heidegger, as I will grant even though I believe that in many ways his influence on twentieth-century thought, also on twentieth-century Christian theology, has been deleterious (Macquarrie 1955). It was chiefly out of his close reading and *explication de texte* of the fragments of Heraclitus in Greek – as that close reading over many decades makes itself palpable in the 60 pages of "Fragments and Translations" from Heraclitus that have now been pulled together from various of his writings (Maly and Emad 1986: 9–68) – that Heidegger was driven not only to deal, as the ground rules of German *Wissenschaft* required, with this particular problem of philosophy or that specific issue of ontology, but to raise the very question of philosophy as such: "Was ist das – die Philosophie?" (Davis 1986). Without getting into a discussion here of Martin Heidegger's political stance during the Nazi period, specifically of his notorious *Rektoratsrede* of May 27, 1933 at the University of Freiburg im Breisgau, under the title *The Self-Assertion of the German University* (Heidegger 1934), nor into an examination of his personal character as revealed in his recently published love letters to the young and impressionable Hannah Arendt, it must nevertheless be acknowledged that, probably more than any other philosophical work of the twentieth century, his *Being and Time [Sein und Zeit]*, which first came out in 1927 and which went through many editions and reprints, having been translated into English in 1962 (Heidegger 1962), has set many of the parameters for how all of us are forced to formulate the question of time – and therefore, to an often unacknowledged degree, for how at least the philosophically sophisticated even among the readers of Albert Einstein (and then of John Archibald Wheeler) have perceived them to be speaking in their speculations about time. For that matter and on a more personal note, I found that at least some awareness of Heidegger's speculations about time was unavoidable even for me when, in my Richard Lectures on St Augustine at the University of Virginia entitled *The Mystery of Continuity*, and then in my Gifford Lectures of 1992/1993 on Greek Christian thinkers of the fourth century at the University of Aberdeen entitled *Christianity and Classical Culture*, I examined the conceptions of time in Late Antiquity and Early Christianity (Pelikan 1986, 1993). It is evident, as Parvis Emad has suggested, that Heraclitus and other pre-Socratics were shaping Heidegger's speculative methods already in his *Beiträge zur Philosophie*, which was "written between 1936–1938 and published posthumously in 1989," and were moving him "from the perspective of fundamental ontology to that of the nonhistoriographical historicality of being" (Emad 1999: 56–7).

And therefore when Professor Wheeler (1994) – declaring that "of all obstacles to a thoroughly penetrating account of existence, none looms up more dismayingly than 'time.' Explain time? Not without explaining existence. Explain existence? Not without explaining time" – projects it as a major task for the future "to uncover the deep and hidden connection between time and existence," he is, from a perspective inspired at that point by Hermann Weyl, asking Heidegger's question about that deceptively simple little word "und" in Heidegger's title *Sein und Zeit* – and thus, yes, about "the nonhistoriographical historicality of being." He is also, if it is not presumptuous of me to add this footnote, echoing St Augustine's preface to one of the most profound examinations of this "obstacle" of time in the entire history of Western thought in Book Eleven of his *Confessions* (XI.xiii.17): "What then is time? Provided that no one asks me, I know. If I want to explain it to an inquirer, I do not know."

It was Heraclitus and the other pre-Socratics, with their speculations about *archē* and about *aiōn*, who taught the scientists and philosophers of the West, as Wheeler in his "itch to speculate" has done again for the scientists of the current generation, the primacy of these questions about time and first things. Heinz Ambronn (1996), in a dissertation at the University of Bremen – for which, he explains, he took an intensive two-semester course in classical Greek (not quite the legendary time in which J. Robert Oppenheimer was said to have learned Greek (and without a teacher) but, considering the complexity of these pre-Socratic texts, a remarkable achievement nonetheless!) – has examined these questions in the pre-Socratics Anaximander, Parmenides, and Leucippus. But Heraclitus, too, was engaged with Ambronn's three fundamental questions about *archē*: "(1) Has this 'first' always already existed, or did it have a beginning? (2) What was this 'first'? (3) What brought about the development from this 'first' to the present-day physical world?" (Ambronn 1996: 231). And when Wheeler, with strikingly similar questions before him, speaks about the need to "uncover the deep and hidden connection between time and existence," he is putting himself into that classical and pre-Socratic (and Augustinian) apostolic succession (Wheeler 1994: 190–1).

A (restrained) reliance on intuition

In a passage that has been preserved by one of my favorite early Christian thinkers, Clement of Alexandria, Heraclitus warned: "If you do not expect the unexpected, you will not find it, since it is trackless and unexplored" (Diels 1956: 22B78; Waterfield 2000: F9). That fragment provided the text for Roger Von Oech (2001) to write a somewhat off-the-wall book under the title *Expect the Unexpected (or You Won't Find It): A Creativity Tool Based on the Ancient Wisdom of Heraclitus*. In this provocative challenge, Heraclitus identifies an issue of methodology that runs across the boundaries not only of the 25 centuries between him and ourselves, but of all the various scholarly and scientific disciplines represented here in this volume, even including my own. But first I take it on myself to be quite precise (and a bit polemical) in response to those self-styled "postmoderns" who use his authority to assert that "nature" or even "reality" is a social construct. This admonition of Heraclitus, "If you do not expect the unexpected, you will not find it" does *not* mean, just as I am assured that Werner Heisenberg's principle of indeterminacy does not mean, that every scholarly or scientific (or, for that matter, theological) question already contains its own answer and that therefore our investigations and Denkexperimente are a solipsistic exercise in tautology and self-deception, where observers (whether they be scientists or historians or philosophers or theologians) end up essentially studying themselves. But the admonition does include a curiosity about analogy based on what is already known (Jüngel 1964); and it does mean that an inquiry informed by experience, both the experience of the researcher and the cumulative experience of the great company of those who have gone before, with all their many mistakes and their occasional triumphs, has a better chance of asking the right initial questions – and then of revising the questions, as well as the answers, as the inquiry proceeds. In John Wheeler's apt formula, which applies to my research no less than to his: "No question, no answer . . . . In brief, the choice of question asked, and choice of when it is asked, play a part – not *the whole part, but a part* – in deciding what we have the right to say" (Wheeler 1990: 14; italics added). *Fortēs fortuna adiuvat*, the Roman poet Terence said (*Phormic* 203), a Latin saying that (for once!) has been improved in the translation by becoming a maxim of conventional wisdom for various scientists, including for example John Wheeler (1998: 14) in speaking about Niels Bohr, "Fortune favors the mind prepared."

That sense of the great company comes through, with intellectual power as well as with emotional force, in Professor Wheeler's charming miniature essays about the international group whom he humbly calls "More Greats," meaning Maria Sklodowska Curie, Hermann Weyl, Hendrik Anthony Kramers, and Hideki Yukawa (Wheeler 1994: 161–98). There is a mysterious relation – or, to quote Wheeler's phrase again, a "deep and hidden" relation – here between question and answer, as well as between heritage and task, celebrated in the axiom of Goethe (*Faust* 682–3) that is the melody of my personal and religious life and of my lifelong scholarly study of tradition (Pelikan 1984):

> *Was du ererbt von deinen Vätern hast,*
> *Erwirb es, um es zu besitzen:*
> What you have as heritage, now take as task;
> For thus you will make it your own.

Another word that could be used for this mysterious process, in which both Heraclitus and Wheeler have participated, is *intuition* (Soumia 1982), which entails a curious blending of humility and self-confidence about the riddle of Ultimate Reality, which addresses puzzling challenges to us but is not out to get us: as Einstein (n.d.) said in a famous epigram, "Raffiniert ist der Herr Gott, aber boshaft ist er nicht (God may be cunning, but He is not malicious)." Speaking in his autobiography about his early collaboration with Tullio Regge, Professor Wheeler tells us that "my *intuition* told me that the Schwarzschild singularity should be stable, but I had not yet been able to prove it" (Wheeler 1998: 266; italics added). Of the many examples of this blending of humility and self-confidence that Professor Wheeler's colleagues remember about him, let me cite one amusing anecdote, which comes from our mutual friend, the late Crawford Greenewalt, with whom I was closely associated when he was President of the American Philosophical Society. Speaking of their collaboration in the war, beginning in 1943, Greenewalt writes (see Klauder 1972: 3; italics added):

> John never used a pencil, but favored a very free-flowing fountain pen, with which he set forth the problem on a lined sheet of paper while saying, in his quiet tone, "Let's look at it this way." (There has been speculation among John's colleagues of those days that he used a pen because it prohibited erasure, and helped to bring home the fact that trial-and-error was not permissible in the Manhattan Project. Perhaps that was his motive; perhaps not. The fact remains that he was *an extraordinarily self-disciplined thinker*.)

The search for links: the social conception of *aretē*

Similarly, Wheeler's "search for links" (Wheeler 1991) manifests a distinctly Heraclitean movement of thought: from information to knowledge, but beyond that, if possible, from knowledge to wisdom (Huber 1996). "I have heard a lot of people speak," Heraclitus is reported to have said, "but not one has reached the point of realizing that the wise is different from everything else (Diels 1956: 22B108; Waterfield 2000: F11)." In his edition of the fragments of Heraclitus, entitled *The Art and Thought of Heraclitus*, C. H. Kahn (1979) has suggested that "for Heraclitus there will be no conflict between the selfish [i.e., personal] and the social conceptions of *aretē* [virtue]." But the great importance of his thought for the history of scientific and metaphysical speculation seems to have caused many modem scholars to neglect the social and political thought of Heraclitus. According to the doxographer Diogenes Laertius, the work of Heraclitus was divided into three books: on the cosmos, on politics, and on theology (though perhaps not in so neat a Dewey Decimal System arrangement as that). Even from the surviving fragments, however, it is clear that the social–political dimension of his thought deserves closer attention. The twentieth-century Greek philosopher Kōstas Axelos, who is perhaps best known for his book of 1961 interpreting Karl Marx as the one philosopher that has dealt the most cogently with the issues of technology, including alienation (Axelos 1961), has, in a study that, significantly, bears two quotations from Marx as its epigraph, also looked at the political philosophy of Heraclitus, especially *la cité et la loi* (Axelos 1968). The result of that examination by Axelos is to confirm the impression that comes from the surviving fragments: that for Heraclitus the two major political themes are war and law. "War," he said a century or so before the celebrated Funeral Oration of Pericles on this theme (Thucydides *The Peloponnesian War* II.vi.35–46), "is father of all and king of all. Some he reveals as gods, others as men; some he makes slaves, others free" (Diels 1956; 22B53; Waterfield 2000: F23). And of law he said: "Those who speak with intelligence must stand firm by that which is common to all, as a state stands by the law, and even more firmly" (Diels 1956: 22B114; Waterfield 2000: F12). His "social conception of *aretē*," to use Kahn's phrase, was shaped by this understanding of war as "father of all and king of all."

Although I was, of course, always aware in general of the great contribution that Professor Wheeler made to the Manhattan Project and thus to the scientific war effort that helped to win the Second World War, I had not realized before just how deeply and specifically it was the fact of the war itself – and, of course, in retrospect the tragic loss of his brother in the war – that shaped his own "social conception of *aretē*." As he himself has said near the very beginning of his autobiography (Wheeler 1998: 20), candidly and certainly more movingly than anyone else could have, "For more than fifty years I have lived with the fact of my brother's death. I cannot easily untangle all of the influences of that event on my life, but one is clear: my obligation to accept government service when called upon to render it." Again, near the very end of that autobiography, he speaks of "the derision with which a few of my colleagues had greeted my decision to work on the hydrogen bomb in the early 1950s, and . . . the fact that my old-fashioned brand of patriotism was in short supply in the 1960s and scarcely respected on college campuses" (Wheeler 1998: 303).

The One and the Many between quantum and cosmos: *ti to on*

In another fragment preserved by an early Christian writer, this time by Hippolytus of Rome rather than by Clement of Alexandria, Heraclitus admonishes: "It is wise for those who listen not to me but to the principle *[logos]* to agree in principle that everything is one *[hen panta einai]*" (Diels 1956: 22B50; Waterfield 2000: F10). Having spent most of my scholarly time for the past six years preparing a comprehensive edition (the first for 125 years) of Christian creeds and confessions of faith from the beginnings of the Church to the end of the twentieth century, I cannot restrain myself from noting that Heraclitus's Greek verb *homologein*, which Waterfield translates as "to agree in principle," was eventually to become in fact the technical term for "to make a confession of faith," and *homologia* the technical word for a confession or creed (Lampe 1961: 957–8), so that Heraclitus's words could be translated, somewhat anachronistically: "to make a confession of the faith that everything is one." In fact, according to Philo of Alexandria (quoted in Hammer 1991: 123), the celebrated Jewish philosopher and contemporary of Jesus and Paul, it was this insight into the One and the Many "which the Greeks say that their great and celebrated Heraclitus set up as the high-point of his philosophy and paraded as a new discovery."

Kōstas Axelos (1968: 124), whom I cited earlier on the political thought of Heraclitus, has also probed the relation of Heraclitus to the recurring question of his so-called "pantheism," suggesting that this is something of an oversimplification (as Einstein also showed when he responded to a query from the synagogue of Amsterdam about whether he believed in God by declaring "I believe in Spinoza's God, who reveals himself in the harmony of all being," a response that was addressed to the same faith community that had excommunicated Spinoza on the grounds of his alleged pantheism (Torrance 1998)). Other scholars who deal with the pre-Socratics continue to debate about just what Heraclitus may have meant by this doctrine "that everything is one," and whether the standard interpretation of it, which comes to us from Aristotle, "may have got Heraclitus wrong," and may even be reading a later problematic by hindsight into what are, we must never forget, scraps and fragments from the fifth or sixth century BCE (Hammer 1991). By entitling his Festschrift of 1988 *Between Quantum and Cosmos*, Wheeler's colleagues and students have put him into the succession of these speculations of Heraclitus about the One and the Many. In this context, there is probably no Greek treatise, not even Aristotle's *Physics* or *Metaphysics*, that stands in a more provocative relation to the thought of Heraclitus than the *Timaeus* of Plato. Not only is this treatise the source (*Timaeus* 24E–25D) from which subsequent thought, all the way down to the Walt Disney Studios, has derived the myth of the lost island of Atlantis (Kurtti 2001), which, as Francis Cornford (1957) says, "serious scholars now agree . . . probably owed its existence entirely to Plato's imagination" (although I must add that nowadays one does have to wonder about that alleged "agreement" among the truly "serious scholars"); but especially when, as part of the Septuagint version produced by Hellenistic Jews in Alexandria a century or two BCE, the Book of Genesis had been translated from Hebrew into a Greek that seems to carry echoes of *Timaeus*, this work of Plato was a major force in shaping cosmology and theology in both the Jewish and the Christian traditions, Latin as well as Greek (Pelikan 1997). As Plato made clear in *Timaeus* in continuity with Heraclitus, three Greek monosyllables of two letters each, *ti to on*, ask the first and the last question of all, the nature of Being: in Cornford's translation of the passage from *Timaeus* 27D, "What is that which is always real *[ti to on aei]* and has no becoming, and what is that which is always becoming and is never real?" (Cornford 1957: 22). Therefore, as Axelos (1968) makes clear, it is in Heraclitus that we find *la première saisie de l'être en devenir de la totalité.*

But the genealogy of this *saisie* comes down from Heraclitus, not only through Plato but also through Socrates, to Aristotle. In his attempt to reconstruct the original state of Aristotle's *Metaphysics* (a later title, which was due to the sequence in which the works of Aristotle were eventually arranged by scribes, with the *Metaphysics* coming after *[meta]* the *Physics*), Werner Jaeger (1948: 185) says: "Plato had asserted the unreality of phenomenal things, because he had been led by Heraclitus to the view that all particular things, all sensible particulars, are in continual flux and have no permanent existence. On the other hand, the ethical inquiries of Socrates had indirectly given rise to the new and important discovery that science is of the universal only, though he himself had not abstracted conceptions from real objects nor declared them separate. Plato then went further – according to Aristotle's retrospective account – and hypostasized universal conceptions as true being *(ousia)*." But in *Metaphysics* Book M, Aristotle, in a sense, reaches across this Platonic hypostasizing of the universal as *ousia* back to Heraclitus, in order to be able to take Becoming seriously again but without abandoning the quest for Being. From at least four ancient sources, including both Plato and Plutarch, we know that Heraclitus formulated this itch to speculate not only about Being but about Becoming in the famous aphorism, "It is impossible to step twice into the same river" (Diels 1956: 22B91; Waterfield 2000: F34).

On that basis, the itch to speculate about *ti to on* can be seen as not only permissible, but even obligatory for any serious scientific inquirer into Becoming, also in what the subtitle of this volume calls "a new century of discovery," no less than it is for the philosophical–theological architect of doctrines of Being. For, as Jaeger (1943–5: I.179) says in his *Paideia* about Heraclitus, "The physicists' conception of reality, the cosmos, the incessant rise and fall of coming-to-be and passing-away, the inexhaustible first principle from which everything arises and to which everything returns, the circle of changing appearances through which Being passes – these fundamentally physical ideas were the basis of his philosophy." It would not require a huge adjustment of language, nor a ransacking of his writings, to attribute to John Archibald Wheeler these five themes from Heraclitus as Jaeger summarizes him – "[1] the physicists' conception of reality, [2] the cosmos, [3] the incessant rise and fall of coming-to-be and passing-away, [4] the inexhaustible first principle from which everything arises and to which everything returns, [and 5] the circle of changing appearances through which Being passes." For in a poignant and almost haunting reminder of these lifelong preoccupations of his, he has dedicated his fascinating account of what his subtitle calls A *Life in Physics* "to the still unknown person(s) who will further illuminate the magic of this strange and beautiful world by discovering How come the quantum? How come existence? [And then in italics:] *We will first understand how simple the universe is when we recognize how strange it is*." (Wheeler 1998; italics original) Heraclitus couldn't have said it better, even in his elegant Greek!

At home in the universe

As those of you who know me and my work will not be altogether surprised to hear, these reflections about the significance of Heraclitus for science, philosophy, and theology, and for our understanding of John Archibald Wheeler, lead me, almost inevitably, to Johann Wolfgang von Goethe and his important debt to Heraclitus, which several scholars have studied (Bapp 1921). Goethe was the author not only of *Faust* but of various scientific works, including a monument in the history of speculation about optics, flawed though it may be scientifically, *Die Farbenlehre* (Martin 1979; Amrine *et al.* 1987). In a letter to Fritz Jacobi of January 6, 1813, and then in revised form as number 807 of his delightful collection of *Maxims and Reflections*, Goethe proposed the following taxonomy of our responses to the itch to speculate: "When we do natural science, we are pantheists; when we do poetry, we are polytheists; when we moralize, we are monotheists." I have used this taxonomy, which does have the advantage of coming from the author himself rather than from Marx or Freud, as a kind of Wagnerian leitmotiv to interpret the plot and the character of Goethe's *Faust*, as the protagonist moves from the scientific or theosophic pantheism of his dealings with the *Weltgeist* to the rhapsodically poetic but also self-indulgent aesthetic experience of the *Klassische Walpurgisnacht* (which was Goethe's own creation, by the way) to the purity of moral vision in which he dies and is saved (Pelikan 1995). But with this classification Goethe was also asking, though by no means answering, the question running throughout *Faust* all the way from the Prologue in Heaven to its exalted close, *Bergschluchten*, which is unforgettably set to music in Part Two of Gustav Mahler's Eighth Symphony, of what it could possibly mean to be, in John Wheeler's luminous phrase, "at home in the universe" (Wheeler 1994) Which is also the ultimate question, simultaneously ontological and existential, being raised by the chapters in this volume on *Science and Ultimate Reality* – and by the scientific and the speculative œuvre of our admired friend and mentor, John Archibald Wheeler.

References

Ambronn, H (1996) *Apeiron-eon-kenon: Zum Archē-Begriff bei den Vorsokratikern.* Frankfurt am Main: P. Lang, Verlag.

Amrine, F, Zucker, F J, and Wheeler, H (eds.) (1987) *Goethe and the Sciences: A Reappraisal*. Dordrecht: D. Reidel.

Axelos, K (1961) *Marx penseur de la technique: de l'aliénation de l'homme à la conquête du monde*. Paris: Editions de Minuit.

(1968) *Héraclite et la philosophie: la première saisie de l'être en devenir de la totalité.* Paris: Editions de Minuit.

Bapp, K (1921) *Aus Goethes griechischer Gedankenwelt: Goethe und Heraklit*. Leipzig: Dieterich.

Cornford, F M (1957) *Plato's Cosmology: The* Timaeus *of Plato Translated with a Running Commentary*. New York: Liberal Arts Press.

(1965) *Principium Sapientiae: A Study of the Origins of Greek Philosophical Thought*, ed. W. K. C. Guthrie. New York: Harper and Row.

Davis, S (1986) Philosophy in its originary pre-metaphysical sense. In *Heidegger on Heraclitus: A New Reading*, ed. K. Maly and P. Emad, pp. 155–66. Lewiston, NY: E. Mellen Press.

Diels, H (ed.) (1956) *Die Fragmente der Vorsokratiker*, 8th edn, ed. W. Kranz, 3 vols. Berlin: Weidmannsche Verlagsbuchhandlung.

Einstein (n.d.) Inscription on a fireplace at Princeton.

Emad, P (1999) Chapter title. In *The Presocratics after Heidegger*, ed. David C. Jacobs, pp. 55–71. Albany, NY: State University of New York Press.

Gilson, E (1944) *L'Esprit de la philosophie médiévale*, 2nd edn. Paris: J. Vrin.

Greenberger, D M and Zeilinger, A (eds.) (1995) *Fundamental Problems in Quantum Theory: A Conference Held in Honor of Professor John A. Wheeler*. New York: New York Academy of Sciences.

Hammer, T (1991) *Einheit und Vielheit bei Heraklit von Ephesus*. Würzburg: Königshausen und Neumann.

Heidegger, M (1934) *Die Selbstbehauptung der deutschen Universität: Rede, gehalten bei der feierlichen Übernahme des Rektorats der Universität Freiburg im Breisgau am 27.5.1933*. Breisgau: Verlag Wilh. Gottl. Korn.
(1962) *Sein und Zeit*, transl. as *Being and Time*. New York: Harper.
Huber, M S (1996) *Heraklit: Der Werdegang des Weisen*. Amsterdam: B. R. Grüner.
Jaeger, W (1943–5) *Paideia: The Ideals of Greek Culture*, transl. G. Highet, 3 vols. New York: Oxford University Press.
(1948) *Aristotle: Fundamentals of the History of His Development*, transl. R. Robinson, 2nd edn. Oxford: Clarendon Press.
Jüngel, E (1964) *Zum Ursprung der Analogie bei Parmenides und Heraklit*. Berlin: DeGruyter.
Kahn, C H (1979) *The Art and Thought of Heraclitus*. Cambridge: Cambridge University Press.
Klauder, J R (ed.) (1972) *Magic Without Magic: John Archibald Wheeler: A Collection of Essays in Honor of His Sixtieth Birthday*. San Francisco, CA: W. H. Freeman.
Kurtti, J (2001) *The Mythical World of Atlantis: Theories of the Lost Empire from Plato to Disney*. London: Turnaround.
Lampe, G (1961) *A Patristic Greek Lexicon*. Oxford: Clarendon Press.
Macquarrie, J (1955) *An Existentialist Theology: A Comparison of Heidegger and Bultmann*. New York: Macmillan.
Maly, K and Emad, P (eds.) (1986) *Heidegger on Heraclitus: A New Reading*. Lewiston, NY: E. Mellen Press.
Marcovich, M (1982) Heraclitus: some characteristics. *Illinois Class. Stud.* 7, 171.
Martin, M (1979) *Die Kontroverse um die Farbenlehre: anschauliche Darstellung der Forschungswege von Newton und Goethe*. Schaffhausen: Novalis-Verlag.
Mill, J S (1859) *On Liberty*. In *The English Philosophers from Bacon to Mill*, ed. Edwin A. Burtt. New York: Modern Library.
Pelikan, J (1984) *The Vindication of Tradition: The 1983 Jefferson Lecture in the Humanities*. New Haven, CT: Yale University Press.
(1993) *Christianity and Classical Culture: The Metamorphosis of Natural Theology in the Christian Encounter with Hellenism*. New Haven, CT: Yale University Press.
(1995) *Faust the Theologian*. New Haven, CT: Yale University Press.
(1997) *What Has Athens to Do with Jerusalem?* Timaeus *and* Genesis *in Counterpoint*. Ann Arbor, MI: University of Michigan Press.
(2001) Athens and/or Jerusalem: cosmology and/or creation. In *Cosmic Questions*, ed. J. B. Miller, pp. 17–27. New York: New York Academy of Sciences.
Soumia, A (1982) *Héraclite, ou, L'Intuition de la science*. Paris: A. Soumia.
Spengler, O (1904) *Der metaphysische Grundgedanke der heraklitischen Philosophie*. Halle: C. A. Kaernmerer.
Torrance, T (1998) Einstein and God. *CTI Reflections* **1**, 10.
Von Oech, R (2001) *Expect the Unexpected (or You Won't Find It): A Creativity Tool Based on the Ancient Wisdom of Heraclitus*, illus. G. Willett. New York: Free Press.
Waterfield, R (ed.) (2000) *The First Philosophers: The Presocratics and Sophists Translated with Commentary*. Oxford: Oxford University Press.
Weber, F J (1976) *Fragmente der Vorsokratiker: Text und Kommentar*. Paderborn: F. Schöningh.
Weisskopf, V F (1991) *The Joy of Insight: Passions of a Physicist*. New York: Basic Books.
(1998) (with K. Ford) *Geons, Black Holes, and Quantum Foam: A Life in Physics*. New York: W. W. Norton.

Wheeler, J A (1990) *Information, Physics, Quantum: The Search for Links*. Princeton, NJ: Department of Physics, Princeton University.
(1994) *At Home in the Universe*. New York: American Institute of Physics.
Zurek, W H, van der Merwe, A, and Miller, W A (eds.) (1988) *Between Quantum and Cosmos: Studies and Essays in Honor of John Archibald Wheeler*. Princeton, NJ: Princeton University Press.

Part III

Quantum reality: theory

3

Why is nature described by quantum theory?

Lucien Hardy

Perimeter Institute for Theoretical Physics, Waterloo, Canada

John A. Wheeler's two favorite questions are: "How come existence?" and "How come the quantum?" (Wheeler 1998). It is difficult to know how to go about answering the first question. What shape would an answer take? This article is concerned instead with the second question which I will expand as: "Why is nature described by quantum theory?"

What shape would an answer to this question take? We can get a handle on this by considering some historical examples. In the seventeenth century physicists were confronted by Kepler's laws of planetary motion. These laws were empirically adequate for predicting planetary motion and yet sufficiently abstract and ad hoc that they cannot really have been regarded as an explanation of "why" planets move the way they do. Later, Newton was able to show that Kepler's laws followed from a set of reasonable laws for the mechanics (his three laws) plus his law for gravitational forces. At this stage physicists could begin to assert with some degree of confidence that they understood "why" the planets move the way they do. Of course there were still mysteries. Newton was particularly bothered by the action at a distance of his inverse square law. What mediated the force? It was not until Einstein's theory of general relativity that an answer to this question became available.

Another historical example is the Lorentz transformations. Lorentz had shown that Maxwell's equations for electromagnetism were invariant under these transformations rather than the more intuitive Galilean transformations of Newtonian physics. However, taken by themselves, they have a similar status to Kepler's laws. They work, but it is not clear "why" nature would choose these transformations over the simpler Galilean transformations. The Lorentz transformations imply all sorts of counterintuitive effects such as length contraction. Einstein was able to show that these transformations are the consequence of two very natural principles:

Science and Ultimate Reality, eds. J. D. Barrow, P. C. W. Davies and C. L. Harper, Jr. Published by Cambridge University Press.

that the laws of physics are the same in all inertial frames of reference and that the speed of light is constant (independent of the source). Thus, he answered the "why" question. In so doing he made the counterintuitive features of the transformations more acceptable.

There is something deeply satisfying about this kind of physics. By finding deeper and more reasonable principles we have the sense that we have explained otherwise mysterious behavior. Note that this requires something more basic than merely positing mathematical rules. Rather, we require principles or axioms which express some deeply reasonable expectation about the universe.

Physics is not just about giving us a warm cosy feeling of having understood something (though this is an important part of it). It is also about making progress in developing new theories which have an extended domain and are more accurate. Past experience has shown that finding the deep principles behind the physics of the day helps us to move on to newer theories. The passage from Kepler's laws to Newtonian physics to special relativity and then to general relativity illustrates this very well. This provides added motivation for searching for the deep principles behind quantum theory.

In this chapter we are interested in quantum theory. In its usual formulation it consists of a set of formal rules. The state is represented by a positive operator on a complex Hilbert space, its evolution is given by the action of a unitary operator, and it can be used to calculate probabilities using the trace rule. This mathematical formalism is both abstract and ad hoc in the same way that Kepler's laws of planetary motion and Lorentz's transformations are. What is required is a set of deep principles from which the usual formalism can be obtained. This might offer the same benefits as earlier historical examples. First, we might find a simple principle that makes the various counterintuitive properties of quantum theory seem reasonable. Second, we may see the way forward to going beyond quantum theory (for example to a theory of quantum gravity). Further, if we can find such a set of deeper principles then we can say that we have answered Wheeler's question "How come the quantum?"

Before addressing such matters let us inquire into what type of theory we are dealing with. Quantum theory can be applied to many different situations. At its root, however, it is simply a way of calculating probabilities. It might better be called "quantum probability theory". Its natural predecessor is not classical mechanics as so often stated but rather classical probability theory. Like quantum theory, classical probability theory can be applied to many different situations (coins, dice, mechanics, optics, ...). What makes one classical and the other quantum is not to do with the type of system (mechanical, optical, magnetic, ...) but something deeper. To underline this point we will see that quantum theory and classical probability theory are the same type of mathematical theory (they have the same type of underlying mathematical structure). Classical probability theory is often regarded and taught as a part of mathematics (rather than physics). With the advent of quantum physics we find systems for which the classical probability calculus does not work and it becomes necessary to invent a new calculus – namely quantum theory. And, with this, probability calculus became part of physics. However, quantum theory's status with respect to quantum physics is actually the same as classical probability theory's status with respect to classical physics – they are both meta-theories. By identifying the differences between these two theories we might hope to obtain the deep insight we need to answer Wheeler's question. This is exactly the approach that will be taken here.

In this chapter we will give an answer to the question "How come the quantum?" We will do this by putting forward five principles or axioms from which quantum theory, for discrete dimensional Hilbert spaces, can be derived. These principles appear to be quite reasonable and easily motivated. Furthermore they have the property that four of them are obviously true in both classical probability theory and quantum theory. It is the remaining principle which gives us quantum theory. This principle states that there should be a continuous reversible transformation between any pair of pure states. It is the single word "continuous" which gives quantum theory. The work in this paper first appeared in Hardy (2001; see also Hardy 2002). In this chapter the key ideas will be presented. The details of the proofs are left out (for those the reader is referred to (Hardy (2001)).

Various papers have investigated the structure and origin of quantum theory (Birkhoff and von Neumann 1936; Mackey 1963; Jauch and Piron 1963; Piron 1964; Ludwig 1968, 1983; Mielnik 1968; Lande 1974; Fivel 1994; Accardi 1995; Landsman 1998; Coecke *et al.* 2000). Much of this work is from a quantum logic point of view. In such approaches various logical relations between measurements (regarded as logical propositions) are posited and, from this, an attempt is made to recover the structure of quantum theory.

More recently, and at least partly inspired by Wheeler's philosophy, other people have been looking at ways of deriving quantum theory from reasonable principles. Particularly interesting papers that discuss this idea are Fuchs (2002) and Zeilinger (1999).

The approach taken in this paper has many advantages. First, the principles or axioms presented can be easily understood even by the nonspecialist and, second, the mathematical techniques involved are, for the most part, rather simple.

Setting the scene

Physical theories apply to a particular type of situation. We are interested in situations in which a physical system is prepared in some initial state, it is subject

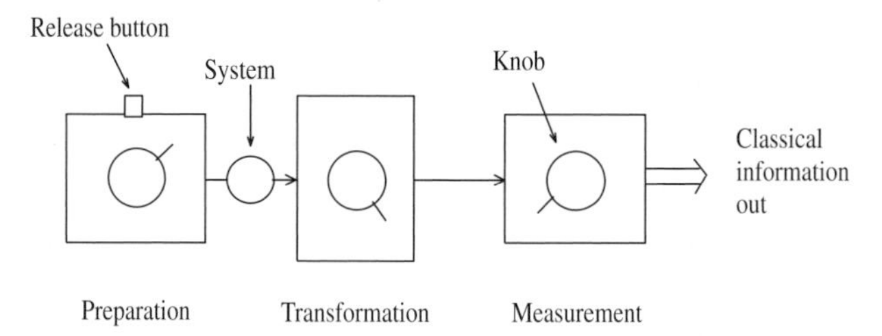

Figure 3.1. The situation considered consists of a preparation device with a knob for varying the state of the system produced and a release button for releasing the system, a transformation device for transforming the state (and a knob to vary this transformation), and a measuring apparatus for measuring the state (with a knob to vary what is measured) which outputs classical information.

to various transformations, and measurements are made on it. This language of preparation/transformation/measurement is quite familiar to us in quantum theory. However, it applies equally well to classical theories. Consider a ball when it is kicked. It is prepared with some initial velocity, then subject to forces which transform its state and at any time measurements may be made on it by simply looking at it, for example. We might even describe this situation as arising from the human condition. Our place in the world forces us to act as preparers, transformers, and measurers. This point of view will be discussed further later.

Given that this picture of preparation/transformation/measurement is fundamental, let us describe it more precisely. Consider the situation shown in Fig. 3.1. We have three types of device. First, we have a preparation device which prepares systems in some particular state. This device has a knob on it to vary the particular state being prepared. It also has a release button whose function will be described in a moment. Second, we have a transformation device which transforms the state of the system. This has a knob on it to vary the transformation effected. And third, we have a measurement device which implements a measurement on the system. This device has a knob to vary the particular measurement being implemented. It also has a measurement outcome readout. There is a null outcome and there are the non-null outcomes the latter being labeled by $l = 1$ to L. We require that when the transformation device is absent (or implements the identity transformation) then, if the release button on the preparation device is pressed, we will certainly see one of the non-null outcomes and, if the release button is not pressed, we will definitely record the null outcome. This requirement does not amount to assuming something additional since it can always be arranged by appealing to the subspace axiom to be introduced later. A typical example of this is when we have an array of detectors. If no particle is incident none of the detectors will fire and we can regard this as the null outcome. The fact that we allow null outcomes means that states, introduced later, need not be normalized. This turns out to be very useful.

It is worth making a philosophical point here. The everyday language usage of the words "state" and "measurement" have an ontological overtone. Thus we talk about measuring some property of the state. For example, we might make a measurement to see where a ball is in its trajectory. We believe that the measurement reveals some ontological property of the state which is independent of the act of measurement. However, here we do not need to commit ourselves to such notions. We think of a measurement as being simply that operation which is implemented by a device of the type illustrated in Fig. 3.1. Likewise, we will give a definition of "state" which carries no ontological commitment.

In using the axioms to derive quantum theory we will implicitly assume that the devices depicted in Fig. 3.1 can be freely manipulated. The experimentalists are allowed to combine them in any way they wish. This is a standard implicit assumption in physics. Einstein, for example, could not have derived special relativity if he had to assume that all the clocks and rulers in the world were tied together with unbreakable string.

General probability theories

We will be interested in building probability theories. Probabilities are problematic even at the stage of defining what a probability is. There are various approaches. In the frequency approach probabilities are defined as limiting relative frequencies, in the propensity approach they are defined as intrinsic propensities resident in a system, and in the Bayesian approach they are defined as degrees of belief. In the laboratory probabilities are measured by taking relative frequencies (though not limiting relative frequencies since it is never possible to repeat an experiment an infinite number of times). All of these approaches to probability have problems that we need not go into here. For pragmatic reasons we will define probabilities to be limiting relative frequencies. We do this because these are conceptually closest to what is actually measured in the laboratory and, furthermore, all approaches to probability must ultimately attempt to account for the fact that we expect relative frequencies to converge on a definite value. Given this pragmatic approach we are free to concentrate on the real topic of this paper which is deriving the structure of quantum theory. However, a proper understanding of probability is likely to be essential in coming to a full understanding of quantum theory. All this having been said, we will now state the first axiom.

Axiom 1 *Probabilities.* Relative frequencies (measured by taking the proportion of times a particular outcome is observed) tend to the same value (which we call the probability) for any case where a given measurement is performed on an ensemble of n systems prepared by some given preparation in the limit as n becomes infinite.

This axiom is sufficient to enable us to construct the form of general probability theories. Were Axiom 1 not true then the concept of probability would not be stable and we could not begin to construct any sort of probability theory. This axiom is sufficient to construct the Kolmogorov axioms for probability theory (see Gillies 2000: 112) modulo technical problems associated with infinite outcome sets (note that in our context the Kolmogorov axioms do not constitute all of classical probability theory since they do not take into account the effects of transformations – rather they apply only to probability measures defined on measurement outcomes).

All measurements amount to probability measurements. For example, an expectation value is a probability weighted sum. Hence, it is sufficient to consider only probability measurements.

In this paper we will use the terminology *measurement* and *probability measurement* to refer to the situation where we measure the probability of a given non-null outcome or set of outcomes with a given knob setting.

For example, we can measure the probability that the outcome is $[l = 1 \text{ or } l = 2]$. This terminology will serve us well but it is perhaps a little nonstandard since the word "measurement" usually refers to all of the possible outcomes.

Associated with each preparation is a state defined as follows:

The *state* of a system is defined to be that thing represented by any mathematical object which can be used to predict the probability associated with every measurement that may be performed on the system.

This definition only makes sense given Axiom 1 (otherwise we could not be sure that the probability is always the same for a given type of measurement). Given this definition it immediately follows that one way of representing the state is simply to list the probability associated with every outcome of every conceivable measurement (labeled by α):

$$\text{STATE} \equiv \begin{pmatrix} \vdots \\ p_\alpha \\ \vdots \end{pmatrix}. \tag{3.1}$$

This mathematical object certainly specifies the state. However, most physical theories have some structure relating the probabilities associated with different measurements and so this object would represent too much information. In general, it will be possible to represent the state by listing the probabilities associated with a smaller set of measurement outcomes. Thus, consider the smallest set of measurements which are just sufficient to fix the state. This set may not be unique but we can fix on one such set and call these the *fiducial measurements*. We will label the fiducial measurements by $k = 1$ to K. The state can now be represented by

$$\mathbf{p} = \begin{pmatrix} p_1 \\ p_2 \\ p_3 \\ \cdot \\ \cdot \\ \cdot \\ p_K \end{pmatrix}. \tag{3.2}$$

We will look at examples of this later in the context of classical probability theory and quantum theory. The integer K, which we will call *the number of degrees of freedom*, will play a very important role in this work.

The *number of degrees of freedom*, K, is defined as the smallest number of measurements of probability required to fix the state.

There is another important integer in this work. This is the maximum number of states which can be distinguished in a single shot measurement. To explain this note we can prepare certain sets of states for which each member gives rise to disjoint outcomes (we do not include the null outcome here). Thus, if Alice randomly prepares one member of this set and sends it to Bob then Bob can make a measurement to determine which state was sent. The largest number of states in any such set is the maximum number of states that can be distinguished in a single shot measurement and we will call this number the *dimension* and denote it by N.

The *dimension*, N, is defined as the maximum number of states that can be distinguished in a single run.

(In this definition we only consider states that have zero probability of activating the null outcome.) We use the terminology "dimension" because, in quantum theory, N is equal to the dimension the Hilbert space associated with the system. We will see that in classical probability theory $K = N$ and in quantum theory $K = N^2$.

One state we can prepare is the null state. We prepare this by never pressing the release button. In this case all the fiducial measurements yield 0 and hence the null state is given by

$$\mathbf{p}_{\text{null}} = \begin{pmatrix} 0 \\ 0 \\ 0 \\ \cdot \\ \cdot \\ \cdot \\ 0 \end{pmatrix}. \tag{3.3}$$

Now imagine we have two preparation apparatuses. Preparation apparatus A prepares state $\mathbf{p}_A$ and preparation apparatus B prepares state $\mathbf{p}_B$. We can use these preparations to construct a new preparation C by tossing a coin. If the coin comes up heads then we prepare $\mathbf{p}_A$ and if it comes up tails we prepare $\mathbf{p}_B$. The new state we prepare is

$$\mathbf{p}_C = \lambda\mathbf{p}_A + (1-\lambda)\mathbf{p}_B \tag{3.4}$$

where λ is the probability of the coin coming up heads. It is clear this is the state because the probability for the kth fiducial measurement will be $\lambda p_k^A + (1-\lambda)p_k^B$ (since probabilities are defined as relative frequencies). There is an implicit use of Axiom 1 here since we need it to be the case that preparations A and B yield the same probabilities whether they are part of preparation C or not. Any state that can be written in this form with $0 < \lambda < 1$ and $\mathbf{p}_A \neq \mathbf{p}_B$ will be called a mixed state.

A *pure state* is any state, except the null state, that is not a mixed state.

Pure states are clearly special in some way. They will play an important role in the axioms.

The probability associated with a general measurement will be given by some function of the state:

$$\text{prob} = f(\mathbf{p}). \tag{3.5}$$

This function will, in general, be different for each measurement. Since $\mathbf{p}_C$ is the mixed state prepared when state $\mathbf{p}_A$ is prepared with probability λ and state $\mathbf{p}_B$ is prepared with probability $1-\lambda$ it follows that

$$f(\mathbf{p}_C) = \lambda f(\mathbf{p}_A) + (1-\lambda)f(\mathbf{p}_B). \tag{3.6}$$

Hence using eqn (3.4) we have

$$f(\lambda\mathbf{p}_A + (1-\lambda)\mathbf{p}_B) = \lambda f(\mathbf{p}_A) + (1-\lambda)f(\mathbf{p}_B). \tag{3.7}$$

This can be used to prove that the function f is linear in $\mathbf{p}$. Hence, we can write

$$\text{prob} = \mathbf{r}\cdot\mathbf{p} \tag{3.8}$$

where $\mathbf{r}$ is a vector associated with the measurement.

We have not yet considered the effect of the transformation device. Clearly its effect must be to take the state $\mathbf{p}$ to some new state $\mathbf{g}(\mathbf{p})$. The argument for linearity above can be applied to each element in the vector $\mathbf{g}(\mathbf{p})$ and hence the transformation must be linear. Thus the transformation is

$$\mathbf{p} \longrightarrow Z\mathbf{p} \tag{3.9}$$

where Z is a $K \times K$ real matrix. The allowed transformations Z will belong to some set Γ. An interesting subclass of transformations are the reversible transformations $\Gamma_{\text{reversible}}$. These are transformations for which there exists another transformation which reverses the effect of that transformation regardless of the initial state. In other words for which Z^{-1} both exists and belongs to Γ. The full set of reversible transformations clearly form a group. It can easily be shown that a reversible transformation acting on a pure state will always output a pure state.

We have discussed the structure of general probability theories. A probability theory is completely characterized for a given system when we know the set S of allowed states $\mathbf{p}$, the set R of allowed measurements $\mathbf{r}$, and the set Γ of allowed transformations Z.

We will now see how classical probability theory and quantum theory fit into this general scheme.

Classical probability theory

As we emphasized in the introduction, the natural predecessor to quantum theory is classical probability theory. We are generally comfortable with classical probability theory because it is possible to give it an ontological underpinning consistent with our usual intuition about the world.

Consider a ball which can be in one of N boxes or may be missing. Associated with each box is a probability p_1 to p_N. This information can be represented as a column vector

$$\mathbf{p} = \begin{pmatrix} p_1 \\ p_2 \\ p_3 \\ \cdot \\ \cdot \\ \cdot \\ p_N \end{pmatrix}. \tag{3.10}$$

This completely specifies the state in this case. Note, since the ball may be missing we require only that $\sum_n p_n \leq 1$. In classical probability theory N probabilities are required to specify the state. Hence $K = N$.

There are some special states in which the ball is always in a particular box. These are the states

$$\mathbf{p}_1 = \begin{pmatrix} 1 \\ 0 \\ 0 \\ \cdot \\ \cdot \\ \cdot \\ 0 \end{pmatrix} \quad \mathbf{p}_2 = \begin{pmatrix} 0 \\ 1 \\ 0 \\ \cdot \\ \cdot \\ \cdot \\ 0 \end{pmatrix} \quad \mathbf{p}_3 = \begin{pmatrix} 0 \\ 0 \\ 1 \\ \cdot \\ \cdot \\ \cdot \\ 0 \end{pmatrix} \quad \text{etc.} \tag{3.11}$$

representing the ball being in the first box, the second box, the third box, etc. These states cannot be regarded as mixtures and are therefore pure states.

We wish to formulate classical probability theory in a similar language to quantum theory. Hence, let us consider measurements on this state. One measurement we might perform is to look and see if the ball is in box 1. The probability for this measurement is p_1. This can be written

$$p_1 = \begin{pmatrix} 1 \\ 0 \\ 0 \\ \cdot \\ \cdot \\ \cdot \\ 0 \end{pmatrix} \cdot \begin{pmatrix} p_1 \\ p_2 \\ p_3 \\ \cdot \\ \cdot \\ \cdot \\ p_N \end{pmatrix} = \mathbf{r}_1 \cdot \mathbf{p}. \tag{3.12}$$

We can identify the vector $\mathbf{r}_1$, defined as

$$\mathbf{r}_1 = \begin{pmatrix} 1 \\ 0 \\ 0 \\ \cdot \\ \cdot \\ \cdot \\ 0 \end{pmatrix}, \tag{3.13}$$

as representing the measurement where we look to see if the ball is in box 1. Similar vectors can be associated with the other boxes. We can consider more general measurements. For example, we could toss a coin and, if the coin comes up heads, look in box 1 and, if it comes up tails, look in box 2. The probability of finding the ball is given by

$$[\mu \mathbf{r}_1 + (1-\mu)\mathbf{r}_2] \cdot \mathbf{p} \tag{3.14}$$

where μ is the probability of getting heads. In this case the measurement is associated with the vector $\mu \mathbf{r}_1 + (1-\mu)\mathbf{r}_2$. Any measurement can be associated with some vector $\mathbf{r}$. In general the probability we measure is given by

$$\text{prob} = \mathbf{r} \cdot \mathbf{p}. \tag{3.15}$$

To illustrate classical probability theory consider the case $N = 2$. Then the state is given by

$$\mathbf{p} = \begin{pmatrix} p_1 \\ p_2 \end{pmatrix}. \tag{3.16}$$

We require that $p_1 + p_2 \leq 1$. Thus, the allowed states are those represented by points in the triangle in Fig. 3.2. The vertices represent the two pure states

$$\mathbf{p}_1 = \begin{pmatrix} 1 \\ 0 \end{pmatrix} \quad \text{and} \quad \mathbf{p}_2 = \begin{pmatrix} 0 \\ 1 \end{pmatrix} \tag{3.17}$$

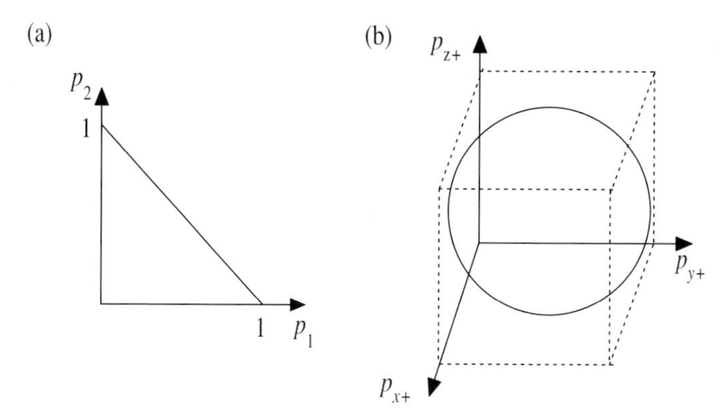

Figure 3.2. (a) Allowed states for classical bit are inside the triangle. States on the hypotenuse are normalized. (b) Normalized states for a qubit are in the ball inside the unit cube as shown.

and the null state

$$\mathbf{p}_{\text{null}} = \begin{pmatrix} 0 \\ 0 \end{pmatrix}. \tag{3.18}$$

Here we learn an essential fact about classical probability theory with finite or countably infinite N:

In classical probability theory the pure states form a countable set.

This is true for general countable N. It implies that if the system is to go from being in one pure state to another it must "jump." As we will see, this contrasts sharply with the situation in quantum theory. Also note that the pure states are distinguishable from one another (not true in quantum theory). For general N the set of allowed states will have $N + 1$ vertices. The set of allowed measurements can easily be determined using some basic considerations.

The reversible transformations will form a group. Since reversible transformations always take pure states to pure states and since the pure states form a countable set this group must be discrete (in fact equal to the permutation group). That is

In classical probability theory the group of reversible transformations is discrete.

To summarize, classical probability theory is characterized by $K = N$, states $\mathbf{p}$ belonging to some allowed set $S_{\text{classical}}$ for which the pure states are both distinguishable and countable, by measurements $\mathbf{r}$ which belong to some allowed set of measurements $R_{\text{classical}}$, and by transformations Z belonging to some set of allowed transformations $\Gamma_{\text{classical}}$ for which the reversible transformations form a discrete group.

Quantum theory

Quantum theory is, at root, a theory of probabilities. In this section we will make this stark by formulating quantum theory in a very similar way to classical probability theory. We will begin with a simple example. Consider a two-level quantum system such as a spin-half particle. In this case $N = 2$. The general state of a two-level system is given by the density matrix

$$\rho = \begin{pmatrix} p_{z+} & a \\ a^* & p_{z-} \end{pmatrix} \tag{3.19}$$

where

$$a = p_{x+} - p_{y+} - \frac{1-i}{2}(p_{z+} + p_{z-}). \tag{3.20}$$

Here, p_{z+} is the probability the particle has spin up along the $+z$ direction and the other probabilities are defined similarly. This density matrix contains the same information as

$$\mathbf{p} = \begin{pmatrix} p_{z+} \\ p_{z-} \\ p_{x+} \\ p_{y+} \end{pmatrix} \tag{3.21}$$

and hence this mathematical object serves as an equally good way of representing the state. Here we have $N = 2$ and $K = 4$. The allowed states will belong to some set. Since $K = 4$ this will be in four dimensions and so a little difficult to picture. However, to simplify matters we can impose normalization $p_{z+} + p_{z-} = 1$ leaving only three parameters (p_{x+}, p_{y+}, p_{z+}). We require that the density matrix be positive. This imposes the constraint

$$\left(p_{x+} - \frac{1}{2}\right)^2 + \left(p_{y+} - \frac{1}{2}\right)^2 + \left(p_{z+} - \frac{1}{2}\right)^2 \leq \left(\frac{1}{2}\right)^2. \tag{3.22}$$

Hence, the allowed states are those represented by points inside this sphere (which fits just in the unit cube in the first octant). It is clear that points on the surface of the sphere cannot be written as mixtures (in the form of eqn (3.4)) and hence they must represent pure states. Here we learn an essential fact about quantum theory:

In quantum theory pure states form a continuous set.

This is true for general N. This means that, unlike the situation in classical probability theory, in quantum theory we do not need to jump when we go from one pure state to another, but rather there exists a continuous path through the pure states. This will turn out to be the key difference between classical and quantum theory.

The z spin-up and z spin-down states represented as $\mathbf{p}$ vectors are

$$\mathbf{p}_1 = \begin{pmatrix} 1 \\ 0 \\ \frac{1}{2} \\ \frac{1}{2} \end{pmatrix} \quad \text{and} \quad \mathbf{p}_2 = \begin{pmatrix} 0 \\ 1 \\ \frac{1}{2} \\ \frac{1}{2} \end{pmatrix}. \tag{3.23}$$

Contrast this with the classical case eqn (3.17).

For general N the quantum state is represented by an $N \times N$ density matrix. The number of real parameters in this Hermitian matrix is N^2 (N real numbers along the diagonal and $N(N-1)/2$ complex numbers above the diagonal). In fact we can find N^2 probabilities p_k which are linearly related to these real parameters as we did in the $N = 2$ case. Therefore the state can be represented by a vector

$$\mathbf{p} = \begin{pmatrix} p_1 \\ p_2 \\ p_3 \\ \cdot \\ \cdot \\ \cdot \\ p_{N^2} \end{pmatrix} \tag{3.24}$$

and hence $K = N^2$. Various authors have noticed that the state can be represented by probabilities (Prugovecki 1977; Wootters 1986, 1990; Busch *et al.* 1995; Weigert 2000).

In quantum theory probabilities are given by the trace formula

$$\text{prob} = \text{tr}(A\rho) \tag{3.25}$$

where ρ is the density matrix and A is a positive matrix associated with the measurement being performed. Since the trace operation is linear and since the parameters in ρ are linearly related to the parameters in $\mathbf{p}$ we can write

$$\text{prob} = \mathbf{r} \cdot \mathbf{p} \tag{3.26}$$

where the vector $\mathbf{r}$ is determined from A and is associated with the measurement.

The effect of the second device in Fig. 3.1 is to transform the system. Textbooks on quantum theory routinely discuss only two types of transformations: unitary evolution which is reversible, and von Neumann projection which happens during a von Neumann type measurement and is irreversible. In fact these two types of evolution are special cases of a much more general type of transformation. In general, quantum states transform under completely positive linear maps. We need not elaborate on what these are but they have two important properties (Nielsen and Chuang 2000). First, they transform allowed states to allowed states even when the system being transformed is part of a composite system and, second, they are linear. This means that $\mathbf{p}$ transforms linearly when acted on by a transformation device. That is

$$\mathbf{p} \longrightarrow Z\mathbf{p} \tag{3.27}$$

where Z is a $K \times K$ real matrix. (A similar statement holds for transformations in the case of classical probability theory.)

The fact that the pure states form a continuous set rather than a discrete set corresponds to the fact that the group of reversible transformations are continuous in quantum theory. Thus,

In quantum theory the reversible transformations form a continuous group.

Quantum theory is characterized by $K = N^2$, by states represented by $\mathbf{p}$ belonging to some set of allowed states S_{quantum} having the property that the pure states form a continuous set, by measurements represented by $\mathbf{r}$ belonging to some set of allowed measurements R_{quantum}, and by transformations Z belonging to some set of allowed transformations Γ_{quantum}.

What is the difference between classical probability theory and quantum theory?

Actually, what is more striking than the differences between classical probability theory and quantum theory are the similarities. In both theories we have states represented by $\mathbf{p}$, measurements represented by $\mathbf{r}$, probabilities calculated by taking the dot product, and transformations represented by Z and given by $Z\mathbf{p}$. The differences between these theories lie in the nature of the sets S, R, and Γ of allowed states, measurements, and transformations respectively. There are a number of further similarities.

The first similarity concerns what we will call subspaces (the terminology being borrowed from quantum theory). Consider a ball that can be in one of five boxes. Now imagine the state is constrained such that ball is never found in the last two boxes – that is it is only ever found in the first three boxes or is missing. In this case the system will have the same properties as a system having $N = 3$. A similar statement holds in quantum theory. If the state is confined to a lower dimensional subspace then the system behaves like one having the dimension of that subspace. In general we can say that

A state is confined to an M-dimensional *subspace* when, with a measurement apparatus is set to distinguish some set of N distinguishable states (where N is the dimension), only outcomes associated with some subset of M of the distinguishable states or the null outcome have a non-zero probability of being observed.

We notice that, in both classical probability theory and quantum theory, the following is true:

Similarity 1 Systems of dimension N or of a higher dimension but where the state is confined to an N dimensional subspace have the same properties.

By saying that two systems have the same properties we mean that there exists a choice of fiducial measurements for each system such that the two systems have the same sets S, R, and Γ.

When we consider composite systems we will see that there are further similarities between the two theories. Consider a composite system consisting of systems A and B. Then

Similarity 2 The number of distinguishable states for the composite system is

$$N = N_A N_B. \tag{3.28}$$

Similarity 3 The number of degrees of freedom for the composite system is

$$K = K_A K_B. \tag{3.29}$$

Similarity 4 If one (or both) of the systems is in a pure state then it is uncorrelated with the other system so that any joint probabilities measured between the two systems factorize

$$p_{AB} = p_A p_B. \tag{3.30}$$

Similarity 5 The number of degrees of freedom associated with a composite system is equal to the number of degrees of freedom associated with the separable states (those states that can be regarded as mixtures of states for which the probabilities factorize as in Similarity 4).

Similarities 1 to 3 will actually form part of the axioms. Similarities 4 and 5 will not but it can be shown that these two properties imply $K = K_A K_B$.

Another striking similarity between classical probability theory and quantum theory which we will not use in our axiom set follows from the fact that, in quantum theory, we can diagonalize the density matrix.

Similarity 6 Any state can be regarded as a mixture of some set of distinguishable states.

It is this property that makes it possible to define entropy in both theories. Of course, the set of distinguishable states is fixed in classical probability theory but not in quantum theory.

The main differences from our present point of view are:

Difference 1 In classical probability theory we have $K = N$ whereas in quantum theory we have $K = N^2$.

Difference 2 In classical probability theory the pure states form a discrete set whereas in quantum theory they form a continuous set. Also, the reversible transformations form a discrete group in classical probability theory and a continuous group in quantum theory.

Difference 3 In classical probability theory the pure states form a distinguishable set whereas in quantum theory there are pure states that cannot be distinguished from one another.

It is interesting to contrast Difference 1 with Similarities 2 and 3. It can be shown that the only strictly increasing functions $K(N)$ which have the properties in Similarities 2 and 3 are of the form $K = N^r$ where r is a positive integer. Hence, classical probability theory is the simplest theory of this kind. However, if we impose that there should exist a continuous set of pure states then it is not possible to have $K = N$ and hence $K = N^2$ becomes the simplest theory of this kind. Thus, classical probability theory and quantum theory have a further similarity of a sort – namely that they each represent the simplest theory consistent with the above considerations. We will use this simplicity property as an axiom.

The axioms

These similarities and differences motivate a set of axioms from which quantum theory can be derived. These axioms are:

Axiom 1 *Probabilities*. Relative frequencies (measured by taking the proportion of times a particular outcome is observed) tend to the same value (which we call the probability) for any case where a given measurement is performed on an ensemble of n systems prepared by some given preparation in the limit as n becomes infinite.

Axiom 2 *Subspaces*. There exist systems for which $N = 1, 2, \ldots,$ and, furthermore, all systems of dimension N, or systems of higher dimension but where the state is constrained to an N dimensional subspace, have the same properties.

Axiom 3 *Composite systems*. A composite system consisting of subsystems A and B satisfies $N = N_A N_B$ and $K = K_A K_B$.

Axiom 4 *Continuity*. There exists a continuous reversible transformation of a system between any two pure states of that system for systems of any dimension N.

Axiom 5 *Simplicity*. For each given N, K takes the minimum value consistent with the other axioms.

If the single word "continuous" is dropped from Axiom 4 then, by virtue of the simplicity axiom, we obtain classical probability theory instead of quantum theory. Hence, the difference between the two theories is enforced by this continuity property.

We claim that, unlike the usual axioms for quantum theory, these are reasonable principles to adopt. A few comments might be appropriate here

Axiom 1 was discussed in section "General probability theories." If one wishes to build up a probability theory then we need to assume something like this so that probabilities are stable. It was shown that the basic structure for general probability theories follows once we have this axiom in place. We could replace this axiom by a different axiom motivated by a Bayesian or propensity-based view of probability and that may be more reasonable for proponents of those viewpoints (Schack (R. Schack, unpublished data) has taken a Bayesian approach). It is the remaining axioms that fix the particular structure of quantum theory.

It has already been shown that Axiom 2 is consistent with both classical probability theory and quantum theory. We maintain that this is a reasonable principle because it imposes a certain fungibility property. We should expect that, in a given type of theory, systems of a given N have the same properties (meaning that there exists a choice of fiducial measurements for which they have the same sets S, R, Γ). This means, for example, that a state of a given N-dimensional system can be represented on any other N-dimensional system. This is a property that we are used to from classical and quantum information theory.

Axiom 3 has two parts. The first part, that $N = N_A N_B$, is quite reasonable. Consider two dice. Then we have $N_A = N_B = 6$ and 36 possibilities in total. However, the second part, $K = K_A K_B$, is a little harder to motivate. We are not used to thinking in terms of the quantity K. However, it can be shown that $K = K_A K_B$ follows from two quite reasonable assumptions.

Assumption A If a system is in a pure state then it should be uncorrelated with any other system so that joint probabilities factorize.

This is Similarity 4 discussed above. It seems reasonable since we expect pure states to be definite states of a system which should therefore be uncorrelated with other systems. It is easy to show that this implies that $K \geq K_A K_B$.

Assumption B The number of degrees of freedom associated with a composite system should be equal to the number of degrees of freedom associated with only the separable states.

This is Similarity 5. It is a reasonable assumption since it implies that new properties do not come into existence when two systems come together. A counterexample is the following. Imagine that Alice and Bob are very simple systems that are each individually fully described when probabilities are given for whether they are happy or not happy. However, when they come together a new property arises. They may be in love. Thus, to fully describe the composite system of Alice and Bob, we need to specify probabilities for [happy, happy], [happy, not happy], [not happy, happy], [not happy, not happy] and [in love]. In this case $K_A = K_B = 2$ but $K = 5$. It can be shown that, given Assumption A, Assumption B implies that $K = K_A K_B$. It is interesting to note that quantum theory formulated with pure states represented by vectors in real Hilbert space violates assumption B (there is some "love" here) while quaternionic quantum theory violates assumption A.

The central axiom in this set is Axiom 4. This has various components. First, that there should exist a transformation between *any* pair of pure states. This is regarded as reasonable since otherwise there would be states that cannot be reached. Second, that there should exist reversible transformations. This is regarded as reasonable since it should be possible to manipulate a system without extracting information from it. In this case we expect to be able to undo whatever changes we made to the system. Third, and most importantly, that the transformations should be continuous. We will try to motivate this more carefully later. But note that transformations of a system are normally implemented in a continuous way perhaps, for example, by subjecting the system to some continuous fields. Hence, we expect the state to change in a continuous way.

Simplicity principles are common in physics and, as such, Axiom 5 may be considered reasonable. However, it would be better if a more direct reason could be found to rule out the higher-order theories. It may be that the axiom is simply unnecessary – that theories with $K = N^r$ for $r \geq 3$ are not possible. Or it may turn out that such theories violate some reasonable principle like Similarity 4, 5, or 6. Alternatively, it may actually be possible to build these higher-order theories. Such theories would be fascinating since they would provide new testing ground for thinking about nonclassical properties. They may lead to faster quantum computing. Furthermore, there is the remote possibility that quantum theory can be embedded into some of these higher-order theories in the same way that classical probability theory can be embedded in quantum theory. In this case, it could be that the world really is described by a higher-order theory but that we have not yet performed sufficiently sophisticated measurements to reveal this (just as we did not reveal quantum theory until the twentieth century).

How the proof works

The proof that quantum theory follows from these axioms, while calling on simple mathematics (mostly just linear algebra) is rather lengthy. Here we will content ourselves with simply outlining how the proof works. The reader is referred to Hardy (2001) for the details. The proof proceeds in a number of stages.

Stage 1 First we need to prove that states can be represented by vectors like $\mathbf{p}$ and that probabilities are given by $\mathbf{r} \cdot \mathbf{p}$. This has already been outlined (pp. 49–52).

Stage 2 We need find the form of the function $K(N)$. To do this we note that Axiom 3 implies

$$K(N_A N_B) = K(N_A)K(N_B). \tag{3.31}$$

Such functions are known in number theory as *completely multiplicative functions*. We expect that

$$K(N+1) > K(N) \tag{3.32}$$

(i.e., that $K(N)$ is strictly increasing) and it can indeed be shown that this follows from the subspace axiom. It can then be shown that all strictly increasing completely multiplicative functions are of the form $K = N^\alpha$ where $\alpha > 0$. Then, since K must be an integer for all N, we have

$$K = N^r \tag{3.33}$$

where $r = 1, 2, 3, \ldots$. Now, by the simplicity axiom we require that r takes the smallest value consistent with the other axioms. It can be shown that the $r = 1$ case is ruled out by the continuity axiom. Thus, the simplest case is when $r = 2$ and hence

$$K = N^2 \tag{3.34}$$

which is the quantum case. Employing an equation like eqn (3.31) Wootters (1986, 1990) also arrived at eqn (3.33) as a possible relationship between K and N, though, since he did not use eqn (3.32), he was not able to show that this is unique.

Stage 3 Next we consider the simplest nontrivial case, namely where $N = 2$ and $K = 4$. If we impose normalization then we have only three degrees of freedom. The pure states can be transformed into one another by a continuous group. It can be shown that, with a suitable choice of fiducial measurements, these pure states correspond to points on a sphere. This is the Bloch sphere of quantum theory and, hence, for $N = 2$ we obtain the space of states of quantum theory.

Stage 4 We can use the $N = 2$ case and the subspace axiom to construct the general N case simply by demanding that every two-dimensional subspace has the properties of the $N = 2$ case.

Stage 5 We show that the most general measurements consistent with these axioms are those of quantum theory (so-called positive operator-valued measures, or POVMs (Krauss 1983; Nielsen and Chuang 2000)).

Stage 6 We show that composite systems can be described by taking the tensor product of the Hilbert spaces of the subsystems.

Stage 7 We show that the most general evolution consistent with the axioms are those of quantum theory (so-called completely positive linear maps).

Stage 8 We show that the most general update rule for the state after a measurement is that of quantum theory (corresponding to the Krauss operator formalism (Krauss 1983; Nielsen and Chuang 2000) which has von Neumann projection as a special case.

Continuity versus discreteness

The classical theory considered here has a countable set of distinguishable states. However, most classical theories actually have a continuum of distinguishable states, for example, a particle moving along a line or an electromagnetic field. Hence, one objection to the approach taken in this paper might be that rather than going to quantum theory, we could fix the problems with the discrete classical probability theory considered here by simply embedding it in a continuous classical probability theory. In this case, when a ball goes from being in one box to another, it does so in a continuous way simply by taking a continuous trajectory between the two boxes. And, indeed, this is what happens during a bit flip in a classical computer (however, this violates the continuity axiom above since it is not true, even with this embedding, that the state $\mathbf{p}$ can evolve in a continuous way for any N). That is to say, we usually employ some coarse graining in describing a classical system as a discrete system. However, we may want to ask what happens at the fundamental level. There are two possibilities within the classical framework. Either we continue to impose that there exists a continuous theory all the way down or, at some level, we hit a discrete domain. The usual approach in classical theories is the former. However, it has certain disadvantages. It implies that there can exist structure on any scale no matter how microscopic. It implies that a classical computer performing a finite calculation uses infinite memory resources (since one can store an infinite number of bits in a continuous variable) which might be regarded as very wasteful. The reason the classical mind imposed this continuity was so that systems could evolve in a continuous way. However, apart from this consideration, it is very natural to suppose that, at some level, we hit a discrete domain. If we still wish to have continuous evolution then it is clear that classical theories are inadequate. By imposing continuity on this discrete classical picture we obtain quantum theory. Quantum theory offers us a way to have the advantages of discreteness and continuity at the same time. We can have our cake and eat it.

What's it all about?

In this section I want to go beyond the technical details and ask more interpretational questions. The comments in this section represent my own take on the axioms and should not be regarded as the only way of understanding their consequences.

The axioms listed above are fairly simple to state. If we are to take these axioms as our starting point then we should consider what kind of philosophical position they correspond to. What picture of the world do they most naturally fit in with? The most striking aspect of this approach is that the preparation/transformation/measurement picture is taken as fundamental. The axioms refer to this macroscopic level of description of the world rather than a microscopic level of description. This can perhaps best be visualized by making some modifications to one of Wheeler's diagrams. Wheeler used the image of a U (representing the universe) with an eye as shown in Fig. 3.3a. An arrow is drawn between to illustrate that the observer can look at the universe. The observer is also taken to be part of the universe this being represented by the fact that the eye is part of the U. However, this picture

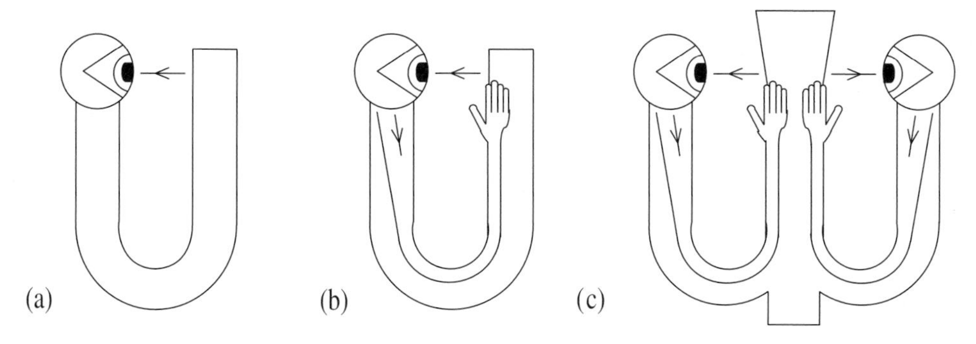

Figure 3.3. (a) Shows an eye observing the universe like Wheeler's original diagram, (b) shows an observer/actor both observing the universe and acting on it with his hands, and (c) shows two observer/actors in the shape of a Ψ.

only brings out the measurement aspect. The notion of humans as observers is deeply ingrained in physics and is culturally related to the habit of presenting science as an objective yet passive way of looking at the universe. In fact, humans (and, indeed, other animals, robots, ...) are observer/actors. They are capable of changing things in the world by reaching out with their hands (or feet, ...). This can be made clear by modifying Wheeler's picture as shown in Fig. 3.3b. Here a hand is shown reaching out to change things in the world. The hand is shown as part of the universe since it is itself made out of the stuff of the universe. This diagram now illustrates humans as observer/actors capable of implementing preparations/transformations/measurements. However, even this diagram is not really sufficient. Physics is the activity of a number of people. Much of our picture of the world comes from interacting with other people. I am not making the point that physics is a societal construct but rather the deeper point that physics is constructed in the context of a world which has many actor/observers. This is illustrated in Fig. 3.3c which shows two observer/actors interacting with the universe. Further, we coincidentally obtain the pleasing visual pun that this diagram resembles Ψ, the preferred Greek letter for representing the state of a quantum system.

The arrows represent classical information going in two directions. First, when we make an observation we "see" something and we can label these different perceptions with words ("I perceive red") which can be converted into bit strings (hence classical information). Likewise, when we form an intention to do something we can reach out and implement this intention. The set of possible intentions can also be labeled by words and hence constitutes classical information. There is the interesting philosophical question of whether we come to know of our actions because we form an intention and implement it or whether we simply subsequently observe the action – for example, is it the case that I intended to pick up the cup or is it the case that I simply subsequently noticed that I picked up the cup? This diagram asserts the former since otherwise there would be no outward-pointing arrow. It seems that this notion of our having kinematical freedom is quite important. Physical equations represent a set of counterfactual possibilities and it is up to the experimentalist to decide which one to implement. When we say that a system is in a particular state we mean this with respect to a set of counterfactual possible states the system could be in. That is to say the notion of state only makes sense with respect to the idea that the same system *could* be in some other state. The motivation for the continuity axiom to be given below is in terms of the hand causing a system to evolve from one state to another and makes most sense when we think in terms of our having the kinematical freedom to implement these transformations.

This point of view runs into some difficulties in astronomy. There we can choose which star to look at but we cannot reach out and change its properties. In this case we cannot set the initial conditions for a particular system but we can look around and hope that we will find one star with the initial conditions we want to test. However, we can argue, first, that choosing where to look constitutes an action on the world and, second, that we could, in principle, change the properties of stars.

It is instructive to identify how the key axioms are motivated by this diagram (Fig. 3.3c). First, the diagram lends itself to an instrumentalist approach by making the interaction of the human with the rest of the world the fundamental aspect of the viewpoint and this is very much in the spirit of the axioms and leads naturally to thinking in terms of probabilities as in Axiom 1. Axiom 2 is best motivated from an informational point of view – that systems of a given N should have the same informational properties. Information is a difficult concept to pin down but the arrows from the world to the eye and from the human to the hands capture something of what information is about. By elevating these arrows to a fundamental role we make information central to physics and then something like Axiom 2 is reasonable. Axiom 3 takes the idea that some systems can be regarded as composite as fundamental. This relates in Fig. 3.3c to the fact that there are two observers/actors each potentially interacting with their own bit of the world (of course, we do not need two observer/actors in order to have two subsystems – a single observer can interact with two separate subsystems).

Axiom 4 is the crucial axiom and so I want to be clear about how it can be motivated from the diagram. Consider one of the observer/actors interacting with a small system. By "small" we mean a system with small N. The observer/actor can use his hand to transform the state of this small system. Imagine that he transforms the state from one pure state to another. In doing this he will move his hand from one position to another. Now, his hand is a big system meaning that it has a large number of distinguishable states (large N). Hence, in moving his hand from one position to another and taking the state from one pure state to another pure state, his hand will go through many intermediate positions. In fact, since his hand is big, these many intermediate positions will approximate a continuum. At each of these intermediate positions the system must be in some state. If no information is being extracted about the state then we expect the transformation to be reversible. Hence, we expect that there will exist a continuous reversible transformation between these two pure states (thus taking the state along a continuous trajectory through the pure states) as stated in Axiom 4.

Axiom 5 is purely a simplicity assumption and, as such, does not have any obvious relation with the diagram.

Taking the picture in Fig. 3.3c as fundamental represents, potentially, a rather radical modification of the standard purely reductionist picture of physics. The reductionist approach entails positing some fundamental laws which apply at the microscopic level from which the macroscopic world can be explained. The present point of view does not contradict this bottom–up picture. Rather it asserts that this is not the full story. In order for this bottom–up picture to be complete we need to say where the microscopic laws come from. Why do atoms obey the laws of quantum theory? Without an answer to this question the bottom–up picture is not a complete explanation of why the world behaves as it does. But, at least so far as the work presented in this paper is concerned, to explain where these quantum laws come from we need a top–down explanation. That is, we need to take the picture represented in Fig. 3.3c which includes macroscopic objects like hands as fundamental. In this way we arrive at a friendly circle of explanation. The top–down aspect of the explanation accounts for why the laws of physics take the particular form they do and, in turn, the bottom–up aspect of the explanation accounts for the existence of macroscopic objects such as hands in terms of smaller systems. If either part of the story is missing then the explanation is incomplete. The picture in Fig. 3.3c effectively represents the human condition so far as our relationship with the physical world is concerned (it makes no reference though to pain and suffering and other Dostoevsky-type themes). That the nature of the human condition should apparently play so fundamental a role in determining the nature of the laws of physics is both surprising and perhaps a little disturbing. It is worth examining alternative points of view. One such counterview is that ultimately there exists an explanation of the world which is purely reductionist in spirit, providing a purely bottom–up explanation of why the world is the way it is. Since we are so firmly embedded in this world, one could assert that all our intuitions derive from our experience of the world and, as such, are not fundamental in themselves. From this point of view apparently reasonable axioms like those in this chapter might not be regarded as fundamental. The problem with this approach is that it cannot easily account for the power of the top–down approach taken in this paper. Nevertheless, both points of view are useful and have played a role in directing physics.

Going beyond quantum theory

We might expect that, like all physical theories before it, quantum theory will turn out to be empirically inadequate at some level. Indeed, the difficulty of incorporating general relativity and quantum theory into a single theory already suggests this. If quantum theory is wrong then how might we find a new theory which goes beyond it? These axioms suggest a number of approaches.

The most obvious is to consider different functions $K(N)$ such as $K = N^3$. If it is the case that quantum theory can be regarded as a restriction of such a higher-order theory then we might expect that nature really corresponds to such a higher order theory. I have tried – so far unsuccessfully – to construct such theories. Certain technical difficulties appear once one goes beyond $K = N^2$. It might be that such theories cannot actually be constructed.

Another way of modifying the axioms would be to relax the subspace axiom to allow the existence of different types of system with different functions $K(N)$ and different laws of composition for composite systems. The study of more general situations of this type may involve some interesting number theory.

Yet another approach is to drop the continuity axiom. The motivation for this axiom was in the context of a big system (large N) interacting with a small system (small N). However, there may be situations in which this is reversed. Further, the state is represented by a list of probabilities. Ultimately, we actually measure relative frequencies without taking the limit of an infinite number of runs and so such empirical probabilities are rational numbers. Hence, the empirical state cannot evolve in a continuous way. We might expect that the evolution should be slightly "jumpy." A small change of this nature to Axiom 4 might lead to some second- or higher-order modifications of quantum theory which could be detected in a sufficiently sophisticated experiment.

Developing this theme a little further, it is interesting to point out that all the empirical data we have to date and are ever likely to have are finite in extent. We can represent the data with integers – we do not need the real numbers. This suggests a move away from using continuous parameters in physical theories. We might hope that many of the fundamental proofs in physics, perhaps all of them, ultimately boil down to number theory (like the derivation of $K = N^r$ above).

Conclusions

We have given an answer to Wheeler's question "How come the quantum?" In so doing we have identified a weakness in classical probability theory. Namely that, for systems having a countable number of distinguishable states, we must have jumps as we evolve from one pure state to another. The reason for quantum theory is that we need it to get rid of those "damned classical jumps!" It is quite remarkable and perhaps a little ironic that, one word – the word "continuous" – marks out the difference between the classical and the quantum.

It is surprising that simply adding this continuity requirement to a list of principles otherwise consistent with classical probability theory gives us the full structure of quantum theory for discrete Hilbert spaces. Hence, by demanding this continuity property, we can account for the counterintuitive properties of quantum theory in a similar way to Einstein when he accounted for the counterintuitive properties of the Lorentz transformations with his light postulate. It is this continuity property alone which forces the difference between classical probability theory and quantum theory.

Acknowledgments

I am very grateful to numerous people for discussions. In particular, I would like to thank Chris Fuchs for getting me interested in the Wheeler-type questions addressed in this paper. This work was funded by a Royal Society University Research Fellowship held at Oxford University.

References

Accardi, L (1995) *Il Nuovo Cimento* **110B**, 685.
Birkhoff, G and von Neumann, J (1936) *Ann. Math.* **37**, 743.
Busch, P, Grabowski, M, and Lahti, PJ (1995) *Operational Quantum Physics*. Berlin: Springer-Verlag.
Coecke, B, Moore, D, and Wilce, A (2000) *Current Research in Operational Quantum Logic: Algebras, Categories, Languages*. quant-ph/0008019.
Fivel, DI (1994) *Phys. Rev. A* **50**, 2108.
Fuchs, CA (2002) *Quantum Mechanics as Quantum Information (and Only a Little More)*. quant-ph/0205039.
Gillies, D (2000) *Philosophical Theories of Probability*. London: Routledge.
Hardy, L (2001) *Quantum Theory from Five Reasonable Axioms*. quant-ph/0101012.
(2002) Why Quantum theory? In *Proceedings of the NATO Advanced Research workshop on Modality, Probability, and Bell's theorem*, ed. J. Butterfield and T. Placek, pp. 61–73. Dordrecht: Kluwer. (also available as quant-ph/0111068)
Jauch, JM and Piron, C (1963) *Helv. Phys. Acta* **36**, 837.
Krauss, K (1983) *States, Effects, and Operations: Fundamental Notions of Quantum Theory*. Berlin: Springer-Verlag.
Lande, A (1974) *Am. J. Phys.* **42**, 459.
Landsman, NP (1998) *Mathematical Topics between Classical and Quantum Mechanics*. New York: Springer-Verlag.
Ludwig, G (1968) *Commun. Math. Phys.* **9**, 1.
(1983) *Foundations of Quantum Mechanics*, vols. 1 and 2. New York: Springer-Verlag.
Mackey, GW (1963) *The Mathematical Foundations of Quantum Mechanics*. New York: W. A. Benjamin.
Mielnik, B (1968) *Commun. Math. Phys.* **9**, 55.
Nielsen, MA and Chuang, IL (2000) *Quantum Information and Quantum Information*. New York: Cambridge University Press.
Piron, C (1964) *Helv. Phys. Acta* **37**, 439.
Prugovecki, E (1977) *Int. J. Theor. Phys.* **16**, 321.
Weigert, S (2000) *Phys. Rev. Lett.* **84**, 802.

Wheeler, JA (1998) (with Kenneth Ford) *Geons, Black Holes, and Quantum Foam: A Life in Physics*. New York: W. W. Norton.
Wootters, WK (1986) *Found. Phys.* **16**, 319.
(1990) Local accessibility of quantum states. In *Complexity, Entropy and the Physics of Information*, ed. W. H. Zurek, pp. 39–46. New York: Addison-Wesley.
Zeilinger, A (1999) *Found. Phys.* **29**, 631.

4

Thought-experiments in honor of John Archibald Wheeler

Freeman J. Dyson

Institute for Advanced Study, Princeton

Beyond the black hole

In 1979 we held a symposium at the Institute for Advanced Study to celebrate the hundredth birthday of Albert Einstein. Unfortunately Einstein could not be there, but John Wheeler made up for Einstein's absence. Wheeler gave a marvelous talk with the title "Beyond the black hole," sketching with poetic prose and Wheelerian pictures his grand design for the future of science. Wheeler's philosophy of science is much more truly relativistic than Einstein's. Wheeler would make all physical law dependent on the participation of observers. He has us creating physical laws by our existence. This is a radical departure from the objective reality in which Einstein believed so firmly. Einstein thought of nature and nature's laws as transcendent, standing altogether above and beyond us, infinitely higher than human machinery and human will.

One of the questions that has always puzzled me is this. Why was Einstein so little interested in black holes? To physicists of my age and younger, black holes are the most exciting consequence of general relativity. With this judgment the man-in-the-street and the television commentators and journalists agree. How could Einstein have been so indifferent to the promise of his brightest brainchild? I suspect that the reason may have been that Einstein had some inkling of the road along which John Wheeler was traveling, a road profoundly alien to Einstein's philosophical preconceptions. Black holes make the laws of nature contingent on the mechanical accident of stellar collapse. John Wheeler embraces black holes because they show most sharply the contingent and provisory character of physical law. Perhaps Einstein rejected them for the same reason.

Science and Ultimate Reality, eds. J. D. Barrow, P. C. W. Davies and C. L. Harper, Jr. Published by Cambridge University Press.

Let me quote a few sentences from Wheeler's Varenna lectures, published in 1979 with the title *Frontiers of Time* (Wheeler 1979). His talk at the Einstein symposium was a condensed version of these lectures. Here is Wheeler:

> Law without law. It is difficult to see what else than that can be the plan of physics. It is preposterous to think of the laws of physics as installed by a Swiss watchmaker to endure from everlasting to everlasting when we know that the universe began with a big bang. The laws must have come into being. Therefore they could not have been always a hundred per cent accurate. That means that they are derivative, not primary . . . Events beyond law. Events so numerous and so uncoordinated that, flaunting their freedom from formula, they yet fabricate firm form . . . The universe is a self-excited circuit. As it expands, cools and develops, it gives rise to observer-participancy. Observer-participancy in turn gives what we call tangible reality to the universe . . . Of all strange features of the universe, none are stranger than these: time is transcended, laws are mutable, and observer-participancy matters.

Wheeler unified two streams of thought which had before been separate. On the one hand, in the domain of cosmology, the anthropic principle of Bob Dicke and Brandon Carter constrains the structure of the universe. On the other hand, in the domain of quantum physics, atomic systems cannot be described independently of the experimental apparatus by which they are observed. The Einstein–Podolsky–Rosen paradox showed once and for all that it is not possible in quantum mechanics to give an objective meaning to the state of a particle, independent of the state of other particles with which it may be entangled. Wheeler has made an interpolation over the enormous gap between the domains of cosmology and atomic physics. He conjectures that the role of the observer is crucial to the laws of physics, not only at the two extremes where it has hitherto been noticeable, but over the whole range. He conjectures that the requirement of observability will ultimately be sufficient to determine the laws completely. He may be right, but it will take a little while for particle physicists and astronomers and string-theorists to fill in the details in his grand picture of the cosmos.

There are two kinds of science, known to historians as Baconian and Cartesian. Baconian science is interested in details, Cartesian science is interested in ideas. Bacon said, "All depends on keeping the eye steadily fixed on the facts of nature, and so receiving their images as they are. For God forbid that we should give out a dream of our own imagination for a pattern of the world." Descartes said, "I showed what the laws of nature were, and without basing my arguments on any principle other than the infinite perfections of God, I tried to demonstrate all those laws about which we could have any doubt, and to show that they are such that, even if God created many worlds, there could not be any in which they failed to be observed." Modern science leapt ahead in the seventeenth century as a result of fruitful competition between Baconian and Cartesian viewpoints. The relation between Baconian science and Cartesian science is complementary, where I use the word complementary as Niels Bohr used it. Both viewpoints are true, and both are necessary, but they cannot be seen simultaneously. We need Baconian scientists to explore the universe and find out what is there to be explained. We need Cartesian scientists to explain and unify what we have found. Wheeler, as you can tell from the passage that I quoted, is a Cartesian. I am a Baconian. I admire the majestic Cartesian style of his thinking, but I cannot share it. I cannot think the way he thinks. I cannot debate with Wheeler the big questions that he is raising, whether science is based on logic or on circumstances, whether the laws of nature are necessary or contingent. In this chapter I do not try to answer the big questions. I write about details, about particles traveling through detectors, about clocks in boxes, about black holes evaporating, about the concrete objects that are the subject matter of Baconian science. Only intermittently, in honor of John Wheeler, I interrupt the discussion of details with a few Cartesian remarks about ideas.

The subject of this chapter is a set of four thought-experiments that are intended to set limits to the scope of quantum mechanics. Each of the experiments explores a situation where the hypothesis, that quantum mechanics can describe everything that happens, leads to an absurdity. The conclusion that I draw from these examples is that quantum mechanics cannot be a complete description of nature. This conclusion is, of course, controversial. I do not expect everyone, or even a majority, to agree with me. The purpose of writing about a controversial subject is not to compel agreement but to provoke discussion. Being myself a Baconian, I am more interested in the details of the thought-experiments than in the philosophical inferences that may be drawn from them. The details are as solid as the classical apparatus with which the experiments are done. The philosophy, like quantum mechanics, is always a little fuzzy.

I have observed in teaching quantum mechanics, and also in learning it, that students go through an experience that divides itself into three distinct stages. The students begin by learning the tricks of the trade. They learn how to make calculations in quantum mechanics and get the right answers, how to calculate the scattering of neutrons by protons or the spectrum of a rigidly rotating molecule. To learn the mathematics of the subject and to learn how to use it takes about six months. This is the first stage in learning quantum mechanics, and it is comparatively painless. The second stage comes when the students begin to worry because they do not understand what they have been doing. They worry because they have no clear physical picture in their heads. They get confused in trying to arrive at a physical explanation for each of the mathematical tricks they have been taught. They work very hard and get discouraged because they do not seem to be able to think clearly. This second stage often lasts six months or longer. It is strenuous and unpleasant. Then, unexpectedly, the third stage begins. The students suddenly say to themselves, "I understand quantum mechanics," or rather they say, "I understand now that there isn't anything there to be understood." The difficulties which seemed so formidable have vanished. What has happened is that they have learned to think directly and unconsciously in quantum-mechanical language.

The duration and severity of the second stage are decreasing as the years go by. Each new generation of students learns quantum mechanics more easily than their teachers learned it. There is less resistance to be broken down before the students feel at home with quantum ideas. Ultimately the second stage will disappear entirely. Quantum mechanics will be accepted by students from the beginning as a simple and natural way of thinking, because we shall all have grown used to it. I believe the process of getting used to quantum mechanics will become quicker and easier, if the students are aware that quantum mechanics has limited scope. Much of the difficulty of the second stage resulted from misguided attempts to find quantum-mechanical descriptions of situations to which quantum mechanics does not apply.

Unfortunately, while the students have been growing wiser, some of the older physicists of my generation have been growing more foolish. Some of us have regressed mentally to the second stage, the stage which should only be a disease of adolescence. We tend then to get stuck in the second stage. If you are a real adolescent, you may spend six months floundering in the second stage, but then you grow up fast and move on to the third stage. If you are an old-timer returning to your adolescence, you don't grow up any more. Meanwhile, we may hope that the students of today are moving ahead to the fourth stage, which is the new world of ideas explored by John Wheeler. In the fourth stage, you are at home in the quantum world, and you are also at home in the brave new world of black holes and mutable laws that Wheeler has imagined.

Complementarity and decoherence

Roughly speaking, there are two schools of thought about the meaning of quantum mechanics. I call them broad and strict. I use the words in the same way they are used in American constitutional law, where the broad interpretation says the constitution means whatever you want it to mean, while the strict interpretation says the constitution means exactly what it says and no more. The broad school says that quantum mechanics applies to all physical processes equally, while the strict school says that quantum mechanics covers only a small part of physics, namely the part dealing with events on a local or limited scale. Speaking again roughly, one may say that the extreme exponent of the broad view is Stephen Hawking, who is trying to create a theory of quantum cosmology with a single wave function for the whole universe. If a wave function for the whole universe makes sense, then any restriction on the scope of quantum mechanics must be nonsense. The historic exponent of the strict view of quantum mechanics was Niels Bohr, who maintained that quantum mechanics can only describe processes occurring within a larger framework that must be defined classically. According to Bohr, a wave function can only exist for a piece of the world that is isolated in space and time from the rest of the world. In Bohr's view, the notion of a wave function for the whole universe is an extreme form of nonsense.

As often happens in the history of religions or philosophies, the disciples of the founder established a code of orthodox doctrine that is more dogmatic and elaborate than the founder intended. Bohr's pragmatic view of quantum mechanics was elaborated by his disciples into a rigid doctrine, the so-called "Copenhagen interpretation." When I use the word strict to describe Bohr's view, I have in mind the orthodox Copenhagen dogma rather than Bohr himself. If you read what Bohr himself wrote, you find that he is much more tentative and broad-minded than his disciples. Bohr's approach to science is based on the principle of complementarity, which says that nature is too subtle to be described adequately by any single viewpoint. To obtain a true description of nature you have to use several alternative viewpoints that cannot be seen simultaneously. The different viewpoints are complementary in the sense that they are all needed to tell a complete story, but they are mutually exclusive in the sense that you can only see them one at a time. In Bohr's view quantum mechanics and classical physics are complementary aspects of nature. You cannot describe what is going on in the world without using both. Quantum language deals with probabilities while classical language deals with facts. Our knowledge of the world consists of an inseparable mixture of probabilities and facts. So our description of the world must be an inseparable mixture of quantum and classical pictures.

Against this dualistic philosophy of Bohr, putting strict limits to the scope of quantum descriptions, the quantum cosmologists take a hard line. They say the quantum picture must include everything and explain everything. In particular, the classical picture must be built out of the quantum picture by a process which they call decoherence. Decoherence is the large-scale elimination of the wave-interference effects that are seen in quantum systems but not in classical systems. For the benefit of any literary scholars who may be among my readers, decoherence is to science as deconstruction is to literature, a fashionable buzzword that is used by different people to mean different things. I quote a few sentences from a classic article by Bryce DeWitt (1992), explaining decoherence, from the point of view of the quantum cosmologists, with unusual clarity:

In the old Copenhagen days one seldom worried about decoherence. The classical realm existed a priori and was needed as a basis for making sense of quantum mechanics. With the emergence of quantum cosmology, it became important to understand how the classical realm emerges from quantum mechanics itself . . . The formalism is able to generate its own interpretation.

After some simple mathematics describing a particular quantum system that first decoheres and thereafter exhibits classical behavior, DeWitt goes on:

The above results have the following implications for decoherence in quantum cosmology: (1) Although complexity (metastability, chaos, thermal baths, wave packets) can only help in driving massive bodies to localized states, it is massiveness, not complexity, that is the key to decoherence. (2) Given the fact that the elementary particles of matter tend, upon cooling, to form stable bound states consisting of massive agglomerations, decoherence at the classical level is a natural phenomenon of the quantum cosmos. (3) Given the fact that the interaction described here is a simple scattering interaction and not at all specially designed like a . . . measurement interaction, the universe is likely to display decoherence in almost all states that it may find itself in. (4) An arrow of time has no basic role to play in decoherence.

I have tried to give you a fair and balanced statement of the two points of view, Bohr on one side and DeWitt on the other. Personally, I find both of them entirely reasonable. As usual, when people are engaged in philosophical argument, what they do is more reasonable than what they say. DeWitt rejects with scorn what he calls "the old Copenhagen days," but his mathematical analysis of decoherence does not contradict the analysis of quantum processes made 60 years earlier by Bohr. From Bohr's point of view, decoherence is just another example of complementarity, showing in detail how quantum and classical descriptions give complementary pictures of events in the early universe. From DeWitt's point of view, complementarity is just a complicated way of talking about decoherence, giving a spurious importance to the classical description which is only an approximation to the true quantum universe. My Princeton colleague Stephen Adler has written a paper with the title "Why decoherence has not solved the measurement problem" (Adler 2003), which I recommend as a clear statement of what decoherence can and cannot do.

The first of my four thought-experiments is an old one, invented long ago by Schrödinger and not by me. To sharpen the issues between Bohr and DeWitt, I look again at the experiment known to experts as Schrödinger's cat. Schrödinger's cat is imprisoned in a cage with a bottle of hydrogen cyanide, arranged with a quantum-mechanical device connected to a hammer so that a single atom decides whether the hammer falls and breaks the bottle. The atom is prepared in a coherent state with equal probabilities for its spin to be up or down. If the spin is up, the hammer falls, and if the spin is down, the hammer stays still. The cat is then in a coherent state, with equal probabilities of being dead or alive. Two questions then arise. What does it mean to be in a coherent state of life and death? What does the cat think about the experiment?

From the point of view of Bohr the answers are simple. When you open the cage and examine the cat, or when the cat inside the cage examines itself to see whether it is alive, the experiment is over and the result can only be stated in classical terms. The coherent state lasts only as long as the examination of the cat is in the future. The cat cannot be aware of the coherent state, because as soon as the cat is aware the state is a matter of fact and not a matter of probability. From the point of view of DeWitt, the cat itself, just because it is a massive object with complicated interactions, achieves its own decoherence. It destroys the paradoxical coherence automatically, as a consequence of Schrödinger's equation.

I like to remain neutral in this philosophical debate. If I were forced to make a choice, I would choose to follow Bohr rather than DeWitt, because I find the idea of complementarity more illuminating than the idea of decoherence. Complementarity is a principle that has wide applications extending beyond physics into biology and anthropology and ethics, wherever problems exist that can be understood in depth only by going outside the limits of a single viewpoint or a single culture. Decoherence, so far as I know, has not yet been adopted by anthropologists as a slogan, although one might consider the loss of traditional family and tribal loyalties, when people migrate from farms and villages into city slums, to be the cultural equivalent of decoherence.

Two more thought-experiments

That is enough about philosophy. I now move on to technical issues which are to me more interesting than philosophy. The next two thought-experiments could be carried out with real apparatus if anybody found them worth the money and time that they would require. The results of the experiments are clear and simple. They seem to show that in some sense Bohr is right, that limits exist to the scope of quantum descriptions of events. This does not mean that DeWitt is necessarily wrong. According to Bohr, there are two kinds of truth, ordinary truth and deep truth. You can tell the difference between the two kinds of truth by looking at their opposites. The opposite of an ordinary truth is a falsehood, but the opposite of a deep truth is another deep truth. The essence of Bohr's idea of complementarity is that you need deep truths to describe nature correctly. So my thought-experiments do not prove that DeWitt is wrong, only that he is not telling us an ordinary truth. His picture of decoherence may be correct as far as it goes, but it can only be a deep truth, giving a partial view of the way nature works.

The second thought-experiment consists of two small particle counters separated by a distance D with empty space in between. An electron is fired through the first counter at time T_1 and hits the second counter at time T_2. The positions of the counters and the times of arrival of the electron are measured. First I give you a simple qualitative argument and then a more careful quantitative argument. The qualitative argument goes like this. Suppose the positions and times are known precisely. Then the velocity of the electron between the counters is also known precisely. If we assume that the mass of the electron is known precisely, the momentum is also known precisely. This contradicts the uncertainty principle, which says that the position and momentum of an electron cannot both be known precisely. The contradiction means that it is not legitimate to use a quantum description of the motion of the electron between the two counters.

To make the conclusion firmer, I now give you a quantitative argument, which takes account of the inevitable inaccuracy of the measurements. This argument is an exercise in elementary quantum mechanics, and the experiment is just an old-fashioned two-slit experiment with the slits arranged in series instead of in parallel. The time interval between the two measurements does not need to be measured accurately. We assume only that the time interval is known to be greater than T. Suppose that there are two parallel slits of width L, one placed at the exit from the first counter and the other at the entrance to the second counter. When an electron is counted in both counters, we know that it has passed through both slits. Let x be the coordinate of the electron perpendicular to the plane containing the slits. The positions of the slits are measured with sufficient accuracy, so that the uncertainty of x as the electron passes through either slit is less than L. Let p be the momentum of the electron conjugate to x. We consider the mean-square dispersions

$$D(t) = \langle |\Delta x|^2 \rangle, \quad K = \langle |\Delta p|^2 \rangle, \tag{4.1}$$

as a function of time t as the electron travels between the slits. Since the electron is traveling freely, K is independent of time. According to the virial theorem, which holds for a free electron in nonrelativistic quantum mechanics,

$$(d^2 D/dt^2) = (2/m^2)K, \tag{4.2}$$

where m is the electron mass. The right-hand side of eqn (4.2) is independent of time, so that

$$D(t) = D(t_0) + K(t - t_0)^2/m^2, \tag{4.3}$$

where t_0 is the time when D is smallest. But the Heisenberg uncertainty principle says

$$D(t_0)K \geq (1/4)\hbar^2, \tag{4.4}$$

which together with eqn (4.3) gives

$$D(t) \geq (\hbar/m)|t - t_0|. \tag{4.5}$$

The value of D at either counter is less than L^2, and the values of $|t - t_0|$ at the two counters add to at least T. Therefore eqn (4.5) implies

$$(2L^2/T) \geq (\hbar/m), \tag{4.6}$$

which must be valid if the electron is described by a wave function satisfying the Schrödinger equation. But it is easy to choose L and T so that eqn (4.6) is violated. Then the uncertainty principle (4.4) will also be violated, and a wave function describing the passage of the electron through the two slits cannot exist.

Let us put in some numbers to show that the violation of the uncertainty principle could be achieved with apparatus small enough to sit on a table-top. The numbers are not absurd. The right-hand side of eqn (4.6) is about 1 square centimeter per second. Without stretching the state of the art, we may take the width L of the slits to be 1 micron or 10^{-4} cm. Then eqn (4.6) will be violated for any time interval T longer than 20 nanoseconds. For an electron with energy 1 kV, a travel time of 20 nanoseconds corresponds to a travel distance of 20 centimeters between the two counters. We can easily imagine doing the experiment on a table-top with a travel distance longer than this. To make the purpose of the thought-experiment clear, I hasten to add that it does not prove quantum mechanics wrong. It only proves that quantum mechanics is wrongly applied to this particular situation.

Before discussing the meaning of this second thought-experiment, I go on to the third experiment. The third experiment uses the Einstein box (Fig. 4.1), a device invented by Einstein for the purpose of violating the uncertainty principle. Einstein wanted to use his box to prove that quantum mechanics was inconsistent, since he didn't believe that quantum mechanics was true. Einstein confronted Bohr with this box at a public meeting in 1930 (Bohr 1949). Bohr won the argument with a dramatic counterattack, pointing out that Einstein had forgotten to take into account his own theory of general relativity when he discussed the behavior of the box. When Bohr included the gravitational effects that follow from general relativity, it turned out that the uncertainty principle was not violated after all. So Einstein was defeated and the box became a victory trophy for Bohr.

The idea of the Einstein box was that you hang it from a spring balance and measure its mass by measuring its weight in a known gravitational field. You measure its weight by measuring the momentum p transferred to the balance by the spring in a given time T. The uncertainty in the mass is then

$$\Delta m = \Delta p/gT, \tag{4.7}$$

where g is the gravitational field. The box has a window with a shutter that can be opened and closed from the inside, and a clock that measures the times when the

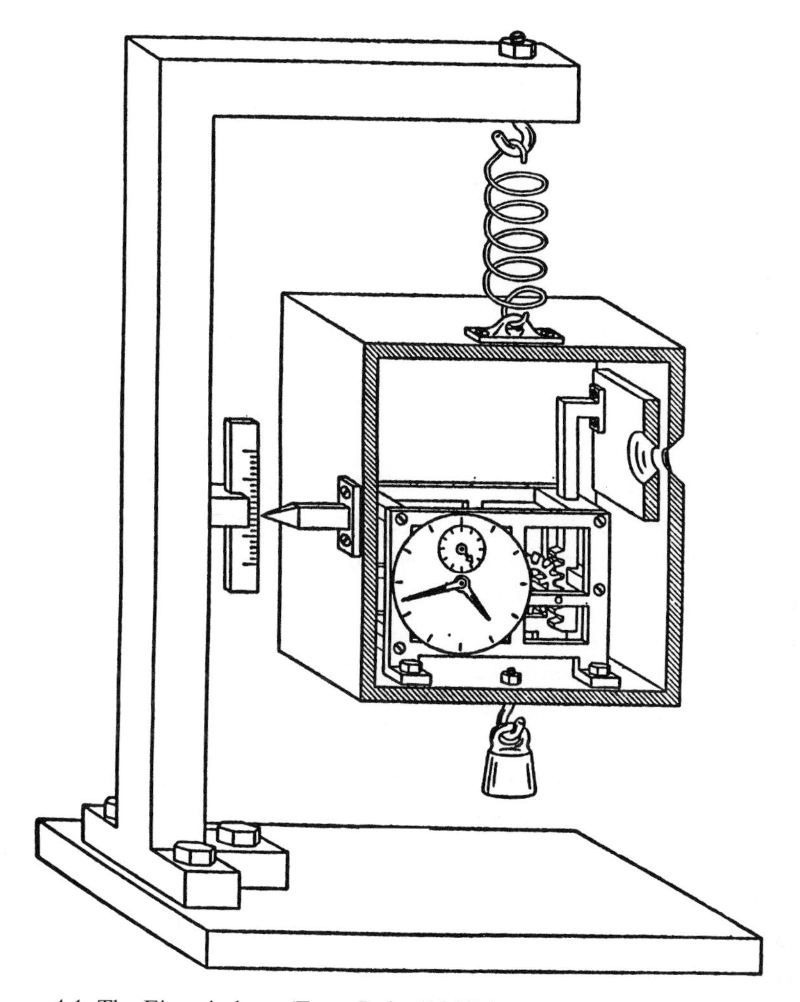

Figure 4.1. The Einstein box. (From Bohr (1949).)

shutter is opened and closed. It is important that the clock sits inside the box, so that the weighing is not disturbed by time signals coming into the box from the outside. At the time when the shutter is open, a photon leaves the box and carries away with it a mass proportional to its energy E. The weighing of the box before and after the emission determines the energy of the photon with an uncertainty

$$\Delta E = c^2 \Delta m = c^2 \Delta p / gT. \tag{4.8}$$

Einstein thought he could violate the uncertainty principle between energy and time,

$$\Delta E \cdot \Delta t \geq (1/2)\hbar, \tag{4.9}$$

because he thought he could set the internal clock to make the uncertainty Δt in the time of emission of the photon as small as he pleased. Bohr defeated this scheme by pointing out that the rate of the clock would be affected by the position of the box in the gravitational potential according to general relativity. If the uncertainty in the position of the box is Δx, then the resulting uncertainty in the clock-time during the weighing is

$$\Delta t = Tg\Delta x/c^2. \tag{4.10}$$

Putting together eqns (4.8) and (4.10), we see that the uncertainty relation (4.9) between time and energy for the photon follows immediately from the ordinary uncertainty relation between position and momentum for the box. Point, set, and match to Bohr.

My third thought-experiment is nothing more than a repetition of the Einstein box experiment with one measurement added. You arrange a photon detector with an accurate clock outside the box, and measure the time at which the photon arrives at the detector. The uncertainty in the arrival time is then independent of the movement of the box. The uncertainty in the emission time is determined by the uncertainty in the travel time of the photon. The travel time is uncertain, according to general relativity again, because the route of the photon in the gravitational potential is uncertain. However, we can arrange an optical system with f-number f that will focus the photon onto a fixed point at a distance

$$l = f\Delta x \tag{4.11}$$

from the window of the box, no matter where the box happens to be at the moment of emission. The travel time of the photon from the window to the focus will be (l/c), with an uncertainty introduced by the gravitational potential as before. The travel of the photon from the focus to the detector is along a known path and introduces no additional uncertainty. The travel time uncertainty is then

$$\Delta t = (l/c)(g\Delta x/c^2) = (fg/c^3)(\Delta x)^2, \tag{4.12}$$

which with eqn (4.8) gives

$$\Delta E \cdot \Delta t = (f/cT)(\Delta x)^2 \cdot \Delta p. \tag{4.13}$$

We can now choose the ratio $(\Delta x/cT)$ to be as small as we like, so that the uncertainty relation (4.9) will be violated for the photon even when the usual relation between position and momentum is valid for the box. In this way we achieve the violation that Einstein intended when he introduced his box. In this third experiment, just like the second, the violation is easily achieved with apparatus of desktop size. The optical system that focuses the photon can be a simple telescope with a length of a few centimeters. After 70 years, Einstein is finally vindicated.

Bohr would not have been disturbed for a moment by these two thought-experiments. Both of them only violate the uncertainty principle by violating the rules that Bohr laid down for a legitimate use of quantum mechanics. Bohr's rules say that a quantum-mechanical description of an object can only be used to predict the probabilities that the object will behave in specified ways when it has been prepared in a specified quantum state. The quantum-mechanical description cannot be used to say what actually happened after the experiment is finished. The two thought-experiments merely confirm that this restriction of the use of quantum mechanics is necessary. If, in the second experiment, it were possible to define a wave function for the electron traveling between the two counters at the observed times, this wave-function could be proved to satisfy the uncertainty principle by the usual mathematical argument. But we saw that the uncertainty principle was violated, and therefore no such wave function can exist. Similarly, in the third thought-experiment, there can be no wave function describing the travel of the photon from the box to the detector. A wave function can only say that a photon has a certain probability of arriving, not that a photon has arrived. Although Bohr would say that the two experiments only confirm the correctness of his strict interpretation of quantum mechanics, Einstein might also claim that they justify his distrust. They prove in a simple and convincing fashion the contention of Einstein that quantum mechanics is not a complete description of nature. Perhaps Einstein would be happy to learn that his box is still alive and well after 70 years, and still making trouble for believers in quantum mechanics.

Let me now summarize the results of these two thought-experiments. They lead to two conclusions. First, statements about the past cannot in general be made in quantum-mechanical language. For example, we can describe a uranium nucleus by a wave function including an outgoing alpha particle wave which determines the probability that the nucleus will decay tomorrow. But we cannot describe by means of a wave function the statement, "This nucleus decayed yesterday at 9 a.m. Greenwich time." As a general rule, knowledge about the past can only be expressed in classical terms. Lawrence Bragg, a shrewd observer of the birth of quantum mechanics, summed up the situation in a few words, "Everything in the future is a wave, everything in the past is a particle." Since a large fraction of science, including most of geology and astronomy as well as the whole of paleontology, is knowledge of the past, quantum mechanics must always remain a small part of science. The second conclusion is that the "role of the observer" in quantum mechanics is solely to make the distinction between past and future. Since quantum-mechanical statements can be made only about the future, all such statements require a precise definition of the boundary separating the future from the past. Every quantum-mechanical statement is relative, in the sense that it describes possible futures predicted from a particular past–future boundary. Only in a classical description can the universe be viewed as an absolute spacetime continuum without distinction between past and future. All quantum-mechanical descriptions are partial. They refer only to particular regions of spacetime, separated from other regions within which the description is classical.

These conclusions of mine contradict both the extreme DeWitt view and the extreme Copenhagen view of quantum mechanics. I contradict DeWitt when he says it makes sense to speak about a wave function for the universe as a whole. I contradict the orthodox Copenhagen view, that the role of the observer in quantum mechanics is to cause an abrupt "reduction of the wave-packet" so that the state of the system appears to jump discontinuously at the instant when it is observed. This picture of the observer interrupting the course of natural events is unnecessary and misleading. What really happens is that the quantum-mechanical description of an event ceases to be meaningful as the observer changes the point of reference from before the event to after it. We do not need a human observer to make quantum mechanics work. All we need is a point of reference, to separate past from future, to separate what has happened from what may happen, to separate facts from probabilities.

Black holes and quantum causality

Twenty-six years ago, Stephen Hawking published a remarkable paper with the title "Breakdown of predictability in gravitational collapse" (Hawking 1976). He had then recently discovered the phenomenon of Hawking radiation, which led him to the prediction that black holes should slowly evaporate and finally vanish into a puff of gamma-rays. Some years earlier, gamma-ray bursts had been detected by orbiting satellites, but we then had no clue concerning how and where the bursts originated. We only knew that they came from some violent process occurring at a great distance from the Earth. When Hawking made his prediction, we hoped at first that the bursts might be the final display of fireworks giving direct evidence of the evaporation of small black holes by the Hawking process. To agree with the observed brightness of the bursts, the evaporating black holes would have to be at distances of the order of 1 light-year, far beyond the planets but still loosely attached to the gravitational field of the Sun. Such a population of small black holes in our neighborhood would have been a wonderful laboratory for studying black hole physics. Unfortunately, the Hawking process would give a final burst of gamma-rays of much higher energy than the observed bursts. Most of the bursts are certainly not produced by the Hawking process. We now know that the bursts are events of stupendous violence occurring in remote galaxies at cosmological distances. The nature of the sources is one of the outstanding problems of astronomy.

I return now to Hawking's 1976 paper. Hawking asked three important questions about the process of black-hole evaporation. None of his questions is yet definitively answered. First, when the horizon around the black hole disappears at the end of the evaporation process, does a naked spacetime singularity exist at the point of disappearance, and is the singularity exposed to view from the outside? Second, when a real star composed of ordinary matter collapses into a black hole that later evaporates, what happens to the law of conservation of baryons? Third, is the process of black-hole evaporation consistent with quantum-mechanical causality? Each of these three questions has given rise to important progress in our understanding of the universe. The first question led to improved understanding of the geometrical structure of horizons. The second question led to a gradual abandonment of the belief that baryon conservation could be an exact law of nature. The third question placed a new limit on the validity of quantum mechanics. Since the validity of quantum mechanics is the subject of this chapter, I am mainly concerned with the third question, but I will have something to say about all three.

Following the good example of Bohr and Einstein, Hawking clarified his questions by means of a thought-experiment. This is the last of the four experiments that I am discussing. The experiment is a simple one, although it requires a rather large laboratory and a long-lived experimenter. The experimenter prepares a massive object in a pure quantum state at zero temperature, and keeps it isolated from all contact with the rest of the universe. The Schrödinger equation, being a linear equation, predicts that the object will remain in a pure state so long as it is not disturbed from the outside. The object is assumed to be so massive that its internal pressure is insufficient to keep it from collapsing into a black hole. After the collapse, the black hole slowly evaporates into thermal radiation as predicted by Hawking's theory. Then the thermal radiation should also be in a pure quantum state. But this is a contradiction in terms. Thermal radiation is in a state of maximum entropy, as far removed as possible from a pure state. Therefore, Hawking concludes, the Schrödinger equation cannot be correct. At some stage in the process of collapse and evaporation, the Schrödinger equation must fail. The failure of the Schrödinger equation is what Hawking means by the phrase "violation of quantum causality." This was a highly unwelcome conclusion for Hawking, who believed that quantum mechanics should be universally valid. With his habitual honesty and open-mindedness, he did not conceal his discomfort but published his disagreable conclusion for all of us to ponder.

What do we learn from Hawking's thought-experiment? It throws light on all three of his questions. First, the question of naked singularities. There exists a tentative model of the evaporation of a black hole which avoids the appearance of a naked singularity. The model consists of a sequence of black hole configurations (see Fig. 4.2), described by John Wheeler nearly 40 years ago (Harrison *et al.* 1965). Wheeler introduced these configurations as a model of rapid gravitational collapse, but they work equally well as a model of slow evaporation. The configurations are described by an exterior Schwarzschild metric with mass M joined onto an interior

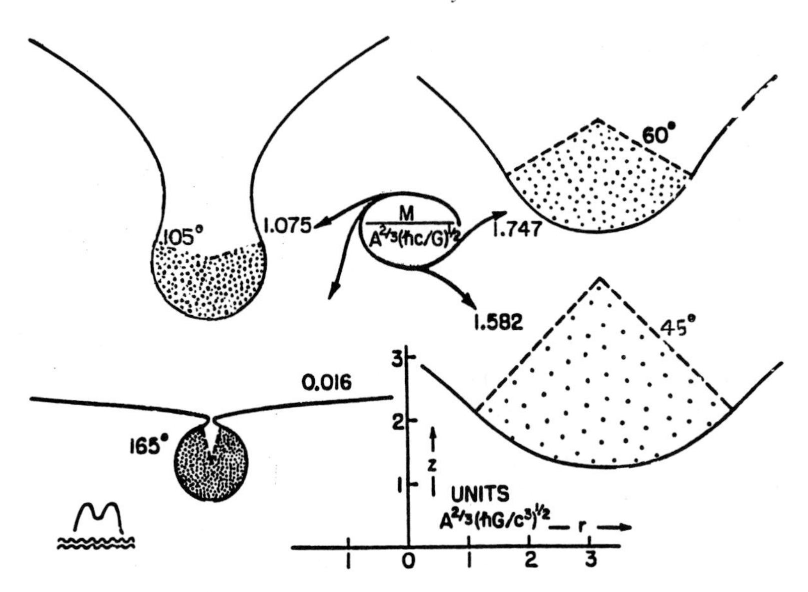

Figure 4.2. The Wheeler model of gravitational collapse, borrowed here as a model of black-hole evaporation. (From Harrison *et al.* (1965).)

uniform-curvature metric with mass density ρ. The uniform-curvature part of the spacetime is a polar cap, a piece of a hypersphere containing all points within an angular distance x from one pole. The relations connecting M, ρ and x are

$$M = M_0(2/3)(\sin x)^3(x - \sin x \cos x)^{-1}, \tag{4.14}$$

$$\rho = \rho_0(x - \sin x \cos x)^2, \tag{4.15}$$

with the initial mass M_0 and the density coefficient ρ_0 constant. The model of the evaporation process has the angle x slowly increasing from 0 to π as evaporation proceeds. The interior metric describes a distribution of cold matter with zero pressure and with local conservation of mass. The total amount of matter in the interior does not vary with x. But M, the gravitational mass of the object as seen from the outside, decreases steadily to zero as x approaches π. When x is close to π, the Schwarzschild radius of the exterior metric becomes small and there is only a narrow neck connecting the interior spacetime with the world outside. At the instant when $x = \pi$, the neck is pinched off and evaporation of the black hole is complete. We are then left with two disconnected pieces of spacetime, a flat piece containing no black hole and only outgoing waves of radiation from the evaporation process, and a completely closed hypersphere containing the original matter of mass M_0 in splendid isolation. There is no naked singularity remaining. A discontinuity in the spacetime curvature only occurs momentarily at the instant of separation, when the interior metric floats away into nothingness. The discontinuity is actually hidden from any possibility of observation from the outside, because the radiation emitted immediately before the end of the evaporation forms an opaque fireball. The size and energy content of the fireball can be roughly calculated. The energy content turns out to be of the order of a few megatons of TNT, comparable to a large hydrogen bomb. The fireball thus provides a highly effective screen, enforcing the rule of "cosmic censorship" which forbids the observation of naked singularities.

Second, we come to the question of baryon conservation. In Wheeler's simple model of the evaporation process, the baryon number is everywhere locally conserved. The baryon-number density is proportional to the mass density ρ of the interior metric. At the end of the evaporation, the baryons contained in the black hole have disappeared from the part of spacetime that is connected with the outside universe. Whether they still "exist" is a question of words and not a question of physics. The question would only become physically meaningful if the detached hypersphere should for some reason reattach itself to our universe at some other point of spacetime. If this happened, we would observe it as a "white hole," and we would be able to verify that the baryons contained in it had been conserved. In the absence of white holes, the model says that baryons are conserved locally but may disappear globally.

The evaporation model is purely schematic and not based upon dynamical calculations. The Wheeler configurations are treated as if they were static or slowly varying. In reality, these configurations are subject to dynamical collapse with a rapid timescale. In order to justify the model as a quantitatively correct description of Hawking's thought-experiment, the evolution of the evaporating black hole should be calculated with a consistent dynamical treatment of the spacetime geometry and of the matter. Since I am not an expert in the solution of the Einstein equations for black-hole dynamics, I have not tried to make the model dynamically consistent.

Lastly I come to Hawking's third question, the question that is directly relevant to this chapter. What does his thought-experiment tell us about quantum causality? I claim that it gives strong support to my main thesis, that quantum-mechanical description can be consistently applied only to the future and not to the past. More precisely, I am saying that quantum-mechanical description is limited to parts of the universe that are confined within the future of a conceivable observer. Hawking's experiment illustrates this principle, because no global separation between past and future is possible in any region of spacetime containing a black hole. The observer who defines the separation between past and future has to make a choice, either to plunge into the black hole or to stay outside it. It is impossible to include the interior of a black hole in the past or the future of an observer outside. This impossibility implies as a corollary that no complete description of a black hole together with its outside environment can be made in quantum-mechanical language. So we confirm Hawking's conclusion that black holes are not subject to quantum causality as he defines it. The Schrödinger equation is not violated locally. The equation remains valid wherever it can be meaningfully stated. It breaks down in Hawking's experiment only because the notion of quantum-mechanical coherence between events occurring inside and outside a black hole has no physical meaning. The processes of formation and evaporation of a black hole provide us with a good model for DeWitt's concept of decoherence. That is my answer to Hawking's third question. His discovery of the laws of radiation from black holes does not imply a breakdown of quantum theory. The Hawking process is, on the contrary, entirely consistent with the known limitations of the classical and quantum-mechanical descriptions of nature.

Concluding remarks

I hope this chapter has left you as confused as it leaves me. Once, when Bohr was accused of confusing people with his convoluted sentences, he replied that one should not speak more clearly than one can think. This wise remark applies particularly to speaking about quantum theory. I am usually reluctant to engage in discussions about the meaning of quantum theory, because I find that the experts in this area have a tendency to speak with dogmatic certainty, each of them convinced that one particular solution to the problem has a unique claim to be the final truth. I have the impression that they are less wise than Bohr. They tend to speak more clearly than they think. Each of them presents to us one particular version of quantum theory as the definitive description of the way nature works. Their efforts do not convince me, because I am a working physicist. As a physicist, I am much more impressed by our ignorance than by our knowledge. During the last hundred years we have made tremendous progress in our understanding of nature, but there is no reason to fear that our progress is coming close to an end. During the twenty-first century we shall probably meet with as many rude surprises as we have met in the twentieth. The discovery of Hawking radiation was the most recent big surprise in theoretical physics, but there is no reason to expect that it will be the last. The structure of theoretical physics as a whole, and of quantum theory in particular, looks to me like a makeshift agglomeration of bits and pieces, not like a finished design. If the structure of science is still provisional, still growing and changing as the years go by, it makes no sense to impose on the structure a spurious philosophical coherence. That is why I am skeptical of all attempts to squeeze quantum theory into a clean and tidy philosophical doctrine. I prefer to leave you with the feeling that we still have a lot to learn.

One question that I have not discussed in this chapter is the existence of gravitons. Theoretical physicists have almost unanimously assumed that the gravitational field must be a quantum field, with associated particles called gravitons. The statement that gravitons exist can only have meaning if one can devise a thought-experiment to demonstrate their existence. I have searched in vain for such a thought-experiment. For example, if one tries to imagine an experiment to detect the emission of a single graviton in a high-energy particle collision, one needs a detector of such astronomical dimensions that it cannot be prevented from collapsing into a black hole. Detection of single gravitons appears always to be frustrated by the extraordinary weakness of the gravitational interaction. Feasible detectors can detect gravitational waves only when the source is massive and the waves are classical. If it turned out to be true that no conceivable thought-experiment could detect effects of quantum gravity, then quantum gravity would have no physical meaning. In that case, the gravitational field would be a purely classical field, and efforts to quantize it would be illusory. Particles with the properties of gravitons would not exist. This conclusion is a hypothesis to be tested, not a statement of fact. To decide whether it is true, a careful and complete analysis of possible thought-experiments is needed. I leave this task to the physicists of the twenty-first century.

References

Adler, S (2003) Why decoherence has not solved the measurement problem: a response to P. W. Anderson. *Stud. Hist. Philos. Modern Phys.*, **34**, 135–42.

Bohr, N (1949) Discussion with Einstein on epistemological problems in atomic physics. In *Albert Einstein, Philosopher–Scientist*, ed. P. A. Schilpp. Evanston, IL: Library of Living Philosophers.

DeWitt, B (1992) Decoherence without complexity and without an arrow of time. Talk given at the Workshop on Time Asymmetry, Mazagón, Huelva, Spain, September 1991. Austin, TX: University of Texas Center for Relativity.

Harrison, K, Thorne, K, Wakano, M, and Wheeler, JA (1965) *Gravitation Theory and Gravitational Collapse*. Chicago, IL: University of Chicago Press.

Hawking, S (1976) Breakdown of predictability in gravitational collapse. *Phys. Rev.* **D(3)14**, 2460.

Wheeler, J A (1979) *Frontiers of Time*. Amsterdam: North-Holland.

5

It from qubit

David Deutsch
University of Oxford

Introduction

Of John Wheeler's "Really Big Questions," the one on which the most progress has been made is "It from bit?" – does information play a significant role at the foundations of physics? It is perhaps less ambitious than some of the other questions, such as "How come existence?", because it does not necessarily require a metaphysical answer. And unlike, say, "Why the quantum?", it does not require the discovery of new laws of nature: there was room for hope that it might be answered through a better understanding of the laws as we currently know them, particularly those of quantum physics. And this is what has happened: the better understanding is the quantum theory of information and computation.

How might our conception of the quantum physical world have been different if "It from bit" had been a motivation from the outset? No one knows how to derive *it* (the nature of the physical world) from *bit* (the idea that information plays a significant role at the foundations of physics), and I shall argue that this will never be possible. But we can do the next best thing: we can start from the qubit.

Qubits

To a classical information theorist, a bit is an abstraction: a certain amount of information. To a programmer, a bit is a Boolean variable. To an engineer, a bit is a "flip-flop" – a piece of hardware that is stable in either of two physical states. And to a physicist? Quantum information theory differs in many ways from its classical predecessor. One reason is that quantum theory provides a new answer to the ancient dispute, dating back to the Stoics and the Epicureans and even earlier, about whether the world is discrete or continuous.

Science and Ultimate Reality, eds. J. D. Barrow, P. C. W. Davies and C. L. Harper, Jr. Published by Cambridge University Press.

Logic is discrete: it forbids any "middle" between true and false. Yet in classical physics, discrete information processing is a derivative and rather awkward concept. The fundamental classical observables vary continuously with time and, if they are fields, with space too, and they obey differential equations. When classical physicists spoke of discrete observable quantities, such as how many moons a planet had, they were referring to an idealization, for in reality there would have been a continuum of possible states of affairs between a particular moon's being "in orbit" around the planet and "just passing by," each designated by a different real number or numbers. Any two such sets of real numbers, however close, would refer to physically different states which would evolve differently over time and have different physical effects. (Indeed the differences between them would typically grow exponentially with time because of the instability of classical dynamics known as *chaos*.) Thus, since even one real variable is equivalent to an infinity of independent discrete variables – say, the infinite sequence of zeros and ones in its binary expansion – an infinite amount of in-principle-observable information would be present in any classical object.

Despite this ontological extravagance, the continuum is a very natural idea. But then, so is the idea (which is the essence of information processing and therefore of "It from bit") that complicated processes can be analysed as combinations of simple ones. These two ideas have not been easy to reconcile. With the benefit of hindsight, I think that this is what Zeno's paradox of the impossibility of motion was really about. Had he been familiar with classical physics and the concept of information processing, he might have put it like this: consider the flight of an arrow as described in classical physics. To understand what happens during the flight, we could try to regard the real-valued position coordinates of the arrow as pieces of information, and the flight as a computation that processes that information, and we could try to analyze that computation as a sequence of elementary computations. But in that case, what is the "elementary" operation in question? If we regard the flight as consisting of a *finite* number of shorter flights, then each of them is, by any straightforward measure, exactly as complicated as the whole: it comprises exactly as many substeps, and the positions that the arrow takes during it are in one-to-one correspondence with those of the whole flight. Yet if, alternatively, we regard the flight as consisting of a literally infinite number of infinitesimal steps, what exactly is the effect of such a step? Since there is no such thing as a real number infinitesimally greater than another, we cannot characterize the effect of this infinitesimal operation as the transformation of one real number into another, and so we cannot characterize it as an elementary computation performed on what we are trying to regard as information.

For this sort of reason, "It from bit" would be a nonstarter in classical physics. It is noteworthy that the black-body problem, which drove Max Planck unwillingly to formulate the first quantum theory, was also a consequence of the infinite information-carrying capacity of the classical continuum.

In quantum theory, it is continuous observables that do not fit naturally into the formalism (hence the name *quantum* theory). And that raises another paradox – in a sense the converse of Zeno's: if the spectrum of an observable quantity (the set of possible outcomes of measuring it) is not a continuous range but a discrete set of values, how does the system ever make the transition from one of those values to another? The remarkable answer given by quantum theory is that it makes it continuously. It can do that because a quantum observable – the basic descriptor of quantum reality – is neither a real variable, like a classical degree of freedom, nor a discrete variable like a classical bit, but a more complicated object that has both discrete and continuous aspects.

When investigating the foundations of quantum theory, and especially the role of information, it is best to use the Heisenberg picture, in which quantum observables (which I shall mark with a caret, as in $\hat{X}(t)$) change with time, and the quantum state $|\Psi\rangle$ is constant. Though the Schrödinger picture is equivalent for all predictive purposes, and more efficient for most calculations, it is very bad at representing information flow and has given rise to widespread misconceptions (see Deutsch and Hayden 2000).

Apart from the trivial observables that are multiples of the unit observable $\hat{1}$, and hence have only one eigenvalue, the simplest type of quantum observable is a *Boolean observable* – defined as one with exactly two eigenvalues. This is the closest thing that quantum physics has to the classical programmer's idea of a Boolean variable. But the engineer's flip-flop is not just an observable: it is a whole physical system. The simplest quantum system that contains a Boolean observable is a *qubit*. Equivalently, a qubit can be defined as any system all of whose nontrivial observables are Boolean. Qubits are also known as "quantum two-state systems" (though this is a rather misleading term because, like all quantum systems, a qubit has a continuum of physical states available to it). The spin of a spin-$\frac{1}{2}$ particle, such as an electron, is an example. The fact that a qubit is a type of physical system, rather than a pure abstraction, is another important conceptual difference between the classical and quantum theories of information.

We can describe a qubit Q at time t elegantly in the Heisenberg picture (Gottesman 1999) using a triple $\hat{\mathbf{q}}(t) = (\hat{q}_x(t), \hat{q}_y(t), \hat{q}_z(t))$ of Boolean observables of Q, satisfying

$$\begin{aligned} \hat{q}_x(t)\hat{q}_y(t) &= i\hat{q}_z(t) \\ \hat{q}_x(t)^2 &= \hat{1} \qquad \text{(and cyclic permutations over } (x, y, z)). \end{aligned} \tag{5.1}$$

All observables of Q are linear combinations, with constant coefficients, of the unit observable $\hat{1}$ and the three components of $\hat{\mathbf{q}}(t)$. Each Boolean observable of Q

changes continuously with time, and yet, because of eqn (5.1), retains its fixed pair of eigenvalues which are the only two possible outcomes of measuring it.

Although this means that the classical information storage capacity of a qubit is exactly one bit, there is no elementary entity in nature corresponding to a bit. It is qubits that occur in nature. Bits, Boolean variables, and classical computation are all emergent or approximate properties of qubits, manifested mainly when they undergo decoherence (see Deutsch 2002a).

The standard model of quantum computation is the *quantum computational network* (Deutsch 1989). This contains some fixed number N of qubits

$$Q_a \quad (1 \leq a \leq N), \text{ with } [\hat{\mathbf{q}}_a(t), \hat{\mathbf{q}}_b(t)] = 0 \quad (a \neq b), \tag{5.2}$$

where $\hat{\mathbf{q}}_a(t) = (\hat{q}_{ax}(t), \hat{q}_{ay}(t), \hat{q}_{az}(t))$.

In physical implementations, qubits are always subsystems of other quantum systems – such as photons or electrons – which are themselves manipulated via a larger apparatus in order to give the quantum computational network its defining properties. However, one of those properties is that the network is *causally autonomous*: that is to say, the law of motion of each qubit depends only on its own observables and those of other qubits of the network, and the motion required of the external apparatus is independent of that of the qubits. Hence, all the external paraphernalia can be abstracted away when we study the properties of quantum computational networks.

Furthermore, we restrict our attention to networks that perform their computations in a sequence of *computational steps*, and we measure the time in units of these steps. The *computational state* of the network at integer times t is completely specified by all the observables $\hat{\mathbf{q}}_a(t)$. Although any real network would interpolate smoothly between computational states during the computational step, we are not interested in the computational state at noninteger times. The network at integer times is itself a causally autonomous system, and so, just as we abstract away the external apparatus, we also abstract away the network itself at noninteger times.

The computational state is not to be confused with the Heisenberg state $|\Psi\rangle$ of the network, which is constant, and can always be taken to be the state in which

$$\langle\Psi|\hat{q}_{az}(0)|\Psi\rangle = 1, \tag{5.3}$$

so that all the $\hat{q}_{az}$ observables are initially sharp with values $+1$. (In this convention, the network starts in a standard, "blank" state at $t = 0$, and we regard the process of providing the computation with its input as being a preliminary computation performed by the network itself.)

During any one step, the qubits of the network are separated (dynamically, not necessarily spatially) into nonoverlapping subsets such that the qubits of each subset interact with each other, but with no other qubits, during that step. We call this process "passing through a quantum gate" – a gate being any means of isolating a set of qubits and causing them to interact with each other for a fixed period. Because we are interested only in integer times, the relevant effect of a gate is its net effect over the whole computational step. The effect of an n-qubit quantum gate may be characterized by a set of $3n$ functions, each expressing one of the observables in the set $\{\hat{\mathbf{q}}_a(t+1)\}$ (where a now ranges over the indices of qubits passing through the gate between times t and $t+1$) in terms of the $3n$ observables $\{\hat{\mathbf{q}}_a(t)\}$, subject to the constraint that the relations are preserved. Every such set of functions describes a possible quantum gate. For examples see Deutsch and Hayden (2000).

Between these interactions, the qubits are computationally inert (none of their observables change); they merely move (logically, not necessarily spatially) from the output of one gate to the input of the next. Thus the dynamics of a quantum computational network can be defined by specifying a network of gates linked by "wires."

It might seem from this description that the study of quantum computational networks is a narrow subspeciality of physics. Qubits are special physical systems, and are often realized as subsystems of what are normally considered "elementary" systems (such as elementary particles). In quantum gates, qubits interact in a rather unusual way: they strongly affect each other while remaining isolated from the environment; their periods of interaction are synchronized, alternating with periods of inertness; and so on. We even assume that all the qubits of the network start out with their spins pointing in the $+z$-direction (or whatever the initial condition (3) means for qubits that are not spin-$\frac{1}{2}$ systems). None of these attributes is common in nature, and none can ever be realized perfectly in the laboratory. At the present state of technology, realizing them well enough to perform any useful computation is still a tremendously challenging, unattained target.

Yet quantum computational networks have another property which makes them far more worthy of both scientific and philosophical study than this way of describing them might suggest. The property is *computational universality*.

Universality

Universality has several interrelated aspects, including:

- the fact that a single, standard type of quantum gate suffices to build quantum computational networks with arbitrary functionality;
- the fact that quantum computational networks are a universal model for computation;
- the fact that a universal quantum computer can simulate, with arbitrary accuracy, the behaviour of arbitrary physical systems;
- the fact (not yet verified) that such computers can be constructed in practice.

The first of those concerns *universal gates*. One of the ways in which the theory of quantum computation lives up to the "It from bit" intuition is that in the most natural sense, the computation performed by the component gates of a network can indeed be simpler than that performed by the network as a whole. The possible motions of one or two qubits through a gate, though continuous, are not isomorphic to the possible motions of a larger network; but by composing multiple instances of only a single type of gate that performs a fixed, elementary operation, it is possible to construct networks performing arbitrary quantum computations. Any gate with this property is known as a *universal quantum gate*. It turns out that not only do there exist universal gates operating on only two qubits, but in the manifold of all possible two-qubit gates, only a set of measure zero is *not* universal (Deutsch *et al.* 1995).

Thus, computational universality is a generic property of the simplest type of gate, which itself involves interactions between just two instances of the simplest type of quantum system. There are also other ways of expressing gate-universality: for instance, the set of all single-qubit gates, together with the controlled-not operation (measurement of one qubit by another) also suffice to perform arbitrary computations. Alternatively, so do single-qubit gates together with the uniquely quantum operation of "teleportation" (Gottesman and Chuang 1999). All this constitutes a strikingly close connection between quantum computation and quantum physics – of which there were only hints in classical computation and classical physics. Models of classical computation based on idealized classical systems such as "billiard balls" have been constructed in theory (Fredkin and Toffoli 1982), but they are unrealistic in several ways, and unstable because of "chaos," and no approximation to such a model could ever be a practical computer. Constructing a universal classical computer (such as Babbage's analytical engine) from "elementary" components that are well described in a classical approximation (such as cogs and levers) requires those components to be highly composite, precision-engineered objects that would fail in their function if they had an even slightly different shape.

The same is true of the individual transistors on the microchips that are used to build today's classical computers. But it is not true, for instance, of the ions in an ion trap (Cirac and Zoller 1995; Steane 1997) – one of many quantum systems that are currently being investigated for possible use as quantum computers. In an ion trap, a group of ions is held in place in a straight line by an ingeniously shaped oscillating electric field. In each ion, one electron forms a two-state system (the states being its ground state and one of its excited states) which constitutes a qubit. The ions interact with each other via a combination of the Coulomb force and an external electromagnetic field in the form of laser light – which is capable of causing the observables of any pair of the qubits to change continuously when the laser is on. *The engineering problem ends there*. Once an arrangement of that general description is realized, the specific form of the interaction does not matter. Because of the generic universality of quantum gates, there is bound to exist *some* sequence of laser pulses – each pulse constituting a gate affecting two of the qubits – that will cause an N-ion trap to perform any desired N-qubit quantum computation.

The same sort of thing applies in all the other physical systems – nuclear spins, superconducting loops, trapped electrons, and many more exotic possibilities – that serve, or might one day serve, as the elementary components of quantum computers. S. Lloyd has summed this up in the aphorism: "Almost any physical system becomes a quantum computer if you shine the right sort of light on it." There is no classical analog of this aphorism.

Quantum computers are far harder to engineer than classical computers, of course, but not for the same reason. Indeed the problem is almost the opposite: it is not to engineer precisely defined composite systems for use as components, but rather, to isolate the physically simplest systems that already exist in nature, from the complex systems in their environment. That done, we have to find a way of allowing arbitrary pairs of them to interact – in some way – with each other. But once that is achieved in a given type of physical system, no shaping or machining is necessary, because the interactions that quantum systems undergo as a matter of course are already computationally universal.

The second aspect of universality is that quantum networks are a universal model for computation. That is to say, consider any technology that could, one day, be used to perform computations – whether quantum or classical, and whether based on gates or anything else. For any computer C built using that technology, there exists a quantum computational network, composed entirely of simple gates (such as instances of a single two-qubit universal gate), that has at least the same repertoire of computations as C. Here we mean 'the same repertoire' in quite a strong sense:

- Given a computational task (say, factorization) and an input (say, an integer), the network could produce the same output as C does (say, the factors of the integer).
- The resources (number of gates, time, energy, mass of raw materials, or whatever) required by the network to perform a given computation would be bounded by a low power of those required by C. I conjecture that this power can be 1. That is to say, there exists a technology for implementing quantum computational networks under which they can emulate computers built under any other technology, using only a constant multiple of the resources required under that technology.
- The network could emulate more than just the relationship between the output of C and its input. It could produce the output *by the same method* – using the same quantum algorithm – as C.

The upshot is that the abstract study of quantum computations (as distinct from the study of how to implement them technologically) is effectively the same as the study of one particular class of quantum computational networks (which need only contain one type of universal quantum gate). This universality is the quantum generalization of that which exists in classical computation, where the study of all computations is effectively the same as the study of any one universal model, such as logic networks built of NAND gates or Toffoli gates, or the universal Turing machine.

However, quantum universality has a further aspect which was only guessed at – and turned out to be lacking – in the case of classical computation: quantum computational networks can simulate, with arbitrary accuracy, the behavior of arbitrary physical systems; and they can do so using resources that are at most polynomial in the complexity of the system being simulated. The most general way of describing quantum systems (of which we are at all confident) is as *quantum fields*. For instance, a scalar quantum field $\hat{\varphi}(\mathbf{x}, t)$ consists of an observable for every point $(\mathbf{x}, t)$ of spacetime, satisfying a differential equation of motion. There are many possible approximation schemes for computing the behavior of such a system by approximating the continuous spacetime fields with continuous spectra as finite sets of observables with finite spectra, on a spacetime lattice. Such approximation schemes would be suitable for quantum computation too, where, for instance, a finite number of qubits would simulate the behavior of the field $\hat{\varphi}$ in the vicinity of each of a set of spatial grid points.

However, suppose that we had come upon quantum field theory from the other direction, convinced from the outset that "it" (a quantum field) is made of qubits. A quantum field can certainly be expressed in terms of fields of Boolean observables. For instance, the set of all Boolean observables "whether the average value of the field over a spacetime region R exceeds a given value ϕ," as R ranges over all regions of non-zero volume and duration, and ϕ ranges over all real numbers, contains the same information as the quantum field $\hat{\varphi}(\mathbf{x}, t)$ itself (albeit redundantly). For each of these Boolean observables, we can construct a "simplest" quantum system containing it, and that will be a qubit.

Local interactions could be simulated using gates in which qubits interact with close neighbors only. In this way, quantum networks could simulate arbitrary physical systems not merely in the bottom-line sense of being able to reproduce the same output (observable behavior), but again, in the strong sense of mimicking the physical events, locally and in arbitrary detail, that bring the outcome about.

In most practical computations, we should only be interested in the output for a given input and not (unless we are the programmer) in how it was brought about. But there are exceptions. An amusing example is given in the science-fiction novel *Permutation City* by Greg Egan (1994). In it, technology has reached the point where the computational states of human brains can be uploaded into a computer, and simulations of those brains, starting from those states, interact there with each other and with a virtual-reality environment – a self-contained world of the clients' choice. Because these computations are expensive, the people who run the service are continually seeking ways to optimize the program that performs this simulation. They run an optimization algorithm which systematically examines the program, replacing pieces of code or data with other pieces that achieve the identical effect in fewer steps. The simulated people cannot of course perceive the effect of such optimizations – and yet . . . eventually the optimization program halts, having deleted the entire simulation with all its data, and reports "This program generates no output."

By the way, there is no reason to believe that a universal *quantum* computer would be required for such simulations (see Tegmark 2000). There is every reason to believe that the brain is a universal classical computer. Nevertheless this strong form of universality of quantum computation assures us that such a technology, and artificial intelligence in general, must be possible, and tractable, regardless of how the brain works.

Provided, that is, that *universal quantum computers can be built in practice.* This is yet another aspect of universality, perhaps the most significant for the "It from qubit?" question. Indeed, universality itself may not be considered quite as significant by many physicists and philosophers if it turns out that qubits cannot, in reality, be composed into networks with universal simulating capabilities.

The world is not "made of information"

Let us suppose that universality does hold in all four of the above senses. Then, since every physical system can be fully described as a collection of qubits,[1] it is natural to wonder whether this can be taken further. Might it have been possible to *start* with such qubit fields and to interpret traditional quantum fields as emergent properties of them? The fact that all quantum systems that are known to occur in nature obey equations that look fairly simple in the language of fields on spacetime, is perhaps evidence against such a naive "qubits-are-fundamental" view of reality. On the other hand, we have some evidence in its favor too. One of the few things that we think we know about the quantum theory of gravity is expressed in the so-called Bekenstein bound: the entropy of any region of space cannot exceed a fixed constant times the surface area of the region (Bekenstein 1981). This strongly suggests that the complete state space of any spatially finite quantum system is finite, so that, in fact, it would contain only a finite number of independent qubits.

But even if this most optimistic quantum-computation-centered view of physics turned out to be true, it would not support the most ambitious ideas that have been

[1] Note added in proof: See also Zizzi (2000) who has used the term "It from qubit" to describe an approach to quantum gravity which does that explicitly.

suggested about the role that information might play at the foundations of physics. The most straightforward such idea, and also the most extreme, is that the whole of what we usually think of as reality is merely a program running on a gigantic computer – a Great Simulator. On the face of it, this might seem a promising approach to explaining the connections between physics and computation: perhaps the reason why the laws of physics are expressible in terms of computer programs is that they are in fact computer programs; perhaps the existence of computers in nature is a special case of the ability of computers (in this case the Great Simulator) to emulate other computers; the locality of the laws of physics is natural because complex computations are composed of elementary computations – perhaps the Great Simulator is a (quantum?) cellular automaton – and so on. But in fact this whole line of speculation is a chimera.

It entails giving up on explanation in science. It is in the very nature of computational universality that if we and our world were composed of software, we should have no means of understanding the real physics – the physics underlying the hardware of the Great Simulator itself. Of course, no one can prove that we are not software. Like all conspiracy theories, this one is untestable. But if we are to adopt the methodology of believing such theories, we may as well save ourselves the trouble of all that algebra and all those experiments, and go back to explaining the world in terms of the sex lives of Greek gods.

An apparently very different way of putting computation at the heart of physics is to postulate that "all possible laws of physics" (in some sense) are realized in nature, and then to try to explain the ones that we see, entirely as a selection effect (see, e.g., Smolin 1997). But selection effects, by their very nature, can never be the whole explanation for the apparent regularities in the world. That is because making predictions about an ensemble of worlds (say, with different laws of physics, or different initial conditions) depends on the existence of a measure on the ensemble, making it meaningful to say things like "admittedly, most of them do not have property X, but most of the ones in which anyone exists to ask the question, do." But there can be no a priori measure over "all possible laws." Tegmark (1997) and others have proposed that the complexity of the law, when it is expressed as a computer program, might be this elusive measure. But that merely raises the question: *complexity according to which theory of computation?* Classical and quantum computation, for instance, have very different complexity theories. Indeed, the very notion of "complexity" is irretrievably rooted in physics, so in this sense physics is necessarily prior to any concept of computation. "It" cannot possibly come from "bit," or from qubit, by this route. (See also my criticism of Wheeler's "Law without law" idea – Deutsch 1986.)

Both these approaches fail because they attempt to reverse the direction of the explanations that the real connections between physics and computation provide.

They seem plausible only because they rely on a common misconception about the status of computation *within mathematics*. The misconception is that the set of computable functions (or the set of quantum-computational tasks) has some a priori privileged status within mathematics. But it does not. The only thing that privileges that set of operations is that it is instantiated in the computationally universal laws of physics. It is only through our knowledge of physics that we know of the distinction between computable and noncomputable (see Deutsch *et al.* 2000), or between simple and complex.

The world is made of qubits

So, what does that leave us with? Not "something for nothing": information does not create the world *ex nihilo*. Nor a world whose laws are really just fiction, so that physics is just a form of literary criticism. But a world in which the stuff we call information, and the processes we call computations, really do have a special status. The world contains – or at least, is ready to contain – universal computers. This idea is illuminating in a way that its mirror-image – that a universal computer contains the world – could never be.

The world is made of qubits. Every answer to a question about whether something that could be observed in nature is so or not, is in reality a Boolean observable. Each Boolean observable is part of an entity, the qubit, that is fundamental to physical reality but very alien to our everyday experience. It is the simplest possible quantum system and yet, like all quantum systems, it is literally not of this universe. If we prepare it carefully so that one of its Boolean observables is sharp – has the same value in all the universes in which we prepare it – then according to the uncertainty principle, its other Boolean observables cease to be sharp: there is no way we can make the qubit as a whole homogeneous across universes. Qubits are unequivocally multiversal objects. This is how they are able to undergo continuous changes even though the outcome of measuring – or being – them is only ever one of a discrete set of possibilities.

What we perceive to some degree of approximation as a world of single-valued variables is actually part of a larger reality in which the full answer to a yes–no question is never just yes or no, nor even both yes and no in parallel, but a quantum observable – something that can be represented as a large Hermitian matrix. Is it really possible to conceive of the world, including ourselves, as being "made of matrices" in this sense? Zeno was in effect asking the same question about real numbers in classical physics: how can we be made of real numbers? To answer that question we have to do as Zeno did, and analyze the *flow of information* – the information processing – that would occur if this conception of reality were true. Whether we could be "made of matrices" comes down to this: what sort of experiences would an observer composed entirely of matrices, living in a world of matrices, have? The theories of decoherence (Zurek 1981) and consistent histories (Hartle 1991) have answered that question in some detail (see also Deutsch 2002a): at a coarse-grained level the world looks as though classical physics is true; and as though the classical theories of information and computation were true too. But where coherent quantum processes are under way – particularly quantum computations – there is no such appearance, and an exponentially richer structure comes into play.

As Karl Popper noted, the outcome of solving a problem is never just a new theory but always a new problem as well. In fundamental science this means, paradoxically, that new discoveries are always disappointing for those who hope for a final answer. But it also means that they are doubly exhilarating for those who seek ever more, and ever deeper, knowledge.

The argument that I used above to rule out Great-Simulator-type explanations has implications for genuine physics too: although in one sense the quantum theory of computation contains the whole of physics (with the possible exception of quantum gravity), the very power of the principle of the universality of computation inherently limits the theory's scope. Universality means that computations, and the laws of computation, are independent of the underlying hardware. And therefore, the quantum theory of computation cannot explain hardware. It cannot, by itself, explain why some things are technologically possible and others are not. For example, steam engines are, perpetual motion machines are not, and yet the quantum theory of computation knows nothing of the second law of thermodynamics: if a physical process can be simulated by a universal quantum computer, then so can its time reverse. An example closer to home is that of quantum computers themselves: the last aspect of universality that I mentioned above – that universal quantum computers can be built in practice – has not yet been verified. Indeed, there are physicists who doubt that it is true. At the present state of physics, this controversy, which is a very fundamental one from the "It from qubit" point of view, cannot be addressed from first principles. But if there is any truth in the "It from qubit" conception of physics that I have sketched here, then the quantum theory of computation as we know it must be a special case of a wider theory.

Quantum constructor theory (Deutsch 2002b) is the theory that predicts which objects can (or cannot) be constructed, and using what resources. It is currently in its infancy: we have only fragmentary knowledge of this type – such as the laws of thermodynamics, which can be interpreted as saying that certain types of machine (perpetual motion machines of the first and second kind) cannot be constructed, while others – heat engines with efficiencies approaching that of the Carnot cycle arbitrarily closely – can. One day, quantum constructor theory will likewise embody principles of nature which express the fact that certain types of information processing (say, the computation of non-Turing-computable functions of integers) cannot be realized in any technology while others (the construction of universal quantum computers with arbitrary accuracy) can. Just as the quantum theory of computation is now *the* theory of computation – the previous theory developed by Turing and others being merely a limiting case – so the present theory of computation will one day be understood as a special case of quantum constructor theory, valid in the limit where we ignore all issues of hardware practicability. As Einstein (1920) said, "There could be no fairer destiny for any physical theory than that it should point the way to a more comprehensive theory in which it lives on as a limiting case."

References

Bekenstein, J D (1981) *Phys. Rev.* **D23(2)**, 287–98.
Cirac, J I and Zoller, P (1995) *Phys. Rev. Lett.* **74**, 4091.
Deutsch, D (1986) *Found. Phys.* **16**(6), 565.
(1989) *Proc. R. Soc. Lond.* **A425**, 1868.
(2002a) *Proc. R. Soc. Lond.* **A458**, 2911.
(2002b) In *Proceedings of the Sixth International Conference on Quantum Communication, Measurement and Computing*, ed. J. H. Shapiro and O. Hirota, pp. 419–25. Princeton, NJ: Rinton Press.
Deutsch, D and Hayden, P (2000) *Proc. R. Soc. Lond.* **A456**, 1759.
Deutsch, D, Barenco, A, and Ekert, A (1995) *Proc. R. Soc. Lond.* **A449**, 669.
Deutsch, D, Ekert, A, and Luppachini, R (2000) *Bull. Symb. Logic* **3**, 3.
Egan, G (1994) *Permutation City*. London: Orion.
Einstein, A (1920) *Relativity: The Special and General Theory*. New York: Henry Holt.
Fredkin, E and Toffoli, T (1982) *Int. J. Theor. Phys.* **21**, 219.
Gottesman, D (1999) in *Group22: Proceedings of the 22nd International Colloquium on Group Theoretical Methods in Physics*, ed. S. P. Corney, R. Delbourgo, and P. D. Jarvis, pp. 32–43. Cambridge, MA: International Press.
Gottesman, D and Chuang, IL (1999) *Nature* **402**, 390.
Hartle, J B (1991) *Phys. Rev.* **D44**(10) 3173.
Smolin, L (1997) *The Life of the Cosmos*. New York: Oxford University Press.
Steane, A (1997) *Appl. Physics* **B64**, 623.
Tegmark, M (1997) Preprint gr-qc/9704009.
(2000) *Phys. Rev.* **E61**, 4194.
Zizzi, P (2000) *Entropy* **2**, 39.
Zurek, W H (1981) *Phys. Rev.* **D24**, 1516.

6

The wave function: it or bit?

H. Dieter Zeh
Universität Heidelberg

Introduction

Does Schrödinger's wave function describe physical reality ("it" in John Wheeler's terminology (Wheeler 1994)) or some kind of information ("bit")? The answer to this question must crucially depend on the definition of these terms. Is it then merely a matter of words? Not quite – I feel. Inappropriate words may be misleading, while reasonably chosen terms are helpful.

A *bit* is usually understood as the binary unit of information, which can be *physically realized* in (classical) computers, but also by neuronal states of having fired or not. This traditional physical (in particular, thermodynamical) realization of information ("bit from it") has proven essential in order to avoid paradoxes otherwise arising from situations related to Maxwell's demon. On the other hand, the concept of a bit has a typical quantum aspect: the very word quantum refers to discreteness, while, paradoxically, the *quantum bit* is represented by a continuum (the unit sphere in a two-dimensional Hilbert space) – more similar to an analog computer. If this quantum state describes "mere information," how can there be *real* quantum computers that are based on such superpositions of classical bits?

The problematic choice of words characterizing the nature of the wave function (or a general "quantum state") seems to reflect the common uneasiness of physicists, including the founders of quantum theory, about its fundamental meaning. However, it may also express a certain prejudice. So let me first recall some historical developments, most of which are discussed in Max Jammer's informative books (Jammer 1966, 1974), where you will also find the relevant "classic" references that I have here omitted.

Science and Ultimate Reality, eds. J. D. Barrow, P. C. W. Davies and C. L. Harper, Jr. Published by Cambridge University Press.

Historical remarks about the wave function

When Schrödinger first invented the wave function, he was convinced that it described real electrons, even though the construction of his wave equation from Hamiltonian mechanics readily led to wave mechanics on *configuration* space. As far as he applied his theory to single electron states, this had no immediate consequences. Therefore, he tried to explain the apparently observed "corpuscles" in terms of wave packets in space (such as coherent oscillator states). This attempt failed, but I will apply it to the configuration space of bounded ("local") systems (below, pp. 110 ff). Since Schrödinger firmly believed that reality must be described in space and time, he proposed nonlinear corrections to his single-electron wave equation, thus temporarily abandoning his own many-particle wave function.

When Born later proposed his probability interpretation, he initially postulated probabilities for spontaneous transitions *of a wave function into a new one*, since at this time he "was inclined to regard it [wave mechanics] as the most profound formalism of the quantum laws" (as he later explained). These new wave functions were either assumed to be bound states (resulting from spontaneous transitions within atoms) or plane waves (resulting from scattering or decay). In both cases the final (and mostly also the initial) states were thus stationary eigenstates of certain subsystem Hamiltonians, which replaced Bohr's semi-quantized electron orbits in the hydrogen atom.[1] Born "associated" plane waves with particle momenta according to de Broglie's mysterious proposal, although this had already been incorporated into wave mechanics in the form of differential momentum operators. Only after Heisenberg had formulated his uncertainty relations did Pauli introduce the *general* interpretation of the wave function as a "probability amplitude" for particle positions *or* momenta (or functions thereof) – cf. Beller (1999). It *seemed* to resemble a statistical distribution representing incomplete knowledge – although not simultaneously about position *and* momentum. This would then allow the *entanglement* contained in a many-particle wave function to be understood as a statistical correlation, and the reduction of the wave function as a "normal increase of information."

However, Pauli concluded (correctly, I think, although this has also been debated) that the potentially observable classical properties (particle position or momentum) are not just unknown, they are unknowable because they do not exist prior to measurement. As he later said in a letter to Born (see Born 1991), "the appearance of a definite position of an electron during an observation is a *creation* outside

[1] The idea of probabilistically changing (collapsing) wave functions was generalized and formalized as applying to measurements by von Neumann in what Wigner later called the "orthodox interpretation" of quantum mechanics. (By this term he did *not* mean the Copenhagen interpretation.) Its historical roots may explain why von Neumann regarded quantum jumps as the *first* kind of dynamics, while calling the Schrödinger equation a *second* "Eingriff" (intervention).

the laws of nature" (my translation and italics). Heisenberg had similarly claimed that "the particle trajectory is created by our act of observing it." In accordance with Born's original ideas (and also with von Neumann's orthodox interpretation), such spontaneous "events" are thus understood as dynamics (in contrast to a mere increase of information), while the process of observation or measurement is not further dynamically analyzed.

According to Heisenberg and the early Niels Bohr, these individual *events* occur in the atoms, but this interpretation had soon to be abandoned because of the existence of larger quantum systems. Bohr later placed them into the irreversible detection process. Others (such as London and Bauer (1939) or Wigner (1962)) suggested that the ultimate events occur in the observer, or that the "Heisenberg cut," where the probability interpretation is applied in the observational chain between observer and observed, is quite arbitrary. Ulfbeck and Aage Bohr (Ulfbeck and Bohr 2001) recently wrote that "the click in the counter occurs out of the blue, and without an event in the source itself as a precursor to the click." Note that there would then be no particle or other real object any more that dynamically connects the source with the counter! These authors add that simultaneously with the occurrence of the click the wave function "loses its meaning." This is indeed the way the wave function is often *used* – though *not* whenever the "particle" is measured repeatedly (such as when giving rise to a track in a bubble chamber). Even if it is absorbed when being measured for the first time, the state thereafter is described by a state vector that represents the corresponding vacuum (which is evidently an *individually meaningful* quantum state). The quantum state changes rather than losing its meaning.

The picture that spontaneous *events* are real, while the wave function merely describes their deterministically evolving probabilities (as Born formulated it), became general quantum folklore. It *could* represent an objective description if these events were consistently described by a fundamental stochastic process "in nature" – for example in terms of stochastic electron trajectories. A physical state at time t_0 would then *incompletely* determine that at another time t_1, say. The former could be said to contain "incomplete information" about the latter in an objective dynamical sense (in contrast to Heisenberg's concept of "human knowledge," or to the formal information processed in a computer). This indeterminism would be described by the spreading of a probability distribution, representing the decay of "objective information" about the later state, contained in an initial one. Unfortunately, this dynamical interpretation fails. It is in conflict with coherent effects in extended systems, or with Wheeler's delayed choice experiment (Wheeler 1979). Therefore, attempts were made to reduce the concept of trajectories to one of "consistent histories," that is, *partially* defined trajectories (Griffiths 1984). Roughly, these histories consist of successions of discrete stochastic events that occur in situations being equivalent to the aforementioned "counters." However, what circumstances let a physical system qualify as a counter in this sense?

Can a wave function that *affects* real events, or that keeps solid bodies from collapsing, itself be unreal? In principle, this is indeed a matter of definition. For example, electromagnetic fields were originally regarded as abstract auxiliary concepts, merely useful to calculate forces between the ("really existing") charged elements of matter. Bohm's quantum theory (Bohm 1952) demonstrates that electron trajectories can be consistently *assumed* to exist, and even to be deterministic under the guidance of a global wave function. Their unpredictability is then due to unknown (and unknowable) initial conditions. John Bell (1981) argued that the assumed global wave function would have to be regarded as real, too, in this theory: it "kicks" the electron (while it is here not being kicked back). Evidently this wave function cannot *merely* represent a statistical ensemble, although it dynamically *determines* an ensemble of potential events (of which but one is supposed to *become* real in each case – note the presumed direction of time!).

In particular, any entanglement of the wave function is *transformed* into statistical correlations whenever (local) events *occur* without being observed. Even when Schrödinger (1935) later called entanglement the greatest mystery of quantum theory, he used the insufficient phrase "probability relations in separated systems" in the title of his important paper. In the same year, Einstein, Podolsky, and Rosen concluded, also by using entangled states, that quantum theory must be incomplete. The importance of entanglement for the (evidently real!) binding energy of the helium atom was well known by then, total angular momentum eigenstates were known to require *superpositions* of products of subsystem states regardless of their distance, while von Neumann, in his book, had discussed the specific entanglement that arises from quantum measurements. Nonetheless, none of these great physicists was ready to dismiss the requirement that reality must be local (that is, defined in space and time). It is this requirement that led Niels Bohr to abandon microscopic reality entirely (while he preserved this concept for the apparently classical realm of events).

The reality of superpositions

There seems to be more to the wave function than its statistical and dynamical aspects. Dirac's general *kinematical* concept of "quantum states" (described by his ket vectors in Hilbert space) is based on the superposition principle. It requires, for example, that the superposition of spin-up and spin-down defines a new *individual* physical state, and does not just lead to interference fringes in the statistics of certain events. For every such spin superposition of a neutron, say, there exists a certain orientation of a Stern–Gerlach device, such that the path of the neutron can be *predicted with certainty* (to use an argument from Einstein, Podolsky, and Rosen). This spinor would not be correctly described as a vector with two unknown components. Other spin components have to be *created* (outside the laws of nature according to Pauli) in *measurements* with different orientations of the Stern–Gerlach magnet.

Superpositions of a neutron and a proton in the isotopic spin formalism are formally analogous to spin, although the SU(2) symmetry is dynamically broken in this case. Since these superpositions do not occur as free nucleons in nature (they may form quasi-particles within nuclei), the validity of the superposition principle has been restricted by postulating a "charge superselection rule." We can now *explain* the nonoccurrence of these and many other conceivable but never observed superpositions by means of environmental decoherence, while *neutral* particles, such as K mesons and their antiparticles, or various kinds of neutrinos, *can* be superposed to form new bosons or fermions, respectively, with novel individual and observable properties.

Two-state superpositions *in space* can be formed from partial waves which escape from two slits of a screen. Since the complete wave can hardly be refocused onto one point, we have to rely on statistical interference experiments (using the probability interpretation) to confirm its existence. (The outcome of the required *series* of events, such as a set of spots on a photographic plate, has quantum mechanically to be described by a tensor product of local states describing such spots – not by an ensemble of possible states.) *General* one-particle wave functions can themselves be understood as superpositions of all possible "particle" positions (space points). They define "real" physical properties, such as energy, momentum, or angular momentum, only as a whole.

Superpositions of different particle numbers form another application of this basic principle, essential to describe quasi-classical fields. If free fields are treated as continua of coupled oscillators, boson numbers appear as the corresponding oscillator quantum numbers. Coherent states (which were first used in Schrödinger's attempt to describe corpuscles as wave packets) may represent spatial fields. Conversely, *quantum* superpositions of classical fields define field functionals, that is, wave functions over a configurations space for classical field amplitudes.

These field functionals (generalized wave functions) were used by Dyson (1949) to derive path integrals and Feynman diagrams for perturbation theory of QED. All particle lines in these diagrams are no more than an intuitive short hand notation for plane waves appearing in the integrals that are actually *used*. The misinterpretation of the wave function as a probability distribution for classical configurations (from which a subensemble could be "picked out" by an increase of information) is often carried over to the path integral. In particular, quantum cosmologists are using the uncertainty relations to justify an *ensemble* of initial states (an initial indeterminacy) for presumed trajectories of the universe. Everett's relative state interpretation (based on the assumption of a universal wave function) is then easily misunderstood as a *many-classical-worlds* interpretation. However, Heisenberg's uncertainty relations refer to *classical* variables. They are valid for *given* quantum states, and so do not require the latters' (initial, in this case) indeterminacy as well. Ensembles of quantum states would again have to be *created* from an initial superposition (outside the laws or by means of new laws), while *apparent* ensembles of cosmic fluctuations may readily form by means of decoherence (Kiefer *et al.* 1998).

Superpositions of different states which are generated from one asymmetric state by applying a symmetry group (rotations, for example) are particularly useful. They define irreducible representations (eigenstates of the corresponding Casimir operators) as new individual physical states, which may give rise to various kinds of families of states (or "particles").

During recent decades, more and more superpositions have been confirmed to *exist* by clever experimentalists. We have learned about SQUIDs (superconducting quantum interference devices), mesoscopic Schrödinger cats, Bose condensates, and even superpositions of a macroscopic current running in opposite directions (very different from two currents canceling each other). Microscopic elements for quantum computers (which would perform different calculations simultaneously as components of one superposition) have been successfully designed. All these superpositions occur and act as individual physical states. Hence, their components "exist" simultaneously. As long as no unpredictable events have occurred (that is, have indeterministically changed the physical state), the components do *not* form ensembles of *possible* states (incomplete information).

A typical example for the *appearance* of probabilistic quantum events is the decay of an unstable state through tunneling through a potential barrier. The decay products (which in quantum cosmology may even represent universes) are here assumed to enter existence, or to leave the potential well, at a certain though unpredictable time. That this description is not generally applicable to quantum tunneling has been demonstrated by experiments in cavities (Rempe *et al.* 1987; Fearn *et al.* 1995), where different decay times may interfere with one another in an individual situation. Many narrow wave packets, approximately representing definite decay times, would have to be superposed in order to form a unitarily evolving wave function that may approximately decay exponentially (described by a complex energy eigenvalue) in a large but limited spacetime region. This evolution according to a Schrödinger equation requires furthermore that exponential tails of an energy eigenstate need time to form. In this way, it excludes superluminal effects that would result (though with very small probability) if exact eigenstates were created in discontinuous quantum jumps (Hegerfeldt 1994).

The conventional quantization rules, which are applied in situations where a "corresponding" classical theory is known or postulated, define the wave function $\psi(q)$ as a continuum of coefficients in the superposition $\int dq\, \psi(q)|q\rangle$ of all classical configurations q.[2] With the exception of single-particle states, $q \equiv \mathbf{r}$, this procedure leads directly to the infamous *nonlocal states*. Their nonlocality is very different from a (classical) *extension* in space, which would quantum mechanically be described by a *product* of local states. Only *superpositions* of such products of subsystem states may be nonlocal in the quantum mechanical sense. In order to prevent reality from being nonlocal, superpositions were therefore usually regarded as states of information – in contrast to the conclusion arrived at above. Even hypothetical "baby" or "bubble universes" are defined to be somewhere else *in space*, and thus far more conventional concepts than the "many worlds" of quantum theory. However, Bell's inequality – and even more so its nonstatistical generalizations (Greenberger *et al.* 1989; Hardy 1992) – have allowed experimentalists to demonstrate by operational means that *reality* is nonlocal. So why not simply accept the reality of the wave function?

As explained above, there are *two*, apparently unrelated, aspects which seem to support an interpretation of the wave function as a state of "information": classical *configuration* space instead of normal space as the new "stage for dynamics" (thus leading to quantum nonlocality), and the probability interpretation. Therefore, this picture and terminology appear quite appropriate for practical purposes. I am using it myself – although preferentially in quotation marks whenever questions of interpretation may arise.

While the general superposition principle, from which nonlocality is derived, requires nonlocal *states* (that is, a *kinematical* nonlocality), most physicists seem to regard as conceivable only a *dynamical* nonlocality (such as Einstein's spooky action at a distance). The latter would even have to include superluminal actions. In contrast, nonlocal entanglement must already "exist" before any crucially related local but spatially separated events would occur. For example, in so-called quantum teleportation experiments, a nonlocal state would have to be carefully prepared initially – so nothing has to be *ported* any more. After this preparation, a global state "exists but is not there" (Joos and Zeh 1985). Or in similar words: the real physical state is *ou topos* (at no place) – although this situation is *not utopic* according to quantum theory. A generic quantum state is not simply *composed* of local properties

[2] Permutation symmetries and "quantum statistics" demonstrate that the *correct* classical states q to be used as a basis are always spatial fields – never the apparent particles, which are a mere consequence of decoherence. This has recently been nicely illustrated by various experiments with Bose condensates. The equidistant energy levels of free field modes (oscillators) mimic particle numbers, and – if robust under decoherence – give rise to the nonrelativistic approximation by means of "N-particle wave functions." This conclusion eliminates any need for a "second quantization," while numbers distinguishing particles are meaningless. *There are no particles even before quantization* (while a fundamental unified theory might not be "based" on classical variables at all)!

(such as an extended object or a spatial field). If nonlocality is thus readily described by the formalism (if just taken seriously), how about the probability interpretation?

The role of decoherence

Most nonlocal superpositions discussed in the literature describe controllable (or usable) entanglement. This is the reason why they are being investigated. In an operationalist approach, this usable part is often exclusively *defined* as entanglement, while uncontrollable entanglement is regarded as "distortion" or "noise." However, if the Schrödinger equation is assumed to be universally valid, the wave function must contain far more entanglement (or "quantum correlations") than can ever be used (Zeh 1970). In contrast to entanglement, uncontrollable noise, such as phases fluctuating in time, would *not* destroy (or dislocalize) an individual superposition at any time. It may at most wash out an interference pattern in the *statistics* of many events – cf. Joos *et al.* (2003). Therefore, entanglement, which leads to decoherence in bounded systems even for the *individual global* quantum states, has to be distinguished from phase averaging in ensembles or with respect to a fluctuating Hamiltonian ("dephasing").

John von Neumann discussed the entanglement that arises when a quantum system is measured by an appropriate device. It leads to the consequence that the relative phases which characterize a superposition are now neither in the object nor in the apparatus, but only in their (shared) total state. These phases cannot affect measurements performed at one or the other of these two subsystems any more. The latter by themselves are then conveniently described by their reduced density matrices, which can be formally *represented by ensembles* of subsystem wave functions with certain formal probabilities. If the interaction dynamics between system and apparatus is (according to von Neumann) reasonably chosen, the resulting density matrix of the apparatus by itself can be represented by an ensemble of slightly overlapping wave packets which describe different pointer positions with precisely Born's probabilities. Does it therefore explain the required ensemble of measurement outcomes? That is, have the quantum jumps into these new wave functions (the unpredictable events) already occurred according to von Neumann's unitary interaction?

Clearly not. Von Neumann's model interaction, which leads to this entanglement, can in principle be reversed in order to reproduce a local superposition that depends on the initial phase relation. For a microscopic pointer variable this can be experimentally confirmed. For this reason, d'Espagnat (1966) distinguished conceptually between proper mixtures (which describe ensembles) and improper mixtures (which are defined to describe entanglement with another system). This difference is of utmost importance when the problem of measurement is being discussed. The density matrix is a formal tool that is sufficient for all practical purposes which *presume* the probability interpretation and neglect the possibility of local phase revivals (recoherence). Measurements by *microscopic* pointers can be regarded as "virtual measurements" (leading to "virtual decoherence") – in the same sense as virtual particle emission or virtual excitation. Similarly, scattering "events" cannot be treated probabilistically as far as phase relations, described by the scattering matrix, remain relevant or *usable*. Nothing can be assumed to have irreversibly happened in virtual measurements (as it would be required for real events).

The concept of a reduced density matrix obtains its pragmatic justification from the fact that all potential measurements are local, that is, described by local interactions with a local apparatus. Classically, dynamical locality means that an object can *directly* affect the state of another one only if it *is* at the same place. However, we have seen that quantum states are at no place, in general. So what does dynamical locality *mean* in quantum theory?

This locality (which is, in particular, a prerequisite for quantum field theory) is based on an important structure that goes beyond the mere Hilbert space structure of quantum theory. It requires (1) that there is a Hilbert space *basis* consisting of local states (usually a "classical configuration space"), and (2) that the Hamiltonian is a sum or spatial integral over corresponding local operators. (The first condition may require the inclusion of gauge degrees of freedom.) For example, the configuration space of a fundamental quantum field theory is expected to consist of the totality of certain classical field configurations on three- (or more) dimensional space, while its Hamiltonian is an integral over products of these field operators and their derivatives. This form warrants dynamical locality (in relativistic and nonrelativistic form) in spite of the nonlocal kinematics of the generic quantum states.

So let us come back to the question why events and measurement results appear actual rather than virtual. In order to answer it we must first understand the difference between reversible and irreversible (uncontrollable) entanglement. For this purpose we have to take into account the realistic environment of a quantum system. We may then convince ourselves by means of explicit estimates that a macroscopic pointer cannot avoid becoming strongly entangled with its environment through an uncontrollable avalanche of interactions, while the quantum state of a microscopic variable may remain almost unaffected in many cases.

This situation has been studied in much detail in the theory of decoherence[3] (Joos *et al.* 2003; Tegmark and Wheeler 2001; Zurek 2003), while many important applications remain to be investigated – for example in chemistry. It turns out that all phase relations between macroscopically different pointer positions become

[3] The concept of decoherence became known and popular through the "causal" chain Wigner–Wheeler–Zurek.

irreversibly nonlocal within very short times and for all practical purposes – similar to the rapid and irreversible formation of *statistical* correlations in Boltzmann's molecular collisions. These chaotic correlations as well as the quantum phases become inaccessible and irrelevant for the future evolution, while they still *exist* according to the assumed deterministic dynamics. If the wave function did "lose meaning," we would *not* have been able to derive decoherence from universal quantum dynamics in a consistent manner.

The asymmetry in time of this dissipation of correlations requires special initial conditions for the state of the universe – in quantum theory for its wave function (Zeh 2001). However, in contrast to classical statistical correlations, the arising entanglement ("quantum correlations") is part of the *individual* state: it represents a formal "plus" rather than an "or" that would characterize an ensemble of incomplete knowledge.

Two conclusions have to be drawn at this point: (1) decoherence occurs according to the reversible dynamical law (the Schrödinger equation) by means of an in-practice irreversible process, and precisely where events *seem* to occur, but (2) even this success does *not* lead to an ensemble representing incomplete information. The improper mixture does *not* become a proper one. We can neither justify the choice of a specific ensemble that would "unravel" the reduced density matrix, nor that of a subsystem to which the latter belongs.

From a *fundamental* point of view it would, therefore, be misleading in a twofold way to regard the entangled wave function as representing "quantum information." This terminology suggests incorrectly the presence of a (local) reality that is incompletely described or predicted by the wave function *and* the irrelevance of environmental decoherence for the measurement process (even though it has been experimentally confirmed (Brune *et al.* 1996)).

A further dynamical consequence of decoherence is essential for the pragmatic characterization of the observed classical physical world. Consider a two-state system with states $|L\rangle$ and $|R\rangle$ which are "continually measured" by their environment, and assume that they have exactly the same diagonal elements in a density matrix. Then this density matrix would be diagonal in *any* basis after complete decoherence. While a very small deviation from this degeneracy would resolve this deadlock, an exact equality *could* arise from a symmetry eigenstate, $|\pm\rangle = (|R\rangle \pm |L\rangle)/\sqrt{2}$. However, if we then measured $|R\rangle$, say, a second measurement would confirm this result, while a measurement of $|+\rangle$ (if possible) would give $|+\rangle$ or $|-\rangle$ with equal probabilities when repeated after a short decoherence time. It is the "robustness" of a certain basis under decoherence (a "predictability sieve" in Zurek's language) that gives rise to its classical appearance. In the case of a measurement apparatus it is called a "pointer basis."

This problem of degenerate probabilities also affects quasi-degenerate *continua* of states. For sufficiently massive particles (or macroscopic pointer variables), narrow wave packets may be robust even though they do not form an orthogonal basis diagonalizing the density matrix. Their precise shapes and sizes may even change under decoherence without violating their robustness. Collective variables (such as the amplitude of a surface vibration) are adiabatically "felt" (or "measured") by the individual particles. In microscopic systems this may represent a mere dressing of the collective mode. (My original work on decoherence was indeed influenced by John Wheeler's work (Griffin and Wheeler 1957) on collective nuclear vibrations by means of *generator coordinates*.) However, real decoherence is an ever-present irreversible *process*. Even the germs of all cosmic inhomogeneities were irreversibly "created" by the power of decoherence in breaking the initial homogeneity during early inflation of the universe (Kiefer *et al.* 1998). In other cases, such as a gas under normal conditions, molecules *appear* as "particles" because of the localizing power of decoherence, while the lack of robustness prevents the formation of extended trajectories for them (their collisions appear stochastic).

Nonetheless, something is still missing in the theory in order to arrive at *definite* events or outcomes of measurements, since the *global* superposition still exists according to the Schrödinger equation. The most conventional way out of this dilemma would be to postulate an appropriate collapse of the wave function as a fundamental modification of unitary dynamics. Several models have been proposed in the literature – see Pearle (1976), Ghirardi *et al.* (1986). They (quite unnecessarily) attempt to mimic precisely the observed environmental decoherence. However, since superpositions have now been confirmed far in the macroscopic realm, a Heisenberg cut for the application of the collapse may be placed *anywhere* between the counter (where decoherence first occurs in the observational chain) and the observer – although it would eventually have to be experimentally confirmed. The definition of subsystems in the intervening medium is entirely arbitrary, while the diagonalization of their reduced density matrices (the choice of their "pointer bases") may be convenient, but is actually irrelevant for this purpose. An individual observer may "solipsistically" assume this border line to exist in the observational chain even after another human observer (who is usually referred to as "Wigner's friend," since Eugene Wigner first discussed this situation in the role of the final observer).

It would in fact not help very much to postulate a collapse to occur only in counters. The physical systems which carry the information from the counter to the observer, including his sensorium and even his brain, must all be described by quantum mechanics. As far as we know, quantum theory applies everywhere, even where decoherence allows it to be approximately replaced by stochastic dynamics in terms of quasi-classical concepts ("consistent histories"). In an important paper, Max Tegmark (2000) estimated that neuronal networks and even smaller subsystems of the brain are strongly affected by decoherence. While this result does allow (or even requires) probabilistic quantum effects, it excludes extended controlled superpositions in the brain, which might represent some kind of quantum computing. However, postulating a probability interpretation at this point would eliminate the need for postulating it anywhere else in the observational chain. It is the (local) *classical* world that seems to be an illusion!

Nobody knows as yet where precisely (and in fact whether) consciousness may be located as the "ultimate observer system." Without any novel empirical evidence there is no way to decide where a collapse really occurs, or whether we have indeed to assume a superposition of many classical worlds – including "many minds" (Zeh 2000) for each observer – in accordance with a *universal* Schrödinger equation. It is sufficient for all practical purposes to know that, due to the irreversibility of decoherence, these different minds are dynamically autonomous (independent of each other) after an observation has been completed. Therefore, Tegmark's quasi-digitalization of the neuronal system (similar to the $|R, L>$ system discussed above) may even allow us to define this subjective Everett branching by means of the states diagonalizing the observer subsystem density matrix and their *relative states*.

A genuine collapse (in the counter or elsewhere) would produce an unpredictable result (described by *one component* of the wave function prior to the collapse). The state of ignorance after a collapse with unobserved outcome is, therefore, described by the *ensemble* of all these components with corresponding probabilities. In order to reduce this ensemble by an increase of information, the observer would have to interact with the detector in a classical process of observation. In the many-minds interpretation, in contrast, there is an objective process of decoherence that does *not* produce an ensemble. (The reduced density matrix resulting from decoherence can be treated for all practical purposes *as though* it represented one. This describes the *apparently observed events*.) The superposition of all resulting many minds forms *one* quantum state of the universe. Only from a subjective (though objectivizable by entanglement) point of view would there be a transition into *one* of these many "minds" (without any intermediary ensemble in this case). This interpretation is reminiscent of *Anaxagoras' doctrine*, proposed to split Anaximander's *apeiron* (a state of complete symmetry): "The things that are in a single world are not parted from one another, not cut away with an axe, neither the warm from the cold nor the cold from the warm. When Mind began to set things in motion, separation took place from each thing that was being moved, and all that Mind moved was separated." (Quoted from Jammer (1974: 482).) Although according to quantum description the role of "Mind" remains that of a passive (though essential) epiphenomenon (that can never be *explained* in terms of physical concepts), we will see in the next section how Anaxagoras' "doctrine" would even apply to the concepts of motion and time themselves.

In this specific sense one might introduce the *terminology* (though *not* as an explanation) that the global wave function represents "quantum information." While decoherence transforms the formal "plus" of a superposition into an effective "and" (an *apparent* ensemble of new wave functions), this "and" becomes an "or" only with respect to a subjective observer. An *additional* assumption has still to be made in order to justify Born's probabilities (which are meaningful to an individual mind in the form of frequencies in *series* of measurements): one has to assume that "our" (quantum correlated) minds are located in a component of the universal wave function that has non-negligible norm (Graham 1970). (Note that this is a *probable* assumption only after it has been made.) It is even conceivable that observers may not have been able to evolve at all in other branches, where Born's rules would not hold (Saunders 1993).

The Wheeler–DeWitt wave function

The essential lesson of decoherence is that the whole universe must be strongly entangled. This is an unavoidable consequence of quantum dynamics under realistic assumptions (Zeh 1970). In principle, we would have to know the whole wave function of the universe in order to make local predictions. Fortunately, there are useful local approximations, and most things may be neglected in most applications that are relevant for us local observers. (Very few systems, such as the hydrogen atom, are sufficiently closed and simple to allow precision tests of the theory itself.)

For example, Einstein's metric tensor defines space and time – concepts that are always relevant. Erich Joos (Joos 1986) first argued that the quantized metric field is strongly decohered by matter, and may *therefore* usually be treated classically. However, some aspects of quantum gravity are essential from a fundamental and cosmological point of view.

General relativity (or any unified theory containing it) is invariant under reparametrization of the (physically meaningless) time coordinate t that is used to describe the dynamics of the metric tensor. This invariance of the classical theory requires trajectories (in the corresponding configuration space) for which the Hamiltonian vanishes. This *Hamiltonian constraint*, $H = 0$, can thus classically be understood as a conserved initial condition (a conserved "law of the instant") for the time-dependent states. Upon quantization it assumes the form of the *Wheeler–DeWitt equation* (WDWE),

$$H\Psi = 0,$$

as the ultimate Schrödinger equation (DeWitt 1967; Wheeler 1968). This wave function Ψ depends on all variables of the universe (matter and geometry, or any unified fields instead). Since now $\partial\Psi/\partial t = 0$, the static constraint is all that remains of dynamics. While the classical law of the instant is compatible with time dependent states (trajectories), time is entirely lost on a fundamental level according to the WDWE. For a wave function that describes reality, this result cannot be regarded as just formal. "Time is not primordial!" (Wheeler 1979).

Dynamical aspects are still present, however, since the Wheeler–DeWitt wave function Ψ describes entanglement between all variables of the universe, including those representing appropriate clocks. Time dependence is thus replaced by quantum correlations (Page and Wootters 1983). Among these variables is the spatial metric ("three-geometry"), which defines time as a "many-fingered controller of motion" for matter (Baierlein *et al.* 1962) just as Newton's time would control motion in an absolute sense – another deep conceptual insight of John Wheeler.

The general solution of this WDWE requires cosmic boundary conditions in its configuration space. They may not appear very relevant for "us", since Ψ describes the superposition of very "many worlds." Surprisingly, for Friedmann-type universes, this static equation is of hyperbolic type after gauge degrees of freedom have been removed: the boundary value problem becomes an "initial" value problem with respect to the cosmic expansion parameter a (Zeh 1986). For appropriate boundary conditions at $a = 0$, this allows one to deduce a cosmic arrow of time (identical with that of cosmic expansion) (Zeh 2001). However, in the absence of external time t, there is neither any justification for interpreting the wave function of the whole universe in a classically forbidden region as describing a tunneling *process* for a (or the probability for an *event* to "occur"), nor to distinguish between its expansion and contraction according to the phase of $e^{\pm ia}$ (see Vilenkin (2002) for a recent misuse of this phase). In contrast, an α-"particle" tunneling out of a potential well according to the *time-dependent* Schrödinger equation is described by an outgoing wave *after* its metastable state has been *prepared* (with respect to external time). Similar arguments as for tunneling apply to the concept of a "slow roll" of the universe along a descending potential well (Steinhardt and Turok 2002).

Since the Wheeler–DeWitt wave function represents a superposition of all three-geometries (entangled with matter), it does not describe quasi-classical histories (defined as one-dimensional successions of states, or instants). Kiefer was able to show (Kiefer 1994) that such histories (which define spacetimes) can be approximately recovered by means of decoherence along WKB trajectories that arise according to a Born–Oppenheimer approximation with respect to the Planck mass. This leads to an effective time-dependent Schrödinger equation along each WKB trajectory in superspace (Wheeler's term for the configuration space of three-geometries). Complex branch wave functions emerge thereby from the real Wheeler–DeWitt wave function by an intrinsic breaking of the symmetry of the WDWE under complex conjugation (cf. Sect. 9.6 of Joos *et al.* (2003)). Each WKB trajectory then describes a whole (further branching) Everett universe for matter.

Claus Kiefer and I have been discussing the problem of timelessness with Julian Barbour (who wrote a popular book about it: Barbour (1999)) since the mid-1980s. Although we agree with him that time can only have emerged as an approximate concept from a fundamental timeless quantum world that is described by the WDWE, our initial approach and even our present understandings differ. While Barbour regards a *classical* general-relativistic world as timeless, Kiefer and I prefer the interpretation that timelessness is a specific quantum aspect (since there *are no* parametrizable trajectories in quantum theory). In classical general relativity, only *absolute* time (a *preferred* time parameter) is missing, while the concept of one-dimensional successions of states remains valid.

In particular, Barbour regards the *classical* configuration space (in contradistinction to the corresponding momentum space or to phase space) as a space of global actualities or "Nows." *Presuming* that time does not exist (on the basis that there is no absolute time), he then *concludes* that trajectories, of which but one would be real in conventional classical description, must be replaced by the multidimensional continuum of *all* potential Nows (his "Platonia"). He assumes this continuum to be "dynamically" controlled by the WDWE. After furthermore presuming a probability interpretation of the Wheeler–DeWitt wave function for his global Nows (in what may be regarded as a *Bohm theory without trajectories*), he is able to show along the lines of Mott's theory of α-particle tracks, and by using Kiefer's results, that classical configurations which are considerably "off-track" (and thus without memory of an *apparent* history) are extremely improbable. Thus come memories without a history.

One might say that according to this interpretation the Wheeler–DeWitt wave function is a *multidimensional generalization of one-dimensional time* as a controller of causal relationships (Zeh 2001). Julian Barbour does not agree with this terminology, since he insists on the complete absence of time (although this may be a matter of words). I do not like this picture too much for other reasons, since I feel that global Nows (simultaneities) are not required, and that the Hamiltonian symmetry between configuration space and momentum space is only dynamically – not conceptually – broken (by dynamical locality). Nonetheless, this is a neat and novel idea that I feel is worth being mentioned on this occasion.

That itsy bitsy wave function

Reality became a problem in quantum theory when physicists desperately tried to understand whether the electron and the photon "really" are particles or waves (in space). This quest aimed at no more than a *conceptually consistent description* that may have to be guessed rather than operationally construed, but would then have to be confirmed by all experiments. This concept of reality was dismissed in the Copenhagen interpretation according to the program of complementarity (which has consequently been called a "nonconcept"). I have here neither argued for particles nor for spatial waves, but instead for Everett's (nonlocal) universal wave function(al). It may serve as a consistent kinematical concept (once supported also by John Wheeler (Wheeler 1957)), and in this sense as a description of reality. The price may appear high: a vast multitude of separately observed (and thus with one exception unobservable to us) quasi-classical universes in one huge superposition. However, in the same way as Everett I have no more than extrapolated those concepts which are successfully *used* by quantum physicists. If this extrapolation is valid, the price would turn into an enormous dividend of grown knowledge about an operationally inaccessible reality.

The concept of reality has alternatively been based on operationalism. Its elements are then defined by means of operations (performed by what Wheeler called "observer–participators" (Wheeler 1979)), while these operations are themselves described in nontechnical "everyday" terms *in space and time*. In classical physics, this approach led successfully to physical concepts that proved consistently applicable. An example is the electric field, which was defined by means of the force on (real or hypothetical) test particles. The required operational means (apparata) can afterwards be self-consistently described in terms of these new concepts themselves ("partial reductionism"). This approach fails in quantum theory, since quantum states of fields would be strongly affected (decohered, for example) by test particles.

The investigation of quantum objects thus required various, mutually incompatible, operational means. This led to mutually incompatible (or "complementary") concepts, in conflict with microscopic reality. Niels Bohr's ingenuity allowed him to recognize this situation very early. Unfortunately, his enormous influence (together with the dogma that the concept of reality must be confined to objects in space and time) seems to have prevented his contemporaries from *explaining* it in terms of a more general (nonlocal) concept that is successfully *used* but not directly accessible by means of operations: the universal wave function. In terms of this hypothetical reality we may now understand why certain ("classical") properties are robust even when being observed, while microscopic objects may interact with mutually exclusive quasi-classical devices under the control of clever experimentalists. However, this does *not* mean that these quantum objects have to be *fundamentally* described by varying and mutually incompatible concepts (waves and particles, for example).

If *it* (reality) is understood in the operationalist sense, while the wave function is regarded as *bit* (incomplete knowledge about the outcome of potential operations), then one or the other kind of *it* may indeed emerge *from bit* – depending on the "very conditions" of the operational situation. I expect that this will remain the pragmatic language for physicists to describe their experiments for some time to come. However, if *it* is required to be described in terms of not necessarily operationally accessible but instead universally valid concepts, then the wave function remains as the only available candidate for *it*. In this case, *bit* (information as a dynamical functional form, as usual) may emerge *from it*, provided an appropriate (though as yet incompletely defined) version of a psychophysical parallelism is postulated in terms of this nonlocal *it*. If quantum theory appears as a "smokey dragon" (Wheeler 1979), the dragon itself may now be recognized as the universal wave function, greatly veiled to us local beings by the "smoke" represented by our own entanglement with the rest of the world.

However you turn it: *In the beginning was the wave function*. We may have to declare victory of the Schrödinger over the Heisenberg picture.

References

Baierlein, R F, Sharp, D H, and Wheeler, J A (1962) *Phys. Rev.* **126**, 1864.
Barbour, J B (1999) *The End of Time*. London: Weidenfeld and Nicolson.
Bell, J S (1981) In *Quantum Gravity 2*, ed. C. Isham, R. Penrose, and D. Sciama, pp. 611–37. Oxford: Clarendon Press.
Beller, M (1999) *Quantum Dialogue*. Chicago, IL: University of Chicago Press.
Bohm, D (1952) *Phys. Rev.* **85**, 166.
Born, M (1991) *A. Einstein, H. and M. Born, Briefwechsel 1916–1956*. München: Nymphenburger Verlagshandlung.
Brune, M, Hagley, E, Dreyer, J, *et al.* (1996) *Phys. Rev. Lett.* **77**, 4887.
d'Espagnat, B (1966) In *Preludes in Theoretical Physics*, ed. A. De-Shalit, H. Feshbach, and L. v. Hove, pp. 185–91. Amsterdam: North-Holland.
DeWitt, B S (1967) *Phys. Rev.* **160**, 1113.
Dyson, F (1949) *Phys. Rev.* **75**, 1736.
Fearn, H, Cook, R J, and Milonni, P W (1995) *Phys. Rev. Lett.* **74**, 1327.
Ghirardi, G C, Rimini, A, and Weber, T (1986) *Phys. Rev.* **D34**, 470.
Graham, N (1970) *The Everett Interpretation of Quantum Mechanics*. Chapel Hill, NC: University of North Carolina Press.
Greenberger, D M, Horne, M A, and Zeilinger, A (1989) In *Bell's Theorem, Quantum Theory, and Conceptions of the Universe*, ed. M. Kafatos, pp. Dordrecht: Kluwer.
Griffin, J J and Wheeler, J A (1957) *Phys. Rev.* **108**, 311.
Griffiths, R B (1984) *J. Stat. Phys.* **36**, 219.
Hardy, L (1992) *Phys. Rev. Lett.* **68**, 2981.
Hegerfeldt, G (1994) *Phys. Rev. Lett.* **72**, 596.

Jammer, M (1966) *The Conceptual Development of Quantum Mechanics*. New York: McGraw-Hill.
(1974) *The Philosophy of Quantum Mechanics*. New York: John Wiley.
Joos, E (1986) *Phys. Lett.* **A116**, 6.
Joos, E and Zeh, H D (1985) *Zeits. Phys.* **B59**, 223.
Joos, E, Zeh, H D, Kiefer, C, *et al.* (2003) *Decoherence and the Appearance of a Classical World in Quantum Theory*. Berlin: Springer-Verlag.
Kiefer, C (1994) In *Canonical Gravity: From Classical to Quantum*, ed. J. Ehlers and H. Friedrich, pp. 170–212. Berlin: Springer-Verlag.
Kiefer, C, Polarski, D, and Starobinsky, A A (1998) *Int. Journ. Mod. Phys.* **D7**, 455.
London, F and Bauer, E (1939) *La Théorie d'observation en méchanic quantique*. Paris: Hermann.
Page, D N and Wootters, W K (1983) *Phys. Rev.* **D27**, 2885.
Pearle, P (1976) *Phys. Rev.* **D13**, 857.
Rempe, G, Walther, H, and Klein, N (1987) *Phys. Rev. Lett.* **58**, 353.
Saunders, S (1993) *Found. Phys.* **23**, 1553.
Schrödinger, E (1935) *Proc. Cambridge Phil. Soc.* **31**, 555.
Steinhardt, P J and Turok, N (2002) report astro-ph/0204479.
Tegmark, M (2000) *Phys. Rev.* **E61**, 4194. quant-ph/9907009.
Tegmark, M and Wheeler, J A (2001) *Scient. Am.* **284**, 54. quant-ph/0101077.
Ulfbeck, O and Bohr, A (2001) *Found. Phys.* **31**, 757.
Vilenkin, A (2002) gr-qc/0204061.
Wheeler, J A (1957) *Rev. Mod. Phys.* **29**, 463.
(1968), In *Battelle recontres*, ed. B. S. DeWitt and J. A. Wheeler, pp. 242–307. New York: Benjamin.
(1979) In *Problems in the Foundations of Physics*, ed. G. Toraldi die Francia, pp. 395–497. Amsterdam: North-Holland.
(1994) In *Physical Origins of Time Asymmetry*, ed. J. J. Halliwell, J. Pères-Mercader, and W. H. Zurek, pp. 1–29. Cambridge: Cambridge University Press.
Wigner, E (1962) In *The Scientist Speculates*, ed. L. J. Good, pp. 284–302. London: Heinemann.
Zeh, H D (1970) *Found. Phys.* **1**, 69.
(1986) *Phys. Lett.* **A116**, 9.
(2000) *Found. Phys. Lett.* **13**, 221. quant-ph/9908084.
(2001) *The Physical Basis of the Direction of Time*, 4th edn. Berlin: Springer-Verlag.
Zurek, W H (2003) *Rev. Mod. Phys.* **75**, 715.

7

Quantum Darwinism and envariance

Wojciech H. Zurek
Los Alamos National Laboratory

Introduction

Quantum measurement problem is a technical euphemism for a much deeper and less well-defined question: How do we, "the observers," fit within the physical universe? The problem is especially apparent in quantum physics because, for the first time in the history of science a majority of (but not all) physicists seriously entertains the possibility that the framework for the ultimate universal physical theory provided by quantum mechanics is here to stay.

Quantum physics relevant for this discussion is (contrary to the common prejudice) relatively simple. By this I mean that some of the key features of its predictions can be arrived at on the basis of overarching principles of quantum theory and without reference to the minutiae of other specific ingredients (such as the details of the forces).

Quantum superposition principle is such an overarching principle of quantum theory. It leads to predictions that seem difficult to reconcile with our perception of the familiar classical universe of everyday experience. The aim of this paper is to show that appearance of the classical reality can be viewed as a result of the emergence of the preferred states from within the quantum substrate through the Darwinian paradigm, once the *survival of the fittest quantum states and selective proliferation* of the information about them throughout the universe are properly taken into account.

Measurement problem has been the focus of discussions of the interpretation of quantum theory since its inception in its present form in the 1920s. Two new ideas that are the focus of this paper – quantum Darwinism (Zurek 2000, 2003a) and envariance (Zurek 2003a, b) – have been introduced very recently. Exploration of their consequences has only started. This presentation provides a somewhat

Science and Ultimate Reality, eds. J. D. Barrow, P. C. W. Davies and C. L. Harper, Jr. Published by Cambridge University Press.

premature (and, consequently, rather speculative) "preview" of their implications. We shall start with the von Neumann model of quantum measurements (von Neumann 1932). It has provided the standard setting for the exploration of the role of observers and information transfer since it was introduced in 1932. We shall then go on and describe how von Neumann's model is modified by the introduction of the environment in the more modern treatments, and briefly summarize consequences of decoherence and of the *environment-induced superselection* or *einselection* that settle some of the issues.

Quantum Darwinism and envariance capitalise on the introduction of the environment into the picture. They explore a similar set of issues as the theory of decoherence and einselection (Zurek 1993, 1998, 2003a; Giulini *et al.* 1996; Paz and Zurek 2001), but from a very different vantage point: rather than limit attention to the consequences of the immersion of the system $\mathcal{S}$ or of the apparatus $\mathcal{A}$ in the environment $\mathcal{E}$ on the state of $\mathcal{SA}$, the focus shifts to the effect of the state of $\mathcal{SA}$ (or more precisely, the to-be-classical observables of that object, including in particular the apparatus pointer $\mathcal{A}$) on state of the environment.

The study of decoherence already calls for a modification of von Neumann's model, for the addition of the environment. Quantum Darwinism is a far more radical change – a change of focus, of the subject of discourse. It is based on the realization that almost without exception we – the observers – acquire information about "measured systems" or the "apparatus pointers" indirectly – by monitoring the environment.

Interaction with the environment is responsible for the negative selection, for destabilization of the vast majority of the states in the relevant Hilbert spaces of the open systems. What is left are the preferred pointer states. This, in essence, is the einselection. Quantum Darwinism is based on the observation that intercepting such "second-hand" information about the system by measuring fragments of the environment makes only some of the states of the system of interest accessible. These states happen to be the preferred pointer states of $\mathcal{S}$. It is the reason for their selection that is "Darwinian": pointer states are not only best in *surviving* the hostile environment, but are also best in *proliferating* – throughout the rest of the universe, and using environment as the medium – the information about themselves. This allows many observers to find out about the pointer states indirectly, and therefore, without perturbing them. Objective existence of pointer states of quantum systems can be accounted for in this way (Zurek 2000, 2003a). Hence, analogs of the Darwinian criterion of "fitness" can be seen in the (ein-)selection of "the classical."

Envariance focuses on the origins of ignorance (and, hence, information) in the quantum universe. It leads to the definition of probabilities – to the completely quantum derivation (Zurek 2003a, b) of Born's rule. Again, introduction of the environment is essential in this argument. In its presence one can delineate what aspects of the state of a system (that is correlated with the environment) cannot be known to the observer. In this way – by starting from a quantum definition of ignorance – the operational definition of probabilities can be obtained as a consequence of a quintessentially quantum sort of a correlation–quantum entanglement. It is interesting to note that analogous derivation cannot be repeated classically. This is because in classical physics information about the state can be "dissociated" from that state, while in quantum physics what is known about the state cannot be treated separately from the state. Consequently, in quantum physics it is possible to know precisely the joint state of two systems, but be provably ignorant about the states of the component subsystems.

Both of these themes – quantum natural selection and envariance – have benefited from the inspiration and support of John Archibald Wheeler. To begin with, one of the two portraits displayed prominently in John's office in Austin, Texas, was of Charles Darwin (the other one was of Abraham Lincoln). This was symptomatic of the role the theme of evolution played in his thinking about physics (see, e.g., Wheeler's (1983) ideas on the evolutionary origin of physical laws). While I was always fond of thinking about the "natural world" in Darwinian terms, this tendency was very much encouraged by John's influence: I was emboldened by his example to look at the emergence of the classical as a consequence of a quantum analog of natural selection. Last but not least (and on a lighter note), while my wife Anna and I were visiting John on his High Island summer estate in Maine, we were put up in a cottage in which – I was told – James Watson wrote *The Double Helix*.

While quantum Darwinism benefited from Wheeler's boldness and encouragement, envariance bears a direct Wheeler imprimatur: late in the year 1981 John and I were putting finishing touches on *Quantum Theory and Measurement* (Wheeler and Zurek 1983), and that included writing a section on "Further Literature." At that time I was fascinated with the idea that quantum states of entangled systems are in a sense relative – defined with respect to one another. Thus, John has caught me speculating:

> Zurek notes that "Nothing can keep one from thinking about [the two spins in a singlet] as the measured system and . . . a quantum apparatus. [In that language] . . . spin-system always points in the direction which is opposite to the direction of the . . . spin-apparatus. This is a definite, "coordinate – independent" statement.

John overcame my reluctance and included these musings about "the relativity of quantum observables" (see Wheeler and Zurek 1983: 772). Yet, these very same ideas have – after a long gestation period – recently begun to mature into a new way of looking at information and ignorance in the quantum context. Derivation of Born's rule based on the symmetries anticipated in that 20-years-old passage is presented in this paper in the sections on envariance. It is my hope that this result is just a "tip of the iceberg," and that envariance will prove to be a useful way of looking at various quantum issues of both fundamental and practical significance.

Quantum measurement: von Neumann's model

The traditional statement of the measurement problem goes back to von Neumann (1932), who has analyzed unitary evolutions that take initial state $|\psi_S\rangle$ *of the system and* $|A_0\rangle$ *of the apparatus into and entangled joint state* $|\Psi_{SA}\rangle$:

$$|\psi_S\rangle|A_0\rangle = \left(\sum_k a_k|s_k\rangle\right)|A_0\rangle \longrightarrow \sum_k a_k|s_k\rangle|A_k\rangle = |\Psi_{SA}\rangle. \qquad (7.1)$$

Von Neumann has realized that while $|\Psi_{SA}\rangle$ exhibits the desired correlation between $\mathcal{S}$ and $\mathcal{A}$, the unitary pre-measurement (as the "conditional dynamics" step described by eqn (7.1) is often called) does not provide a satisfactory account of "real world" measurements. There are two reasons why eqn (7.1) falls short: they are respectively identified as "basis ambiguity" and "collapse of the wave packet."

Basis ambiguity (Zurek 1981) is a reflection of the superposition principle: according to it, one can rewrite an entangled bipartite state such as $|\Psi_{SA}\rangle$ of eqn (7.1) in an arbitrary basis of one of the two subsystems (say, $\mathcal{S}$) and then identify the corresponding basis of the other (i.e., the apparatus $\mathcal{A}$). That is:

$$|\Psi_{SA}\rangle = \sum_k a_k|s_k\rangle|A_k\rangle = \sum_k b_k|r_k\rangle|B_k\rangle = \ldots, \qquad (7.2)$$

where $\{|s_k\rangle\}$ and $\{|r_k\rangle\}$ (as well as $\{|A_k\rangle\}$ and $\{|B_k\rangle\}$) span the same Hilbert space $\mathcal{H}_S$ ($\mathcal{H}_A$), while a_k (and b_k) are complex coefficients.

Basis ambiguity is also a consequence of entanglement. It is troubling, as it seems to imply that not just the outcome of the measurement, but also the set of states that describe the apparatus is arbitrary. Hence, any conceivable superposition (including the counterintuitive "Schrödinger cat" states (Schrödinger 1935a, b, 1936)) should have an equal right to be a description of a real apparatus (or a real cat) in a quantum universe. This is blatantly at odds with our experience of the macroscopic objects (including, for instance, states of the pointers of measuring devices) which explore only a very limited subset of the Hilbert space of the system restricted to the familiar, localized, effectively classical states.

The problem with the "collapse of the wave packet" would persist even if one were to somehow identify the preferred basis in the Hilbert space of the apparatus, so that prior to observer's contact with $\mathcal{A}$ one could be at least certain of the "menu" of the possible outcome states of the apparatus, and the basis ambiguity would disappear. For, in the end, we perceive only one of the possibilities, the actual outcome of the measurement. "Collapse" is the (apparently random) selection of just one of the positions on the "menu" of the potential outcomes with the probability given by Born's rule (Born 1926).

Von Neumann discussed two processes that address the two aspects of the "quantum measurement process" described above. While his investigation preceded the famos EPR paper (Einstein *et al.* 1935), and, hence, appreciation of the role of entanglement (which is behind the basis ambiguity), he has nevertheless postulated ad hoc a nonunitary "reduction" from a pure state into a mixture:

$$|\Psi_{\mathcal{SA}}\rangle\langle\Psi_{\mathcal{SA}}| \longrightarrow \sum_k |a_k|^2 |s_k\rangle\langle s_k||A_k\rangle\langle A_k| = \rho_{\mathcal{SA}}. \tag{7.3}$$

This process would have (obviously) selected the preferred basis. Moreover, von Neumann has also speculated about the nature of the next step – the collapse, i.e., the perception, by the observer, of a unique outcome. This could be represented by another nonunitary transition, e.g.:

$$\sum_k |a_k|^2 |s_k\rangle\langle s_k||A_k\rangle\langle A_k| \longrightarrow |s_{17}\rangle\langle s_{17}||A_{17}\rangle\langle A_{17}|. \tag{7.4}$$

In the collapse, the probability of any given outcome is given by Born's rule, $p_k = |a_k|^2$. Von Neumann has even considered the possibility that collapse may be precipitated by the conscious observers. This "anthropic" theme was later taken up by the others, including London and Bauer (1939) and Wigner (1963).

The aim of this chapter is to investigate and – where possible – to settle open questions within the unitary quantum theory per se, without invoking any nonunitary or anthropic *deus ex machina*.

Decoherence and einselection

The contemporary view (dubbed even "the new orthodoxy" (Bub 1997)) is that the solution of the measurement problem – and, in particular, the resolution of the issues described above in the context of von Neumann's original model – requires a more realistic account of what actually happens during a measurement: while von Neumann has treated the $\mathcal{SA}$ pair as isolated from the rest of the universe, the discussions over the past two decades have paid a lot of attention to the consequences of the immersion of the apparatus (and, more generally, of all the macroscopic objects) in their environments (Zurek 1981, 1982, 1991, 2003a, b; Giulini *et al.* 1996; Paz and Zurek 2001).

When the impossibility of perfect isolation of $\mathcal{A}$ is recognized, the solution of the basis ambiguity problem can be obtained (Zurek 1981, 1982, 2003a, b). A preferred basis – a candidate for the classical basis in the Hilbert space of the system coupled to the environment – is induced by the interaction with the environment. The egalitarian principle of superposition – the cornerstone of quantum mechanics – is grossly violated in such "open" quantum systems. Different quantum states exhibit very different degrees of resilience in the presence of the interaction with the outside. Thus, the question about effective classicality is answered by the study of stability. This is one of the tenets of the existential interpretation (Zurek 1993, 1998): *states that exist are the states that persist.*

Preferred pointer states are – in contrast to arbitrary superpositions, which, in accord with the superposition principle, have equal right to inhabit Hilbert space of an *isolated* system – resilient to the entangling interaction with the environment. Hence, they maintain their identity – their ability to faithfully represent the system. Selection of the preferred set of resilient pointer states in the Hilbert space is the essence of the environment-induced superselection (einselection). It is caused by a (pre-)measurement – like unitary evolution in which the environment $\mathcal{E}$ becomes entangled with the apparatus:

$$|\Psi_{\mathcal{SA}}\rangle|e_0\rangle = \left(\sum_k a_k|s_k\rangle|A_k\rangle\right)|e_0\rangle \longrightarrow \sum_k a_k|s_k\rangle|A_k\rangle|e_k\rangle = |\Phi_{\mathcal{SAE}}\rangle. \quad (7.5)$$

When the state of the environment contains an accurate record of the outcome, so that $|\langle e_k|e_l\rangle|^2 = \delta_{kl}$, the density matrix of the apparatus–system pair acquires the desired form, as can be seen by tracing out the environment:

$$\rho_{\mathcal{SA}} = \mathrm{Tr}_{\mathcal{E}}|\Phi_{\mathcal{SAE}}\rangle\langle\Phi_{\mathcal{SAE}}| = \sum_k |a_k|^2|s_k\rangle\langle s_k||A_k\rangle\langle A_k|. \quad (7.6)$$

This is clearly what is needed to solve the basis ambiguity problem (compare with eqn (7.3) above). Moreover, it has by now been confirmed in model calculations and corroborated by experiments that the preferred pointer basis will habitually appear on the diagonal of the density matrix describing $\mathcal{A}$. The question, however, can be raised about the justification of the trace operation: the form of the density matrix relies on Born's rule (von Neumann 1932). Moreover, eqn (7.6) gets only half of the job done: eqn (7.3) – the collapse – still needs to be understood.

Within the context of decoherence and einselection both of these questions – basis ambiguity and collapse – can be (albeit to a different degree) addressed. It is by now largely accepted (as a result of extensive studies of specific models) that under a reasonable set of realistic assumptions a preferred basis of an apparatus pointer (or of selected observables of any open macroscopic system) does indeed emerge. Thus, quantum entanglement (present after the pre-measurement, eqn (7.1)) will give way to a classical correlation between $\mathcal{S}$ and $\mathcal{A}$, with the same preferred pointer basis $\{|A_k\rangle\}$ habitually appearing on the diagonal of $\rho_{\mathcal{SA}}$. This takes care of the basis ambiguity.

This conclusion, however, crucially depends on the trace operation, which is justified by employing Born's rule – an important part of the quantum foundations, that is often regarded as an independent axiom of quantum theory intimately tied with the process of measurement. One may (as many have) simply accept Born's rule as one of the axioms. But it would be clearly much more satisfying to derive it. This will be our aim in the discussion of envariance.

The other outstanding issue is the apparent collapse and – in particular – the *objectivity* of effectively classical (but, presumably, ultimately quantum) states. That is, classical states can be simply "found out" by an observer who is initially completely ignorant. This is not the case for quantum states. Ideal measurement will always yield an eigenvalue of the measured observable. Hence, it will select its (possibly degenerate) eigenstate. When the system does not happen to be in one of the eigenstates of the observable selected by the observer (or when the pointer basis of eqn (7.6) does not commute with the measured observable) its measurement will perturb the state of the system by resetting it to one of the eigenstates of what is being measured. Yet, in our everyday experience we never have to face this problem: somehow, at the macroscopic level of classical reality, we find out about the rest of the universe at will, without having to worry about what does (and what does not) *exist*. We start by addressing this second issue of the emergence of *objectivity*.

Quantum Darwinism

A part of the paradigm of "quantum measurements" that is shared not just by von Neumann's model, but by most of the other approaches to the interpretation of quantum mechanics is the belief that we – the observers – acquire information about quantum systems directly, i.e., by interacting with them. As was pointed out some time ago (Zurek 1993, 1998), this is never the case. For instance, a vast majority of our information is acquired visually. The information we obtain in this way does not concern photons, although our eyes act as photon detectors: rather, photons play the role of carriers of information about objects that emitted or scattered them. Moreover, we obtain all the information by intercepting only a small fraction of photons emitted by or scattered from the object of interest with our eyes. Thus, many more copies of the same information must be carried away by this photon environment. Upon reflection one is led to conclude that essentially the same scheme (but involving different carriers of information) is the rule rather than exception. Measurements carried out on the macroscopic objects are invariably indirect, and carriers of information always "fan out" most of the copies of the "data," spreading it indiscriminately throughout the universe. Observers use the same environment that causes decoherence to obtain information.

This distinction between direct and indirect acquisition of information may seem inconsequential. After all, replacing a direct measurement with an indirect one only extends the "von Neumann chain" (von Neumann 1932). The overall state has a form of eqn (7.5) and is still pure, with all of the potential outcomes present, superficially with no evidence of either eqn (7.3) or the "collapse" of eqn (7.4). Still, we shall show that when this situation is analyzed from the point of view of the observer, most (and perhaps all) of the symptoms of classicality emerge in this setting.

To investigate a simple model of this situation we consider an obvious generalization of eqn (7.5) we have used to describe decoherence:

$$\begin{aligned} |\Psi_{\mathcal{SA}}\rangle \otimes_{n=1}^{\mathcal{N}} \left|e_0^{(n)}\right\rangle &= \left(\sum_k a_k |s_k\rangle |A_k\rangle\right) \otimes_{n=1}^{\mathcal{N}} \left|e_0^{(n)}\right\rangle \\ &\longrightarrow \sum_k a_k |s_k\rangle |A_k\rangle \otimes_{n=1}^{\mathcal{N}} \left|e_k^{(n)}\right\rangle = |\Phi_{\mathcal{SAE}^{\mathcal{N}}}\rangle. \end{aligned} \tag{7.7}$$

There are $\mathcal{N}$ environment subsystems here. The assumption is that they exist, and that they can be (like photons) accessed one at a time.

We first note that enlarging this composite environment $\mathcal{E}^{\mathcal{N}}$ of eqn (7.7) is absolutely irrelevant from the point of view of its effect on the density matrix of the "object of interest," $\rho_{\mathcal{SA}}$. For, when either a simple environment of eqn (7.5) or the multiple environment of eqn (7.7) is traced out, the same $\rho_{\mathcal{SA}}$ of eqn (7.6) will obtain. So what (if anything) have we gained by complicating the model? Whatever it is obviously cannot be inferred from the state of $\mathcal{SA}$ alone. Yet, in classical physics the state of "the object of interest" was all that mattered! So where should we look now?

The inability to appreciate the implications of the difference between these two situations is indeed firmly rooted in the "classical prejudice" that the information about the system is synonymous with its state, but that the presence of that information is physically irrelevant for that state. This belief in the analog of the "separation of church from state" is untenable in the quantum setting. For starters, there is "no information without representation" (Zurek 1994)!

Guided by our previous considerations, we shift attention from the state of the object of interest (the $\mathcal{SA}$ pair) to the record of its state in the environment. Now there is our difference! Instead of a single (fragile) record of the state of the system we now have many identical copies. How many? The preferred states $\{|A_k\rangle\}$ of the apparatus have left $\mathcal{N}$ imprints on the environment. This is easily seen in the example above, and can be quantified by one of the versions of the *redundancy ratio* (Zurek 2000, 2003a), which in effect count the number of copies of the information about the object of interest spread throughout the environment.

One definition of the redundancy ratio is based on the mutual information between the fragment of the environment $\mathcal{E}^{(n)}$ and the object of interest (Zurek 2000, 2003a). This leads to:

$$I\left(\mathcal{S}:\mathcal{E}^{(n)}\right) = H(\mathcal{S}) + H\left(\mathcal{E}^{(n)}\right) - H\left(\mathcal{S},\mathcal{E}^{(n)}\right). \tag{7.8}$$

Above, we have replaced $\mathcal{SA}$ of eqn (7.5) by a single object to simplify discussion, and to emphasize that this approach applies in general – and not just in measurement situations. Various entropies can be defined in several ways using obvious reduced density matrices of the relevant subsystems of the whole (Zurek 2000, 2003a, 2003c; Ollivier and Zurek 2002). Redundancy can be then estimated as:

$$\mathcal{I}^{(\mathcal{N})} = \sum_{k=1}^{\mathcal{N}} I\left(\mathcal{SA}:\mathcal{E}^{(k)}\right). \tag{7.9}$$

The physical significance of redundancy in the context of our discussion is similar to its import in the classical information theory (Cover and Thomas 1991): redundancy protects information about the object of interest. From the point of view of the interpretation of quantum theory, this implies, for example, that many different observers can find out the state of the object of interest independently – by measuring different fragments of the environment. This is how – I believe – states of the ultimately quantum but macroscopic objects in the world of our everyday experience acquire their *objective existence* (Zurek 2000, 2003a).

However, viewed in a Darwinian fashion, redundancy has also a rather different significance: it provides, in effect, a measure of the number of "offspring" of the state in question. Thus, in the ideal case we have considered above proliferation of information has led to $\mathcal{N}$ descendants of the original state of the apparatus. The redundancy ratio in the example given above is:

$$\mathcal{R} = \mathcal{I}^{(\mathcal{N})}/H(\mathcal{S}) = \mathcal{N}. \tag{7.10}$$

Both the preconditions for and the consequences of high redundancy have significance that is best appreciated by invoking analogies with the "survival of the fittest." To begin with, a state that manages to spread many imprints of its "genetic information" throughout the environment must be resistant to the perturbations caused by the environment. This points immediately to the connection with the "pointer states" (Zurek 1981) – they remain unperturbed by decoherence. But this is in a sense just a different view of selection of the preferred states, which does not capitalize on the measure of their fecundity we have introduced above.

Thus, Darwinian analogy focuses on the fact that proliferation of certain information throughout the environment makes its further proliferation more likely. This is best seen in a still more realistic extension of the models of the environments we have considered so far: suppose that in addition to the immediate environments $\mathcal{E}^{(k)}$ there are also distant environments $\varepsilon^{(l)}$, which do not interact directly with $\mathcal{S}$ but interact with the immediate environments through interaction that is local – i.e., that allows individual subsystems of the immediate environment to become correlated with individual subsystems of the distant environment. Then it is easy to argue that the only information about $\mathcal{S}$ that can be passed along from $\mathcal{E}$'s to ε's will have to do with the preferred pointer states: Only locally accessible information (Ollivier and Zurek 2002; Zurek 2003c) can be passed along by such local interactions. Indeed, this connection between the selection of the preferred basis and redundancy was already noted some time ago (Zurek 1983).

We note in passing that there is an intimate relation between this necessity to make a selection of preferred states in the setting that involves "fan-out" of the information and the *no-cloning theorem*, which, in effect, says that copying implies a selection of a preferred set of states that are copied. We also note that all of the above considerations depend on the ability to split the universe as a whole into subsystems. This – as was already noted in the past – is a prerequisite of decoherence. Moreover, problems of interpretation of quantum physics do not arise in a universe that does not consist of subsystems (Zurek 1993, 1998, 2003a).

Environment-assisted invariance

Envariance is an abbreviation for environment-assisted invariance, the peculiarly quantum symmetry exhibited by the states of entangled quantum systems. Our first step is to consider a state vector describing system $\mathcal{S}$ entangled (but no longer interacting) with the environment $\mathcal{E}$. The joint state can be – for our purpose – always written in the Schmidt basis:

$$|\psi_{\mathcal{S}\mathcal{E}}\rangle = \sum_{k}^{N} \alpha_k |s_k\rangle |\varepsilon_k\rangle. \tag{7.11a}$$

For, even when the initial joint state is mixed, one can always imagine purifying it by enlarging the environment. As the environment no longer interacts with the system, probabilities of various states of the system cannot be – on physical grounds – influenced by such purification. In writing eqn (7.11*a*) we assumed that such purification was either unnecessary or was already carried out.

Environment-assisted invariance refers to the fact that there is a family of unitary quantum transformations $U_{\mathcal{S}}$ that act on a system alone, and that are nontrivial, so that $U_{\mathcal{S}}|\psi_{\mathcal{S}\mathcal{E}}\rangle \neq |\psi_{\mathcal{S}\mathcal{E}}\rangle$, but their effect can be undone by acting solely on $\mathcal{E}$. Thus, for any $U_{\mathcal{S}}$ that has Schmidt states as eigenstates one can always find $U_{\mathcal{E}}$ such that:

$$U_{\mathcal{E}}(U_{\mathcal{S}}|\psi_{\mathcal{S}\mathcal{E}}\rangle) = |\psi_{\mathcal{S}\mathcal{E}}\rangle\cdot \tag{7.12}$$

This is evident, as unitaries with Schmidt eigenstates acting on $\mathcal{S}$ will only rotate the phases of the coefficients $\psi_{\mathcal{S}\mathcal{E}}$. But these phases can be also rotated by acting on $\mathcal{E}$ alone. Hence, transformations of this kind are envariant. It turns out that envariant transformations always have Schmidt eigenstates (Zurek 2003a).

In the spirit of decoherence we now focus on the system alone. Clearly, for an observer with no access to $\mathcal{E}$, the system must be completely characterized by the set of pairs $\{|\alpha_k|, |s_k\rangle\}$: only the absolute values of the coefficients can matter since phases of α_k can be altered by acting on $\mathcal{E}$ alone, and $\mathcal{E}$ is causally disconnected from $\mathcal{S}$. Thus, in the case when all $|\alpha_k|$ are equal:

$$|\bar{\psi}_{\mathcal{S}\mathcal{E}}\rangle = \sum_k^N |\alpha|e^{-i\varphi_k}|s_k\rangle|\varepsilon_k\rangle, \qquad (7.11b)$$

any orthonormal basis is obviously Schmidt, and we can use envariance to reassign the coefficients to different states $\alpha_k \rightarrow \alpha_l, \alpha_l \rightarrow \alpha_k$, etc. Such swapping leaves the description of the system invariant: the coefficients can differ only by the phase, and we have proved above that phases of the Schmidt coefficients cannot influence probabilities of the system alone (Zurek 2003a, b). (Indeed, if this was possible, faster-than-light communication would be also possible, as the reader can easily establish by extending the above argument.)

It is now evident that the probabilities of all k's must be equal. Hence, assuming the obvious normalization, they are given by:

$$p_k = 1/N. \qquad (7.13a)$$

Moreover, a collection of a subset of n amongst N mutually exclusive events (orthogonal states) has the probability:

$$p_{k_1 \vee k_2 \vee \ldots \vee\, k_n} = n/N. \qquad (7.13b)$$

These results were easy to arrive at, but we have started with very strong assumption about the coefficients.

The case when $|\alpha_k|$ are *not* equal is of course of interest. We shall reduce it to the case of equal coefficients by extending the Hilbert space of the environment. In the process we shall recover Born's rule $p_k = |\alpha_k|^2$. This will also provide a firmer foundation for the decoherence approach which until now had used Born's rule to justify its reliance on reduced density matrices. We note that we have, in a sense, already gone half way in that direction: phases in the Schmidt decomposition have been already shown to be irrelevant, so the probabilities must depend on the absolute values of the coefficients. We still do not know in what specific function is this dependence embodied.

To illustrate the general strategy we start with an example involving a two-dimensional Hilbert space of the system spanned by states $\{|0\rangle, |2\rangle\}$ and (at least) a three-dimensional Hilbert space of the environment. The correlated state of $\mathcal{SE}$ is:

$$|\psi_{\mathcal{SE}}\rangle = (\sqrt{2}|0\rangle|+\rangle + |2\rangle|2\rangle)/\sqrt{3}. \tag{7.14a}$$

The state of the system is on the left, and $|+\rangle = (|0\rangle + |1\rangle)/\sqrt{2}$ exists in the (at least two-dimensional) subspace of $\mathcal{E}$ orthogonal to the environment state $|2\rangle$, so that $\langle 0|1\rangle = \langle 0|2\rangle = \langle 1|2\rangle = \langle +|2\rangle = 0$. To reduce this case to the case of eqn (11b) we extend $|\psi_{\mathcal{SE}}\rangle$ above to a state $|\phi_{\mathcal{SEE}'}\rangle$ with equal coefficients by acting only on the causally disconnected $\mathcal{E}$ (which implies that probabilities we shall infer for $\mathcal{S}$ could not have changed). This can be done by allowing a c-shift act between $\mathcal{E}$ and $\mathcal{E}'$ so that (in the obvious notation) $|k\rangle|0'\rangle \Rightarrow |k\rangle|k'\rangle$, and:

$$\begin{aligned} |\psi_{\mathcal{SE}}\rangle|0\rangle &= \frac{\sqrt{2}|0\rangle|+\rangle|0'\rangle + |2\rangle|2\rangle|0'\rangle}{\sqrt{3}} \\ &\Longrightarrow \left(\sqrt{2}|0\rangle\frac{|0\rangle|0'\rangle + |1\rangle|1'\rangle}{\sqrt{2}} + |2\rangle|2\rangle|0'\rangle\right)\Big/\sqrt{3}. \end{aligned} \tag{7.15a}$$

The cancellation of $\sqrt{2}$ leads to:

$$|\phi_{\mathcal{SEE}'}\rangle = (|0\rangle|0\rangle|0'\rangle + |0\rangle|1\rangle|1'\rangle + |2\rangle|2\rangle|2'\rangle)/\sqrt{3}. \tag{7.16a}$$

The phases are again irrelevant as they can be altered by manipulating $\mathcal{E}'$ alone. Clearly, for the bipartite combination of $\mathcal{S}$ and $\mathcal{E}$ the three orthonormal product states have coefficients with the same absolute values and can be swapped. Hence, all of them have the same probability. Moreover, two of them involve state $|0\rangle$ of the system. Thus, by eqn (7.13a), probabilities of $|0\rangle|0\rangle$, $|0\rangle|1\rangle$, and $|2\rangle|2\rangle$ are all equal. Now, by eqn (7.13b), the probability of $|0\rangle$ state of the system is twice the probability of $|2\rangle$. Consequently:

$$p_0 = 2/3; \quad p_2 = 1/3. \tag{7.17a}$$

Hence, in this special case – but using ideas that are generally applicable – I have derived Born's rule, i.e., demonstrated that entanglement leads to envariance and this implies $p_k = |\alpha_k|^2$. It is straightforward (if a bit notationally cumbersome) to generalize this derivation, and we shall do so in a moment. But the basic idea is already apparent and worth contemplating before we proceed with a general case (where the main point is somewhat obscured by notation).

Envariance, ignorance, and information

The above derivation of probabilities in quantum physics is very much in the spirit of the "ignorance interpretation," but in the quantum context it can be carried out with an important advantage: in the classical case observers assume that an unknown state they are about to discover exists objectively prior to the measurement, and that the ignorance allowing for various swappings reflects their "subjective lack of knowledge." Indeed, the clash between this subjectivity of information on one hand and its obvious physical significance on the other has been a source of a long-standing confusion distilled into the Maxwell's demon paradox. *In quantum theory ignorance can be demonstrated in an objective fashion, as a consequence of envariance of a state perfectly known as a whole.* Above, $\mathcal{SEE}'$ is pure. Quantum complementarity enforces ignorance of the states of the parts as the price that must be paid for the perfect knowledge of the state of the whole.

It seems ironic that a natural (and a very powerful) strategy to justify probabilities rests – in quantum physics – on a more objective and secure foundation of perfectly known *entangled* pure states than in the deterministic classical physics: When the state of the observer's memory $|\mu\rangle$ is not correlated with the system,

$$|\Psi_{\mu\mathcal{SE}}\rangle \sim |\mu\rangle \sum_k |s_k\rangle|\epsilon_k\rangle \tag{7.18}$$

and the absolute values of the coefficients in the Schmidt decomposition of the entangled state describing $\mathcal{SE}$ are all equal, and $\mathcal{E}$ cannot be accessed, the resulting state of $\mathcal{S}$ is *objectively invariant* under all *local* measure-preserving transformations. Thus, with no need for further excuses, probabilities of events $\{|s_k\rangle\}$ must be – prior to measurement – equal.

By contrast, after an observer (pre-)measures the system, the overall state;

$$|\Phi_{\mu\mathcal{SE}}\rangle \sim \sum_k |\mu_k\rangle|s_k\rangle|\epsilon_k\rangle \tag{7.19}$$

obtains, with the correlation between his record $|\mu_k\rangle$ and the system state $|s_k\rangle$ allowing him to infer the state of the system from his record state. The invariance we have appealed to before is substantially restricted: correlated pairs $|\mu_k\rangle|s_k\rangle$ can be no longer separated and have to be permuted together. Thus, to a friend of the observer, all outcomes remain equiprobable, but to the "owner of the memory μ" his state is in part described by what he has found out about the system. Consequently, $|\mu_k\rangle$ implies $|s_k\rangle$ and the probability conditioned on the observer's own state in the wake of the perfect measurement is simply $p_{s_l|\mu_k} = \delta_{lk}$. Conditional probability in quantum theory emerges as an objective consequence of the relationship between the state of the observer and the rest of the universe, as the combined state under consideration (and not just the ill-defined and dangerously subjective "state of observer's knowledge" about a "definite but unknown classical state") is invariant in a manner that allows one to deduce equality of probabilities much more rigorously, directly, and without the copious apologies required in the classical setting.

We note that the above discussion of the acquisition of information owes a great deal to Everett (1957a, b). The apparent collapse occurrs on the way from eqn (7.18) to eqn (7.19). Envariance has given us a new insight into the nature of collapse: it is the extent of the correlations – the proliferation of information – that is of essential in determining what states of the quantum systems can be perceived by observers. When an envariant swap can be carried out on the $\mathcal{SE}$ pair, without involving the state of the observer (see eqn (7.18)), he is obviously ignorant of the state of $\mathcal{S}$. By contrast, a swap in eqn (7.19) would have to involve the state of the observer. This is because the information he has acquired is inscribed in the state of his own memory. There is no information without representation (Zurek 1994). In a sense, envariance extends the *existential interpretation* (Zurek 1993, 1998) introduced some time ago to deal with the issue of collapse.

Born's rule from envariance: general case

To discuss the general case we start with the state:

$$|\Psi_{\mathcal{SE}}\rangle = \sum_{k=1}^{N} \sqrt{\frac{m_k}{M}} |s_k\rangle|\epsilon_k\rangle, \tag{7.14b}$$

where $M = \sum_{k=1}^{N} m_k$ assures normalization. As the coefficients are commensurate, and as we assume that the Hilbert subspaces of $\mathcal{E}$ corresponding to different k are at least m_k dimensional, appropriate c-shift (Zurek 2000, 2003a):

$$|\epsilon_k\rangle|\varepsilon'\rangle = \frac{1}{\sqrt{m_k}}\left(\sum_{l_k}^{m_k} |\varepsilon_{l_k}\rangle\right)|\varepsilon'\rangle \Longrightarrow \frac{1}{\sqrt{m_k}}\sum_{l_k}^{m_k} |\varepsilon_k\rangle|\varepsilon'_k\rangle$$

that couples $\mathcal{E}$ with at least as large $\mathcal{E}'$ yields:

$$|\Psi_{\mathcal{SE}}\rangle|\varepsilon'\rangle \Longrightarrow M^{-1}\sum_{k=1}^{N} |s_k\rangle\left(\sum_{l_k}^{m_k} |\varepsilon_{l_k}\rangle|\varepsilon'_{l_k}\rangle\right). \tag{7.15b}$$

Here, in contrast to eqn (7.15a), we have immediately carried out the obvious cancellation, ($\sqrt{m_k}|\epsilon_k\rangle = \sum_{l_k}^{m_k} |\varepsilon_{l_k}\rangle$). It follows as a direct consequence of the relation between the states $|\epsilon_k\rangle$ and their Fourier–Hadamard transforms $|\varepsilon_k\rangle$.

The resulting state can be rewritten in a simpler and more obviously invariant form:

$$|\Phi_{\mathcal{S}\mathcal{E}\mathcal{E}'}\rangle = M^{-1} \sum_{j=1}^{M} |s_{k(j)}\rangle |\varepsilon_j\rangle |\varepsilon'_j\rangle \tag{7.16b}$$

where the environmental states are orthonormal, and the system state is the same within different m_k-sized blocks (so that the same state $|s_{k(j)}\rangle$ appears for m_k different values of j, and $\sum_{k=1}^{N} m_k = M$).

As before (see eqn (7.16a)) phases are irrelevant because of envariance. Hence, terms corresponding to different values of j can be swapped, and – by eqns (7.13) – their probabilities are all equal to $1/M$. It follows that:

$$p_k = p(|s_k\rangle) = m_k/M = |\alpha_k|^2 \tag{7.17b}$$

in obvious notation. This, as promised, is Born's rule. When $|\alpha_k|^2$ are not commensurate, one can easily produce sequences of states that set up convergent bounds on p_{s_k} so that – when the probabilities are assumed to be continuous in the amplitudes – the interval containing p_k shrinks in proportion to $1/M$ for large M.

We emphasize again that one could not carry out the basic step of our argument – the proof of the independence of the probabilities from the phases of the Schmidt expansion coefficients, eqn (7.12) and below – for an equal amplitude pure state of a single, isolated system. The problem with:

$$|\psi\rangle \;=\; N^{-\frac{1}{2}} \sum_{k}^{N} \exp(i\phi_k)|k\rangle$$

is the accessibility of the phases. Consider, for instance:

$$|\psi\rangle = (|0\rangle + |1\rangle - |2\rangle)/\sqrt{3} \quad \text{and} \quad |\psi'\rangle = (|2\rangle + |1\rangle - |0\rangle)/\sqrt{3}.$$

In the absence of entanglement there is no envariance and swapping of states correspondint to various k's is detectable: interference measurements (i.e., measurements of the observables with phase-dependent eigenstates $|1\rangle + |2\rangle$; $|1\rangle - |2\rangle$, etc.) would have revealed the difference between $|\psi\rangle$ and $|\psi'\rangle$. Indeed, given an ensemble of identical *pure* states a skilled observer should be able to confirm that they are pure and find out what they are. Loss of phase coherence is needed to allow for the shuffling of the states and coefficients.

Note that in our derivation environment and einselection play an additional, more subtle role: Once a measurement has taken place – i.e., a correlation with the apparatus or with the memory of the observer was established (e.g., eqns (7.18) and (7.19)) – one would hope that records will retain validity over a long time, well beyond the decoherence timescale. Thus, a "collapse" from a multitude of possibilities to a single reality (implied by eqn (7.19) above) can be confirmed by subsequent measurements only in the einselected pointer basis.

With this in mind, it is easy to see that – especially on the macroscopic level – the einselected states are the only sensible choice as outcomes: other sets of states lose correlation with the apparatus (or with the memory of the observer) far too rapidly – on the decoherence timescale – to serve as candidate events in the sample space.

We close this part of our discussion by calling reader's attention to the fact that the above derivation did not rely on – or even invoke – reduced density matrices, which are at the very foundation of the decoherence program. Indeed, we have used envariance to *derive* Born's rule, and, hence, in a sense, to justify the form and the uses of the reduced density matrices. More extensive discussion of this point shall be given elsewhere (Zurek, in preparation).

Summary and conclusions: quantum facts

In spite of the preliminary nature of much of the above (which would seem to make "Conclusions" premature) we point out that if one were forced to attach a single label to the topics explored above, *quantum facts* would be a possible choice. Quantum Darwinism approaches this theme directly: quantum states, by their very nature, share epistemological and ontological role – they are simultaneously a description of the state, and "the dream stuff is made of." One might say that they are *epiontic*. These two aspects may seem contradictory, but, at least in quantum setting, there is a union of these two functions.

Quantum Darwinism puts forward a specific theory of how the ontic aspect – reliable classical existence of the states – can emerge from the quantum substrate. We shall not repeat the arguments already given in detail. But one might sum up the key idea by pointing to the role of the redundancy: tenuous quantum facts acquire objective existence when the information they encode is amplified and widely disseminated (and therefore easily accessible). Approximate (exact) classicality obtains in the limit of a large (infinite) redundancy. Redundancy is a measure of classicality.

Envariance is, by contrast, a way to capture the most tenuous aspect of the quantum – the ignorance (and, hence, the essence of what is epistemic: the information). Quantum facts are the opposite of envariant properties. Quantum facts remain invariant under envariance. Thus, in a sense, what we have accomplished is to "corral" the problem of the emergence of the classical from quantum states between two extremes: the case – exploited by quantum Darwinism – where quantum facts become solid and reliable, and the opposite, when some candidate properties of these states are envariant, and, therefore, demonstrably inconsequential. Investigation, in terms of envariance and quantum Darwinism, of what lies in between these two extremes is still in its early stages.

References

Born, M (1926) *Zeits. Physik*, **37**, 863.

Bub, J (1997) *Interpreting the Quantum World*. Cambridge: Cambridge University Press.

Cover, T M and Thomas, J A (1991) *Elements of Information Theory*. New York: John Wiley.

Einstein, A, Podolsky, B, and Rosen, N (1935) *Phys. Rev.* **47**, 777. (Reprinted in Wheeler and Zurek (1983).)

Everett H, III. (1957a) *Rev. Mod. Phys.* **29**, 454. (Reprinted in Wheeler and Zurek (1983).)

(1957b) The theory of the universal wave function. Ph.D. thesis, Princeton University. (Reprinted in DeWitt B S, and Graham N (1973) *The Many-Worlds Interpretation of Quantum Mechanics*. Princeton, NJ: Princeton University Press).)

Giulini, D, Joos, E, Kiefer, C, *et al.* (1996) *Decoherence and the Appearance of a Classical World in Quantum Theory*. Berlin: Springer-Verlag.

London, F and Bauer, E (1939) *La Théorie de l'observation en méchanique quantique*. Paris: Hermann. (English translation in Wheeler and Zurek (1983).)

Ollivier, H and Zurek, W H (2002) *Phys. Rev. Lett.* **88**, 017901.

(2003c) Quantum discord and Maxwell's demons. *Phys. Rev. A.*, **67**, 012320.

Paz, J-P and Zurek, W H (2001) In *Coherent Atomic Matter Waves, Les Houches Lectures*, ed. R. Kaiser, C. Westbrook, and F. David, pp. 533–614. Berlin: Springer-Verlag.

Schrödinger, E (1935a) *Nautürwiss.* **23**, 807–812, 823–828, 844–849. (English translation in Wheeler and Zurek (1983).)

(1935b) *Proc. Cambridge Phil. Soc.* **31**, 555.

(1936) *Proc. Cambridge Phil. Soc.* **32**, 446.

von Neumann, J (1932) *Mathematische Grundlagen der Quantenmechanik*. Berlin: Springer-Verlag. English translation by R T Beyer (1955) *Mathematical Foundations of Quantum Mechanics*. Princeton, NJ: Princeton University Press.

Wheeler, J A (1983) In *Quantum Theory and Measurement*, ed. J. A. Wheeler and W. H. Zurek, p. 772. Princeton, NJ: Princeton University Press.

Wheeler, J A and Zurek, W H (eds.) (1983) *Quantum Theory and Measurement*. Princeton, NJ: Princeton University Press.

Wigner, E P (1963) *Am. J. Phys.* **31**, 6. (Reprinted in Wheeler and Zurek (1983).)

Zurek, W H (1981) *Phys. Rev.* **D24**, 1516.

(1982) *Phys. Rev.* **D26**, 1862.

(1983) In *Quantum Optics, Experimental Gravitation, and Measurement Theory*, ed. P. Meystre and M. O. Scully, p. 87. New York: Plenum Press.

(1991) *Physics Today* **44**, 36.

(1993) *Progr. Theor. Phys.* **89**, 281.

(1994) In *From Statistical Physics to Statistical Inference and Back*, ed. P. Grassberger and J.-P. Nadal, pp. 341–350. Dordrecht: Plenum Press.

(1998) *Phil. Trans. Roy. Soc. Lond.* **A356**, 1793.

(2000) *Ann. der Physik* (*Leipzig*) **9**, 855.

(2003a) Decoherence, einselection, and the quantum origins of the classical. *Rev. Mod. Phys.*, **75**, 715.

(2003b) Environment-assisted invariance, ignorance, and probabilities in quantum physics. *Phys. Rev. Lett.*, **90**, 120404.

(2003c) Quantum discord and Maxwell's demons. *Phys. Rev. A.*, **67**, 012320.

8

Using qubits to learn about "it"

Juan Pablo Paz
University of Buenos Aires

Introduction

Almost a century after its birth, quantum theory remains odd and counterintuitive. As Richard Feynman wrote, it seems that nobody really understands it. This is indeed true if by understanding we mean being able to explain it using our common sense and everyday experience. The development of some kind of "quantum common sense" has been very slow even though our everyday life is being continuously influenced by technologies whose roots lie in quantum laws. In recent years the growing field of quantum information and quantum computation became a fruitful playground for physicists, mathematicians, computer scientists, and researchers from other fields who developed new interesting ways of storing, transmitting, and processing information using quantum mechanics at its best. Thus, both theoretical and experimental research on multiparticle entanglement, on the manipulation of individual quantum systems, on decoherence, and on the transition from quantum to classical are subjects of interest not only for their basic relevance but also for their potential practical significance as they might be of help for the development of a real quantum computer. Even though this technology may be far in the future, it is interesting to speculate about what lessons on quantum reality could be learned from quantum computation (eg., what would happen if one could operate a quantum computer). From a physicist's perspective this is much more interesting than, for example, being able efficiently to factor large numbers or rapidly to search a giant database (two of the most important killer applications known today). In this chapter we will give a simple presentation of some results connected to the general idea of using quantum computers for physics simulations. Among the applications that would enable us to learn about quantum properties of natural systems we will focus on the simulation of spectroscopy and state tomography,

Science and Ultimate Reality, eds. J. D. Barrow, P. C. W. Davies and C. L. Harper, Jr. Published by Cambridge University Press.

which can be viewed as dual forms of quantum computation. Lessons learned from these studies will hopefully help us develop our quantum intuition and, maybe, enable us finally to understand the quantum. The paper is organized as follows. In the first section we will present an introduction to the use of quantum circuits. We will do this using a simple example of a quantum circuit describing a process that closely resembles a scattering experiment. Next we will use this tool to establish an analogy between spectroscopy and tomography, showing the reason why these two tasks can be viewed as dual forms of the same quantum computation. We will describe in some detail the way in which the scattering circuit can be adapted to perform a class of tomographic experiments with the goal of efficiently measuring a phase space distribution characterizing the quantum state of a system. We will go on to analyze how the same scattering circuit can be generalized to build a set of programmable devices that can be used to measure expectation values of a class of operators. Finally, we will discuss how it is possible that, by applying the same strategies discussed in the previous sections, quantum computers could be used to address problems that cannot even be formulated classically. Such problems have input data which are inherently quantum and are encoded, for example, in the quantum state of a system. Some of these problems can also be addressed using variants of the quantum scattering circuit.

A quantum circuit representing a scattering experiment

The evolution of a quantum system can be pictorially represented using a quantum circuit (Nielsen and Chuang 2000). In this approach, the initial and final states of the system are respectively the input and output states of the circuit. The temporal evolution, which in quantum mechanics is enforced by unitary operators, is represented by a sequence of black boxes that map the input state into its output. Each of these black boxes represents the evolution of the system under conditions that can be controlled by the experimenter. Clearly, the theory of quantum circuits is of interest in the context of the study of quantum computers. In such a case, the quantum circuit defines the program that the quantum computer is executing. However, the use of quantum circuits to describe the evolution of a system is not restricted to quantum computation. Thus, quantum circuits provide an abstract representation of temporal evolution and, as such, they can be used to describe a wide variety of physically interesting problems. Most importantly, they can be used to establish analogies between different physical processes. In fact, we believe that it would be very fruitful for a variety of areas of physics to incorporate the language and framework provided by the theory of quantum circuits. Below, we will discuss an interesting example of possible analogies between different physical processes that can be most naturally exhibited using the language of

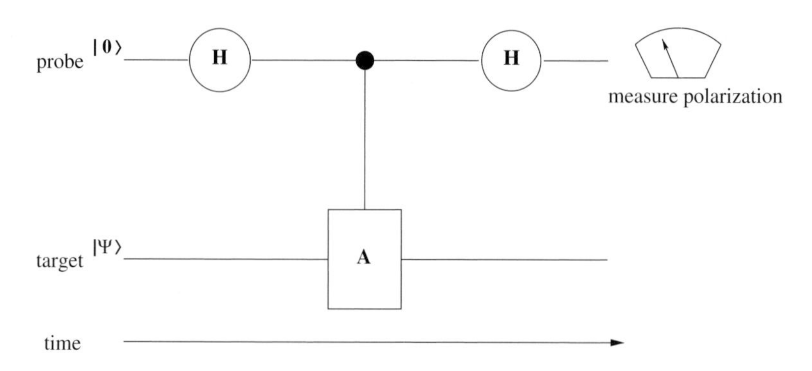

Figure 8.1. The scattering circuit. The auxiliary qubit, initially in state $|0\rangle$, plays the role of a probe particle in a scattering experiment. The target system is initially in state $|\Psi\rangle$. After the interaction, the polarization of the probe provides information either about the state of the target or about the interaction

quantum circuits. We start by describing one of the simplest such circuits that plays an important role in many known quantum algorithms. This circuit will enable us to establish a close connection between quantum computation, spectroscopy, and tomography.

Let us consider in some detail the so-called quantum "scattering" circuit shown in Fig. 8.1. In the circuit time runs from left to right and each cable represents a quantum system with a space of states of arbitrary dimensionality (it is often convenient, but not necessary, to use quantum circuits where lines represent qubits, i.e., quantum systems with a two dimensional space of states). To describe the behavior of the scattering circuit we will follow the temporal evolution of the quantum state along it. For this purpose we define two intermediate times (t' and t'') between the initial and final ones which will be denoted $t_{\rm i}$ and $t_{\rm f}$ respectively. In the scattering circuit a system, initially in the state $|\Psi\rangle$, is brought in contact with an ancillary qubit prepared in the state $|0\rangle$. This ancilla acts as a "probe particle" in a scattering experiment where the target is initially in the quantum state $|\Psi\rangle$. Therefore, the initial state of the complete system is:

$$|\Phi(t_{\rm i})\rangle_{12} = |0\rangle_1 \otimes |\Psi\rangle_2. \tag{8.1}$$

The process described by the quantum circuit consist of the following steps: apply a Hadamard transform H to the ancillary qubit. This is a unitary operator whose action in the computational basis is $H|0\rangle = (|0\rangle + |1\rangle)/\sqrt{2}$, $H|1\rangle = (|0\rangle - |1\rangle)/\sqrt{2}$. Therefore, the new state of the combined system at the intermediate time t' is:

$$|\Phi(t')\rangle_{12} = \frac{1}{\sqrt{2}}(|0\rangle_1 + |1\rangle_1) \otimes |\Psi\rangle_2. \tag{8.2}$$

The second step of the algorithm is to apply a "controlled-A" operator. This is a unitary operator corresponding to a special type of interaction between the probe and the target system. Such operator does not change the state of the target if the state of the ancilla is $|0\rangle$. However, if the ancilla is in state $|1\rangle$ then the system evolves with the unitary operator A. In this sense, the evolution of the target is controlled by the probe. Therefore, the complete state after this step is:

$$|\Phi(t'')\rangle_{12} = \frac{1}{\sqrt{2}}(|0\rangle_1 \otimes |\Psi\rangle_2 + |1\rangle_1 \otimes A|\Psi\rangle_2). \tag{8.3}$$

The fourth step of the algorithm is to apply another Hadamard gate to the ancilla. The complete state is:

$$|\Phi(t_{\rm f})\rangle_{12} = \frac{1}{2}(|0\rangle_1 \otimes (|\Psi\rangle_2 + A|\Psi\rangle_2) + |1\rangle_1 \otimes (|\Psi\rangle_2 - A|\Psi\rangle_2)). \tag{8.4}$$

After this sequence of operations one performs a measurement of the probe detecting this qubit either in state $|0\rangle$ or in state $|1\rangle$. It is simple to compute the probabilities of these two events. More interestingly, one can show that the spin polarizations of this qubit along the z- and y-axes are $\langle\sigma_z\rangle = \mathrm{Re}[\langle\Psi|A|\Psi\rangle]$, $\langle\sigma_y\rangle = -\mathrm{Im}[\langle\Psi|A|\Psi\rangle]$ (notice that these polarizations should be measured using sufficiently many instances of the experiment). More generally, if the quantum state of the system is described by a density matrix ρ instead of a pure state $|\Psi\rangle$ the polarization of the probe measures the real and imaginary parts of $\mathrm{Tr}(\rho A)$, i.e.,

$$\langle\sigma_z\rangle = \mathrm{Re}[\mathrm{Tr}(\rho A)],\ \langle\sigma_y\rangle = -\mathrm{Im}[\mathrm{Tr}(\rho A)]. \tag{8.5}$$

What is the relevance and utility of the scattering circuit? From a physicist's point of view, the connection between the quantum algorithm and a scattering experiment is quite interesting. In physics, one uses this kind of experiment at least in two different ways. On the one hand, one can use a scattering experiment to learn about the interaction between probe and target provided one has some information about the state of the latter. On the other hand, one can use the same kind of experiment to learn about the state of the target provided one has information about the interaction. In the scattering circuit this duality is explicit: this is because the state ρ and the unitary operator A appear symmetrically in the right-hand side of eqn (8.5). As a consequence, the final polarization measurement in the scattering circuit can be used to reveal properties of either A or ρ. For this, we should use known instances of ρ or A, respectively. This is the reason why the circuit can be used to show the duality between two relevant tasks whose use is widespread in physics: spectroscopy and tomography. In the next section we will explore some of the practical consequences we can derive from this analogy. Here, we would like to point out that versions of the scattering circuit play an important role in many quantum algorithms. For pure input states, it occurs in the phase estimation algorithm, which is the core of Kitaev's (1997) solution to the Abelian stabilizer problem and is itself closely related to Shor's factoring algorithm (see Nielsen and Chuang 2000). This circuit was later adapted by Cleve *et al.* (1998) to revisit most quantum algorithms. Moreover Abrams and Lloyd (1999) gave another presentation of the algorithm as a tool for finding random, approximate eigenvalues of certain Hamiltonians. The simple extension to mixed states is described in Knill and Laflamme (1999) as an example illustrating the potential power of a single qubit in a pure state. More recently, the scattering circuit was used to establish and explore the duality between tomography and spectroscopy (Miquel *et al.* 2002a).

Two dual faces of quantum computation

Tomography and spectroscopy are techniques that are closely related to two of the most important problems a physicist faces: determining the state of a system and measuring properties of its evolution. For the first purpose one uses state tomography (Smithey *et al.* 1993; Breitenbach *et al.* 1995). This is a methodology that after subjecting the system to a number of experiments completely determines its quantum state For the second task, one can use spectroscopy, a set of techniques used to determine the spectrum of eigenvalues of the evolution operator. In Miquel *et al.* (2002a), we showed that tomography and spectroscopy can be naturally interpreted as dual forms of quantum computation. Moreover, we showed how to use the scattering circuit for both tasks constructing a general tomographer and a spectrometer. It is worth pointing out that the above framework can be used in practice provided one has enough control over the physical systems involved. Thus, the circuit describes a quantum simulation run by a quantum computer that could be used to determine properties of the spectrum of an unknown but implementable evolution A. For example, consider the case in which the initial state of the target is completely mixed, i.e., $\rho = I/N$. Then, using eqn. (8.5) we can show that the final polarization measurement turns out to be $\langle \sigma_z \rangle = \text{Tr}(A)/N$. Thus, in such way we directly measure the trace of a unitary operator, which carries relevant information about its spectrum. In Miquel *et al.* (2002a) we showed how to adapt the scattering circuit to perform a deterministic measurement of the spectral density of the unitary operator A. For this purpose one needs to add an extra register to the circuit whose state determines the value of the energy E where one wants to evaluate the spectral density. Potential applications and generalizations of this method could turn out to be useful in the future for measuring quantities characterizing fluctuations and statistical correlations in the spectrum of an operator. In fact, these techniques could provide a way in which quantum computers could be used to perform novel physics simulations aimed at finding information on properties of physical models characterized by some unitary operators with the final goal of contrasting them with experimental predictions. It is also worth pointing out that it is quite straightforward to generalize these ideas to analyze spectral properties of nonunitary (i.e., dissipative) operators (Miquel *et al.* 2002a).

As mentioned above, the scattering circuit can also be viewed as a tomographer. The purpose of tomography is to completely determine an unknown but preparable state ρ. Let us briefly review how to use the scattering circuit as a primitive to design a tomographer. As a consequence of eqn (8.5) we know that every time we run the algorithm for a known operator A, we extract information about the state ρ. Doing so for a complete basis of operators $\{A(\alpha)\}$ one gets complete information and determines the full density matrix. Different tomographic schemes are characterized by the basis of operators $A(\alpha)$ they use. Of course, completely determining the quantum state requires an exponential amount of resources. In fact, if the dimensionality of the Hilbert space of the system is N then the complete determination of the quantum state involves running the scattering circuit for a complete basis of N^2 operators $A(\alpha)$. However, evaluating any coefficient of the decomposition of ρ in a given basis can be done efficiently provided that the operators $A(\alpha)$ can be implemented by efficient networks. A convenient basis set that could be used for this purpose (Miquel *et al.* 2002a, b) is defined as:

$$A(\alpha) = A(q, p) = U^q R V^{-p} \exp(i2\pi pq/2N). \qquad (8.6)$$

Here, both q and p are integers between 0 and $N - 1$, U is a shift operator in the computational basis ($U|n\rangle = |n + 1\rangle$), V is the shift operator in the basis related to the computational one via the discrete Fourier transform, and R is the reflection operator ($R|n\rangle = |N - n\rangle$). It is straightforward to show that the operators $A(\alpha)$ are hermitian, unitary and form a complete orthonormal basis of the space of operators satisfying

$$\mathrm{Tr}[\hat{A}(\alpha)\hat{A}(\alpha')] = N\delta_N(q' - q)\delta_N(p' - p), \qquad (8.7)$$

where $\delta_N(x)$ denotes the periodic delta function that is nonvanishing if $x = 0$ (modulo N). With this specific choice for $A(\alpha)$ the scattering circuit directly evaluates the discrete Wigner function (Hannay and Berry 1980; Wooters 1987; Leonhardt 1996; Miquel *et al.* 2002a, b). This function enables us to represent the state of a quantum system in phase space and its properties are worth analyzing in some detail.

The use of phase space methods to analyze the state and evolution of quantum systems is widespread in other areas of physics such as quantum optics (see Davidovich (1999) and Hillery *et al.* (1984) for a review), but only recently has been applied in the context of quantum computation (Bianucci *et al.* 2002; Miquel *et al.* 2002b; Paz 2002). Let us mention some of the most relevant properties of the Wigner function, which is the basic tool on which phase space methods are based. This function is the

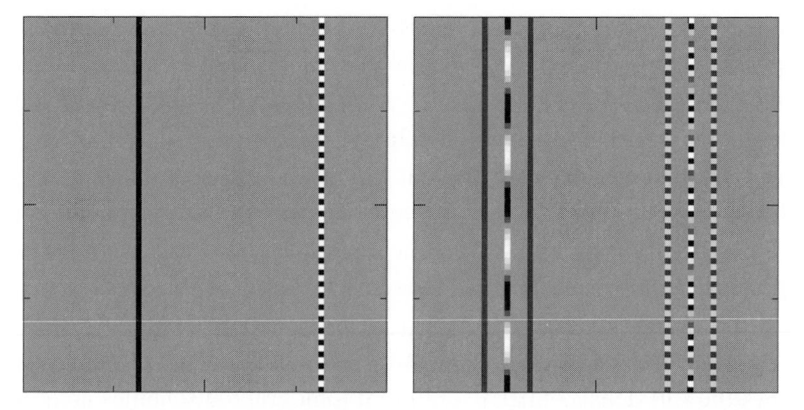

Figure 8.2. The discrete Wigner function for a computational state (left) and a superposition of two computational states (right). Black (white) areas correspond to positive (negative) values of the Wigner function. Thus, a computational state is characterized by a positive vertical strip and an oscillatory partner. This oscillatory component comes from the interference between the main peak and the mirror images created by the periodic boundary conditions. For a superposition state there are two positive vertical strips with interference fringes in between (and three oscillatory partners).

closest one can get to a phase space distribution function in quantum mechanics. In fact, it is a real function which uniquely determines the state of a quantum system. As it can be negative, it cannot be interpreted as a true probability density in phase space. In this sense, Wigner functions clearly display the key ingredients of quantum mechanics: quantum coherence and interference. These functions can be used to compute observable quantities as if they were probability densities since they relate traces of products of operators to phase space averages through an expression like $\mathrm{Tr}(\rho_1\rho_2) = N\sum_{q,p} W_1(q, p)W_2(q, p)$. Moreover, they have a crucial property: adding the value of the Wigner function along any line in phase space one computes the probability distribution for the measurement of an observable (the generator of translations along the line). Discrete and continuous Wigner functions share many similarities but the main difference between them is that for a system with an N-dimensional Hilbert space the Wigner function is defined on a grid of $2N \times 2N$ points, with periodic boundary conditions (and the topology of a torus). In Fig. 8.2 we display the discrete Wigner function for two different states (a computational state and a superposition of two computational states). The main difference between the discrete and the continuous cases is due to the periodic boundary conditions imposed in the discrete case. As a consequence of them, for every positive peak in the discrete Wigner function, there is a mirror image located a distance $2N$ away (mirror images appear both in position and momentum directions). Therefore, in between the peak and its mirror image there are regions with interference fringes where the discrete Wigner function oscillates (these regions move towards infinity in the large N limit and are absent in the continuous case).

One can ask if there are potential advantages in using a phase space representation for a quantum computer. Phase space methods have been fruitful for example, in analyzing issues concerning the classical limit of quantum mechanics (Paz *et al.* 1992; Paz and Zurek 2001). For quantum computers, the semiclassical regime corresponds to the limit of a large number of qubits since $1/N$ plays the role of an effective Planck constant (since $2\pi\hbar = 1/N$, where $log(N)$ is the number of qubits of the quantum computer; in this paper, the logarithm is always taken in base 2). Therefore, one may speculate that some quantum algorithms display interesting properties in such limit when represented as a quantum map in phase space (indeed, a quantum algorithm is just a sequence of unitary operators that are applied successively). Of course, whether this representation will be useful or not will depend on properties of the algorithm and we do not expect this to be true in every case. In Miquel *et al.* (2002b) we applied this idea to analyze Grover's quantum search algorithm in phase space showing an interesting analogy between this algorithm a classical phase space flow (a map with a fixed point). Here, we will not discuss the phase space representation of quantum algorithms but concentrate on discussing the tomographic scheme that can be used to measure the discrete Wigner function. This was discussed in detail in Miquel *et al.* (2002a, b) where it was shown that the evaluation of the Wigner function in any phase space point $\alpha = (q, p)$ can be done efficiently. The reason for this is that in such case the resulting scattering circuit involves the use of controlled displacements, reflections, and the Fourier transform which can all be built using efficient networks consisting of a number of gates scaling that grows polynomially with the number of qubits.

Measuring directly the Wigner function has been the goal of a series of experiments in various areas of physics, all dealing with continuous systems (Dunn *et al.* 1995; Leibfried *et al.* 1996, 1998). However, it is worth pointing out that all those experiments are highly inefficient in the following sense: they involve the complete determination of the quantum state by measuring marginal distributions for a complete family of observables. After those measurements the Wigner function is reconstructed by using a Radon-like transform. However, our above discussion shows that a general efficient scheme for the measurement of the Wigner function exists and is represented by the scattering circuit. Most remarkably some of the most recent experiments that have been performed to determine $W(\alpha)$ for the electromagnetic field in a cavity QED setup (Lutterbach and Davidovich 1997, 1998; Maitre *et al.* 1997; Nogues *et al.* 2000) turn out to be concrete realizations of the scattering circuit discussed above. In such case the system is the mode of the field stored in a high-Q cavity and the ancillary qubit is a two-level atom. The Wigner function for this continuous system can be defined as $W(\alpha) = \mathrm{Tr}(B(\alpha)\rho)/\pi\hbar$ where $B(\alpha) = D(\alpha)RD^\dagger(\alpha)$, and $D(\alpha)$ is a phase space displacement operator and R a reflection. Therefore, the measurement of the continuous Wigner function can be done by first preparing the state $D(\alpha)\rho D^\dagger(\alpha)$ and using later the scattering circuit with $A = R$. (The initial state should be prepared by displacing the quantum state ρ. It is worth noticing that this scheme is slightly simpler than applying the scattering circuit for the operator $A = B(\alpha)$. The experiment that was originally proposed in Lutterbach and Davidovich (1997, 1998) consists of the following sequence, which closely follows the scheme described in Fig. 8.1.)

1. The atom goes through a Ramsey zone that has the effect of implementing an Hadamard transform. An r.f. source is connected to the cavity displacing the field (by an amount parametrized by α).
2. The atom goes through the cavity interacting dispersively with the field. The interaction is tuned in such a way that only if the atom is in state $|e\rangle$ does it acquire a phase shift of π per each photon in the cavity (i.e., this interaction is a controlled-$\exp(-i\pi\hat{N})$ gate, where $\hat{N}$ is the photon number, which is nothing but a controlled reflection).
3. The atom leaves the cavity entering a new Ramsey zone and is finally detected in a counter either in the $|g\rangle$ or $|e\rangle$ state.

The Wigner function is measured as the difference between both probabilities: $W(\alpha) = 2(P(e) - P(g))/\hbar$. As we see, this cavity-QED experiment is a concrete realization of the general tomographic scheme described above. Recently, the measurement of the discrete Wigner function characterizing the quantum state of a system of two qubits was measured using NMR quantum computation techniques in a liquid sample (Miquel *et al.* 2002a). The remarkable analogy between this and other kind of experiment mentioned above is possible thanks to the use of a common language and framework provided by quantum circuits which, as it is quite clear from the above discussion, are of interest not only in the context of quantum computers.

Quantum programs

It is important to understand the differences and similarities between quantum and classical computers. A well-known feature of classical computers is that they can be programmed. That is to say, a fixed universal device can perform different tasks depending on the state of some input registers (that define the program the computer is executing). Quantum computers, by contrast, have a different property: in fact, a general-purpose quantum computer cannot be programmed using quantum software. Thus, Nielsen and Chuang (1997) established that it is not possible to build a fixed quantum gate array which takes as input a quantum state, specifying a quantum program, and a data register to which the unitary operator corresponding to the quantum program is applied. The existence of nondeterministic programmable gate arrays was also established in Nielsen and Chuang (1997) and analyzed later in a variety of interesting examples (Vidal, and Cirac 2000; Huelga *et al.* 2001; Hillery *et al.* 2002). Other programmable quantum devices were studied more recently: quantum multimeters were introduced and discussed first in Dusek and Buzek (2002). Such a device is defined as a fixed gate array acting on a data register and an ancillary system (the program register) together with a final fixed projective measurement on the composite system. They are programmable quantum measurement devices (see Dusek and Buzek 2002) that act either nondeterministically or in an approximate way (see Fiurasek *et al.* (2002) for a proposal of a quantum multimeter that approximates any projective measurement on a qubit). Here, we will show that the scattering circuit can be easily adapted to describe a different kind of programmable quantum gate array. The array will have a data register, a program register, and an auxiliary qubit. The circuit is such that the final measurement of the polarization of the auxiliary qubit is equal to the expectation value of an operator specified by the program register. The expectation value is evaluated in the quantum state of the data register. We will discuss what kind of operators we can evaluate with this method and exhibit a way in which this scheme could be used to evaluate probabilities associated with measurements of a family of observables on a quantum system. Remarkably, the gate arrays we will use are simple application of the scattering circuit introduced above.

The procedure to construct a programmable circuit that evaluates the expectation value of a family of operators O is rather simple. To do this we can first generalize the scattering circuit used to measure the discrete Wigner function. In such a circuit q and p enter as classical information that determines the quantum network (i.e., the hardware). However, it is simple to realize that one can construct an equivalent circuit where both q and p are stored in the quantum state of a program register. Let us denote the state of such register as $|\Psi\rangle_P$. The universal network that computes the discrete Wigner function at a point $\alpha = (q, p)$ has a program register that should be prepared in the state $|\Psi\rangle_P = |q\rangle \otimes |p\rangle$. The quantum circuit is such that the qubits of the program register control the application of a displacement operator either in the computational or in the conjugate basis. The network obtained in this way requires a number of elementary gates which scales polynomially with the number of qubits (i.e., with $log(N)$). As we mentioned above, the Wigner function is nothing but the expectation value of the operator $A(q, p)$. Therefore, the circuit we have just described has an obvious property: different states $|q\rangle|p\rangle$ of the program register are used to evaluate the expectation value of orthogonal operators $A(q, p)$. However, the same circuit can be used with arbitrary states of the program register. In particular, if the program state is $|\Psi\rangle_P = \sum_{q,p} c(q, p)|q\rangle|p\rangle$ then the programmable gate array would evaluate the expectation value of a linear combination of the operators $A(\alpha)$ given by:

$$\langle \sigma_z \rangle = Re\left(\sum_{q,p} |c(q,p)|^2 A(q,p)\right). \tag{8.8}$$

Therefore, it is clear that using this method we can evaluate the expectation value of any operator O that can be written as a convex sum of the basis set $A(q,p)$. To do this we should proceed as follows: one should expand O in the basis $A(q,p)$ writing $O = \sum_{q,p} O(q,p)A(q,p)/N$ (note that the coefficients $O(q,p)$ are simply given by $O(q,p) = Tr(O\,A(q,p))$). (2) To measure O one needs to prepare the program state $|\Psi\rangle_P = \sum_{q,p} c(q,p)|q\rangle|p\rangle$ where $O(q,p) = |c(q,p)|^2$. Thus, the coefficients of the expansion of the operator O in the basis $A(q,p)$ determine the program state $|\Psi\rangle_P$ required to measure the expectation value of O. Due to eqn (8.8) this method is limited to evaluating operators that can be expressed as a convex sum of the basis $A(q,p)$. The basis one uses defines the quantum multimeter. Thus, for any operator one can assure that there is a quantum multimeter that evaluates it but a given multimeter is necessarily limited in scope by the above considerations. The use of the basis $A(\alpha)$ given in eqn (8.6) is merely a possible option that may be convenient in some cases (see below). The choice of such basis determines the way in which the gate array is designed (the hardware) and the way in which the quantum program should be written (the software). For this method to have a chance at being efficient the operators chosen as a basis set should be implementable by efficient networks (a requirement that is satisfied by eqn (8.6)). There are two implicit requirements for this scheme to be practical. First, one needs to be able to express the operator to be measured as a linear combination of the basis $A(q,p)$. This is of course in principle possible for any operator but the task of finding the coefficients $O(q,p)$ can be hard for most cases. The second implicit requirement that could affect the practicality of the method is that the program state $|\Psi\rangle_P$ given in eqn (8.8) should be efficiently preparable.

An application of the above ideas is the design of programmable circuits to measure sums of Wigner functions over regions of phase space. Let us consider a line in phase space, which is defined as the set of points (q,p) such that $ap - bq = c$ (mod $2N$) where a, b, and c are integers between 0 and $2N$. As we mentioned above, a crucial property of the Wigner function is that by adding them along a line $ap - bq = c$ one obtains the probability to find a certain result in the measurement of a specific observable (both the result and the observable are associated with the line). This is a consequence of the fact that when adding the operators $A(\alpha)$ along the line one obtains a projection operator along an eigenstate of the translation operator $T(b,a) = U^a V^b \exp(i\pi ab/N)$ with eigenvalue $\exp(i\pi c/N)$.

More precisely,

$$\sum_{aq-bp=c} W(q,p) = \mathrm{Tr}(\rho|\Phi(a,b,c)\rangle\langle\Phi(a,b,c)|) \tag{8.9}$$

where $|\Phi(a,b,c)\rangle$ is an eigenstate of $T(b,a)$ with eigenvalue $\exp(i\pi c/N)$ (if $\exp(i\pi c/N)$ is not in the spectrum of $T(b,a)$ then $|\Phi(a,b,c)\rangle$ is the null vector). It is interesting to notice that a basis of the Hilbert space is associated with a foliation of the phase space with lines defined by fixed values of the parameters a and b. A line within the foliation specifies a particular state within the basis set (parametrized by the integer c). It is clear that by choosing the state of the program register as $|\Psi\rangle_P = \sum_q |q\rangle|p_0\rangle/\sqrt{N}$ the final polarization measurement of the ancillary qubit in the scattering circuit turns out to be equal to $\langle\sigma_x\rangle = \sum_q W(q,p_0)/N$. Therefore, in this way we measure the sum of values of the Wigner function along a vertical line defined by the equation $p = p_0$. A similar result can be used to measure the sum of values of the Wigner function along horizontal lines. More interestingly, lines with an arbitrary slope can be handled in this way. In fact, this can be done by first noticing that a vertical line can be mapped into a tilted one by applying a linear, area preserving, transformation (a linear homeomorphism on the discrete phase space torus). Unitary operators corresponding to quantizations of such linear transformations have been extensively studied and are known as "quantum cat maps" (Arnold's cat is a notorious member of this family). As shown in Miquel *et al.* (2002a) such cat maps have a very important property: when the evolution of the system is analyzed in phase space by using the discrete Wigner function one discovers that the action of the cat map is completely classical. Thus, the Wigner function flows from one point to another following a classical path determined by the classical linear transformation. Using this observation it is easy to see how to compute the sum of values of the Wigner function along a tilted line. One first transforms the state using an appropriately chosen cat map and later adds the Wigner function along a vertical or horizontal line (Roncaglia 2002). The existence of entanglement between subsystems can be related to some properties of the discrete Wigner function. In fact, for a composite system formed by two parts A and B, the discrete Wigner function can be defined as $W(\alpha_1,\alpha_2) = \mathrm{Tr}(\rho_{AB}A(\alpha_1)\otimes A(\alpha_2))/N^2$. In Paz (2002) we showed the existence of a relation between nonseparable states and nonseparable lines in the phase space. An example of a foliation of phase space with nonseparable lines is defined by the equations $q_1 - q_2 = q_0$ and $p_1 + p_2 = p_0$. Such foliation defines a basis of the complete Hilbert space. A state within this basis corresponds to a given line, which is associated with a given value of the relative coordinate q_0 and the total momentum p_0. Such a basis consists of nonseparable Bell states. Therefore, adding the total Wigner function over a nonseparable line such as the above one obtains the probabilities for the results of a Bell measurement on the system.

Quantum questions

Quantum computers are usually thought of as devices being useful for efficiently answering questions motivated in computational problems. Factoring integers, searching databases, etc., are very important problems with many practical applications. However, this is a limited view of quantum computation. In fact, Feynman himself emphasized from the very early days of quantum computation that performing interesting physics simulations could be one of the most relevant uses of quantum computers. In the previous sections we discussed a few interesting examples of quantum computations that are aimed at analyzing properties of physical systems and are related to well-known physical processes. Here, we would like to conclude by pointing out that in the future, one of the most relevant uses of quantum computers may be to answer questions that cannot even be formulated in classical terms. In fact, a quantum computer would have the unique property of being useful to address an entirely new kind of "quantum decision problems." These kinds of problems can be thought of as ordinary decision problems where the input data is inherently quantum. A simple example of this kind is the following: given (many copies of) a quantum state ρ, we may be interested to determine if the state ρ is pure or not (indeed, a physically meaningful question). Similarly, we could ask if the purity $\chi = \mathrm{Tr}\rho^2$ is less than some specified value. More generally, quantum decision problems can be thought of as problems where the input data are given in terms of a quantum state. It is clear that the question cannot even be formulated classically! The answer to the questions posed above is indeed a piece of classical information (a classical bit corresponding to the answer yes or no). This information can later be used to make a decision that would depend on the quantum state satisfying the desired criterion or not.

Remarkably, this type of quantum decision problem can be addressed using a minor variations of the scattering circuit we have described above. For example, the quantum decision problem concerning the purity of a quantum state can be solved as follows: the basic idea is to use the fact that the quantum scattering circuit measures $\mathrm{Tr}(\rho A)$ for a unitary operator A. So, we just need to find a convenient initial state and unitary operator well-adapted to answer the question we are interested in. For this, let us consider ρ as the quantum state of a composite system formed by two identical, and initially uncorrelated, subsystems A and B: $\rho = \rho_A \otimes \rho_B$. Consider also the unitary operator $A = U_{swap}$ where U_{swap} is the operator swapping the state of the two subsystems (i.e., $U_{swap}|\Psi\rangle_A \otimes |\phi\rangle_B = |\phi\rangle_A \otimes |\Psi\rangle_B$). In such case the scattering circuit directly measures the overlap between the states ρ_A and ρ_B. In fact, $\mathrm{Tr}(\rho_A \otimes \rho_B U_{swap}) = \mathrm{Tr}(\rho_A \rho_B)$. As the control-swap operation can be implemented by an efficient network, the scattering circuit can be used to efficiently measure the overlap between two quantum states. Of course, for this to be possible we need to have a supply of many identically prepared copies of the above states. In the particular case in which we consider $\rho_A = \rho_B$ the circuit measures the purity of the state. As discussed in Horodecki and Ekert (2001), this method can be generalized to measure the von Neumann entropy of a quantum state. For this purpose one would need to measure traces of powers of ρ where the power ranges up to the dimensionality of the Hilbert space of the system (which is a highly inefficient scheme). It is worth stressing again that the techniques required to formulate this problem in terms of quantum circuits are esentially the same as those used in the previous sections and only require minor generalizations of the scattering circuit. In fact, if the data of a quantum decision problem are given in terms of a quantum state, the problem can be viewed as an ordinary problem with a tomographic subroutine (it is also clear that there are dual problem with spectroscopic subroutines). A class of quantum decision problems have been recently considered by P. Horodecki, A. Ekert and others (Ekert *et al.* 2001; Horodecki and Ekert 2001). In particular, the question addressed by those authors as a decision problem was the detection of entanglement (i.e., to determine whether a given quantum state is separable or not). Answering these kinds of questions, which cannot even be formulated in classical terms, is indeed a unique feature of quantum computers. It is clear that these problems, whose relevance and importance have not been yet fully explored, are of interest from the point of view of a physicist (but they are not directly connected to classical computational problems). The search for other classes of quantum problems, formulated in terms of input data with quantum structure, seems to be a rather interesting avenue where one could find novel uses and application for quantum computers. In helping to solve these problems, quantum computers will enable us to develop new intuition into the nature of the quantum. In this way the qubit will help us to learn about it.

Acknowledgments

This work was partially supported with grants from Fundación Antorchas, UBACyT and ANPCyT. The author acknowledges useful discussions with Marcos Saraceno and Augusto Roncaglia.

References

Abrams, D and Lloyd, S (1999) *Phys. Rev. Lett.* **83**, 5162.
Bianucci, P, Miquel, C, Paz J P, *et al.* (2002) *Phys. Lett.* **A299**, 353.
Breitenbach, G, *et al.* (1995) *J. Opt. Soc. Am.* **B12**, 2304.
Cleve, R, Ekert, A, Macchiavello, C, and Mosca, M (1998) *Proc. Roy. Soc. Lond.* **A454**, 339.
Davidovich, L (1999) Quantum optics in cavities, phase space representations and the classical limit of quantum mechanics. In *New Perspectives on Quantum Mechanics*, ed. S. Hacyan *et al.* pp. New York: AIP.; http://kaiken.df.uba.ar.
Dunn, T J, Welmsley, I A, and Mukamel, S (1995) *Phys. Rev. Lett.* **74**, 884.

Dusek, M and Buzek, V (2002) Quantum multimeters: a programmable state discriminator. quant-ph/0201097.
Ekert, A, Moura-Alves, C, Oi, D, *et al.* (2001) Universal quantum estimator. quant-ph/0112073.
Fiurasek, J, Dusek, M, and Filip, R (2002) Universal measurement apparatus controlled by quantum software. quant-ph/0202152.
Hannay, J H and Berry, M V (1980) *Physica* **1D**, 267.
Hillery, M, O'Connell, R F, Scully, M O, *et al.* (1984) *Phys. Rep.* **106**, 121.
Hillery, M, Buzek, V, and Ziman, M (2002) *Phys. Rev.* **A65**, 012301.
Horodecki, P and Ekert, A (2001) Direct detection of quantum entanglement. quant-ph/0111064.
Huelga, S F, Vaccaro, J A, Cheflies, A, *et al.* (2001) *Phys. Rev.* **A63**, 042303.
Kitaev, A Y (1997) Quantum measurements and the Abelian stabilizer problem. quant-ph/9511026.
Knill, E and Laflamme, R (1999) *Phys. Rev. Lett.* **81**, 5672.
Leibfried, D, Meekhof, D M, King, B E, *et al.* (1996) *Phys. Rev. Lett.* **77**, 4281.
(1998) *Physics Today* **51**(4), 22.
Leonhardt, U (1996) *Phys. Rev.* **A53**, 2998.
Lutterbach, L G and Davidovich, L (1997) *Phys. Rev. Lett.* **78**, 2547.
(1998) *Optics Express* **3**, 147.
Maitre, X, Hagley, E, Nogues, G, *et al.* (1997) *Phys. Rev. Lett.* **79**, 769.
Miquel, C, Paz, J P, Saraceno, M, *et al.* (2002a) *Nature* **418**, 59.
(2002b) *Phys. Rev.* **A65**, 062309.
Nielsen, M and Chuang, I (1997) *Phys. Rev. Lett.* **79**, 321.
(2000) *Quantum Information and Computation*. Cambridge: Cambridge University Press.
Nogues, G, Rauschenbeutel, A, Osnaghi, S, *et al.* (2000) *Phys. Rev.* **A**.
Paz, J P (2002) *Phys. Rev.* **A65**, 062311.
Paz, J P and Zurek, W H (2001) Environment induced superselection and the transition from quantum to classical. In *Coherent Matter Waves*, ed. R. Kaiser, C. Westbrook, and F. David, pp. 533–614. Berlin: Springer-Verlag.
Paz, J P, Habib, S, and Zurek, W H (1992) *Phys. Rev.* **D47**, 488.
Roncaglia, A (2002) M.Sc. thesis, University of Buenos Aires.
Smithey, D T, Beck, M, Reymer, M G, *et al.* (1993) *Phys. Rev. Lett.* **70**, 1244.
Vidal, G and Cirac, J I (2000) quant-ph/0012067.
Wooters, W K (1987) *Ann. Phys. (N.Y.)* **176**, 1.

9

Quantum gravity as an ordinary gauge theory

Juan M. Maldacena
Institute for Advanced Study, Princeton

Why quantum gravity?

In the twentieth century we gained an enormous amount of knowledge about the basic fundamental laws that govern the physical world. We can summarize this knowledge by saying that particles experience four kinds of forces: electromagnetic, weak, strong, and gravitational. For the first three we have a quantum mechanical description but for gravity we have Einstein's theory, which is rather difficult to quantize. It is not logically consistent to describe particles with quantum mechanics but spacetime with classical physics since matter causes spacetime curvature. So we should be able to consider a particle which is in a quantum mechanical superposition of two states with different positions. These particles should create a gravitational field which contains a similar superposition. This is possible only if the gravitational field itself is quantized. Finding a theory of quantum gravity is not just a question of mathematical consistency, there are physical processes that we cannot describe with current theories. The most notable of these is the beginning of the universe, the initial moments of the Big Bang. We need a quantum gravity theory to be able to understand that moment. The moment is very important since it sets the initial conditions for the subsequent classical evolution of spacetime. Quantum gravity is important when the typical energies of the particles involved are very high. We know from the form of Einstein's action that quantum gravity must be important when particle energies are close to 10^{19} GeV, which is called the Planck energy. One should note, however, that quantum gravity could appear at much lower energies in some theories, such as the large extra dimensions scenario (Arkani-Hamed *et al.* 1998), where quantum gravity is important at a TeV scale. Quantum gravity is also important in the interior of black holes. Inside a black hole there is a singularity where the curvature diverges. When the curvature becomes of Planck size the classical approximation becomes

Science and Ultimate Reality, eds. J. D. Barrow, P. C. W. Davies and C. L. Harper, Jr. Published by Cambridge University Press.

invalid and should be replaced by something else. For an observer who is sitting outside the black hole the singularity is surrounded by a horizon which seems to shield the outside observer from the singularity. Nevertheless, quantum effects bring back the problem since black holes emit Hawking radiation which seems thermal in the semi-classical approximation. Quantum mechanics, on the other hand, implies that there should exist subtle correlations between the radiation coming out and the matter that formed the black hole. In order to compute these correlations we need a quantum gravity theory.

String theory is a theory of quantum gravity (see Green *et al.* 1987; Polchinski 1998). It is a theory under construction. Many new theoretical advances have been made recently which clarify various aspects of the theory. In this chapter we will describe in detail one of those advances. Namely the relation between quantum gravity with asymptotically anti-de-Sitter boundary conditions and conformal field theories. We will emphasize the fact that this relationship enables us to *define* a consistent theory of quantum gravity when spacetime is asymptotically anti-de-Sitter.

Quantum gravity with fixed asymptotic boundary conditions

In Einstein's theory gravity is due to the curvature of spacetime and spacetime is a dynamical variable. As John A. Wheeler emphasized, we should study geometrodynamics, the geometry of space as a dynamical variable. In a quantum theory one should assign probabilities to various geometries, or alternatively one should devise some formulation such that, in a low energy limit, one is effectively assigning probabilities to various geometries. In a path integral formulation one would expect to have a sum over all spacetime geometries. How to formulate this sum in a general setting and how to interpret it is a great challenge. One important point is that when one has suitable asymptotic boundary conditions some aspects of the problem simplify. What is an asymptotic boundary condition? Suppose that instead of summing over all geometries one only sums over geometries that have infinite size and are such that they asymptotically, for large distances, become a fixed prescribed geometry. The simplest example is asymptotically flat boundary conditions. So we only sum over geometries that at large distances become flat Minkowski space and have finite action relative to flat space. Of course we should first understand why it is consistent to restrict the sum in this fashion. The reason is that at long distances gravity becomes very weakly coupled, it becomes classical so that it is consistent to impose a classical condition such as the condition that the geometry asymptotes to a prescribed geometry. The fact that we have a boundary can be used to define the Hilbert space and the observables of the theory. For example, with asymptotically flat space boundary conditions an observable is the

S-matrix. Our main focus will be the case of asymptotically anti-de-Sitter boundary conditions. Anti-de-Sitter (AdS) space is the simplest negatively curved solution of Einstein's equations with a negative cosmological constant. By simplest we mean the one with the largest number of isometries. In fact AdS has as many isometries as flat space but forming a different group, the SO(2, d) group for a $d+1$ dimensional AdS space. Though we live in a spacetime that has $3+1$ large dimensions, let us consider for the moment $d+1$ dimensional spacetimes. The metric is

$$ds^2 = R^2 \frac{-dt^2 + d\vec{x}_{d-1}^2 + dz^2}{z^2} = R^2\left(-\cosh^2\rho d\tau^2 + d\rho^2 + \sinh^2\rho d\Omega_{d-1}^2\right). \tag{9.1}$$

We have written the metric in two coordinate systems. The first is called Poincaré coordinates where AdS is sliced in flat $R^{1,d-1}$ slices, each of which has Poincaré symmetry. The coordinates cover only a portion of AdS space, the surface at $z=\infty$ is a horizon which particles can cross in finite proper time. The second coordinate system in eqn (9.1) is called "global" coordinates and they have the virtue that they cover the whole AdS space. The causal structure of AdS space is most simply exhibited by its Penrose diagram which we obtain by taking out an overall factor of $\cosh^2\rho$ in eqn (9.1). In other words, we define the rescaled metric

$$d\bar{s}^2 = \frac{ds^2}{\cosh^2\rho}. \tag{9.2}$$

This rescaled metric describes a solid cylinder of finite radius which has a boundary at $\rho=\infty$. In the original metric (9.1), $\rho=\infty$ is infinitely far in proper distance, while in the rescaled metric it is at a finite distance. This Penrose diagram shows that AdS space has a boundary of the form $S^{d-1}\times R$ where R is the time dimension (see Fig. 9.1).

A spacetime that is asymptotically anti-de-Sitter is a spacetime whose metric approaches (9.1) at long distances. We will need boundary conditions for all fields at the boundary of AdS space. Depending on the choice of these boundary conditions we will have different physics in the interior. The fact that we can choose a definite boundary condition is related to the fact that at long distances gravity becomes classical and therefore fixes the values of some fields at long distances.

Physics of anti-de-Sitter spacetimes

Let us start by considering the trajectory of a massless particle in AdS space. We can compute it either in the original metric (9.1) or the rescaled metric (9.2). In the rescaled metric it is clear that a light ray can go from the center at $\rho=0$ to the boundary and back in finite time, where time is measured by an observer at rest at the center. This makes it clear that we will need suitable boundary conditions

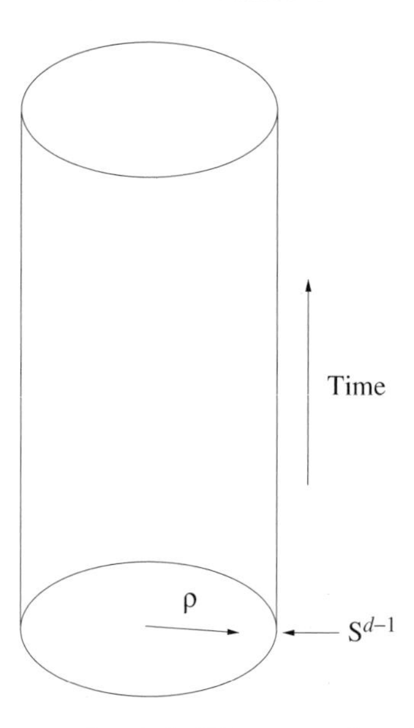

Figure 9.1. Penrose diagram of anti-de-Sitter spacetime. We see a solid cylinder. The boundary is a spatial sphere and a time direction. The radial direction is the coordinate ρ in eqn (9.1).

in order to have a well-defined problem. An important property of AdS space is the fact that there is a large redshift factor $g_{00} \sim 1/z^2$. This means that there is a gravitational force attracting particles to the large z region. The fact that the redshift factor diverges as $z \to 0$ implies that a massive particle will never be able to get to $z = 0$ (or $\rho = \infty$) if it has finite energy (see Fig. 9.2). So AdS is like an infinite gravitational potential well. A very important parameter for AdS spacetimes is their radius of curvature, R. The radius of curvature is some inverse power of the cosmological constant. Depending on the value of the cosmological constant we can have a radius which is as big as the size of the universe or we can have a radius which is microscopic. Clearly the physics in both situations will be very different. It is important to know what the radius is in terms of some other natural scale in the problem. The most natural scale is the Planck scale, so the radius in Planck scale units is very important. In other words we can form the dimensionless quantity

$$c \sim \frac{R^{d-1}}{G_N}. \tag{9.3}$$

In order to have a good semi-classical approximation we need $c \gg 1$ so that the radius of curvature is much bigger than l_p (the Planck length). If we have some other particles propagating in this spacetime, such as electrons of mass m_e, then in order to think of the electron as a localized particle in this space we need that $m_e R \gg 1$. This ensures that the Compton wavelength is much smaller than the

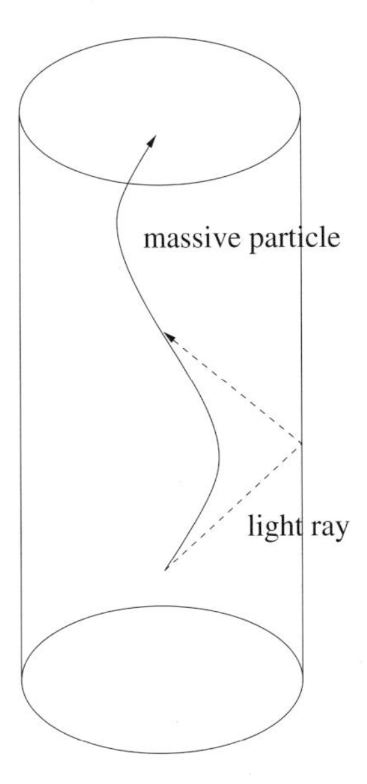

Figure 9.2. The solid line is a trajectory of a massive particle in AdS space. It can never get to the boundary of AdS. A massless particle can get to the boundary and back in finite proper time as measured by proper time along a timelike geodesic.

radius of curvature. Of course we could also consider situations where this is not the case, then it will be very important to think of the electron as a wave using the relativistic theory. In string theory the graviton is not strictly pointlike but it is a small string of size l_s. So we also need that $R \gg l_s$ in order to have a geometrical description of the background. It is only in this case that the intrinsic size of the graviton is much smaller than the radius of curvature of the spacetime.

Anti-de-Sitter has a globally defined timelike Killing vector $\partial/\partial\tau$ (see eqn (9.1)). We can use this Killing vector to compute energies in AdS. Moreover we can use it in any asymptotically AdS spacetime to define the energy of the configuration, or the energy of the particular solution. For example, we can have a black hole in AdS space. In that case the spacetime is topologically and geometrically different from AdS but it is still asymptotically AdS so we can compute the energy of this black hole by comparing this new metric with the metric of empty AdS space near the boundary. Even more simply we can measure the energy of a particle propagating in AdS. Due to quantum mechanics we will need to solve for the wave function of the particle in AdS space. Since we said that AdS space is basically a big gravitational potential well, we will find that the energy levels are quantized. Measured with respect to proper time the energy level spacings are of order $\delta E \sim 1/R$, while measured with respect to to the energy E_τ (which generates translations in τ) they are of the order of $\delta E_\tau \sim 1$. So one question we can ask about a quantum theory of gravity in AdS space is its energy spectrum. Depending on the particular theory we will have different energy spectra. By a "particular theory" we mean the complete specification of all other fields, besides the graviton, that propagate on the spacetime. If the Newton constant is small and other interactions are also small the first-order approximation to this spectrum will be the spectrum of free particles in AdS space. Once we take into account the interactions interesting things can happen. For example, we expect that large number of particles might collapse to form a black hole. So the energy spectrum we are talking about includes all the energy levels of possible black holes.

Finally let us note that AdS space arises as the near-horizon geometry of certain extremal black holes and black branes. For example, the near-horizon geometry of a four-dimensional extremally charged black hole (with $Q = M$) is $AdS_2 \times S^2$. Similarly AdS_{d+1} arises as the near-horizon geometry of black $(d - 1)$ branes. For example, $AdS_5 \times S^5$ arises as the near-horizon geometry of a black three-brane in ten spacetime dimensions. In these cases the horizon is at $z = \infty$, the metric looks like eqn (9.1) up to some small value of z and then it smoothly connects to the metric of flat space.

Geometrodynamics equals chromodynamics

In this section we will explain the conjecture (Gubser *et al.* 1998; Maldacena 1998; Witten 1998) (see Aharony *et al.* (2000) for a review) which states that the physics of AdS space can be equivalently described in terms of an ordinary quantum field theory living on the boundary of AdS space. As we explained above, AdS_{d+1} has a boundary which is $S^{d-1} \times R$, where R is the time direction. The boundary is d-dimensional, one dimension less than AdS space. The quantum field theory is a *conformal* quantum field theory. A conformal theory is a theory that is scale invariant, it is a theory which contains no dimensionful parameters. When we consider such a theory on a sphere we will find that the energy levels are given in terms of the inverse radius of the sphere, which we can set to 1. These energy levels are conjectured to be completely equal to the energy levels that we have in the corresponding AdS theory. The precise definition of the quantum field theory depends on the precise quantum gravity theory that we have in AdS.

We will now proceed to discuss an example of this relationship in order to make the discussion a bit more concrete. We will discuss first an example in which the boundary theory is four-dimensional, so that the AdS spacetime is five-dimensional. An example of a four-dimensional scale-invariant theory is classical electromagnetism. The quantum version of this theory contains free photons. An interesting generalization of electromagnetism is non-Abelian Yang–Mills theory, otherwise known as chromodynamics. This is a theory where we have massless spin-1 particles, similar to the photon, which carry "colors." More precisely, they carry a pair of indices i, j which go from 1 to N where N is the number of colors. The theory describing strong interactions is an example of such a theory with $N = 3$. (In reality we need also to add quarks which are fermions with only one color index; here we will only discuss theories with no quarks.) These Yang–Mills theories are not scale invariant once we include quantum corrections. The coupling depends on the energy: it becomes weak at high energies and strong at low energies. One can add to this theory some fermions and scalar fields, all carrying a pair of color indices in such a way that the resulting theory is scale invariant. An easy way to arrange this is to demand that the theory has enough supersymmetry. Supersymmetry is a symmetry relating bosons and fermions (see Bagger and Wess 1990). Among other things it says that if there is a boson with a certain mass there should be a fermion with the same mass. Supersymmetry relates the couplings of the bosons to the couplings of the fermions. More precisely, it relates various couplings in the Lagrangian. A theory can have various amounts of supersymmetry. A theory with many supersymmetries is a theory where there are many independent relations between the bosons and the fermions. There is a maximum number of supersymmetries that a local quantum field theory can have. In four spacetime dimensions a theory with this maximal amount of supersymmetry is constrained to be a particular supersymmetric version of Yang–Mills theory. This is conventionally called $\mathcal{N} = 4$ Super Yang–Mills, where $\mathcal{N}$ indicates the number of supersymmetries. We can have an arbitrary number of colors N. The theory contains the spin-1 gluons, plus six spin-0 bosonic fields plus four spin-$\frac{1}{2}$ fermions. Its Lagrangian can be found in Aharony *et al.* (2000). The quantum version of the theory is scale invariant. The coupling constant g_{YM} does not depend on the energy and it is a dimensionless number.

After having described the field theory in some detail let us focus on the gravitational theory. We can construct quantum gravity theories using string theory. Again it is simpler to study supersymmetric string theories. A particular string theory that was extensively studied is ten-dimensional superstring theory. At low energies this theory reduces to a ten-dimensional supergravity theory. Flat ten-dimensional spacetime is one solution of this supergravity theory but there are many others. Of particular interest to us is a solution of the form $AdS_5 \times S^5$. This theory contains a generalized "magnetic field" which has some flux over S^5. Here the word "generalized" means that the field strength has five indices $F_{\mu_1 \dots \mu_5}$, while for electromagnetism the field strength has two indices $F_{\mu\nu} = \partial_{[\mu} A_{\nu]}$. The flux over S^5 is quantized. We can consider a solution with N units of magnetic flux over S^5. In other words $N = \int_{S^5} F$. The radii of AdS and S^5 are equal. In ten-dimensional Planck units they are $R/l_p \sim N^{1/4}$. It is also useful to relate the radius in string units and the string coupling constant to the parameters of the quantum field theory on the boundary

$$\frac{R}{l_s} \sim \left(g_{\rm YM}^2 N\right)^{1/4}, \qquad g_s = g_{\rm YM}^2. \tag{9.4}$$

$AdS_5 \times S^5$ has a four-dimensional boundary, when we rescale the metric as in eqn (9.2) the size of S^5 shrinks to zero in the rescaled metric at the boundary. It is on this boundary that the field theory lives.

In conclusion, we find that $\mathcal{N} = 4$ SU(N) Yang–Mills is the same as superstring theory on $AdS_5 \times S^5$ where the relation between the parameters of the two theories is as in eqn (9.4). This relationship is sometimes called a "duality." What this word means is that the two sides are weakly coupled in complementary regions in parameter space. More explicitly, the Yang–Mills theory is weakly coupled when

$$g_{\rm YM}^2 N \ll 1. \tag{9.5}$$

The fact that a factor of N appears here is due to the fact that two gluons can interact by exchanging N other gluons. On the other hand the description of *string* theory in $AdS_5 \times S^5$ in terms of *gravity* is appropriate only if the radius of curvature is much larger than l_s. This means that we need

$$g_{\rm YM}^2 N \gg 1. \tag{9.6}$$

We see that eqns (9.5) and (9.6) describe complementary regions in parameter space. This is good news since weakly coupled Yang–Mills behaves very differently than weakly coupled gravity. This also explains why the relationship is a duality, for small $g_{\rm YM}^2\, N$ the Yang–Mills description is weakly coupled and when $g_{\rm YM}^2\, N$ is large the gravity description is more appropriate. But we should not forget that Yang–Mills theory is a well-defined theory for any value of $g_{\rm YM}^2\, N$. Of course it will be hard to do computations when $g_{\rm YM}^2\, N$ is large, for the same reason that it is hard to compute low energy phenomena (such as the mass of the proton) with quantum chromodynamics. So Yang–Mills theory is well defined and it can be used to *define* gravity and string theory on $AdS_5 \times S^5$. Of course if we want to have an astronomically large space we need an astronomically large value of $g_{\rm YM}^2\, N$. It is interesting also to note that the value of c in eqn (9.3) is proportional to

$$c \sim N^2 \tag{9.7}$$

which is the number of degrees of freedom (the number of fields) in the boundary quantum field theory. More concretely, c is the coefficient of the two point function of the stress tensor in the boundary theory, $T(0)T(x) \sim c/|x|^8$ where the dependence on x is determined by conformal symmetry.

Let us now say a few words about why we expect this duality to be true. One can present an argument based on properties of certain black branes in string theory which relates the Yang–Mills theory to the near-horizon geometry of such branes. This near horizon geometry is $AdS_5 \times S^5$. Instead of following this route (Aharony *et al.* 2000) we will just present some more qualitative reasons why this duality holds. Many years ago t' Hooft (1974) argued that the large N limit of gauge theories should be some kind of string theory. He observed that the gauge theory Feynman diagrams that are dominant in the large N limit are diagrams that can be drawn on the plane (after adopting a double-line notation to represent the two colors that a gluon carries). These diagrams can be thought of as discretizing a two-dimensional worldsheet, so that this leading contribution would be given by a free string theory. Subleading contributions in $1/N$ would take into account string interactions. Originally it was thought that this would be a string theory living in four dimensions, the four spacetime dimensions of the gauge theory. But strings are not consistent in four dimensions. So we need to add more dimensions, at least one more (Polyakov 1998). From the four-dimensional point of view, the extra dimension is roughly the thickness of the string. All the symmetries of the field theory should be present in the corresponding string theory. If the field theory is conformal then the symmetry group is $SO(2,4)$. The spacetime where strings propagate should be invariant under an $SO(2,4)$ isometry group. The only such five-dimensional spacetime is AdS_5. In the particular case of $\mathcal{N} = 4$ Yang–Mills we also have many supersymmetries. Their number is the same as the number of supersymmetries that flat ten dimensional string theory has. So one expects that the string theory corresponding to the t' Hooft large N limit of $\mathcal{N} = 4$ U(N) Yang–Mills should be related to ten-dimensional string theory on some spacetime with an AdS_5 factor. The quantum field theory also has an $SO(6)$ symmetry so that it is natural to think that the space should be $AdS_5 \times S^5$.

One surprising aspect of this duality is the fact that a theory without gravity, the Yang–Mills theory, is dual to a gravitational theory. Where does the graviton come from? The graviton is related to the stress tensor operator in the quantum field theory. When we make an insertion of the stress tensor operator in the boundary quantum field theory we create a state which, from the bulk point of view, contains a graviton excitation (see Fig. 9.3). So the graviton is a composite particle but the particles that make up the graviton do not live in the five or ten dimensions where it is moving, they live in four dimensions. Four-dimensional particles combine to form stringlike excitations which effectively move in ten dimensions. The graviton is one of them. In the t' Hooft limit these are the free strings of ten-dimensional superstring theory on $AdS_5 \times S^5$. If N is finite these are the standard interacting strings of string theory. A puzzling feature is the fact that these strings move in more than four dimensions. This puzzle goes away once we view them as complicated

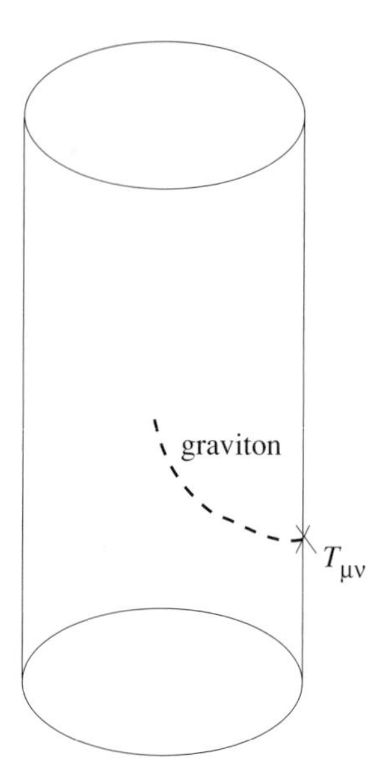

Figure 9.3. An insertion of the stress tensor in the boundary theory creates a state in the boundary theory that is equivalent to a graviton in AdS.

"bound" states of gluons. Since the theory is conformal such a bound state cannot have any particular scale or size. So we can have bound states of all possible sizes. In fact the size is then another dynamical variable we need to specify. This variable becomes the fifth (radial) dimension in AdS_5 (see Fig. 9.4).

In this section we explained in detail one particular example of this relationship. Many related examples are know with theories that have less supersymmetry and that are not conformal. In many cases the S^5 gets replaced by another manifold. If the boundary field theory is not conformal then also AdS gets replaced by another manifold with a boundary. Unfortunately we do not have a general prescription for finding which quantum field theory corresponds to which gravity background, but we can do an analysis on a case-by-case basis. We discussed a relationship between AdS_5 and a four-dimensional quantum field theory. If we have an AdS_4 space the duality relates it to a three-dimensional quantum field theory. There is an example that relates eleven-dimensional supergravity on $AdS_4 \times S^7$ to a particular conformal field theory in $2 + 1$ dimensions.

Some implications for gravity

We do not have a theory of quantum gravity that would work in all circumstances. If we have asymptotically AdS boundary conditions we can define quantum gravity via an ordinary quantum field theory, assuming the conjecture is true. One question

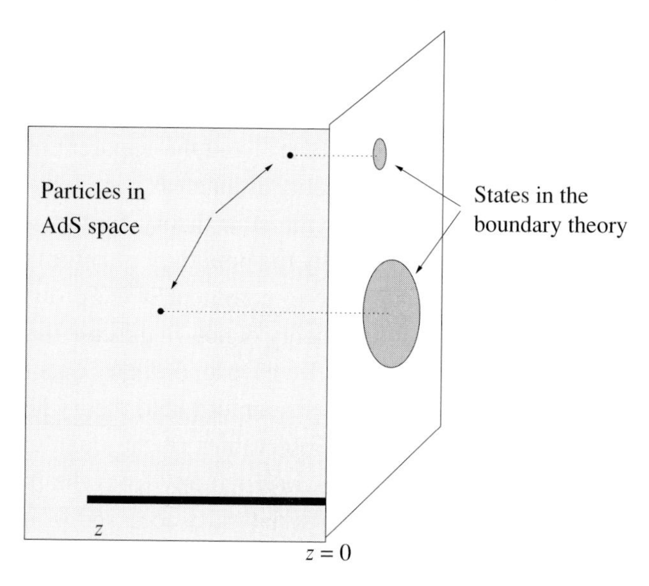

Figure 9.4. Here the two blobs represent two states in the boundary field theory that differ only in their overall scale. They correspond to the same particle in AdS space localized at different positions in the radial direction. The smaller the blob is in the boundary theory the closer the particle is to the boundary of AdS space. Large blobs correspond to particles at large values of z, in the coordinates of eqn (9.1).

of principle that this relationship resolves is the problem of unitarity in black-hole evaporation. The semi-classical analysis of black-hole evaporation suggests that information is lost and that "quantum" gravity would not be an ordinary quantum theory with unitary evolution. In this case the black-hole formation and evaporation process can be described as a process occurring in the boundary quantum field theory, which is certainly a unitary theory. This is a proof of principle that the process is unitary. It is quite difficult to describe the process explicitly since the quantum field theory is strongly coupled. It is even difficult to extract the semi-classical results for small evaporating black holes from the quantum field theory. Big black holes, black holes whose Schwarzschild radius is larger than the radius of AdS, can be thought of as describing a thermal state in the quantum field theory. These black holes do not evaporate because Hawking radiation is confined to the bottom of the gravitational potential and after a short time the black hole begins to absorb as much as it emits. In these cases we can understand rather easily the statistical origin of the Bekenstein–Hawking entropy. In the quantum field theory this is the ordinary statistical entropy of the thermal state. Something that is particularly easy to understand is the inverse dependence on the Newton constant in the Bekenstein–Hawking formula, $S = \text{area}/4G_N$. We have seen from eqn (9.7) that the inverse Newton constant is proportional to the number of degrees of freedom in the quantum field theory. Indeed the entropy of a thermal state in the quantum field theory is proportional to the number of degrees of freedom. Some people might object that most of these computations are done for supersymmetric theories while in nature supersymmetry, if present at all, is broken. This is true, but on the other hand the apparent information loss problem is present in a supersymmetric theory as much as it is in an ordinary theory. Note also that thermal states in a supersymmetric theory are not supersymmetric themselves and can indeed be described by the boundary quantum field theory.

Another issue on which we would like to comment is the problem of whether quantum gravity is an ordinary quantum theory or not. In the case that the spacetime is asymptotically AdS we clearly see that we have an ordinary quantum theory. We have a Hamiltonian, which is the boundary quantum field theory Hamiltonian and we have observables which are all the observables of the quantum field theory. On the other hand these observables look rather funny from the point of view of the bulk AdS theory since they are all most naturally accessed from the boundary and seem to be related to the long-distance behavior of the state. In particular it seems hard to define observables, even approximate ones, that are localized in AdS space. Of course, these might not exist in a quantum theory of gravity but there should be an approximation where they emerge. One lesson is that the variables in terms of which the theory is an ordinary quantum field theory are intimately related to the spacetime asymptotics. The same is true in flat space where there are prescriptions (for some supersymmetric gravity theories) for the computation of the exact S-matrix (Banks *et al.* 1997). Of course, we would like to understand as much as possible the description of the quantum gravity theory from the point of view of an observer which is part of the system, such as we are. It seems that by studying this duality further one might be able to isolate the approximately local description of physics inside and we might learn some lessons that we can generalize to other cases.

Another issue we would like to discuss is that of wormholes and their effects on the coupling constants (Coleman 1988). The picture was that microscopic wormholes would shift the values of the coupling constants. A natural consequence of those ideas was that the value of the cosmological constant could change and depend on the particular superselection sector that one is in. These superselection sectors were supposed to be related to the different possible wave functions for the baby universes. In the $AdS_5 \times S^5$ case there is a unique vacuum which is the unique vacuum of the quantum field theory. There does not seem to be room for choosing a large number of wave functions for baby universes which would change the parameters of the bulk theory since there is no such freedom in the boundary theory either. In this case the structure of the boundary theory is largely fixed by supersymmetry, so the relevance of wormholes for less supersymmetric theories is still an open question.

Our own universe does not look like AdS so why should we care that one can define quantum gravity in AdS? We have already seen the fact that this description clarifies the black hole entropy and unitarity problems. Suppose for a moment that we did indeed live in a universe which was AdS_4 at long distances. Then in order to describe nature we would need to figure out which conformal field theory describes it and we would need to understand some features of the state in which we are. It might be possible that our expanding universe is embedded in a much bigger AdS space and that it is some particular state. Since the radius of AdS_4 would have to be bigger than the size of the universe, we can see from eqns (9.3) and (9.7) that the effective number of degrees of freedom in the $2 + 1$ dimensional conformal field theory living on the boundary should be very large, $c \sim R^2/l_p^2 > 10^{122}$.

One lesson that we learnt through this discussion is that in cases where quantum gravity has some simple asymptotic directions one can have a rather simple dual description. A de Sitter space has a simple behavior in the far future, and moreover, our universe might be asymptotically de Sitter in the future if there is a positive cosmological constant as current observations suggest. In that case one might hope that there is a simple description of the physics of this de Sitter space in terms of a dual theory that lives in the boundary of de Sitter. This theory would have to be a Euclidean quantum field theory, since the boundary of de Sitter space is Euclidean (it is S^3 for dS_{3+1}) (Strominger 2001; Witten 2001). Unfortunately nobody has found yet a concrete example where we know both the quantum field theory and the gravity theory. So we do not know if such a description is possible or not.

To conclude let us emphasize that we now have one concrete definition of quantum gravity. We are just beginning to explore what it means for quantum gravity in general. We think that there is still a lot to learn from this duality which will illuminate many aspects of quantum gravity and might probably help us understand how to define quantum gravity in more general and realistic situations such as the universe where we live.

Acknowledgments

I would like to thank the organizers of the John A. Wheeler ninetieth birthday conference for a very nice and stimulating conference.

I would also like to thank the John Templeton Foundation for support.

References

Aharony, O, Gubser, S S, Maldacena, J M, *et al.* (2000), *Phys. Rept.* **323**, 183. arXiv:hep-th/9905111.

Arkani-Hamed, N, Dimopoulos, S, and Dvali, G R (1998) *Phys. Lett.* **B429**, 263. arXiv:hep-ph/9803315.

Bagger, J and Wess, J (1990) *Supersymmetry and Supergravity*. Princeton, NJ: Princeton University Press.
Banks, T, Fischler, W, Shenker, S H, *et al.* (1997) *Phys. Rev.* **D55**, 5112. arXiv:hep-th/9610043.
Coleman, S R (1988) *Nucl. Phys.* **B310**, 643.
Green, M B, Schwarz, J H, and Witten, E (1987) *Superstring Theory*, vols. 1 and 2. Cambridge: Cambridge University Press.
Gubser, S S, Klebanov, I R, and Polyakov, A M (1998) *Phys. Lett.* **B428**, 105. arXiv:hep-th/9802109.
Maldacena, J M (1998) *Adv. Theor. Math. Phys.* **2**, 231.
Polchinski, J (1998) *String Theory*, vols. 1 and 2. Cambridge: Cambridge University Press.
Polyakov, A M (1998) *Nucl. Phys. Proc.* **Suppl. 68**, 1. arXiv:hep-th/9711002.
Strominger, A (2001) *J. High Energy Phys.* **0110**, 034. arXiv:hep-th/0106113.
t' Hooft, G (1974) *Nucl. Phys.* **B72**, 461.
Witten, E (1998) *Adv. Theor. Math. Phys.* **2**, 253. arXiv:hep-th/9802150.
(2001) arXiv:hep-th/0106109.

10

The Everett interpretation of quantum mechanics

Bryce S. DeWitt
The University of Texas at Austin

Introduction

In the July 1957 issue of the *Reviews of Modern Physics* Hugh Everett III put forward a new interpretation of quantum mechanics (Everett 1957). John Wheeler, Everett's thesis adviser, published, in the same issue, an accompanying paper supporting Everett's views (Wheeler 1957). Everett's aim was to cut through the fuzzy thinking displayed by many authors, some of them quite prominent, who in previous years had written incredibly dull papers on how they understood quantum mechanics. Everett's idea was simply to assume that quantum mechanics provides a description of reality in exactly the same sense as classical mechanics was once thought to do.

This is a shocking idea, for it leads to a multiplicity of "realities." Few physicists in 1957 were prepared to accept it. And yet it can be shown to work. It is the purpose of this article to expand on Everett's original demonstration and to reveal the courage John Wheeler displayed in betting that his student was right.

Classical theory of measurement

System, apparatus, coupling

Our starting point is the standard theory of measurement, which we shall first examine classically. In its simplest form a measurement involves just two dynamical entities: a *system* and an *apparatus*. It is the role of the apparatus to record the value of some system observable s. For this purpose system and apparatus must be coupled together for a certain period of time, which will be taken to be finite. Since couplings occurring in nature cannot be switched on and off at will the finiteness of the interval

This article contains compressed and rearranged portions of material included in Chapters 8, 9, and 12 of DeWitt (2003) and in *Int. J. Mod. Phys.* **A13**, 1881 (1998).

Science and Ultimate Reality, eds. J. D. Barrow, P. C. W. Davies and C. L. Harper, Jr. Published by Cambridge University Press.

will usually be an *effective* finiteness, determined by a carefully chosen apparatus trajectory or *history*. It is the change in this history, brought about by the coupling, that yields information about the system, in particular about the observable s.

Let S and Σ be the action functionals for system and apparatus respectively. The action for the two together, when uncoupled, is the sum $S + \Sigma$. The measurement is brought about by a change in the total action, of the form

$$S + \Sigma \longrightarrow S + \Sigma + g\mathfrak{X}, \tag{10.1}$$

where $g\mathfrak{X}$ describes the coupling, g being an adjustable "coupling constant" and $\mathfrak{X}$ a functional of the combined system–apparatus histories. The disturbance in some chosen apparatus observable P, brought about by the coupling, is what constitutes the measurement. The system also gets disturbed by the coupling, and this disturbance may change the observable s, complicating the measurement one is trying to make. For this reason one imagines, in classical physics, that the coupling can be made as weak as desired. However, the weaker the coupling the harder it is to detect the disturbance in the apparatus.

The disturbances: apparatus inertia

For discussing disturbances a notation due to Peierls (1952) is helpful. Let the action functional of any system be changed by the addition to it of a term ϵA, where A is some observable and ϵ is an infinitesimal real number. Because infinitesimal disturbances can be superposed linearly, the resulting change in any other observable B will be proportional to ϵ and can be written in the form

$$\delta B = \epsilon D_A B, \tag{10.2}$$

where $D_A B$, which is Peierls's symbol, is in principle calculable (e.g., by means of Green's functions). Here it will be assumed that the initial data are the same both before and after the addition of ϵA. One is then dealing with retarded boundary conditions, although because of the time-reversibility of classical systems the analysis could equally well be carried out with advanced boundary conditions.

If the dynamical variables out of which B is constructed are taken from an interval of time that is *earlier* than that out of which the variables composing A are taken, a situation described symbolically by

$$B \longleftarrow A, \tag{10.3}$$

then $D_A B = 0$. In the case of the coupled system $S + \Sigma + g\mathfrak{X}$, if $P \longleftarrow \mathfrak{X}$ then the disturbance in P vanishes and no measurement takes place. One therefore assumes

$$\mathfrak{X} \longleftarrow P, \tag{10.4}$$

$\mathfrak{X}$ being limited by the *effective* time interval specified by the chosen apparatus history.

To lowest order in an expansion in powers of g the disturbance in P is given simply by

$$\delta P = g D_{\mathfrak{X}} P. \tag{10.5}$$

To second order in g the expression for δP is more complicated. However, it simplifies if one invokes the dominant feature of any good apparatus: *its inertia is large compared to that of the system to which it is coupled.* Technically this means that certain terms involving the retarded Green's function of the apparatus may be ignored compared to similar terms involving system Green's functions. As a result the dominant second-order correction to expression (10.5) is just the change in $g D_{\mathfrak{X}} P$ brought about by its dependence, *through* $\mathfrak{X}$, on the disturbed system trajectory, and this (again by the dominant inertia condition) is adequately approximated by $g^2 D_{\mathfrak{X}} D_{\mathfrak{X}} P$, yielding for the total value of δP, correct to second order,

$$\delta P = g(1 + g D_{\mathfrak{X}}) D_{\mathfrak{X}} P, \tag{10.6}$$

all quantities being evaluated at the undisturbed histories.

Peierls's bracket: design of the coupling: uncertainties

Peierls originally introduced his notation to give an alternative (noncanonical, covariant, global) definition of the Poisson bracket. The Peierls bracket of any two observables A and B is defined to be

$$(A, B) := D_A B - D_B A. \tag{10.7}$$

It can be shown (DeWitt 1965) to satisfy the Jacobi identity and to be identical to the conventional Poisson bracket in the case of standard canonical systems. In view of relation (10.4) one may rewrite expression (10.6) in the form

$$\delta P = g(\mathfrak{X}, P) + g^2 D_{\mathfrak{X}}(\mathfrak{X}, P). \tag{10.8}$$

The purpose of the disturbance δP is to reveal the value of the system observable s, at least to lowest order in g. This is achieved by choosing $\mathfrak{X}$ to satisfy

$$(\mathfrak{X}, P) = s, \qquad (\mathfrak{X}, s) = 0. \tag{10.9}$$

These conditions introduce an asymmetry between system and apparatus additional to the inertial asymmetry already noted. But *it is the role of the apparatus to observe the system, not vice versa.* Whether couplings satisfying eqns (10.9) can be achieved in practice is a question of the experimenter's art, on which no comment will be made here. We note that an easy way to secure eqns (10.9) *on paper* is simply to choose

$$\mathfrak{X} = sX, \tag{10.10}$$

where X is an apparatus variable conjugate to P:

$$(X, P) = 1. \tag{10.11}$$

In practice the history of the apparatus will be rigid enough (owing to the apparatus's inertia) that the simple product (10.10) will be replaced by the *time integral* of a product, over the coupling interval.

Equations (10.8) and (10.9) yield the following expression for the disturbed apparatus observable:

$$\bar{P} := P + \delta P = P + gs + g^2 D_{\mathfrak{X}} s. \tag{10.12}$$

The "experimental" value of s to which δP corresponds is

$$s = \frac{\delta P}{g} - g D_{\mathfrak{X}} s. \tag{10.13}$$

The accuracy with which eqn (10.13) determines s depends on the accuracy with which P and $D_{\mathfrak{X}} s$ are known for the undisturbed histories. Denote by ΔP and $\Delta D_{\mathfrak{X}} s$ the uncertainties in these quantities. The mean square error in the experimental value of s that these uncertainties generate is

$$(\Delta s)^2 = \frac{(\Delta P)^2}{g^2} + g^2 (\Delta D_{\mathfrak{X}} s)^2. \tag{10.14}$$

This expression displays at once how the error Δs behaves as the coupling constant is varied. When g is large Δs is large due to the uncertainty $g \Delta D_{\mathfrak{X}} s$ in the disturbance produced in the system. (Note that it is the *uncertainty* in the disturbance that is important here and not the disturbance itself, which could in principle be allowed for.) When g is small, on the other hand, Δs again becomes large because of the difficulty of obtaining a meaningful value for δP. (It gets swamped by the uncertainty ΔP.) The minimum value of Δs occurs for $g^2 = \Delta P / \Delta D_{\mathfrak{X}} s$, at which coupling strength one has

$$\Delta s = (2 \Delta P \Delta D_{\mathfrak{X}} s)^{\frac{1}{2}}. \tag{10.15}$$

When $\mathfrak{X}$ has the form (10.10) then

$$D_{\mathfrak{X}} s = X D_s s, \tag{10.16}$$

and if, as is often the case, $D_s s$ varies slowly under changes in the (at least partially unknown) trajectory of the system S, eqn (10.15) becomes

$$\Delta s = (2\Delta P \Delta X |D_s s|)^{\frac{1}{2}} . \tag{10.17}$$

Compensation term

In the classical theory ΔP and $\Delta D_{\mathfrak{X}} s$ (or ΔX) can in principle be made as small as desired. In the quantum theory, however, they are limited by uncertainty relations. Equation (10.17) therefore suggests that in the quantum theory there is a fundamental limit to the accuracy with which the value of any *single* observable can be measured. Equation (10.17), in the context of electrodynamics, was thrown down as a challenge to Bohr by two cocky youngsters, Landau and Peierls. The equation contradicts the well-established principle that the value of any single observable should be determinable with arbitrary precision even in quantum mechanics.

The answer to the challenge came in a famous paper by Bohr and Rosenfeld (1933). It is not an easy paper to read, but the key point is that the Landau–Peierls contradiction can be overcome by modifying the coupling between system and apparatus through the insertion of a *compensation term*. Bohr and Rosenfeld went through elaborate steps in constructing (on paper) the device that would yield the effect of such a term. In the end it turns out that the term is generically just $-\frac{1}{2}g^2 D_{\mathfrak{X}}\mathfrak{X}$, so that eqn (10.1) gets replaced by

$$S + \Sigma \longrightarrow S + \Sigma + g\mathfrak{X} - \frac{1}{2}g^2 D_{\mathfrak{X}}\mathfrak{X}. \tag{10.18}$$

When the dominant inertia condition holds this leads to

$$\delta P = gs + g^2 D_{\mathfrak{X}} s - \frac{1}{2}g^2 (D_{\mathfrak{X}}\mathfrak{X}, P). \tag{10.19}$$

Using the dominant inertia condition one can show that the last two terms on the right-hand side of eqn (10.19) cancel each other. This is most easily seen when $\mathfrak{X}$ has the form (10.10). In that case one has

$$D_{\mathfrak{X}}\mathfrak{X} = (D_s s)\, X^2 \tag{10.20}$$

which, together with eqn (10.16), yields

$$\delta P = gs \tag{10.21}$$

and

$$\Delta s = \frac{\Delta P}{g}. \tag{10.22}$$

This uncertainty can be made as small as desired.

Quantum theory of measurement

Quantum operators and state vectors

In the formalism of quantum mechanics the observables s, P, A, B are replaced by self-adjoint operators $\boldsymbol{s}$, $\boldsymbol{P}$, $\boldsymbol{A}$, $\boldsymbol{B}$, and one defines

$$\langle \boldsymbol{A} \rangle := \langle \psi | \boldsymbol{A} | \psi \rangle, \qquad \langle \psi | \psi \rangle = 1, \tag{10.23}$$

$$(\Delta \boldsymbol{A})^2 := \langle (\boldsymbol{A} - \langle \boldsymbol{A} \rangle)^2 \rangle = \langle \boldsymbol{A}^2 \rangle - \langle \boldsymbol{A} \rangle^2, \tag{10.24}$$

$|\psi\rangle$ being Dirac's symbol for the normalized state vector of the quantum system and $\langle\psi|$ its dual. The state-vector space on which the operators act is not given a priori but is constructed in such a way as to yield a minimal representation of the operator algebra of the system. This algebra is always determined in some way by the *heuristic quantization rule*

$$[\boldsymbol{A}, \boldsymbol{B}] = i\,(\boldsymbol{A}, \boldsymbol{B}) \quad \text{(in units with } \hbar = 1), \tag{10.25}$$

which tries to identify, up to a factor i, each Peierls bracket with a commutator. We ignore here problems of factor ordering and turn at once, with this rule in hand, to examine how the measurement process described in the previous section looks in the quantum theory.

Equations (10.9) get replaced by

$$[\mathfrak{X}, \boldsymbol{P}] = i\boldsymbol{s}, \qquad [\mathfrak{X}, \boldsymbol{s}] = 0. \tag{10.26}$$

It is not difficult to see that, to order g^2, the modified coupling (10.18) produces a change in the apparatus observable $\boldsymbol{P}$ equivalent to that generated by the unitary operator $e^{ig\mathfrak{X}}$:

$$\boldsymbol{P} \longrightarrow \bar{\boldsymbol{P}} = \boldsymbol{P} + \delta\boldsymbol{P} = \boldsymbol{P} + g\boldsymbol{s} = e^{-ig\mathfrak{X}}\boldsymbol{P}e^{ig\mathfrak{X}}. \tag{10.27}$$

The quantum mechanical description of the measurement starts with a state in which the system and apparatus, in the absense of coupling, would be *uncorrelated*, which simply means that the state vector of the combination is expressible as a tensor product:

$$|\Psi\rangle = |\psi\rangle|\Phi\rangle, \tag{10.28}$$

$|\psi\rangle$ and $|\Phi\rangle$ being the state vectors of system and apparatus respectively. Note that the meaning of the word "state," whether applied to the vector $|\psi\rangle$, $|\Phi\rangle$, or $|\Psi\rangle$, is still somewhat vague. In the classical theory "state" is synonymous with "history," or with the Cauchy data that determine the history. If the formalism of quantum mechanics is to provide a faithful representation of reality, as Everett proposes, then it must also yield its own interpretation, and a certain effort is going to be required to determine how a vector can represent a state.

A preferred basis

The structure of the coupling operator $\mathfrak{X}$, as expressed in eqns (10.26), defines a *preferred* set of basis vectors:

$$|s, P\rangle = |s\rangle|P\rangle, \tag{10.29}$$

where $|s\rangle$ and $|P\rangle$ are eigenvectors of $\boldsymbol{s}$ and $\boldsymbol{P}$ respectively:

$$\boldsymbol{s}|s\rangle = s|s\rangle, \qquad \boldsymbol{P}|P\rangle = P|P\rangle. \tag{10.30}$$

Other labels beside s and P will, of course, generally be needed for a complete specification of the basis, but they are omitted here because they play no direct role in what follows. Indeed, the validity of eqns (10.26), or of the relation (10.4), often depends on some of the suppressed labels having certain values. (Think of the time constraints on the effective coupling interval, for example.) But the values in question are experimental details and do not bear directly on questions of interpretation of the quantum formalism.

With respect to the basis vectors (10.29) the state vector is represented by the function

$$\langle s, P|\Psi\rangle = c_s\Phi(P), \tag{10.31}$$

where

$$c_s := \langle s|\psi\rangle, \qquad \Phi(P) := \langle P|\Phi\rangle. \tag{10.32}$$

It will be convenient to assume that the eigenvalues of $\boldsymbol{s}$ range over a discrete set while those of $\boldsymbol{P}$ range over a continuum. It is then natural to impose the orthonormality conditions

$$\langle s|s'\rangle = \delta_{ss'}, \qquad \langle P|P'\rangle = \delta(P, P'), \tag{10.33}$$

so that the conditions that $|\psi\rangle$ and $|\Phi\rangle$ be normalized are expressible in the forms

$$\sum_s |c_s|^2 = 1, \qquad \int |\Phi(P)|^2 dP = 1. \tag{10.34}$$

Since it is the disturbed apparatus observable $\bar{P}$ that stores the results of the measurement, the basis that is appropriate for discussing the measurement process consists of the unitarily related vectors

$$|s, \bar{P}\rangle = e^{-ig\mathfrak{X}}|s, P\rangle. \tag{10.35}$$

P and $\bar{P}$ are numerically equal, but when standing in the bracket "$|\rangle$" they represent different vectors. From eqn (10.27) it is easy to see that

$$|s, P\rangle = |s, \bar{P} + gs\rangle \quad \text{or} \quad |s, \bar{P}\rangle = |s, P - gs\rangle, \tag{10.36}$$

$|s, \bar{P} + gs\rangle$ being an eigenvector of $\bar{\boldsymbol{P}}$ corresponding to the eigenvalue $\bar{P} + gs$ and $|s, P - gs\rangle$ being an eigenvector of $\boldsymbol{P}$ corresponding to the eigenvalue $P - gs$.

Relative states: good measurements

The vectors $|s, \bar{P}\rangle$ constitute a kind of mixed basis in that they are eigenvectors of a disturbed apparatus observable and an *un*disturbed system observable. This basis is nevertheless the appropriate one to use for the analysis of the measurement process because the coupling, including the compensation term, is deliberately designed to set up a correlation between the disturbed apparatus and the value that the system observable would have assumed had there been no coupling. This correlation is displayed by decomposing the total state vector (10.28) in terms of the $|s, \bar{P}\rangle$ and using eqns (10.32) and (10.36):

$$\begin{aligned} |\Psi\rangle &= \sum_s \int |s, P\rangle\langle s, P|\Psi\rangle dP \\ &= \sum \int |s, \bar{P} + gs\rangle c_s \Phi(\bar{P}) d\bar{P}. \end{aligned} \tag{10.37}$$

The final form follows from the numerical equality of P and $\bar{P}$.

In view of the fact that

$$[\boldsymbol{s}, \bar{\boldsymbol{P}}] = [\boldsymbol{s}, \boldsymbol{P} + g\boldsymbol{s}] = 0, \tag{10.38}$$

the vectors $|s, \bar{P}\rangle$, like the vectors $|s, P\rangle$, can be expressed as products:

$$|s, \bar{P}\rangle = |s\rangle|\bar{P}\rangle \tag{10.39}$$

Equation (10.37) then takes the form

$$|\Psi\rangle = \sum_s c_s |s\rangle|\Phi[s]\rangle \tag{10.40}$$

where

$$|\Phi[s]\rangle := \int |\bar{P} + gs\rangle\Phi(\bar{P}) d\bar{P}. \tag{10.41}$$

The apparatus state represented by the vector $|\Phi[s]\rangle$ is frequently referred to as a *relative state*, terminology that can be traced back to Everett (1957) and Wheeler

(1957).[1] Equation (10.40) shows that relative to each system state $|s\rangle$ the apparatus, as a result of the coupling, "goes into" a corresponding state $|\Phi[s]\rangle$. All the possible outcomes of the measurement are contained in the superposition (10.40), weighted by coefficients c_s determined by the system state vector $|\psi\rangle$ (eqn (10.32)). Each relative state itself is represented as a superposition of apparatus state vectors $|\bar{P} + gs\rangle$, with the eigenvalues $\bar{P} + gs$, being centered (in the sense of the weights $\Phi(\bar{P})$) at a distance gs from the unperturbed "average"

$$\langle \boldsymbol{P} \rangle = \langle \Phi | \boldsymbol{P} | \Phi \rangle = \int P |\Phi(P)|^2 dP \tag{10.42}$$

and distributed with a "width"

$$\Delta \boldsymbol{P} = (\langle \boldsymbol{P}^2 \rangle - \langle \boldsymbol{P} \rangle^2)^{\frac{1}{2}} \tag{10.43}$$

around the displaced value.

The relative state vectors will be orthogonal to one another if the measurement is *good*, namely if $\Delta \boldsymbol{P}$ satisfies

$$\Delta \boldsymbol{P} << g \Delta \boldsymbol{s} \tag{10.44}$$

where $\Delta \boldsymbol{s}$ is the minimal spacing between those eigenvalues of $\boldsymbol{s}$ that are contained in the support of the function c_s. When the measurement is good then, to high accuracy,

$$\langle \Phi[s] | \Phi[s'] \rangle = \delta_{ss'}. \tag{10.45}$$

Note that the apparatus "state," represented by $|\Phi\rangle$, must be carefully prepared if the measurement is to be good. In this respect quantum measurements are like classical measurements.

Many worlds

In other respects classical and quantum measurements are not at all alike. If the formalism of quantum mechanics truly provides a faithful description of reality then the reality described by the vector (10.40) consists of simultaneous worlds (or *mini-worlds*, since there is only a system and an apparatus) in each of which the system observable $\boldsymbol{s}$ has a certain value s and the apparatus has observed that value. Here is what John Wheeler (1957) has to say about expression (10.40):

[1] The terminology is almost certainly due to Wheeler himself. Wheeler once told the author that he had sat down with Everett and told him precisely what to abstract out of a much larger *Urwerk* that Everett had prepared (Everett 1973). The *Urwerk*, which did not get published until 1973, does not use the terminology. The reader will recognize that the analysis up to this point is simply von Neumann's (1996) standard treatment of measurement theory, but expanded to include a detailed description of the role of the apparatus and the formal structure of its coupling to the system.

It is difficult to make clear how decisively the "relative state" formulation drops classical concepts. One's initial unhappiness at this step can be matched but few times in history: when Newton described gravity by anything so preposterous as action at a distance; when Maxwell described anything as natural as action at a distance in terms as unnatural as field theory; when Einstein denied a privilege character to any coordinate system, and the whole foundations of physical measurement at first sight seemed to collapse. How can one consider seriously a model for nature that follows neither the Newtonian scheme, in which coordinates are functions of time, nor the "external observation" description, where probabilities are ascribed to the possible outcomes of a measurement? Merely to analyze the alternative decompositions of a state function, as in [10.40], without saying what the decomposition means or how to interpret it, is apparently to define a theoretical structure as poorly as possible!

It is not clear to the author precisely what Wheeler means by "alternative decompositions." A given state vector of a composite system can be decomposed in an infinity of ways into products of vectors taken from orthonormal sets, and some writers have worried about the question of finding preferred sets. The author does not believe this is a valid worry. Measurements do not take place in an abstract Hilbert space. The experimenter knows perfectly well the observable he is trying to measure, and he chooses his coupling accordingly. The coupling, together with the choice of observable, determines the preferred basis, as we have seen.

A greater worry is the following: the idea that the world may "split," as a result of couplings, into many worlds, is hard to reconcile with the testimony of our senses, namely that we simply do not split. Those who object to the many-worlds view for this reason Everett (1957) likens to the anti-Copernicans in the time of Galileo, who did not feel the Earth move. To the extent to which we may be regarded as automata, and hence on a par with ordinary measuring apparatus, it is not hard to show that *the laws of quantum mechanics do not allow us to feel ourselves split.* For this purpose it is useful to speak of the apparatus as having a *memory*, represented by the bracket "$[s]$" in the symbol $|\Phi[s]\rangle$, in which the measurement is stored. If the storage is assumed to last for a reasonable period of time one may in fact speak of a *memory bank.* Now what would happen, in the case of the measurement described by the superposition (10.40), if a second apparatus were introduced, which not only "looks at" the memory bank of the first apparatus but also carries out an independent direct check on the value of the system observable? If the splitting into many worlds is to be unobservable then the results had better agree.

Following the methods that have already been outlined it is not difficult to set up the appropriate couplings. It is helpful to assume that the variance in the observable $\boldsymbol{P}$ of the first apparatus satisfies $\Delta \boldsymbol{P} << \Delta \boldsymbol{Q}/\bar{g}$, where $\boldsymbol{Q}$ is the observable (of the second apparatus) that records the value of $\boldsymbol{P}$, and $\bar{g}$ is the corresponding coupling constant. The total state vector is then again revealed as a superposition of vectors, each of which represents the system observable $\boldsymbol{s}$ as having assumed one of its possible values. Although the value varies from one element of the superposition to another, not only do both apparatus within a given element observe the value appropriate to that element, but also the second apparatus "sees" that both have observed that value. (The second apparatus, which may be assumed to have known in advance that the undisturbed average of $\boldsymbol{P}$ was $\langle \boldsymbol{P} \rangle$, "sees" that this average has shifted to $\langle \boldsymbol{P} \rangle + gs$.)

It is not difficult to devise increasingly complicated situations in which, for example, each apparatus can make decisions by switching on various couplings depending on the outcome of other observations. No inconsistencies will ever arise, however, that will permit a given apparatus to be aware of more than one world at a time. Not only is its own memory content self-consistent (think of the two apparatus above as a single apparatus that can communicate with itself) but consistency is always maintained as well in rational discourse with other automata.

Some physicists, while admitting all this, protest against the violation, implied by Everett's insights, of the principle of Occam's razor, which urges scientists to keep entities to a minimum. According to them it is preposterous to assert the "reality" of worlds of which one cannot be aware. A consistent application of this logic would require one to deny the existence of planets in distant galaxies. Someday such planets, as well as the "other worlds," may become observable, at least indirectly. One should remember that in the nineteenth century many physicists denied the existence of atoms.

Other physicists object to the superfluousness of the "other worlds" and to the prodigality of a universe that includes them all, forgetting the prodigal scale of the universe we actually see. These physicists are comfortable with little huge numbers, but not with big ones.

The Everett interpretation *does* satisfy Occam's principle in the sense that it keeps *concepts* to a minimum, taking the mathematical formalism as it stands without adding excess metaphysical baggage in the form of "collapsing wave functions" or probabilities imposed from outside. The implications of this "bare bones" interpretation are admittedly bizarre. But physicists have learned over the years that it is almost always rewarding to push any formalism (Maxwell's electromagnetic theory, Einstein's general relativity theory, quantum field theory) to its extreme logical conclusions.

Lesser objections to Everett's interpretation concern the language used in explaining it. In the last analysis the formalism should be allowed to speak for itself. Words like "splitting" or "many worlds" should not be used as substitutes for the mathematical theory, and if the words offend then one should choose others.[2]

[2] Cf. Wheeler (1957): "To describe this situation one can use if he will the words 'communication in clear terms always demands classical concepts.' However, the kind of physics that goes on does not adjust itself to the available terminology: the terminology has to adjust itself in accordance with the kind of physics that goes on."

Probability

Determinism versus indeterminism

In Everett's view quantum mechanics is a completely deterministic theory. Up to this point its formalism implies the following:

1. An apparatus that measures an observable never records anything but an eigenvalue of the operator that represents the observable, at least if the measurement is good.[3]
2. The operator represents not the value of the observable, but rather all the values that the observable can assume under various conditions, the values themselves being the eigenvalues.
3. The dynamical variables of the system, being operators, do not represent the system other than generically. They represent not the system as it really is, but rather all the situations in which it might find itself.
4. Which situation a system is actually in is specified by the state vector. Reality is therefore described jointly by the dynamical variables and the state vector. This reality is not the reality we customarily experience but is a reality composed of many worlds.

Obviously this list is insufficient to tell us how to apply the quantum formalism to practical problems. The symbols that describe a given system, namely the state vector and the dynamical variables, describe not only the system as it is observed in one of the many worlds comprising reality, but also the system as it is seen in all the other worlds. We, who inhabit only one of these worlds, have no symbols to describe our world alone. Because we ordinarily have no access to the other worlds we are unable to make rigorous predictions about reality as we observe it. Although reality as a whole is completely deterministic, our own corner of it suffers from indeterminism. The interpretation of the quantum formalism is complete only when we show that this indeterminism is nevertheless governed by rigorous statistical laws.

Permutations: equal likelihood

Suppose, in the superposition (10.40), that a particular eigenvalue of $\boldsymbol{s}$ fails to appear, i.e., that $c_s = 0$ for that eigenvalue. Then in none of the many worlds represented by the superposition does the apparatus observe the value in question. It is natural to say that the apparatus will not observe that value, or that the value has *zero likelihood* of being observed. But what significance must one ascribe to the nonvanishing coefficients?

[3] Imperfect measurements are described in DeWitt (2003), from the Everett point of view supplemented with the "consistent histories" viewpoint. (See also Deutsch 1999.)

David Deutsch[4] has given the best answer to this question, an answer that is free from a priori statistical notions and is based solely on factual physical properties of the system. Denote by $\mathcal{S}$ the set of eigenvalues s. Let $\mathcal{F}$ be a finite subset of $\mathcal{S}$ contained in the support of the function $c : \mathcal{S} \longrightarrow \mathbb{C}$. That is

$$s \in \mathcal{F} \Longrightarrow c_s \neq 0 \tag{10.46}$$

Denote by $\mathcal{P}$ the set of all bijective maps

$$\Pi : \mathcal{S} \longrightarrow \mathcal{S} \ \text{ such that } \ \Pi(s) = s \ \text{ if } \ s \notin \mathcal{F}. \tag{10.47}$$

$\mathcal{P}$ is a group, its elements Π being permutations of the eigenvalues in $\mathcal{F}$. Let $\boldsymbol{A}$ be the operator algebra of the system S and let $\boldsymbol{U} : \mathcal{P} \longrightarrow \boldsymbol{A}$ be a mapping that satisfies

$$\boldsymbol{U}(\Pi_1)\boldsymbol{U}(\Pi_2) = \boldsymbol{U}(\Pi_1 \circ \Pi_2), \tag{10.48}$$

$$\boldsymbol{U}(\Pi)|s\rangle = |\Pi(s)\rangle, \tag{10.49}$$

for all Π_1, Π_2, Π in $\mathcal{P}$ and all s in $\mathcal{S}$. It is easy to verify that $\boldsymbol{U}(\Pi)$ is a unitary operator for each Π and hence that

$$\langle s|\boldsymbol{U}(\Pi) = \langle \Pi^{-1}(s)|. \tag{10.50}$$

Being unitary each $\boldsymbol{U}(\Pi)$ is in principle realizable by an external dynamical agent acting on S and inducing a unitary transformation of all its dynamical variables. In particular

$$\boldsymbol{s} \longrightarrow \boldsymbol{s}_\Pi = \boldsymbol{U}(\Pi^{-1})\boldsymbol{s}\boldsymbol{U}(\Pi), \tag{10.51}$$

and any ordering of the eigenvalues of $\boldsymbol{s}_\Pi$ is just a permutation of a corresponding ordering of the eigenvalues of $\boldsymbol{s}$.

Suppose $|\psi\rangle$ has the property that the physical state it describes looks no different from the point of view of the transformed dynamical variables than it does from the point of view of the original variables, no matter which Π is chosen. This implies

$$\boldsymbol{U}(\Pi)|\psi\rangle = e^{i\theta(\Pi)}|\psi\rangle \ \text{ for all } \ \Pi \in \mathcal{P} \tag{10.52}$$

for some $\theta : \mathcal{P} \longrightarrow \mathbb{R}$. It also implies that, as far as $\boldsymbol{s}$ is concerned, the scrambling of its eigenvalues is immaterial, and indeed undetectable, in the state represented by $|\psi\rangle$. This means that if a measurement of $\boldsymbol{s}$ were to be made in this state then all those outcomes s that lie in $\mathcal{F}$ would be *equally likely*. Note that although this statement is probabilistic it concerns a purely factual property of $|\psi\rangle$.

[4] What follows is based on a 1999 Oxford preprint that Deutsch did not publish.

From eqns (10.50) and (10.52) it follows that

$$\langle s|\psi\rangle = e^{-i\theta(\Pi)}\langle s|\boldsymbol{U}(\Pi)|\psi\rangle = e^{-i\theta(\Pi)}\langle \Pi^{-1}(s)|\psi\rangle \tag{10.53}$$

and hence

$$|\langle s|\psi\rangle| = |\langle \Pi^{-1}(s)|\psi\rangle| \quad \text{for all } \Pi \in \mathcal{P}. \tag{10.54}$$

Suppose $\mathcal{F}$ has n members and happens to coincide with the support of the function c. Then no eigenvalues will be observed other than those contained in $\mathcal{F}$, and it is natural to push the probabilistic terminology one step further by saying that each of the eigenvalues in $\mathcal{F}$ has *probability* $1/n$ of being observed. When this happens one has

$$1 = \langle\psi|\psi\rangle = \sum_{s'}\langle\psi|s'\rangle\langle s'|\psi\rangle = n|\langle s|\psi\rangle|^2, \qquad s \in \mathcal{F}, \tag{10.55}$$

so this probability can be expressed in the form

$$1/n = |\langle s|\psi\rangle|^2, \qquad s \in \mathcal{F}. \tag{10.56}$$

The case of degeneracy

It is easy to include other labels, α, in the basis vectors when the spectrum of $\boldsymbol{s}$ is degenerate. Equation (10.49) then gets replaced by

$$\boldsymbol{U}(\Pi)|s, \alpha\rangle = |\Pi(s), \alpha_\Pi\rangle, \tag{10.57}$$

and one conveniently introduces the projection operators

$$\boldsymbol{P}_s := \sum_\alpha |s, \alpha\rangle\langle s, \alpha|. \tag{10.58}$$

Evidently

$$\boldsymbol{U}(\Pi)\boldsymbol{P}_s\boldsymbol{U}(\Pi^{-1}) = \sum_{\alpha_\Pi} |\Pi(s), \alpha_\Pi\rangle\langle\Pi(s), \alpha_\Pi| = \boldsymbol{P}_{\Pi(s)}. \tag{10.59}$$

Undetectability of the permutations in the state $|\psi\rangle$ is still expressed by eqn (10.52), but this now implies

$$\langle\psi|\boldsymbol{P}_{\Pi(s)}|\psi\rangle = \langle\psi|\boldsymbol{P}_s|\psi\rangle, \tag{10.60}$$

and when $\mathcal{F}$ coincides with the s-support of $\langle s, \alpha|\psi\rangle$ eqns (10.55) and (10.56) get replaced by

$$1 = \langle\psi|\psi\rangle = \sum_{s'} \langle\psi|\boldsymbol{P}_{s'}|\psi\rangle = n\langle\psi|\boldsymbol{P}_s|\psi\rangle, \qquad s \in \mathcal{F}, \tag{10.61}$$

$$\langle\psi|\boldsymbol{P}_s|\psi\rangle = 1/n. \tag{10.62}$$

Unequal probabilities

We stress again that although the terminology of probability theory is now being used, the words themselves have no probabilistic antecedents. They are defined not in terms of an a priori metaphysics but in terms of factual physical properties of the state that $|\psi\rangle$ represents. However, once the terminology of probability theory has been introduced there need be no hesitation in using it in exactly the same way as it is used in the standard probability calculus. That is, the probability calculus, in particular the calculus of *conditional* or *joint probabilities*, may be freely used to motivate further definitions.

When $\langle\psi|\boldsymbol{P}_s|\psi\rangle$ is not constant over the s-support of the function c it is convenient to imagine two auxiliary physical systems, Q and R, in addition to S, together with their state vectors $|\phi\rangle$ and $|\chi\rangle$. Let $\mathcal{A}$ be a subset of m distinct eigenvalues q of a nondegenerate observable $\boldsymbol{q}$ of Q and let $\mathcal{B}$ be a subset of $n-m$ distinct eigenvalues r of a nondegenerate observable $\boldsymbol{r}$ of R. Let the r's in $\mathcal{B}$ be all different from the q's in $\mathcal{A}$ so that the set $\mathcal{A}\cup\mathcal{B}$ has n distinct elements.

Suppose $|\phi\rangle$ and $|\chi\rangle$ are such that all the q's in $\mathcal{A}$ are equally likely and all the r's in $\mathcal{B}$ are equally likely. Suppose, furthermore, that $\langle q|\phi\rangle = 0$ when $q \notin \mathcal{A}$ and $\langle r|\chi\rangle = 0$ when $r \notin \mathcal{B}$. Then each q in $\mathcal{A}$ has probability $1/m$ of being observed and each r in $\mathcal{B}$ has probability $1/(n-m)$ of being observed. In mathematical language

$$\left.\begin{aligned} |\langle q|\phi\rangle|^2 &= 1/m, & q &\in \mathcal{A}, \\ |\langle r|\chi\rangle|^2 &= 1/(n-m), & r &\in \mathcal{B}. \end{aligned}\right\} \tag{10.63}$$

Consider the action, on the combined state-vector space of the systems Q, R, and S, of the operator

$$\boldsymbol{u} := \boldsymbol{q}\otimes\boldsymbol{1}\otimes\boldsymbol{P}_s + \boldsymbol{1}\otimes\boldsymbol{r}\otimes(\boldsymbol{1}-\boldsymbol{P}_s) \tag{10.64}$$

This operator, which is an observable of the combined system, can be measured as follows. First measure $\boldsymbol{s}$. If s is obtained then measure $\boldsymbol{q}$. If s is not obtained measure $\boldsymbol{r}$ instead. The final outcome in either case is the measured value of $\boldsymbol{u}$. Note that $\boldsymbol{u}$ has n distinct eigenvalues lying in the set $\mathcal{A}\cup\mathcal{B}$, and these are the only eigenvalues that can turn up when the state vector of the combined system is

$$|\Psi\rangle = |\phi\rangle|\chi\rangle|\psi\rangle. \tag{10.65}$$

Now suppose that the state $|\psi\rangle$ of S is such that, when $\boldsymbol{u}$ is measured, all the outcomes u lying in $\mathcal{A}\cup\mathcal{B}$ are equally likely. Then each u in $\mathcal{A}\cup\mathcal{B}$ has probability $1/n$ of being observed. Given the prescription for measuring $\boldsymbol{u}$, it follows from the calculus of joint probabilities that

$$\frac{1}{n} = \begin{cases} p \times (1/m) \\ \text{or} \\ (1-p) \times [1/(n-m)], \end{cases} \tag{10.66}$$

where p is the probability that s will be observed when $\boldsymbol{s}$ is measured. Note that p is defined *via* the calculus of joint probabilities and that both possibilities in (10.66) lead to

$$p = m/n. \tag{10.67}$$

Although the derivation of this result refers to hypothetical measurements made on the hypothetical auxiliary systems Q and R, it nevertheless refers only to factual properties of S.

To relate p to $\langle\psi|\boldsymbol{P}_s|\psi\rangle$ consider the projection operator on the eigenvalue u of $\boldsymbol{u}$:

$$\boldsymbol{P}_u = \sum_q \delta_{qu}|q\rangle\langle q| \otimes \mathbf{1} \otimes \boldsymbol{P}_s + \sum_r \delta_{ur}\mathbf{1} \otimes |r\rangle\langle r| \otimes (\mathbf{1} - \boldsymbol{P}_s). \tag{10.68}$$

If $u \in \mathcal{A} \cup \mathcal{B}$ then

$$\begin{aligned} \frac{1}{n} &= \langle\Psi|\boldsymbol{P}_u|\Psi\rangle \\ &= \sum_{q\in\mathcal{A}} \delta_{uq}|\langle q|\phi\rangle|^2\langle\psi|\boldsymbol{P}_s|\psi\rangle + \sum_{r\in\mathcal{B}} \delta_{ur}|\langle r|\chi\rangle|^2\,(1 - \langle\psi|\boldsymbol{P}_s|\psi\rangle) \\ &= \begin{cases} \langle\psi|\boldsymbol{P}_s|\psi\rangle \times (1/m) \\ \text{or} \\ (1 - \langle\psi|\boldsymbol{P}_s|\psi\rangle) \times [1/(n-m)]\,, \end{cases} \end{aligned} \tag{10.69}$$

in which (10.63) is used in passing to the final expression. Comparison with (10.66) and (10.67) yields

$$\langle\psi|\boldsymbol{P}_s|\psi\rangle = p = m/n. \tag{10.70}$$

Note that this result does not require the s-support of the function c to be a finite subset of $\mathcal{S}$.

Irrational probabilities

The above analysis gives a factual meaning to all rational probabilities. It can be extended to irrational probabilities (Deutsch 1999) by introducing the notion "at least as likely" and making a kind of Dedekind cut between those states in which the eigenvalue s is at least as likely to be observed as in the given state and those in which it is not. It is essential that one prescribe a specific class of *physical* processes for carrying out the state companion. However, even if one does this the Dedekind cut method of defining irrational probabilities is unphysical (and hence basically meaningless), for the following reasons. First, probabilities themselves are physically measurable only when viewed as frequencies (see below), and physically measured numbers are always rational. Second, the set of comparison states in the Dedekind cut definition is an infinite one, so the definition is untestable. It cannot be checked by any physical process.

Expectation value: single system versus an ensemble

Deutsch's analysis above shows that *the conventional probability interpretation of quantum mechanics emerges from the formalism itself and does not have to be imposed from outside.* This fact is important when one adopts Everett's view that the formalism corresponds directly to reality so that there is no room for a priori probabilistic concepts. The probability that is defined here, in terms of factual physical statements, is precisely the probability that a rational person (or automaton) would use in placing bets about the outcomes of observations.[5] For example, if he could buy a gaming machine that would pay him a dollar amount equal to the outcome of a measurement of the observable $\boldsymbol{s}$ of a system having the state vector $|\psi\rangle$, the maximum he would rationally be willing to pay for the machine (i.e., its *value*) is

$$\langle \boldsymbol{s} \rangle = \sum_{s} s \langle \psi | \boldsymbol{P}_s | \psi \rangle = \sum_{s,\alpha} \langle \psi | s, \alpha \rangle s \langle s, \alpha | \psi \rangle = \langle \psi | \boldsymbol{s} | \psi \rangle. \qquad (10.71)$$

This value is known as the *expectation value* of $\boldsymbol{s}$ in the state represented by $|\psi\rangle$.

If the reasoning up to now has been truly rational and consistent, the expectation value (10.71) should be equal to the average payoff if the game is repeated many times, with the system always in the same initial state. Repetition of a game is equivalent to a single game played on an ensemble of identical systems in identical states. The total state vector of such a game has the form

$$|\Psi\rangle = |\psi_1\rangle|\psi_2\rangle \ldots |\Phi\rangle, \qquad (10.72)$$

where $|\psi_1\rangle$, $|\psi_2\rangle$, ... are the system state vectors and $|\Phi\rangle$ is the apparatus state vector, all assumed normalized. The apparatus may be regarded as measuring consecutively the values of observables $\boldsymbol{s}_1, \boldsymbol{s}_2, \ldots$ by means of couplings $g\mathfrak{X}_1, g\mathfrak{X}_2, \ldots$ that produce unitary transformations on a set of basis vectors $|s_1\rangle|s_2\rangle \ldots |P_1, P_2, \ldots\rangle$

[5] Deutsch (1999) derives this statement from a "theory of values" that he constructs.

singled out by the couplings. Assume for simplicity that the spectra of the $\boldsymbol{s}_n$ are nondegenerate. Then

$$\langle s_n | \psi_n \rangle = c_{s_n} \quad \text{for all } n, \tag{10.73}$$

and after N measurements have taken place the first N of the undisturbed apparatus observables $\boldsymbol{P}_1, \boldsymbol{P}_2, \ldots$ find themselves transformed into disturbed observables $\bar{\boldsymbol{P}}_1, \bar{\boldsymbol{P}}_2, \ldots$ having basis eignenvectors given by

$$\begin{aligned} &|s_1\rangle|s_2\rangle \ldots |\bar{P}_1, \ldots, \bar{P}_N, P_{N+1}, \ldots\rangle \\ &\quad = |s_1\rangle|s_2\rangle \ldots |P_1 - g s_1, \ldots, P_N - g s_N, P_{N+1}, \ldots\rangle \end{aligned} \tag{10.74}$$

(cf. eqn (10.36)). If one decomposes $|\Psi\rangle$ in terms of these vectors one finds

$$|\Psi\rangle = \sum_{s_1, s_2, \ldots} c_{s_1} c_{s_2} \ldots |s_1\rangle|s_2\rangle \ldots |\Phi\,[s_1, \ldots, s_N]\rangle \tag{10.75}$$

where

$$\begin{aligned} |\Phi\,[s_1, \ldots, s_N]\rangle = &\int d\bar{P}_1 \ldots \int d\bar{P}_N \int dP_{N+1} \ldots |\bar{P}_1 + g s_1, \ldots, \bar{P}_N \\ &+ g s_N, P_{N+1}, \ldots\rangle \Phi(\bar{P}_1, \ldots, \bar{P}_N, P_{N+1}, \ldots), \end{aligned} \tag{10.76}$$

$$\Phi(P_1, P_2, \ldots) = \langle P_1, P_2, \ldots | \Phi \rangle. \tag{10.77}$$

It will be observed that although every system is initially in the same state as every other, the apparatus, as represented by the relative state vectors $|\Phi\,[s_1, \ldots, s_N]\rangle$, does not generally record a sequence of identical values for the system observable, even within a single element of the superposition (10.75). Each *memory sequence* $s_1, \ldots, s_N$ yields a distribution of possible values for the system observable. Each of these distributions may be subjected to a statistical analysis. The first and simplest part of such an analysis is the calculation of the *histogram* or *relative frequency function* of the distribution:

$$f(s\,; s_1, \ldots, s_N) := \frac{1}{N} \sum_{n=1}^{N} \delta_{s s_n}. \tag{10.78}$$

In terms of this function one may define

$$\delta\,(s_1, \ldots, s_N) := \sum_{s} [f(s\,; s_1, \ldots, s_N) - |c_s|^2]^2. \tag{10.79}$$

This is the first of a hierarchy of functions that measure the degree to which the sequence $s_1, \ldots, s_N$ deviates from a random sequence with weights $|c_s|^2$.

Let ϵ be an arbitrarily small positive number. Call the sequence *first random* if $\delta(s_1, \ldots, s_N) < \epsilon$ and *non-first-random* otherwise. Denote by $|\chi_N^\epsilon\rangle$ the sum of all those elements of the superposition for which the apparatus memory sequence is non-first-random. Then the total probability that the memory sequence will be non-first-random is

$$
\begin{aligned}
\langle \chi_N^\epsilon | \chi_N^\epsilon \rangle &= \sum_{\substack{s_1, s_2, \ldots \\ \delta(s_1, \ldots, s_N) \geq \epsilon}} |c_{s_1}|^2 |c_{s_2}|^2 \ldots = \sum_{\substack{s_1, \ldots, s_N \\ \delta(s_1, \ldots, s_N) \geq \epsilon}} |c_{s_1}|^2 \ldots |c_{s_N}|^2 \\
&\leq \frac{1}{\epsilon} \sum_{s_1, \ldots, s_N} \delta(s_1, \ldots, s_N) |c_{s_1}|^2 \ldots |c_{s_N}|^2. \qquad (10.80)
\end{aligned}
$$

Through use of the easily verified identities

$$\sum_{s_1, \ldots, s_N} f(s; s_1, \ldots, s_N) |c_{s_1}|^2 \ldots |c_{s_N}|^2 = |c_s|^2, \qquad (10.81)$$

$$\sum_{s_1, \ldots, s_N} [f(s; s_1, \ldots, s_N) - |c_s|^2]^2 |c_{s_1}|^2 \ldots |c_{s_N}|^2 = \frac{1}{N} |c_s|^2 (1 - |c_s|^2), \qquad (10.82)$$

one readily obtains

$$\langle \chi_N^\epsilon | \chi_N^\epsilon \rangle \leq \frac{1}{N\epsilon}, \qquad (10.83)$$

from which it follows that no matter how small one chooses ϵ one can always find an N large enough so that the probability of a non-first-random memory sequence becomes smaller than any positive number.

A similar result is obtained if $|\chi_N^\epsilon\rangle$ is defined by including, in addition, elements of the superposition (10.76) whose memory sequences fail to meet, to a chosen accuracy, any finite combination of the infinity of other requirements for a random sequence. This means that, as N becomes large, any observed sequence of eigenvalues $s_1, \ldots, s_N$ becomes overwhelmingly likely to be a random sequence with weights $|c_s|^2$. This result forms the basis for the *ensemble interpretation of quantum mechanical probabilities*, in which one is willing to bet that the histogram of any experimentally obtained long sequence $s_1, \ldots, s_N$ is close to $|c|^2$ and that the sequence itself approximates, up to some finite order, the conditions for randomness.

Note: All that has been proved here is that the ensemble interpretation of quantum mechanical probability is *consistent* with the factual definition based on invariances under permutation. The ensemble idea cannot be used as a vehicle for showing that the probability interpretation of quantum mechanics emerges from the formalism itself, because one has to *invoke* the probability interpretation of $\langle \chi_N^\epsilon | \chi_N^\epsilon \rangle$ in order to *get* the ensemble interpretation, and hence ensemble arguments are circular. Only Deutsch's arguments do the job. Everett himself left the job unfinished.

Note also that ensemble arguments do not apply to single measurements, which are just as physical as repeated measurements. Deutsch's arguments allow one to place rational bets on the outcome of single measurements. They also allow one to assign relative weights to the many "worlds" in superpositions like (10.40) and, when applied to superpositions like (10.75), to infer the rigorous statistical laws that govern one's own indeterministic corner of a globally deterministic reality.

Density operator for single measurements

Although ensembles are not needed to obtain the probability interpretation of quantum mechanics they *are* needed in order to *measure* probabilities. They also yield insights into the phenomenon of *decoherence*. Consider the superposition (10.75). When N is large the average value of a typical memory sequence in the superposition is approximately equal to the expectation value (10.71). Note that the basis vectors $|s\rangle$ do not appear in the final expression $\langle\psi|\boldsymbol{s}|\psi\rangle$ for this expectation value. It is evident therefore that had one chosen to introduce a different apparatus, designed to measure some other observable $\boldsymbol{r}$, a long sequence of measurements would have yielded an average approximately equal to $\langle\psi|\boldsymbol{r}|\psi\rangle$, in which again no basis vectors appear. One can, if one likes, reintroduce the basis vectors $|s\rangle$, obtaining

$$\langle\boldsymbol{r}\rangle = \langle\psi|\boldsymbol{r}|\psi\rangle = \sum_{s,s'} c_s^*\langle s|\boldsymbol{r}|s'\rangle c_{s'}. \tag{10.84}$$

Now suppose that instead of performing a sequence of identical measurements to obtain an experimental value for $\langle\boldsymbol{r}\rangle$, one first measures $\boldsymbol{s}$ in each case and *then* performs a statistical analysis on $\boldsymbol{r}$. This could be accomplished by introducing a second apparatus which performs a sequence of observations on a set of identical two-component systems, all in identical states given by the vector $|\Psi\rangle$ of eqn (10.40). Each of the latter systems is composed of one of the original systems together with an apparatus that has just measured the observable s. The job of the second apparatus is to make observations of the $\boldsymbol{r}$'s ($\boldsymbol{r}_1$, $\boldsymbol{r}_2$, etc.) of these two-component systems. Because a measurement of the corresponding $\boldsymbol{s}$ has intervened in each case, however, these $\boldsymbol{r}$'s are not the undisturbed $\boldsymbol{r}$'s but the $\boldsymbol{r}$'s resulting from the couplings $g\mathfrak{X}_1$, $g\mathfrak{X}_2$, etc. What the second apparatus is really observing in each case is the observable

$$\bar{\boldsymbol{r}} = e^{-ig\mathfrak{X}}\boldsymbol{r}e^{ig\mathfrak{X}}. \tag{10.85}$$

Within each element of the grand superposition the second apparatus will have observed a sequence $\bar{r}_1, \ldots, \bar{r}_N$ of values for $\bar{\boldsymbol{r}}$. When N is large the average of a typical sequence will be approximately equal to

$$\begin{aligned}
\langle \bar{\boldsymbol{r}} \rangle_s &= \langle \Psi | \bar{\boldsymbol{r}} | \Psi \rangle \\
&= \sum_{s,s'} \int d\bar{P} \int d\bar{P}' \langle \Psi | s, \bar{P} \rangle \langle s, \bar{P} | e^{-ig\mathfrak{X}} \boldsymbol{r} e^{ig\mathfrak{X}} | s', \bar{P}' \rangle \langle s', \bar{P}' | \Psi \rangle \\
&= \sum_{s,s'} \int d\bar{P} \int d\bar{P}' \langle \Psi | s, \bar{P} \rangle \langle s, P | \boldsymbol{r} | s', P' \rangle \langle s', \bar{P}' | \Psi \rangle \\
&= \sum_{s,s'} \int d\bar{P} c_s{}^* \langle s | \boldsymbol{r} | s' \rangle c_{s'} \Phi \left(\bar{P} - gs \right)^* \Phi(\bar{P} - gs')
\end{aligned} \tag{10.86}$$

(see eqns (10.35) and (10.36)). If the measurements of $\boldsymbol{s}$ are good in every case, so that the relative state vectors (10.41) satisfy the orthonormality condition (10.45), then this average reduces to

$$\langle \bar{\boldsymbol{r}} \rangle_s = \sum_s |c_s|^2 \langle s | \boldsymbol{r} | s \rangle = \mathrm{tr}(\boldsymbol{\rho}_s \boldsymbol{r}) \tag{10.87}$$

where $\boldsymbol{\rho}_s$ is the *density operator*

$$\boldsymbol{\rho}_s := \sum_s |s\rangle |c_s|^2 \langle s|. \tag{10.88}$$

The averages (10.84) and (10.87) are generally not equal. In (10.87) the measurement of $\boldsymbol{s}$, which the first apparatus has performed, has destroyed the *quantum interference effects* that are still present in (10.84). The word "decoherence" has come to be applied to this situation. One speaks of the *decoherence* of the state of a low-inertia quantum mechanical system by coupling to a high-inertia apparatus. As a result of decoherence the elements of the superposition (10.40) may, insofar as the quantum behavior of the *system* is concerned, be treated *as if* they were members of a statistical ensemble. In practice what one does, in following the subsequent behavior in one's own corner of reality, is to *collapse* the state vector $|\Psi\rangle$, i.e., to replace it by that single member of the superposition which corresponds to the new information received in the apparatus memory bank.

According to Everett the state vector does not really collapse, of course. Since the system and apparatus become correlated as a result of the first measurement it is not strictly possible to speak of the "state" of the system independently of the apparatus. The state of the total system is always *pure*, and one can only say that the state of the subsystem S has become *effectively* a mixed state, i.e., described by a density operator.

The effective collapsibility of the total state vector $|\Psi\rangle$ of eqn (10.40) is what permits one to introduce and study the quantum behavior of systems having well-defined initial states without at the same time introducing into the mathematical formalism the apparatus that prepared the systems in those states. The irrelevance of the apparatus, after it has performed its function, is expressed in one of the standard ways of obtaining the density operator, namely that of *tracing out* the apparatus:

$$\begin{aligned}\langle s|\boldsymbol{\rho}_s|s'\rangle &= \int \langle s, \bar{P}|\Psi\rangle\langle\Psi|s', \bar{P}\rangle d\bar{P} \\ &= \int c_s c_s{}^* \Phi(\bar{P} - gs)\Phi(\bar{P} - gs')^* d\bar{P} \\ &= |c_s|^2 \delta_{ss'}. \end{aligned} \tag{10.89}$$

Emergence of classical worlds

The question of narrow variances

Of far greater importance in daily life than the decoherence engendered in low-inertia systems by coupling to high-inertia apparatus is an *inverse process*, consisting of the decoherence produced in high-inertia systems through coupling to low-inertia systems. The density operator for this process involves tracing out the low-inertia systems and provides the answer to another important question that Everett left dangling: Why can one assume the ready availability of the narrow variances (as in eqn (10.22)) that allow one to perform accurate measurements in the first place?

It is certainly true that if the apparatus has high enough inertia to be basically classical, i.e., to follow a precise trajectory, then despite the quantum restriction $\Delta\boldsymbol{X}\Delta\boldsymbol{P} \geq 1/2$, variances that are sufficiently narrow to satisfy condition (10.44) are in principle available in almost any convenient apparatus observable $\boldsymbol{P}$. But why should it be so easy to get narrow variances? In practice they fell into our laps. We do not have to sweat to get them. Why is this so? How does the classical behavior emerge spontaneously from the state vector of the universe (or of a suitable isolated part of the universe)?

The germ of the answer was first given by Mott in 1929, in a beautiful paper called "The wave mechanics of α-ray tracks" (Mott 1929). Mott's example involves the decoherence of a high-inertia system (an α-particle) by coupling to low-inertia systems (the electrons in the ambient gas). The principle can be illustrated on a very simple model, which consists of a massive body moving in one dimension in an arbitrary potential V and colliding, *via* a δ-function interaction, with a light body moving in the same one-dimensional space. The Hamiltonian operator of the combined system is

$$H = -\frac{1}{2M}\frac{\partial^2}{\partial X^2} + V(X) - \frac{1}{2m}\frac{\partial^2}{\partial x^2} + g\delta(x - X) \tag{10.90}$$

where M and m are the masses of the two bodies, X and x are their positions, and g is the strength of their interaction. We seek a solution of the Schrödinger equation for this Hamiltonian operator, with the light body in an "incoming" momentum state at momentum p.

The collision with the light body will leave the motion of the massive body virtually undisturbed, provided

$$m \ll M. \tag{10.91}$$

The state of the massive body can be quite arbitrary. We shall assume only that the velocity states into which it can be decomposed correspond to velocities that are small compared to the velocity of the light body. If $p > 0$ and condition (10.91) holds, the time-dependent wave function of the combined system is then given very accurately by

$$\begin{aligned}\langle X, x, t|\Psi\rangle = L^{-1/2}\big\{\theta(X - x)\big[e^{ipx} + Re^{ip(2X-x)}\big]\\ + \theta(x - X)Te^{ipx}\big\}e^{-i(p^2/2m)t}\langle X, t|\psi\rangle,\end{aligned} \tag{10.92}$$

where θ is the step function, L is the length of an effective "box" controlling the normalization of the momentum wave functions, T and R are the *transmission* and *reflection* coefficients for the collision:

$$T = \frac{1}{1 + i\frac{g\mu}{p}}, \qquad R = \frac{-i\frac{g\mu}{p}}{1 + i\frac{g\mu}{p}}, \qquad \mu = \frac{Mm}{M + m} \approx m, \tag{10.93}$$

and $\langle X, t|\psi\rangle$ is the wave function for the massive body in the absence of the light body:

$$\left[-i\frac{\partial}{\partial t} - \frac{1}{2M}\frac{\partial^2}{\partial X^2} + V(X)\right]\langle X, t|\psi\rangle = 0. \tag{10.94}$$

Density operator

The factorization of the wave function (10.92) into a part referring to the light body and a part $\langle X, t|\psi\rangle$ satisfying the Schrödinger equation (10.94) of the massive body alone is entirely due to condition (10.91). Note, however, that the first factor does not refer *solely* to the light body. The term involving the reflection coefficient contains a phase factor e^{2ipX} representing the effect of the momentum transfer to the massive body. Although this momentum transfer has no practical effect on the motion of the massive body, which continues to be described by the wave function $\langle X, t|\psi\rangle$, its role in decoherence is crucial. In order to see this, construct the density operator $\boldsymbol{\rho}$ of the massive body by tracing out the light body:

$$\begin{aligned}\langle X, t|\boldsymbol{\rho}|X', t\rangle &= \int_{-L/2}^{L/2} \langle X, x, t|\Psi\rangle\langle\Psi|X', x, t\rangle dx \\ &\underset{L\to\infty}{\to} \langle X, t|\psi\rangle\langle\psi|X', t\rangle M_p(X - X'),\end{aligned} \tag{10.95}$$

where

$$M_p(X - X') = \left[1 + \left(\frac{g\mu}{p}\right)^2\right]^{-1}\left[1 + \left(\frac{g\mu}{p}\right)^2 e^{ip(X-X')} \cos p(X - X')\right]. \tag{10.96}$$

With the light body traced out, the density operator is no longer that of a pure state. Its matrix representation now includes the *modulation function* M_p. The absolute value of this function has the following general appearance:

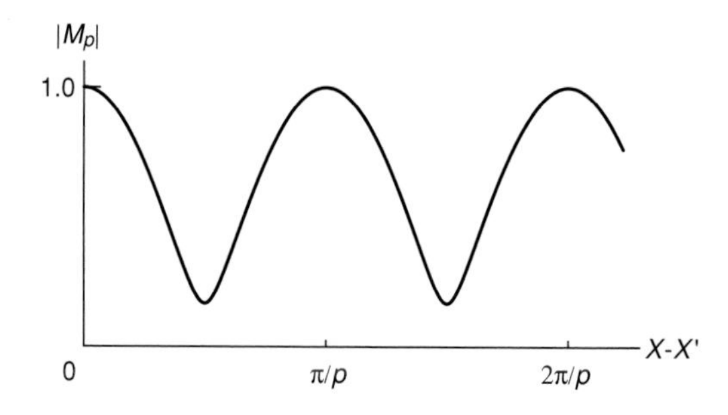

The function $M_p(X - X')$, regarded as a continuous matrix, has several important properties:

1. It is Hermitian and positive definite.
2. If each of its (continuous infinity of) elements is raised to the power N, where N is a fixed positive integer, the result is again a positive definite Hermitian matrix. (Note that this is not the same thing as raising the matrix itself to the Nth power.)
3. Properties 1 and 2 are invariant under time reversal.

Remark: The justification for tracing out the light body is not that one may neglect either the effect it may have (through other couplings) on other systems, or the quantum interference effects its wave function may produce on expectation values (or other matrix elements) of *its own* observables, but that the expectation value of any observable $\mathbf{A}$ of the *massive* body is given, in the state represented by expression (10.92), by

$$\langle\Psi|\mathbf{A}|\Psi\rangle = tr(\boldsymbol{\rho}\mathbf{A}). \tag{10.97}$$

Localization: sharp decoherence

The density matrix (10.95) does not describe the massive body as being in a localized state. However, suppose the massive body is allowed to collide with N identical light bodies, all in the same momentum state and having identical δ-function interactions with the massive body. Then in the wave function for the combined system there will be a factor for each light body, and when all these are traced out the density matrix of the massive body will take the form

$$\langle X, t|\rho|X', t\rangle \underset{L\to\infty}{\to} \langle X, t|\psi\rangle\langle\psi|X', t\rangle E(X - X') \tag{10.98}$$

where

$$E(X - X') = [M_p(X - X')]^N. \tag{10.99}$$

The function E may be called the *environmental modulation function*. Because of property 2 above it, like M_p, is positive definite and Hermitian when viewed as a continuous matrix. For $N = 20$ its absolute value, based on the previous figure, has the appearance

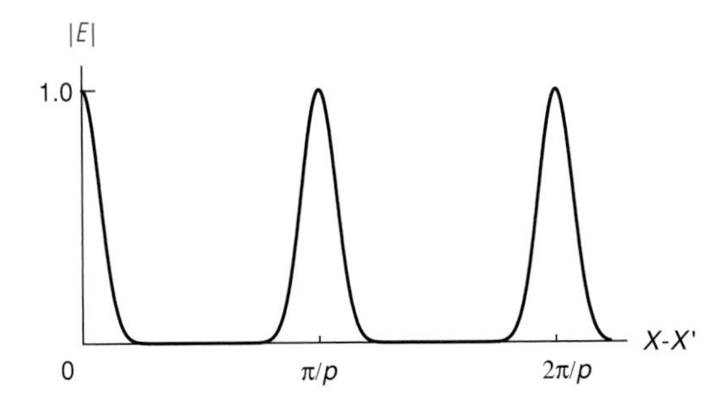

Here localization is beginning to show itself. But it is a localization modulo π/p. To get true localization it is clear what one must do. One must let the massive body collide with N_1 light bodies having momentum p_1, N_2 having momentum p_2, and so on, and choose the p's to be incommensurable. The density matrix takes again the form (10.98), but with an environmental modulation function given by

$$E(X - X') = [M_{p_1}(X - X')]^{N_1}[M_{p_2}(X - X']^{N_2}\dots. \tag{10.100}$$

This function, like all the others, defines a positive definite Hermitian continuous matrix, but *its* absolute value has the appearance

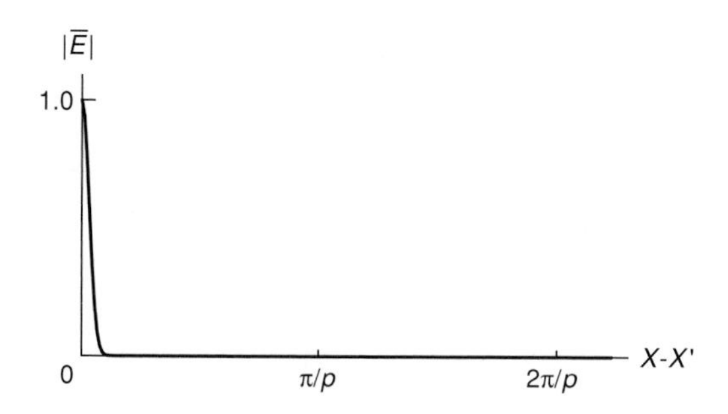

As the N's become large E becomes a very narrow function. If the wave function $\langle X, t|\psi\rangle$ varies negligibly over a distance equal to the width of E then, switching to a time-independent basis $\{|X\rangle\}$, one may write

$$\langle X, t|\rho|X', t\rangle = \langle X|\rho(t)|X'\rangle \tag{10.101}$$

where

$$\rho(t) = \int_{-\infty}^{\infty} |E^{1/2}_{X''}\rangle|\langle X'', t|\psi\rangle|^2\langle E^{1/2}_{X''}|dX'' \tag{10.102}$$

with

$$\langle X|E^{1/2}_{X''}\rangle = E^{1/2}(X - X''), \tag{10.103}$$

the continuous matrix defined by the function $E^{1/2}$ being the positive definite Hermitian square root of that defined by E. Note that

$$\langle E^{1/2}_{X}|E^{1/2}_{X'}\rangle = E(X - X'), \tag{10.104}$$

$$\langle E^{1/2}_{X}|E^{1/2}_{X}\rangle = E(0) = 1. \tag{10.105}$$

The limiting form of the density operator, as the N's become large, evidently describes the massive body as *localized at a point*, the probability, at time t, that the point is in an infinitesimal neighborhood of width dX containing X being $|\langle X, t|\psi\rangle|^2 dX$.

Discussion

The following remarks are in order:

1. The localization occurs no matter what $|\psi\rangle$ is, provided only that (10.91) and the velocity spread condition are satisfied.
2. Only two incommensurable momenta are needed. One does not need a thermal bath.
3. If the light bodies arrive in the form of packets with a momentum distribution function $f(p)$, then M_p is modified to $\int M_p|f(p)|^2 dp$, which tends already to suppress the periodicity in $E(X - X')$ displayed when only a single sharp momentum is present and can

narrow $E(X - X')$ even further. But an arrow of time is then introduced, and the density matrix will not take the form (10.98) until after the collisions have occurred.

The above results have the following implications for decoherence in quantum cosmology, where one attempts to deal, at least schematically, with the wave function of the universe:

1. Although complexity (metastability, chaos, thermal baths, wave packets) can only help in driving massive bodies to localized states, it is *inertia*, not complexity, that is the *key* to localization and sharp decoherence.
2. Given the fact that the elementary particles of nature tend, upon cooling, to form stable bound states consisting of massive agglomerations, localization–decoherence at the classical level is a natural phenomenon of the quantum cosmos.
3. Given the fact that the interaction described above, between the massive body and the light ones, is a simple scattering interaction and not at all specially designed like that of a good measurement, the universe is likely to display localization–decoherence in almost all states that it may find itself in. The initial state of the universe does *not* have to be special.
4. Decoherence does not depend on the existence of an arrow of time. This follows from the time-reversal invariance of the key properties 1 and 2 of the modulation function M_p above.

Coarse graining: decoherence function

There is a more general and, at the same time, more precise approach to decoherence, which accounts not only for localization but also for the emergence of classicality (Griffiths 1984; Omnès 1988a, b; Gell-Mann and Hartle 1990; Hartle 1995). The formalism used in this approach is easily applied to the present model. One first introduces a set of projection operators that defines a *coarse graining* of the possible dynamical histories of the massive body. For example

$$\mathbf{P}_\epsilon(\bar{X}, t) = \int_{\bar{X}-\epsilon/2}^{\bar{X}+\epsilon/2} |X, t\rangle\langle X, t| dX, \tag{10.106}$$

where ϵ determines the coarseness of the graining. If $\bar{X}$ is chosen from a discrete set of points, separated by intervals ϵ from one another, then these projection operators, at a fixed instant of time, are mutually orthogonal. More generally, one introduces projection operators at successive instants of time, and the finest useful graining is controlled by the phenomenon of wave-packet spreading. If the spreading were entirely due to the quantum behavior of the massive body one would choose

$$\epsilon \gg \sqrt{\Delta t_{\max}/M} \tag{10.107}$$

where $\Delta t_{\max}$ is the largest of the successive time intervals. In fact the environment (i.e., the light bodies) causes additional spreading. But in any case the larger M is the finer the graining can be.

Using the projection operators one can define the so-called *decoherence function*:

$$\begin{aligned} D(\bar{X}_n, \ldots, \bar{X}_2, \bar{X}_1 | \bar{X}'_1, X'_2, \ldots, \bar{X}'_n) \\ = \operatorname{tr}[\mathbf{P}_\epsilon(\bar{\mathrm{X}}_n, t_n) \ldots \mathbf{P}_\epsilon(\bar{\mathrm{X}}_2, t_2) \mathbf{P}_\epsilon(\bar{\mathrm{X}}_1, t_1) \rho \mathbf{P}_\epsilon(\bar{\mathrm{X}}'_1, t_1) \mathbf{P}_\epsilon(\bar{\mathrm{X}}'_2, t_2) \ldots \mathbf{P}_\epsilon(\bar{\mathrm{X}}'_n, t_n)]. \end{aligned} \tag{10.108}$$

The times $t_1, t_2, \ldots, t_n$ are assumed to be in chronological order and fixed a priori. The function D, regarded as a matrix, is positive definite and Hermitian. It is not difficult to show that its positive real diagonal elements have a simple interpretation. $D(\bar{X}_n, \ldots, \bar{X}_1 | \bar{X}_1, \ldots, \bar{X}_n)$ is the (joint) probability that the massive body will be observed (by measurements of $\mathbf{P}_\epsilon(\bar{X}_1, t_1), \mathbf{P}_\epsilon(\bar{X}_2, t_2) \ldots$, for example) to pass within intervals of width ϵ about the points $\bar{X}_1, \bar{X}_2, \ldots$ at the successive times $t_1, t_2, \ldots$ respectively.

Emergence of classicality

Suppose there are just three instants of time, t_1, t_2, and t_3. One easily sees from expressions (10.98) and (10.108) that if the width of the environmental modulation function $E(X - X')$ is small compared to ϵ, the decoherence function will vanish if $\bar{X}'_1$ differs from $\bar{X}_1$. Because of the cyclic invariance of the trace, the same will be true if $\bar{X}'_3$ differs from $\bar{X}_3$. One is therefore led to study

$$\begin{aligned} D(\bar{X}_3, \bar{X}_2, \bar{X}_1 | \bar{X}_1, \bar{X}'_2, \bar{X}_3) \\ = \int_{\bar{X}_1-\epsilon/2}^{\bar{X}_1+\epsilon/2} dX_1 \int_{\bar{X}_1-\epsilon/2}^{\bar{X}_1+\epsilon/2} dX'_1 \int_{\bar{X}_2-\epsilon/2}^{\bar{X}_2+\epsilon/2} dX_2 \int_{\bar{X}'_2-\epsilon/2}^{\bar{X}'_2+\epsilon/2} dX'_2 \int_{\bar{X}_3-\epsilon/2}^{\bar{X}_3+\epsilon/2} dX_3 \\ \times \langle X_3, t_3 | X_2, t_2 \rangle \langle X_2, t_2 | X_1, t_1 \rangle \langle X_1, t_1 | \rho | X'_1, t_1 \rangle \langle X'_1, t_1 | X'_2, t_2 \rangle \\ \times \langle X'_2, t_2 | X_3, t_3 \rangle. \end{aligned} \tag{10.109}$$

Consider the integration over X_2. If it were extended to the whole real line then the two factors in which X_2 appears would combine to yield $\langle X_3, t_3 | X_1, t_1 \rangle$, which can be expressed as a Feynman functional integral over all histories of the massive body connecting the spacetime points (X_1, t_1) and (X_3, t_3). Constructive interference between the contributions that the functional integral receives from its integrand occurs for those histories that lie close to a classical trajectory (stationary point of the action) between (X_1, t_1) and (X_3, t_3). Suppose there is only one such trajectory. Then as long as the massive body satisfies the Schrödinger equation (10.94) the dominant contributions to the functional integral will come from trajectories that, at time t_2, pass well within a distance ϵ of this trajectory. Contributions from other trajectories will destructively interfere.

This means that when the integration over X_2 is constrained to the interval $\bar{X}_2 - \epsilon/2$ to $\bar{X}_2 + \epsilon/2$, as in (10.109), the integral will vanish unless $\bar{X}_2$ lies, at time t_2, within a distance of order ϵ from the classical trajectory between $(\bar{X}_1, t_1)$ and $(\bar{X}_3, t_3)$. By exactly the same kind of argument one sees that (10.109) will vanish unless $\bar{X}'_2$ too lies, at time t_2, within a distance of order ϵ from this trajectory. Since the points of the set from which the $\bar{X}$'s are chosen are separated by intervals ϵ from one another, it is clear that (10.109) will vanish unless $\bar{X}_2 = \bar{X}'_2$.

These results are easily generalized. When condition (10.107) is satisfied and when the environmental modulation function is sufficiently narrow, the matrix defined by the decoherence function has the following properties:

1. It is diagonal.
2. Even its diagonal elements will vanish unless the points $(\bar{X}_1, t_1), (\bar{X}_2, t_2), \ldots, (\bar{X}_n, t_n)$ in spacetime lie within a distance ϵ of a classical trajectory.
3. The diagonal elements will also vanish unless $(\bar{X}_1, t_1)$ lies in the support of the function $|\langle X, t|\psi\rangle|^2$.

The most sophisticated modern investigations are those that turn the problem around and try to discover, in more realistic contexts, the kinds of coarse graining that will lead to decoherence functions having the above properties. The coarse graining may involve projection operators of a more general kind than those defined in eqn (10.106), in which observables other than position are bracketed within certain limits. Or it may involve projection operators that place limits on observables that are themselves averages over regions of space*time*. In quantum cosmology, where one is dealing with the wave function of a whole universe and where it may be meaningless to introduce a space of possible state vectors for this universe, one may attempt to define the decoherence function (and its associated coarse graining) as a double functional integral (one for the rows and one for the columns of D) over sets of histories that are restricted in even more general ways (Hartle 1995). Any definition is useful to the extent that it yields a decoherence function that (1) is often diagonal and (2) satisfies the identities

$$\sum_{\bar{X}_k, \bar{X}'_k} D(\bar{X}_n, \ldots \bar{X}_1 | \bar{X}'_1, \ldots \bar{X}'_n)$$
$$= D(\bar{X}_n, \ldots \bar{X}_{k+1}, \bar{X}_{k-1}, \ldots \bar{X}_1 | \bar{X}'_1, \bar{X}'_{k-1}, \bar{X}'_{k+1}, \ldots \bar{X}'_n), \quad (10.110)$$

$$\sum_{\bar{X}_1, \ldots \bar{X}_n, \bar{X}'_1, \ldots \bar{X}'_n} D(\bar{X}_n, \ldots \bar{X}_1 | \bar{X}'_1, \ldots \bar{X}'_n) = 1, \quad (10.111)$$

where the $\bar{X}$'s and $\bar{X}'$'s are the labels relevant to the coarse graining.

Many worlds again: probability as an emergent concept

Quite generally decoherence is said to occur whenever the decoherence function, regarded as a matrix, is diagonal. The nonvanishing diagonal elements represent alternative "realities" or alternative histories. These histories, which are known as "consistent histories" (Griffiths 1984; Hartle 1995), do not quantum-mechanically interfere with each other. From the Everett viewpoint they constitute a new class of *many worlds*, in addition to those that arise in the course of good measurements. These worlds, like those induced by measurement situations, are unaware of one another (no interference), but they arise from processes (e.g., scatterings) that are much more common than those found in a laboratory setting. They are therefore likely to be ubiquitous in the states of any sufficiently complicated system.

It is a corollary of eqns (10.110) and (10.111) that whenever decoherence occurs, the diagonal elements of the decoherence function satisfy

$$\sum_{\bar{X}_k} D(\bar{X}_n, \ldots, \bar{X}_1 | \bar{X}_1, \ldots, \bar{X}_n)$$
$$= D(\bar{X}_n, \ldots, \bar{X}_{k+1}, \bar{X}_{k-1}, \ldots, \bar{X}_1 | \bar{X}_1, \ldots, \bar{X}_{k-1}, \bar{X}_{k+1}, \ldots, \bar{X}_n), \tag{10.112}$$

$$\sum_{\bar{X}_1, \ldots, \bar{X}_n} D(\bar{X}_n, \ldots, \bar{X}_1 | \bar{X}_1, \ldots, \bar{X}_n) = 1. \tag{10.113}$$

These are exactly the identities that express the laws of joint probability in the probability calculus. It is natural, therefore, whenever a coarse graining can be found leading to a decoherence function satisfying (10.110) and (10.111), and whenever decoherence based on this coarse graining occurs, to identify

$$P(\bar{X}_1, \ldots, \bar{X}_n) := D(\bar{X}_n, \ldots, \bar{X}_1 | \bar{X}_1, \ldots, \bar{X}_n) \tag{10.114}$$

as the probability of occurrence of the history described by the $\bar{X}$'s. Such an extension of the probability concept is important for two reasons:

1. It is applicable even in the absence of a meaningful state-vector space, e.g., in quantum cosmology.
2. It brings into sharper focus the fact that probability, in the last analysis, is an *emergent* concept, depending on the phenomenology of the state of the universe, and is not necessarily a useful concept for all universes. One can easily construct (on paper) a universe that is so tiny, or so simple, or in such a special state, that decoherence does not occur for *any* coarse graining. One may be at a loss to know what such a universe would be good for, or how to interpret it, but one cannot say that it could not exist.

It should finally be stressed that decoherence does not depend on a pre-existing arrow of time. To be sure, if a universe is sufficiently complicated, the imprecision (due to coarse graining) in the alternative histories of a decohering set, combined with the classical phenomenon of chaos, may, *in each history that is not already in thermal equilibrium with respect to the coarse graining*, quickly generate an arrow of time (Boltzmann's view), at least over periods of time short compared to a classical Poincaré cycle. But this does not mean that the state vector of the universe as a whole has an arrow of time. For let $|\Phi\rangle$ be such a possible state vector. If it has an arrow of time replace it by $\frac{1}{\sqrt{2}}(|\Phi\rangle + |\Phi\rangle^T)$, "$T$" denoting time reversal. Each decohering world, or history, in the latter vector is paired with a time-reversed world. The two arrows of time do not conflict, for the two worlds are unaware of one another.

Problems for the future

There are still loose ends to be tied up to complete Everett's vision. Here are three:

1. A good full analysis is needed of the Einstein, Podolsky, Rosen experiment as seen from Everett's viewpoint. The excitement over so-called "nonlocality" will almost certainly turn out to be a red herring. Such an analysis has in good measure already been provided by Deutsch and Hayden (2000). One hopes that this will jog physicists loose from their traditional mindset.
2. This mindset is responsible (among other things) for the use of the word "entanglement" in preference to "interacting worlds." If quantum computers, of sufficient strength to factor products of huge prime numbers, ever become a reality, the entanglement will be so severe as to make "other worlds" seem a more convenient concept. A textbook needs to be written setting forth this conceptual framework.
3. The universe itself is the prime example of an isolated system, uncoupled to any outside "observer" who could collapse its wave function. A proper many-worlds analysis needs to be made of this wave function, starting, for example, from the Wheeler–DeWitt equation. It is astonishing that cosmologists today are ready to entertain all sorts of ill-conceived notions about "many universes" while ignoring Everett's solidly grounded ideas. A partial antidote to this will be found in Deutsch (2002).

References

Bohr, N and Rosenfeld, L (1933) *Kgl. Danske Videnskab. Selskab, Mat.-fys. Med.* **12**, 8.
Deutsch, D (1999) *Proc. Roy. Soc. (Lond.)* **A455**, 3129.
(2002) *Proc. Roy. Soc. (Lond.)* **A458**, 2028, 2911.
Deutsch, D and Hayden, P (2000) *Proc. Roy. Soc. (Lond.)* **A456**, 1759.
DeWitt, B (1965) *Dynamical Theory of Groups and Fields*. New York: Gordon and Breach.
(2003) *The Global Approach to Quantum Field Theory*. Oxford: Oxford University Press.
Everett, H, III (1957) *Rev. Mod. Phys.* **29**, 454.

(1973) In *The Many Worlds Interpretation of Quantum Mechanics*, ed. B. DeWitt and N. Graham, p. 3. Princeton, NJ: Princeton University Press.

Gell-Mann, M and Hartle, J B (1990) In *Complexity, Entrophy and the Physics of Information*, ed. W. H. Zurek. New York: Addison Wesley.

Griffiths, R (1984) *J. Stat. Phys.* **36**, 219.

Hartle, J B (1995) In *Gravitation and Quantizations*, ed. B. Julia and J. Zinn-Justin, p. 285. Dordrecht: North Holland.

Mott, N F (1929) *Proc. Roy. Soc. (Lond.)* **126**, 79.

Omnès, R (1988a) *J. Stat. Phys.* **53**, 893, 933, 957.

(1988b) *J. Stat. Phys.* **57**, 357.

Peierls, R E (1952) *Proc. Roy. Soc. (Lond.)* **A214**, 143.

von Neumann, J (1996) *Mathematical Foundations of Quantum Mechanics*. Princeton, NJ: Princeton University Press.

Wheeler, J A (1957) *Rev. Mod. Phys.* **29**, 459.

Part IV

Quantum reality: experiment

11

Why the quantum? "It" from "bit"? A participatory universe? Three far-reaching challenges from John Archibald Wheeler and their relation to experiment

Anton Zeilinger
University of Vienna

Introduction

First a word of thanks. When I first came across the papers of John Archibald Wheeler on the foundations of quantum mechanics, most of them reprinted in Wheeler and Zurek (1983), I could not believe what I read. Finally here was a colleague of worldwide reputation, given his many contributions to theoretical physics, who was not afraid to discuss openly the conceptual problems of quantum mechanics. The outstanding feature of Professor Wheeler's viewpoint is his realization that the implications of quantum mechanics are so far-reaching that they require a completely novel approach in our view of reality and in the way we see our role in the universe. This distinguishes him from many others who in one way or another tried to save pre-quantum viewpoints, particularly the obviously wrong notion of a reality independent of us.

Particularly remarkable is Professor Wheeler's austerity in thinking. He tries to use as few concepts as possible and to build on this the whole of physics. A fascinating case in point is the title of one of his papers "Law without law," the attempt to arrive at the laws of nature without assuming any law a priori.

For me personally his work on fundamental issues in quantum mechanics has been particularly inspiring. The questions he raises are exceptionally far-reaching and some of his concepts in the foundations of physics are so radical that calling them revolutionary would not do them justice. Such radically new concepts are certainly needed in view of such challenges as the measurement problem, the Schrödinger cat paradox, the conceptual nature of quantum entanglement, or the transition from quantum to classical.

Science and Ultimate Reality, eds. J. D. Barrow, P. C. W. Davies and C. L. Harper, Jr. Published by Cambridge University Press.

In his discussions of the foundations of quantum mechanics Wheeler uses thought-experiments a number of times. In this way he continues the beautiful tradition which was set in quantum mechanics from the very beginning, for example by Heisenberg's gamma microscope and culminating in the Bohr–Einstein dialogue, showing that thought-experiments are the vehicle of choice to demonstrate counterintuitive features of quantum theory or even to challenge it. In the last two to three decades technological progress has made it possible for many of these thought-experiments to be realized in the laboratory, and this has led to a perfect confirmation of all the counterintuitive predictions of quantum mechanics. It has also led to the invention of novel experiments which the forefathers did not even dream of in their gedanken version, and, most recently, this work on the foundations of quantum mechanics has brought into existence a new field of information science signified by such interesting topics as quantum cryptography, quantum teleportation, and quantum computation.

This experimental development is now giving rise to new thinking about the foundations of quantum mechanics, having increasingly freed the minds of physicists, particularly of young physicists, from the prejudices about how the world ought to work, prejudices that are based on pre-quantum classical concepts. In search of the final understanding of quantum mechanics, John Archibald Wheeler's far-reaching questions provide bright beacons for illuminating the abysses of prejudice, of preconceived notions, and of complacency with seemingly satisfactory yet immature partial solutions.

A participatory universe?

Quantum physics has raised the question of the role of the observer in a novel way, at least for physics. In classical physics the observer has a role that is essentially passive. It is certainly legitimate within that world-view to assume reality as existing prior to and independently of our observation. The situation might be compared with that of actors on a stage, in the sense that the stage with its objects and features, including the other actors, is essentially present and we just move through it. There is clearly some influence by the observer on the world even in classical physics; for example the actor can certainly move objects around on stage, but this influence can be understood, at least in principle, on the basis of an unbroken causal chain. The most essential point here is the view that we are dealing with features of an outside world, a world in which, while it might be changed somehow by the observer through the act of observation or through other acts, any such change is a change of features pre-existing before observation.

Not so in quantum physics. Already in the famous double-slit experiment it depends on which question we ask whether the particle passing through the

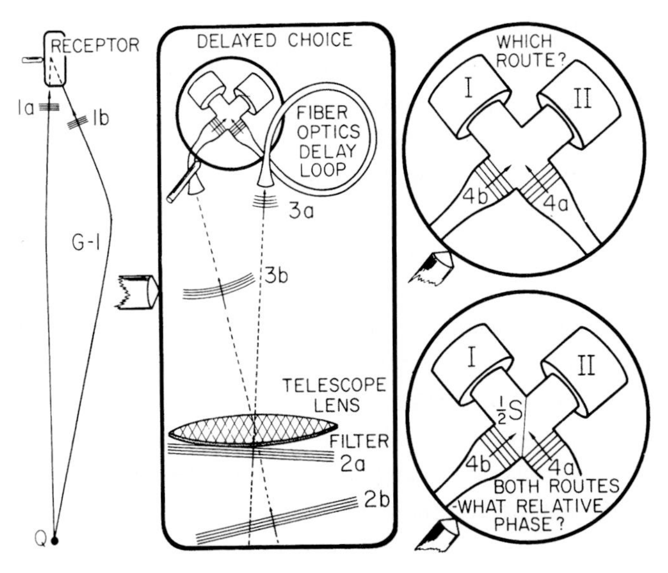

Figure 11.1. Proposed delayed-choice experiment extending over a cosmological reach of space and time. Left, quasar Q recorded at receptor as two quasars by reason of the gravitational lens action of the intervening galaxy G-1. Middle, schematic design of receptor for delayed-choice experiment: (a) filter to pass only wavelengths in a narrow interval, corresponding to a long wave train, suitable for interference experiments; (b) lens to focus the two apparent sources on to the acceptor faces of the optic fibers; (c) delay loop in one of these fibers of such length, and of such rate of change of length with time, as to bring together the waves traveling the two very different routes with the same, or close to the same, phase. Right, the choice. Upper diagram, nothing is interposed in the path of the two waves at the crossing of the optic fibers. Wave 4a goes into counter I, and wave 4b into counter II. Whichever of these photodetectors goes off, that – in a bad way of speaking – signals "by *which* route, a or b, the photon in question traveled from the quasar to the receptor." Lower diagram, a half-silvered mirror, $\frac{1}{2}$S, is interposed as indicated at the crossing of the two fibers. Let the delay loop be so adjusted that the two arriving waves have the same phase. Then there is never a count in I. All photons are recorded in II. This result, again in a misleading phraseology, says that "the photons in question come by both routes." However, at the time the choice was made whether to put in $\frac{1}{2}$S or leave it out, the photon in question had already been on its way for billions of years. It is not right to attribute to it a route. No elementary phenomenon is a phenomenon until it is a registered phenomenon.

apparatus can be viewed as a particle or as a wave. This has been brought into focus by Wheeler's proposal of a delayed-choice experiment (Fig. 11.1). Wheeler considers the ultimate interferometer, which is of the size of the universe. The essential starting point is the observation of more than one image of one and the same quasar at two spots in the sky which are close to each other. The explanation is that light from these quasars is deflected in some way by an intervening galaxy which is placed along the path of the light from the quasar to us. Wheeler then argues that the light which has come along the two (or more) routes must be coherent as it comes from the same source, so it should be possible to bring the light which has come along the two routes to interference, as is shown by the middle part of Fig. 11.1. In order to achieve this, Wheeler chooses to couple the light into optical fibers and to bring it to interference at a fiber-optic coupler. The two routes are certainly not of equal length because the geometrical arrangement of quasar, galaxy, and our position is rarely a symmetric one, so the light along one of the two routes will have arrived earlier, meaning that we have to store it until the light from the other route also arrives. Considering the cosmic differences this could be a very long time and thus beyond any practical feasibility, but that is not the point here. Then Wheeler makes the interesting suggestion that it is up to the observer to decide at the last instance just before the photon is measured whether it behaved like a particle or like a wave. The observer is free to decide to either detect the photons having propagated on their separate paths separately or to insert a semi-reflecting beam-splitter (the right-hand part of Fig. 11.1), in which case the waves which have come from the two routes are coherently superposed. It is clear that it is a decision at the disposal of the experimentalist whether or not to insert the semi-reflecting mirror at the time after the light has already propagated to us. This choice then decides whether or not the light has come to us as a particle or as a wave. In Wheeler's own words, "One decides whether the photon should have come by one route or by both routes after it has already done its traveling."

In an experiment a few years ago in my group we brought Wheeler's thought-experiment into the laboratory and carried it a step further (Dopfer 1998; Zeilinger 1999a). The idea was to demonstrate that it can be decided after the photon has been registered already whether the phenomenon observed can be understood as a particle or as a wave. Let us first contemplate the relationship of path information and interference pattern in the two-slit experiment.

Consider a double-slit experiment with electrons (Fig. 11.2). We have an electron gun which emits electrons at such low intensity that they come one by one. The electrons then pass through a diaphragm with two slit openings and are collected on an observation screen. On the observation screen we will observe an interference pattern consisting of bright and dark stripes. This fringe pattern can easily be understood on the basis of waves which came through both slits and which interfere constructively at the maxima of intensity on the observation screen and destructively at the minima. Evidently the interference pattern only forms because of the wave having come along two routes. Let us then also consider some light source which produces photons with energy $h\nu$. These photons may be scattered by the electrons and we view the scattered photons using a Heisenberg microscope. This microscope assembly had been invented by Heisenberg (1927) in order to demonstrate the position-momentum uncertainty relation for electrons.

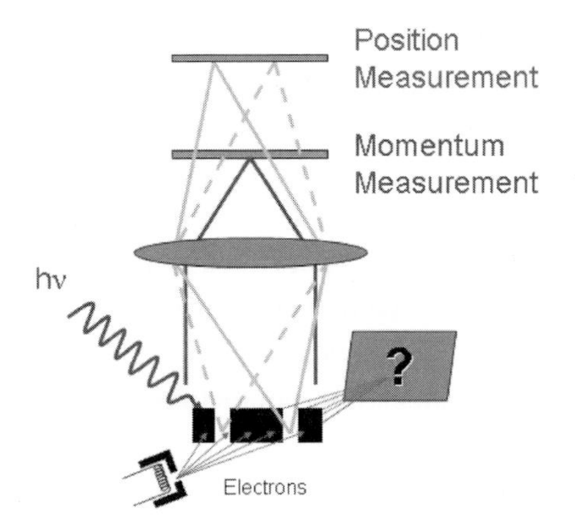

Figure 11.2. The Heisenberg microscope in a double-slit experiment for electrons. Photons with energy $h\nu$ are scattered off electrons passing one by one through a double-slit assembly. The scattered photons are then imaged using a Heisenberg microscope. The experimentalist has the choice to position the observation screen for the scattered photons at any distance behind the Heisenberg lens. If the observation screen is placed in the focal plane, detection of a photon there collapses the incoming wave of the photon onto a momentum eigenstate containing no position information, hence double-slit electron interference should appear in that case. If the observation screen is placed at the image plane of the microscope lens, the position where the scattering took place can be determined and thus the slit through which the electron passed. In that case no double-slit electron interference should appear.

There are clearly various choices in which observation plane we detect our photons behind the microscope. Let us assume that we use a position-sensitive photon detector which for each photon gives us the position where it arrives. At first we consider the detector being placed at the image plane of the microscope. Each position on the detector plane corresponds to a unique position in the plane of the two-slit assembly. Therefore, by registering where the photon arrives we can find out which path the electron took through the assembly. Therefore no interference pattern can arise as it is well known that any interference pattern disappears as soon as we have path information.

Yet we also have other alternative choices available as to where we place our single-photon detector. For example, if we place our detector in the focal plane of the lens then each incident direction is imaged onto one spot of the detector. Therefore detection of the photon now gives us the momentum of the photon after it has been scattered by the electrons and no path information for the electrons whatsoever, which therefore must show an interference pattern. The conceptual problem now arises that we can easily consider a situation where the electrons are detected in the observation plane earlier than the photons are detected. Therefore, we could consider the choice as to whether a momentum measurement or a position measurement is made through the photon to be done at the last instant after the electron has already been measured. This possibility has actually been remarked upon for the Heisenberg microscope by C. F. von Weizsäcker (1931).

So what does the poor electron then do, when it arrives on the observation screen? Does it behave like a wave forming an interference pattern, as would be necessary if a momentum measurement is made on the photon, or does it behave like a particle arriving randomly somewhere on the observation plane, allowing for the possibility that the detector be placed at a location for position measurement? Let us therefore refer to the real experiment.

In the experiment, instead of using an electron and a photon, we use entangled photon pairs created in the process of type-I parametric down-conversion, in a $LiIO_3$ crystal, with an optical nonlinearity, pumped by a UV laser beam (Fig. 11.3). This results in rare spontaneous creation of entangled photon pairs. These pairs are entangled in the sense that neither of the two photons carries any well-defined momentum or well-defined energy on its own but all that is defined is that their momenta and their energies have to add up to the momentum and energy of the incident photon. Experimentally this means that as soon as one of the two photons is measured it spontaneously assumes some energy and momentum and then the other photon, no matter how far away it is, immediately assumes the corresponding energy and momentum such that they add up to the energy and momentum of the original photon. Here we only consider the momentum entanglement. One of the two photons is then sent to a double-slit set-up and detected behind its double-slit using a movable single-photon detector. The other photon passes through the Heisenberg lens and is detected in the single-photon Heisenberg detector D1. The Heisenberg detector may be placed at any position behind the lens including the two positions considered above. As we will now see, photon 1 passing through the Heisenberg lens plays exactly the same role as the photon in our previous considerations.

We expect that if the detector is placed at the focal distance f behind the lens, the incoming photon is projected on a well-defined momentum state, and thus it cannot carry any position information. Therefore, we have no information where photon 2 passes through the two-slit assembly, as it also is projected onto a momentum eigenstate and thus it should exhibit a two-slit interference pattern. On the other hand, if we place the detector at the distance 2f, the plane with the two-slit assembly is exactly imaged, as this is 2f in front of the lens measuring the distance from the lens via the crystal to the slits. Then we can get position information and no interference pattern should arise for the second photon. This is exactly what we have seen in the experiment (Fig. 11.4). Does this now mean that the distribution of the photons in the observation plane behind the two-slit assembly changes depending on what we

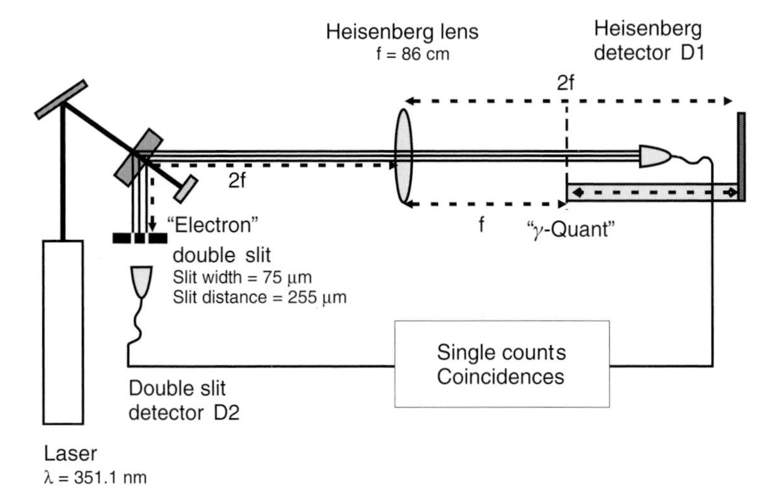

Figure 11.3. Experimental realization of the Heisenberg microscope double-slit experiment. In the LiO_3 crystal an incoming UV photon may spontaneously convert into two red photons which are momentum-entangled. One of the two photons plays the same role as the photon in the thought-experiment of Fig. 11.2. It passes through a Heisenberg lens and then the Heisenberg detector can be placed at any position behind the lens. The other photon plays the role of the electron of the thought-experiment passing through a double slit. Using proper electronics, one can determine both the individual counts in the detectors and the coincidences.

do with photon 1? Obviously this is impossible, as photon 1 is detected at a time after photon 2 has been registered already. The solution is that we have to register the two photons in coincidence. Thus whether we obtain the two-slit pattern or not depends on whether the possible position information carried by the other photon has been irrevocably erased or not.

The important conclusion here is that the distribution of events in the observation plane behind the two-slit assembly is independent of what we do with photon 1. Yet the interpretation of that distribution is crucially dependent on whether we place the detector for photon 1 in the focal plane or at the distance 2f. In the first case we can consider each photon that has already passed through the two-slit assembly and has already been registered as a wave having passed through both slits. In the second case we have to consider each photon as a particle having passed through only one slit. The important conclusion is that, while individual events just happen, their physical interpretation in terms of wave or particle might depend on the future; it might particularly depend on decisions we might make in the future concerning the measurement performed at some distant spacetime location in the future. It is also evident that the relative spacetime arrangement of the two observations does not matter at all. We could carry out the two registrations in a spacelike separated manner, in which case the relative time ordering of the two events is not

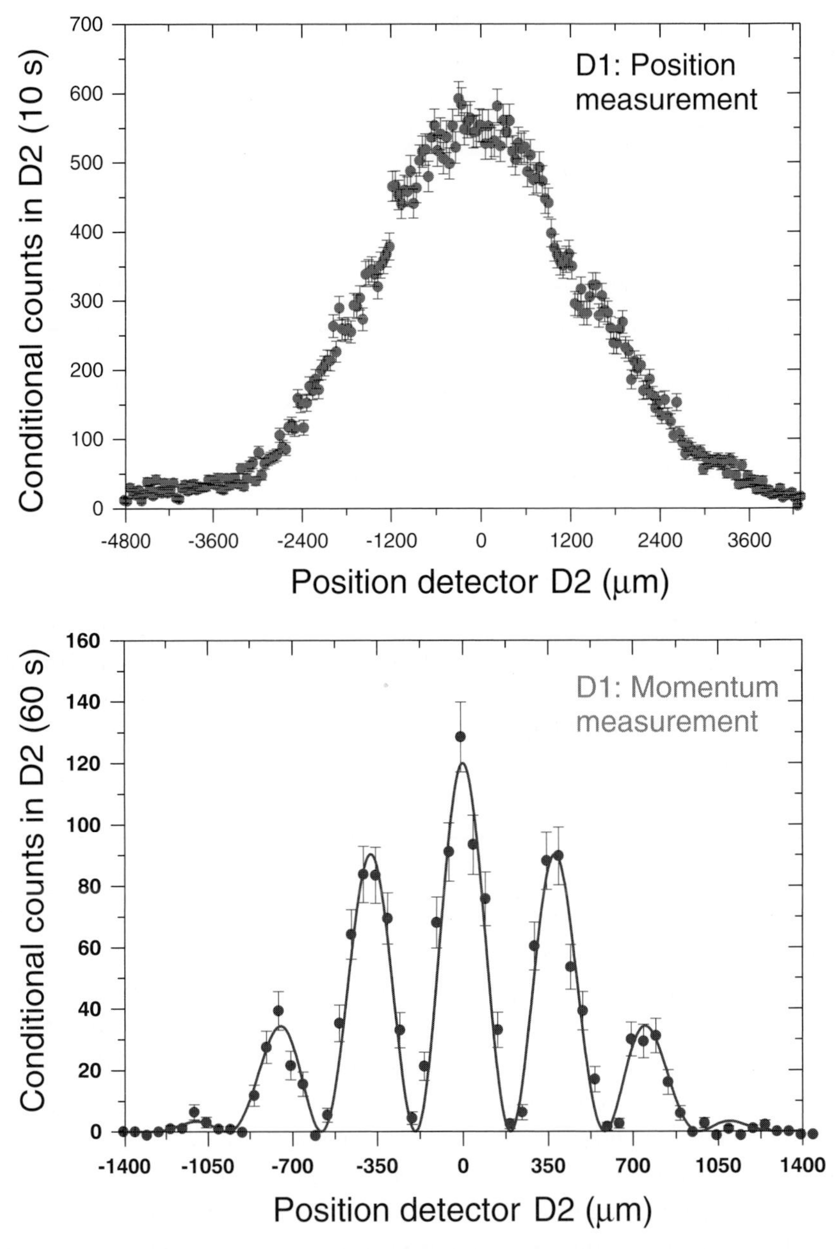

Figure 11.4. Experimental result of the experiment of Fig. 11.3. If the Heisenberg detector is placed at a distance suitable for position measurement (top), no interference fringe results as path information for the second photon passing through the double slit is available. If the detector for the first photon is placed in the focal plane behind the Heisenberg lens, no position information is available and hence for those photons arriving behind the double slit in coincidence, with the other photon being registered behind the lens, beautiful interference fringes result (bottom). The reader should also check the intensity, which clearly demonstrates that we have single-photon interference.

well defined. Depending on the relative motion of an observer and the apparatus, either one might precede the other or they might appear to be simultaneous. Or we might arrange the two detections in a timelike separated manner, having a clear temporal sequence between the two. In any case this experiment, besides being a manifestation of Wheeler's delayed-choice proposal, can also be viewed as supporting Niels Bohr's famous dictum, "No phenomenon is a phenomenon unless it is an observed phenomenon." Here it means that we are not allowed to talk about photon 2 as a particle or as a wave, even at a time when it has been registered already, unless the respective experiment has actually been carried out by also registering photon 1.

The experiment just discussed also provides a clear illustration of the role of the experimentalist. By choosing the apparatus the experimentalist determines whether the phenomenon observed can be seen as a wave or as a particle phenomenon and once the observer has made this choice, Nature gives the respective answer and the other possibility is forever lost. Thus, we conclude, by choosing the apparatus the experimentalist can determine which quality can become reality in the experiment. In that sense, the experimentalist's choice is constitutive to reality, yet one should be warned strongly against a subjective interpretation of the role of the experimentalist or of the observer. It is clear that the consciousness of the observer does not influence the particle at all, in contradiction to a widespread but unfortunate interpretation of the quantum situation.

"It" from "bit"?

This is the second far-reaching question raised by John Archibald Wheeler which we will discuss here, the question concerning the role of information. What is the relation between material existence and knowledge, between reality and information?

As scientists, indeed as human beings, we look at the world and, from the information streaming in on us, we construct some kind of reality. Probably science began when the first person looking up to the sky and wondering at its beauty asked the question of how to interpret the small bright points up there. Prehistoric humans had very little information at their disposal to answer this question and thus they had to invent additional information in order to construct a consistent picture. Therefore we have scores of different explanations as to what the stars really are. Today, due to modern technology, we have much more information available and therefore we have very refined yet in general less romantic pictures of the stars, of the galaxies, and of the universe.

We would now like to address an important question, namely that of the relation between the size of a system and the amount of information it can carry. Clearly a huge system like a galaxy needs an immense number of bits of information in order to be characterized completely. But how do we expect the information to scale with the size of a system? Apparently, if we split a system into two, it is reasonable to assume that each half needs about half the information to be characterized on its own. So we continue to split our system into smaller ones and smaller ones and smaller ones, and therefore the number of bits necessary to characterize one of these partial systems will be further and further reduced. Evidently we will arrive at a fundamental limit if we keep continuing in this way, and the limit is reached when one system carries only one bit of information. Less is obviously not possible (Zeilinger 1999b). So it is suggestive to define the most elementary system in the following way: the most elementary system carries one bit of information.

As a word of caution we point out that an elementary particle in physics might in general not be a most elementary system in every sense, as it might carry electrical charge, spin, position information, energy, etc. In that sense the definition of a most elementary system pertains to the observation in the specific experimental context.

Our observation that the most elementary system carries only one bit of information simply means that it can carry only the answer to one question or the truth value of one proposition only. We can now show how this simple, innocuous observation leads to an understanding of such basic notions as complementarity, of the randomness of individual quantum events, and of entanglement. Complementarity is one of the most fundamental conceptual notions in quantum mechanics. We might quote Niels Bohr here: "Phenomena under different experimental conditions must be termed complementary in the sense that each is well defined and that together they exhaust all definable knowledge about the object concerned." The most basic situation where complementarity arises is the one between the path and the interference pattern (Fig. 11.5). In the most simple version we have two paths available, a and b, which are superposed at a semi-reflecting beam-splitter, and finally two detectors, I and II. If we consider our most elementary system passing through this set-up, how would we use the one bit of information available? Clearly there are at least two different possibilities. On the one hand we can use the one bit of information to define whether the particle passes along path a or path b. This is done by preparing the particle in the appropriate quantum state. Or, alternatively, we can prepare the state such that the system represents the information defining whether detector I or II will fire. In either case we have completely exhausted the one bit of information available and therefore there is no information present at all to define the other quantity. Therefore, once the one bit of information is used to define which-path information, no information is available any more to determine if detector I or II will fire. Alternatively, once the one bit of information is used to define whether detector I or II will fire, no information is available to define the particle's path, a or b. In both cases, the property for which no information is

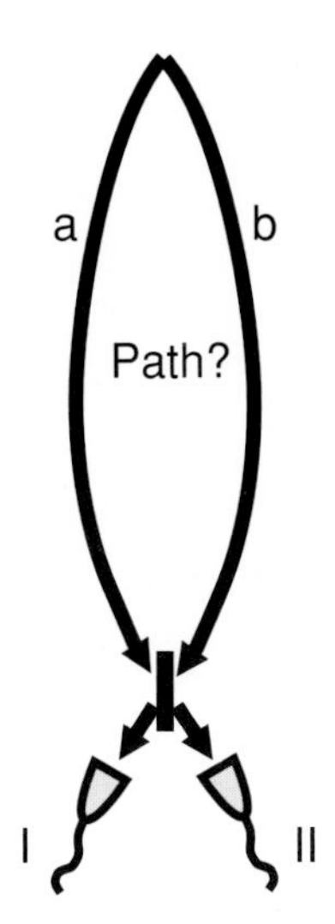

Figure 11.5. Complementarity and information in quantum interference. An incoming particle can propagate along path a or path b to a semi-reflecting mirror. Behind the mirror detectors I and II can observe the particle: either which-path (a or b) information can be defined or which-detector (I or II) information.

available any more must therefore be completely undefined, so such a quantity must be objectively undefined. Therefore, for that very simple reason, there is no room for considerations about hidden variables.

For completeness we point out that one can also choose to define either information partly, so it is possible (see, e.g., Wooters and Zurek (1979)) to have both partial information about the path taken and partial information about which detector will fire. But this can only be done in such a way as not to exhaust the total one bit of information available. It is interesting that this alone already points to a measure of information different from Shannon's (Brukner and Zeilinger 2001).

As stated above, the definition of the most elementary system pertains only to a specific experimental context. Therefore there is no limit in principle to the internal complexity of a system to show quantum interference. All that is needed is an experimental set-up where the way of reasoning just exposed can be applied. In that sense quantum interference has been realized with many different kinds of particles, the largest ones being the fullerenes C_{60} and C_{70} (Arndt *et al.* 1999). These molecules (Fig. 11.6) are extremely complex systems, containing a huge amount of information. Not only do they consist of a number of individual atoms, each atom already being a complex arrangement by itself. In the experiments performed so far the fullerenes are at high temperatures, typically around 900 K. This means that they are highly excited in many internal quantum states. Nevertheless, with respect to external motion they clearly exhibit an interference pattern, as seen in Fig. 11.7.

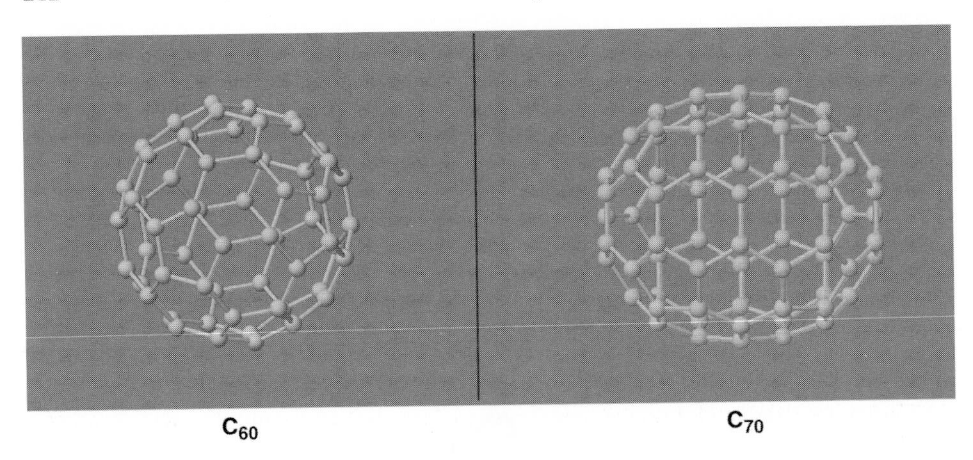

Figure 11.6. The fullerenes C_{60} and C_{70}, the largest individual objects for which quantum interference has been demonstrated hitherto.

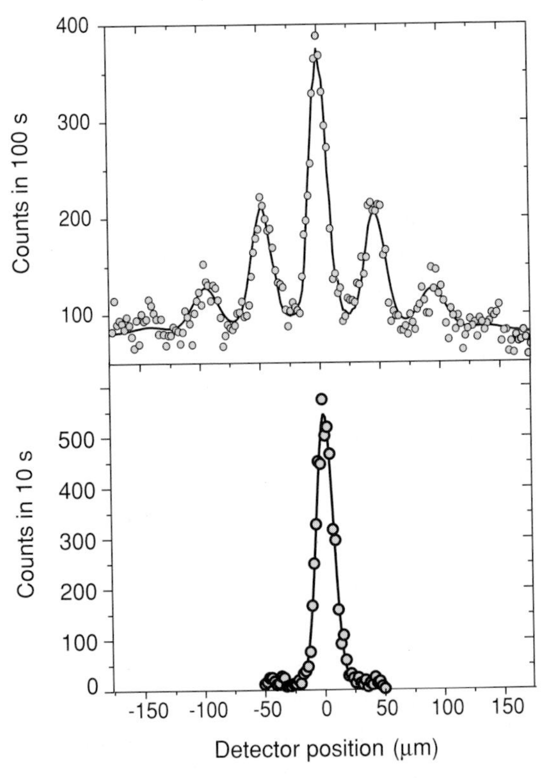

Figure 11.7. Interference pattern of C_{60} molecules after passage through a multi-slit assembly (Arndt *et al.* 1999) (top). The bottom shows the fullerene distribution without the diffraction grating present.

We note that in principle nothing in quantum physics limits the size of objects for which such interference phenomena might be observed some day. It is a safe bet that no limit for the validity of quantum superposition will ever be found in experiments. Therefore it is just an experimental challenge to further develop and refine these techniques in order to extend the realm of systems for which quantum interference has been experimentally observed to larger and larger systems, perhaps one day all the way to small viruses or maybe even larger living systems. Clearly in that case the challenge of isolation of the system from the environment becomes more and more serious. Yet we note that, already in our present experiments, the fullerene molecules were not completely isolated from the environment, as at these temperatures the fullerene molecules can already be viewed as small objects emitting black-body radiation (Mitzner and Campbell 1995). The reason why interference was observed in our experiments is simply the fact that the photons emitted have such a long wavelength that observation of the photon does not reveal any which-path information. Therefore for biological systems, perhaps tiny bacteria, one might hope that they emit such long-wave radiation that the coupling to the environment does not deteriorate quantum interference. Yet even if that were so, one could even contemplate to provide such small bacteria with a micro-life-support system, thus sufficiently isolating it from the environment. In any case, there is ample space for fantasy and creativity for experimentalists.

Another consequence of our observation that the most elementary system carries only one bit of information is an immediate understanding of the nature of quantum randomness. Let us consider again our basic interference set-up in Fig. 11.5. Suppose we use up the one bit to define which-path information. Thus, the answer to one question we might ask the system, namely the question as to which path is taken, *a* or *b*, is well defined. Then, by the mere fact that information is limited to one bit, no information is left for the particle to "know what to do" when it meets the detectors I and II, and therefore by necessity the click at detectors I and II must be random and they must be irreducibly random with no hidden possibility of explanation. This randomness therefore is an objective randomness, as opposed to the subjective randomness in classical physics and in everyday life, where we assume that any random event has an explanation in terms of its individual causal chain, where we assume that such an interpretation is at least in principle possible and not in contradiction with any other concepts. Since that randomness is subjective, it is the ignorance of the subject describing the situation that leads to apparent randomness. Not so in the quantum situation. It is not just subjective ignorance but there is objectively no information present to define which detector will fire in the situation just discussed.

This randomness of individual events in quantum mechanics has been used to create physical random number generators. A specific example is our random-number generator (Jennewein *et al.* 2000a), which is based on the randomness of the path taken by photons after meeting a semi-reflecting beam-splitter, exactly the situation just discussed.

Our point of view that the most elementary system carries one bit of information only also leads to a natural understanding of entanglement. The notion of entanglement was coined by Erwin Schrödinger (1935a) (in German *Verschränkung* (Schrödinger 1935b)) and he called it the most essential feature of quantum physics. A quintessential entangled state is

$$|\Psi^-\rangle = \frac{1}{\sqrt{2}}(|0\rangle_1|1\rangle_2 - |1\rangle_1|0\rangle_2) \tag{11.1}$$

where we have two quantum bits, or qubits, carrying the bit value "0" or "1". The entangled state presented above means that if qubit 1 (the first ket in either product state) carries the bit value "0" or "1", then the other qubit, the second one, carries the other bit value, "1" or "0", so there is perfect correlation between the two. Physically a qubit could be any dichotomic, that is, two-valued, observable, for example an electron's spin, a photon's polarization, or a particle's path taken in an interferometer (Horne and Zeilinger 1985). Most importantly, the state (11.1) represents a coherent superposition of the two possibilities and not just a statistical mixture. This implies that interference takes place. For the state of eqn (11.1) it means that it has the same mathematical form in any basis, whichever one might choose. This would not be the case for a statistical mixture.

To see the relation of eqn (11.1) to information it is suggestive to assume that two elementary systems just carry two bits of information. One way to view this is simply by assuming that each bit of information represents a possible measurement result for each elementary system on its own. This we would like to call local coding. In that case any relations between the possible measurement results on both sides are just a consequence of the information carried by each individual system. For example, should we elect to use the two bits, one each to define the spin along the z-axis, then we also definitely know how the spin measurements along that axis relate to each other. This apparently is one further bit of information but it is not independent information, it is a direct consequence of how the information is encoded into the two systems on their own.

But there could also be completely different situations. Instead of defining the information carried by each system separately, we could use up both bits of information to represent just how measurement results on the two systems relate to each other. For example, state (1) is uniquely defined by the two statements "the two qubits are orthogonal in the basis chosen" and by "the two qubits are orthogonal in a conjugate basis," where a conjugate basis is defined as

$$|0'\rangle = \frac{1}{\sqrt{2}}(|0\rangle + |1\rangle) \quad \text{and} \quad |1'\rangle = \frac{1}{\sqrt{2}}(|0\rangle - |1\rangle). \tag{11.2}$$

Thus we used up the two bits of information, the propositions are clearly independent of each other, and there is no information left to define measurement results on the individual systems on their own. As there is no information left to define the properties of the systems on their own, the measurement result on each individual system on its own must be completely random, as prescribed by quantum mechanics. This is the puzzle of entanglement exactly as expressed by Schrödinger. How can it be that measurement results are perfectly correlated without individuals carrying any information whatsoever? We just saw that our principle of finiteness of information, together with the new way of distributing the information between two systems, leads to a direct intuitive understanding of entanglement.

We have thus seen that three of the most fundamental conceptual notions or consequences of quantum mechanics can readily be understood on the basis of our identification of a most fundamental system being the basic element of information, the bit. We have, finally, to analyze the notion of "system" used so far. One might be tempted to assume that a system in the sense we are talking about is something which exists with all its features in its own right independent of observation. Yet if we take our notion of elementary system carefully it cannot be more than, in the concrete experimental situation, that which is characterized by information. Therefore the system is not anything more than that to which the information relates; in other words, there is no more than this information. To ascribe to a system more reality would mean to assign it more information, in contradiction with our fundamental assumption.

While entanglement is one of the most counterintuitive notions in quantum mechanics and while it has been investigated for this very reason with increasing intensity over the last three decades (Freedman and Clauser 1972), a surprising new development has set in, namely applications of entanglement in novel quantum information protocols. All these are just based on the property that entangled systems can carry information in a nonlocal way (Zeilinger 1998).

As one example let us consider first quantum teleportation. There Alice would like to teleport a qubit in a state unknown to her over to Bob. It has been well known for a long time that the most basic procedure is not possible, namely that Alice simply measures her qubit, determines its quantum state, sends all the information obtained over to Bob, and he reconstructs the original system. The problem here is that no measurement is possible to reveal the quantum state of an individual system, yet

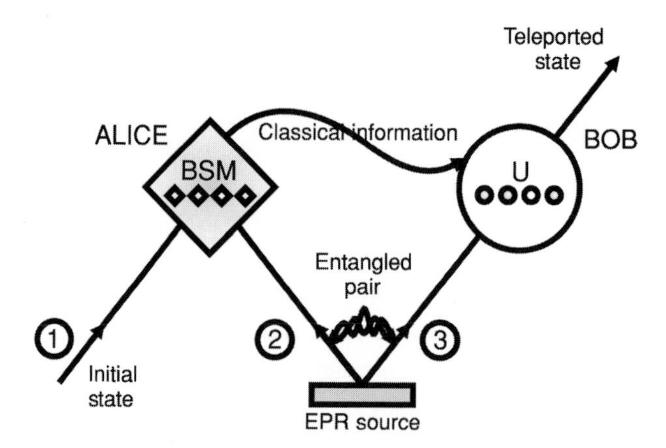

Figure 11.8. Principle of quantum teleportation of qubits. A qubit in an arbitrary initial state unknown to Alice is teleported to Bob. Alice performs a joint Bell-state measurement (BSM) on her initial qubit and on one of the two qubits of an auxiliary entangled pair emerging from an Einstein–Podolsky–Rosen source. Bob, after receiving the classical information about the result of the Bell-state measurement through a simple rotation can transform his qubit into an exact replica of the original.

we wish to teleport individual systems. Therefore it seems that quantum mechanics puts a fundamental limitation on all aspirations to achieve quantum teleportation some day.

Yet quantum mechanics itself comes to the rescue (Bennett *et al.* 1993). The basic idea there is to use entanglement to transfer quantum information over large distances. Alice and Bob (Fig. 11.8), anticipating the need for teleportation, share an entangled pair of qubits. Alice then performs a Bell-state measurement on her qubit to be teleported and on her member of the entangled pair. A Bell-state measurement projects the two qubits into an entangled state. For qubits there are four different Bell states, the state of eqn (11.1) being one of them. All these Bell states can be understood as representing any combination of the two possible truth-values of the two propositions mentioned above. We thus by a simple chain of logical reasoning know exactly how, after the Bell-state measurement, the quantum state of Bob's particle relates to Alice's original. In one of the four cases, Alice obtains a Bell-state measurement with exactly the result that corresponds to the state originally shared by Alice and Bob. Then Bob's particle is immediately projected into the state of the original qubit. In the other three cases this is not the case and Bob has to perform a rotation on his qubit, depending on the specific result Alice has obtained. This rotation is completely independent of the state of the original qubit to be teleported. In particular, the results for the Bell-state measurement occur randomly, each with 25% probability, completely independent of what the initial incoming state was.

It has been conjectured that quantum teleportation might beat the speed of light limit imposed on us by Einstein's special theory of relativity. True, it is necessary that Alice transmits to Bob which of the four Bell states was obtained and these two classical bits can at most travel at the speed of light. Yet, as we noticed above, the possibility arises that Bob's particle is immediately projected into an exact replica of the original whenever by chance a specific one of the four Bell states results appears in Alice's measurement. So, we are faced with the possibility that in one of the four cases, Bob's particle instantly becomes identical with the original. Can this now be implied to demonstrate a violation of Einstein locality? The answer, which evidently is in the negative, rests on a very subtle curiosity. This is that Bob cannot know immediately whether or not the system he receives has been projected into a state such that it is an exact replica of the original. Thus, while it might very well happen that Bob's particle is instantly projected into the same state as Alice's original, this cannot be used to transfer information faster than the speed of light. Bob has to wait for the classical message from Alice to arrive at his location and that can only happen at the speed of light. In a sense, to put it more succinctly, it can very well be argued that while quantum systems appear to be able to communicate faster than the speed of light, this cannot be utilized in a practical way by humans.

We will now briefly analyze the teleportation experiment (Bouwmeester *et al.* 1997) from our information theoretical approach to quantum mechanics. The situation is rather simple. From the initial preparation of the auxiliary entangled state we know how its two qubits relate to each other, should they be measured. Then Alice's Bell-state measurement does nothing else than provide us with the two bits of information necessary to tell us how the photon to be teleported and Alice's member of the entangled pair relate to each other. Therefore we know how both the photon to be teleported relates to Alice's entangled photon and how Alice's entangled photon relates to Bob's entangled photon, and thus finally by a simple logical chain of reasoning we know how the original relates to Bob's. This is a very simple chain of reasoning and gives us a unique state for Bob.

There are other important applications of quantum entanglement in the science and technology of information. Technically most advanced is quantum cryptography (Jennewein *et al.* 2000b; Naik *et al.* 2000; Tittel *et al.* 2000), where entanglement is used to circumvent a standard problem in conventional cryptography, namely the necessity to transfer the key for encryption from one place to another. Using entanglement, the key is generated at two distant locations at the same time. Finally quantum computation relies on the superposition of very complex states consisting of many qubits. This, evidently, immediately leads to the entanglement of information just discussed above.

Why the quantum?

In the beginning was the word.

(The Gospel according to John 1.1)

The quest for the reason for quantum mechanics is one of the most fundamental ones advocated by John Archibald Wheeler. He simply asks whether there is any possibility to arrive at the fundamental understanding of why we have quantum mechanics at all. What is the simple, basic reason for the existence of quantum physics? What is the underlying principle? Thus the point of view and context of these questions simply is that while the counterintuitive properties of quantum mechanics, such as for example Schrödinger's cat paradox (Schrödinger 1935b), most likely will stay with us forever, we would at least like to have an understanding why we are forced to accept these counterintuitive properties. We will now attempt such an explanation.

A guide in our consideration is again Niels Bohr, who, according to J. P. Petersen once remarked, "There is no quantum world. There is only an abstract quantum physical description. It is wrong to think that the task of physics is to find out what Nature is. Physics concerns what we can say about Nature." It is suggestive to assume that this implies that what can be said at all limits our possible knowledge about the world. So what we are doing both as scientists and in our daily lives is that we collect information about the world, information which always can be structured as a series of answers to questions or a series of truth-values of propositions. The way one constructs the world out of such a series of propositions has been beautifully illustrated by John Archibald Wheeler in his version of the game of "Twenty Questions." In the standard way of playing the game, one person leaves a room and the remaining persons then agree on some object or concept. The other person then comes back and has to find out by successively questioning the others through questions which can only be answered by "yes" or "no" what the object or concept the others agreed upon is. Usually, and interestingly, this can often be found out in less than 20 questions. John Archibald Wheeler's version is an amusing one. He suggests that the persons remaining in the room do not agree at all upon the object. Indeed, all they agree upon is that everyone is free to give whatever answer she or he wants but any answers have to be consistent with previous ones. So, the object or concept is then constructed together by all persons present following the course of questioning. This is a beautiful example of how we construct reality out of nothing.

But still, one may be tempted to assume that whenever we ask questions of nature, of the world there outside, there is reality existing independently of what can be said about it. We will now claim that such a position is void of any meaning. It is obvious that any property or feature of reality "out there" can only be based on information we receive. There cannot be any statement whatsoever about the world or about reality that is not based on such information. It therefore follows that the concept of a reality without at least the ability in principle to make statements about it to obtain information about its features is devoid of any possibility of confirmation or proof. This implies that the distinction between information, that is knowledge, and reality is devoid of any meaning. Evidently what we are talking about is again a unification of very different concepts. The reader might recall that unification is one of the main themes of the development of modern science. One of the first unifications was the discovery by Newton that the same laws apply to bodies falling on earth and to the motion of heavenly bodies. Other well-known unifications concern the unification of electricity and magnetism by Maxwell or the later unification of electromagnetism and the weak force.

In other words, it is impossible to distinguish operationally in any way reality and information. Therefore, following Occam's razor, the notion of the two being distinct should be abandoned, as the assumption of the existence of such a difference does not add anything that could not also be obtained without it.

Therefore, if we now investigate fundamental elements of information, we automatically investigate fundamental elements of the world. We have already seen earlier that any representation of information is based on bits. Any object is representing a huge number of bits. If we go to smaller and smaller objects we necessarily arrive at the fact that such objects can be characterized by one bit, two bits, three bits, etc., that is, information is quantized in truth-values of propositions. In view of our proposal that information and reality are basically the same, it follows that reality also has to be quantized. In other words, the quantization in physics is the same as the quantization of information. To conclude, it is worth mentioning that this idea can be turned into a research program developing the structure of quantum physics from first principles (Brukner and Zeilinger 1999, 2001, 2003; Baeyer 2001).

Acknowledgment

This work was supported by the Austrian Science Foundation FWF through SFB 015 "Control and Measurement of Coherent Quantum Systems."

References

Arndt, M, Nairz, O, Voss-Andreae, J, *et al.* (1999) *Nature* **401**, 680.
Baeyer, H C von (2001) *New Scientist* **169,** 26.
Bennett, C H, Brassard, G, Crépeau, C, *et al.* (1993) *Phys. Rev. Lett.* **70**, 1895.
Bouwmeester, D, Pan, J-W, Mattle, K, *et al.* (1997) *Nature* **390**, 575.
Brukner, Č and Zeilinger, A (1999) *Phys. Rev. Lett.* **83**, 335.

(2001) *Phys. Rev.* **A63**, 022113.
(2003) In *Time, Quantum and Information: The Festschrift for C. F. von Weizsäcker*, ed. L. Castell and O. Ischebeck. Berlin: Springer-Verlag. quant-ph/0212084.
Dopfer, B (1998) *Zwei Experimente zur Interferenz von Zwei-Photonen Zuständen: Ein Heisenbergmikroskop und Pendellösung*. Ph.D. thesis, University of Vienna.
Freedman, S J and Clauser, J S (1972) *Phys. Rev. Lett.* **28**, 938.
Heisenberg, W (1927) *Zeits. Phys. (Leipzig)* **43**, 172. (English translation in Wheeler and Zurek (1983).)
Horne, M A and Zeilinger, A (1985) In *Proc. Symposium "Foundations of Modern Physics*," ed. P. Lahti and P. Mittelstaedt, p. 435. Singapore: World Scientific Press.
Jennewein, T, Achleitner, U, Weihs, G, *et al.* (2000a) *Rev. Sci. Inst.* **71**, 1675.
Jennewein, T, Simon, C, Weihs, G, *et al.* (2000b) *Phys. Rev. Lett.* **84**, 4729.
Mitzner, R and Campbell, EEB (1995) *J. Chem. Phys.* **103**, 2445.
Naik, D S, Peterson, C G, White, A G, *et al.* (2000) *Phys. Rev. Lett.* **84**, 4733.
Schrödinger, E (1935a) *Proc. Cambridge Phil. Soc.* **31**, 555.
(1935b) *Naturwiss.* **23**, 807, 823, 844. (English translation in Wheeler and Zurek (1983).)
Tittel, W, Brendel, J, Zbinden, H, *et al.* (2000) *Phys. Rev. Lett.* **84**, 4737.
Weizsäcker, C F von (1931) *Zeits. Phys.* **40**, 114.
Wheeler, J A and Zurek, W H (eds.) (1983) *Quantum Theory and Measurement*. Princeton, NJ: Princeton University Press.
Wooters, W and Zurek, W (1979) *Phys. Rev.* **D19**, 2.
Zeilinger, A (1998) *Physica Scripta* **T76**, 203.
(1999a) *Rev. Mod. Phys.* **71**, S288–S297.
(1999b) *Found. Phys.* **29**, 631.

12

Speakable and unspeakable, past and future

Aephraim M. Steinberg

University of Toronto

Introduction

A volume in honor of a visionary thinker such as John Archibald Wheeler is a rare license to exercise in the kind of speculation and exploration for which Wheeler is famous, but which most of the rest of us usually feel we had better keep to ourselves. We have all – even those of us who never had the fortune to work directly with him – been inspired and motivated by Wheeler's creativity and open-mindedness. For all of our apparent understanding of quantum mechanics, our ability to calculate remarkable things using this theory, and the regularity with which experiment has borne out these predictions, at the turn of the twenty-first century it seems there are as many puzzles on the road to a true *understanding* of quantum theory as there were at the start of the previous century. Then, at least, one could hope to be guided by the mysteries of unexplained experiment. Now, by contrast, we may seem to have lost our way, as even though our experiments are all "explained" (in some narrow sense which can only be deemed satisfactory out of fear to leap beyond the comfortable realm of formalism), the theory itself is mysterious. Further explorations, without the anchor of experiment, certainly run the risk of becoming mere flights of metaphysical fancy, giving rise to factions characterized less by intellectual rigor than by fundamentalist zeal. Yet it would be premature to give up the journey before at least trying to establish a foothold on the terrain ahead. Following Wheeler's example, we can invent new experiments to help us speak about some of the unspeakable aspects of our theory, and to venture forward.

I have therefore decided to use this occasion to describe a number of loosely connected ideas we have been thinking about and experiments we have been working on in my group, which I believe relate to deep questions about how one should understand quantum mechanics. In keeping with the best tradition, I provide no

Science and Ultimate Reality, eds. J. D. Barrow, P. C. W. Davies and C. L. Harper, Jr. Published by Cambridge University Press.

answers to these questions, but I hope that I can show how a variety of questions are related to one another, and related to experiments both *gedanken* and real. Everything which follows takes place in the setting of standard quantum theory, and therefore even the most surprising predictions or observations I discuss are of course unambiguous, and implicit in every quantum textbook. Why then are they surprising? Clearly, we are not surprised only by results which contradict our theories; as is obvious when one discusses classical physics with students learning it for the first time, we are surprised by results that contradict what we *understand* of these theories. Over and over again in the past decade or two, experiments in fields such as quantum optics have revealed phenomena that surprise even those of us who ought by now to know quantum theory reasonably well. While many thinkers seem to consider such experiments mere parlor tricks, does not the ability of these experiments to evoke continued surprise demonstrate that we still do not *understand* quantum theory the way we understand classical theory? This simple observation is so clichéd as to bear repeating, for too many physicists have fallen prey to the reassuring but nihilistic thesis that since so many before us have failed, we would be wasting our time to seek any deeper understanding of quantum theory than is contained in our beautiful equations.

Past and future, particle and wave, locality and nonlocality

"Prediction is difficult, especially of the future."

This famous phrase is generally attributed to Yogi Berra, although among scientists one hears the credit given to Niels Bohr with some frequency. While the latter attribution has a certain comforting believability to it, one wonders whether Bohr's theory would make the past any more amenable to analysis than the future. A moment's thought suffices to realize that as difficult as prediction of the future may be, prediction of the past is not necessarily any easier (even aside from the semantic issue, which leads us to adopt the term "retrodiction" for inferences about the past). Neither is more or less the domain of science, although physics has traditionally concentrated on prediction while fields such as archaeology and cosmology have dealt with retrodiction. In classical mechanics, nevertheless, time-reversal symmetry guarantees that retrodiction is *precisely* the same task as prediction.[1] But in quantum mechanics as it is generally taught, despite the time-reversibility of the Schrödinger equation, retrodiction appears particularly mysterious. If I fire a photon towards a double slit, quantum mechanics unambiguously tells me what the state of the photon is after passing through the slits, although this state only gives probabilities for individual measurement outcomes. But when I see the

[1] For closed systems, at any rate – the thermodynamic arrow of time breaks the symmetry in the case of open systems.

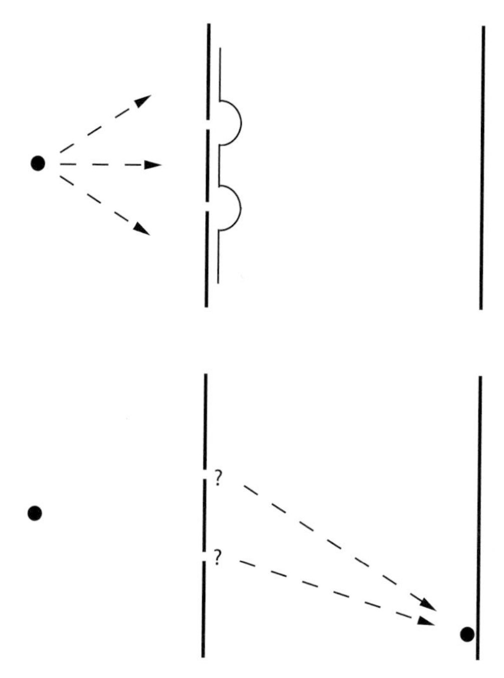

Figure 12.1. A two-slit experiment. When a particle is fired from a source towards the double slit, we can use Schrödinger's equation to predict its state as it passes the two slits: a symmetric wave function localized equally behind both slits. But when a particle appears at one point on the screen, what can we conclude about its history? As we all know, we cannot state it went through one slit or the other. Shall we say it went through both with equal likelihoods, as determined by the state preparation? Or from the location of the spot on the screen, can we construct some more accurate wave function? Can we just use Schrödinger's equation to propagate the electron backwards in time? This would discard all information about the state preparation, which seems extreme. Yet to discard all information about the future may also be unnecessary – for instance, even the claim of a symmetric double peaked wave function only made sense given the knowledge that the particle did make it through the double slit to eventually reach the screen, knowledge only obtained via postselection.

photon land at a particular point on the screen (see Fig. 12.1), what can I conclude about which slit it went through? The usual approach to measurement, involving an uncontrollable, irreversible disturbance, effectively decouples the "collapsed" state from what came before, except insofar as the probabilities for the measurement results are determined by the initial state. This is quite different from the usual treatment of state preparation, which sets up a well-defined initial condition and allows unitary evolution to take over.

The orthodox view of quantum mechanics holds that what has been measured can be known, and what has not is "unspeakable." If a particle is prepared in a certain wave packet, that function is to be considered a complete description, and any additional questions about where the particle "is"[2] are deemed uncouth, at least until such a measurement is made. The absence of trajectories in quantum mechanics means that one supposedly has no right to discuss where the particle "was" prior to that measurement. Yet the fundamental laws of quantum mechanics are as time-reversible as those of Newton, and one quite reasonably wonders why it is any less valid to use a measurement to draw inferences about a particle's history than to make predictions as to its future behavior. Such considerations led Yakir Aharonov and his coworkers to a formalism of "weak measurements" which allows one to discuss the state of evolving quantum systems in a fundamentally time-symmetric way. This chapter draws heavily on their ideas, whose main elements I will introduce below. I will analyze how weak measurements can be applied to several experimentally interesting situations. Consider, for one example, the problem of a tunneling particle. What can we know about where a particle was before it appeared on the far side of a forbidden barrier? Is it ever localized in the "forbidden" region? Can we obtain more information about the particle's history from the state preparation, or from the observation that it was transmitted?

These new ideas about measurement naturally lead one to think about epistemology. Is the wave function the fullest description of what we can know about a system? Is there then a real sense in which a particle may be in two places at the same time? Can we sometimes have more information than is encoded in a single wave function, by utilizing preselection and postselection simultaneously? Or, on the contrary, is it impossible even to know as much as a wave function, and are we limited to knowing the outcomes of the specific measurements we have performed? Can we have anything more than statistical knowledge about the outcomes of future measurements? Some experiments we plan to perform are designed to touch on these issues. In addition, they make one question whether even our probabilistic description of reality is complete, or whether exotic entities such as negative or complex probabilites may actually be meaningful.

The explosive growth of the field of quantum information, with its potential applications and headline-making buzzwords, has surprised many by turning "philosophical" research programs into timely, relevant, and some suspect even lucrative projects. These questions about past and future are no exception. Some of our recent work has involved the development of a quantum "switch," in which a single photon may be transmitted or not, depending on whether or not a single other photon is present. The thorn is that it is impossible to know whether either photon was ever present in the first place . . . as in many quantum optics experiments, the outcome depends on conditions which can only be measured after the fact. On this new work,

[2] Indeed, I once received an anonymous referee report which read, in essence, "This work is interesting, but I am unsure what the author means by the word 'is'."

I have no philosophical conclusions to draw: only a cautionary tale about how tricky these quantum conundrums remain even for those building the experiments, and a hope that others will help us learn how to think about our own experiments in new ways.

To come full circle, our first planned application of this "switch" is to carry out an experimental investigation of quantum reality first proposed by Lucien Hardy, extending ideas due to Elitzur and Vaidman. This experiment allows one to demonstrate that what at first glance appears to be perfectly airtight reasoning about the history of particles once they have been detected can lead to a seeming contradiction. More recently, it has been recognized that this contradiction can be eliminated if one applies the formalism of weak measurements and accepts these "exotic" probabilities as a correct description of reality. We believe that most if not all of these ideas are now accessible in the laboratory.

Weak measurements

The question of what measurement is is of course one of those which has haunted quantum theory from the start. Why does one thing occur and not another (let alone more than one)? When is a measurement? How does this relate to the arrow of time? By thinking carefully about retrodiction as well as prediction, some of these issues can be, if not resolved, then perhaps at least brought into starker relief. Aharonov *et al.* have led the way in generalizing concepts of measurement in this direction (Aharonov *et al.* 1988; Aharonov and Vaidman 1990), with their formalism of "weak measurement." In particular, weak measurements allow one to put past and future on an equal footing – and, better yet, to do something which is commonplace to any experimentalist and yet seemingly at odds with the usual machinery of quantum theory: to use one's knowledge of the initial *and* the final conditions of a system together to draw conclusions about what came in between.

If the task of deducing what happened *before* a measurement was made based on the result of that single measurement seems to conflict with the standard prescriptions of quantum theory, this is because the measurement is postulated to irrevocably change the state of the system. But what is the origin of this disturbance? Let us leave aside any considerations of "collapse" for the time being, and think only about the effect of an interaction between some system to be studied, and some other quantum mechanical system which will serve as a "pointer," or measuring device. Amplification of the state of this pointer to the macroscopic realm, so that a human observer might take note of it, can happen at some later stage if necessary; for our purposes, the important questions about measurement can all be treated simply by considering the effects of this quantum mechanical interaction.

In the standard approach due to von Neumann (1955, 1983), a measurement of a system observable A_s can be effected via an interaction Hamiltonian

$$\mathcal{H} = g(t)A_s \cdot P_p, \tag{12.1}$$

where the time-dependence $g(t)$ allows the measurement to take place during a finite interval of time, and where P_p is the canonical momentum of the pointer. Since the momentum is the generator of spatial translations, the effect of this interaction is to displace the pointer position by an amount proportional to the value of A_s. In particular, for suitably normalized $g(t)$, the expectation value of the pointer position will change by an amount that is proportional to the expectation value of A_s, and thus serves as a record of this measured value. Naturally, the requirement for a "good" measurement is that the pointer position be sufficiently well-defined that for different eigenvalues of A_s, the final state of the pointer is measurably different.[3] In this case, the pointer and the system become *entangled*, and the irreversibility of the measurement can be seen as arising from the effective decoherence of the system wave function when one traces over the state of the pointer.

The back-action on the system can be seen in another way, which is that the above Hamiltonian exerts an uncertain force on the system, to the extent that P_p is uncertain. If the pointer were in an eigenstate of momentum, then the measurement interaction would be an entirely predictable, unitary evolution of the system, $\mathcal{H} \propto A_s$; no irreversibility would thereby be introduced. Of course, if the pointer momentum were perfectly well defined, the pointer position would be entirely uncertain, and it would be impossible to observe a translation of the pointer. No measurement would have occurred.

Aharonov *et al.* argue that it is reasonable to consider an intermediate regime, where *some* information is captured during a measurement interaction, yet where the disturbance on the system is limited. Although this is not the textbook model of a quantum measurement, it is in fact a good model of how countless experiments are actually performed. Frequently, measurements on individual systems have such large uncertainties that only by averaging over thousands or millions of trials can statistical information be extracted.

The theoretical idea of a "weak measurement" is then to carry out a von Neumann interaction, but with an initial pointer state which is so delocalized in position that no single measurement can determine with certainty the value of A_s. On the flip side, this pointer may have such small uncertainty in momentum that the back-action on the system can be made arbitrarily small. It is in fact straightforward to

[3] Clearly, in the case of an observable with a continuous spectrum at least, one must be more cautious in defining precisely which eigenvalues ought to be distinguishable.

verify that under these conditions, instead of entangling the system and pointer according to

$$|\Psi\rangle_s \phi_p(x) \to \sum_i c_i |\psi_i\rangle_s \phi_p(x - ga_i) \quad (12.2)$$

(where the ψ_i and c_i are the eigenkets of A_s and their corresponding amplitudes, and ga_i is the shift in the pointer wavefunction ϕ_p which corresponds to an eigenvalue a_i), the system and pointer remain to lowest order unentangled:

$$|\Psi\rangle_s \phi_p(x) \to |\Psi\rangle_s \phi_p(x - g\langle A_s\rangle). \quad (12.3)$$

On average, the pointer is displaced by an amount related to the expectation value of A_s, but since this shift is too small to significantly modify the pointer state, the system is unaffected.

Importantly, this means that the original evolution of the particle may continue, and one may ask not only about the correlations between the pointer position and the initial state of the system, but equally well about correlations between the pointer position and the state the system is *later* observed to be in. One may quite generally ask what will happen to the pointer on those occasions where the system was prepared in state $|i\rangle$ before the measurement interaction, and later measured to be in some final state $|f\rangle$. Using standard quantum theory, Aharonov and coworkers showed that the mean shift of the pointer position for this subensemble corresponds to a "weak value" of A_s given by

$$\langle A_s\rangle_{\text{wk}} = \frac{\langle f|A_s|i\rangle}{\langle f|i\rangle}. \quad (12.4)$$

Clearly, for the trivial case $f = i$, this reduces to the usual expression for an expectation value. But for the more general case, it is heartening to note that the initial and final states have equal importance for the measured value of A_s; one can learn as much about a particle's state by observing its future as by knowing its past.

There are many other striking properties of weak measurements which suggest that they are a powerful tool for analyzing a broad variety of physical situations, and also that there may be some deep physical meaning to these quantities themselves. I will not go over these in detail, but Reznik and Aharonov (1995) and Aharonov and Vaidman (2002a) provide a deep analysis. In many ways, these values can be seen as a natural application of Bayesian probability theory to quantum mechanics (Steinberg 1995a, b), satisfying many of the natural axioms of probability theory. More important, they describe the outcomes of *any* measurements which can be described using the (modified) von Neumann formalism, and therefore show a clear connection to physical observables, not to mention a unifying framework within which a broad class of experiments may be treated. At the same time, they display a number of troubling features. Notably, the measured weak value need not be consistent with any physically plausible values of A_s; it need not even fall within this operator's eigenvalue spectrum. More shocking still, some positive–definite quantities such as energy (or even probability) may be measured to be negative (Aharonov *et al.* 1993). In fact, weak values are in general complex numbers rather than reals. As explained in some of the above references, this is not an entirely untenable state of affairs, and the physical significance of the real and imaginary parts of the weak value may be clearly identified. Roughly speaking, the real part indicates the size of the physical shift in pointer position, the measurement result one expected classically from such a device. The imaginary part indicates how much the *momentum* of the pointer will change as an unintended consequence of the measurement interaction, and consequently, how large the back-action of the measurement on the system.

One of the truly exciting features of weak measurements is that simultaneous weak measurements may be made on noncommuting observables, and do not render each other impossible, or even modify each other's results. For instance, if a particle is prepared in an eigenstate of some operator B with eigenvalue b_j, then a weak measurement of B is guaranteed to yield the value b_j, regardless of the postselection. Similarly, if it is *postselected* to have an eigenvalue c_j of some operator C, then a weak measurement of C is certain to yield c_j, regardless of the preparation. If both B and C are measured weakly between the preparation and the postselection, both of these values will be observed (albeit as *average* shifts of a very uncertain pointer position) – even if B and C do not commute. For that matter, if $B + C$ is measured, the result will be $b_j + c_j$, something which makes intuitive "classical" sense, but which one could never hope for in the context of strong quantum measurements. Such properties clearly hold out the tantalizing possibility of making more of reality "speakable" (in John Bell's term (Bell 1987)) than we are usually led to believe. When we think about a particular system which survived from state preparation through postselection, should we merely think of the initial state evolving in a unitary fashion until the postselection induced a collapse, or should we think about its properties as depending on both pre- and postmeasurements? While the orthodox view may be that if no measurement is performed between preparation and postselection, the question is meaningless, it is thought-provoking that *any* von Neumann-style interaction that takes place at intermediate times, provided that it is not so strong as to irreversibly modify the system dynamics, will produce an effect whose magnitude is defined by this new formalism. Such observations led to a variety of speculations about the "reality of the wave function" (Aharonov and Anandan 1993) and to a general formulation of quantum mechanics via "two-time wave functions" (Reznik and Aharonov 1995).

Recently, a connection has been drawn between weak measurements and more widespread techniques for dealing with the quantum evolution of open systems (Wiseman 2002), and this has proved useful for explaining the "negative-time correlations" in a cavity QED experiment (Foster *et al.* 2000). Specifically, an experiment in Luis Orozco's group designed to observe the evolution of an electromagnetic field after the detection of one photon also found interesting dynamics in the evolution of the field *before* the detection of a photon. Howard Wiseman pointed out that when the photodetection event is treated as a postselection, an extension of weak measurement theory can be fruitfully applied to understand this negative-time evolution, which had not previously been fully explained.

A quantum-mechanical shell game

While it was recognized from the outset that weak measurements could yield anomalously large values, and the first (intentional!) experimental implementation of weak measurements was a linear-optics experiment to demonstrate how a spin measurement could yield an apparently nonsensical value (Ritchie *et al.* 1991), it was pointed out (Steinberg 1995b) that there is a striking mathematical relationship between weak measurements and classical probability theory. In fact, the result of eqn (12.4) can be obtained quite generally by summing over the "conditional probabilities" for each of the eigenstates of the operator

$$\langle A \rangle_{\mathrm{wk}} = \sum_j a_j P(j|i, f), \tag{12.5}$$

where the probability of being in an eigenstate $|\psi_j\rangle$ is defined as the expectation value of the projector $|\psi_j\rangle\langle\psi_j|$; the "conditional probability" is the natural generalization based on the weak-measurement prediction for the shift experienced by a pointer which couples to this projection operator, conditioned on the appropriate postselection:

$$P_{\mathrm{wk}}(j|i, f) = \frac{\langle f|\psi_j\rangle\langle\psi_j|i\rangle}{\langle f|i\rangle}. \tag{12.6}$$

Of course, these conditional probabilities sometimes prove to have values greater than 1, less than 0, or even with imaginary components. It is on the one hand unclear if it is meaningful in any real sense to interpret these as probabilities, while on the other hand the weak-value expressions for probability are defined in clear analogy to classical probabilities, and satisfy the same axioms. Furthermore, the experiments which are predicted to yield negative or complex "probabilities" are designed in precisely the fashion one would choose classically to measure the conditional probabilities, and they would correctly measure these probabilities when used in the classical regime; is this not the operational prescription for developing the quantum mechanical formalism for a given observable?

I do not possess the hubris to attempt to pronounce a final verdict on how seriously one should take these probabilities, or on whether one would be better to avoid such a loaded term at all. Nevertheless, the expressions derived in this fashion have clear physical significance for a wide-ranging class of experiments. Suffice it to note that there are a number of other contexts (such as "rescuing" locality, in the context of Bell's theorem) in which other authors have suggested taking seriously the concept of negative probability in quantum mechanics (Pitowski 1982; Muckenheim *et al.* 1983; Feynman 1987; Scully *et al.* 1994), not to mention the negative quasi-probabilities which are familiar in the context of the Wigner function and other phase–space distributions (Wigner 1932; Liebfried *et al.* 1996).

Let us for now accept this terminology of probabilities, with all its caveats, and examine some striking examples of what weak-measurement theory predicts. In 1991, Aharonov and Vaidman applied the formalism to the following toy problem (Aharonov and Vaidman 1991). Consider a particle which can be in any of three boxes, which we will denote as three orthogonal states $|A\rangle$, $|B\rangle$, and $|C\rangle$. Let us prepare the particle in an initial state

$$|i\rangle = \frac{|A\rangle + |B\rangle + |C\rangle}{\sqrt{3}}, \tag{12.7}$$

i.e., a symmetric equal superposition of being in each of the three boxes. Suppose that some time later we choose another basis, and measure whether or not the particle is in the final state

$$|f\rangle = \frac{|A\rangle + |B\rangle - |C\rangle}{\sqrt{3}}, \tag{12.8}$$

where the sign in front of box C has been changed. Note that there is some probability for this postselection to succeed, without any need for the particle to change its state between the measurements: $|\langle f|i\rangle|^2 = 1/9$.

Obviously, the question of interest is how we should describe the state of the particle between the state-preparation and a successful postselection. Should we evolve $|i\rangle$ forward in time under the free Hamiltonian, the particle remaining symmetrically distributed among the three boxes, until the final measurement disturbs its phase? Or should we instead evolve $|f\rangle$ backwards in time? Clearly, orthodox quantum mechanics says there is no meaning to the question of *at what time C* stopped being in phase with A or B, and began being out of phase with them; Bohr would tell us that the value of this phase during a period when nothing in the apparatus is sensitive to it is meaningless. Similarly, we cannot ask which of the three boxes the particle was in before it was detected in $|f\rangle$, although it seems quite natural to suppose it had equal probabilities to be found in any of them.

One can conceive of measuring such probabilities, by using a large ensemble of particles. For instance, a test charge held near box A may experience a slight momentum shift if and only if the particle is in box A. If this shift is arranged to be far smaller than the uncertainty in the test charge's momentum, then it may be possible to carry out such measurements without any appreciable effect on the evolution of the particle. If no postselection is performed, the magnitude of this shift will be proportional to the probability that the particle was indeed in A, i.e., the expectation value of the projection operator $|A\rangle\langle A|$. For the state $|i\rangle$, for instance, this probability is one third: the impulse imparted to the test charge after N particles go through the boxes will be precisely what one would expect if $N/3$ had been in box A ... or, equivalently, if one third of *each* of the N particles had been in box A.

What if the momentum shift on the test charge is recorded (including its large uncertainty) each time a particle passes, but is discarded unless the postselection fails ? Then the sum of the momentum shifts for all the test charges which interacted with particles eventually detected in $|f\rangle$ will describe the *conditional* probability that those particles had been in box A:

$$P_{\text{wk}}(A|i, f) = \frac{\langle f|\text{Proj(A)}|i\rangle}{\langle f|i\rangle} = \frac{\langle f|A\rangle\,\langle A|i\rangle}{\langle f|i\rangle}. \tag{12.9}$$

It is easy to verify that this probability is unity. The postselected test charges will display precisely the same mean momentum shift as they would for a particle prepared with 100% certainty in box A. Similarly, the weak (or conditional) probability for the particle to be in box B is 100%. And the axioms of probability? Must not the probabilities of all the exclusive possibilities add up to 1? Indeed – it is equally easy to verify that $P_{\text{wk}}(C|i, f)$, the conditional probability for a particle to have been in box C between its preparation in $|i\rangle$ and its detection in $|f\rangle$, is -1. Meaningless? Not at all. If the mean momentum shift of test charges which interact with particles eventually detected in state $|f\rangle$ is measured, it will be found to have the "wrong" sign – that is, if the particle and the test particle have charges of like signs and ought to repel each other, the test charge will be found to have a mean momentum *towards* box C. Perhaps it is risky to interpret this by saying the particle truly had a negative probability to be in that box – yet physically, its effect was equal and opposite to the effect of a particle in box C.

Perhaps more striking yet is the observation that the particle was "definitely" in box A, but also in box B. We are quite accustomed to saying that a particle must go through "both slits at once" in Young's interferometer, but how many of us truly mean it? The wave function, of course, traverses both slits, but we know full well that to talk of "the" position of the particle, we must introduce some position measurement, in which case the particle will be observed at one slit or the other. Weak measurements show us that this is not necessarily always the case, so long as no "collapse" (or decoherence, more precisely) is introduced during the measurement. Aharonov *et al.* have used these features of the theory to argue in favor of the ontological "reality of the wave function" (Aharonov and Anandan 1993), while these arguments have incited a great deal of controversy (Unruh 1994). More recently, Aharonov and Vaidman have tried to respond to some objections to their shell-game paper by introducing a *strong* measurement – they show that if this particle is a "shutter," then a photon heading towards either box A or box B, or indeed any superposition of the two, is guaranteed to be intercepted by the shutter (in cases where the shutter is postselected to be in $|f\rangle$, as always) (Aharonov and Vaidman 2002b). This suggests that the nonlocality of quantum mechanics may be even deeper than usually recognized, in that a given particle could actually have measurable effects in two places at the same time.

We are currently setting up an experiment, shown schematically in Fig. 12.2, designed to test some of the features of this quantum conundrum. Photons are prepared in a symmetric superposition of the three "boxes" A, B, and C, by the use of beam-splitters; each box is in fact one path in an interferometer. By carefully adjusting the relative phases of the paths (specifically, by introducing an extra π phase shift along path C before symmetrically recombining the three beams at another beam-splitter), it is possible to project out light in the state $[|A\rangle + |B\rangle - |C\rangle]/\sqrt{3}$. Several varieties of weak measurement may be performed. In particular, a small piece of glass can introduce a spatial shift in one of the three beams, smaller than the width of the beam (i.e., the uncertainty in the photon's transverse position). Alternatively, a waveplate can rotate the polarization of one of the paths by a small angle. It is an optics problem left for the reader to show that the deviations to be expected are precisely those predicted by weak measurement theory: if beam A or B is displaced by δx, then the output will be displaced by $\delta x \ldots$ on the other hand, if beam C is displaced by the same amount, the displacement at the output will be $-\delta x$. (In the optics context, it is not difficult to understand this as an interference effect related to the π phase shift introduced in arm C.) We plan not only to confirm the weak-measurement predictions, but also to study the *correlations* between the different probabilities. In particular, we are interested in the question of nonlocality. If we can say with certainty that the particle was in A and that it was in B, can we also say that it was simultaneously in A and B? This may seem obvious, but again, with weak measurements one must be careful.

In their paper on "How one shutter can close N slits" Aharonov and Vaidman (2002b) note that a *pair* of test particles, one heading to shutter position A and the

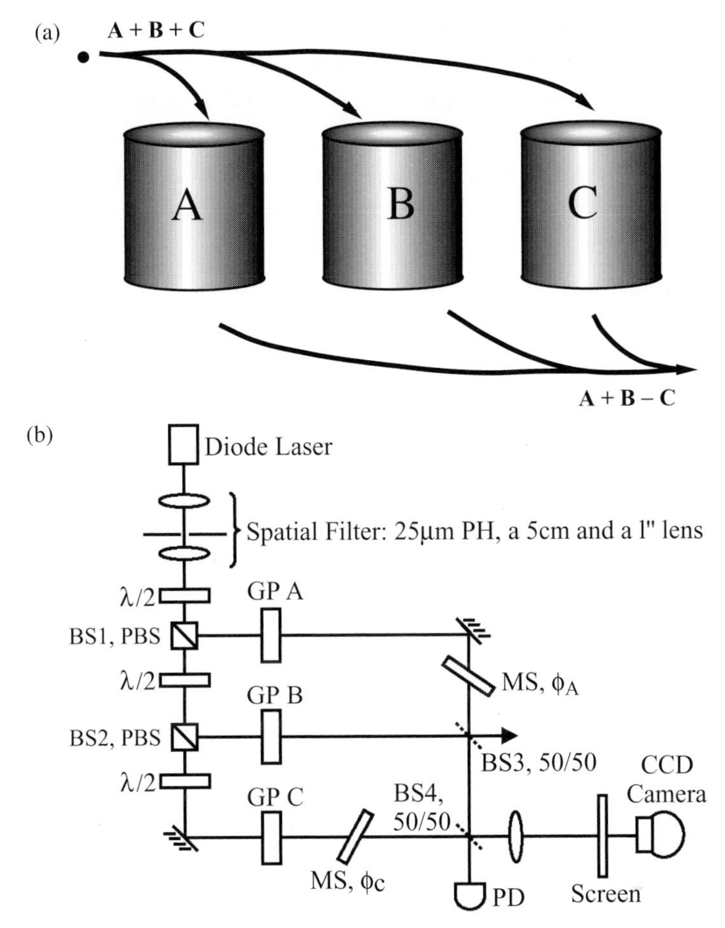

Figure 12.2. (a) The quantum three-box problem. If a particle is hidden in three boxes in a superposition $(A + B + C)/\sqrt{3}$, but is subsequently found to be in the (different but nonorthogonal) superposition $(A + B - C)/\sqrt{3}$, what can one say about the state of the particle while in the box? (b) Experimental schematic for an optical implementation of the three-box problem. Photons are sent into a three-rail interferometer, with the three rails playing the roles of boxes A, B, and C. A π phase shift is introduced in rail C, such that detection at the camera postselects a superposition $(A + B - C)/\sqrt{3}$. To weakly "measure" the particle in one or another of the boxes, small transverse displacements are induced in each of the rails, and an image of the postselected photon distribution is taken to determine the size of the effects of displacements in each of the boxes.

other to shutter position B, could *not* both be reflected by a single shutter (although they make interesting observations about the case of multiple slits, multiple shutters, and multiple incident particles). In essence, the reflection of a particle heading towards A is a strong measurement, and prevents the slit from stopping a second particle heading towards B. However, one can put this even more succinctly if one

Table 12.1. *Summary of the probabilities and joint probabilities of finding the particle in or out of box* A *and in or out of box* B*, demonstrating how a negative probability in one column can allow the joint probability of two "certain" events to vanish*

Probabilities	A	not A	A or not A
B	0	1	1
not B	1	−1	0
B or not B	1	0	

accepts the definition

$$P(A\&B) = \langle \text{Proj}(A) \cdot \text{Proj}(B) \rangle. \tag{12.10}$$

Although this definition has certain pathologies associated with it (Steinberg 1995b) (notably, this product of two projectors need not be a Hermitian operator, and therefore could yield complex "joint probabilities" even in non-postselected systems), it seems the most natural way of describing joint probabilities. It generalizes easily to the case of weak (conditional) measurements. However, if A and B are orthogonal, as in the present case, then the product of their projectors

$$\begin{aligned} \text{Proj}(A) \cdot \text{Proj}(B) &= |A\rangle \quad \langle A|B\rangle \quad \langle B| \\ &= |A\rangle \quad\; 0 \quad\; \langle B| = 0. \end{aligned} \tag{12.11}$$

Under no circumstances is there a nonzero joint probability, conditional or otherwise, to be in box A *and* to be in box B. As discussed in Aharonov *et al.* (2002), weak measurements do not allow one to conclude that because $P(A) = P(B) = 1$, then $P(A\&B)$ must also be 1; this is because the probabilities themselves are not bounded by 0 and 1. The probability of "A and B" may vanish, in spite of the certainty of A and B individually, for the probability of "A and not B" is 1. If this seems strange, given that the probability of "not B" is zero, no worries: for the probability of "not A and not B" is negative 1. This odd state of affairs is summarized in Table 12.1.

Tunneling

Another problem where nonlocality has been a topic of discussion in recent years is that of tunneling through a barrier. It has been well known since early in the century (MacColl 1932; Wigner 1955; Büttiker and Landauer 1982; Hauge and Støvneng 1989) that the group delay (stationary phase time) for a wave-packet incident on an opaque barrier of thickness d to appear on the far side saturates to a finite value as d tends to infinity. For large enough d, this implies superluminal propagation speeds for the peak of the wave packet, which naturally provoked much skepticism. A number of experiments, including one I performed along with Paul Kwiat in Ray Chiao's group at Berkeley (Steinberg *et al.* 1993), demonstrated that this prediction is indeed correct (Enders and Nimtz 1993; Spielmann *et al.* 1994), although no violation of causality is implied (Chiao and Steinberg 1997). Due to the difficulty of timing the arrival of matter particles through any reasonable tunnel barrier, and the problems of reaching the relativistic regime with massive particles, these experiments were carried out with photons. We are now building at Toronto a series of experiments designed to observe the tunneling of laser-cooled atoms through micron-scale barriers formed by focused beams of light (Steinberg 1998a; Steinberg *et al.* 1998). Although the experiments are complex, this should open up a broad new vista of phenomena to study. In particular, it becomes possible to *probe* the particles while they are traversing the "forbidden" region, and also to study the effects of decoherence on the tunneling process (Steinberg 1999).

While it is certainly strange that a wave-packet peak should arrive in less time than if the original peak had traveled at the speed of light, it was pointed out comparatively early in the (latest bout of the) tunneling time controversy that no physical law guarantees any direct causal connection (let alone identity) between an incoming peak and an outgoing peak (Büttiker and Landauer 1982). We generally interpret these effects as remarkable but entirely causal "pulse reshaping" phenomena, in which the leading edge of a pulse is preferentially transmitted, while its trailing edge is preferentially reflected, thus biasing the peak towards earlier times. Similar effects had been observed in the 1980s in the context of propagation through absorbing media (Garrett and McCumber 1970; Chu and Wong 1982), and much excitement has recently been created by the analogous observation of faster-than-light propagation in transparent (but active) media (Steinberg and Chiao 1994; Steinberg 2000; Wang *et al.* 2000). A review of superluminality and causality in optics is given in Chiao and Steinberg (1997).

These counterintuitive effects occur only when the tunneling probability is relatively small. In other words, like many weak-measurement paradoxes, the anomalies are dependent on the success of a postselection which occurs only rarely. If one tracks the center of mass of a wave packet incident on a tunneling barrier, it never moves faster than light – only when one projects out the transmitted portion alone does the peak abruptly appear to have traveled superluminally. In this sense, one may well argue that the superluminality is not a function of propagation through the tunnel barrier, but only of this mysterious "collapse" event whereby a particle previously spread out across two peaks may choose to localize itself on one.

Nevertheless, it seems reasonable to apply the formalism of weak measurements to the tunneling problem, in order to see whether this can shed light on the counterintuitive aspects of the situation. For instance, can one verify that the tunneling particles originated predominantly near the peak of the wave packet? Can one, alternatively, determine the length of time a particle spends (on average) under the barrier? This "sojourn" or "dwell" time is a quantity which had been of much interest to the condensed-matter community, as it would allow one to describe the importance of interactions between a tunneling particle and the surrounding environment, and the validity of approximations such as adiabatic following. Even those who were not troubled by the superluminal *peak delay* presumed that the physical time spent in a given region of space would have to be greater than or equal to d/c; a number of models of the interaction between a tunneling particle and the environment were used to support this conjecture and yield "interaction times" for the tunneling problem (Büttiker and Landauer 1985).

In Steinberg (1995a, b), I applied the ideas of weak measurement to this question, and was surprised. On the one hand, no weak measurement would show the supposed "bias" towards the leading edge of the incident wave packet. Furthermore, one could rewrite the tunneling "interaction time" as a time-integral of the probability to be in the barrier, which in turn decomposed into a probability density at each position and time:

$$\begin{aligned}\tau &\equiv \int_{-\infty}^{\infty} dt\, P_{\text{bar}}(t) \\ &= \int_{-\infty}^{\infty} dt \int_{0}^{d} dx |\Psi(x,t)|^2. \end{aligned} \tag{12.12}$$

By generalizing this to the case of postselected subensembles (i.e., calculating the weak values of the projector $\delta(\hat{X} - x)$ for various positions x), it proved possible to derive a "conditional probability distribution" for a particle to be at position x, *given* that it was prepared in a state $|i\rangle$ (incident on the barrier from the left, in a given wave packet) and detected in a final state $|f\rangle$ (transmitted to the far side of the barrier). The time τ turned out to be in general complex, but its real part – that part which describes the position shift of a pointer coupled to the particle's presence in the barrier region – is of the same order of magnitude as the group delay, and exhibits the same "superluminal" features. A plot of the evolving conditional probability distribution is shown in Fig. 12.3.

One of the striking things about this figure is that the particle appears to spend essentially no "time" (in the sense of the real part of a weak value) near the center of the barrier. A reflected particle only spends time within an exponential decay length of the input facet; while a transmitted particle spends roughly equal amounts of time near the entrance and near the exit (as one might have surmised from the

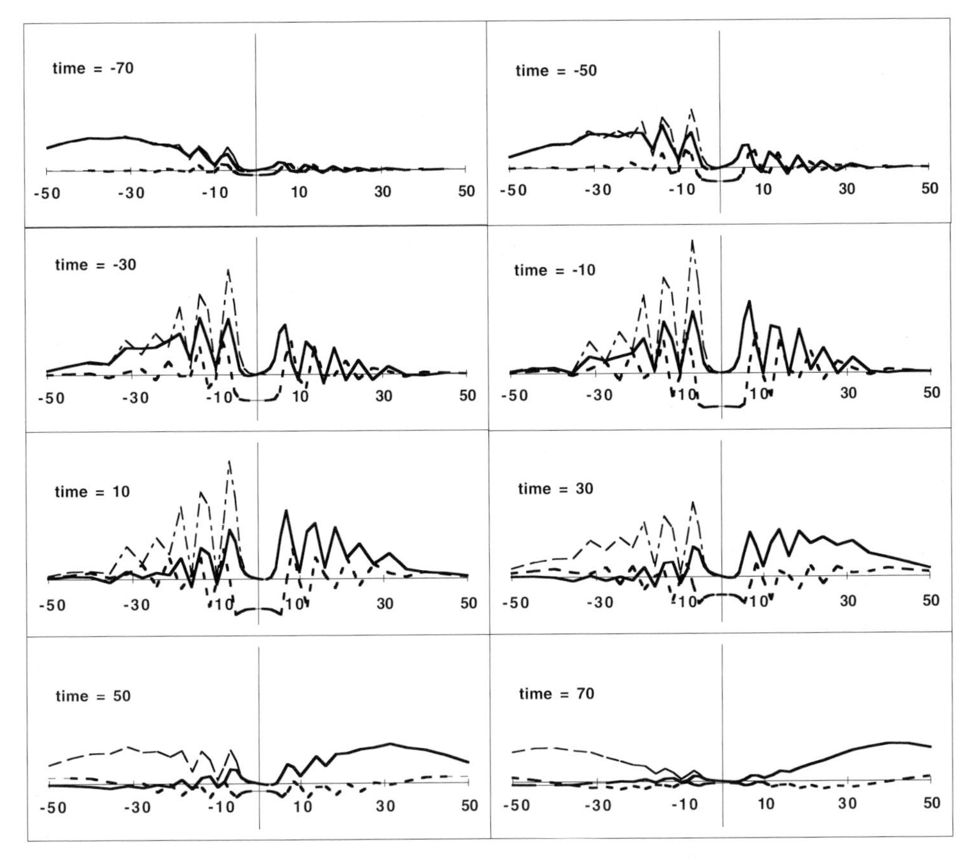

Figure 12.3. The time-evolution of the "weak" conditional probability distribution for a particle's position as it tunnels through a barrier. The heavy curve shows the real part of this distribution (the magnitude of the expected measurement result), while the dashed curve shows its imaginary value (the "back-action" due to measurement), and the light curve shows the distribution for reflected particles (essentially equal to $|\Psi|^2$). Note that at early and late times, the weak distribution mimics the full incident or transmitted wave packet, while at intermediate times it has an exponentially small magnitude inside the forbidden region.

symmetry of the experimental arrangement, or of the formula for weak values). Figure 12.4 presents a thought-experiment to elucidate the physical meaning of these curves. Consider a proton constrained to tunnel in one dimension. It tunnels along a series of holes in parallel conducting sheets, which serve to break the tunnel barrier up into a sequence of electrically shielded regions. As described in the context of the three-box problem above, one way to measure the weak value of a "probability" (or of its time-integral, a dwell time) is to study the momentum shift of a test charge which interacts with the particle in question. Here we imagine an electron, initially at rest, between each pair of conducting plates. We measure the

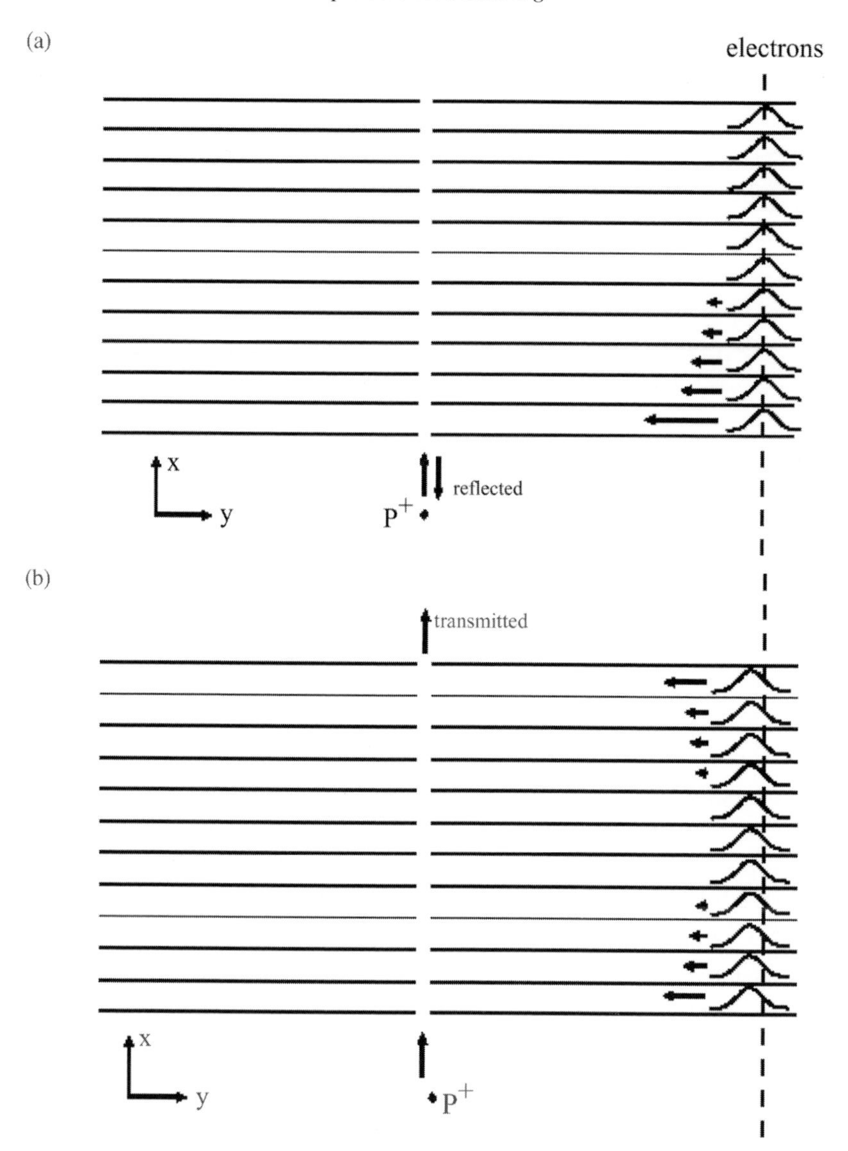

Figure 12.4. A thought-experiment using distant electrons to measure how much time a tunneling proton spends in each of several shielded regions of space. While the proton is between a given pair of conducting plates, only the corresponding electron feels a significant force. After the tunneling event, the momentum shift of each electron thus records the amount of time spent by the proton between the plates in question. The implication of weak-measurement theory is that reflected protons only transfer momentum to electrons near the entrance (a), while transmitted protons affect electrons near both edges of the barrier (b). Electrons in the center only undergo a position shift, related to the back-action of the measurement.

final momentum of each electron after the passage (reflection or transmission) of the proton, sorting according to whether the proton was transmitted or reflected. On each event, by definition of a weak measurement, the electrons' momenta are far too uncertain to draw any conclusions (or else the presence of the electrons would so perturb the motion of the proton that there would be no sense in discussing it as a tunneling problem; see Steinberg (1999)). After averaging over the momenta found for numerous transmitted protons, however, one would find the symmetric distribution indicated in the figures.

In keeping with our intuitions, but not with the standard (time-asymmetric) recipe for dealing with quantum evolution and measurement, we see that in addition to concluding from the initial condition (a particle approaching the barrier from the left) that the wave packet penetrates roughly one exponential decay length into the left side of the barrier, one may conclude from the final condition (a particle exiting the barrier on the right, for instance) that it had penetrated one decay length into the right side of the barrier as well. Weak measurements allow us to discuss the behavior of "to-be-transmitted" particles and "to-be-reflected" particles separately, and observe that even when described by the same initial wave function, they may have different physical effects on weakly coupled environments.

It turns out that one of the popular approaches to tunneling times, the Larmor time (Büttiker 1983), is in essence nothing but a weak value. This time has two different components, whose individual physical meanings were obscure, however, until reinterpreted in the light of this new formalism. It is now clear that they correspond to the real and imaginary parts of the weak measurement, and that the former corresponds to the pointer shift (the measurement result as extrapolated from the classical limit), while the latter indicates a necessary back-action of the particle due to the measurement, which can be made arbitrarily small by using a sufficiently weak measurement.

One question raised by the evolving conditional probability distributions plotted above is whether, in the superluminal-tunneling regime, the particle really does move from a wave packet on the left of the barrier to one on the right in a time shorter than d/c, without spending significant time in the center of the barrier. While we all know that a cause cannot have any measurable effect at a spacelike separated point, is it perhaps possible for a single particle to have an effect at *two* points spacelike separated from one another (but not from the source of the particle) (Steinberg 1998b)? Clearly, it suffices for two people on opposite sides of a radio transmitter to listen to the same broadcast, for a cause to have two spacelike-separated effects. But is a single quantum particle truly as nonlocal as this radio wave? We all know that if a strong measurement is made of the position of a photon, it can no longer be found in a different position. But since repeated weak measurements can be made on the same wave function, and are not modified by the action of other weak

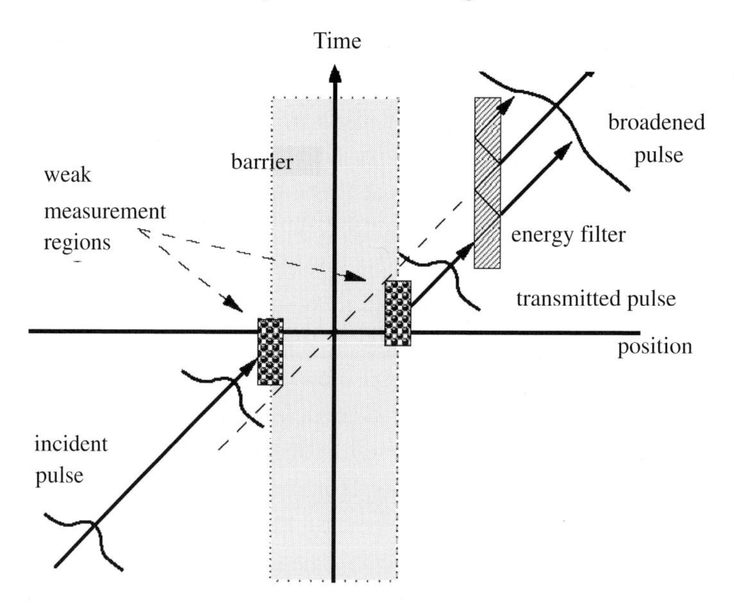

Figure 12.5. A thought-experiment to investigate whether or not a subset of tunneling particles may truly prove to have "been" in two places at the same time, due to the superluminal group velocity in tunneling. The peak of the transmitted Gaussian may emerge at a point spacelike separated from the peak of the incident Gaussian. An energy filter is necessary to "erase" any timing information which would preclude the detected particle from having been present at the incident peak; once a particle is transmitted through a narrowband filter, information about its time of origin is smeared out.

measurements made at the same time, I was led to suspect that it should be possible to weakly measure the probability of a tunneling particle passing through a region of spacetime which contains the bulk of the incident wave packet, as well as the probability of the same particle passing through the (spacelike-separated) region which contains the bulk of the transmitted wave packet. If conditioned on eventual transmission of the tunneling particle, both of these would be close to unity – on average, each individual particle would have had an effect on two spacelike-separated detectors. Figure 12.5 shows a spacetime diagram for the experiment under consideration (Steinberg 1998a; Steinberg *et al.* 1998). An energy filter is added after transmission, to "erase" (Scully *et al.* 1991; Kwiat *et al.* 1992) any information about the time of arrival of the transmitted peak; without this filter, the possibility of a *strong* measurement of the time of arrival of the particle would preclude any possibility that it had come from the initial peak.

We have been setting up an experiment (Steinberg *et al.* 1998, 1999) to observe laser-cooled atoms tunneling through an optical barrier, wherein probes interacting with atoms at various positions and various points in time should allow us to study the weak-measurement predictions. In parallel, we have been thinking about the theoretical approach necessary to determine whether each single particle had actually affected two measurement apparatus at spacelike separation, or whether despite this appearance on average, each particle could be thought of as being at only one device at a time. If the wave function is not merely a measure of our ignorance, but in some deeper sense "real," then one ought perhaps not to be surprised by a particle having a (weak) effect in two places at the same time, so long as no "collapse" occurs. Nevertheless, I believe that most physicists still have an underlying intuition about the indivisibility of particles which would lead them to predict such effects could not occur. Amusingly, when I have tried to explain our proposed experiments, most of the physicists *I* know, who are willing to discuss such things, had the opposite reaction: of course a particle can be in two places at the same time, and of course both pointers may shift simultaneously!

Our initial proposal was to build on the following idea. Consider pointers P_1 and P_2 at spacelike-separated positions. We would like to demonstrate that even though each picks up only a small shift on a single event, it is possible to show that individual particles interacted with *both* pointers. Let us therefore assume the opposite, the corpuscular hypothesis that on a given event, either P_1 or P_2 was affected, but not both. Nevertheless, weak measurements will show that both P_1 and P_2 are shifted on average by an amount roughly equal to unity (a measurement that the particle was almost certainly in a given region). This must imply that on some occasions, P_2 is unshifted, while on other occasions, it is shifted by an amount greater than unity; and the same for P_1. Due to the anticorrelation of these shifts, we expect the distribution of the *difference* $P_1 - P_2$ to develop a larger uncertainty. If the uncertainty of $P_1 - P_2$ did not grow, we would conclude that the shifts of P_1 and P_2 were not anticorrelated, and that each individual particle must really have interacted with both.

While some work has been started on higher-moment weak values (Iannaccone 1996), this field is far from mature. We decided that a simple approach would be to use the same measuring device at P_1 and P_2, but with equal and opposite signs. For instance, using the Larmor-clock approach, a magnetic field along $+z$ at region P_1 could couple to the electron's spin so long as the particle was in that region, while a magnetic field along $-z$ at region P_2 could couple to the same spin with the opposite sign. The rotation of the spin in the $x - y$ plane would automatically record the difference between P_1 and P_2. It is straightforward to show, in the limit of very weak measurements and narrow-band energy filters, that the effects of the two magnetic fields should cancel perfectly. All the transmitted particles should have their spin unaffected, implying that they were affected equally by the two interaction regions. This would, I thought, support the hypothesis that quantum particles can truly be in two places (and have measurable effects there) at the same time.

More recently, consideration of the three-box problem described above led me to carry out the same calculation in that situation. Spin rotations of opposite sign in arms A and B would also cancel out, implying that the particle was really in *both* A and B simultaneously. Yet we saw earlier that the joint probability for being in A and B was in fact zero. One can go through the same argument in the tunneling case. Even though the conditional probability distribution does fill both regions P_1 and P_2, the product of projection operators onto two spacelike-separated regions automatically vanishes (in the Heisenberg picture), because these regions constitute orthogonal subspaces of Hilbert space. It now seems that even in the case of superluminal tunneling, a true weak measurement of the joint probability of being in two places at once is always guaranteed to yield zero. Thus even though $\langle P_1 \rangle_{\mathrm{wk}} = \langle P_2 \rangle_{\mathrm{wk}}$ and $\langle P_1 - P_2 \rangle_{\mathrm{wk}} = 0$, one can show

$$\begin{aligned} \langle (P_1 - P_2)^2 \rangle_{\mathrm{wk}} = & \\ &= \langle P_1^2 \rangle_{\mathrm{wk}} + \langle P_2^2 \rangle_{\mathrm{wk}} - \langle P_1 P_2 \rangle_{\mathrm{wk}} - \langle P_2 P_1 \rangle_{\mathrm{wk}} \\ &= = 1 + 1 + 0 + 0 = 2. \end{aligned} \tag{12.13}$$

If one treats this as the definition of the uncertainty in a weak value, one certainly finds anticorrelations: P_1 and P_2 only shift by unity at the expense of their difference growing uncertain by $\sqrt{2}$, just as though they had shifted in an entirely uncorrelated fashion. On the other hand, if one simply calculates the final state of a transmitted spin which was subject to equal and opposite interactions at P_1 and P_2, one finds no increase in the uncertainty of its orientation. Further work will be necessary to determine what weak values can really teach us about nonlocality, and how best to define the uncertainties and correlations of these probabilities which are not bounded by the usual classical rules. Nevertheless, it is apparent that weak measurements allow us to discuss postselected systems (such as tunneling particles) in a much more powerful way than was possible in the more conventional language of evolving and collapsing wave functions. In the meantime, we continue to build our laser-cooling experiment, to verify these predictions, and to study generalizations which occur when "real" measurements (i.e., decoherence or dissipation) are introduced, and when the "weakness" of an interaction is varied.

Quantum information and postselection

In the burgeoning field of quantum information (Nielsen and Chuang 2000), it is well known that photons are excellent carriers of quantum information, easily produced, manipulated, and detected, and relatively immune to "decoherence" and undesired interactions with the surrounding environment. This has led to their widespread application in quantum communications (Bennett and Brassard 1984;

Brendel *et al.* 1999; Buttler *et al.* 2000). Unfortunately, the superposition principle of linear optics implies that different photons behave independently of one another – without some nonlinearity, it is impossible for one photon to influence the evolution of another photon, and this has long made it seem that optics would be an unsuitable platform for designing a quantum *computer*. Even certain straightforward projective measurements, such as the determination of which of the four Bell states[4] a photon pair is in, prove to be intractable without significantly stronger nonlinearities than exist in practice (Mattle *et al.* 1996; Bouwmeester *et al.* 1997; Calsamiglia and Lütkenhaus 2001). Much work has focused on developing exotic systems such as cavity-QED experiments (Turchette *et al.* 1995; Nogues *et al.* 1999) in which enhanced nonlinearities allow for the design of effective quantum logic gates, while most of quantum-computation research has instead focused on using atoms, ions, or solids to store and manipulate "qubits" (Cirac and Zoller 1995; Monroe *et al.* 1995; Kane 1998). Recently, it was noted that detection *itself* is a nonlinear process, and that appropriately chosen postselection may be used to "mimic" the kinds of optical nonlinearity one would desire for the construction of an optical quantum logic gate (Knill *et al.* 2001; Pittman *et al.* 2001). In parallel, work has continued on searches for systems in which true optical nonlinearities might be enhanced by factors on the order of 10^9 or 10^{10}, as would be necessary for the construction of fundamental logic gates (Franson 1997; Harris and Hau 1999; Kash *et al.* 1999).

We recently showed that it is possible to use quantum interference between photon pairs to effectively enhance nonlinearities by a similar order of magnitude. Using a crystal of beta-barium borate (BBO), it is possible to frequency-double a beam of light, converting two photons at ω into one photon at 2ω with some small ($\mathcal{O}(10^{-10})$) probability, or alternatively to "down-convert" a photon at 2ω into a pair of photons around ω, with equally low probability (Steinberg *et al.* 1996). These effects are extremely common and extremely important in modern nonlinear optics, but rely on high-intensity beams to generate significant effects; two individual photons entering such a crystal would have a negligible interaction. For this reason, one experiment which purported to perform "100% efficient" quantum teleportation by using a nonlinear interaction to carry out the necessary Bell-state determination actually needed to replace one of the incident photons with a beam containing billions of identical copies (Kim *et al.* 2001). By contrast, we discovered that adding an additional pump beam (with billions of photons) to the system leads to a quantum-interference effect which can enhance the interaction between two single-photon-level beams by many orders of magnitude. In Resch *et al.* (2003), we show that this can lead to >50%-efficient frequency-doubling of photon pairs.

[4] The maximally entangled polarization states of two particles: $|HV\rangle \pm |VH\rangle$ and $|HH\rangle \pm |VV\rangle$ in the case of photons, or equivalently, $|J=0, m=0\rangle$, $|J=1, m=0\rangle$, and $|J=1, m=1\rangle \pm |J=1, m=-1\rangle$ for a pair of spin-$\frac{1}{2}$ particles.

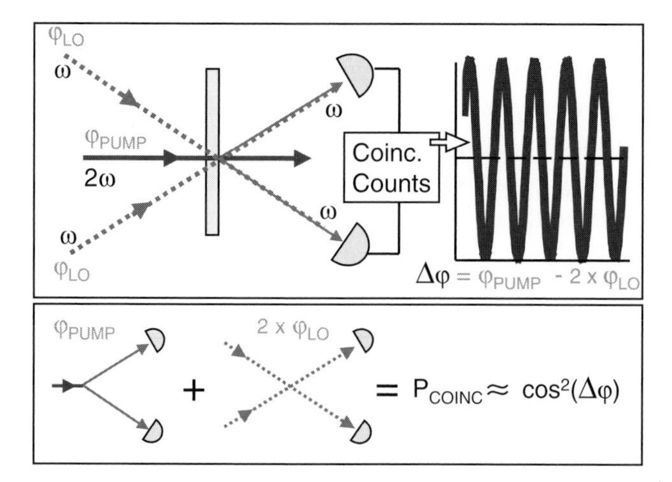

Figure 12.6. The two-photon "switch" experiment: quantum interference between photon pairs being generated through down-conversion and already being present in two laser beams can lead to nearly unit-efficiency up-conversion of photon pairs from classical beams.

This effect is closely related to earlier work on quantum suppression of parametric down-conversion by Anton Zeilinger's group (Herzog *et al.* 1994).

The basic scheme is shown in Fig. 12.6. Two beams at ω, each containing less than 1 photon on average, enter a nonlinear crystal; these beams are conventionally known as "signal" and "idler." Simultaneously, a strong pump beam at 2ω pumps the crystal in a mode which couples to signal and idler via the interaction Hamiltonian

$$\mathcal{H} = g a_p^{\dagger} a_s a_i + \text{h.c.}. \tag{12.14}$$

This can convert a single pump photon into a signal–idler pair, or vice versa, albeit with vanishingly small efficiency. The three input beams are in coherent states, and thus the initial state of the system may be written $|\Psi\rangle = |\alpha_p\rangle_p|\alpha_s\rangle_s|\alpha_i\rangle_i$. For weak inputs $|\alpha_s|, |\alpha_i| \ll 1$, but a strong classical pump ($|\alpha_p|^2 \sim 10^{10}$), the interaction can be controlled such that to lowest order, all photon pairs are removed from the signal and idler beams (i.e., they are up-converted into the pump mode, although this effect is too weak to be directly observed). This occurs due to destructive interference between the amplitude for a photon pair to be present in s and i, and the amplitude for a pump photon to down-convert into the same modes (Resch *et al.* 2002a). Importantly, this interference effect depends on the relative *phase* of the three beams, which means that it cannot work if any of the beams has a well-defined photon number, since the optical phase and the photon number are incompatible observables (roughly speaking – on the same order of roughness as the time-energy uncertainty principle – $\Delta n \Delta\varphi \geq 1/2$).

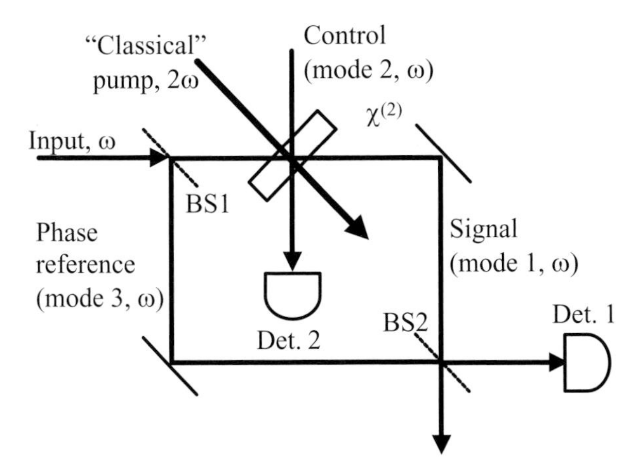

Figure 12.7. The two-photon switch incorporated into a Mach–Zehnder interferometer serves to demonstrate a conditional-phase gate, i.e., cross-phase modulation at the single-photon level.

This up-conversion effect can be thought of as a highly efficient switch – if a photon happens to be present in the signal mode, than no photon in the idler mode can be transmitted; and vice versa. Unfortunately, this is only true if it is fundamentally *unknown* whether the signal mode possessed a photon or not. By observing the absence of coincidence counts after the device, we may conclude that any photon pairs which had been present disappeared . . . but on no individual occasion did we know a photon pair actually existed!

We extended this work to a geometry more closely related to one of the standard logic gates of quantum information theory, the controlled-phase gate (Nielsen and Chuang 2000; Resch *et al.* 2002b). Still relying on interference between incoming photon pairs and the down-conversion process, we altered the relative phase so that the *probability* of a photon pair emerging was not significantly altered, but its quantum phase would be shifted relative to that of the vacuum or a single photon in either beam alone. To measure this, we built the homodyne set-up in Fig. 12.7. This can be thought of as a simple Mach–Zehnder interferometer for a signal photon (really a signal beam with an average photon number per pulse much less than 1). Into one arm of the interferometer, our pumped crystal is inserted. At the same time, a "control" beam is sent through the crystal's idler mode. If a control photon is present, then a phase shift is impressed on any passing signal photon; this is observed as a shift in the Mach–Zehnder interference pattern (see Fig. 12.8). We were able to observe shifts as large as $\pm 180°$, or very small phase shifts with little effect on the probability itself, depending on the strength of the pump beam relative to that of the signal and control beams. Once more, however, to operate this gate, we had to operate in a condition of ignorance. We send in beams which may or

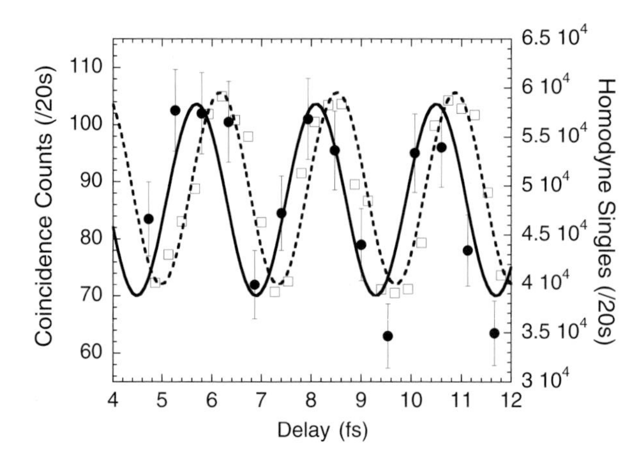

Figure 12.8. Fringe patterns observed at the output of the Mach–Zehnder interferometer when a trigger photon was detected (black circles; solid line), versus when no trigger photon was detected (white squares; dashed line). A significant phase shift is observed on the signal beam due to the presence of a single photon in the trigger mode.

may not have photons, but when we observe a "control" photon leaving the crystal, we find the desired effect on the signal. Can we conclude that this postselection determined that there had been a control photon there all along, and that the logic gate performed the correct operation for an input of logical "1"? To understand the operation of the gate – the phase shift imprinted on the signal beam – it is necessary to take into account both the state preparation (the well-defined phase differences between the beams) and the postselection (the presence of a control photon).

In a manner somewhat reminiscent of the KLM scheme (Knill *et al.* 2001), this requires a fundamental change in the way one thinks about logic operations, with inputs being determined not by preparing the appropriate state, but by postselecting the desired value of the input (Resch *et al.* 2002c). So far, it remains unclear how widely such effects could be applied in quantum information; we do not presently know of a way to incorporate them into the standard paradigm of quantum computing. On the other hand, we have shown (Resch *et al.*, 2001b) that despite its eccentricity and potential pitfalls, this "conditional-phase switch" can indeed be used to implement the Bell measurements which were previously impossible for individual photon pairs, provided only that the photon pairs are produced in the appropriate superposition with vacuum. For subtle but important reasons, this means our technique *cannot* be used for unconditional quantum teleportation; but it can be used to improve earlier experiments on subjects such as quantum dense coding (Mattle *et al.* 1996).

While it is not possible to have a well-defined phase *and* a well-defined photon number in a quantum state, it is possible to prepare one and postselect the other: and weak measurements show us that at intermediate times, the system possesses some characteristics of both the initial and final states. It seems that weak measurement may be precisely the formalism needed for describing such enhanced nonlinearities, and probably a broader range of "nondeterministic" operations currently being investigated in quantum logic.

Having your cake and eating it too

There is another example of a possible application for these enhanced nonlinearities, and we are presently setting up an experiment to demonstrate this. In 1992, Lucien Hardy proposed an ingenious quantum paradox which involved intersecting electron and positron interferometers, wherein colliding electrons and positrons would undergo certain annihilation (Hardy 1992a, b). Of course, this scheme was quickly recognized to be something of a stretch experimentally, and it was hoped that the experiment could be performed with optical interferometers instead. Unfortunately, as mentioned several times already, the interaction between different photons is so weak in practical systems that the equivalent of an "annihilation" event – an up-conversion event, for instance – was exceedingly rare. A mathematically equivalent paradox was eventually tested optically (Torgerson *et al.* 1995; White *et al.* 1999), but no direct demonstration of the original conundrum has been possible to date.

Hardy's paradox relies on the concept of "interaction-free measurements" introduced by Elitzur and Vaidman (1993). Briefly, it is possible to set up an interferometer as in Fig. 12.9 to transmit all the input light out of one port, known as the "bright" port. Ideally, no photon should ever be detected at the "dark" port. However, any object that blocks one of the paths of the interferometer will destroy the interference, and therefore generate some probability of a photon exiting the dark port. Clearly, in the cases in which a photon is observed at this port, one can conclude that (a) it was not blocked by the object; but (b) the object must have been in place (since without the object, interference prevents any counts from being observed there). In the original example, this made it possible to achieve the surprising feat of confirming that an infinitely sensitive bomb was functioning – without setting it off. In later work (Kwiat *et al.* 1995), it was shown that this task could be accomplished with arbitrarily high efficiency, through ingenious modifications to the interferometer.

Although such measurements are popularly referred to as "interaction-free," in some quantum mechanical sense, they clearly do involve an interaction: a "bomb" initially in an uncertain position may be collapsed into the interferometer arm, through the detection of a photon at the dark port. Such considerations motivated the extension of the problem to two overlapping interaction-free-measurement (IFM)

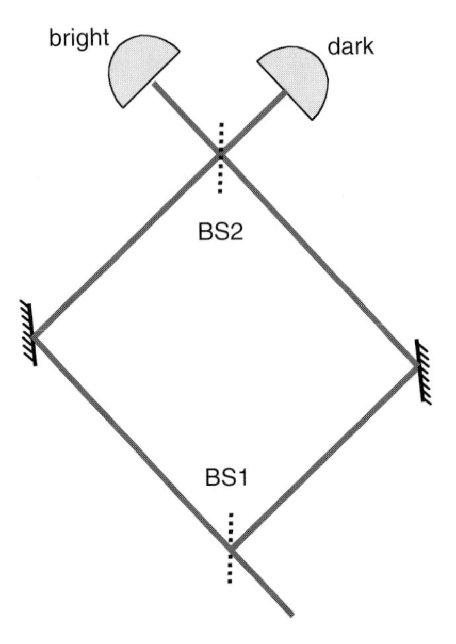

Figure 12.9. A Mach–Zehnder interferometer as proposed by Elitzur and Vaidman (1993) for performing "interaction-free measurements." When the path lengths are balanced, all photons reach the "bright" port and none the "dark" port. An absorbing object placed inside the interferometer may cause photons to reach the dark port, indicating the presence of the object even though those photons could (in some sense) never have interacted with the object directly.

interferometers, shown in Fig. 12.10, each of which can be thought of as measuring whether or not the *other* interferometer's particle is in the "in" path. The reasoning now is simple. If an electron interferometer and a positron interferometer overlap at "in," in such a way that the electron and positron are certain to annihilate if they meet there, then each particle may serve to "block" the other particle if and only if it takes the "in" path. If each interferometer is aligned so that all electrons reach B_- and all positrons B_+, then these two interferometers are IFMs. An electron will only be detected at D_- if the positron was in the way. Similarly, a positron can only reach D_+ if an electron is in the way. Naturally, if both the electron and the positron are at "in," then they annihilate, and cannot be observed. For this reason, one should never observe an electron at D_- *and* a positron at D_+ at the same time.

Yet this is not the case. Quantum mechanically, there is a finite probability for both the electron and the positron to reach their dark ports. How do we interpret this? The conventional answer is that we have learned the error of our classical ways. While the IFM was able to tell us whether or not a classical particle was blocking one arm of an interferometer, we transgressed by drawing counterfactual conclusions about a quantum particle which was not directly observed. Clearly, this

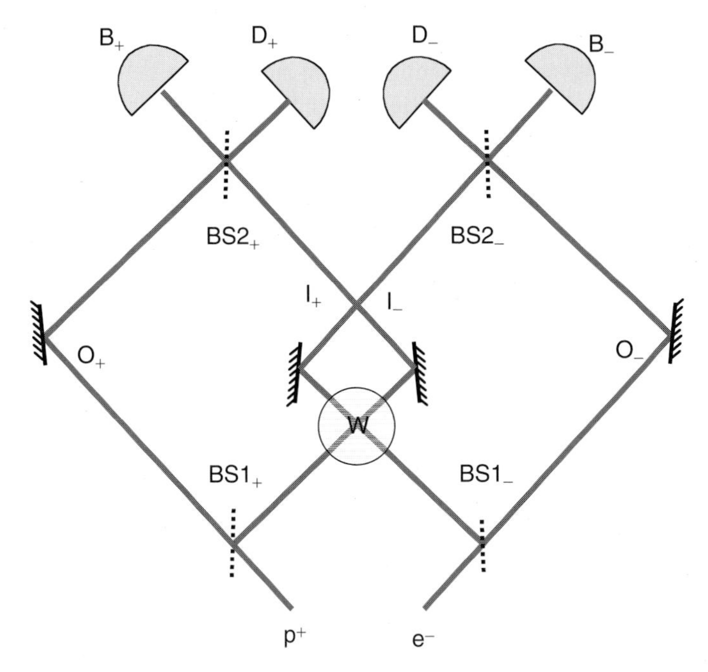

Figure 12.10. Two overlapping interaction-free measurement devices ("IFMs," in the jargon) implement Hardy's paradox. One device is an electron interferometer, and the other a positron interferometer. They overlap at W, where it is supposed that an electron and a positron arriving simultaneously will annihilate with certainty. If one can truly conclude from electron detection at D_- that the positron was in the interaction region W, and from positron detection at D_+ that the electron was in W, then one should never see coincident detections between the two dark detectors, since the particles would have been annihilated at W. Quantum mechanics shows that this is not the case.

is not a very satisfying state of affairs, but perhaps it is true that quantum mechanics does not allow us to make "retrodictions" of the sort we rely on to construct this paradox.

Despite the clear contradiction with classical reasoning, the astute reader may recall that at least in the case of weak measurements, classical intuition often works surprisingly well, albeit at the expense of certain other intuitions, such as the positive–definiteness of probabilities. Indeed, it was recently pointed out (Aharonov *et al.* 2002) that weak measurements can "resolve" the paradox raised by Hardy. How is this? Consider weak measurements of the probabilities for the various particles to be in the various arms of the interferometer, and of the corresponding joint probabilities. From where does the apparent paradox arise? If we postselect on cases where both photons reach the dark port, we want to conclude that the probability of the electron having followed the "in" path, $P(e^{-}\text{in}) = 1$; and also that $P(e^{+}\text{in}) = 1$. So far so good, except that we also believe that $P(e^{-}\ \text{in}\ \textit{and}\ e^{+}\ \text{in})$

must = 0, since both particles would have annihilated had they met along the "in" path. Of course, we have already seen a similar situation in Table 12.1. Just because A and B both happen with certainty (in a weak-measurement sense) does *not* imply that A and B ever happen simultaneously. Aharonov *et al.* (2002) calculate that the above probabilities do in fact hold, and that to satisfy the various sum rules, the probability of one particle being "in" and the other being "out" is 100%, and that the probability of *both* particles being "out" is -100%. In this sense, there is no more paradox. All the paths can be measured simultaneously and in arbitrary combinations, so long as the measurements are all weak. And given this proviso, all our expectations from intuitive analysis of the IFMs should prove to be correct. The price we need to pay for this resolution is to accept that, at least in situations of postselection, certain probabilities may turn out to be negative.

Although we still do not know how to turn our "switch" into a quantum computer, recall that it allows us to cause photon pairs to up-convert with nearly unit efficiency. This is the analog of the e^+e^- annihilation in Hardy's original formulation, and we can now hope to observe his paradox directly, using a coherently driven nonlinear crystal as the interaction region for "annihilating" our photon pairs. Now this switch, which had the disturbing property of working only in a "nondeterministic," after-the-fact manner, becomes the ideal tool for studying the difficult situations one gets into when trying to make retrodictions about quantum-mechanical systems.

Conclusion

In this rapid tour of a variety of recent (and future) experiments and theoretical investigations, I have tried to focus some attention on the new trend towards attempting to talk about *history* in quantum mechanics, and in particular to talk about the history of specific subensembles defined by both state preparation and postselection. The formalism of weak measurements addresses such problems in a very natural fashion, but yields all manner of counterintuitive predictions. At the same time, it has an unshakable connection to real measurements which could be (and often are) performed in the laboratory; I describe certain experiments now in progress which should further demonstrate the fruitfulness of this formalism. The relationship between weak measurements and generalized probability theories appears to be particularly strong, but more work remains to be done to elucidate the meaning of these exotic (negative, or even complex) quantities which obey many of the axioms of probability theory. In particular, weak measurements provide one with a little more leeway than orthodox quantum mechanics when it comes to describing what the state of a system "really was" between preparation and detection, but in so doing, raises a variety of difficult questions, especially relating to the reality of the wave function, and the nonlocality of individual quantum particles. It is interesting to note that a variety of experiments, ranging from new concepts for quantum computation to cavity-QED studies of open-system quantum dynamics, have recently provoked increased interest in the mathematical description of post-selected subensembles. Perhaps the time is finally right for mainstream quantum physicists to attack these problems, and in the process develop a better understanding of the nature of space, time, and measurement in quantum mechanics.

Acknowledgments

I would like to acknowledge my coworkers at Toronto – Kevin Resch, Jeff Lundeen, Stefan Myrskog, Jalani Fox, Ana Jofre, Chris Ellenor, Masoud Mohseni, and Mirco Siercke – both for their efforts in the lab and for their ideas, many of which have found their way into this chapter. I would also like to thank Ray Chiao and Paul Kwiat for their collaboration and useful discussions over many years. Finally, for thought-provoking conversations about weak measurements I would like to thank Howard Wiseman, Jeff Tollaksen, Sandu Popescu, Lev Vaidman, Avshalom Elitzur, Gonzalo Muga, Markus Büttiker, and Jeeva Anandan, in addition of course to Yakir Aharonov, whose seminars first introduced me to the concept. Some of the work described in this chapter was supported by NSERC, by Photonics Research Ontario, by the Canadian Foundation for Innovation, and by the US Air Force Office of Scientific Research under the QuIST programme (F49620-01-1-0468).

References

Aharonov, Y and Anandan, J (1993) *Phys. Rev.* **A47**, 4616.
Aharonov, Y and Vaidman, L (1990) *Phys. Rev.* **A41**, 11.
(1991) *J. Phys.* **A24**, 2315.
(2002a) In *Time in Quantum Mechanics*, ed. J. G. Muga, R. Sala Mayato, and I. L. Egusquiza, pp. 369–412. New York: Springer-Verlag.
(2002b) quant-ph/0206074.
Aharonov, Y, Albert, D Z, and Vaidman, L (1988) *Phys. Rev. Lett.* **60**, 1351.
Aharonov, Y, Popescu, S, Rohrlich, D, *et al.* (1993) *Phys. Rev.* **A48**, 4084.
Aharonov, Y, Botero, A, Popescu, S, *et al.* (2002) *Phys. Lett.* **A 301**, 130.
Bell, J S (1987) *Speakable and Unspeakable in Quantuum Mechanics*. Princeton, NJ: Princeton University Press.
Bennett, C H and Brassard, G (1984) In *Proceedings IEEE Int. Conf. on Computing Systems and Signal Processing*, Bangalore, India, p. 175.
Bouwmeester, D, Pan, J-W, Mattle, K, *et al.* (1997) *Nature* **390**, 575.
Brendel, J, Gisin, N, Tittel, W, *et al.* (1999) *Phys. Rev. Lett.* **82**, 2594.
Büttiker, M (1983) *Phys. Rev.* **B27**, 6178.
Büttiker, M and Landauer, R (1982) *Phys. Rev. Lett.* **49**, 1739.
(1985) *Phys. Scr.* **32**, 429.
Buttler, W T, Hughes, R J, Lamoreaux, S K, *et al.* (2000) *Phys. Rev. Lett.* **84**, 5652.
Calsamiglia, J and Lütkenhaus, N (2001) *Appl. Phys.* **B72**, 67.

Chiao, R Y and Steinberg, A M (1997) In *Progress in Optics*, vol. 37, ed. E. Wolf, pp. 375–405. New York: Elsevier.
Chu, S and Wong, S (1982) *Phys. Rev. Lett.* **48**, 738.
Cirac, J I and Zoller, P (1995) *Phys. Rev. Lett.* **74**, 4091.
Elitzur, A C and Vaidman, L (1993) *Found. Phys.* **23**, 987.
Enders, A and Nimtz, G (1993) *J. Phys. I France* **3**, 1089.
Feynman, R P (1987) In *Quantum Implications*, ed. B. J. Hiley and F. D. Peat. London: Routledge and Kegan Paul.
Foster, G T, Orozco, L A , Castro-Beltran, H M, *et al.* (2000) *Phys. Rev. Lett.* **85**, 3149.
Franson, J D (1997) *Phys. Rev. Lett.* **78**, 3852.
Garrett, C G B and McCumber, D E (1970). *Phys. Rev.* **A1**, 305.
Hardy, L (1992a) *Phys. Lett.* **A167**, 17.
(1992b) *Phys. Rev. Lett.* **68**, 2981.
Harris, S E and Hau, L V (1999) *Phys. Rev. Lett.* **82**, 4611.
Hauge, E H and Støvneng, J A (1989) *Rev. Mod. Phys.* **61**, 917.
Herzog, T J, Rarity, J G, Weinfurter, H, *et al.* (1994) *Phys. Rev. Lett.* **72**, 629.
Iannaccone, G (1996) quant-ph/9611018.
Kane, B E (1998) *Nature* **393**, 133.
Kash, M M, Sautenkov, V A, Zibrov, A S, *et al.* (1999) *Phys. Rev. Lett.* **82**, 5229.
Kim, Y-H, Kulik, S P, and Shih, Y (2001) *Phys. Rev. Lett.* **86**, 1370.
Knill, E, Laflamme, R, and Milburn, G (2001) *Nature* **409**, 46.
Kwiat, P G, Steinberg, A M, and Chiao, R Y (1992) *Phys. Rev.* **A45**, 7729.
Kwiat, P G, Weinfurter, H, Herzog, T, *et al.* (1995) *Phys. Rev. Lett.* **74**, 4763.
Liebfried, D, Meekhof, D M, King, B E (1996) *Phys. Rev. Lett.* **77**, 4281.
MacColl, L A (1932) *Phys. Rev.* **40**, 621.
Mattle, K, Weinfurter, H, Kwiat, P G, *et al.* (1996) *Phys. Rev. Lett.* **76**, 4656.
Monroe, C, Meekhof, D M, King, B E, *et al.* (1995) *Phys. Rev. Lett.* **75**, 4714.
Muckenheim, W, Ludwig, G, Dewdney, C, *et al.* (1983). *Phys. Rep.* **133**, 339.
von Neumann, J (1955) *Mathematical Foundations of Quantum Mechanics*. Princeton, NJ: Princeton University Press.
(1983) In *Quantum Theory and Measurement*, ed. J. A. Wheeler and W. H. Zurek. Princeton, NJ: Princeton University Press.
Nielsen, M A and Chuang, I L (2000) *Quantum Computation and Information*. Cambridge: Cambridge University Press.
Nogues, G, *et al.* (1999) *Nature* **400**, 239.
Pitowski, I (1982) *Phys. Rev. Lett.* **48**, 1299.
Pittman, T B, Jacobs, B C, and Franson, J D (2001) *Phys. Rev.* **A64**, 062311.
Resch, K J, Lundeen, J S, and Steinberg, A M (2001a) *Phys. Rev. Lett.* **87**, 123603.
(2002a) *J. Mod. Opt.* **49**, 487.
(2002b) *Phys. Rev. Lett.* **89**, 037904.
(2002c) *Phys. Rev. Lett.* **88**, 113601.
(2003) Practical creation and detection of polarization Bell states using parametric down-conversion. In *The Physics of Communication*, Proceedings of the XXII Solvay Conference on Physics, Antoniou, Sadovnichy, and Walther, eds. World Scientific, pp. 437–451. (also available as quant-ph/0204034).
Reznik, B and Aharonov, Y (1995) *Phys. Rev.* **A52**, 2538.
Ritchie, N W M, Story, J G, and Hulet, R G (1991) *Phys. Rev. Lett.* **66**, 1107.
Scully, M O, Englert, B-G, and Walther, H (1991) *Nature* **351**, 111.
Scully, M O, Walther, H, and Schleich, W (1994) *Phys. Rev.* **A49**, 1562.
Spielmann, C, Szipöcs, R, Stingl, A, *et al.* (1994) *Phys. Rev. Lett.* **73**, 2308.

Steinberg, A M (1995a) *Phys. Rev. Lett.* **74**, 2405.
(1995b) *Phys. Rev.* **A52**, 32.
(1998a) *Superlattices and Microstructures* **23**, 823.
(1998b) *Found. Phys.* **28**, 385.
(1999) *J. Kor. Phys. Soc.* **35**, 122.
(2000) *Phys. World* **13**, 21.
Steinberg, A M and Chiao, R Y (1994) *Phys. Rev.* **A49**, 2071.
Steinberg, A M, Kwiat, P G, and Chiao, R Y (1993) *Phys. Rev. Lett.* **71**, 708.
(1996) In *American Institute of Physics Atomic, Molecular, and Optical Physics Handbook*, ed. G. W. F. Drake. Woodbury, NY: AIP Press.
Steinberg, A M, Myrskog, S, Moon, H S, *et al.* (1998) *Ann. Phys. (Leipzig)* **7**, 593.
(1999) *Am. Inst. Phys. Conf. Proc.* **461**, 36.
Torgerson, J R, Branning, D, Monken, C H, *et al.* (1995) *Phys. Lett.* **A204**, 323.
Turchette, Q A, Hood, C J, Lange, W, *et al.* (1995) *Phys. Rev. Lett.* **75**, 4710.
Unruh, W G (1994) *Phys. Rev.* **A50**, 882.
Wang, L J, Kuzmich, A, and Dogariu, A (2000) *Nature* **406**, 277.
White, A G, James, D F V, Eberhard, P H, *et al.* (1999) *Phys. Rev. Lett.* **83**, 3103.
Wigner, E P (1932) *Phys. Rev.* **40**, 749.
(1995) *Phys. Rev.* **98**, 145.
Wiseman, H M (2002) *Phys. Rev.* **A65**, 032111.

13

Conceptual tensions between quantum mechanics and general relativity: are there experimental consequences?

Raymond Y. Chiao
University of California at Berkeley

Introduction

Mercy and truth are met together; righteousness and peace have kissed each other.
(Psalm 85:10)

In this volume honoring John Archibald Wheeler, I would like to take a fresh look at the intersection between two fields to which he devoted much of his research life: general relativity (GR) and quantum mechanics (QM). As evidence of his keen interest in these two subjects, I would cite two examples from my own experience. When I was an undergraduate at Princeton University during the years from 1957 to 1961, he was my adviser. One of his duties was to assign me topics for my junior paper and for my senior thesis. For my junior paper, I was assigned the topic: "Compare the complementarity and the uncertainty principles of quantum mechanics: Which is more fundamental?" For my senior thesis, I was assigned the topic: "How to quantize general relativity?" As Wheeler taught me, more than half of science is devoted to the asking of the right question, while often less than half is devoted to the obtaining of the correct answer, but not always!

In the same spirit, I would like to offer up here some questions concerning conceptual tensions between GR and QM, which hopefully can be answered in the course of time by experiments, with a view towards probing the tension between the concepts of *locality* in GR and *nonlocality* in QM. I hope that it would be appropriate and permissible to ask some questions here concerning this tension. It is not the purpose of this chapter to present demonstrated results, but to suggest heuristically some interesting avenues of research which might lead to future experimental discoveries.

Science and Ultimate Reality, eds. J. D. Barrow, P. C. W. Davies and C. L. Harper, Jr. Published by Cambridge University Press.

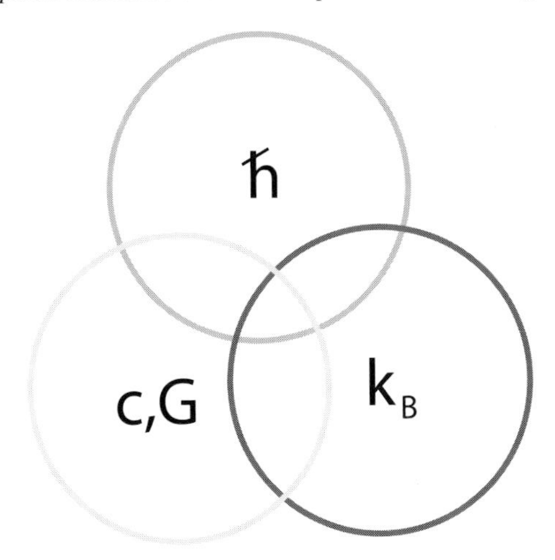

Figure 13.1. Three intersecting circles in a Venn-like diagram represent the three main pillars of physics at the beginning of the twenty-first century. The top circle represents quantum mechanics, and is labeled by Planck's constant $\hbar$. The left circle represents relativity, and is labeled by the two constants c, the speed of light, and G, Newton's constant. The right circle represents statistical mechanics and thermodynamics, and is labeled by Boltzmann's constant k_B . Conceptual tensions exist at the intersections of these three circles, which may lead to fruitful experimental consequences.

One question that naturally arises at the border between GR and QM is the following: are there novel experimental or observational ways of studying quantized fields coupled to curved spacetime? This question has already arisen in the context of the vacuum embedded in curved spacetime (Birrell and Davies 1982), but I would like to extend this to possible experimental studies of the ground state of a nonrelativistic quantum many-body system with off-diagonal long-range order, i.e., a "quantum fluid," viewed as a quantized field, coupled to curved spacetime. As we shall see, this will naturally lead to the further question: are there *quantum* methods to detect gravitational radiation other than the *classical* ones presently being used in the Weber bar and LIGO (i.e., the "Laser Interferometer Gravitational Wave Observatory") (Misner *et al.*1973; Tyson and Gifford 1978; Taylor and Weisberg 1982)?

As I see it, the three main pillars of physics at the beginning of the twenty-first century are quantum mechanics, relativity, and statistical mechanics, which correspond to Einstein's three papers of 1905. There exist conceptual tensions at the intersections of these three fields of physics (see Fig. 13.1). It seems worthwhile re-examining these tensions, since they may entail important experimental consequences. In this introduction, I shall only briefly mention three conceptual tensions between these three fields: *locality* versus *nonlocality* of physical systems, *objectivity* versus *subjectivity* of probabilities in quantum and statistical mechanics (the problem of the nature of information), and *reversibility* versus *irreversibility* of time (the problem of the arrows of time). Others in this volume will discuss the second and the third of these tensions in detail. I shall limit myself to a discussion of the first conceptual tension concerning locality versus nonlocality, mainly in the context of GR and QM. (However, in my Solvay lecture (Chiao 2003a), I have discussed the other two tensions in more detail. See also my Rome lecture (Chiao 2001) for a discussion of three different kinds of quantum nonlocalities.)

Why examine conceptual tensions? A brief answer is that they often lead to new experimental discoveries. It suffices to give just one example from late nineteenth- and early twentieth-century physics: the clash between the venerable concepts of *continuity* and *discreteness*. The concept of continuity, which goes back to the Greek philosopher Heraclitus ("everything flows"), clashed with the concept of discreteness, which goes back to Democritus ("everything is composed of atoms"). Eventually, Heraclitus's concept of continuity, or more specifically that of the *continuum*, was embodied in the idea of *field* in the classical field theory associated with Maxwell's equations. The atomic hypothesis of Democritus was eventually embodied in the kinetic theory of gases in statistical mechanics.

Conceptual tensions, or what Wheeler calls the "clash of ideas," need not lead to a complete victory of one conflicting idea over the other, so as to eliminate the opposing idea completely, as seemed to be the case in the nineteenth century, when Newton's idea of "corpuscles of light" was apparently completely eliminated in favor of the wave theory of light. Rather, there may result a reconciliation of the two conflicting ideas, which then often leads to many fruitful experimental consequences.

Experiments on black-body radiation in the nineteenth century were exploring the intersection, or borderline, between Maxwell's theory of electromagnetism and statistical mechanics, where the conceptual tension between continuity and discreteness was most acute, and eventually led to the discovery of quantum mechanics through the work of Planck. The concept of *discreteness* metamorphosed into the concept of the *quantum*. This led in turn to the concept of *discontinuity* embodied in Bohr's *quantum jump* hypothesis, which was necessitated by the indivisibility of the quantum. Many experiments, such as Millikan's measurements of *h/e*, were in turn motivated by Einstein's heuristic theory of the photoelectric effect based on the "light quantum" hypothesis. Newton's idea of "corpuscles of light" metamorphosed into the concept of the *photon*. This is a striking example showing how many fruitful experimental consequences can come out of one particular conceptual tension.

Within a broader cultural context, there have been many acute conceptual tensions between science and faith, which have lasted over many centuries. Perhaps the above examples of the fruitfulness of the resolution of conceptual tensions within physics itself may serve as a parable concerning the possibility of a peaceful reconciliation of these great cultural tensions, which may eventually lead to the further growth of both science and faith. Hence we should not shy away from conceptual tensions, but rather explore them with an honest, bold, and open spirit.

Three conceptual tensions between quantum mechanics and general relativity

Here I shall focus my attention on some specific conceptual tensions at the intersection between QM and GR. A commonly held viewpoint within the physics community today is that the only place where conceptual tensions between these two fields can arise is at the microscopic Planck length scale (1.6×10^{-33} cm), where quantum fluctuations of spacetime ("quantum foam") occur. Hence manifestations of these tensions would be expected to occur only in conjunction with extremely high-energy phenomena, accessible presumably only in astrophysical settings, such as the early Big Bang.

However, I believe that this point of view is too narrow. There exist other conceptual tensions at macroscopic, non-Planckian distance scales ($\gg 1.6 \times 10^{-33}$ cm), which should be accessible in low-energy laboratory experiments involving macroscopic QM phenomena. It should be kept in mind that QM not only describes *microscopic* phenomena, but also *macroscopic* phenomena, such as superconductivity. Specifically, I would like to point out the following three conceptual tensions:

(I) The *spatial nonseparability* of physical systems due to entangled states in QM, versus the complete *spatial separability* of all physical systems in GR.
(II) The *equivalence principle* of GR, versus the *uncertainty principle* of QM.
(III) The *mixed state* (e.g., of an entangled bipartite system, one part of which falls into a black hole; the other of which flies off to infinity) in GR, versus the *pure state* of such a system in QM.

Conceptual tension (III) concerns the problem of the natures of information and entropy in QM and GR. Since others will discuss this tension in detail in this volume, I shall limit myself only to a discussion of the first two of these tensions.

These conceptual tensions originate from the *superposition principle* of QM, which finds its most dramatic expression in the *entangled state* of two or more spatially separated particles of a single physical system, which in turn leads to Einstein–Podolsky–Rosen (EPR) effects. It should be emphasized here that it is necessary to consider *two or more* particles for observing EPR phenomena, since only then does the *configuration space* of these particles no longer coincide with that of ordinary spacetime. For example, consider the entangled state of two spin-$\frac{1}{2}$

particles in a singlet state initially prepared in the total spin-0 state

$$|S = 0\rangle = \frac{1}{\sqrt{2}} \{|\uparrow\rangle_1 |\downarrow\rangle_2 - |\downarrow\rangle_1 |\uparrow\rangle_2\}, \tag{13.1}$$

in which the two particles in a spontaneous decay process fly arbitrarily far away from each other into two spacelike separated regions of spacetime, where measurements on spin by means of two Stern–Gerlach apparatus are performed separately on these two particles.

As a result of the quantum entanglement arising from the *superposition* of *product* states, such as in the above singlet state suggested by Bohm in connection with the EPR "paradox," it is in general impossible to factorize this state into products of probability amplitudes. Hence it is impossible to factorize the *joint* probabilities in the measurements of spin of this two-particle system. This mathematical *nonfactorizability* implies a physical *nonseparability* of the system, and leads to instantaneous, spacelike correlations-at-a-distance in the joint measurements of the properties (e.g., spin) of discrete events, such as in the coincidence detection of "clicks" in Geiger counters placed behind the two distant Stern–Gerlach apparatus. Bell's inequalities place an upper limit the amount of angular correlations possible for these two-particle decays, based on the *independence* (and hence *factorizability*) of the joint probabilities of spatially separated measurements in all local realistic theories, such as those envisioned by Einstein.

Violations of Bell's inequalities have been extensively experimentally demonstrated (Stefanov *et al.* (2002) and references therein). Therefore these observations cannot be explained on the basis of any *local realistic* world view; however, they were predicted by QM. If we assume a *realistic* world view, i.e., that the "clicks" of the Geiger counters really happened, then we must conclude that we have observed *nonlocal* features of the world. Therefore a fundamental *spatial nonseparability* of physical systems has been revealed by these Bell-inequalities-violating EPR experiments (Chiao and Garrison 1999). It should be emphasized that the observed spacelike EPR correlations occur on macroscopic, non-Planckian distance scales, where the conceptual tension (I) between QM and GR becomes most acute.

Although some of these same issues arise in the conceptual tensions between quantum mechanics and *special* relativity, there are new issues that crop up due to the long-range nature of the gravitational force, which are absent in special relativity, but present in *general* relativity. The problem of quantum fields in *curved* spacetime can be more interesting than in *flat* spacetime.

Gravity is a *long-range* force. It is therefore natural to expect that experimental consequences of conceptual tension (I) should manifest themselves most dramatically in the interaction of macroscopically coherent quantum matter, which exhibit *long-range* EPR correlations, with *long-range* gravitational fields. In particular, the question naturally arises: how do entangled states, such as the above singlet state, interact with tidal fields, such as those in gravitational radiation? Stated more generally: how do quantum many-body systems with entangled ground states possessing off-diagonal long-range order couple to curved spacetime? ("Off-diagonal long-range order" means that the off-diagonal elements of the reduced density matrix in a coordinate space representation of the system are nonvanishing and possess long-range order, i.e., macroscopic quantum phase coherence.) It is therefore natural to look to the realm of *macroscopic* phenomena associated with quantum fluids, rather than phenomena at *microscopic*, Planck-length scales, in our search for these experimental consequences.

Already a decade or so before Bell's ground-breaking work on his inequality, Einstein himself was clearly worried by the radical, spatial nonseparability of physical systems in quantum mechanics. Einstein (1971: 168–73) wrote:

> Let us consider a physical system S_{12}, which consists of two part-systems S_1 and S_2. These two part-systems may have been in a state of mutual physical interaction at an earlier time. We are, however, considering them at a time when this interaction is at an end. Let the entire system be completely described in the quantum mechanical sense by a ψ-function ψ_{12} of the coordinates $q_1, \ldots$ and $q_2, \ldots$ of the two part-systems (ψ_{12} cannot be represented as a product of the form $\psi_1 \psi_2$ but only as a sum of such products [*i.e., as an entangled state*]). At time t let the two part-systems be separated from each other in space, in such a way that ψ_{12} only differs from zero when $q_1, \ldots$ belong to a limited part R_1 of space and $q_2, \ldots$ belong to a part R_2 separated from R_1. . . .
>
> There seems to me no doubt that those physicists who regard the descriptive methods of quantum mechanics as definitive in principle would react to this line of thought in the following way: they would drop the requirement for *the independent existence of the physical reality present in different parts of space*; they would be justified in pointing out that the quantum theory nowhere makes explicit use of this requirement. [*Italics mine.*]

This radical, *spatial nonseparability* of a physical system consisting of two or more entangled particles in QM, which seems to undermine the very possibility of the concept of *field* in physics, is in an obvious conceptual tension with the complete spatial separability of any physical system into its separate parts in GR, which is a *local realistic* field theory.

However, I should hasten to add immediately that the battle-tested concept of *field* has of course been extremely fruitful not only at the classical but also at the quantum level. Relativistic quantum field theories have been very well validated, at least in an approximate, correspondence-principle sense in which spacetime itself is treated classically, i.e., as being describable by a rigidly flat, Minkowskian metric, which has no possibility of any quantum dynamics. There have been tremendous successes of quantum electrodynamics and electroweak gauge field theory (and, to a lesser extent, quantum chromodynamics) in passing all known high-energy experimental tests. Thus the conceptual tension between *continuity* (used in the concept of the spacetime *continuum*) and *discreteness* (used in the concept of *quantized excitations* of a field in classical spacetime) seems to have been successfully reconciled in these relativistic quantum field theories. Nevertheless, the problem of a satisfactory relativistic treatment of quantum measurement within these theories remains an open one (Gingrich and Adami 2002; Peres 2002; Peres and Scudo 2002; Peres *et al.* 2002; Peres and Terno 2003).

Is there any difference between the response of classical and quantum fluids to tidal gravitational fields?

Motivated by the above discussion, a more specific question arises: is there any difference between classical and quantum matter when it is embedded in curved spacetime, for instance, in the *linear* response to the gravitational tidal field of the Earth of a *classical* liquid drop, as compared to that of a *quantum* one, such as a liquid drop of superfluid helium? In order to answer this question, consider a thought-experiment to observe the shape of a freely floating liquid drop placed at the center of the Space Station sketched in Fig. 13.2.

At first glance, the answer to this question would seem to be "no," since the equivalence principle would seem to imply that all freely falling bodies, whether

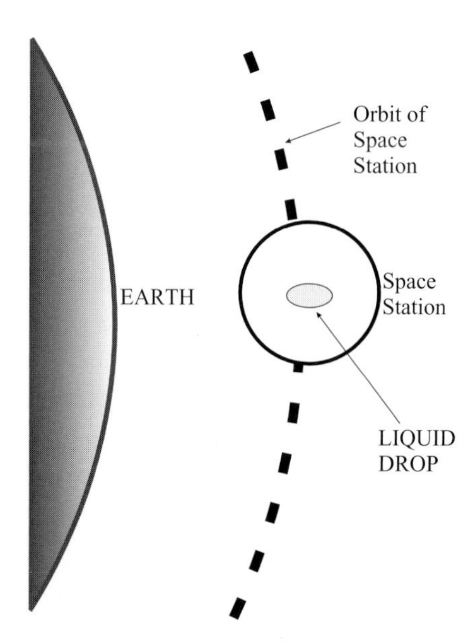

Figure 13.2. Liquid drop placed at the center of a not-to-scale sketch of the Space Station, where it is subjected to the tidal force due to the Earth's gravity. Is there any difference between the shape of a classical and a quantum liquid drop, for example, between a drop of water and one composed of superfluid helium?

classical or quantum, must respond to gravitation, e.g., Earth's gravity, in a *mass-independent*, or more generally, in a *composition-independent* way. Thus whether the internal dynamics of the particles composing the liquid drop obeys classical mechanics or quantum mechanics would seem to make no difference in the response of this body to gravity. Just as in the case of the response of the tides of the Earth's oceans to the Moon's gravity, the shape of the surface of a liquid of any mass or composition would be determined by the equipotential surfaces of the total gravitational field, and should be independent of the mass or composition of the liquid, provided that the fluid particles can move *freely* inside the fluid, and provided that the surface tension of the liquid can be neglected.

However, one must carefully distinguish between the response of the *center of mass* of the liquid drop inside the Space Station to Earth's gravity, and the response of the *relative motions* of particles within the drop to Earth's tidal gravitational field. Whereas the former clearly obeys the mass- and composition-independence of the equivalence principle, one must examine the latter with more care. First, one must define what one means by "classical" and "quantum" bodies. By a "classical body," we shall mean here a body whose particles have undergone *decoherence* in the sense of Zurek (see Chapter 7 in this volume), so that no macroscopic, Schrödinger-cat-like states for widely spatially separated subsystems (i.e., the fluid elements inside the classical liquid drop) can survive the rapid decoherence arising from the environment. This is true for the vast majority of bodies typically encountered in the laboratory. It is the rapid decoherence of the spatially separated subsystems of a classical body that makes the *spatial separability* of a system into its parts, and hence *locality*, a valid concept.

Nevertheless, there exist exceptions. For example, a macroscopically coherent quantum system, e.g., a quantum fluid such as the electron pairs inside a superconductor, usually possesses an energy gap which separates the ground state of the system from all possible excited states of the system. Cooper pairs of electrons in a Bardeen, Cooper, and Schrieffer (BCS) ground state are in the entangled spin singlet states given by eqn (13.1). At sufficiently low temperatures, such a quantum fluid develops a macroscopic quantum coherence, as is manifested by a macroscopic quantum phase which becomes well defined at each point inside the fluid. The resulting macroscopic wave function must remain single valued, in spite of small perturbations, such as those due to weak external fields.

The energy gap, such as the BCS gap, protects spatially separated, but entangled, particles within the body, such as the electrons that are members of Cooper pairs inside a superconductor, against decoherence. Therefore, these quantum fluids are *protectively entangled*, in the sense that the existence of some sort of energy gap separates the nondegenerate ground state of the system from all excited states, and hence prevents any rapid decoherence due to the environment. Under these

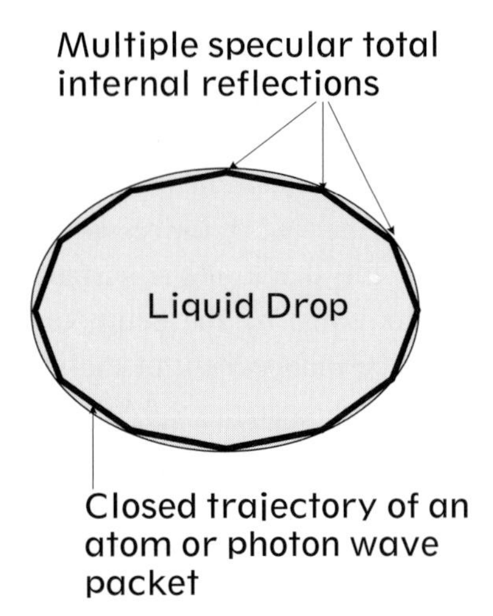

Figure 13.3. Whispering gallery modes of a liquid drop arise in the correspondence principle limit, when an atom or a photon wave packet bounces at grazing incidence off the inner surface of the drop in multiple specular internal reflections, to form a closed polygonal trajectory. The Bohr–Sommerfeld quantization rule leads to a discrete set of such modes.

circumstances, the macroscopically entangled ground state of a quantum fluid, becomes a meaningful *global* concept, and the notion of *nonlocality*, that is, the *spatial nonseparability* of a system into its parts, enters in an intrinsic way into the problem of the interaction of matter with gravitational fields.

For example, imagine a liquid drop consisting of superfluid helium at zero Kelvin, which is in a pure quantum state, floating at the center of the Space Station, as pictured in Fig 13.3. Although the microscopic many-body problem for this superfluid has not been completely solved, there exists a successful macroscopic, phenomenological description based on the Gross–Pitaevskii equation

$$-\frac{\hbar^2}{2m}\nabla^2\Psi + V(x, y, z)\Psi + \beta\,|\Psi|^2\,\Psi = -\alpha\Psi, \tag{13.2}$$

where Ψ is the macroscopic complex order parameter, and the potential $V(x, y, z)$ describes Earth's gravity (including its tidal gravitational potential, but neglecting for the moment the frame-dragging term coupled to superfluid currents), along with the surface tension effects which enters into the determination of the free boundary of the liquid drop. Macroscopic quantum entanglement is contained in the nonlinear term $\beta\,|\Psi|^2\,\Psi$, which arises microscopically from atom–atom S-wave scattering events, just as in the case of the recently observed atomic Bose–Einstein condensates (BECs). (The parameter β is directly proportional to the S-wave scattering length a; the interaction between two atoms in a individual scattering event entangles the two scattering atoms together, so that a measurement of the momentum of one atom immediately determines the momentum of the other atom which participated in the scattering event.) As in the case of the BECs, where this equation has been successfully applied to predict many observed phenomena, the physical meaning of Ψ is that it is the condensate wavefunction.

There should exist near the inside surface of the superfluid liquid drop, closed trajectories for helium atom wave packets propagating at grazing incidence, which, in the correspondence-principle limit, should lead to the atomic analog of the "whispering gallery modes" of light, such as those observed inside microspheres immersed in superfluid helium (Braginsky *et al.* 1989; Treussart *et al.* 1998). In the case of light, these modes can possess extremely high Q's (of the order of 10^9), so that the quadrupolar distortion from a spherical shape due to tidal forces can thereby be very sensitively measured optically (the degeneracy of these modes has been observed to be split by nontidal quadrupolar distortions (Hartings *et al.* 1998)). The atomic wave packets propagating at grazing incidence near the surface are actually those of individual helium atoms dressed by the collective excitations of the superfluid, such as phonons, rotons, and ripplons (Sridhar 1979).[1] Application of the Bohr–Sommerfeld quantization rule to the closed trajectories which correspond to the whispering gallery modes for atoms should lead to a *quantization* of the sizes and shapes of the superfluid drop. For a classical liquid drop, no such quantization occurs because of the decoherence of an atom after it has propagated around these large, polygonal closed trajectories. Hence there should exist a *difference* between classical and quantum matter in their respective responses to gravitational tidal fields. At a fundamental level, this difference arises from the quantum phase shift which is observable in the shift of the interference fringe pattern that results from an atom travelling coherently along two nearby, but intersecting, geodesics in the presence of spacetime curvature (Chiao and Speliotopoulos 2003).

Another difference between a classical and a quantum liquid drop is the possibility of the presence of *quantized* vortices in the latter, along with their associated persistent, macroscopic quantum flows. These quantum flows possess quantized vorticities of $\pm h/m$, where m is the mass of the superfluid atom. The question naturally arises: how do two such vortices placed symmetrically around the center of mass of a superfluid liquid drop react to the presence of tidal forces associated with gravitational radiation? I suspect that these vortices will move *at right angles* in response to these forces in accordance with the Magnus force law, which is a Lorentz-like force law for vortex motion in superfluids. The perpendicularity of this kind of motion is manifestly different from that of a test particle of a classical "perfect" fluid.

[1] The critical angle for total internal reflection of quasi-particles in superfluid helium is 15° with respect to the inward normal of a flat and free surface. These quasi-particles, i.e, dressed helium-atom wave packets, can undergo multiple specular total internal reflections at grazing incidence in accordance with the Gross–Pitaevskii equation, with a negligible nonlinear term near the surface where Ψ becomes vanishingly small, so that it reduces to the standard, linear nonrelativistic Schrödinger equation.

Such *differences* in the linear response between classical and quantum matter in the induced quadrupole moment ΔQ_{ij} of the liquid drop can be characterized by a linear equation relating ΔQ_{ij} to the metric deviations from flat spacetime h_{kl} by means of a phenomenological susceptibility tensor $\Delta \chi_{ij}{}^{kl}$, viz.,

$$\Delta Q_{ij} = \Delta \chi_{ij}{}^{kl} h_{kl}, \tag{13.3}$$

where i, j, k, l are spatial indices (repeated indices are summed). The susceptibility tensor $\Delta \chi_{ij}{}^{kl}$ should in principle be calculable from the many-body current–current correlation function in the linear-response theory of superfluid helium (Forster 1975).

Here, however, I shall limit myself only to some general remarks concerning $\Delta \chi_{ij}{}^{kl}$ based on the Kramers–Kronig relations. Since the response of the liquid drop to weak tidal gravitational fields is *linear* and *causal*, it follows that

$$\text{Re } \Delta \chi_{ij}{}^{kl}(\omega) = \frac{1}{\pi} P \int_{-\infty}^{\infty} d\omega' \frac{\text{Im } \Delta \chi_{ij}{}^{kl}(\omega')}{\omega' - \omega} \tag{13.4}$$

$$\text{Im } \Delta \chi_{ij}{}^{kl}(\omega) = -\frac{1}{\pi} P \int_{-\infty}^{\infty} d\omega' \frac{\text{Re } \Delta \chi_{ij}{}^{kl}(\omega')}{\omega' - \omega}, \tag{13.5}$$

where P denotes Cauchy's principal value. From the first of these relations, there follows the zero-frequency sum rule

$$\text{Re } \Delta \chi_{ij}{}^{kl}(\omega \to 0) = \frac{2}{\pi} \int_{0}^{\infty} d\omega' \frac{\text{Im } \Delta \chi_{ij}{}^{kl}(\omega')}{\omega'}. \tag{13.6}$$

This equation tells us that *if* there should exist a difference in the linear response between classical and quantum matter to tidal fields near DC (i.e., $\omega \to 0$) in the quadrupolar shape of the liquid drop, *then* there must also exist a difference in the rate of absorption or emission of gravitational radiation due to the imaginary part of the susceptibility $\text{Im } \Delta \chi_{ij}{}^{kl}(\omega')$ between classical and quantum matter. The purpose here is not to calculate how big this difference is, but merely to point out that such a difference exists. The above considerations also apply equally well to an atomic BEC, indeed, to any quantum fluid, in its linear response to tidal fields.

Quantum fluids versus perfect fluids

At this point, I would like to return to the more general question: where to look for experimental consequences of conceptual tension (I)? The above discussion suggests the following answer: look at *macroscopically entangled*, and thus *radically delocalized*, quantum states encountered, for example, in superconductors, superfluids, atomic BECs, and quantum Hall fluids, i.e., in what I shall henceforth call "quantum fluids." Again it should be stressed that since gravity is a *long-range* force, it should be possible to perform *low-energy* experiments to probe the interaction between gravity and these kinds of quantum matter on large, non-Planckian distance scales, without the necessity of performing high-energy experiments, as is required for probing the short-range weak and strong forces on very short distance scales. The quantum many-body problem, even in its nonrelativistic limit, may lead to nontrivial interactions with weak, long-range gravitational fields, as the above example suggests. One is thereby strongly motivated to study the interaction of these quantum fluids with weak gravity, in particular, with gravitational radiation.

One manifestation of this conceptual tension is that the way one views a quantum fluid in QM is conceptually radically different from the way that one views a perfect fluid in GR, where only the *local* properties of the fluid, which can conceptually always be spatially separated into independent, infinitesimal fluid elements, are to be considered. For example, interstellar dust particles can be thought of as being a perfect fluid in GR, provided that we can neglect all interactions between such particles.[2] At a fundamental level, the spatial separability of the perfect fluid in GR arises from the rapid decoherence of quantum superposition states (i.e., Schrödinger-cat-like states) of various interstellar dust particles at widely separated spatial positions within a dust cloud, due to interactions with the environment. Hence the notion of *locality* is valid here. The response of these dust particles in the resulting *classical* many-body system to a gravitational wave passing over it is characterized by the local, classical, free-fall motion of each individual dust particle.

In contrast to the classical case, due to their radical delocalization, particles in a macroscopically coherent quantum many-body system, i.e., a quantum fluid, are entangled with each other in such a way that there arises an unusual "quantum rigidity" of the system, closely associated with what London (1964) called "the rigidity of the macroscopic wavefunction." One example of such a rigid quantum fluid is the "incompressible quantum fluid" in both the integer and the fractional quantum Hall effects (Laughlin 1983). This rigidity arises from the fact that there exists an energy gap (for example, the quantum Hall gap) which separates the ground state from all the low-lying excitations of the system. This gap, as pointed out above, also serves to protect the quantum entanglement present in the ground state from decoherence due to the environment, provided that the temperature of these quantum systems is sufficiently low. Thus these quantum fluids exhibit a kind of "gap-protected quantum entanglement." Furthermore, the gap leads to an evolution in accordance with the quantum adiabatic theorem: the system stays adiabatically in a rigidly unaltered ground state, which leads in first-order perturbation

[2] I thank my graduate student Colin McCormick for stimulating discussions concerning the perfect fluid in GR.

theory to quantum diamagnetic effects. Examples of consequences of this "rigidity of the wave function" are the Meissner effect in the case of superconductors, in which the magnetic field is expelled from their interiors, and the Chern–Simons effect in the quantum Hall fluid, in which the photon acquires a mass inside the fluid.

Spontaneous symmetry breaking, off-diagonal long-range order, and superluminality

The unusual states of matter in these quantum fluids usually possess *spontaneous symmetry breaking*, in which the ground state, or the "vacuum" state, of the quantum many-body system breaks the symmetry present in the free energy of the system. The physical vacuum, which is in an intrinsically nonlocal ground state of relativistic quantum field theories, possesses certain similarities to the ground state of a superconductor, for example. Weinberg (1986) has argued that in superconductivity, the spontaneous symmetry breaking process results in a broken *gauge* invariance, an idea which traces back to the early work of Nambu (Nambu and Jona-Lasino 1961a, b).

The Meissner effect in a superconductor is closely analogous to the Higgs mechanism of high-energy physics, in which the physical vacuum also spontaneously breaks local gauge invariance, and can also be viewed as forming a condensate which possesses a single-valued complex order parameter with a well-defined local phase. From this viewpoint, the appearance of the London penetration depth for a superconductor is analogous in an inverse manner to the appearance of a mass for a gauge boson, such as that of the W or Z boson. Thus, the photon, viewed as a gauge boson, acquires a mass inside the superconductor, such that its Compton wavelength becomes the London penetration depth. Similar considerations apply to the effect of the Chern–Simons term in the quantum Hall fluid.

Closely related to this spontaneous symmetry breaking process is the appearance of Yang's (1962) off-diagonal long-range order (ODLRO) of the reduced density matrix in the coordinate-space representation for most of these macroscopically coherent quantum systems. In particular, there seems to be no limit on how far apart Cooper pairs can be inside a single superconductor before they lose their quantum coherence. ODLRO and spontaneous symmetry breaking are both purely quantum concepts with no classical analogs.

Within a quantum fluid, there should arise both the phenomenon of instantaneous EPR correlations-at-a-distance, and the phenomenon of London's "rigidity of the wave function," i.e., a Meissner-like response to radiation fields. Both phenomena involve at the microscopic level *interactions* of entangled particles with an external environment, either through local *measurements*, such as in Bell-type measurements, or through local *perturbations*, such as those arising from radiation fields interacting locally with these particles.

Although at first sight the notion of "infinite quantum rigidity" would seem to imply infinite velocities, and hence would seem to violate relativity, there are in fact no violations of relativistic causality here, since the instantaneous EPR *correlations*-at-a-distance (as seen by an observer in the center-of-mass frame) are not instantaneous *signals*-at-a-distance, which would instantaneously connect causes to effects (Chiao and Kwiat 2002). Also, experiments have verified the existence of superluminal wave packet propagations, i.e., faster-than-c, infinite, and even negative group velocities, for finite-bandwidth, analytic wave packets in the excitations of a wide range of physical systems (Chiao and Steinberg 1997; Nimtz and Heitmann 1997; Boyd and Gauthier 2002; Chiao *et al.* 2004). An analytic function, e.g., a Gaussian wave packet, contains sufficient information in its early tail such that a causal medium can, during its propagation, reconstruct the entire wave packet with a superluminal pulse advancement, and with little distortion. Relativistic causality forbids only the *front* velocity, i.e., the velocity of *discontinuities* which connect causes to their effects, from exceeding the speed of light c, but does not forbid a wave packet's *group* velocity from being superluminal. One example is the observed superluminal tunneling of single-photon wave packets (Steinberg *et al.* 1993; Steinberg and Chiao 1995). Thus the notion of "infinite quantum rigidity," although counterintuitive, does not in fact violate relativistic causality.

The equivalence versus the uncertainty principle

Concerning conceptual tension (II), the equivalence principle is formulated at its outset using the concept of "trajectory," or equivalently, "geodesic." By contrast, Bohr has taught us that the very *concept* of trajectory must be abandoned at fundamental level, because of the uncertainty principle. Thus the equivalence and the uncertainty principles are in a fundamental conceptual tension. The equivalence principle is based on the notion of locality, since it requires that the region of space, inside which two trajectories of two nearby freely falling objects of different masses, compositions, or thermodynamic states are to be compared, go to zero volume, before the principle becomes exact. This limiting procedure is in a conceptual tension with the uncertainty principle, since taking the limit of the volume of space going to zero, within which these objects are to be measured, makes their momenta infinitely uncertain. However, whenever the correspondence principle holds, the *center of mass* of a quantum wave packet (for a single particle or for an entire quantum object) moves according to Ehrenfest's theorem along a classical trajectory, and *then* it is possible to reconcile these two principles.

Davies (P. C. W. Davies, pers. comm.) has come up with a simple example of a quantum violation of the equivalence principle (Viola and Onofrio 1997; Adunas *et al.* 2001; Herdegen and Wawrzycki 2002). Consider two perfectly elastic balls, e.g., one made out of rubber, and one made out of steel, bouncing against a perfectly elastic table. If we drop the two balls from the same height above the table, their classical trajectories, and hence their classical periods of oscillation will be identical, and independent of the mass or composition of the balls. This is a consequence of the equivalence principle. However, quantum mechanically, there will be the phenomenon of tunneling, in which the two balls can penetrate into the classically forbidden region *above* their turning points. The extra time spent by the balls in the classically forbidden region due to tunneling will depend on their mass (and thus on their composition). Thus there will in principle be *mass-dependent* quantum corrections of the classical periods of the bouncing motion of these balls, which will lead to quantum violations of the equivalence principle.

There might exist macroscopic situations in which Ehrenfest's form of the correspondence principle fails. Imagine that one is inside a macroscopic quantum fluid, such as a big piece of superconconductor. Even in the limit of a very large size and a very large number of particles inside this object (i.e., in the thermodynamic limit), there exists no correspondence-principle limit in which classical trajectories or geodesics for the *relative motion* of electrons which are members of Cooper pairs in Bohm singlet states within the superconductor, make any sense. This is due to the superposition principle and the entanglement of a macroscopic number of identical particles inside these quantum fluids. Nevertheless, the *motion of the center of mass* of the superconductor may obey perfectly the equivalence principle, and may therefore be conceptualized in terms of a geodesic.

Quantum fluids as antennas for gravitational radiation

Can the quantum rigidity arising from the energy gap of a quantum fluid circumvent the problem of the tiny rigidity of classical matter, such as that of the normal metals used in Weber bars, in their feeble responses to gravitational radiation? One consequence of the tiny rigidity of classical matter is the fact that the speed of sound in a Weber bar is typically five orders of magnitude less than the speed of light. In order to transfer energy coherently from a gravitational wave by classical means, for example, by acoustical modes inside the bar to some local detector, e.g., a piezoelectric crystal glued to the middle of the bar, the length scale of the Weber bar L is limited to a distance scale on the order of the speed of sound times the period of the gravitational wave, i.e., an acoustical wavelength λ_{sound}, which is typically five orders of magnitude smaller than the gravitational radiation wavelength λ to be detected. This makes the Weber bar, which is thereby limited in its length to $L \simeq \lambda_{\rm sound}$, much too short an antenna to couple efficiently to free space.

However, rigid quantum objects, such as a two-dimensional electron gas in a strong magnetic field which exhibits the quantum Hall effect, in what Laughlin (1983) has called an "incompressible quantum fluid", are not limited by these classical considerations, but can have macroscopic quantum phase coherence on a length scale L on the same order as (or even much greater than) the gravitational radiation wavelength λ. Since the radiation efficiency of a quadrupole antenna scales as the length of the antenna L to the fourth power when $L \ll \lambda$, such quantum antennas should be much more efficient in coupling to free space than classical ones like the Weber bar by at least a factor of $(\lambda/\lambda_{\rm sound})^4$.

Weinberg gives a measure of the radiative coupling efficiency $\eta_{\rm rad}$ of a Weber bar of mass M, length L, and velocity of sound $v_{\rm sound}$, in terms of a branching ratio for the emission of gravitational radiation by the Weber bar, relative to the emission of heat, i.e., the ratio of the *rate* of emission of gravitational radiation $\Gamma_{\rm grav}$ relative to the *rate* of the decay of the acoustical oscillations into heat $\Gamma_{\rm heat}$, which is given by (Weinberg 1972):

$$\eta_{\rm rad} \equiv \frac{\Gamma_{\rm grav}}{\Gamma_{\rm heat}} = \frac{64GMv_{\rm sound}^4}{15L^2c^5\Gamma_{\rm heat}} \simeq 3 \times 10^{-34}, \tag{13.7}$$

where G is Newton's constant. The quartic power dependence of the efficiency $\eta_{\rm rad}$ on the velocity of sound $v_{\rm sound}$ arises from the quartic dependence of the coupling efficiency to free space of a quadrupole antenna upon its length L, when $L \ll \lambda$.

The long-range quantum phase coherence of a quantum fluid allows the typical size L of a quantum antenna to be comparable to the wavelength λ. Thus the phase rigidity of the quantum fluid allows us in principle to replace the velocity of sound $v_{\rm sound}$ by the speed of light c. Therefore, quantum fluids can be more efficient than Weber bars, based on the $v_{\rm sound}^4$ factor alone, by 20 orders of magnitude, i.e.,

$$\left(\frac{c}{v_{\rm sound}}\right)^4 \simeq 10^{20}. \tag{13.8}$$

Hence quantum fluids could be much more efficient receivers of this radiation than Weber bars for detecting astrophysical sources of gravitational radiation. This has previously been suggested to be the case for superfluids and superconductors (Anandan 1981, 1984, 1985; Anandan and Chiao 1982; Chiao 1982; Peng and Torr 1990; Peng *et al.* 1991).

Another important property of quantum fluids lies in the fact that they can possess an extremely low dissipation coefficient $\Gamma_{\rm heat}$, as can be inferred, for example, by the existence of persistent currents in superfluids that can last for indefinitely long periods of time. Thus the impedance matching of the quantum antenna to free space,[3] or equivalently, the branching ratio of energy emitted into the gravitational radiation channel rather than into the heat channel can be much larger than that calculated above for the classical Weber bar.

Minimal-coupling rule for a quantum Hall fluid

The electron, which possesses charge e, rest mass m, and spin $s = 1/2$, obeys the Dirac equation. The nonrelativistic, interacting, fermionic many-body system, such as that in the quantum Hall fluid, should obey the minimal-coupling rule which originates from the covariant-derivative coupling of the Dirac electron to curved spacetime, viz. (Weinberg 1972; Birrel and Davies 1982):

$$p_\mu \to p_\mu - eA_\mu - \frac{1}{2}\Sigma_{AB}\omega_\mu^{AB} \tag{13.9}$$

where p_μ is the electron's four-momentum, A_μ is the electromagnetic four-potential, Σ_{AB} are the Dirac γ matrices in curved spacetime with tetrad (or vierbein) A, B indices, and ω_μ^{AB} are the components of the spin connection

$$\omega_\mu^{AB} = e^{A\nu}\nabla_\mu \, e^B{}_\nu \tag{13.10}$$

where $e^{A\nu}$ and $e^B{}_\nu$ are tetrad four-vectors, which are sets of four orthogonal unit vectors of spacetime, such as those corresponding to a local inertial frame.

Spacetime curvature directly affects the phase of the wave function, leading to fringe shifts of quantum-mechanical interference patterns within atomic interferometers (Chiao and Speliotopoulos 2003). Moreover, it is well known that the vector potential A_μ will also lead to a quantum interference effect, in which the gauge-invariant Aharonov–Bohm phase becomes observable. Similarly, the spin connection ω_μ^{AB}, in its Abelian holonomy, should also lead to a quantum interference effect, in which the gauge-invariant Berry phase (Berry 1984; Chiao and Wu 1986; Tomita and Chiao 1986; Chiao and Jordan 1988) becomes observable.[4] The following Berry-phase picture of a spin coupled to curved spacetime leads to an intuitive way of understanding why there could exist a coupling between a classical GR wave and a classical electromagnetic (EM) wave mediated by a quantum fluid with charge and spin, such as the quantum Hall fluid.

Due to its gyroscopic nature, the spin vector of an electron undergoes *parallel transport* during the passage of a GR wave. The spin of the electron is constrained

[3] In linearized GR, the impedance of free space for GR plane waves is $Z_G = 16\pi G/c = 1.12 \times 10^{-17}$ m^2 s^{-1} kg^{-1} in SI units. For details concerning the concept of "impedance matching to free space" of a GR plane wave in its near field to a thin dissipative film, see Chiao (2003b).

[4] I thank Dung-Hai Lee and Jon Magne Leinaas for discussions on Berry's phase and the quantum Hall effect, and Robert Littlejohn and Neal Snyderman for discussions on Thomas precession (http://bohr.physics.berkeley.edu/209.htm).

to lie inside the spacelike submanifold of curved spacetime. This is due to the fact that we can always transform to a co-moving frame, such that the electron is at rest at the origin of this frame. In this frame, the spin of the electron must be purely a spacelike vector with no timelike component. This imposes an important *constraint* on the motion of the electron's spin, such that whenever the spacelike submanifold of spacetime is disturbed by the passage of a gravitational wave, the spin must remain at all times *perpendicular* to the local time axis. If the spin vector is constrained to follow a conical trajectory during the passage of the gravitational wave, the electron picks up a Berry phase proportional to the solid angle subtended by this conical trajectory after one period of the GR wave.

In a manner similar to the persistent currents induced by the Berry phase in systems with ODLRO (Stern 1992; Lyanda-Geller and Goldbart 2000), such a Berry phase induces an electrical current in the quantum Hall fluid, which is in a macroscopically coherent ground state (Girvin and MacDonald 1987; Zhang *et al.* 1989). This macroscopic current generates an EM wave. Thus a GR wave can be converted into an EM wave. By reciprocity, the time-reversed process of the conversion from an EM wave to a GR wave must also be possible.

In the nonrelativistic limit, the four-component Dirac spinor is reduced to a two-component spinor. While the precise form of the nonrelativistic Hamiltonian is not known for the many-body system in a weakly curved spacetime consisting of electrons in a strong magnetic field, I conjecture that it will have the form

$$H = \frac{1}{2m}\left(p_i - eA_i - \frac{1}{2}\sigma_{ab}\Omega_i^{ab}\right)^2 + V \tag{13.11}$$

where i is a spatial index, a, b are spatial tetrad indices, σ_{ab} is a two-by-two matrix-valued tensor representing the spin,[5] and $\sigma_{ab}\Omega_i^{ab}$ is the nonrelativistic form of $\Sigma_{AB}\omega_\mu^{AB}$. Here H and V are two-by-two matrix operators on the two-component spinor electron wave function in the nonrelativistic limit. The potential energy V includes the Coulomb interactions between the electrons in the quantum Hall fluid. This nonrelativistic Hamiltonian has the form

$$H = \frac{1}{2m}(\mathbf{p} - \mathbf{a} - \mathbf{b})^2 + V, \tag{13.12}$$

where the particle index, the spin, and the tetrad indices have all been suppressed. Upon expanding the square, it follows that for a quantum Hall fluid of uniform

[5] The spin connection couples the spin s, and not the mass m, to curved spacetime. Spin can be fundamentally different from mass as a source for spacetime curvature: spin can also be a source for torsion and nonmetricity (Adak *et al.* 2002). Hence the dimensionless ratio $Gm^2 \cdot 4\pi\varepsilon_0/e^2 = 2.4 \times 10^{-43}$ need not in principle apply to the efficiency for the conversion of GR to electromagnetic waves (and vice versa) mediated by the quantum Hall fluid. Furthermore, there may exist macroscopically coherent, many-body enhancements of this coupling.

density, there exists a cross-coupling or interaction Hamiltonian term of the form

$$H_{\text{int}} \sim \mathbf{a} \cdot \mathbf{b}, \tag{13.13}$$

which couples the electromagnetic **a** field to the gravitational **b** field. In the case of time-varying fields, $\mathbf{a}(t)$ and $\mathbf{b}(t)$ represent EM and GR radiation, respectively.

In first-order perturbation theory, the quantum adiabatic theorem predicts that there will arise the cross-coupling energy between the two radiation fields mediated by the quantum fluid

$$\Delta E \sim \langle \Psi_0 | \mathbf{a} \cdot \mathbf{b} | \Psi_0 \rangle \tag{13.14}$$

where $|\Psi_0\rangle$ is the unperturbed ground state of the system. For the adiabatic theorem to hold, there must exist an energy gap E_{gap} (e.g., the quantum Hall energy gap) separating the ground state from all excited states, in conjunction with the approximation that the time variation of the radiation fields must be slow compared to the gap time $\hbar/E_{\text{gap}}$. This suggests that under these conditions, there might exist an interconversion process between these two kinds of classical radiation fields mediated by this quantum fluid, as indicated in Fig. 13.4.

The question immediately arises: EM radiation is fundamentally a spin-1 (photon) field, but GR radiation is fundamentally a spin-2 (graviton) field. How is it possible to convert one kind of radiation into the other, and not violate the conservation of angular momentum? The answer: the EM wave converts to the GR wave *through a medium*. Here specifically, the medium of conversion consists of a strong DC magnetic field applied to a system of electrons. This system possesses an axis of symmetry pointing along the magnetic field direction, and therefore transforms like a spin-1 object. When coupled to a spin-1 (circularly polarized) EM radiation field, the total system can in principle produce a spin-2 (circularly polarized) GR radiation field, by the addition of angular momentum. However, it remains an open question as to how strong this interconversion process is between EM and GR radiation. Most importantly, the size of the conversion efficiency of this transduction process needs to be determined by experiment.

We can see more clearly the physical significance of the interaction Hamiltonian $H_{\text{int}} \sim \mathbf{a} \cdot \mathbf{b}$ once we convert it into second quantized form and express it in terms of the creation and annihilation operators for the electromagnetic and gravitational radiation fields, as in the theory of quantum optics, so that in the rotating-wave approximation

$$H_{\text{int}} \sim a^{\dagger} b + b^{\dagger} a, \tag{13.15}$$

where the annihilation operator a and the creation operator $a^{\dagger}$ of the single classical mode of the plane wave EM radiation field corresponding to the **a** term obey the commutation relation $[a, a^{\dagger}] = 1$, and where the annihilation operator b and the

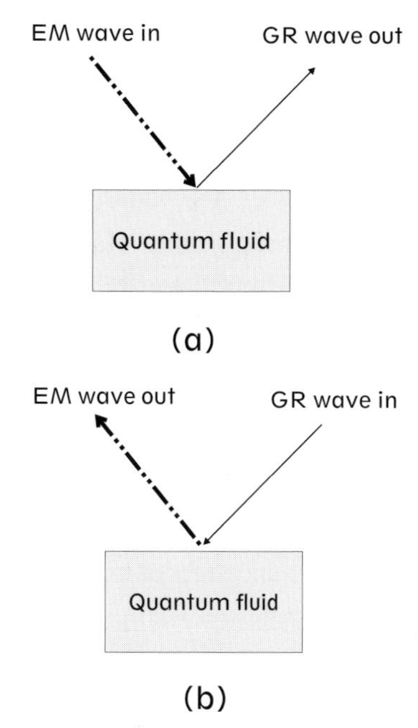

Figure 13.4. Quantum transducer between electromagnetic (EM) and gravitational (GR) radiation, consisting of a quantum fluid with charge and spin, such as the quantum Hall fluid. The minimal-coupling rule for an electron coupled to curved spacetime via its charge and spin results in two processes. In process (a) an EM plane wave is converted upon reflection from the quantum fluid into a GR plane wave; in process (b), which is the reciprocal or time-reversed process, a GR plane wave is converted upon reflection from the quantum fluid into an EM plane wave. Transducer interconversion between these two kinds of waves may also occur upon *transmission* through the quantum fluid, as well as upon *reflection*.

creation operator $b^\dagger$ of the single classical mode of the plane-wave GR radiation field corresponding to the **b** term obey the commutation relation $[b, b^\dagger] = 1$. (This represents a crude, first attempt at quantizing the gravitational field, which applies only in the case of weak, linearized gravity.) The first term $a^\dagger b$ then corresponds to the process in which a graviton is annihilated and a photon is created inside the quantum fluid, and similarly the second term $b^\dagger a$ corresponds to the reciprocal process, in which a photon is annihilated and a graviton is created inside the quantum fluid.

Let us return once again to the question of whether there exists *any* difference in the response of quantum fluids to tidal fields in gravitational radiation, and the response of classical matter, such as the lattice of ions in a superconductor, for example, to such fields. The essential difference between quantum fluids and classical matter is the presence or absence of macroscopic quantum phase coherence. In quantum matter, there exist quantum interference effects, whereas in classical matter, such as in the lattice of ions of a superconductor, decoherence arising from the environment destroys any such interference. As argued earlier (pp. 260 ff), the response of quantum fluids and of classical matter to these fields will therefore differ from each other.

In the case of superconductors, Cooper pairs of electrons possess a macroscopic phase coherence, which can lead to an Aharonov–Bohm-type interference absent in the ionic lattice. Similarly, in the quantum Hall fluid, the electrons will also possess macroscopic phase coherence (Girvin and MacDonald 1987; Zhang *et al.* 1989), which can lead to Berry-phase-type interference absent in the lattice. Furthermore, there exist ferromagnetic superfluids with intrinsic spin, in which an ionic lattice is completely absent, such as in spin-polarized atomic BECs (Cornell *et al.* 1998; Hall *et al.* 1998) and in superfluid helium 3 (Osheroff *et al.* 1972a, b).[6] In such ferromagnetic quantum fluids, there exists no ionic lattice to give rise to any classical response which could prevent a quantum response to tidal gravitational radiation fields. The Berry-phase-induced response of the ferromagnetic superfluid arises from the spin connection (see the above minimal-coupling rule, which can be generalized from an electron spin to a nuclear spin coupled to the curved spacetime associated with gravitational radiation), and leads to a purely quantum response to this radiation. The Berry phase induces time-varying macroscopic quantum flows in this ferromagnetic ODLRO system (Stern 1992, Lyanda-Geller and Goldbart 2000), which transports time-varying orientations of the nuclear magnetic moments. This ferromagnetic superfluid can therefore also in principle convert gravitational into electromagnetic radiation, and vice versa, in a manner similar to the case discussed above for the ferromagnetic quantum Hall fluid.

Thus we expect there to exist differences between classical and quantum fluids in their respective linear responses to weak external perturbations associated with gravitational radiation. Like superfluids, the quantum Hall fluid is an example of a quantum fluid which differs from a classical fluid in its current–current correlation function (Forster 1975) in the presence of GR waves. In particular, GR waves can induce a transition of the quantum Hall fluid out of its ground state *only* by exciting a quantized, collective excitation across the quantum Hall energy gap. This collective excitation would involve the correlated motions of a macroscopic number of electrons in this coherent quantum system. Hence the quantum Hall fluid is effectively incompressible and dissipationless, and is thus a good candidate for a quantum antenna.

There exist other situations in which a minimal-coupling rule similar to the one above, arises for *scalar* quantum fields in curved spacetime. DeWitt (1966) suggested such a coupling in the case of superconductors.[7] Speliotopoulos (1995) noted

[6] I thank Joel Moore for helpful discussions on ferromagnetic superfluids.

[7] I thank my graduate student Daniel Solli for pointing out to me the term $H_{int} \sim \mathbf{a} \cdot \mathbf{b}$ in the case of DeWitt's Hamiltonian.

that a cross-coupling term of the form $H_{\text{int}} \sim \mathbf{a} \cdot \mathbf{b}$ arose in the long-wavelength limit of a certain quantum Hamiltonian derived from the geodesic deviation equations of motion using the transverse-traceless gauge for GR waves.[8]

Speliotopoulos and I have been working on the problem of the coupling of a scalar quantum field to curved spacetime in a general laboratory frame, which avoids the use of the long-wavelength approximation (Speliotopoulos and Chiao 2003). In general relativity, there exists in general no global time coordinate that can apply throughout a large system, since for nonstationary metrics, such as those associated with gravitational radiation, the local time axis varies from place to place in the system. It is therefore necessary to set up operationally a general laboratory frame by which an observer can measure the motion of slowly moving test particles in the presence of weak, time-varying gravitational radiation fields.

For either a classical or quantum test particle, the result is that its mass m should enter into the Hamiltonian through the replacement of $\mathbf{p} - e\mathbf{A}$ by $\mathbf{p} - e\mathbf{A} - m\mathbf{N}$, where $\mathbf{N}$ is the small, local tidal velocity field induced by gravitational radiation on a test particle located at X_a relative to the observer at the origin (i.e., the center of mass) of this frame, where, for the small deviations h_{ab} of the metric from that of flat spacetime,

$$N_a = \frac{1}{2} \int_0^{X_a} \frac{\partial h_{ab}}{\partial t} dX^b. \tag{13.16}$$

Due to the quadrupolar nature of gravitational tidal fields, the velocity field $\mathbf{N}$ for a plane wave grows linearly in magnitude with the distance of the test particle from the center of mass, as seen by the observer located at the center of mass of the system. Therefore, in order to recover the standard result of classical GR that only *tidal* gravitational fields enter into the coupling of radiation and matter, one expects in general that a new characteristic length scale L corresponding to the typical size of the distance X_a separating the test particle from the observer, must enter into the determination of the coupling constant between radiation and matter. For example, L can be the typical size of the detection apparatus (e.g., the length of the arms of the Michelson interferometer used in LIGO), or of the transverse Gaussian wave packet size of the gravitational radiation, so that the coupling constant associated with the Feynman vertex for a graviton–particle interaction becomes proportional to the *extensive* quantity $\sqrt{GL}$, instead of an *intensive* quantity involving only $\sqrt{G}$.

For the case of superconductors, treating Cooper pairs of electrons as bosons, we would expect the above arguments would carry over with the charge e replaced by $2e$ and the mass m replaced by $2m$. For quantum fluids which possess an order

[8] I thank Achilles Speliotopoulos for many helpful discussions.

parameter Ψ obeying the Ginzburg–Landau equation, the above minimal-coupling rule suggests that this equation be generalized as follows:

$$\frac{1}{2m}\left(\frac{\hbar}{i}\nabla - \mathbf{a} - \mathbf{b}\right)^2 \Psi + \beta|\Psi|^2\Psi = -\alpha\Psi, \tag{13.17}$$

where $\mathbf{b} \propto \mathbf{N}$.

Quantum transducers between EM and GR waves?

Returning to the general problem of quantum fields embedded in curved spacetime, we recall that the ground state of a superconductor, which possesses spontaneous symmetry breaking, and therefore ODLRO, is very similar to that of the physical vacuum, which is believed also to possess spontanous symmetry breaking through the Higgs mechanism. In this sense, therefore, the vacuum is "superconducting." The question thus arises: how does a ground or "vacuum" state of a superconductor, and that of the other quantum fluids viewed as ground states of nonrelativistic quantum field theories with ODLRO, interact with dynamically changing spacetimes, in particular with a GR wave? I believe that this question needs both theoretical and experimental investigation.

In particular, motivated by the discussion in the previous section, I suspect that there might exist superconductors, viewed as quantum fluids, which are transducers between EM and GR waves based on the cross-coupling Hamiltonian $H_{\text{int}} \sim \mathbf{a} \cdot \mathbf{b}$. One possible geometry for an experiment is shown in Fig. 13.4. An EM wave impinges on the quantum fluid, which converts it into a GR wave in process (a). In the time-reversed process (b), a GR wave impinges on the quantum fluid, which converts it back into an EM wave. It is an open question at this point as to what the conversion efficiency of such quantum transducers will be.[9] This question is best settled by an experiment to measure this efficiency by means of a Hertz-type apparatus, in which process (a) is used for generating gravitational radiation, and process (b), inside a separate quantum transducer, is used to detect this radiation.

[9] One might expect a transducer power conversion efficiency of $Gm^2 \cdot 4\pi\varepsilon_0/e^2 = 2.4 \times 10^{-43}$ based on a naive classical picture in which each individual electron follows a deterministic, Newtonian trajectory. If this classical picture had been correct, there would have been no hope of actually observing this conversion process, based on the limited sensitivity of existing experimental techniques such as those described in Chiao and Fitelson (2003). However, superconductivity is fundamentally a quantum mechanical phenomenon. Due to the macroscopic coherence of the ground state with ODLRO, and the existence of a non-zero energy gap, there may exist quantum many-body enhancements to this classical conversion efficiency. In addition to these enhancements, there must exist additional enhancements due to the fact that the intensive coupling constant $\sqrt{G}$ of the Feynman graviton-matter vertex should be replaced by the extensive coupling constant $\sqrt{G}L$, in order to account correctly for the *tidal* nature of GR waves (Speliotopoulos and Chiao 2003). Furthermore, it is difficult to calculate the branching ratio for converting EM waves into GR waves versus into heat, since it is difficult to predict theoretically the rate of dissipation into heat Γ_{heat} in eqn (13.7) for quantum fluids. Hence an experiment is needed to measure this conversion efficiency.

If the quantum transducer conversion efficiency turns out to be high, this will lead to an avenue of research which could be called "gravity radio." I have performed a preliminary version of this Hertz-type experiment with Walt Fitelson using the high T_c superconductor yttrium barium copper oxide (YBCO) to measure its transducer efficiency at microwave frequencies. (Faraday cages were used to block EM couplings). We have obtained an upper limit on the conversion efficiency for YBCO at liquid nitrogen temperature of 1.6×10^{-5}. Details of this experiment will be reported elsewhere (Chiao and Fitelson 2003; Chiao *et al.* 2003).[10]

Conclusions

The conceptual tensions between QM and GR, the two main fields of interest of John Archibald Wheeler, could indeed lead to important experimental consequences, much like the conceptual tensions of the past. I have covered here in detail only one of these conceptual tensions, namely, the tension between the concept of *spatial nonseparability* of physical systems due to the notion of nonlocality embedded in the superposition principle, in particular, in the entangled states of QM, and the concept of *spatial separability* of all physical systems due to the notion of locality embedded in the equivalence principle in GR. This has led to the idea of antennas and transducers using quantum fluids as potentially practical devices, which could possibly open up a door for exciting discoveries. Quantum transducers, if sufficiently efficient, would allow us to directly observe for the first time the CMB (Cosmic Microwave Background) in GR radiation, which would tell us much about the very early universe.

Acknowledgments

I dedicate this paper to my teacher, John Archibald Wheeler, whose vision helped inspire this paper. I am grateful to the John Templeton Foundation for the invitation to contribute to this volume, and would like to thank my father-in-law, the late Yi-Fan Chiao, for his financial and moral support of this work. This work was supported also by the Office of Naval Research.

NOTE ADDED IN PROOF: I thank Joseph Orenstein and Andrew Mackenzie for pointing out to me that the direct spin–spacetime coupling mechanism described above for the quantum Hall fluid might also apply to spin–triplet superconductors, such as Sr_2RuO_4, and also to ferromagnetic superconductors, such as URhGe (Mackenzie and Maeno 2003; Aoki *et al.* 2001).

[10] I thank Sander Weinreb and Richard Packard for helpful comments.

References

Adak, M, Dereli, T, and Ryder, L H (2002) gr-qc/0205042.
Adunas, G Z, Rodriguez-Milla, E, and Ahluwalia, D V, (2001) *Gen. Rel. Grav.* **33**, 183.
Anandan, J (1981) *Phys. Rev. Lett.* **47**, 463.
(1984) *Phys. Rev. Lett.* **52**, 401.
(1985) *Phys. Lett.* **110A**, 446.
Anandan, J and Chiao, R Y (1982) *Gen. Rel. Grav.* **14**, 515.
Aoki D, Huxley A, Ressouche E, *et al.* (2001) *Nature* **413**, 613.
Berry, M V (1984) *Proc. Roy. Soc. Lond.* **A392**, 45.
Birrell, N D and Davies, P C W (1982) *Quantum Fields in Curved Space*. Cambridge: Cambridge University Press.
Boyd, R W and Gauthier, D J (2002) *Prog. Optics* **43**, 497.
Braginsky, V B Gorodetsky, M L, and Ilchenko, V S (1989) *Phys. Lett.* **137**, 393.
Chiao, R Y (1982) *Phys. Rev.* **B25**, 1655.
(2001) In *Quantum Mechanics: Scientific Perspectives on Divine Action*, ed. R. J. Russell, P. Clayton, K. Wegter-McNelly, and J. Polkinghorne, p. 17. Vatican City State: Vatican Observatory Publications.
(2003a) In *Proc. XXII Int. Solvay Conf. Physics*, ed. I. Antoniou, V. A. Sadovnichy, and H. Walther, p. 287. Singapore: World Scientific Press.
(2003b) gr-qc/0208024.
Chiao, R Y and Fitelson, W J (2003) In *Proc. "Time and Matter" Conf.* Venice 2002. gr-qc/0303089.
Chiao, R Y, Fitelson, W J, and Speliotopoulos, A D (2003) gr-qc/0304026.
Chiao, R Y and Garrison, J C (1999) *Found. Phys.* **29**, 553.
Chiao, R Y and Jordan, T F (1988) *Phys. Lett.* **A132**, 77.
Chiao, R Y and Kwiat, P G (2002) *Fortschritte Phys.* **50**, 5. quant-ph/0201036.
Chiao, R Y, Ropers, C, Solli, D, *et al.* (2004) In *Coherence and Quantum Optics*, vol. 8, ed. N.P. Bigelow, J.H. Eberly, C.R. Stroud, and I.A. Walmsley p. 109. New York: Plenum Press.
Chiao, R Y and Speliotopoulos, A D (2003) gr-qc/0304027.
Chiao, R Y and Steinberg, A M (1997) *Prog. Optics* **37**, 347.
Chiao, R Y and Wu, Y S (1986) *Phys. Rev. Lett.* **57**, 933.
Cornell, E A, Hall, D S, and Wieman, C E, (1998) cond-mat/9808105.
DeWitt, B S (1966) *Phys. Rev. Lett.* **16**, 1092.
Einstein, A (1971) *The Born–Einstein Letters.* English translation by Irene Born. New York: Walker.
Forster, D (1975) *Hydrodynamic Fluctuations, Broken Symmetry, and Correlation Functions*. New York: Addison Wesley.
Gingrich, R M and Adami, C (2002) *Phys. Rev. Lett.* **89**, 270402.
Girvin, S M and MacDonald, A H (1987) *Phys. Rev. Lett.* **58**, 1252.
Hall, D S, Matthews, M R, Wieman, C E, *et al.* (1998) *Phys. Rev. Lett.* **81**, 1543.
Hartings, J M, Cheung, J L, and Chang, R K (1998) *Appl. Opt.* **37**, 3306.
Herdegen, A and Wawrzycki, J (2002) *Phys. Rev.* **D66**, 044007.
Laughlin, R B (1983) *Phys. Rev. Lett.* **50**, 1395.
London, F (1964) *Superfluids*, vols. 1 and 2. New York: Dover.
Lyanda-Geller, Y and Goldbart, P M (2000) *Phys. Rev.* **A61**, 043609.
Mackenzie, A P and Maeno, Y (2003) *Rev. Mod. Phys.* **75**, 657.
Misner, C W, Thorne, K S, and Wheeler, J A (1973) *Gravitation*. San Francisco, CA: W. H. Freeman.

Nambu, Y and Jona-Lasino, G (1961a) *Phys. Rev.* **122**, 345.
(1961b) *Phys. Rev.* **124**, 246.
Nimtz, G and Heitmann, W (1997) *Prog. Qu. Electr.* **21**, 81.
Osheroff, D D, Gully, W J, Richardson, R C, *et al.* (1972b) *Phys. Rev. Lett.* **29**, 920.
Osheroff, D D, Richardson, R C, and Lee, D M (1972a) *Phys. Rev. Lett.* **28**, 885.
Peng, H, and Torr, D G (1990) *Gen. Rel. Grav.* **22**, 53.
Peng, H, Torr, D G, Hu, E K, *et al.* (1991) *Phys. Rev.* **B43**, 2700.
Peres, A (1995) *Quantum Theory: Concepts and Methods*. Dordrecht: Kluwer.
Peres, A and Scudo, P F (2002) *J. Mod. Opt.* **49**, 1235.
Peres, A Scudo, P F, and Terno, D R (2002) *Phys. Rev. Lett.* **88**, 230402.
Peres, A and Terno, D R (2003) quant-ph/0212023.
Speliotopoulos, A D (1995) *Phys. Rev.* **D51**, 1701.
Speliotopoulos, A D and Chiao, R Y (2003) gr-qc/0302045 (to be published in *Phys. Rev. D.*).
Sridhar, R (1997) *Lectures on Some Surface Phenomena in Superfluid Helium*. Madras: Institute of Mathematical Sciences.
Stefanov, A, Zbinden, H, Gisin, N, *et al.* (2002) *Phys. Rev. Lett.* **88**, 120404.
Steinberg, A M, and Chiao, R Y (1995) *Phys. Rev.* **A51**, 3525.
Steinberg, A M, Kwiat, P G, and Chiao, R Y (1993) *Phys. Rev. Lett.* **71**, 708.
Stern, A (1992) *Phys. Rev. Lett.* **68**, 1022.
Taylor, J H and Weisberg, J M (1982) *Astrophys. J.* **253**, 908.
Tomita A and Chiao, R Y (1986) *Phys. Rev. Lett.* **57**, 937.
Treussart, F, Ilchenko, V S, Roch, J-F, *et al.* (1998) *Eur. Phys. J.* **D1**, 235.
Tyson, J A and Giffard, G P (1978) *Ann. Rev. Astron. Astrophys.* **16**, 521.
Viola, L, and Onofrio, R (1997) *Phys. Rev.* **D55**, 455.
Weinberg, S (1972) *Gravitation and Cosmology: Principles and Applications of the General Theory of Relativity*. New York: John Wiley.
(1986) *Prog. Theor. Phys. (Suppl.)* **86**, 43.
Yang, C N (1962) *Rev. Mod. Phys.* **34**, 694.
Zhang, S C, Hansson, T H, and Kivelson, S (1989) *Phys. Rev. Lett.* **62**, 82.

14

Breeding nonlocal Schrödinger cats: a thought-experiment to explore the quantum–classical boundary

Serge Haroche

Collège de France and Ecole Normale Supérieure, Paris

Introduction: about quanta, atoms, photons, and cats

Experiments which manipulate and study isolated quantum systems have come of age. We can now trap single atoms or photons in a box, entangle them together, observe directly their quantum jumps, and realize in this way some of the thought-experiments imagined by the founding fathers of quantum physics. Schrödinger, who believed that observing an atom so to speak *in vivo* would remain forever impossible (Schrödinger 1952), would have been amazed, could he have seen what experimenters now achieve by manipulating atoms with lasers. These experiments are not just textbook illustrations of quantum concepts. They are considered by many as first steps towards harnessing the quantum world and realizing classically impossible tasks A quantum computer, for instance, would be a machine using quantum interference effects at a macroscopic scale in order to perform massive parallelism in computation (Nielsen and Chuang 2000). It would achieve an exponential speed-up to solve some problems such as the factoring of large numbers (Shor 1994). Such a machine would manipulate large ensembles of "quantum bits" made of atoms, molecules, or photons. Each bit would evolve in a superposition of two states labeled as "0" and "1". These bits would be entangled together by quantum gates exploiting electromagnetic interactions between them. The behavior of this machine would be strange and counterintuitive. It would be a system made of thousands of two-level particles following during the calculation a huge number of different routes among which it remains coherently suspended. The formidable enemy to defeat in order to build such a device is decoherence, which tends to destroy with a remarkable efficiency large quantum superpositions, transforming them into mundane classical mixtures of states (Zurek 1991). Experimenters are given the daunting task to isolate quantum bits from their environment and to correct

Science and Ultimate Reality, eds. J. D. Barrow, P. C. W. Davies and C. L. Harper, Jr. Published by Cambridge University Press.

efficiently the effects of decoherence in complex entangled systems. Whether they will succeed to build a practical quantum computer is debatable, to say the least, but it is clear that, by probing into quantum superpositions of increasing complexity, we will learn more about the quantum.

There are two ways a quantum superposition can be considered as "macroscopic." If two particles are in an entangled state, separated by a large distance, quantum effects manifest themselves – in a sense – at a macroscopic scale. What is done to one particle is immediately correlated to what happens to the other, even if they are very far apart, in a way that cannot be described in classical terms. This is the famous *nonlocality* aspect of quantum physics, first discussed by Einstein, Podolsky, and Rosen (Einstein *et al.* 1935), then by Bohm (1951) and Bell (1964), and tested over the last 30 years in beautiful experiments involving "twin photons" (Clauser and Friedman 1972; Aspect *et al.* 1982; Ghosh and Mandel 1987, Shih and Alley 1988; Rarity and Tapster 1990; Zeilinger 1998). A quantum superposition can also be considered as macroscopic, in a different sense, if it is made of a large number of particles or quanta. Such a situation is usually referred to as a "Schrödinger cat," since it recalls the fate of the mythical feline that Schrödinger had imagined to be suspended in a superposition of "dead" and "alive" states (Schrödinger 1935a, b, c). A quantum computer would be, in a sense, a Schrödinger cat tamed to compute faster than classical machines.

The experimental investigation of these states has developed very fast over the last few years. In quantum optics, Schrödinger-cat-like states (or rather "Schrödinger kitten" since they involve so far only a few quanta) have been realized with photons in cavities (Brune *et al.* 1996) and ions in traps (Monroe *et al.* 1996). In my research group at l'Ecole Normale Supérieure in Paris, we have produced photonic "cats" in which a field of several photons is suspended in a superposition of two states corresponding to different classical phases (Brune *et al.* 1996; Haroche 1998). The coherent nature of the superpositions has been probed by detecting the interferences produced by the separation and subsequent recombination of the Schrödinger cat parts. The gradual disappearance of these interferences, signaling the onset of decoherence, has also been observed. Recently, the discovery of the Bose–Einstein condensation of very cold atoms (Anderson *et al.* 1995; Davis *et al.* 1995) and the very fast development of spectacular experiments in this domain have given a new impetus to the Schrödinger cat breeding industry. Bose–Einstein condensates (BEC) are ensembles of identical "bosonic" atoms all in the same quantum state, exhibiting strong matter–wave features. The collective behavior of atoms in BEC is very similar to the one of identical photons in a laser beam. Many recent proposals (Cirac *et al.* 1998; Gordon and Savage 1999; Dalvit *et al.* 2000; Montina and Arecchi 2002) envision the preparation of "cat states" in which a collection of such atoms would be in a coherent superposition of two distinct matter waves, each containing a large number of particles. There are clear indications that such states (at least their few particle "kitten" version) can be generated with present technology (Greiner *et al.* 2002).

The experiments in quantum optics usually start with isolated atoms, or small numbers of them, and try to build large objects by adding progressively more and more particles to the system. Schrödinger cats are so to speak built "from the bottom up." Solid-state physicists are following the opposite route. Starting from bulk systems, they try to miniaturize them and get to the quantum regime "from the top down" (Leggett 1987). A very promising domain of "mesoscopic physics" is developing, in which one tries to generate "Schrödinger cat" states made of superconducting electron pairs in small circuits (Friedman *et al.* 2000; van der Wal *et al.* 2000). The number of electrons involved in these systems is in the million to the billion range, much larger than the numbers of photons or atoms in quantum optics experiments, but the coherence of these superpositions, an essential feature of the "Schrödinger cat" systems, has not been demonstrated so far.

We can now even consider states which possess the two kinds of macroscopic features I have just listed (spatial extension *and* large number of particles). One can think of Schrödinger cat states which contain many particles *and* are delocalized at two different points in space. In each location, the system could also differ by some other parameter, such as its energy, momentum, or polarization. It would be – so to speak – a cat simultaneously *dead* in one box and *alive* in another, combining two kinds of weirdness at once. Intuitively, we have the feeling that such systems are possible if they are made of at most a few photons or atoms, but that they will become increasingly difficult to build if we try to make them with large ensembles of particles. As John Archibald Wheeler pointed out in his recent autobiography (Wheeler 1998), the cloud of probability attached to a single photon or a single atom can take two routes at once, but the cloud of a baseball never behaves in that strange way. This raises directly the issue about the "quantum–classical boundary." At which scale do the quantum features vanish and why? How does the classical world where interferences are banished emerge from the underlying quantum laws? These issues are deeply connected with the famous question raised by Wheeler about existence: *what does it mean "to be"*? Clearly the meaning of life and death would be quite special for a cat coherently suspended between the two! I will not attempt to discuss here in general terms these fundamental questions which are not fully resolved yet and which are addressed in other chapters of this book dealing with the interpretation of the quantum theory. Part of the answer lies in the theory of decoherence (Giulini *et al.* 1996), which I will however evoke at the end of this chapter.

My goal here is more modest. I will describe a realistic thought-experiment for preparing and probing "delocalized" systems of N particles in two boxes. The system

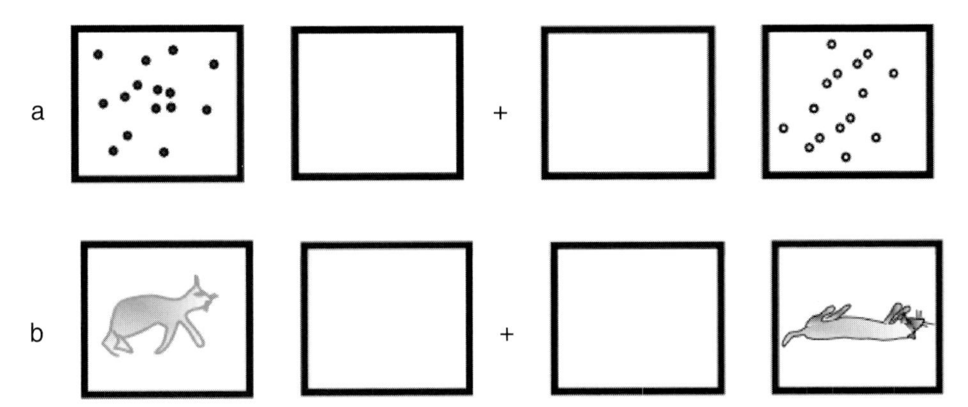

Figure 14.1. (a) Schematic representation of the quantum superposition of N particles delocalized in two separate boxes C_1 and C_2. The left and the right parts of the drawing describe the two classical situations between which the system is "suspended." The particles may have different "attributes" in the two boxes, (e.g., different energies) symbolized by different black and white colors. The + sign between the two parts of the figure means that the two classical states can "interfere" under appropriate experimental conditions. (b) Comparison with the situation of a "Schrödinger cat" suspended between a "live" state in a box and a "dead" state in the other.

is made of identical photons and is manipulated by single atoms used to prepare the coherent superposition and to probe it. This experiment belongs to a field of quantum optics known as cavity quantum electrodynamics (Haroche 1992; Berman 1994). The physics involved – the laws of light–matter interaction – are simple and well understood from first principles. The system's evolution can be completely calculated, which is not the case of more complex situations, in solid-state physics for example. Most importantly for my purpose, the physics behind the preparation of these cat states can be simply understood in intuitive terms, requiring only a basic knowledge of quantum physics. The emergence of Schrödinger cats, and the processes by which one can monitor their decoherence, can thus be discussed in a rather transparent way. This is not so easy for other proposals, in BEC or mesoscopic physics, which exploit subtle quantum effects more difficult to explain by hand-waving arguments.

The situation I will consider is illustrated (as far as a classical picture can do it) in Fig. 14.1a. Two cavities C_1 and C_2 can store identical photons. The system is suspended in a superposition of the two states where all the particles are either in C_1 or in C_2. The left and right parts of the figure represent the two classical components of this superposition. As we will see later, the + sign between the two parts means that a probability amplitude is attached to each part of this quantum alternative and that these amplitudes can give rise to interferences. All the particles in a cavity are in the same quantum state, but this state can differ by some physical attribute from one cavity to the other. Each photon in C_2 can for instance have an energy slightly different from the ones in C_1. The situation thus depicted, although of course much simpler, bears indeed some analogy with a "cat" simultaneously "alive" in one box and "dead" in the other, as shown in Fig. 14.1b.

By considering a "realistic" system, I will also be able to estimate the upper limit to the number of photons one can pack in such a delocalized quantum superposition. A theoretically oriented mind might be tempted to say that this number is limited by mundane technical difficulties and that increasing N to arbitrary values is "only" a matter of hard work and money (a not so subtle argument to ask for unlimited funding in an experiment proposal!). Such a view is naive and overoptimistic. Technical limitations have a fundamental origin. The "noise" in an experiment is, deep down, a quantum phenomenon which cannot be suppressed altogether. The dimensions of atoms and cavities and the wavelengths of photons, ultimately depend on the size of Nature's fundamental fine structure constant which cannot be tuned arbitrarily as mathematicians would do. Even a thought-experiment must take these physical restrictions into account. I will show that in my cat experiment they limit N to about 1000. How general is this limit? Are other kinds of cats made of different particles subjected to similar constraints? Can we go beyond it by actively correcting for the effects of decoherence? What are the implications for quantum computing? Many questions remain to be solved, which I will evoke at the end of this chapter. In any case, I hope that actual experiments of the kind I am describing here will soon investigate the fascinating properties of photonic or atomic delocalized "cats," helping us answering some of the deep questions that Wheeler has asked about the quantum.

Can an ordinary beam-splitter produce a Schrödinger cat? Single- versus many-particle interferences

The separation of fields between two spatially distinct modes is a general feature of optical interferometers. Beam-splitters are usually employed, which are made of dielectric layers of transparent material obeying the rules of ordinary linear optics (their "response" to an incident field is proportional to this field). Figure 14.2a shows how a beam-splitter B can be used to distribute photons between two cavities C_1 and C_2. A photon impinging on B is, with equal probabilities, transmitted into C_1 or reflected into C_2. Instead of storing the photon in a cavity, one can alternatively, as shown in Fig. 14.2b, let it propagate freely and reflect it on two mirrors (M, M'), before recombining its two routes (1 and 2) with a second beam-splitter B'. One realizes in this way a Mach–Zehnder interferometer. The probability to detect finally the photon in either one of the output channels of B' varies periodically when

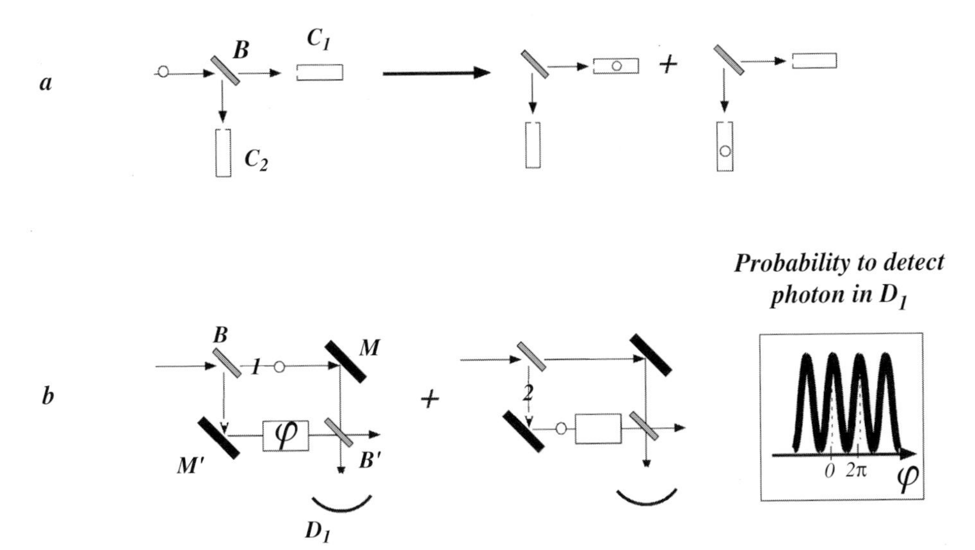

Figure 14.2. (a) A normal beam-splitter B channels with 50% probability an incident photon in either one of its two output modes. Using it to feed two cavities would result in a delocalized trapped photon. (b) The combination of two such beam splitters B and B' with folding mirrors M, M' constitutes a Mach–Zehnder interferometer. The probability to detect the photon in one output mode of B' exhibits interference fringes when the phase φ between the interferometer arms is swept.

the phase difference between the two routes is modified with a phase-shifter device introduced into one of the interferometer arms.

Experiments of this kind are usually performed by sending many photons through the apparatus and accumulating statistical data from successive photon detections. Although large ensembles of particles are involved, the photon state one would prepare by sending many photons in the set-up of Fig. 14.2a is fundamentally different from the one depicted in Fig. 14.1. Each impinging photon is dispatched by B with equal probabilities in channels 1 and 2, but different photons are randomly channeled, resulting in a binomial probability law for the global photon number distribution in the beam-splitter's two output modes. If one tried to use such a device to fill two cavities with a stream of impinging photons, one would get the situation shown in Fig. 14.3. The two cavities would be in a superposition of states containing each about $N/2$ photons with fluctuations (on the order of $\sqrt{N}$) between the two cavities. These fluctuations are identical to the one observed between the number of molecules in two equal volumes of a gas at thermal equilibrium. Such a distribution is quite different from the macroscopic superposition described by Fig. 14.1, in which the number of particles in C_1 (and C_2) obeys a bimodal distribution sharply peaked at two values (0 and N).

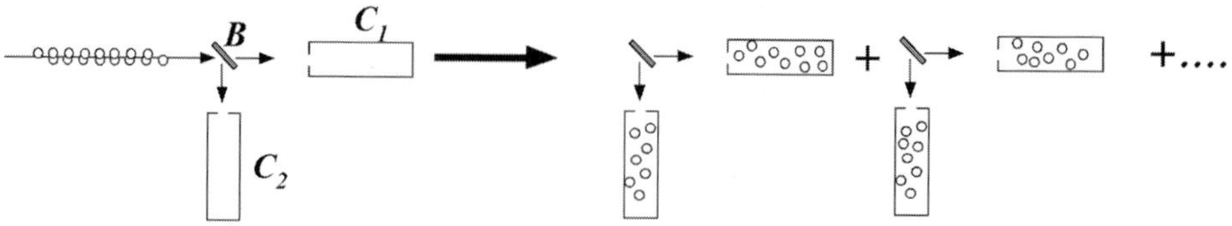

Figure 14.3. Successive photons impinging on a beam-splitter are channeled independently in the two output modes, resulting in a balanced partition of the photons, with only small photon number fluctuations. A beam-splitter coupled to the two cavities does not realize a "Schrödinger cat."

Similarly, the interferometer experiment of Fig. 14.2b reflects the existence of single particle superpositions, not macroscopic ones. Each photon propagating between B and B' is indeed in a superposition state, being at the same time transmitted and reflected by B, but this superposition is independent from one photon to the next (different photon states are not entangled). The ensemble of photons in the light beam can be described as a mere collection of single photons in this superposition state and the modulated signal observed when measuring the photon final detection probability results from the quantum interference occurring for the probability amplitude of each photon, independently from the others. Interferometers of the Mach–Zehnder kind using linear beam-splitters thus appear essentially as devices sensitive to single particle state superpositions and not to macroscopic superpositions of the kind depicted in Fig. 14.1. This has been forcefully expressed by Dirac when he stated that "photons *only* interfere with themselves".

For a long time this restrictive property was accepted as a general law of Nature. We now know that it is not general at all. In fact, the development of quantum optics over the last 40 years has been marked by the discovery, time and again, of subtle and amazing (because largely counterintuitive) interference effects in which more than one photon takes part (for a review of early experiments of this kind, see for example Greenberger *et al.* 1993). The only general rule imposed by quantum physics is that an interference occurs between quantum amplitudes associated to different "routes" followed by the system, provided the experimental apparatus does not permit one to "tell" which path has been taken. A large class of optical experiments involves interference between single photons following several routes, but some also involve two or more photons taking collectively different undistinguishable paths.

The observation of multi-photon interference generally requires an experimental set-up more complex than the simple Mach–Zehnder interferometer. In some cases a subtle combination of linear beam-splitters and detectors will work (Greenberger *et al.* 1993). In general, the breeding of Schrödinger cats requires ingredients of a different nature. Ingenious early proposals of cat state preparation have suggested to exploit the optical properties of a nonlinear optical medium (of the kind used to double the frequency of light or to mix light beams together). A slab made of such a medium, introduced on the path of a coherent light beam made of many photons could in principle "split" this beam into a superposition of two multi-photon coherent beams of opposite phases (Walls and Milburn 1985; Yurke and Stoler 1986). This situation, which has no classical counterpart, bears a strong analogy with the one I have sketched in Fig. 14.1 (although the macroscopic splitting occurs in phase space rather than in real space). In the same way as the beam has been split, it could be recombined, giving then rise to multi-photon interferences violating the Dirac rule. These clever proposals have not been implemented in the laboratory, due to technical difficulties. Other proposals have described a beam-splitting scheme involving two atoms that get entangled with the field (Brune *et al.* 1992; Davidovich *et al.* 1996). The features of entanglement and the fundamental properties of a quantum measurement performed on the atoms are then exploited in order to produce and detect the desired multi-particle field state superposition. This general method is the one we have used to generate and study the superpositions of field states with different phases I mentioned above. The thought experiment I discuss below to prepare nonlocal field superpositions also exploits the amazing properties of these atomic quantum beam splitters. Before describing the "realistic" experiment, I will start by discussing a simple model which will have the merit of introducing the basic ingredients of the method.

A quantum tap to prepare nonlocal Schrödinger cats

I wish to realize the situation shown in Fig. 14.4. Our two cavities are now connected to a large reservoir of particles S through a tap T which can send the particles either in C_1 or in C_2. We can (somewhat naively) think of T as made of a valve rotating in a cylinder pierced with three holes connecting it to S, C_1, and C_2. It has an "off" position when the valve is horizontal and two "on" positions corresponding to the filling of C_1 (tap in state T_1) or C_2 (tap in state T_2). Classically, we can store N particles in C_1, leaving C_2 empty, or put them in C_2, leaving C_1 empty, just by setting for some time t_0 the tap to the T_1 or T_2 positions. Note that classically, there is no intermediate position sending particles in both cavities at once, unless of course the tap is defective.

Let us consider now a quantum tap which could be genuinely in a linear superposition of the T_1 and T_2 states, dispatching *at the same time all the particles in one cavity and in the other*. Having the tap operating for time t_0 in this state, we would obtain a quantum superposition of the two classical situations representing a macroscopic system delocalized between two boxes separated by a macroscopic distance. The particles could be photons stored in two cavities (macroscopic superpositions

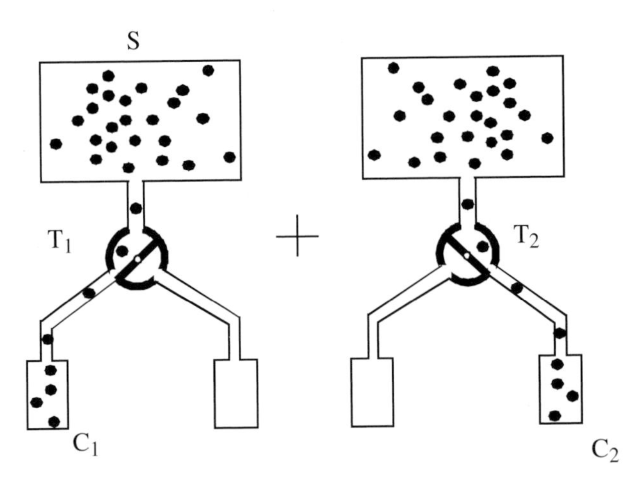

Figure 14.4. A quantum tap connecting a source S to two cavities C_1 and C_2 can be used to prepare a "Schrödinger cat" made of photons delocalized in the two cavities. The tap must be itself in a superposition of two operating states T_1 and T_2, filling respectively C_1 and C_2.

of electromagnetic field states) or atoms at a very low temperature, in a BEC delocalized between two atom traps (macroscopic superpositions of matter waves).

Without being too mathematical, a few equations will be useful to make us understand the physics. I will only assume here that the reader is familiar with the general features of quantum physics, such as the concept of state superpositions and the basic notions of measurement theory. I will find it handy to use the Dirac formalism for representing quantum states. I will write between brackets ($|\rangle$) the symbols representing the system's states of interest. Quite generally, the kind of quantum superposition prepared with our quantum tap can be written as:

$$|\Psi\rangle_{\text{cat}}^{N} = \left(\frac{1}{\sqrt{2}}\right)|N, 0\rangle \pm \left(\frac{1}{\sqrt{2}}\right)|0, N\rangle \tag{14.1}$$

where the first (second) symbol inside each bracket represents the number of field quanta (or particles) in C_1 (C_2) respectively. As we will see later, the sign ($\pm$) in eqn (14.1) corresponds to a quantum phase which depends upon the result of a final manipulation and measurement performed on the tap.

Equation (14.1) has a very rich and subtle meaning. Each state in the superposition is multiplied by its "amplitude," a number equal here to $\pm 1/\sqrt{2}$. The squares of these amplitudes represent the probabilities for finding all the particles in one box or the other, if we perform a measurement on the system. Here, these probabilities are equal. There is a 50% chance of finding all the particles in the left-hand box and the same chance of finding all of them in the right-hand one. Up to this point the situation can be understood classically. Suppose for instance you have been told that a fortune of N coins has been deposited in either one of two identical sealed coffers. Until you have opened them to check inside, your knowledge about the system would be exactly the same as the one I have described so far about the particles in C_1 and C_2.

The quantum situation is however much richer than the classical one because state superpositions can be manipulated and combined as waves in classical physics, using the linear equations of quantum theory. The amplitudes, which in eqn (14.1) are real numbers, usually become complex numbers when the system evolves in time and these c-numbers can be combined in various ways, leading to interference effects. As an example, suppose that the particles in C_2 are given, for a time interval τ, a small excess energy δE per particle, the energy being finally reset to its initial value. This can be done, for instance, by slightly deforming the C_2 cavity walls during time τ. The photons bouncing on the moving boundaries will experience a small Doppler effect which will change by a small amount their frequency and hence their energy. This happens of course only if the photons are in C_2. We have thus, during the time interval τ, the spooky situation I have evoked above, where the N particles are at the same time in the left-hand box with a given energy *and* in the right-hand one with a different energy. At the end of this operation, each photon in C_2 has experienced a phase shift $\varphi = \partial E\tau/\hbar$ where $\hbar$ is Planck's constant. The Schrödinger equation which rules the evolution of this quantum system tells us that the system evolves at time τ into the state:

$$|\Psi\rangle_{\Phi}^{N} = \left(\frac{1}{\sqrt{2}}\right)(|N,0\rangle \pm \exp(-iN\varphi)|0,N\rangle). \tag{14.2}$$

The two probability amplitudes of finding the field in the left-hand or right-hand box have thus acquired a phase difference $N\varphi$.

I have assumed so far that the quantum tap channels a well-defined number of particles from the source S to C_1 or C_2. Due to the wavy nature of the particles, however, there is an uncertainty about this number. As we will see later in a realistic situation, the tap, instead of transmitting exactly N photons, introduces in each cavity a "coherent" field described by a superposition of photon number states (Glauber 1963a, b):

$$|\alpha\rangle = \sum_N C_N(\alpha)|N\rangle \tag{14.3}$$

with:

$$C_N(\alpha) = \exp(-|\alpha|^2/2)\frac{\alpha^N}{\sqrt{N!}}. \tag{14.4}$$

This field is completely defined by its "classical amplitude" $\alpha = \sqrt{\bar{N}}$. The photon number probability distribution $P(N) = |C_N(\alpha)|^2$ obeys a Poisson law peaked around $\bar{N} = \alpha^2$, with a fluctuation $\Delta N = \alpha = \sqrt{\bar{N}}$ and a relative photon number dispersion $\Delta N/\bar{N} = 1/\sqrt{\bar{N}}$. Taking into account this fluctuation, the state prepared by the tap operation writes:

$$|\Psi\rangle^{\alpha}_{\text{cat}} = \left(\frac{1}{\sqrt{2}}\right)(|\alpha, 0\rangle \pm |0, \alpha\rangle)\,. \tag{14.5}$$

After the phase shift, each N amplitude in the C_2 field coherent state superposition acquires a phase $N\varphi$. As long as φ remains smaller than $1/\sqrt{\bar{N}}$ the phase dispersion can be neglected and, to a good approximation, the field evolves into the state:

$$|\Psi\rangle^{\alpha}_{\varphi} = \left(\frac{1}{\sqrt{2}}\right)(|\alpha, 0\rangle \pm \exp(-i\bar{N}\varphi)|0, \alpha\rangle). \tag{14.6}$$

At this stage, we have prepared the system in a well-defined "Schrödinger cat" state. The challenge is now to probe this superposition and to demonstrate that we have indeed a macroscopic quantum coherence in our system. This can be done, as I will show below, by opening the tap for a second time and letting the photons flow again in the cavities for an additional time t_0. This will lead to a quantum interference involving the two parts of the cat state prepared by the first tap operation.

Probing the Schrödinger cat: multi-particle interferences and collective de Broglie wavelength

Assume now that the tap is put again, for another time t_0, in the superposition of the T_1 and T_2 states. During this second operation, the coherent superposition of photon number states in S is supposed to have the same phase as during the first one. Let us see what happens to the two states of the superposition given by eqn (14.6). Starting from the $|\alpha, 0\rangle$ state, the open tap will either keep adding particles in C_1, or start filling C_2. In fact, the system will again evolve into a superposition of these two situations, corresponding to the transformation:

$$|\alpha, 0\rangle \rightarrow \left(\frac{1}{\sqrt{2}}\right)(|2\alpha, 0\rangle \pm |\alpha, \alpha\rangle)\,. \tag{14.7}$$

Note that when the tap adds a field into the first cavity, it doubles its amplitude and quadruples the photon number. This may appear strange if we think about photons in terms of independent classical particles being added in a box. In fact we are dealing here with gregarious bosons which "like" to accumulate in the same quantum state. The photons present in the box after the first tap operation "stimulate" the arrival of more photons when the tap is opened again, explaining why the final mean particle number is larger than $2\bar{N}$. Note also that here again, the relative phases of the two states in the superposition of eqn (14.7) depend on the result of a final manipulation and measurement of the tap, a point to which we will come back later. Starting from the $|0, \alpha\rangle$ state, the second tap operation would similarly induce the transformation:

$$|0, \alpha\rangle \rightarrow \left(\frac{1}{\sqrt{2}}\right)(|\alpha, \alpha\rangle \pm |0, 2\alpha\rangle)\,. \tag{14.8}$$

Since the preparation stage had in fact left the system in the superposition described by eqn (14.6), we can invoke the linearity of quantum physics to obtain, from eqns (14.6), (14.7), and (14.8) the system's final state as:

$$\begin{aligned}|\Psi\rangle_{\text{final}} \approx &\left(\frac{1}{A(\varphi)}\right)[(|2\alpha, 0\rangle + \varepsilon|\alpha, \alpha\rangle) \\ &+ \varepsilon' \exp(-i\bar{N}\varphi)(|\alpha, \alpha\rangle + \varepsilon|0, 2\alpha\rangle)]\end{aligned} \tag{14.9}$$

where $\varepsilon,\varepsilon' = \pm 1$ and $A(\varphi) \approx \sqrt{4 + 2\varepsilon\varepsilon' \cos(\bar{N}\varphi)}$ is a normalization constant insuring that the total probability of all possible outcomes of the system's final measurement is equal to 1.

There are obviously two "classical" ways to put photons in both cavities with successive tap operations: we can fill first C_1, then C_2 or vice versa. These two classical routes result in the same final state $|\alpha, \alpha\rangle$ which appears twice in the field expression, with two amplitudes having different phases. The probability $P_{\bar{N},\bar{N}}$ to find (on average) $\bar{N}$ particles in *both* C_1 and C_2 is the squared sum of these two amplitudes. It exhibits a quantum interference term sensitive to this phase difference. When $\varepsilon\varepsilon' = 1$, we find:

$$P_{\bar{N},\bar{N}} \approx \left(\frac{1}{A^2(\varphi)}\right)|1 + \exp(-i\bar{N}\varphi)|^2 = \frac{1 + \cos(\bar{N}\varphi)}{2 + \cos(\bar{N}\varphi)}. \tag{14.10}$$

This probability exhibits a modulation with 100% contrast when φ is varied. This modulation cannot be observed with coins in coffers. There is definitely something more to the superposition of states than the mere statistical uncertainty about where the particles are. Quantum phase differences between the states do matter and can be physically observed. Most strikingly, for $\varphi = \pi/\bar{N}, 3\pi/\bar{N}, \ldots,$ $P_{\bar{N},\bar{N}}$ vanishes altogether. It is then impossible to fill both cavities at once, in spite of the fact that the tap operations, taken separately, do fill both of them. This kind of negative interference is one of the weirdest manifestations of quantum physics.

The discussion so far has neglected the fluctuation of the photon number, an approximation legitimate only if $\varphi < 1/\sqrt{\bar{N}}$. For larger φ values, the phase dispersion becomes significant resulting in a blurring of the "side fringes" of the interference pattern. A complete calculation taking this phase dispersion into account is a bit more technical than the one we have just made. It leads to the following exact

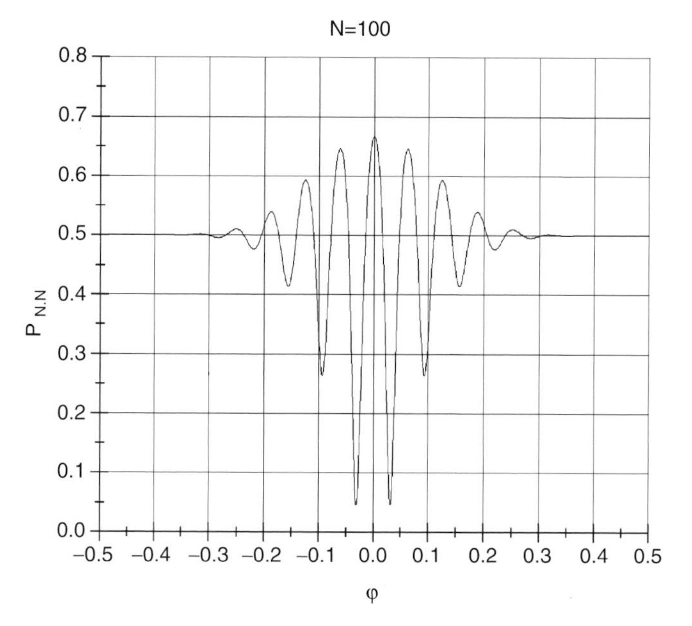

Figure 14.5. Multi-particle interference fringes signaling the transient existence of a Schrödinger cat "suspended" between the two cavities. The average photon number is $\bar{N} = 100$. The signal represents the probability to find both cavities filled with photons after a succession of two quantum tap operations, the first preparing the cat state and the second "reading" it out. A phase shift φ is applied to one of the cavities between the two quantum tap openings and the fringes are recorded as a function of φ. The fringe spacing $2\pi/\bar{N}$ is inversely proportional to the mean particle number. Only a few "central fringes" are visible, due to the fluctuation of the photon number.

expression for $P_{\bar{N},\bar{N}}$:

$$P_{\bar{N}\,\bar{N}} = \frac{1 + e^{-\bar{N}(1-\cos\varphi)}\cos(\bar{N}\sin\varphi)}{2 + e^{-\bar{N}(1-\cos\varphi)}\cos(\bar{N}\sin\varphi)}. \tag{14.11}$$

This probability is plotted in Fig.14.5 for $\bar{N} = 100$. We clearly see a central fringe, with a width $\pi/100$, flanked by smaller lateral fringes, the fringe contrast going to zero for $\varphi > 0.2$. The situation is similar to classical interference fringes observed with a broadband light containing a continuum of wavelengths. The dispersion of the fringe spacing, proportional to the wavelength, then washes out all but a few fringes around the central one. In the quantum situation considered here, the field is monochromatic, but the experiment is sensitive to a different kind of interference, with a fringe spacing inversely proportional to the particle number. Since this number fluctuates, only a few central fringes are visible.

The period $2\pi/\bar{N}$ of this quantum interference is a clear signature of its multi-particle character. It is because each tap opening channels *all the particles together* in one cavity *and* the other that the phase difference between the two interfering paths is equal to $\bar{N}\varphi$. The spacing of the "fringes" becomes smaller and smaller when $\bar{N}$ increases. By measuring this spacing, one could directly determine the mean number of particles in the state superposition.

One way of interpreting the interference pattern with its $2\pi/\bar{N}$ fringe spacing is to introduce for N photons of wavelength λ the notion of a collective wavelength λ/N (Jacobson *et al.* 1995). Collective de Broglie wavelengths of multi-particle systems have in a way already been measured in atomic (Pfau *et al.* 1994; Chapman *et al.* 1995) or molecular (Arndt *et al.* 1999) interferometers, where the fringe spacing is inversely proportional to the total mass M of the atoms or molecules involved. The existence of these $1/M$ de Broglie wavelengths just reflects the fact that all the components (nucleons, quarks, electrons) of these composite systems are collectively sent into one arm or the other of the interferometer by beam-splitters which do not split the atom or the molecule. The $1/\bar{N}$ collective wavelength considered here is quite different, though, because the photons which make up our macroscopic superposition are not bound together. We thus need a very special kind of quantum beam-splitter to channel all of them at the same time into one arm or the other of the interferometer, without splitting apart our unbound and thus very fragile composite system. Note that the collective de Broglie wavelength of a "two-photon" state has already been observed in a recent quantum optics experiment (Fonseca *et al.* 1999). Interference fringes with $1/N$ spacing (N up to 4) have also been recorded in quantum interference experiments with trapped ions (Sackett *et al.* 2000) Observing the very narrow fringes produced by a large cat remains a challenging goal.

A realistic Schrödinger cat prepared and read out by two atoms

The experiment described so far is of course unrealistic. A mechanical tap of the kind shown in Fig. 14.4 would be a macroscopic object made of a huge number of atoms. Putting it in a state superposition would be, by itself, preparing a large Schrödinger cat, thus taking for granted what we try to achieve. Instead of a macroscopic gate, we need in fact a truly microscopic tap which could be prepared in a superposition of two quantum states controlling the flow of photons in one cavity or the other. A single atom crossing the cavities successively makes it possible to realize such a device, by using basic properties of cavity quantum electrodynamics.

In fact, several variants of the experiment are possible. One of them has been described by my colleagues and myself a few years ago (Davidovich *et al.* 1993). It involved an atom acting as a small piece of dielectric medium with a tunable refractive index crossing the cavities. I will present here a somewhat simpler version, illustrated by the set-up sketched in Fig. 14.6. The two cavities are made of two

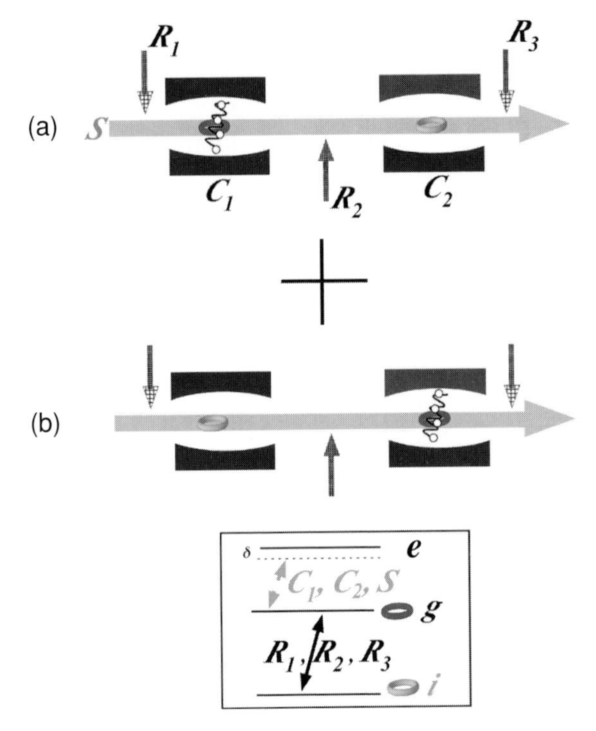

Figure 14.6. A "realistic" implementation of the quantum tap with a single atom in a cavity quantum electrodynamics experiment. The source mode propagates transversally across the cavities made of mirrors facing each other. The atom flies along the source beam and deflects some photons in either C_1 or C_2. Auxiliary pulses R_1, R_2, and R_3 are used to manipulate the internal states of the atom (whose relevant energy levels e, g, and i are represented in the inset). Figures 14.6a and 14.6b represent the two "routes" which the system can follow. In Fig. 14.6a it crosses C_1 in level g and C_2 in level i, thus filling only C_1. In Fig. 14.6b it crosses C_1 in level i and C_2 in level g, thus filling only C_2. Sending the atom in a superposition of g and i makes it follow the two routes at once and results in the preparation of a cat state of the field.

highly reflecting spherical mirrors facing each other. They can store for a time T_C photons bouncing vertically up and down before they get lost by absorption in the mirrors or by escaping on the side due to scattering on mirrors imperfections. A "source" field S, resonant with the cavities, propagates horizontally across the cavities at mid distance between the mirrors. This field is coherent and has an amplitude α_S. In a quantum description, it corresponds to a superposition of states with different photon numbers, according to a Poisson statistic. The source photons do not hit the mirrors and are thus normally uncoupled to the cavities.

A single atom flying along the direction of the source field crosses successively C_1 and C_2 and provides a mechanism for photon exchange between S and the cavities. Three atomic energy levels (shown in the insert of Fig. 14.6) play a role in the process. The source S as well as C_1 and C_2 are slightly off resonant with the atomic transition between levels g and e (with a small frequency mismatch δ). An atom in level g will thus be weakly excited by the source field S while it flies along with the S photons. A small electric dipole will develop on the atom, which will scatter some light in all directions, in the same way as a speck of dust scatters the light around from a laser beam. A third level i, well separated from g, is not coupled at all to the S, C_1, and C_2 fields by allowed transitions. An atom in this level remains thus completely insensitive to the source photons. In other words, a piece of "atomic dust" in level i remains totally invisible.

Auxiliary electromagnetic field pulses, propagating vertically, can be applied to the atom before C_1 (R_1 pulse), between the cavities (R_2 pulse), and after C_2 (R_3 pulse). These fields induce transitions between the levels g and i, allowing to exchange them or to mix them coherently in well-defined proportions. We can for example transform an atom scattering light into an invisible one (and vice versa) by applying a pulse exchanging levels g and i. This is technically called a π pulse, realizing the following transformations:

$$|g\rangle \rightarrow |i\rangle; \qquad |i\rangle \rightarrow -|g\rangle. \tag{14.12}$$

Between C_1 and C_2, such a π pulse will be applied on the atom in R_2. One can also, in R_1 and R_3, mix coherently with equal amplitudes the scattering state g and the invisible state i, according to the transformations that are called "$\pi/2$" pulses:

$$|g\rangle \rightarrow \left(\frac{1}{\sqrt{2}}\right)(|g\rangle + |i\rangle); \qquad |i\rangle \rightarrow \left(\frac{1}{\sqrt{2}}\right)(|g\rangle - |i\rangle). \tag{14.13}$$

Forgetting for the time being R_1 and R_3, let us assume that the atom enters C_1 in level g. While it crosses C_1 (which takes a time t_0) the tiny field it scatters gets coupled into the cavity mode (see Fig. 14.6a). The light re-emitted towards the mirrors gets reflected back and forth between them, undergoing a huge number of bounces before the atom leaves the cavity. Since the cavity is resonant with this light, all the partial waves produced by successive reflections interfere constructively, resulting in a very effective build-up of the field in the cavity. The atom thus plays during this time the role of a small flying mirror transferring the field from the source mode into the first cavity. The efficiency of this process can be easily estimated. It depends on three parameters, all expressed as frequencies. The first one, Ω_S, proportional to α_S, measures the strength of the atom coupling to the source mode S; the second frequency, Ω_C, measures the strength of the atom coupling to the cavity mode. It depends on the characteristics of the atomic electronic state and on the geometry of the cavity. We will assume that the product $\Omega_C T_C$ is much larger than 1. This expresses the condition of "strong coupling regime" in cavity quantum electrodynamics (Haroche 1992), whose importance will appear clearly below. Finally, the amplitude of the driven dipole, for a given source amplitude, is inversely proportional to the frequency mismatch δ between the atom and S (the closer the source frequency to the atomic one, the larger the induced dipole). To sum up, the amplitude α of the coherent field scattered into C is proportional to Ω_S, Ω_C, and to the time the atom stays in C. It is also inversely proportional to δ. It simply writes in dimensionless units:

$$\alpha = (\Omega_S \cdot \Omega_C/\delta)t_0. \tag{14.14}$$

This field is a superposition of photon number states, given by eqns (14.3) and (14.4). The average number of injected photons $\bar{N} = \alpha^2$ varies as t_0^2. This means that the photon number increases faster and faster when considering successive time intervals of the same duration. As discussed above, this can be seen as an expression of the bosonic character of photons, which "like" to accumulate in the same mode.

This simple analysis seems to imply that a field of arbitrarily large amplitude (and thus arbitrarily large $\bar{N}$) can be produced, by just having the atom stay a time t_0 long enough inside the cavity. This is deceptive for at least two reasons. First, t_0 cannot exceed T_C, the cavity photon damping time. Second, the validity of the above model requires that the atom must be driven gently enough, without saturating its electric dipole. This means that the ratio Ω_S/δ must be at most on the order of $1/10$. Combining these conditions leads to $\alpha < \Omega_C T_C/10$. We have already noticed that we must have $\Omega_C T_C \gg 1$ (strong atom cavity coupling regime). We now see that this condition is indeed required for a single atom to be able to channel several photons into C.

Let us now come back to the description of our atom evolution. When it leaves C_1, it is transferred by R_2 from level g to i so that, when crossing C_2 it no longer scatters light. The second cavity thus stays empty. As a result, the atom$+C_1+C_2$ system undergoes the global transformation $|g, 0, 0\rangle \to |i, \alpha, 0\rangle$ where the first, second, and third symbols refer to the three parts of the system. Similarly, if the atom is sent into C_1 in level i, this cavity stays empty and C_2 is filled, since the atom, switching from i to g between the cavities, acts as a reflecting mirror for C_2 only. The atom$+C_1+C_2$ system then undergoes the transformation $|i, 0, 0\rangle \to -|g, 0, \alpha\rangle$.

Assume now that we send through the apparatus an atom prepared in level g and that we activate the first auxiliary $\pi/2$ pulse R_1. The system will then follow two routes at once, one in which the atom enters C_1 in level g, the other in which it does so in level i. Then, by linearity of quantum mechanics, we expect the system to evolve into the state $(1/\sqrt{2})(|i, \alpha, 0\rangle - |g, 0, \alpha\rangle)$. The single atom has behaved as a multi-photon switch, and we have achieved our goal to channel all the photons in a collective superposition state.

In order to do this, we have paid a price, however, which is to produce entanglement not only between the cavities but also between the cavities and the atom. The atom+C_1+C_2 superposition cannot indeed be separated as the product of a field state by an independent atom's state. The entanglement of this combined global state entails strong quantum correlations between the atom and the field. If we are interested in observing the state of the field alone, its entanglement with the atom constitutes a cause of decoherence. Observing (really or virtually) the atom's state would indeed result in projecting the field into a well-defined component of the initial superposition (in C_1 if the atom is found in i, or in C_2 if the atom is found in g). The atom entangled with the field thus plays the role of a kind of "spy," potentially able to reveal the position of the field. We will come back to this important notion when we discuss decoherence in the next section.

In order to avoid this field localization, we will play a final trick by applying to the atom the $\pi/2$ pulse R_3 mixing again g and i after C_2. We thus "erase" the information about the field state carried by the atom, since a subsequent measurement on the atom cannot reveal any more what were the atom's states when it crossed the cavities. The final atom+field state then becomes:

$$|\Psi\rangle_{\text{final}} = \frac{1}{\sqrt{2}}\left[(|g\rangle - |i\rangle)\,|\alpha,\,0\rangle - (|g\rangle + |i\rangle)\,|0,\,\alpha\rangle\right]. \tag{14.15}$$

After the atom's detection, the field is left in one of the two states given by eqn (14.5). We find the field in the state:

$$|\Psi_+\rangle^{\alpha}_{\text{cat}} = \left(\frac{1}{\sqrt{2}}\right)(|\alpha,\,0\rangle + |0,\,\alpha\rangle) \tag{14.16}$$

if the atom has been found in level i and in the state

$$|\Psi_-\rangle^{\alpha}_{\text{cat}} = \left(\frac{1}{\sqrt{2}}\right)(|\alpha,0\rangle - |0,\alpha\rangle) \tag{14.17}$$

if the atom has been detected in level g.

The efficiency of the whole process relies on the existence of a strong coupling between a single atom and the field of a cavity. This condition is satisfied in the cavity quantum electrodynamics experiments we perform at Ecole Normale Supérieure, in which we manipulate Rydberg atoms interacting with microwave superconducting cavities (Raimond *et al.* 2001). A Rydberg state is prepared by promoting, with the help of laser beams and radio-frequency fields, the outer electron of an atom into an excited level having a very large spatial extension. Such an atom is very weakly bound and very sensitive to all kinds of electric and magnetic interactions. It must be kept isolated in a very dilute atomic beam and manipulated in a good vacuum at a very low temperature, to avoid the perturbation of thermal photons. The size of the atom can be tuned almost at will by choosing, in the preparation process, the outer electron energy. The closer this energy is to the atomic ionization, the larger the electron orbit.

Among all the Rydberg states, the "circular" ones, whose excited electron orbits on a circle around the atomic nucleus, are particularly well suited for cavity quantum electrodynamics experiments, because they are very stable, decaying only very slowly by spontaneous radiation. Another big asset of these atoms is their strong intrinsic coupling to microwaves. The large electron orbit behaves as a very sensitive antenna for fields resonant or nearly resonant with a transition between nearby Rydberg levels. The circular Rydberg states are simply labeled by the value of their principal quantum number n. The electron orbit radius r_a is equal to $n^2 a_0$ where $a_0 = 0.5\ 10^{-10}\ m$ is the atomic length unit. The transition frequency between nearby Rydberg levels scales as n^{-3}. Their coupling Ω_C to a resonant cavity scales as n^{-4} and their radiative life time varies as n^{-5}. For our atomic quantum tap experiment, we could employ the circular Rydberg atoms we are typically manipulating in our present cavity quantum electrodynamics studies. They correspond to $n \approx 50$ ($\boldsymbol{r}_a \approx 2500 a_0 = 1.25\ 10^{-7}\ m$). The two circular states with $n = 50$ and 51 would play the role of the g and e states in the scheme we have just described, while the $n = 49$ circular state would be the i level. The e–g transition frequency would then be 51 GHz (6 mm wavelength radiation).

The cavities we are using to store photons at this frequency are made of polished niobium mirrors. The intermirror distance is on the order of 3 cm and the transverse size w (waist) of the cavities is 6 mm. Two such cavities, placed side by side, have their center separated by a distance $D = 5$ cm. The typical damping time of the cavities in our present set-up is $T_C = 1$ ms, and their resonant coupling to a circular Rydberg atom on the $n = 50 \rightarrow n = 51$ transition is $\Omega_C = 3.10^5\ \mathrm{s}^{-1}$. The product $\Omega_C T_C = 300$ largely satisfies the strong coupling regime condition. By improving the cavity mirror technology, it is not unrealistic to aim for $T_C = 0.3$ s. Such values have already been achieved with closed cavities used in other kinds of Rydberg atom experiments (Raithel *et al.* 1994). Since we need to prepare the multi-photon "cat state" in a time much shorter than T_C, we could choose an atomic velocity v such that the travel time D/v between C_1 and C_2 is 3 ms ($v = 15\ \mathrm{ms}^{-1}$). We would then have $t_0 = 300\ \mu$s and, setting $\Omega_S/\delta = 1/10$, eqn (14.14) gives $\alpha = 10$ and $\bar{N} = 100$.

This nonlocal cat state could be probed by an interferometer experiment of the kind described above, giving the field in C_2 a phase shift equal to φ per photon, then subjecting the cavity fields to a second tap atom, finally performing a detection of the resulting field in both cavities. When $\varphi < 1/\sqrt{\bar{N}}$ this field is in the state given by eqn (14.9), where the sign (+1 or –1) of ε depends upon the level (i or g) in which the first atom has been detected, the sign of ε' depending similarly on the detection of the second atom. The probability to find photons *in both cavities* is given by eqn (14.11), when the two atoms are detected in the same quantum state ($\varepsilon\varepsilon' = +1$).

The $|\Psi_\pm\rangle_{\text{cat}}^{\alpha}$ state is one among the possible field superpositions one could produce using similar Rydberg atom microwave techniques. Other fields of the form $|\alpha, \beta\rangle \pm |\beta, \alpha\rangle$ could also be generated by simple variants. For instance, by adding a common coherent field $-\alpha/2$ to both cavities prepared in the $|\Psi_+\rangle_{\text{cat}}^{\alpha}$ state, one would obtain the state $|\alpha/2, -\alpha/2\rangle \pm |-\alpha/2, \alpha/2\rangle$. Also, the two field components should not necessarily belong to two distinct cavities. They could be within two modes of the same cavity. Experiments along these lines have recently been started in our laboratory (Rauschenbeutel *et al.* 2001). Note finally that the two-mode versions we are discussing here are strongly related to the localized cat states involving only one mode of the field, of the form $|\alpha\rangle \pm |\beta\rangle$, which I mentioned above (Brune *et al.* 1996; Haroche 1998).

Decoherence of the Schrödinger cat

Suppose we have been able to prepare the cat state described by eqn (14.16) or (14.17), with on the average $\bar{N}$ photons stored in C_1 or C_2. For how long will we be able to keep this state "alive" in a coherent superposition? An upper limit is obviously given by the finite life time T_C of the field in the cavities. But we will not in fact be able to keep our Schrödinger cat state for that long. Decoherence due to an uncontrollable entanglement of the system with its environment will reduce the lifetime of the quantum superposition to a value much shorter than T_C.

The field we are manipulating here is indeed inherently coupled to its surrounding. Since we have – realistically – assumed that the cavity damping time is finite, this means that we have implicitly introduced in our system a reservoir in which the cavity photons are damped. The exact nature of this reservoir does not really matter. We can, for example, suppose that each of our cavities is surrounded by a big "environment box" made of perfectly reflecting walls into which the photons lost by the cavity eventually escape, as shown in Fig. 14.7. In fact, this model is not too unrealistic since it is indeed the escape of photons scattered on the cavity mirror defects which up to now has limited the quality of our cavities.

Let us assume that such a scattering event introduces a photon in the reservoir box surrounding C_1 (as sketched in Fig. 14.6a). In principle, one could detect this photon (the way we would do this is not relevant; it is enough to know that such a detection is in principle possible). If such a detection were made, we would be sure that our initial system could not have been in the state $|0, \alpha\rangle$ since such a state, corresponding to an empty C_1, could not produce a photon in the box surrounding C_1. Hence, the information acquired by watching the environment would "force"

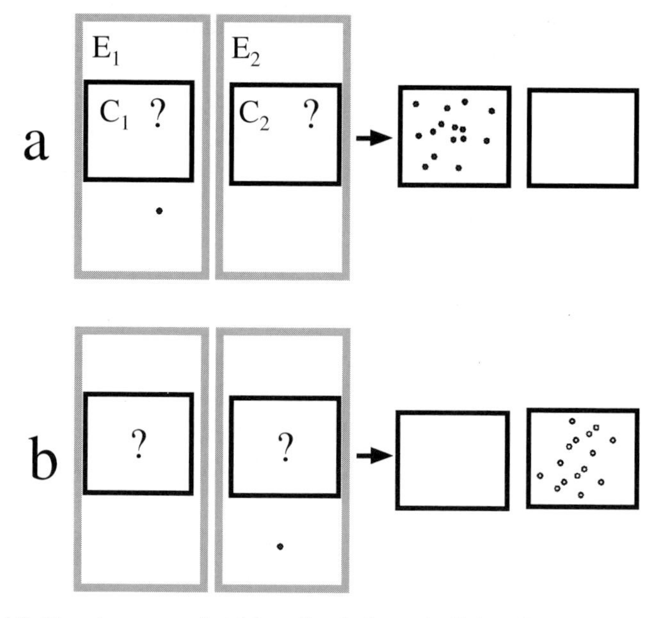

Figure 14.7. Decoherence of a delocalized photonic Schrödinger cat. Each cavity is coupled to its own "environment," represented by the big boxes E_1 and E_2 surrounding cavities C_1 and C_2. As soon as a photon escapes into one of these environment boxes, the "quantum ambiguity" of the cat state (symbolized by the question-marks on the cavity boxes) is lost. The field "collapses" in C_1 if a photon "appears" in E_1 (Fig. 14.7a), in C_2 if a photon appears in E_2 (Fig.14.7b).

the system to collapse in this case in the state $|\alpha, 0\rangle$. In the same way, a photon escaping into the box surrounding C_2 would force the system into the state $|0, \alpha\rangle$.

Of course, we do not look into the environment (the experiment would be too difficult, and moreover, it is not of practical interest). However, even without looking, we know for sure that, after some time, there will be at least one photon in one of the two boxes, and, that, were we looking at it, we would know in which cavity the field is. We know furthermore that there are equal probabilities to find this first photon in either one of the two boxes. This is equivalent to saying that the quantum coherence of the system has disappeared at the time the first photon has been emitted, all our information on the system being from now on that the field is with 50% probability either in the first or in the second box, a situation that has become similar to the two-coffer experiment described in the introduction. The quantum superposition is then transformed into a classical mixture, with a statistical uncertainty about the location of the field. The multi-particle interference signal of Fig. 14.5 is completely washed out.

How long will it take before we can be sure that at least a photon has been scattered out of the system? Since the photons escape independently from each other and since a large fraction of them has escaped after a time T_C, it is safe to assume that the first photon will have left the C_1–C_2 system after a time of the order $T_C/\bar{N}$. In the situation considered above ($T_C = 300$ ms and $\bar{N} = 100$) decoherence would thus occur within 3 ms. This simple qualitative analysis can be confirmed by a more sophisticated calculation of the system's evolution (Raimond *et al.* 1997).

We see, from this simple model, that decoherence becomes faster and faster when the number of particles in the system increases. This result is quite general when the particles are interacting independently from each other with the environment, as our photons scattered independently of each other out of the cavities. Increasing the system's size increases the number of ways it can decay and provides more and more information on the system path, making decoherence all the more efficient. The situation is different if we consider strongly bound composite systems such as a large molecule (Arndt *et al.* 1999). Their parts do not generally interact independently with the environment and decoherence does not appear simply related to the number of particles involved.

Conclusion: how big a Schrödinger cat?

The size of multi-photon state superpositions appears to be limited by the finite damping time of the cavity field. Does this time have an upper bound? The reflectivity of superconducting metals for microwaves tends in principle towards 1 when the temperature decreases to absolute zero, so that it is possible to envision cavities made of nearly perfect ultra-cold mirrors having very long damping times (perhaps on the order of seconds). Does this mean that we could – in principle – build much larger cat states? In fact, in the experiment I have described above, very large T_C values will not be of any help, since the circular Rydberg atoms we are using have a finite lifetime which eventually sets the effective decoherence time of the system. As noted above, the atoms indeed radiate very weakly spontaneous microwave photons in all directions and their life time, $T_a = 30$ ms for the $n = 50$ circular Rydberg state, fixes an upper bound for T_0 approximately equal to $T_a/30 = 1$ ms. The atom must indeed enter into C_1, travel from C_1 to C_2, and be finally detected, which takes a total time of the order 30 t_0, before decaying. Even for infinite cavity damping time, this condition restricts α to being smaller than $\Omega_C T_a/300$ and $\bar{N}$ smaller than 10^{-5} $(\Omega_C \mathrm{T}_a)^2$. With the values of Ω_C and T_a considered above, we find $\bar{N} \leq 1000$.

To increase $\bar{N}$ beyond this value, we could be tempted to choose as a quantum tap a Rydberg atom of larger size, with a longer lifetime (remember it scales as n^{-5}). Such an atom would however have a smaller coupling to the cavity, making the cat's state preparation longer. In order to find out whether we win or lose by changing the atom's size, we need a more precise estimate of the $(\Omega_C T_a)^2$ product. It can in fact be expressed simply in terms of three dimensionless parameters: the fine structure constant $\alpha_{fs} = 1/137$, the Rydberg atom's size in atomic units $l_a = r_a/a_0$, and the cavity intermirror separation in wavelength units l_C:

$$(\Omega_C \cdot T_a)^2 \approx \frac{l_a}{l_C^2} \alpha_{fs}^{-3}. \tag{14.18}$$

Using the smallest possible cavity ($l_C \approx 10$) and Rydberg atoms with a radius $l_a \approx 2.10^3$, we find $\Omega_C T_a$ on the order of 10^4, which is the situation we have considered above. Increasing l_a beyond this size would allow us in principle to prepare cats with more photons, but the situation would rapidly become unrealistic. Dreaming about much larger $\Omega_C T_a$ would mean considering huge atoms tremendously sensitive to all kinds of perturbations and requiring a set-up of unreasonable size. Increasing N by just one order of magnitude, to about 10 000, would for instance mean using a ten times larger atom (with a radius of more than 1 μm), whose coupling to the cavity would be 100 times smaller. The size of the cavity would be 30 times larger, on the order of 1m between mirrors. The preparation time of the cat state would be on the order of 1 second and the cavities should have at least a 3-hour damping time in order for decoherence not to occur during the 1-second preparation time of the cat's state. Just quoting these figures clearly convinces us that the maximum number of particles one can, in practice, put in such a state is on the order of a few thousand, an upper bound ultimately linked to the finite – and not so large – value of $\alpha^{-1}{}_{\text{fs}}$.

It might well be that using unstable Rydberg atoms to build our multi-photon cat state was not such a good idea after all, since its size is ultimately limited by the finite value of T_a. Could we change the method and use as a quantum tap a stable atom in its electronic ground state, whose lifetime is in principle infinite? Ground state atoms interact most strongly with visible photons, so that we should then store short wavelength radiation in two small optical cavities. Such cavities do exist and the field of optical cavity quantum electrodynamics is indeed as active as its microwave counterpart (Münstermann *et al.* 1999; Hood *et al.* 2000). What would be the limit then? Let us again assume that we could design small micron-size cavities of arbitrarily large quality factor (not an easy task!). Would we then have potentially larger cats? In fact, the radiative lifetime issue will again limit us. Exciting an atomic dipole on a transition linking a ground state g to an electronically excited e, even if it is done gently, populates with a small probability the atom in the upper level e. From that level, the atom can radiate by spontaneous emission a photon escaping from the cavity on the side. If a single spontaneous photon is emitted, it induces decoherence since it "tells" which cavity the atom is filling. As in the Rydberg atom microwave experiment, the rate at which the atom is coupled to the cavities is simply related to the rate at which it radiates such spontaneous photons. The ratio of these two rates is expressed by a formula which involves again the value of $\alpha^{-1}{}_{\mathrm{fs}}$. From the finite value of this ratio, it is easy to deduce an upper bound for $\bar{N}$, on the order of a few hundred.

As mentioned above, other schemes can be imagined to build various kinds of multi-particle cat states using cavity quantum electrodynamics, or other quantum optics techniques. One could also think of building "massive" cat states made of bosonic atoms at very low temperature. All the schemes one might think of, in order to be realistic, need to be based on a careful analysis of the system's preparation, which always involves some kind of electromagnetic interaction between particles. The same basic interaction plays also an essential role in the cat's state decoherence. We must *in fine* compare the rate at which the cat state can be prepared using "good interactions" and the rate at which it will lose its coherence due to "bad interactions." In the case of BEC cats, for instance, the preparation of the macroscopic superposition relies on the existence of elastic collisions between the atoms, which provide a nonlinear mechanism to couple the matter waves together. The decoherence processes are primarily due to other kinds of inelastic collisions expelling the atoms from the condensate. It is not possible to increase at will the rate of "good elastic collisions" without affecting the rate of the "bad inelastic" ones. The ratio of these two rates will always be finite and will in turn restrict the size of the cat states one could build and observe. Presently, all "realistic" proposals in BEC physics speak of cats made of hundreds to thousands of particles, not millions or billions. Is the similarity between these realistic photonic and atomic cat sizes just fortuitous or is it due to some fundamental argument?

The above discussion can be rephrased in similar terms when analyzing the feasibility of a quantum computer. As I have already mentioned, the evolution of such a machine would involve the operation of many gates coupling together atomic or photonic bits via electromagnetic interactions. The result would be the emergence in the machine of a kind of Schrödinger cat. In order to be coherent, the whole process, including all the gate operations, should take place before decoherence sets in. One might naively think that this is just a matter of operating the gates fast enough. The problem is that the rate at which the gates operate is not independent of the rate at which the bits are coupled to the environment, also via electromagnetic interactions. In a simple model of a quantum computer using trapped ions as qubits, for instance, the estimation of these rates is relatively easy to make (Plenio and Knight 1996). Comparing them introduces a dimensionless ratio involving some not so high power of $\alpha^{-1}{}_{\mathrm{fs}}$ (Haroche and Raimond 1996). The finite value of this ratio imposes an upper limit to the number of possible gate operations, far below what is required for useful applications.

Is this the end of the game? Just watching the flurry of activity in the field of quantum information processing suggests to the contrary. In fact, I have discussed above situations where quantum systems are protected only passively from decoherence, by just trying to couple them as weakly as possible to their environment. We have seen that there is a limit to what we can do in this way. There is another possible strategy. It consists in watching the system as it gets coupled to its environment and trying to correct the effects of decoherence by an active back-action process on the bits. This is called quantum error correction in the jargon of quantum information. This idea is adapted from classical computer science, where error correction of spurious bit flips is an essential ingredient. Theorists in quantum information have shown that such an active strategy should make it possible to operate a quantum computer realizing an arbitrary number of operation, provided that the fidelity of each gate is close enough to 100% (Steane 1999).

What are the implications for Schrödinger cats? Could we use quantum error correction to maintain their coherence in a system whose size would be beyond the limits we have found above? Some ingenious schemes have already been proposed which should allow us to lengthen somewhat the decoherence time of multi-photon superpositions, using set-ups similar to the one described in this chapter (Fortunato *et al.* 1999). They consist in continuously measuring some atomic or field observable and using the result of this measurement to modify the cat state with the help of additional atoms interacting successively with the cavity fields. To quote a colleague of mine in this cat-taming business, we would give "quantum food" to the cat to keep it in its "healthy" quantum superposition. How far will we be able to go along this route? The question remains open. I don't know whether huge cats will eventually be bred, but I have no doubt that we will find surprises and interesting applications along the way.

References

Anderson, M H, Ensher, J R, Matthews, M R, *et al.* (1995) *Science* **269**, 198.
Arndt, M, Nairz, O, Vos-Andreae, J, *et al.* (1999) *Nature* **401**, 680.
Aspect, A, Dalibard, J, Roger, G (1982) *Phys. Rev. Lett.* **47**, 1804.
Bell, J S (1964) *Physics* **1**, 195.
Berman, P (ed.) (1994) *Cavity Quantum Electrodynamics*. Boston, MA: Academic Press.
Bohm, D (1951) *Quantum Theory*. New York: Prentice Hall.
Brune, M, Haroche, S, Raimond, J M, *et al.* (1992) *Phys. Rev.* **A45**, 5193.
Brune, M, Hagley, E, Dreyer, J, *et al.* (1996) *Phys. Rev. Lett.* **77**, 4887.
Chapman, M S, Hammond, T D, Lenef, A, *et al.* (1995) *Phys. Rev. Lett.* **75**, 3783.
Cirac, J I, Lewenstein, M, Molmer, K, *et al.* (1998) *Phys. Rev.* **A57**, 1208.
Clauser, J, and Friedman, S (1972) *Phys. Rev. Lett.* **28**, 472.
Dalvit, D A R, Dziarmaga, J, and Zurek, W (2000) *Phys. Rev.* **A62**, 013607.
Davidovich, L, Maali, A, Brune, M, *et al.* (1993) *Phys. Rev. Lett.* **71**, 2360.
Davidovich, L, Brune, M, Raimond, J M, *et al.* (1996) *Phys. Rev.* **A53**, 1295.
Davis, K B, Mewes, M O, Andrews, M R, *et al.* (1995) *Phys. Rev. Lett.* **75**, 3969.
Einstein, A, Podolsky, B, and Rosen, N (1935) *Phys. Rev.* **47**, 777.
Fonseca, E J S, Monken, C H, and Padua, S (1999) *Phys. Rev. Lett.* **82**, 2868.

Fortunato, M, Raimond, J M, Tombesi, P, *et al.* (1999) *Phys. Rev.* **A60**, 1687.
Friedman, J R, Patel, V, Chen, W, *et al.* (2000) *Nature* **406**, 43.
Ghosh, R and Mandel, L (1987) *Phys. Rev. Lett.* **59**, 1903.
Giulini, D, Joos, E, Kiefer, C, *et al.* (1996) *Decoherence and the Appearance of a Classical World in Quantum Theory*. Berlin: Springer-Verlag.
Glauber, R (1963a) *Phys. Rev.* **130**, 2529.
(1963b) *Phys. Rev.* **131**, 2766.
Gordon, D and Savage, C M (1999) *Phys. Rev.* **A59**, 4623.
Greenberger, D M, Horne, M A, and Zeilinger, A (1993) *Phys. Today* **Aug.**, 22.
Greiner, M, Mandel, O, Hanschand, T W, *et al.* (2002) *Nature* **419**, 51.
Haroche, S (1992) In *Fundamental Systems in Quantum Optics*, ed. J. Dalibard, J.-M. Raimond and J. Zinn-Justin, p. 767. Amsterdam: North Holland.
Haroche, S (1998) *Phys. Today* **July**, 36.
Haroche, S and Raimond, J-M (1996) *Phys. Today* **Aug.**, 51.
Hood, C J, Lynn, T W, Doherty, A C, *et al.* (2000) *Science* **287**, 1447.
Jacobson, J, Björk, G, Chuang, I (1995) *Phys. Rev. Lett.* **74**, 4835.
Leggett, A J (1987) In *Chance and Matter*, ed. J. Souletie, J. Vannimenus and R. Stora, p. 395. Amsterdam: North Holland.
Monroe, C, Meekhof, D M, King, B E (1996) *Science* **272**, 1131.
Montina, A and Arecchi, F T (2002) *Phys. Rev.* **A66**, 013605.
Münstermann, P, Fisher, T, Maunz, P, *et al.* (1999) *Phys. Rev. Lett.* **82**, 3791.
Nielsen, M A, and Chuang, I (2000) *Quantum Computation and Quantum Information*. Cambridge: Cambridge University Press.
Pfau, T, Spälter, S, Kurtsiefer, C, *et al.* (1994) *Phys. Rev. Lett.* **73**, 1223.
Plenio, M B, and Knight, P L (1996) *Phys. Rev.* **A53**, 2986.
Raimond, J M, Brune, M, and Haroche, S (1997) *Phys. Rev. Lett.* **79**, 1964.
(2001) *Rev. Mod. Phys.* **73**, 565.
Raithel, G, Wagner, C, Walther, H, *et al.* (1994) In *Cavity Quantum Electrodynamics*, ed. P. Berman, p. 57. Boston, MA: Academic Press.
Rarity, J G and Tapster, P R (1990) *Phys. Rev. Lett.* **64**, 2495.
Rauschenbeutel, A, Bertet, P, Osnaghi, S, *et al.* (2001) *Phys. Rev.* **A64**, 050301(R).
Sackett, C A, Kielpinski, D, King, B E, *et al.* (2000) *Nature* **404**, 256.
Schrödinger, E (1935a) *Naturwiss.* **23**, 807.
(1935b) *Naturwiss.* **23**, 823.
(1935c) *Naturwiss.* **23**, 844.
(1952) *Br. J. Philos. Sci.*, **3**.
Shih, Y H and Alley, C O (1988) *Phys. Rev. Lett.* **61**, 2921.
Shor, P W (1994) In *Proc. 35th Ann. Symp. Foundations of Computer Science*, ed. S. Goldwasser, p. 124.
Steane, A (1999) *Nature* **399**, 124.
Wal, C H van der, Haar, A C J ter, Wilhelm, F K, *et al.* (2000) *Science* **290**, 773.
Walls, D F and Milburn, G J (1985) *Phys. Rev.* **A31**, 2403.
Wheeler, J A (1998) (with K. Ford) *Geons, Black Holes, and Quantum Foam: A Life in Physics*. New York: W. W. Norton.
Yurke, B and Stoler, D (1986) *Phys. Rev. Lett.* **57**, 13.
Zeilinger, A (1998) *Rev. Mod. Phys.* **71**, S288.
Zurek, W (1991) *Phys. Today*, **44**, 36.

15

Quantum erasing the nature of reality: or, perhaps, the reality of nature?

Paul G. Kwiat
University of Illinois–Urbana

Berthold-Georg Englert
National University of Singapore

Introduction

WHY must I treat the measuring device classically? What will happen to me if I don't?!
Eugene Wigner

The quantum measurement problem – how does an apparently "classical" definite world arise out of the random world of quantum superpositions – was and continues to be one of the fundamental philosophical issues in quantum mechanics. What we mean by a classical world, for example, is one in which macroscopic objects are not in superposition states of being simultaneously in several locations at once, and cats are never in coherent superpositions of being alive and dead. This lack of coherence is actually a *loss* of the coherence that exists at the level of the isolated quanta, but somehow does not survive the transition to the classical level of measuring apparatus. Such incoherent states are known as mixed states. Therefore, to study the quantum–classical interface, or even to investigate whether such an interface exists at all other than in the minds of "classical sympathizers,"[1] one should look carefully at mixed states, how they arise and how they behave. Here we describe a set of experiments, both real and *gedanken*, investigating the subtleties of quantum interference when mixed states are involved. We start by describing the well-known double-slit experiment, and the loss of interference when which-path information can be had. Next we discuss the principle of the "quantum eraser," by which one can make measurements on the which-path detector in such a way as to remove the path-information; by selecting only a particular set of results one can recover the interference fringes. We will see that this is true even if there was no which-path

[1] (At least) one of us secretly belongs to that camp!

Science and Ultimate Reality, eds. J. D. Barrow, P. C. W. Davies and C. L. Harper, Jr. Published by Cambridge University Press.

information available at the outset because the which-path quantum system was in a mixed state. Because the notion of mixture is strongly correlated to the idea of entanglement with some unobserved system, we describe (pp. 315–20) an analogous experiment performed using entangled photon pairs. We can then show clearly the relationship between the two central quantum mechanical "mysteries" (to quote Feynman): interference and entanglement. We then describe a variety of methods to produced mixed states. Although formally these have the same outcome, physically the underlying interpretations are quite different, and may stimulate some insight into the nature of reality, and the reality of nature. Final thoughts and remarks are gathered in the summary section.

Quantum eraser for pure and mixed states

Interference is at the heart of quantum mechanics. In reality it contains the *only* mystery. We cannot make the mystery go away by 'explaining' how it works.

(*Richard Feynman*)

The complementary nature of wavelike and particlelike behavior is frequently interpreted as follows: as a consequence of the uncertainty principle, any attempt to measure the position (particle aspect) of a quantum leads to an uncontrollable, irreversible disturbance in its momentum, thereby washing out any interference pattern (wave aspect) (Bohr 1983; Feynman *et al.* 1965). This picture is incomplete though; no "state reduction," or "collapse," is necessary to destroy interference, and measurements which do not involve reduction can be reversible. One must view the loss of coherence as arising from an *entanglement* of the system with the which-way marker (WWM), which may be another degree of freedom or another quantum system entirely. Previously interfering alternatives can thereby become distinguishable, such that no interference is observed. Interference may be regained, however, if one manages to "erase" the distinguishing information. This is the physical content of *quantum erasure* (QE) (Scully and Drühl 1982; Hillery and Scully 1983; Scully *et al.* 1991) (though as we shall see below, this simple physical picture cannot explain the results when nonpure WWM states are considered).

The double slit

As the simplest possible example, consider the archetypal double-slit experiment as indicated in Fig. 15.1. A photon is directed from a (relatively) far away source to a wall with two slits, behind which sits (again, at a relatively large spacing) an observation screen, consisting of some sort of classical mechanism for indicating the arrival point of the single photon. As is very well known, the likelihood for the

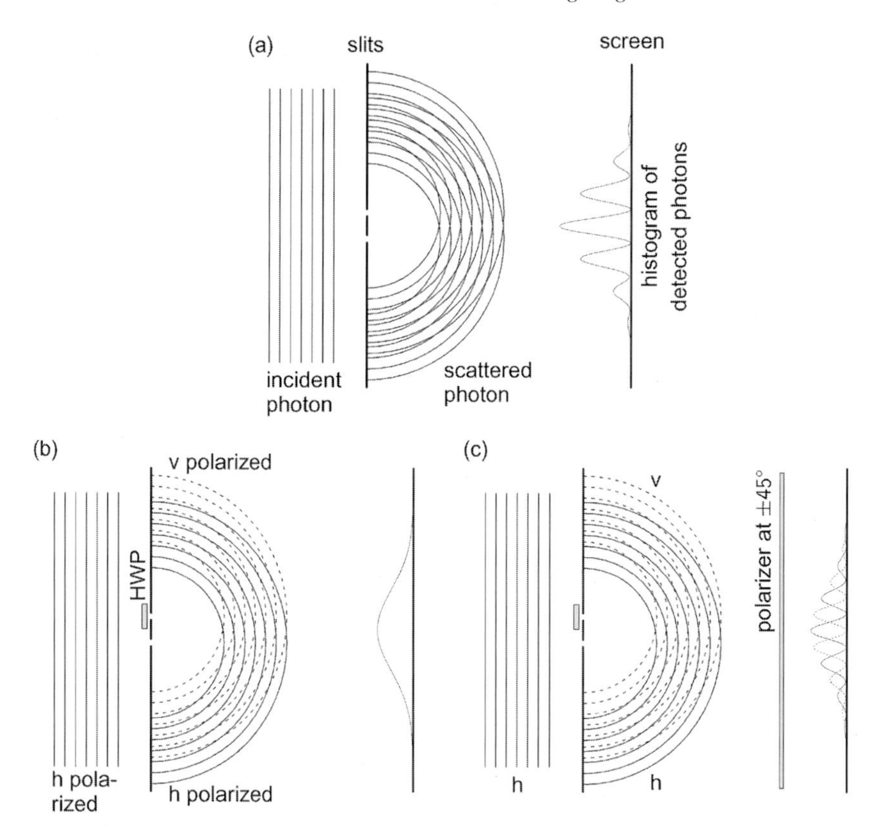

Figure 15.1. The "mysterious" double-slit experiment. (a) Photons are directed at a pair of slits. They arrive with well-defined wavelength and are scattered at the two slits. At the instant under consideration, one photon is approaching the slits, and its immediate predecessor has just been scattered but has not reached the screen as yet. The histogram built up by photons that have been detected on that far screen already shows that they are observed to be more likely to arrive in some places, and completely unlikely to arrive at others. This interference results from the indistinguishability – the in-principle unknowability – of which slit a photon traversed. (b) A half-wave plate (HWP) in front of the top slit rotates the initial horizontal polarization (h) to vertical (v). Accordingly, the scattered spherical wave centered at the top slit is v polarized (dashed half-circles), whereas the one from the bottom slit carries the original h polarization (solid half-circles). In this situation, no fringes are observable on the screen – the which-path information stored in the photon polarization distinguishes the top-slit/bottom-slit processes. (c) The distinguishing which-path information may be erased by analyzing the polarization in a particular way. For example, if we place a polarizer at 45° in front of the screen, we will recover the original fringe pattern of (a), only reduced in intensity by half (solid curve). If instead we use a −45° polarizer, we will again see fringes, but now shifted in phase by 180° (dashed curve). The sum of the two histograms in (c) equals the fringeless pattern of (b). Likewise, we could set the polarizer such that only v (or only h) polarized photons reach the screen, and they would all have passed through the top slit (or all through the bottom slit, respectively).

photon to arrive at any given point on the screen is determined by the coherent sum of the probability amplitudes for the photon to have traveled through the top slit or the bottom slit. Although it is not necessary for the experiment just described, we will now assume that the incident photon is horizontally polarized, i.e., in the polarization state $|\mathbf{h}\rangle$. Next we use a (very small) half-wave plate (HWP) to rotate the polarization of the photon passing through the upper slit by 90° to $|\mathbf{v}\rangle$, the state of vertical polarization. Again, it is well established that no interference will be observed at the screen (Fig. 15.1b). Of course, according to classical physics, the loss of interference occurs because the electric fields $\vec{E}_{\text{top}} = \vec{E}_{\text{v}}$ and $\vec{E}_{\text{bottom}} = \vec{E}_{\text{h}}$ corresponding to the electromagnetic wave passing through the top and bottom slits, respectively, are orthogonal; therefore, the interference cross term, which is proportional to $\vec{E}_{\text{top}} \cdot \vec{E}_{\text{bottom}}$, is zero.

A more modern and more useful perspective is that the process of passing through the double slits with the wave plate covering one of them results in an entanglement of the path degree of freedom with the polarization degree of freedom of the photon. The polarization acts as the WWM here. The interference is determined by the reduced density matrix ρ_{red} that one obtains by tracing over the quantum state of the WWM, since it is not being measured. The off-diagonal elements of ρ_{red} are proportional to the overlap of the two WWM quantum states associated with the top and bottom slits. If these WWM states are orthogonal, as in our example above, then the off-diagonal elements in ρ_{red} will vanish, and no interference will be observed. Note that it is absolutely not necessary to actually measure the polarization; the mere fact that such which-way information in principle *could* be obtained is enough to destroy the interference.

The interference may be recovered, however, if we can find some way to "erase" the distinguishing information being carried in the WWM quantum state. In this example, we can perform this erasing operation by simply passing the photons through a final analyzer at ±45° before the screen (Fig. 15.1c). This effectively removes the distinguishability of the two paths – any photon which is transmitted through an analyzer at 45° is equally likely to have originally been v-polarized (upper slit) or h-polarized (lower slit). This is an example of QE. Note that the presence of the polarizer at 45° merely serves to partition the final data into two subsets: those photons which were transmitted through the polarizer, and those which were absorbed by it. If, instead, we have the polarizer at −45°, 100%-visibility fringes will again be observed, but now they will be shifted 180° with respect to the fringes obtained with the 45° polarizer. Note that we could have instead decided to partition the ensemble of data in such a way that our polarization measurements yielded directly the information of which path the photon chose.

The primary lesson is that one must consider the total physical state, including any WWM with which the interfering quantum has become entangled. If the coherence of the WWM is maintained, then interference may be recovered by effectively postselecting particular subensembles. In general there are two distinct possibilities – either one can measure the WWM in such a way as to recover the which-way (WW) information, a particlelike characteristic; or one can measure the WWM in such a way as to recover interference, a wavelike characteristic. The familiar phrase "each experiment must be described either in terms of particles or in terms of waves" emphasizes the extreme cases and disregards the intermediate situations, in which particle aspects and wave aspects are present simultaneously.[2] Remarkably, if the WWM is a separate system from the interfering system, one can even make the choice of how to measure the WWM *after* the other quantum has been detected, as we discuss below. This is an extension of the original delayed-choice experiment popularized by Wheeler (1979).[3] In all cases though, it is necessary to correlate the results of the measurements. Only after this postselection do the fringes reappear.[4]

Set-up and procedure

In contrast to many interference situations where the WW information may be inaccessible, for our measurements the WW labels are easily manipulated: we used a photon in a simple Mach–Zehnder interferometer as the interfering system, and its polarization as the WWM (see Fig. 15.2) (Kwiat *et al.* 1999a; Schwindt *et al.* 1999).[5] By changing the polarization in one path of the interferometer, we entangle the spatial mode with the polarization and so partially or completely "label" the path followed by the photon. This is enough to make the paths partially or completely distinguishable, resulting in a reduced visibility *even if the polarization is not measured.*

Photons at 670 nm from the output of a single-mode fiber were directed into our Mach–Zehnder interferometer. The entire interferometer was compressed from the usual rectangular configuration (i.e., the angle of incidence on the beam-splitter was set to 10°) in order to minimize polarization variations in the reflection and

[2] Nature allows compromises between the particle aspects and the wave aspects of a quantum object. In the context of two-path interferometers, there are various inequalities that quantify these compromises. These matters are reviewed in Englert and Bergou (2000) where the interested reader will also find extended remarks on the history of the subject, in which the theoretical developments of Wootters and Zurek (1979), Glauber (1986), Greenberger and Yasin (1988), Mandel (1991), Jaeger *et al.* (1995), and Englert (1996) and the experiments reported in Rauch and Summhammer (1984), Summhammer *et al.* (1987), Mittelstaedt *et al.* (1987), Dürr *et al.* (1998a, b, c), and Schwindt *et al.* (1999) play a central role.

[3] The first discussion of delayed choice was given in von Weizsäcker (1941). The first experiments on Wheeler's conception appeared in Hellmuth *et al.* (1987), Alley *et al.* (1987), and Baldzuhn *et al.* (1989).

[4] It is precisely the requirement of correlating the results – necessarily requiring the transmission of classical information at or below the speed of light – which prevents this procedure from enabling superluminal communication.

[5] A similar experiment was carried out much earlier by Baldzuhn and Martienssen (1991) – they used a fast electrooptic switch to decide between measuring fringes or which-path information. However, they focused exclusively on pure input states.

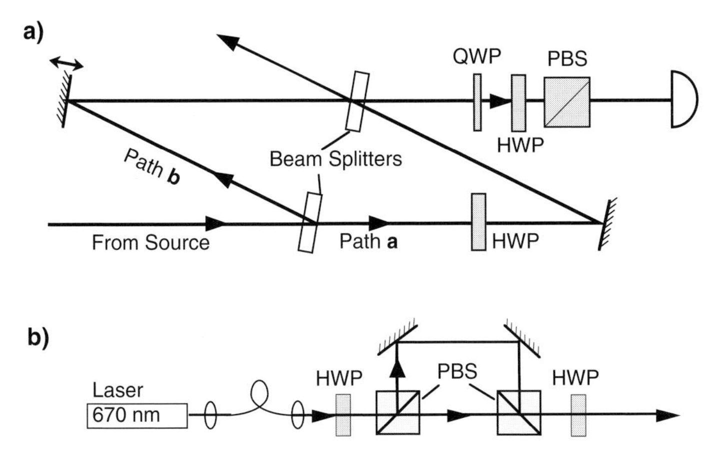

Figure 15.2. Experimental setup to investigate the phenomenon of quantum erasure (QE). (a) The half-wave plate (HWP) in path a of the interferometer generates varying amounts of which-way information, depending on the input polarization state. The quarter-wave plate (QWP), HWP, and Rochon prism (polarizing beam-splitter PBS) in the output port allow analysis in an arbitrary polarization basis, for determining which-way information or to enable QE of it. (b) One method to prepare mixed and partially mixed quantum states. By varying the input polarization to the polarizing Mach–Zehnder interferometer (which is unbalanced by more than the coherence length of the diode laser), one can make an arbitrarily mixed polarization state.

transmission amplitudes. The interferometer arms were adjusted to have equal lengths after the desired polarization-transforming elements were inserted into one or both arms. For the results presented below, an adjustable HWP in path a of the interferometer was used to vary the amount of polarization "labeling" of the photons.

Our analysis system consisted of a Rochon calcite prism preceded by adjustable half- and quarter-wave plates, allowing the polarization of the photons to be measured in any arbitrary polarization basis. The photons were detected using a Geiger-mode avalanche photodiode – a Single Photon Counting Module (EG&G #SPCM-AQ), with detection efficiency ~40%. The input source, described below, was greatly attenuated so that the maximum detection rates were always less than 50 000 s^{-1}; for the interferometer passage time of 1 ns, this means that on average there were fewer than 10^{-4} photons in the interferometer at any time. This one-photon-at-a-time operation is essential to allow sensible discussion of the likely path taken by an individual light quantum.[6] For visibility measurements the

[6] It is, therefore, not true that all features of an experiment of this kind can be understood in terms of classical physics, which is a misplaced objection that has been raised occasionally, most recently by Trifonov *et al.* (2002). They, and others, perhaps have in mind a "semi-classical" description of the phenomenon, consisting of a classical treatment of the electromagnetic field and a quantum treatment of matter (to account for the "clicking" of detectors that indicates the arrival of a single light quantum). It is indeed well known that single-photon interference patterns are identical with the classical intensity patterns (Fermi 1929). But it is equally well known

maximum and minimum count rates at the detector were measured as the length of path b was adjusted using a piezoelectric transducer on the mirror. After subtracting out the separately measured detector background (i.e., the count rate when the input to the interferometer was blocked, typically 100–400 s^{-1}), the fringe visibility is calculated in the standard manner:

$$V = (\text{Max} - \text{Min})/(\text{Max} + \text{Min}).$$

In the limiting case that the paths are completely distinguishable, no interference is observed and the visibility vanishes, $V = 0$. This occurs, for example, when the HWP in path a of the interferometer is used to rotate the polarization in that path by 90°. In fact, with the HWP in place the visibility will be zero *even when the input is in a mixed state of polarization*. This is less intuitive, since there is no WW information to distinguish the paths, but can be understood by examining the behavior of orthogonal *pure* states, with no definite phase relationship between them. In the basis where the HWP rotates the states by 90°, the orthogonal polarizations from paths a and b cannot interfere; in the basis aligned with the HWP's axes, each polarization individually interferes, but the interference patterns are shifted relatively by 180° (due to the birefringence of the HWP), so the sum is a fringeless constant.

Now that we have made interference disappear, we may consider methods to recover it. If we have initially a pure state of the polarization (or, more generally, of the quantum system comprising the WWM), the lack of interference may be directly attributed to the WW information stored in the polarization. By making a suitable analysis of the WWM we can read out this information, e.g., by detecting photons in a state that is orthogonal to the polarization associated with path a, we can know with certainty that they must have traveled via path b, and vice versa. Or we can analyze the polarization so that a detected photon is equally likely to have come from either path in the interferometer. There is actually a whole class of measurements that achieve this: if the polarizations from two paths are θ_{a} and θ_{b}, then analyzing in any basis of the form $(|\theta_{\mathrm{a}}\rangle + e^{i\phi}|\theta_{\mathrm{b}}\rangle)/\sqrt{2}$ will suffice to recover complete interference, i.e., $V = 1$. For example, if the polarizations from the paths a and b are horizontal (h) and vertical (v), respectively, analysis at +45° will recover complete fringes, while analysis at −45° will recover complete *anti*-fringes (i.e., $V = 1$, but shifted by 180°).

that such a semi-classical description has its limitations and is not generally valid, as has been demonstrated, e.g., in Clauser (1974) and Grangier *et al.* (1986). Light really consists of photons, whether or not a given experiment directly proves that fact. Put most simply, the notion of taking this path or that path through an interferometer is utterly meaningless in Maxwellian electromagnetism where there is always intensity in both paths.

In this context it should be mentioned that this erroneous reasoning would also imply that quantum cryptography with faint pulses is somehow completely classical. There are, however, cryptography protocols with attenuated coherent states, rather than true single-photon states, that are *provably* secure (Gisin *et al.* 2002). Somewhat paradoxically, it is the classical "click" of a photon counter that collapses the quantum state of a weak pulse (described by a low-amplitude coherent state) into a very close approximation of a true single-photon state, albeit "posthumously."

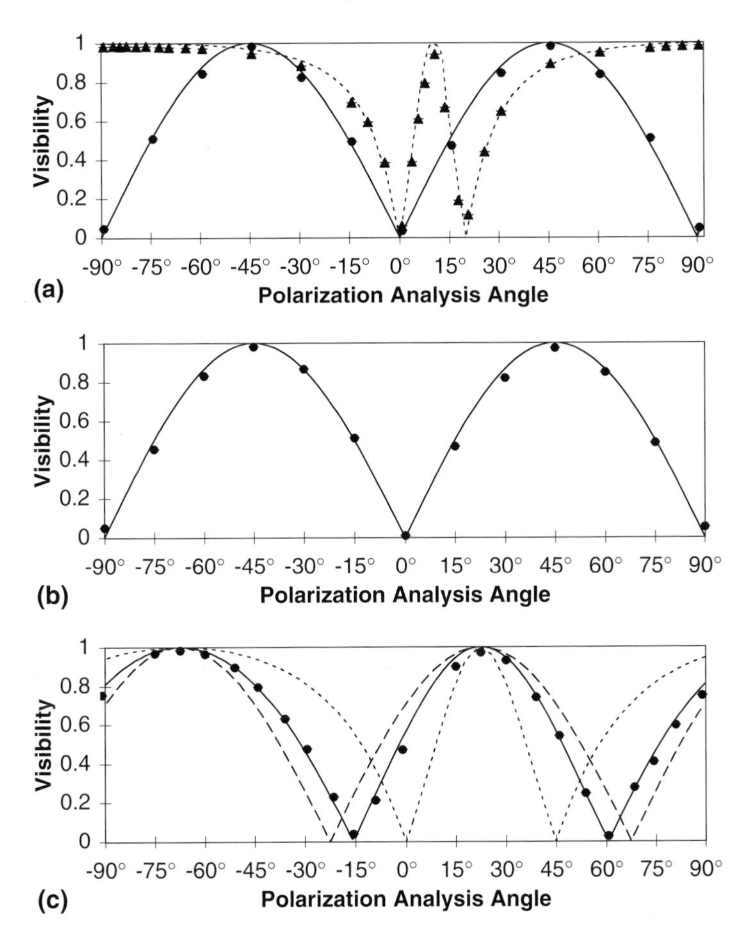

Figure 15.3. QE visibility data and theory curves for various input states, with a HWP in path a of the interferometer: (a) a purely vertically polarized input ($\equiv 90^\circ$), with the polarization rotated by the HWP in path a by 90° (circles, solid line; $\theta_{\text{HWP}} = 45^\circ$) or 20° (triangles, dashed line; $\theta_{\text{HWP}} = 10^\circ$); (b) a completely-mixed state, with $\theta_{\text{HWP}} = 45^\circ$; and (c) a partially mixed state (1:2 pure to mixed; circles, solid line), with the HWP at $\theta_{\text{HWP}} = 22.5^\circ$. For comparison, the dotted and dashed curves show the corresponding theoretical predictions for pure and completely mixed states, respectively.

Pure and mixed states of the which-way marker

Figure 15.3a shows the results when a pure vertical-polarization state was input to the interferometer, and rotated by the HWP in path a by either 90° or 20°, so that the intrinsic visibilities without QE were 0 and 0.94, respectively. We see that the visibility *after* the analyzer can assume any value between 0 and 1. In the latter case, we have a complete QE of WW information. The minima on the curves correspond to analysis that transmits light from only one or the other path; the maxima fall midway between these minima. The maximum measured values of V are slightly lower than 100% because the intrinsic visibility of the interferometer (even without the HWP in path a) is only ∼98%, due to imperfect optics.

Next we investigated QE for non-pure WWM states. If the WWM is initially in a mixed or partially mixed state, formally the key to recovering full fringe visibility is to measure in the basis for which the Hamiltonian coupling the interferometer path to the WWM is diagonal. With polarization as the WWM, this means analyzing along one of the two eigenmodes of the optical element responsible for rotating the polarization in one of the paths.[7] The effect is to select out one of the two perfect-visibility subensembles.

To generate photons in arbitrary mixed and partially mixed states of polarization, we used a "tunable source" (see Fig. 15.2b). It consisted of a 670 nm diode laser – which was spectrally filtered with a narrow-band interference filter (1.5 nm FWHM) and spatially filtered via a single-mode optical fiber – whose horizontal and vertical components were separated by much more than the laser's ∼1-cm coherence length, using an asymmetric Mach–Zehnder interferometer with polarizing beam-splitters (we will discuss below (pp. 321–25) other possible preparation methods). By rotating the (pure linear) polarization input to the first polarizing beam-splitter, one can control the relative contribution of h and v components. For example, for incident photons at 45°, one has equal amplitudes of h and v which are then added together with a random and rapidly varying phase to produce an effectively completely mixed state of polarization. If one has two times more vertical than horizontal, the state is then 1/3 pure to 2/3 completely mixed.[8]

In Fig. 15.3b we show that even for a completely mixed state, it is still possible to recover interference. Since there is no WW information to erase, this *nonerasing QE* may seem quite remarkable at first sight. However, as discussed above, the essential feature of QE is not that it destroys the possibly available WW information,[9] but that it enables us to sort the photons into subensembles, each exhibiting high-visibility fringes. Complete interference is recoverable by analyzing along the eigenmodes of the internal HWP – along one axis we see fringes, and along the other we see anti-fringes, shifted by 180°. The amount of this shift is precisely that imparted to light polarized along the fast axis of the HWP relative to light polarized along the slow axis. Finally, Fig. 15.3c shows the results for a partially mixed state. Note

[7] Throughout we are implicitly assuming a *unitary* transformation of the WWM; if this were not the case, e.g., if a polarizer were placed in one of the interferometer arms, the behavior would be much more complex; see Englert and Bergou (2000) for such complications.

[8] In terms of density matrices (referring to the h/v basis):

$$\frac{1}{3}\begin{pmatrix}1 & 0\\ 0 & 0\end{pmatrix}+\frac{2}{3}\begin{pmatrix}0 & 0\\ 0 & 1\end{pmatrix}=\frac{1}{3}\begin{pmatrix}1 & 0\\ 0 & 2\end{pmatrix}=\frac{1}{3}\begin{pmatrix}0 & 0\\ 0 & 1\end{pmatrix}+\frac{2}{3}\begin{pmatrix}\frac{1}{2} & 0\\ 0 & \frac{1}{2}\end{pmatrix}.$$

[9] In hindsight, that is merely a side effect, but historically it used to be emphasized much; see Englert and Bergou (2000) and pp. 315–20 below.

that the analysis angles yielding zero visibility for the partially mixed state fall between those for pure and mixed states. Specifically, if the fractional purity is s, the angles are at $\theta_{\text{HWP}} \pm 1/2 \arccos[s \cos(2\theta_{\text{HWP}})]$, where θ_{HWP} is the HWP angle. For example, in Fig. 15.3c we have $s = 1/3$ and $\theta_{\text{HWP}} = 22.5°$, so that $V = 0$ is observed at 60.7° and −15.7°.

The WW labeling in the experiment of Fig. 15.2 arose from an *entanglement* between the photon's spatial mode and polarization state (a similar result was obtained in Baldzuhn and Martienssen (1991)). As discussed at length below, it could just as well have been with another photon altogether (as was the case in the experiments reported in Zajonc *et al.* (1991), Kwiat *et al.* (1992), Herzog *et al.* (1995), Monken *et al.* (1995), and Kim *et al.* (2000)), or even with a totally different kind of quantum system (Eichmann *et al.* 1993). Analogous outcomes are predicted – our results are relevant as long as the WWM can be mapped onto a two-state system. For example, instead of the photon's polarization we could use a two-level atom as the WWM. The analogy to unpolarized light would then be an atom in a quantum mechanical mixture of ground and excited state, perhaps because it is entangled to yet another system. More generally, our findings are extendable to analogous experiments with different kinds of interfering quanta, such as interferometers with electrons (Buks *et al.* 1998), neutrons (Badurek *et al.* 1988, 2000; Rauch 1995), or atoms (Dürr *et al.* 1998a, b, c).

Sorting at a distance: what does a quantum eraser erase?

[Entanglement is] the characteristic trait of quantum mechanics, the one that enforces its entire departure from classical lines of thought.

(Erwin Schrödinger)

Consider the experimental set-up sketched in Fig. 15.4.[10] Paired photons are emitted by the source, one of each pair to Alice and the other to Bob. For the moment, assume that the source emits the paired photons in the polarization-entangled state

$$|\Psi_1\rangle \equiv \frac{1}{\sqrt{2}}(|\text{vh}\rangle - |\text{hv}\rangle) = \frac{1}{\sqrt{2}}(|-+\rangle - |+-\rangle), \tag{15.1}$$

where $|\text{vh}\rangle$, for instance, symbolizes the state in which Alice's photon is vertically polarized and Bob's horizontally. The states labeled by ±,

$$|+\rangle \equiv \frac{1}{\sqrt{2}}(|\text{v}\rangle + |\text{h}\rangle) \quad \text{and} \quad |-\rangle \equiv \frac{1}{\sqrt{2}}(|\text{v}\rangle - |\text{h}\rangle), \tag{15.2}$$

are linearly polarized at ±45°.

[10] A simpler version was discussed in problem 9-6 in Ballentine (1998); this extended variant was introduced in Englert (1999).

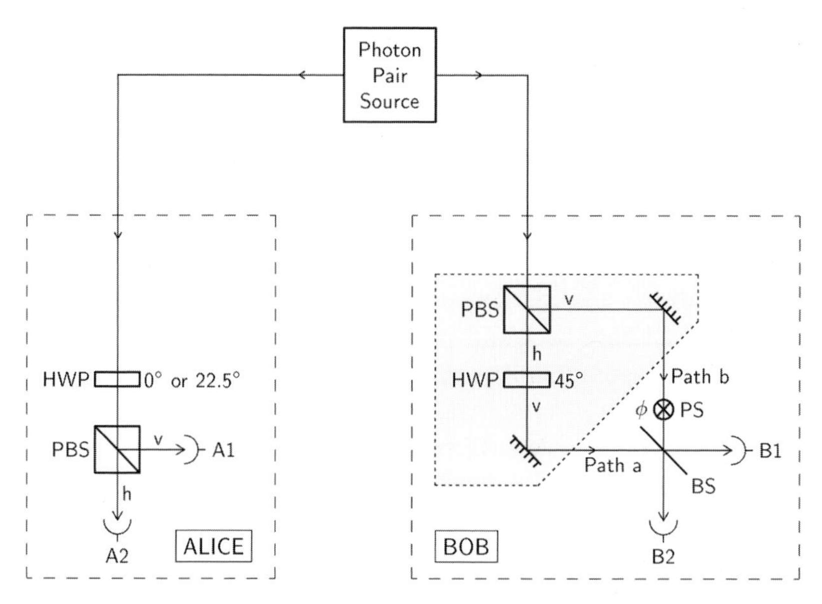

Figure 15.4. Scheme of a photon-pair experiment illustrating various aspects of wave–particle duality. A source emits paired photons whose polarization qubits are entangled. Alice sends her photon through a HWP, set either at 0° or at 22.5°, and then through a PBS to detectors A1 and A2. With the HWP set at 0°, a click of A1 or A2 indicates vertical polarization (v) or horizontal polarization (h), respectively. When the HWP is set at 22.5°, however, a click of A1 means linear polarization at +45° of the incoming photon, and at −45° for a click of A2. Bob uses a PBS in conjunction with a HWP set at 45° to convert the polarization qubit of his photon into a spatial-mode qubit. This conversion happens in the shaded area of Bob's apparatus: photons that arrive h polarized emerge v polarized in path a, and those arriving with v polarization are deflected into path b. An adjustable phase shifter (PS) and a 50–50 beam-splitter (BS) enable Bob to measure the fringe visibility of the Mach–Zehnder interference pattern. Alternatively, Bob can determine the path of the photon by removing BS altogether, so that path a only goes to B1 and path b only goes to B2.

Alice performs one of two polarization measurements on her photons, either distinguishing vertical from horizontal polarization – detecting the polarization states $|\mathsf{v}\rangle$ or $|\mathsf{h}\rangle$ – or distinguishing between $|+\rangle$ and $|-\rangle$. She observes that the outcome is utterly unpredictable: both outcomes are obtained equally frequently in the v/h mode of operation, and also in the $+/-$ mode.

Bob does not perform a polarization measurement (at least, not a direct one). Rather, he converts the polarization qubit of his photon into the spatial-mode qubit of a Mach–Zehnder interferometer.[11] Then he measures the visibility of the

[11] Within the context of the new field of quantum information, Bob's transformations also have another interpretation. Namely, the HWP in path a of the interferometer is actually implementing a controlled-not (CNOT) gate, with the spatial mode as the control qubit and the polarization as the target qubit: only if the control is in state

interferometer fringes, thereby probing for definite phase relations between the amplitudes referring to the two paths. He observes no fringes at all – each of his detectors, B1 and B2, registers approximately 50% of the photons.

The conversion in the shaded area of Bob's apparatus in Fig. 15.4 is summarized by

$$|\mathbf{h}\rangle \to |\mathbf{a}\rangle, \qquad |\mathbf{v}\rangle \to |\mathbf{b}\rangle, \tag{15.3}$$

and its effect on $|\Psi_1\rangle$ is given by

$$|\Psi_1\rangle \to \frac{1}{\sqrt{2}}(|\mathbf{va}\rangle - |\mathbf{hb}\rangle). \tag{15.4}$$

The statistical operator of the reduced state that applies to Alice's polarization qubit is then

$$\rho_1^{(A)} = \frac{1}{2}(|\mathbf{v}\rangle\langle\mathbf{v}| + |\mathbf{h}\rangle\langle\mathbf{h}|) = \frac{1}{2}(|+\rangle\langle+| + |-\rangle\langle-|). \tag{15.5}$$

Bob in turn accounts for the spatial properties of his photon by the statistical operator

$$\rho_1^{(B)} = \frac{1}{2}(|\mathbf{a}\rangle\langle\mathbf{a}| + |\mathbf{b}\rangle\langle\mathbf{b}|). \tag{15.6}$$

Expressions (15.5) and (15.6) are obtained by tracing $|\Psi_1\rangle\langle\Psi_1|$ over Bob's spatial-mode qubit or Alice's polarization qubit, respectively.

In more sophisticated correlation experiments, Alice and Bob can exploit the entanglement between the two qubits. For example, by performing the v/h distinction, Alice can easily determine the trajectory followed by each of Bob's photons: as indicated in (15.4), path a is strictly correlated with v polarization, and path b with h polarization. This is summarized by the conditional density matrices

$$\rho_{1\mathsf{v}}^{(B)} = |\mathsf{a}\rangle\langle\mathsf{a}| \quad \text{and} \quad \rho_{1\mathsf{h}}^{(B)} = |\mathsf{b}\rangle\langle\mathsf{b}| \tag{15.7}$$

which apply to Bob's subensembles that are identified, and labeled, by the polarization detected by Alice.[12] This sorting of Bob's photons is fittingly called *WW sorting*.

a is the target qubit flipped. A polarizing beam-splitter realizes the other CNOT gate, in which the polarization qubit controls the spatial mode qubit. It is now well known that in fact all quantum algorithms can be synthesized using only linear optics, e.g., waveplates, beam-splitters, etc. (Cerf *et al.* 1998); however, there is a price to pay – for more than ~5 qubits, the number of optical elements needed to implement a particular algorithm grows exponentially with the number of qubits required (Kwiat *et al.* 2000). For instance, a 10-qubit algorithm would require interferometers of order 2^{10} spatial paths, or cavities with 2^{10} resolvable modes (Bhattacharya *et al.* 2002). Remarkably, it has recently been shown theoretically how to achieve *scalable* quantum computation using the same linear optics (and true single-photon sources), combined with the implicit effective nonlinearity of the photon detection process itself (Knill *et al.* 2001).

[12] The superscript identifies this as the state of Bob's photon; the first subscript refers to the state $|\Psi_1\rangle$ produced by the source; and the second subscript identifies the results Alice obtained from her measurement.

If Alice instead performs the $+/-$ distinction, the density matrices for the corresponding subensembles of Bob's photons are then given by

$$\rho_{1+}^{(B)} = \frac{1}{2}(|\mathsf{a}\rangle\langle\mathsf{a}| + |\mathsf{b}\rangle\langle\mathsf{b}| - |\mathsf{a}\rangle\langle\mathsf{b}| - |\mathsf{b}\rangle\langle\mathsf{a}|) = \frac{|\mathsf{a}\rangle - |\mathsf{b}\rangle}{\sqrt{2}} \frac{\langle\mathsf{a}| - \langle\mathsf{b}|}{\sqrt{2}} \tag{15.8}$$

and

$$\rho_{1-}^{(B)} = \frac{1}{2}(|\mathsf{a}\rangle\langle\mathsf{a}| + |\mathsf{b}\rangle\langle\mathsf{b}| + |\mathsf{a}\rangle\langle\mathsf{b}| + |\mathsf{b}\rangle\langle\mathsf{a}|) = \frac{|\mathsf{a}\rangle + |\mathsf{b}\rangle}{\sqrt{2}} \frac{\langle\mathsf{a}| + \langle\mathsf{b}|}{\sqrt{2}}. \tag{15.9}$$

The probability that the photon is detected by $\mathsf{B1}$, say, is then

$$\left| \frac{(\langle\mathsf{a}| + e^{i\phi}\langle\mathsf{b}|)}{\sqrt{2}} \frac{(|\mathsf{a}\rangle - |\mathsf{b}\rangle)}{\sqrt{2}} \right|^2 = \frac{1}{2}(1 - \cos\phi) \tag{15.10}$$

if it is in the $\mathsf{1+}$ subensemble (i.e., Alice measured the entangled partner to be polarized at $+45°$), and

$$\left| \frac{(\langle\mathsf{a}| + e^{i\phi}\langle\mathsf{b}|)}{\sqrt{2}} \frac{(|\mathsf{a}\rangle + |\mathsf{b}\rangle)}{\sqrt{2}} \right|^2 = \frac{1}{2}(1 + \cos\phi) \tag{15.11}$$

if it is in the $\mathsf{1-}$ subensemble (Alice's photon polarized at $-45°$). Accordingly, Bob finds unit visibility for the fringes of these individual subensembles, and we note that this is the "fringes and anti-fringes situation" that we encountered already (p. 312), and earlier in Fig. 15.1c. For historical reasons, one speaks of the *QE sorting* here. This terminology derives from the fact that, as soon as she has performed the $+/-$ polarization distinction, Alice can no longer determine the path of Bob's photon by the $\mathsf{v/h}$ distinguishing measurement. She has "erased" the WW information that was latently available before.

But, one could equivalently argue that the WW sorting erases the latently available phase information that could be revealed by QE sorting (whether the relative phase between $|a\rangle$ and $|b\rangle$ is 0 or π as in (9) and (8), respectively). The bias in favor of WW information, i.e., in favor of the photon's particle aspect, is very natural – given the local nature of all physical interactions, the state reduction associated with generic measurements tends to leave the observed system in a well-localized state and thus with attributes that are analogous to those of a classical particle.[13] On a more abstract level, however, one cannot fail to note that in experiments of the kind we are discussing here it is hardly justifiable to regard the photon's particle nature as more fundamental than its wave nature, or vice versa.

We now address the question: what does a quantum eraser erase? Or, in more technical terms, is the availability of WW information a precondition for QE? The

[13] This natural bias in favor of particles rather than waves is ubiquitous. For instance, it makes us prefer the term "elementary particle physics" over "elementary wave physics."

answer is no, which can be seen as follows. First, consider that the source instead of $|\Psi_1\rangle$ emits photon pairs in the entangled state

$$|\Psi_2\rangle \equiv \frac{1}{\sqrt{2}}(|\mathbf{vv}\rangle - |\mathbf{hh}\rangle) = \frac{1}{\sqrt{2}}(|-+\rangle + |+-\rangle). \tag{15.12}$$

After Bob's conversion,

$$|\Psi_2\rangle \rightarrow \frac{1}{\sqrt{2}}(|\mathbf{vb}\rangle - |\mathbf{ha}\rangle), \tag{15.13}$$

so that the which-path sorting results in the subensembles specified by

$$\rho_{2\mathsf{v}}^{(\mathrm{B})} = |\mathsf{b}\rangle\langle\mathsf{b}| \quad \text{and} \quad \rho_{2\mathsf{h}}^{(\mathrm{B})} = |\mathsf{a}\rangle\langle\mathsf{a}| \tag{15.14}$$

and QE produces

$$\rho_{2+}^{(\mathrm{B})} = \frac{1}{2}(|\mathsf{a}\rangle\langle\mathsf{a}| + |\mathsf{b}\rangle\langle\mathsf{b}| - |\mathsf{a}\rangle\langle\mathsf{b}| - |\mathsf{b}\rangle\langle\mathsf{a}|) = \frac{|\mathsf{a}\rangle - |\mathsf{b}\rangle}{\sqrt{2}}\frac{\langle\mathsf{a}| - \langle\mathsf{b}|}{\sqrt{2}} \tag{15.15}$$

and

$$\rho_{2-}^{(\mathrm{B})} = \frac{1}{2}(|\mathsf{a}\rangle\langle\mathsf{a}| + |\mathsf{b}\rangle\langle\mathsf{b}| + |\mathsf{a}\rangle\langle\mathsf{b}| + |\mathsf{b}\rangle\langle\mathsf{a}|) = \frac{|\mathsf{a}\rangle + |\mathsf{b}\rangle}{\sqrt{2}}\frac{\langle\mathsf{a}| + \langle\mathsf{b}|}{\sqrt{2}}. \tag{15.16}$$

In short, we have $\rho_{2\mathsf{v}}^{(\mathrm{B})} = \rho_{1\mathsf{h}}^{(\mathrm{B})}$, $\rho_{2\mathsf{h}}^{(\mathrm{B})} = \rho_{1\mathsf{v}}^{(\mathrm{B})}$ and $\rho_{2+}^{(\mathrm{B})} = \rho_{1+}^{(\mathrm{B})}$, $\rho_{2-}^{(\mathrm{B})} = \rho_{1-}^{(\mathrm{B})}$. Just as before, the WW sorting yields full path knowledge for each photon in question, and the QE sorting gives subensembles with unit fringe visibility.

But now consider the situation in which the source emits with equal probability photon pairs either in state $|\Psi_1\rangle$ or in state $|\Psi_2\rangle$, randomly alternating between the two possibilities. Bob's subensembles are then described by

$$\rho_{3\mathsf{v}}^{(\mathrm{B})} = \rho_{3\mathsf{h}}^{(\mathrm{B})} = \frac{1}{2}(|\mathsf{a}\rangle\langle\mathsf{a}| + |\mathsf{b}\rangle\langle\mathsf{b}|) \tag{15.17}$$

and

$$\begin{aligned}\rho_{3+}^{(\mathrm{B})} &= \frac{1}{2}(|\mathsf{a}\rangle\langle\mathsf{a}| + |\mathsf{b}\rangle\langle\mathsf{b}| - |\mathsf{a}\rangle\langle\mathsf{b}| - |\mathsf{b}\rangle\langle\mathsf{a}|) = \frac{|\mathsf{a}\rangle - |\mathsf{b}\rangle}{\sqrt{2}}\frac{\langle\mathsf{a}| - \langle\mathsf{b}|}{\sqrt{2}},\\ \rho_{3-}^{(\mathrm{B})} &= \frac{1}{2}(|\mathsf{a}\rangle\langle\mathsf{a}| + |\mathsf{b}\rangle\langle\mathsf{b}| + |\mathsf{a}\rangle\langle\mathsf{b}| + |\mathsf{b}\rangle\langle\mathsf{a}|) = \frac{|\mathsf{a}\rangle + |\mathsf{b}\rangle}{\sqrt{2}}\frac{\langle\mathsf{a}| + \langle\mathsf{b}|}{\sqrt{2}}.\end{aligned} \tag{15.18}$$

No WW information is available at all – nothing that Alice can do to her photon would enable her to guess the path of Bob's photon – and yet the QE sorting still identifies subensembles with unit fringe visibility. Clearly, no WW information is erased by Alice's +/− polarization measurement here, because there is simply no such information present to begin with. This is, of course, exactly the situation of Fig. 15.3b, and the oxymoron *nonerasing QE* summarizes the matter quite appropriately.

Similarly, one can easily construct situations in which Alice cannot identify *any* subensembles that display interference fringes. For example, if the source instead emits with equal probability photons in states $|\Psi_1\rangle$ and

$$\frac{1}{\sqrt{2}}(|\mathrm{vh}\rangle + |\mathrm{hv}\rangle) = \frac{1}{\sqrt{2}}(|{+}{+}\rangle - |{-}{-}\rangle) \rightarrow \frac{1}{\sqrt{2}}(|\mathbf{va}\rangle + |\mathbf{hb}\rangle), \quad (15.19)$$

then we have, in this fourth case, $\rho_{4\mathrm{v}}^{(\mathrm{B})} = |\mathrm{a}\rangle\langle\mathrm{a}|$, $\rho_{4\mathrm{h}}^{(\mathrm{B})} = |\mathrm{b}\rangle\langle\mathrm{b}|$ so that full WW information can be acquired, and

$$\rho_{4+}^{(\mathrm{B})} = \rho_{4-}^{(\mathrm{B})} = \frac{1}{2}(|\mathrm{a}\rangle\langle\mathrm{a}| + |\mathrm{b}\rangle\langle\mathrm{b}|), \quad (15.20)$$

so that no fringes can be recovered.

As a final remark concerning the scheme of Fig. 15.4, we note that the temporal order in which Alice and Bob perform their measurements is of no relevance. For instance, one could incorporate delay lines (e.g., many loops of optical fiber guiding the photons) such that Bob completes his measurements before Alice does hers. Thus, in the best Wheelerian tradition (Wheeler 1979), Alice could delay her choice of measuring v vs. h or performing the $+/-$ distinction.[14] Likewise, Bob could delay his choice between the wave experiment (BS in place) and the particle experiment (BS removed) until *after* Alice has finished her polarization measurements. No delay of either kind has any bearing on what is said above; in particular, none of the various conditional density matrices depends on whether Alice measures first, or Bob, or both measurements are made in spacelike separated regions. No superluminal signaling is possible, despite that the correlations concommitant with the state $|\Psi_1\rangle$ (and also with $|\Psi_2\rangle$) can be shown to be nonlocal.

Although the experiment of Fig. 15.4 has not yet been performed, the various scenarios discussed have been realized in spirit in the experiment by Schwindt *et al.* (1999) (see Fig. 15.2), which differs from the set-up sketched in Fig. 15.4 mainly by the circumstance that Alice's polarization qubit and Bob's spatial-mode qubit are carried by one and the same photon, rather than by the two photons of a pair. By contrast, photon pairs are actually used in the experiments of Trifonov *et al.* (2002) or Walborn *et al.* (2002). The latter experiment combines elements described above, inasmuch as the setup is similar to Fig. 15.4 but with the double slit of Fig. 15.1b replacing Bob's Mach–Zehnder interferometer.

[14] There is no paradox here that Bob's photons seem to display interference "after the fact" – no matter how Alice measures (or even whether she measures at all), his results as a whole do *not* display fringes. It is only *after* he receives the classical information from Alice on how to partition his data into subensembles that interference may be recovered.

The four (or five!) types of mixed states

Contraria non contradictoria sed complementa sunt.

(*Niels Bohr*)

As alluded to above, it is certainly well known that *if* system A is entangled with system B, then tracing over system B leaves system A in a mixed or partially mixed state. In fact, this is precisely one method of quantifying the amount of entanglement for bipartite pure states, namely, calculating the entropy of the reduced density matrix after tracing over one of the subsystems: $E = S(\rho^{(A)})$, where $S(\rho) = -\mathrm{Tr}(\rho \ln \rho)$, and $\rho^{(A)} = \mathrm{Tr}_B(\rho^{(A\&B)})$. The reverse question is perhaps more profound, and potentially has very important physical and philosophical implications – if the system A is in a mixed state, is it necessarily the case that it is entangled with another system? Hiding behind this seemingly innocuous question is the much deeper issue of whether or not a classical world actually exists. For, if indeed quantum mechanics does describe every possible interaction[15] between two systems – if quantum mechanics is "the whole story" – then indeed one might believe that the universe is described by one humongous entangled state of all the constituent parts.[16]

To highlight this question, we can identify four or five arguably distinct methods for producing a completely mixed (i.e., unpolarized) photon (see Fig. 15.5).

(a) In Fig. 15.5a we depict a situation where the mixture comes from the entanglement of the polarization to another degree of freedom (e.g., frequency) of the same photon, as was used in the experiment described above (pp. 310–15) (see Fig. 15.2b). The PBS transmits the horizontal component of the light directly, while deflecting the vertical component into a long delay line. The phase accumulated by light in that delay line depends on the frequency ω: $\Delta\phi = \omega L/c$, where L is the optical path length of the delay line, and c is the speed of light. If L is much greater than the coherence length of the light, in turn determined by the spread in frequencies, the resultant reduced density matrix of the polarization will be in a completely mixed state (assuming the incident light had equal horizontal and vertical polarization components).[17]

[15] We are actually restricting ourselves to a fairly limited scope of the word "interaction." For example, we do not necessarily mean to include gravity, as there does not yet exist a quantum theory of gravity. However, it should not be dismissed that such a theory may have tremendous implications for the questions we are addressing, e.g., such a theory could in principle completely resolve the semi-paradox of the emergence of a seemingly classical world from solely quantum mechanical foundations. Interesting discussions on the connection between quantum mechanics and gravity can be read elsewhere in this book.

[16] We can almost hear the Ghost of Bohr, acting as the custodian of the Copenhagen spirit, speak up: "Some of us (including one of you!) do not take kindly to your invoking the 'wave function of the universe': a quantum state is merely a mathematical entity that summarizes the conditions under which we make statistical predictions about the quantum system in question." Of course, there are those who might wish to give more meaning to Schrödinger's wave function, but if they do, then they are responsible for the consequences, as van Kampen (1988) has rightly emphasized.

[17] *Individual* photons thus manipulated are unpolarized; for a discussion of unpolarized multi-photon states see (Lehner *et al.* 1996).

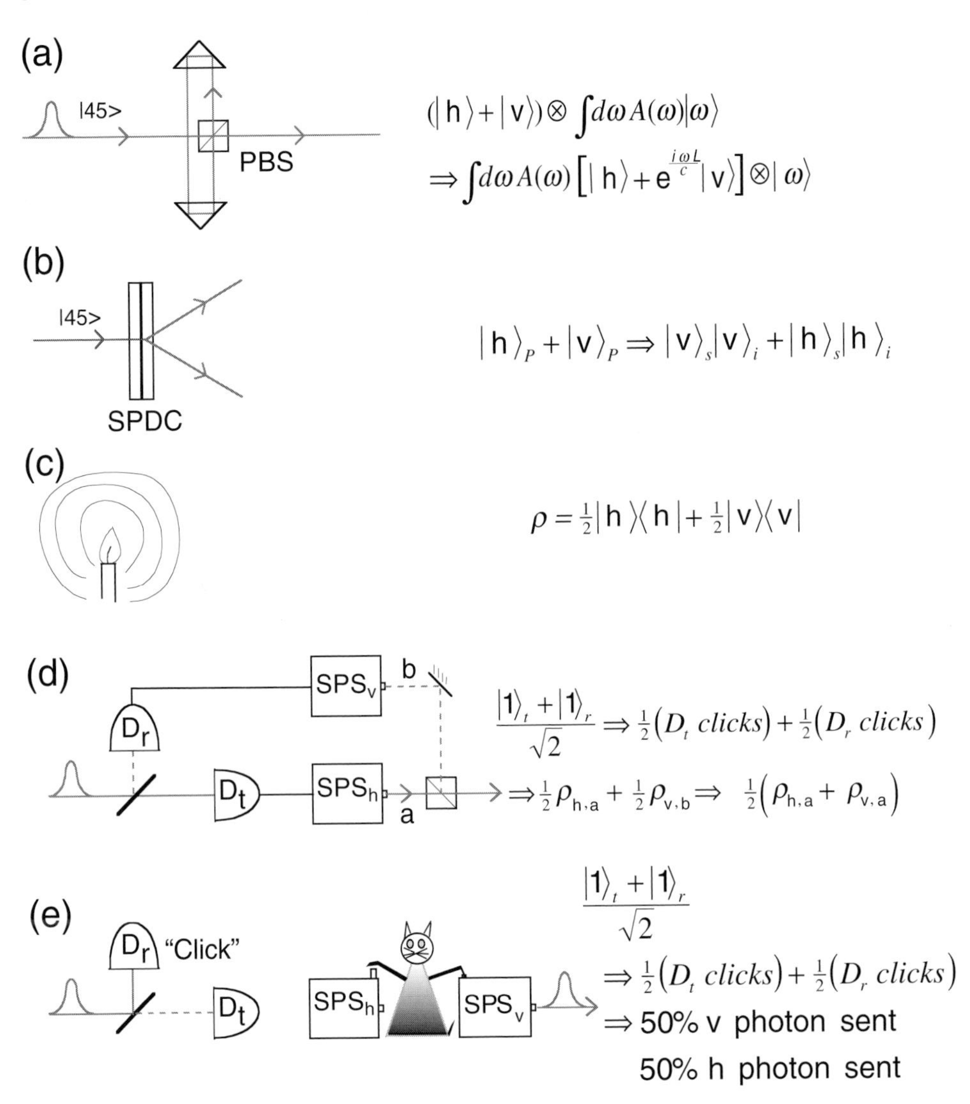

Figure 15.5. Several ways of producing a mixed polarization quantum state of a single photon. The transformations of the incident state are indicated on the right.

(b) Figure 15.5b illustrates one method that has been used to generate polarization entangled pairs of photons (Kwiat *et al.* 1999b): an ultraviolet pump photon is directed into two adjacent nonlinear crystals. Crystal 1 is oriented such that a horizontally polarized pump photon can down-convert into two vertically polarized photons (each with approximately half the energy of the pump), while the second crystal allows for the possibility of a vertically polarized pump photon to down-convert into two horizontally polarized photons. By coherently driving the crystals with the pump polarization of 45°

we arrive at the entangled state of $|\text{vv}\rangle + |\text{hh}\rangle$. As discussed above (p. 317), tracing over one of the photons leaves the other in a completely mixed state.[18]

(c) Figure 15.5c is meant to illustrate some natural random process that emits photons, such as a candle, a star, or simply some fluorescing atoms in a trap. In this case the polarization is seen to be completely random on all timescales long compared to the coherence time of the light. If we look at a classical light source of this type, at any instance in time there is a definite polarization, but it wanders rapidly over all possible states. At the single photon level, the polarization of any given photon is completely undetermined and, perhaps more importantly, *indeterminable*.

(d) Figure 15.5d shows a slightly more sophisticated method of making an unpolarized single photon state. A photon is directed at a 50–50 beam-splitter, with detectors in each output. If the photon is transmitted (and detected), then a SPS (gedanken single-photon source) is automatically fired, producing an h-polarized photon; if the original photon is reflected (and detected), then a different SPS is automatically fired, producing a v-polarized photon. Because it is intrinsically random which way the initial photon travels at the beam-splitter, it is unpredictable which of the two detectors will fire.[19] Finally, the two modes are combined using a polarizing beam-splitter.

(e) Figure 15.5e shows an extension of the previous scheme. The difference is that now a sentient observer is required to actually activate one of the single photon sources, depending on which detector registers the initial photon.[20] Although interpretations of quantum mechanics that rely on consciousness to collapse the wave function are no longer in vogue,[21] by this example we mean to highlight explicitly the role that complexity of the measuring system may play in reducing the state from a pure superposition to a mixed state.

Let us now briefly discuss the five scenarios from the point of view of entanglement and reversibility. The systems are presented in order of increasing size and complexity. In each case, at the very least we start with some entanglement. This is most obvious for the case (b), where we have explicitly written down the entangled state of one photon with the other. In system (a), the entanglement is with another degree of freedom carried by the same quantum particle. Although it is true that such a system could never be used in tests of nonlocality, at least formally there

[18] Of course, one could also use the technique of Fig. 15.5a to prepare the pump itself in a mixed state, which would then prepare the partially mixed, but correlated state $\rho \propto |\text{vv}\rangle\langle\text{vv}| + |\text{hh}\rangle\langle\text{hh}|$. No quantum erasure is possible in this case.

[19] In fact, this very system has been used to generate random numbers for quantum cryptography (Rarity *et al.* 1994; Gisin *et al.* 2002), and for a rapid-switching test of Bell's inequalities (Weihs *et al.* 1998).

[20] We hope the reader will accept this at least as a thought-experiment. We fully admit the unlikelihood that cats, even if sentient, could ever be trained to reliably perform such a task (a dog might be better suited (Bergou and Englert 1998)). On the other hand, a cat's own fickleness is likely to lead to outcomes just as random as the photon at the beam-splitter!

[21] The Ghost of Bohr interjects: "And for good reasons!"

seems to be no difference between entanglement of two degrees of freedom on separate particles and on the same particle.[22]

For scenario (c) we envision an atom in a gas, or part of a much larger ensemble of similar atoms. If there are enough of them – especially if they are more closely spaced than the wavelength of the emitted photon – it becomes impossible ever to determine which emitted the photon. One can further identify two subcategories: (1) it is impossible *in principle* to make the determination; (2) it is only impossible *in practice*. Note that if we reduce the number of emitters in the system to only a few, then it becomes possible in principle and perhaps even in practice to distinguish which emitter emitted the photon, i.e., by observing the motions of the emitters and seeing which one recoiled. In this case we have returned to a situation completely analogous to that in scenario (b).

In variation (d) the superposition of the initial photon is rapidly converted into an entangled state. Assuming, as in Fig. 15.5d, that the path for the photon to reach one detector is longer than the path to reach the other, and that the initial photon arrives at a well-specified time, the state of the whole system when the photon has had time to reach detector D_r but not yet D_t is the entangled state whose two terms are: (photon-in-the-longer-path and all-detector-atoms-quiescent) and (no-photon-in-the-long-path and first-few-detector-electrons-in-an-excited-state), which state will eventually, through various amplification processes, lead to a "click." Note that by this stage it is pretty clear that we have made the transition to the classical realm, apparently somewhere in the multi-step amplification process from the single-photon input to the electrical pulse containing perhaps hundreds or thousands of electrons, each of them interacting with its own local environment. Of course, if one accepts that explanation, then there is hardly a need to list scenario (e). However, if one instead takes the viewpoint that our inability to disentangle a complicated entangled state of the photon and detector is merely a consequence of our *technical* inability to sufficiently isolate the detector from the rest of the world, then it *does* make sense to ask whether or not there might be another natural dividing line at the level of some form of consciousness or life.[23]

We can also consider our various mixed-state creation schemes in terms of reversibility. When does it become impossible to "unscramble" the mixture? For example, the entangling process in system (a) may be very easily reversed simply by sending the photon through another delay line which now delays the horizontal

[22] Actually, the entangled state in system (a) tends to be more robust to decoherence, since the two quantum labels (polarization and frequency), are carried by the same entity, and therefore interact in most scenarios with the same local environment (Remarkably, recent experimental results seem to provide evidence for the spatial separation of the charge and spin wave functions of an electron in a one-dimensional conductor (Lorenz *et al.* 2002).) In contrast, for system (b) the two photons are able to interact with other quantum systems which are not located at the same spacetime coordinates, in general leading to faster decoherence.

[23] Clearly, again, there are numerous levels that one could consider, ranging from a single cell bacterium (which might indeed be physically smaller than the detector element in Fig. 15.5d) all the way up to human consciousness.

component with respect to the vertical, thereby "stitching" the wave function back together in its original productlike form. Scenario (b) is somewhat more difficult to reverse, in that it requires strong nonlinear process to up-convert the two photons.[24] As mentioned above, in the third scenario, as long as we were dealing with only a few atoms, it might be possible to engineer the reversibility of the system. However, it is clear that at some point it will become effectively impossible to achieve this (and maybe even impossible in principle). Apparently, by the time we are at the "click" of the detectors in scenarios (d) and (e), there is no possibility of reversing the procedure and returning to the pure initial state.[25]

Summary and discussion

This isn't right. This isn't even wrong.

(*Wolfgang Pauli*)

It was absolutely marvelous working for Pauli. You could ask him anything. There was no worry that he would think a particular question was stupid, since he thought all questions were stupid.

(*Victor Weisskopf*)

Now let us make a connection to the procedure of QE and harken back to the opening theme of the Introduction. In order for QE to recover interference fringes, the entangled quantum system plus WWM must remain in an entangled state, not irreversibly decohere into a mixed state. More specifically, the quantum phase coherence between the different terms of the entangled state must be maintained if that phase is to be mapped onto the subensemble corresponding to a particular measurement outcome of the WWM. For example, the quantum phase coherence implied by (15.8) depends entirely on the fact that (15.4) also possesses phase coherence between the two terms. If the joint state of Alice's and Bob's photons was instead a mixed state (i.e., $\rho^{(A\&B)} \propto |\mathsf{va}\rangle\langle\mathsf{va}| + |\mathsf{hb}\rangle\langle\mathsf{hb}|$), then no recovery of fringes would be possible, even in principle. We see, then, that QE is in fact a very useful tool for establishing whether or not the system has evolved out of the reversible quantum realm into the classical world of definite reality. As discussed in sections "Quantum eraser for pure and mixed states" and "Sorting at a distance," the sources described by Fig. 15.5a and 15.5b clearly possess this erasable feature.

[24] Although the presence of a bright phase-coherent pump can dramatically improve the efficiency of the up-conversion process (Herzog *et al.* 1994; Resch *et al.* 2001), in this case there is no way to pick out the particular photon that resulted from the pair.

[25] On the macroscopic scale, we are used to the fact that changes are not fully reversible, as witnessed by the folk wisdom of the Humpty Dumpty nursery rhyme: all the king's men cannot put a broken egg together again. This has its quantum analog – a silver atom split in two by a Stern–Gerlach magnet cannot be brought back to its initial spin state (Schwinger *et al.* 1988), and this observation contains a lesson about the reversibility of a quantum evolution (Englert 1997).

In this sense we would not say that the state of the photon prior to measurement was classical.

On the other extreme, to the extent that we believe that cats (Fig. 15.5e), or even photon counters (Fig. 15.5d), do not exist in quantum mechanical superposition states,[26] then there is, even in principle, no possibility of ever undoing the mixture of the polarization state of the finally emitted photon. There could be a record kept at the source and, therefore, each photon is h or v polarized in an objective sense.[27] No such objective reality can be ascribed to a photon emerging from the sources in Figs. 15.5a and 15.5b.

Figure 15.5c in some sense could be thought to cover the transition between the two extremes. If, for example, there are only a few atoms that might be emitting the photon, then – at least in principle, if not in practice – one could envision carefully measuring the atoms before and after emission to see which of them produced the photon. If we do this before the atom has had a chance to interact with other parts of the environment, then we might be able to measure the atom in such a way as to erase the which-polarization information. On the other hand, when sufficient numbers of atoms are involved, with sufficient numbers of degrees of freedom to degrade the quantum coherence, it rapidly becomes impossible – at least in practice, if not in principle – to perform QE.

As we continue to improve our experimental technology, we will undoubtedly be able to recover fringes in still larger systems; but since the difficulties of such an experiment seem to scale exponentially with the number of degrees of freedom we have to keep track of and control, the most likely scenario is that the in-practice constraints will greatly dominate over any possible in-principle constraint like those we alluded to here. In other words, we may never be able to tell whether the unrecoverable nature of the fringes is due to an actual irreversible transition to a classical mixture, or whether no such transition ever occurs but our measurement apparatus nevertheless are not able to reverse the complicated string of interactions. Quantum mechanics with its fundamental indeterminism sets the ultimate limits to how well we can make measurements. But these measurements, curiously, are our window back into the quantum realm: somewhat paradoxically, the only way to experimentally probe the underlying quantum mechanical reality is by making the transition – through amplification-enabled measurement – to the classical world:

No elementary quantum phenomenon is a phenomenon until it is a registered phenomenon.
(*John A. Wheeler*)

[26] There is certainly an *effective* nonexistence of such macroscopic superpositions because there are no known phenomena that are sensitive to the relative phases between the superposed states: how would one distinguish $|\text{dead cat}\rangle + |\text{live cat}\rangle$ from $|\text{dead cat}\rangle - |\text{live cat}\rangle$? And perhaps, as the Ghost of Bohr endorses, this effective nonexistence is all that is needed for the apparent emergence of the classical world.

[27] To the experimenter who has no knowledge of the preparation method, it is, for all purposes, as if the photon were h or v polarized without him knowing which one is the case or, equivalently, + or − polarized. In fact, many such "as if" realities are consistent with the experimenter's data.

References

Alley, C O, Jakubowicz, O G, and Wickes, W C (1987) In *Proc. 2nd Int. Symposium on Foundations of Quantum Mechanics*, Tokyo 1986, ed. M. Namiki, M. *et al.*, p. 36. Tokyo: Physical Society of Japan.

Badurek, G, Rauch, H, and Summhammer, J (1988) *Physica B&C* **151**, 82.

Badurek, G, Buchelt, R J, Englert, B-G, *et al.* (2000) *Nucl. Instr. Meth. Nucl. Res.* **A440**, 562.

Baldzuhn, J, and Martienssen, W (1991) *Zeits. Phys.* **B82**, 309.

Baldzuhn, J, Mohler, E, and Martienssen, W (1989) *Zeits. Phys.* **B77**, 347.

Ballentine, L E (1998) *Quantum Mechanics*, 2nd edn. Singapore: World Scientific Press.

Bergou, J A and Englert, B-G (1998) *J. Mod. Opt.* **45**, 701.

Bhattacharya, N, van Linden van den Heuvell, H B, and Spreeuw, R J C (2002) *Phys. Rev. Lett.* **88**, 137901.

Bohr, N (1983) In *Quantum Theory and Measurement*, ed. J. A. Wheeler and W. H. Zurek, p. 9. Princeton NJ: Princeton University Press.

Buks, E, Schuster, R, Heiblum, M, *et al.* (1998) *Nature* **391**, 871.

Cerf, N J, Adami, C, and Kwiat, P G (1998) *Phys. Rev.* **A57**, R1477.

Clauser, J F (1974) *Phys. Rev.* **D9**, 853.

Dürr, S, Nonn, T, and Rempe, G (1998a) *Nature* **395**, 33.

(1998b) *Phys. Rev. Lett.* **81**, 5705.

(1998c) *Phys. Rev.* **A57**, R1477.

Eichmann, U, Bergquist, J C, Bollinger, J J, *et al.* (1993) *Phys. Rev. Lett.* **70**, 2359.

Englert, B-G (1996) *Phys. Rev. Lett.* **77**, 2154.

(1997) *Zeits. Naturforsch.* **52a**, 13.

(1999) *Zeits. Naturforsch.* **54a**, 11.

Englert, B-G, and Bergou, J A (2000) *Opt. Commun.* **179**, 337.

Fermi, E (1929) *Rend. Lincei* **10**, 72.

Feynman, R P, Leighton, R B, and Sands, M (1965) *The Feynman Lectures on Physics*, Reading, MA: Addison-Wesley.

Gisin, N, Ribordy, G, Tittel, W, *et al.* (2002) *Rev. Mod. Phys.* **74**, 145.

Glauber, R (1986) *Ann. N. Y. Acad. Sci.* **480**, 336.

Grangier, P, Roger, G, and Aspect, A (1986) *Europhys. Lett.* **1**, 173.

Greenberger, D M, and Yasin, A (1988) *Phys. Lett.* **A128**, 391.

Hellmuth, T, Walther, H, Zajonc, A, *et al.* (1987) *Phys. Rev.* **A35**, 2532.

Herzog, T J, Rarity, J G, Weinfurter, H, *et al.* (1994) *Phys. Rev. Lett.* **72**, 629.

Herzog, T J, Kwiat, P G, Weinfurter, H, *et al.* (1995) *Phys. Rev. Lett.* **75**, 3034.

Hillery, M and Scully, M O (1983) In *Quantum Optics, Experimental Gravitation, and Measurement Theory*, ed. P. Meystre and M. O. Scully, p. 65. New York: Plenum Press.

Jaeger, G, Shimony, A, and Vaidman, L (1995) *Phys. Rev.* **A51**, 54.

Kim, Y-H, Yu, R, Kulik, S, *et al.* (2000) *Phys. Rev. Lett.* **84**, 1.

Knill, E, Laflamme, R, and Milburn, G (2001) *Nature* **409**, 46.

Kwiat, P G, Steinberg, A M, and Chiao, R Y (1992) *Phys. Rev.* **A45**, 7729.

Kwiat, P G, Schwindt, P D D, and Englert, B-G (1999a) In *Mysteries, Puzzles, and Paradoxes in Quantum Mechanics*, ed. R. Bonifacio, p. 69. New York: The American Institute of Physics.

Kwiat, P G, Waks, E, White, A G, *et al.* (1999b) *Phys. Rev.* **A60**, R773.

Kwiat, P G, Mitchell, J R, Schwindt, P D D, *et al.* (2000) *J. Mod. Opt.* **47**, 257.

Lehner, J, Leonhardt, U, and Paul, H (1996) *Phys. Rev.* **A53**, 2727.

Lorenz, T, Hofmann, M, Grüninger, M, *et al.* (2002) *Nature* **418**, 614.

Mandel, L (1991) *Opt. Lett.* **16**, 1882.

Mittelstaedt, P, Prieur, A, and Schieder, R (1987) *Found. Phys.* **17**, 891.
Monken, C H, Branning, D, and Mandel, L (1995) In *Coherence and Quantum Optics VII*, ed. J. H. Eberly, L. Mandel, and E. Wolf, p. 701. New York: Plenum Press.
Rarity, J G, Owens, P C M, and Tapster, P R (1994) *J. Mod. Opt.* **41**, 2435.
Rauch, H (1995) *Ann. N. Y. Acad. Sci.* **755**, 263.
Rauch, H and Summhammer, J (1984) *Phys. Lett.* **A104**, 44.
Resch, K J, Lundeen, J S, and Steinberg, A M (2001) *Phys. Rev. Lett.* **87**, 123603.
Schwindt, P D D, Kwiat, P G, and Englert, B-G (1999), *Phys. Rev.* **A60**, 4285.
Schwinger, J, Scully, M O, and Englert, B-G (1988) *Zeits Phys.* **D10**, 135.
Scully, M O, and Drühl, K (1982) *Phys. Rev.* **A25**, 2208.
Scully, M O, Englert, B-G, and Walther, H (1991) *Nature* **351**, 111.
Summhammer, J, Rauch, H, and Tuppinger, D (1987) *Phys. Rev.* **A36**, 4447.
Trifonov, A, Björk, G, Söderholm, J, *et al.* (2002) *Eur. Phys. J.* **D18**, 251.
van Kampen, N G (1988) *Physica* **A153**, 97.
von Weizsäcker, C F (1941) *Zeits. Phys.* **118**, 489.
Walborn, S P, Terra Cunha, M O, Pádua, S, *et al.* (2002) *Phys. Rev.* **A65**, 033818.
Weihs, G, Jennewein, T, Simon, C, *et al.* (1998) *Phys. Rev. Lett.* **81**, 5039.
Wheeler, J A (1979), In *Problems in the Formulation of Physics*, ed. G. T. diFrancia, p. 395. Amsterdam: North-Holland.
Wootters, W K and Zurek, W H (1979) *Phys. Rev.* **D19**, 473.
Zajonc, A G, Wang, L J, Zou, X Y, *et al.* (1991) *Nature* **353**, 507.

16

Quantum feedback and the quantum–classical transition

Hideo Mabuchi
California Institute of Technology

Exploring quantum reality?

Twentieth-century physics bequeaths us an unruly enigma in the equivocal dichotomy between quantum and classical. Mesoscopic systems: which are they, or when? To some this distinction is but a matter of modeling convenience; to others, the partition bears ontological weight. Whichever one's stance, debates on this issue sharpen our introspection on "Why the quantum?" by demanding rigorous justification for choices of calculative consequence, intuitively made on every day in every field of physics.

The limits seem clear. For few particles, left to their own devices, quantum mechanics runs rampant with its nonclassical phenomenology, viz. superposition, tunneling, and entanglement. But for the largish objects of our everyday experience, the sensory familiarity of classical mechanics holds sway: each object has its (singular) place, and every obstacle must be gone round or over. Strange, then, to ponder how big things are made from small! Somehow the assemblage of perceivable matter inevitably converts quantum constituents to classical collective, as if the ordering of the universe were ruled by atoms' aversion to the public embarrassment of quantum behavior writ large. It's a shame in a sense, for the senses slighted of paranormal experience, but superposition . . . what would it look like anyway?

Perhaps in progressing from the classical to the quantum, in our vivisective history of science, we've been subverting teleology. Try it on – wonder not why the quantum rules dictate the dynamics of isolated microsystems (apparently they do); wonder instead at the overwhelming evidence that these simple rules can imply, innately, such dissonant rules for classical behavior that emerges robustly from the selfsame interaction of particles in profusion. Our task would then be to clarify and

Science and Ultimate Reality, eds. J. D. Barrow, P. C. W. Davies and C. L. Harper, Jr. Published by Cambridge University Press.

to elucidate, to verify that this is so. We may presume nothing new is needed, to guide the transition from quantum to classical, but until we have *proof* of sufficiency how can we be sure? A science thus ensues, whose goal lies in the derivation of classical mechanics as the typical behavior of aggregate degrees of freedom, when yet embedded in a web of unheeded quantum variables.

From quantum whole, classical excerpts – such are the *fin de siècle* thoughts of a growing cabal, and at first the program seems clear. But a problem rises quickly, from confounding incompatibility of the very descriptors employed by these theories at hand. In quantum realms we routinely ascribe states that admit no classical equivalents, and generic physical dynamics have an obliging propensity to produce them. What excludes the excess possibility, and on what basis? These are murky waters for our derivational ambition, and we have need of clues.

One finds first guidance in the tenets of decoherence, as environmental entanglement can suffice to suppress the most conspicuous of quantum phenomena, organically. But to go further we must address necessity, and ask, how is it that *all* classically illicit behavior is so categorically shielded from perception (what would it look like anyway)? Can we illumine the mechanism (whence the partial trace)? Vestiges of the quantum world surely trickle through to our own, in the form of statistics or matrix elements, but something insulates us from insensible quantum phenomena per se.

Of course, the need for such was apparent from the outset. Copenhagen's answer's still with us – measurement – but fathom this: what was then conjured and hallowed has of late become tantalizingly tangible; we nearly hear its clockworks. If we are to believe in the sufficiency of quantum for classical, we must demystify the measurement in our physics, which task is twofold with both theoretical and experimental components. In our doings, we should strain the utmost claims of quantum measurement theory, to gain an artisan's easy familiarity with its inner workings. We should put it to work! And in so doing, progress from invocation to engagement – as a great teacher once said, we do not understand what we cannot build.

In our thinking, all the while, we must reorder what we know. The key ideas are around us; we need but complete the reconception of quantum mechanics as a theory of *inference* (not ontology!). What emanates from brains can be but a means of rationalizing concurrence or causality, so the strange states of quantum physics must be an unwitting constraint on our powers of prediction. In the passage from micro to macro we demand a certain definitude, whose price would seem to be certainty. And underneath it all, we feel a universe's irrepressible will to spawn incongruous possibility, too fast – too fast even to be swept under the nebulous rug of inexact stipulations and modeling.

Can we *control* such a nature? Let's see.

Conditional evolution in quantum mechanics

Our aim in the introductory section was to argue that measurement should be viewed as a linchpin of the complex interface between quantum and classical physics. In what follows we will try to sketch a program of research to really put quantum measurement theory through its paces, in the guise of developing viable methods for studying quantum feedback and for utilizing it in technological contexts. Quite a start has been made down this road, in recent years, so the chapter will conclude with a brief survey of important results and proposals.

Measurement theory plays two crucial roles in quantum physics. First, it must predict the statistics of measurement outcomes performed on a system of given preparation. Second, it must provide an "optimal" means for computing the *post-measurement state* of a measured system, given a certain outcome. The superb performance of standard Copenhagen for the first task has been proven time and again, for instance in the remarkable agreement of laboriously calculated QED parameters to the results of painstaking experiments (Kinoshita 1996). Its ultimate adequacy for the second task is hardly in doubt, but in all fairness it must be said that this confidence is founded on far less comprehensive experimental evidence. Some key support has been provided, e.g., by detailed studies of photon antibunching (Kimble *et al.* 1977) and of Bell-inequality violations (Kwiat *et al.* 1999), but validation for the most general scenarios remains a topic of ongoing research.

For the purposes of this exposition, generalized measurement theory may be understood to assert (Nielsen and Chuang 2000) that any experimentally realizable procedure that extracts information about the state of a quantum system can be represented by a set of measurement operators $\{\hat{A}_j\}$ such that

$$\langle\psi|\hat{A}_j^\dagger\hat{A}_j|\psi\rangle \geq 0, \qquad \sum_j \hat{A}_j^\dagger\hat{A}_j = \hat{1}, \tag{16.1}$$

where the first equation holds for all states $|\psi\rangle$ in the Hilbert space of the system being measured, and $\hat{1}$ denotes the identity operator on that space. The index j corresponds to possible outcomes of the measurement,[1] which occur with probabilities

$$\Pr(j) = \mathrm{Tr}[\hat{A}_j\rho\hat{A}_j^\dagger], \tag{16.2}$$

where the density operator ρ represents the premeasurement state of the system. Note that eqn (16.2) satisfies the first "task" of measurement theory as described above. As for the second task, the postmeasurement state *given a particular outcome*

[1] If not all of the j-labeled outcomes can be distinguished by the experimenter, one must use more general versions of eqns (16.2) and (16.3) – see Nielsen and Chuang (2000).

j should be given by

$$\rho \mapsto \frac{\hat{A}_j \rho \hat{A}_j^\dagger}{\mathrm{Tr}[\hat{A}_j \rho \hat{A}_j^\dagger]}. \tag{16.3}$$

Understanding that a laboratory procedure must be very well characterized in order for suitable $\{\hat{A}_j\}$ to be chosen with confidence, our task will be to design experimental methodology for the definitive validation of (16.3) in a broad range of physical systems.

Adaptive measurement and feedback control

Experimentally, the validation of conditional evolution models presents an intriguing methodological challenge. The most straightforward test would consist of an experimental procedure such as the following. First, prepare a physical system in some known quantum state $|\psi_0\rangle$. Second, perform a measurement procedure corresponding to measurement operators $\{\hat{A}_j\}$, obtaining a particular (random) outcome value $j = J$. Invoking the pure-state version of (16.3), theory predicts that the postmeasurement should be given by

$$|\psi_0\rangle \to |\psi_J\rangle = \frac{\hat{A}_J |\psi_0\rangle}{\sqrt{\langle\psi_0|\hat{A}_J^\dagger \hat{A}_J|\psi_0\rangle}}, \tag{16.4}$$

and our final task would seem to be to compare the experimentally produced postmeasurement state with $|\psi_J\rangle$. This would represent a stringent test to the extent that the initial state and measurement operators are chosen such that the $|\psi_j\rangle$ corresponding to various obtainable outcomes are distinct. The problem of course lies in the fact that the state of a quantum system cannot be measured in a single realization, which would seem to imply that this comparison between the theoretical state $|\psi_J\rangle$ and the actual experimental product can only be made at an ensemble level.

The thought along these lines would be to appeal to quantum state tomography (Caves *et al.* 2002); if the set of possible measurement outcomes j is finite, we could in principle repeat the preparation of $|\psi_0\rangle$ and performance of $\{\hat{A}_j\}$ many times and focus on each subset of trials in which the same outcome J is obtained. Given that the evolution from $|\psi_0\rangle$ to the postmeasurement state is in fact identical in every trial in which outcome J is obtained, we would then have an ensemble of preparations of $|\psi_J\rangle$ and could implement a tomographic procedure to characterize this quantum state. While essentially valid, this type of "conditional tomography" approach lacks a certain elegance and will generally not be practicable. Serious difficulties arise when the typical set of postmeasurement states $|\psi_j\rangle$ span a high-dimensional space, as for instance could be the case when j is a continuous parameter. Also, it will commonly be the case that for some given physical system of interest, the reliable preparation of an appropriate $|\psi_0\rangle$ and performance of a well-characterized $\{\hat{A}_j\}$ are within experimental capabilities, but execution of a full quantum state tomography procedure is not. One is therefore motivated to seek more efficient and intrinsic methodology for validating conditional evolution models in general experimental scenarios.

One indication of a way to proceed can be gleaned from recent investigations of quantum nondemolition (QND) measurement (Braginsky *et al.* 1995). A number of authors have pointed out that QND measurement of a discrete variable (such as atomic spin) can be verified by comparing the results of measurements performed successively on one and the same quantum system (see Brune *et al.* (1992) for a particularly insightful discussion). Making formal connections with our previous section, if $\hat{O}$ is a discrete observable we may write its spectral decomposition

$$\hat{O} = \sum_j \lambda_j \hat{\Pi}_j, \tag{16.5}$$

where $\hat{\Pi}_j = |\phi_j\rangle\langle\phi_j|$ and

$$\hat{O}|\phi_j\rangle = \lambda_j|\phi_j\rangle. \tag{16.6}$$

The projectors $\{\hat{A}_j = \hat{\Pi}_j\}$ may be taken as a set of measurement operators since projectors are positive and hermiticity of $\hat{O}$ guarantees

$$\sum_j \hat{\Pi}_j^\dagger \hat{\Pi}_j = \sum_j \hat{\Pi} = \hat{1}. \tag{16.7}$$

In this case our conditional evolution rule (16.4) becomes

$$\begin{aligned} |\psi_0\rangle &\mapsto \frac{\hat{\Pi}_j|\psi_0\rangle}{\sqrt{\langle\psi_0|\hat{\Pi}_j^\dagger\hat{\Pi}_j|\psi_0\rangle}} \\ &= |\phi_j\rangle, \end{aligned} \tag{16.8}$$

for any outcome j that can actually occur ($\langle\phi_j|\psi_0\rangle \neq 0$). This being the case, we are guaranteed that in an ideal series of measurements corresponding to $\{\hat{A}_j = \hat{\Pi}_j\}$ (performed on one and the same quantum system, without anything else happening to it in the meantime) we *must* obtain the same outcome J in every measurement. In traditional language, this is just the idea that "measurement of an observable" collapses the system into an eigenstate corresponding to the obtained eigenvalue. But here we have written out the longhand in anticipation of the more general scenario.

The preceding discussion establishes a central idea, that details of the conditional evolution of a measured quantum system are visible in correlations among the results of subsequent measurements. Note that for a given initial state $|\psi_0\rangle$, the sequence of results obtained in repeated measurements of a given $\{\hat{A}_j\}$ is generally random. As long as more than one outcome is possible in the first measurement, the first postmeasurement will depend on the first random outcome, thus determining the statistics of the next measurement, *et cetera*. In the case of a discrete projection-valued measurement $\{\hat{A}_j = \hat{\Pi}_j\}$, it so happens that the first random outcome strictly determines all subsequent outcomes, but this is clearly a special case.

For future use let us refer (loosely) to the path through Hilbert space, which starts at a given initial state $|\psi_0\rangle$ and proceeds under the conditional evolution rule according to a *particular* sequence of measurement results, as a "quantum trajectory." Thus, each individual experimental "trial" comprising preparation of a physical system in $|\psi_0\rangle$ and performance of a sequence of measurements generates a quantum trajectory. Up to normalization, we may write

$$|\psi_0\rangle \rightarrow \hat{A}_{j_1}|\psi_0\rangle \rightarrow \hat{A}_{j_2}\hat{A}_{j_1}|\psi_0\rangle \rightarrow \hat{A}_{j_n}\cdots\hat{A}_{j_2}\hat{A}_{j_1}|\psi_0\rangle, \tag{16.9}$$

where j_i indicates the ith measurement outcome. In the general case of continuous measurements performed on an open quantum system (Peres and Wootters 1985; Caves 1987; Caves and Milburn 1987; Wiseman and Milburn 1993), the sequence generalizes to a stochastic differential equation driven by a continuous (in time) measurement signal $j(t)$.

Returning to our original goal of formulating an experimental methodology for validation of conditional evolution models, we now see that one strategy would be to perform many sequences of measurements (which could all begin, for convenience, with a given initial state $|\psi_0\rangle$) and then to evaluate autocorrelation statistics of the outcome variable j. While this could certainly be done (see for example Mabuchi (1996), Gambetta and Wiseman (2001)), we know from analogous classical contexts (Ljung 1999) that the addition of *real-time feedback* can greatly enhance the discriminatory power of such methods. That is, *within* a single sequence of measurements, we would like to be able to change the parameters of a given measurement based on the outcomes obtained in the previous ones. This procedure (together with its continuous-measurement generalization) introduces the notion of an *adaptive quantum measurement* (Peres and Wootters 1991; Wiseman 1995).

In the case of open quantum systems, further embellishments are possible and desirable. For the current purposes, we can define an open quantum system to be one that evolves according to its own intrinsic Hamiltonian $\hat{\mathcal{H}}_0$ but also interacts with a reservoir. When a Markov approximation applies, the dynamical evolution of the system density operator $\rho(t)$ can be described by a (Lindblad) master equation

(Gardiner and Zoller 2000):

$$\dot{\rho} = \frac{-i}{\hbar}[\hat{\mathcal{H}}_0, \rho(t)] + \sum_j \{2\hat{c}_j \rho(t) \hat{c}_j^\dagger - \hat{c}_j^\dagger \hat{c}_j \rho(t) - \rho(t) \hat{c}_j^\dagger \hat{c}_j\}. \quad (16.10)$$

Here the $\{\hat{c}_j\}$ are operators on the system Hilbert space that represent the decohering effects of interactions with the reservoir. In some contexts, a given $\hat{c}_j$ may be associated with a physical "output channel" that is accessible to experimental measurements. For example, if the system of interest is an eigenmode of a high-finesse optical cavity with the usual $\hat{\mathcal{H}}_0 = \hbar\omega\hat{a}^\dagger\hat{a}$ (where $\hat{a}$ is the annihilation operator for photons), leakage of light through one of the mirrors will be associated with a decay operator $\hat{c} \propto \hat{a}$. In as much as the light leaks out into a well-defined Gaussian optical mode, this output channel is indeed accessible to measurement by a photon counter or a homodyne/heterodyne receiver. As mentioned above, continuous monitoring of the output channel will induce a stochastic conditional evolution (quantum trajectory) for the system state.

While it is clear that one could contemplate "adaptive monitoring" of an open quantum system, additional possibilities arise if we are able to change the system Hamiltonian $\hat{\mathcal{H}}_0$ in real time. In the simplest case where $\hat{\mathcal{H}}_0$ depends on one or several c-number parameters $\vec{\theta}$ that are subject to experimental control, we may envision a process of dynamical *quantum feedback control* (Belavkin 1999; Wiseman 1994; Doherty *et al.* 2000; Lloyd 2000; Rabitz *et al.* 2000; Lloyd and Viola 2001) in which the parameters $\vec{\theta}(t)$ are made to be a function of information (measurement results) obtained by monitoring of output channels. As we know from classical engineering contexts (Doyle *et al.* 1992; Jacobs 1997), the use of real-time feedback can drastically expand our ability to control dynamical systems that are open to environmental perturbation. Hence, within the context of our overall goal of validating conditional evolution models, we see the alluring possibility of designing purposeful feedback control schemes that accomplish some interesting goal if and only if the conditional evolution model is correct. While this type of "closed-loop system identification" (Ljung 1999) has not yet been carefully examined in the quantum context (although (Doherty *et al.* 2001) has some relevance), a number of interesting proposals for quantum feedback control have appeared in the literature (see below).

Proposed and ongoing experiments

In order to perform definitive experiments on quantum feedback, we require a number of challenging technical requirements to be met. While it would be specious to try to dictate a list of necessary and sufficient criteria, the following points must certainly be considered in any proposal or attempted implementation:

1. It should be possible to identify a well-defined Hilbert space in which a quantum state will be estimated and/or controlled in a manner that relies on the use of real-time feedback.
2. It should be possible to verify that the addition of real-time feedback induces a predicted statistical effect on ensembles of trajectories.
3. The signal-to-noise ratio achievable in a continuous measurement or sequence of discrete measurements should be sufficient to allow the observation of quantum fluctuations in the dynamical variable(s) of interest, or of quantum uncertainty in a parameter that is to be estimated. This is essentially a requirement that the quantum features of the system of interest should be observable and relevant to the feedback loop.
4. Entropy production associated with environmental perturbations or model uncertainties should not dominate conditioning of the system state by the continuous measurement or sequence of discrete measurements. This point is meant to ensure that measurement back-action is a significant factor in the system's evolution.

In a sense, we might say that quantum trajectory theory (Carmichael 1991; Wiseman 1993) should be a natural and enabling formalism for the analysis and design of any bona fide experiment on quantum feedback. However, experience has shown that it is often possible to utilize conceptual insights from the quantum trajectory paradigm while getting away with simplified theoretical treatments. In any case an ideal quantum feedback experiment should cover all of the above points, and should additionally demonstrate the use of quantum feedback to accomplish "phenomenological" goals that would be more difficult or impossible to achieve otherwise. A number of proposals and even a few seminal experiments in line with our desiderata have recently emerged in the general area of atomic and optical physics. In the remainder of this chapter we will survey some highlights of such recent work, indicating key features of the proposals and demonstrations with regard to quantum feedback in the context of the quantum–classical transition.

Quantum optics

Noise rejection and state preparation via real-time feedback have long been a subject of intense investigation in quantum optics. Dating back to such early works as Machida and Yamamoto (1986), developing experimental capabilities in quantum-noise limited measurement and opportunities for real-time feedback have driven accommodative theoretical research (Shapiro *et al.* 1987). Much of the initial and sustained excitement has revolved around the use of real-time feedback to reduce quadrature–amplitude fluctuations of an electromagnetic field mode below the vacuum limit; recent theoretical analyses have greatly clarified the relation of such "in-loop squeezing" schemes to investigations of bona fide optical squeezed states (Taubman *et al.* 1995; Wiseman 1998).

The implications of real-time feedback in quantum optics have also been explored in contexts related to metrology and communication. Going back to early work by Dolinar (1973), there has been a general appreciation that the addition of real-time feedback capability to otherwise standard (photon-counting or homodyne) photodetection apparatus can enable new and interesting quantum measurements. Dolinar proposed that the Helstrom measurement (Helstrom 1976) for optimal discrimination between two nonorthogonal quantum states (here weak coherent states) could be realized by promptly adjusting the amplitude and phase of an interferometric reference beam, triggered by the output of a high-efficiency photon counter. This would be an adaptive quantum measurement according to the discussion in section "Adaptive measurement and feedback control." Subsequent theoretical analyses have further shown that photodetection with real-time adaptation of local oscillator parameters can be used to implement nearly-ideal measurements of optical phase (Yamamoto *et al.* 1990; Wiseman 1995), and unambiguous discrimination among multiple overlapping coherent states (van Enk 2002).

Wiseman's proposed scheme for adaptive homodyne measurement of optical phase (Wiseman 1995) has recently been implemented experimentally by Armen *et al.* (2002). This work demonstrates that for a signal ensemble consisting of weak coherent states, the addition of real-time feedback to an optical homodyne set-up enables phase measurements whose variance approaches closer to the fundamental quantum uncertainty limit than any previous technique. In addition to its purely metrological significance, these results demonstrate an important technical achievement in the quantum feedback paradigm – the quantitative performance of adaptive phase measurement depends in a critical way on having technically "perfect" preparation of the optical signal and reference states, on realizing broadband[2] quantum-noise-limited photodetection with flat gain and phase, and on implementing signal processing with negligible latency. The work of Armen *et al.* also introduced the use of programmable logic devices (Stockton *et al.* 2002) for execution of signal processing algorithms in the feedback loop, demonstrating a promising hardware platform for future quantum feedback experiments involving state estimation (Doherty and Jacobs 1999).

The type of focused activity discussed in this section has engendered keen understanding among quantum optics researchers of how to discriminate between technical noise and intrinsic quantum fluctuations in real experimental scenarios, and of how theoretical analyses can be made to conform as closely as possible to the details of laboratory procedure (see Warszawski *et al.* (2001) for an extreme example). The putative dividing line between semi-classical and "deep quantum" effects has likewise been the object of withering (though still inconclusive) scrutiny in quantum

[2] Here the term "broadband" is meant in the signal processing sense, connoting coverage of multiple decades of spectrum rather than ultrafast absolute timescales.

optics, which in some ways can be taken as a paragon for broader investigations of the quantum–classical interface. For the purposes of this chapter, however, it should be noted that such discussions within quantum optics are unavoidably couched in a historical context of distinguishing between classical and quantum *optics*. This can largely be understood as an investigation of how to characterize optical processes and photocurrent correlation functions that necessarily involve noncoherent states of the electromagnetic field, where "coherent state" here specifically denotes an eigenstate of the field annihilation operator. Quantum–classical debates within quantum optics can thus be rather narrowly focused on issues of squeezing or antibunching, at times.

Recent interest in establishing connections between quantum optics and quantum information science (Braunstein and Kimble 1998) has broadened the perspective, however, and has highlighted the crucial role of *conditional* dynamics. A string of theoretical results such as those by Cirac, Zoller, and coworkers (Duan *et al.* 2000, 2001) has brought attention to the fact that quantum-noise-limited photodetection coupled with real-time feedback (as in Thomsen *et al.* (2002), discussed below) can lead to the generation of entangled states of atomic systems that have coupled to coherent states of light. Likewise, works by Knill, Laflamme, and Milburn (Knill *et al.* 2001) and by Raussendorf and Briegel (2001) have emphasized subtle forms of "exchangeability" among states, dynamics, and measurements in quantum mechanics that have yet to be fully explored. Deep connections between the subject of quantum error correction and quantum feedback control have also been recognized, and are starting to be explored (Ahn *et al.* 2002).

Atomic motion

There has been rapidly growing interest of late in the use of cavity quantum electrodynamics (cavity QED) in the strong coupling regime (Kimble 1998) to implement real-time measurement of the motion of individual atoms. Theoretical analyses (Quadt *et al.* 1995; Rempe 1995) of such measurements were quickly followed by rudimentary experimental demonstrations (Mabuchi *et al.* 1996), which in turn spurred new investigations of the feasibility of real-time feedback control on single atoms (Dunningham *et al.* 1997; Wong *et al.* 1997; Doherty and Jacobs 1999). A primary interest of cavity QED schemes is that they promise an ability to reach the Standard Quantum Limit for continuous position measurements of single atoms (Mabuchi 1998; Verstraete *et al.* 2001), although some formidable technical issues remain to be solved (Mabuchi *et al.* 1999).

The use of real-time feedback to affect atomic motion has already been accomplished experimentally (Hood *et al.* 2000; Fischer *et al.* 2002). Although these demonstrations were quite convincing in the efficacy of the feedback to gain control over an individual atom's motion, in the sense of our above criteria 1 and 2, they did not manage to reach a truly quantum regime for feedback control in the sense of criteria 3 and 4. Of course, work continues in these laboratories and in others (Morrow *et al.* 2002), and it seems certain that true quantum feedback control of atomic motion will be accomplished in the next few years.

Cavity QED

In addition to atomic motion, cavity QED with strong coupling provides a promising paradigm for feedback control of discrete quantum variables. Of particular interest is the system composed of a single two-level atom coupled to photons in a near-resonant optical cavity. The coherent dynamics of such a system are described by the fundamental Jaynes–Cummings model (Thompson *et al.* 1998), and rigorous quantitative treatment of reservoir couplings can easily be achieved with a master equation (Mabuchi *et al.* 1999).

The conditioning of the atom–cavity dynamics on measured output fields can be quite strong in current experimental systems, but until recently the true single-system lifetime (dwell time of any given atom within the cavity mode volume) had been too short to allow for real-time signal processing and feedback. The latter problem has been largely eliminated through the integration of laser cooling techniques with cavity QED (Mabuchi *et al.* 1996), and in addition it has been realized that there are interesting feedback protocols to apply that require minimal complexity of signal processing.

An important step has recently been taken by Smith *et al.* (2002), who succeeded in utilizing real-time feedback to predictably and nontrivially alter the photon statistics of a cavity QED system. The formulation of their scheme relies essentially on insights from an appropriate conditional evolution model, and in a sense satisfies all of the criteria discussed at the beginning of this section. The one shortcoming of their work (from the very single-minded perspective of this chapter) is the lack of any compelling means for *quantitative* verification of conditional evolution equations, which in turn stems from the timescales being somewhat marginal. Nonetheless, these results pioneer a vital new direction for cavity QED and for quantum feedback research.

Along a similar direction, a recent theoretical analysis by Mabuchi and Wiseman (1998) has extended earlier work by Alsing and Carmichael (1991) to show that single-atom cavity QED experiments should be able to observe quantum "analogs" of dynamical phenomena originally studied in the context of optical bistability (Lugiato 1984). The control of such dynamics by feedback would be fascinating, and an initial investigation following Mabuchi and Wiseman (1998) is currently in preparation (J. E. Reiner *et al.*, unpubl. data).

While at present far from experimental tractability, several schemes have been elaborated for the use of quantum feedback to preserve coherence in the quantum states of optical resonators (Mabuchi *et al.* 1996; Fortunato *et al.* 1999), much in the spirit of quantum error correction.

Additional proposals

As briefly mentioned above in connection with quantum information science, a new frontier for applications of quantum feedback is the production of entanglement in atomic ensembles. Beginning with experimental work by Polzik *et al.* (Julsgaard *et al.* 2001) and by Bigelow *et al.* (Kuzmich *et al.* 2000), there has been surging interest in the ability of continuous observation to produce conditional spin squeezing (Bouchoule and Mølmer 2002) and in the use of quantum feedback to produce unconditional spin squeezing (Thomsen *et al.* 2002). Much like optical squeezing, spin squeezing is a promising technique for reduction of quantum noise below the Standard Quantum Limit set by (spin) coherent states; as an experimental phenomenon, it has the additional interest of implying a high degree of entanglement among a large number of atoms.

Also dealing with collective quantum variables, Wiseman and Thomsen (2001) have investigated feedback schemes for narrowing the linewidth of an atom laser. Moving towards even more macroscopic systems, several authors have recently considered the ultimate ability of feedback to reduce fluctuations in the position of an optical mirror (Buchler *et al.* 1999; Vitali *et al.* 2002). In the opposite extreme there have been several highly detailed theoretical works on quantum feedback control of the bare state of a single two-level atom (Hofmann *et al.* 1998; Wang and Wiseman 2001), which provide important examples of how complete an analysis is possible in some cases.

On the purely experimental side, it is worth mentioning that interesting quantum feedback experiments should soon become feasible with electrons in Penning traps (Peil and Gabrielse 1999), atomic ions in rf Paul traps (Meyer *et al.* 2001), degenerate atomic gases (Andrews *et al.* 1996), single-electron transistors (Devoret and Schoelkopf 2000), and micromechanical systems (Armour *et al.* 2002). We should note that appropriate conditional evolution models have not yet been rigorously derived in the latter three cases, although the rough ideas are self-evident.

Concluding remarks

While hardly exhaustive, the preceding survey shows that considerable activity is mounting in quantum feedback research. Much of the excitement comes from the possibility of applying quantum feedback towards important technological problems in metrology and information science. But if conditional measurement models enable the design of feedback schemes that could not properly be formulated otherwise, and if the the use of feedback delivers system performance that beats whatever could be achieved without it, we must surely feel that significant strides are being taken towards demystifying quantum measurement.

Acknowledgments

Christopher A. Fuchs and John C. Doyle have been crucial influences in the development of intellectual perspectives expressed in this chapter, but don't blame them if you disagree with what I wrote. The poetry of Richard Kenney was an inspiration for the introductory section, but don't blame him for anything either. Let me also acknowledge vital research support from the National Science Foundation, the Office of Naval Research, the Army Research Office, and the A. P. Sloan Foundation. Finally, I owe all kinds of debts to the Fellows Program of the John D. and Catherine T. MacArthur Foundation and to programs sponsored by the Jefferson Institute; they and their people have my sincerest gratitude.

References

Ahn, C, Doherty, A C, and Landahl, A (2002) Continuous quantum error correction via quantum feedback control. *Phys. Rev.* **A65**, 042301.

Alsing, P and Carmichael, H J (1991) Spontaneous dressed-state polarization of a coupled atom and cavity mode. *Quantum Opt.* **3**, 13.

Andrews, M R, Mewes, M O, van Druten, N J, *et al.* (1996) Direct, nondestructive observation of a Bose condensate. *Science* **273**, 84.

Armen, M A, Au, J K, Stockton, J K, *et al.* (2002) Adaptive homodyne measurement of optical phase. *Phys. Rev. Lett.* **89**, 133602.

Armour, A D, Blencowe, M P, and Schwab, K C (2002) Entanglement and decoherence of a micromechanical resonator via coupling to a Cooper-pair box. *Phys. Rev. Lett.* **88**, 148301.

Belavkin, V P (1999) Measurement, filtering, and control in quantum open dynamical systems. *Rep. Math. Phys.* **43**, 405.

Bouchoule, I and Mølmer, K (2002) Preparation of spin squeezed atomic states by optical phase shift measurement. quant-ph/0205082.

Braginsky, V B, Khalili, F Y, and Thorne, K S (1995) *Quantum Measurement*. Cambridge: Cambridge University Press.

Braunstein S L, and Kimble, H J (1998) Teleportation of continuous quantum variables. *Phys. Rev. Lett.* **80**, 869.

Brune, M, Haroche, S, Raimond, J M, *et al.* (1992) Manipulation of photons in a cavity by dispersive atom-field coupling: quantum-nondemolition measurements and generation of Schrödinger cat states. *Phys. Rev.* **A45**, 5193.

Buchler, B C, Gray, M B, Shaddock, D A, *et al.* (1999) Suppression of classic and quantum radiation pressure noise by electro-optic feedback. *Opt. Lett.* **24**, 259.

Carmichael, H J (1991) *An Open Systems Approach to Quantum Optics*. Berlin: Springer-Verlag.

Caves, C M (1987) Quantum-mechanics of measurements distributed in time. 2. connections among formulations. *Phys. Rev.* **D35**, 1815.
Caves, C M and Milburn, G J (1987) Quantum mechanical model for continuous position measurements. *Phys. Rev.* **A36**, 5555.
Caves, C M, Fuchs, C A, and Schack, R D (2002) Quantum probabilities as Bayesian probabilities. *Phys. Rev.* **A65**, 022305.
Devoret, M H and Schoelkopf, R J (2000) Amplifying quantum signals with the single-electron transistor. *Nature* **406**, 1039.
Doherty, A C and Jacobs, K (1999) Feedback control of quantum systems using continuous state estimation. *Phys. Rev.* **A60**, 2700.
Doherty, A C, *et al.* (2000) Quantum feedback control and classical control theory. *Phys. Rev.* **A62**, 012105.
Doherty, A C, Jacobs, K, and Jungman, G (2001) Information, disturbance, and Hamiltonian quantum feedback control. *Phys. Rev.* **A63**, 062306.
Dolinar, S (1973) An optimum receiver for the binary coherent state quantum channel. *MIT Res. Lab. Electron. Quart. Prog. Rep.* **111**, 115.
Doyle, J C, Tannenbaum, A, and Francis, B (1992) *Feedback Control Theory*. Englewood Cliffs, NJ: Macmillan.
Duan, L M, Cirac, J I, Zoller, P, *et al.* (2000) Quantum communication between atomic ensembles using coherent states of light. *Phys. Rev. Lett.* **85**, 5643.
Duan, L M, Lukin, M D, Cirac, J I, *et al.* (2001) Long-distance quantum communication with atomic ensembles and linear optics. *Nature* **414**, 413.
Dunningham, J A, Wiseman, H M, and Walls, D F (19997) Manipulating the motion of a single atom in a standing wave via feedback. *Phys. Rev.* **A55**, 1398.
Enk, S J van (2002) Phase measurements with weak reference pulses. quant-ph/0207142.
Fischer, T, Maunz, P, Pinske, P W H, *et al.* (2002) Feedback on the motion of a single atom in an optical cavity. *Phys. Rev. Lett.* **88**, 163002.
Fortunato, M, Raimond, J M, Tombesi, P, *et al.* (1999) Autofeedback scheme for preservation of coherence in microwave cavities. *Phys. Rev.* **A60**, 1687.
Gambetta, J and Wiseman, H M (2001) State and dynamical parameter estimation for open quantum systems. *Phys. Rev.* **A64**, 042105.
Gardiner, C W and Zoller, P (2000) *Quantum Noise*. Berlin: Springer-Verlag.
Helstrom, C W (1976) *Quantum Detection and Estimation Theory*. New York: Academic Press.
Hofmann, H F, Mahler, G, and Hess, O (1998) Quantum control of atomic systems by homodyne detection and feedback. *Phys. Rev.* **A57**, 4877.
Hood, C J, Lynn, T W, Doherty, A C, *et al.* (2000) The atom–cavity microscope: single atoms bound in orbit by single photons. *Science* **287**, 1447.
Jacobs, O L R (1997) *Introduction to Control Theory*. Oxford: Oxford University Press.
Julsgaard, B, Kozhekin, A, and Polzik, E S (2001) Experimental long-lived entanglement of two macroscopic objects. *Nature* **413**, 400.
Kimble, H J (1998) Strong interactions of single atoms and photons in cavity QED. *Phys. Scripta* **T76**, 127.
Kimble, H J, Dagenais, M, and Mandel, L (1977) Photon anti-bunching in resonance fluorescence. *Phys. Rev. Lett.* **39**, 691.
Kinoshita, T (1996) The fine structure constant. *Rep. Prog. Phys.* **59**, 1459.
Knill, E, Laflamme, R, and Milburn, G J (2001) A scheme for efficient quantum computation with linear optics. *Nature* **409**, 46.
Kuzmich, A, Mandel, L, and Bigelow, N P (2000) Generation of spin squeezing via continuous quantum nondemolition measurement. *Phys. Rev. Lett.* **85**, 1594.

Kwiat, P G, Waks, E, White, A G, *et al.* (1999) Ultrabright source of polarization-entangled photons. *Phys. Rev.* **A60**, 773.

Ljung, L (1999) *System Identification: Theory for the User*. Upper Saddle River, NJ: Prentice Hall.

Lloyd, S (2000) Coherent quantum feedback. *Phys. Rev.* **A62**, 022108.

Lloyd, S and Viola, L (2001) Engineering quantum dynamics. *Phys. Rev.* **A65**, 010101.

Lugiato, L A (1984) Theory of optical bistability. *Prog. Optics* **21**, 69.

Mabuchi, H (1996) Dynamical identification of open quantum systems. *Quant. Semicl. Opt.* **8**, 1103.

(1998) Standard Quantum Limits for broadband position measurement. *Phys. Rev.* **A58**, 123.

Mabuchi, H and Wiseman, H M (1998) Retroactive quantum jumps in a strongly coupled atom-field system. *Phys. Rev. Lett.* **81**, 4620.

Mabuchi, H, Turchette, Q A, Chapman, M S, *et al.* (1996) Real-time observation of individual atoms falling through a high-finesse optical cavity. *Opt. Lett.* **21**, 1393.

Mabuchi, H, Ye, J, and Kimble, H J (1999) Full observation of single-atom dynamics in cavity QED. *Appl. Phys.* **B68**, 1095.

Machida, S and Yamamoto, Y (1986) Observation of sub-Poissonian photoelectron statistics in a negative feedback semiconductor laser. *Opt. Comm.* **57**, 290.

Meyer, V, Rowe, M A, Kielpinski, D, *et al.* (2001) Experimental demonstration of entanglement-enhanced rotation angle estimation using trapped ions. *Phys. Rev. Lett.* **86**, 5870.

Morrow, N V, Dutta, S K, and Raithel, G (2002) Feedback control of atomic motion in an optical lattice. *Phys. Rev. Lett.* **88**, 093003.

Nielsen, M A, and Chuang, I L (2000) *Quantum Computation and Quantum Information*. Cambridge: Cambridge University Press.

Peil, S and Gabrielse, G (1999) Observing the quantum limit of an electron cyclotron: QND measurements of quantum jumps between Fock states. *Phys. Rev. Lett.* **83**, 1287.

Peres, A and Wootters, W K (1985) Quantum measurements of finite duration. *Phys. Rev.* **D32**, 1968.

(1991) Optimal detection of quantum information. *Phys. Rev. Lett.* **66**, 1119.

Quadt, R, Collett, M, and Walls, D F (1995) Measurement of atomic motion in a standing light-field by homodyne detection. *Phys. Rev. Lett.* **74**, 351.

Rabitz, H, *et al.* (2000) Whither the future of controlling quantum phenomena? *Science* **288**, 824.

Raussendorf, R and Briegel, H J (2001) A one-way quantum computer. *Phys. Rev. Lett.* **86**, 5188.

Rempe, G (1995) One-atom in an optical cavity: spatial-resolution beyond the standard diffraction limit. *Appl. Phys.* **B60**, 233.

Shapiro, J H, Saplakoglu, G, Ho, S T, *et al.* (1987) Theory of light detection in the presence of feedback. *J. Opt. Soc. Am.* **B4**, 1604.

Smith, W P, Reiner, J E, Orozco, L A, *et al.* (2002) Capture and release of a conditional state of a cavity QED system by quantum feedback. *Phys. Rev. Lett.* **89**, 133601.

Stockton, J, Armen, M, and Mabuchi, H (2002) Programmable logic devices in experimental quantum optics. quant-ph/0203143.

Taubman, M S, Wiseman, H, McClelland, D E, *et al.* (1995) Intensity feedback effects on quantum-limited noise. *J. Opt. Soc. Am.* **B12**, 1792.

Thompson, R J, Turchette, Q A, Carnal, O, *et al.* (1998) Nonlinear spectroscopy in the strong coupling regime of cavity QED. *Phys. Rev.* **A57**, 3084.

Thomsen, L K, Mancini, S, and Wiseman, H M (2002) Spin squeezing via quantum feedback. *Phys. Rev.* **A65**, 061801.

Verstraete, F, Doherty, A C, and Mabuchi, H (2001) Sensitivity optimization in quantum parameter estimation. *Phys. Rev.* **A64**, 032111.

Vitali, D, Mancini, S, Ribichini, L, *et al.* (2002) Mirror quiescence and high-sensitivity position measurements with feedback. *Phys. Rev.* **A65**, 063803.

Wang, J and Wiseman, H M (2001) Feedback-stabilization of an arbitrary state of a two-level atom. *Phys. Rev.* **A64**, 003810.

Warszawski, P, Wiseman, H M, and Mabuchi, H (2001) Quantum trajectories for realistic photodetection. *Phys. Rev.* **A65**, 023802.

Wiseman, H M (1994) Quantum theory of continuous feedback. *Phys. Rev.* **A49**, 2133.

(1995) Adaptive phase measurements of optical modes: going beyond the marginal Q distribution. *Phys. Rev. Lett.* **75**, 4587.

(1998) In-loop squeezing is like real squeezing to an in-loop atom. *Phys. Rev. Lett.* **81**, 3840.

Wiseman, H M and Milburn, G J (1993) Quantum theory of field-quadrature measurements. *Phys. Rev.* **A47**, 642.

Wiseman, H M and Thomsen, L K (2001) Reducing the linewidth of an atom laser by feedback. *Phys. Rev. Lett.* **86**, 1143.

Wong, K S, Collett, M J, and Walls, D F (1997) Atomic juggling using feedback. *Opt. Commun.* **137**, 269.

Yamamoto, Y, Machida, S, Saito, S, *et al.* (1990) Quantum-mechanical limit in optical-precision measurement and communication. *Prog. Optics* **28**, 87.

17

What quantum computers may tell us about quantum mechanics

Christopher R. Monroe
University of Michigan, Ann Arbor

Quantum mechanics occupies a unique position in the history of science. It has survived all experimental tests to date, culminating with the most precise comparison of any measurement to any theory – a 1987 measurement of the electron's magnetic moment, or gyromagnetic ratio $g_e = 2.002\,319\,304\,39$ (Van Dyck *et al.* 1987), agreeing with QED theory to 12 digits. Despite this and other dramatic successes of quantum mechanics, its foundations are often questioned, owing to the glaring difficulties in reconciling quantum physics with the classical laws of physics that govern macroscopic bodies. If quantum mechanics is indeed a complete theory of nature, why does it not apply to everyday life? Even Richard Feynman (1982), a fierce defender of quantum mechanics, memorably stated that:

> We have always had a great deal of difficulty in understanding the world view that quantum mechanics represents . . . Okay, I still get nervous with it . . . It has not yet become obvious to me that there is no real problem. I cannot define the real problem, therefore I suspect there's no real problem, but I'm not sure there's no real problem.

In the dawn of the twenty-first century, John A. Wheeler's big question "Why the quantum?" has returned to the forefront of physics with full steam. Advances in experimental physics are beginning to realize the same thought-experiments that proved helpful to Einstein, Bohr, Heisenberg, Schrödinger, and the other founders of quantum mechanics. The current progression toward nanotechnology, where electronic computing and storage media are being miniaturized to the atomic scale, is beginning to confront quantum-mechanical boundaries, as foreseen in Feynman's early charge, "There's plenty of room at the bottom." While many of these effects are inhibiting the continued miniaturization, new opportunities

Science and Ultimate Reality, eds. J. D. Barrow, P. C. W. Davies and C. L. Harper, Jr. Published by Cambridge University Press.

such as quantum information processing are arising (Nielsen and Chuang 2000), providing a great incentive to build devices that may not only eclipse the performance of current devices, but also may push quantum theory to its limits. From the standpoint of physics, the new field of quantum information science gives us a very useful language with which to revisit the fundamental aspects of quantum mechanics.

Quantum information processing

Information theory began in the mid twentieth century, with Claude Shannon's seminal discovery of how to quantify classical information (Shannon 1948). Shannon's bit, or binary digit, became the fundamental unit, providing a metric for comparing forms of information and optimizing the amount of resources needed to faithfully convey a given amount of information, even in the presence of noise. Shannon's pioneering work led to the experimental representation of bits in nature, from unwieldy vacuum tubes in the mid twentieth century to the modern VLSI semiconductor transistors of under 0.1 μm in size. Under this impressive progression of technology, we have enjoyed an exponential growth in computing power and information processing speed given by the familiar "Moore's law," where computer chips have doubled in density every year or two.

But this growth will not continue indefinitely. As bits continually shrink in size, they will eventually approach the size of individual molecules – by the year 2020 if the current growth continues. At these nanometer-length scales, the laws of quantum mechanics begin to hold sway. Quantum effects are usually thought of as "dirty" in this context, causing unwanted tunneling of electrons across the transistor gates, large fluctuations in electronic signals, and generally adding noise. However, Paul Benioff and Richard Feynman showed in the early 1980s that quantum-mechanical computing elements such as single atoms could, in principle, behave as adequate electronic components not hampered by dirty quantum effects (Benioff 1980, 1982; Feynman 1982). They even discussed using "quantum logic gates" largely following the laws of quantum mechanics, and Feynman became interested in the idea of using model quantum systems to simulate efficiently other intractable quantum systems (Feynman 1982).

Soon after, David Deutsch went a step further by using the full arsenal of quantum mechanical rules. Deutsch proposed that the phenomenon of quantum superposition be harnessed to yield massively parallel processing – computing with multiple inputs at once in a single device (Deutsch 1985). Instead of miniaturizing chip components further, Deutsch posed an end-run around the impending limits of Moore's law by taking advantage of different physical principles underlying these components.

Whereas Shannon's classical bit can be either 0 or 1, the simplest quantum-mechanical unit of information is the quantum bit or *qubit*, which can store superpositions of 0 and 1. A single qubit is represented by the quantum state

$$\Psi_1 = \alpha|0\rangle + \beta|1\rangle, \tag{17.1}$$

where α and β are the complex amplitudes of the superposition. The states $|0\rangle$ and $|1\rangle$ may represent, for example, horizontal and vertical polarization of a single photon, or two particular energy levels within a single atom. The standard (Copenhagen) rules of quantum mechanics dictate that: (a) the time development of amplitudes α and β is described by the Schrödinger wave equation, and (b) when the above quantum bit is measured, it yields either $|0\rangle$ or $|1\rangle$ with probabilities given by $|\alpha|^2$ and $|\beta|^2$, respectively. The measurement of a quantum bit is much like flipping a coin – the results can only be described within the framework of probabilities.

Hints of the power of quantum computing can be seen by considering a register of many qubits. In general, N qubits can store a superposition of all 2^N binary numbers:

$$\Psi_N = \gamma_0|000\cdots 0\rangle + \gamma_1|000\cdots 1\rangle + \cdots + \gamma_{2^N-1}|111\cdots 1\rangle. \tag{17.2}$$

To appreciate the power of this exponential storage capacity, note that with merely $N = 300$ quantum bits, the most general quantum state requires over 10^{90} amplitudes. This is more than the number of fundamental particles in the universe!

When a quantum computation is performed on a quantum superposition, each piece gets processed in superposition. For example, quantum logic operations can shift all the qubits one position to the left, equivalent to multiplying the input by two. When the input state is in superposition, all inputs are simultaneously doubled in one step (see Fig. 17.1a). After this quantum parallel processing, the state of the qubits must ultimately be measured. Herein lies the difficulty in designing useful quantum computing algorithms: according to the laws of quantum mechanics, this measurement yields just one answer out of 2^N possibilities; worse still, there is no way of knowing which answer will appear. Apparently quantum computers cannot compute one-to-one functions (where each input results in a unique output as in the doubling algorithm above) any more efficiently than classical computers.

The trick behind a useful quantum computer algorithm involves the phenomenon of quantum interference. Since the amplitudes $\gamma_0, \gamma_1, \ldots \gamma_{2^N-1}$ in the superposition of eqn (17.2) evolve according to a wave equation, they can be made to interfere with each other. In the end, the parallel inputs are processed with quantum logic gates so that almost all of the amplitudes cancel, leaving only a very small number of answers, or even a single answer, as depicted in Fig. 17.1b. By measuring this answer (or repeating the computation a few times and recording the distribution of

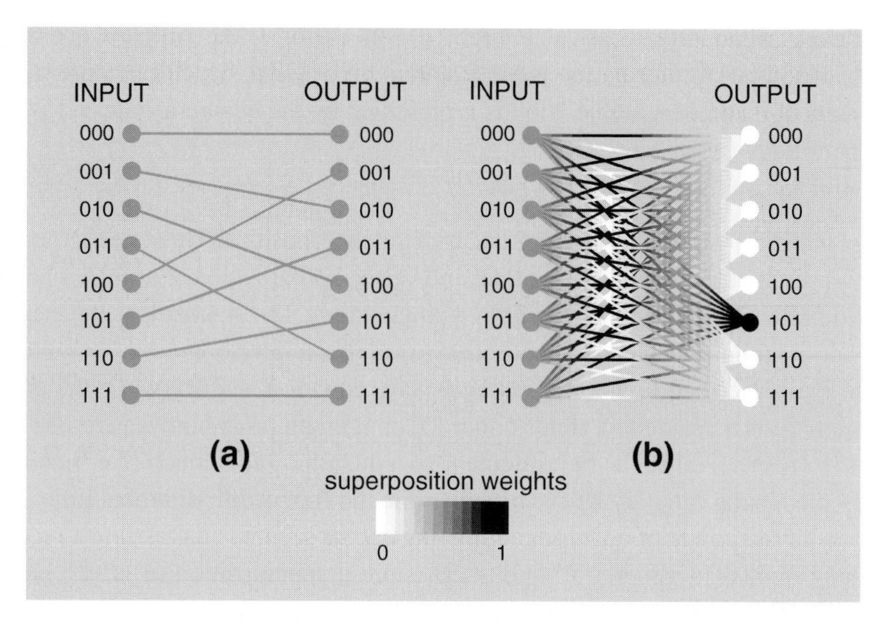

Figure 17.1. Simplified evolution during a $N = 3$ quantum bit quantum algorithm. The inputs are prepared in superposition states of all $2^N = 8$ possible numbers (written in binary). The weights of the superposition are denoted by the grayscale, where black is a large weight and white is a zero weight. (a) Quantum algorithm for simultaneously doubling all input numbers (Modulo 7), by shifting all qubits one position to the left and wrapping around the leftmost bit. The outputs are also in superposition, and a final measurement projects one answer at random. (b) Quantum algorithm involving wavelike interference of weights. Here, quantum logic gates cause the input superposition to interfere, ultimately canceling all of the weights except for one (101 in the figure) which can then be measured. For some algorithms, this lone answer (or the distribution of a few answers after repeated runs) can depend on the weights of all 2^N input states, leading to an exponential speed-up over classical computers.

answers), information can be gained pertaining to all 2^N inputs. In some cases, this implies an exponential speed-up over what can be obtained classically.

In 1994, Peter Shor devised a quantum algorithm to factor numbers into their divisors (Shor 1997). He showed that a quantum computer is able to factorize exponentially faster than any known classical algorithm. This discovery led to a rebirth of interest in quantum computers, in part due to the importance of factoring for cryptography – the security of popular cryptosystems such as those used for internet commerce is derived from the *inability* to factor large numbers (Rivest *et al.* 1978). But perhaps more importantly, Shor's algorithm showed that quantum computers are indeed good for something, spurring physicists, mathematicians, and computer scientists to search for other algorithms amenable to quantum computing. In 1996, for example, Lov Grover proved that a quantum computer can search unsorted databases faster than any search conducted on a classical computer (Grover 1997). The happy result of this flurry of activity is that scientists, mathematicians, engineers, and computer scientists are now studying and learning quantum physics, and their language is quantum information science.

Useful quantum algorithms such as Shor's algorithm are not plentiful, and it is unknown how many classes of problems will ultimately benefit from quantum computation. In pursuit of useful quantum algorithms, it's natural to investigate what makes a quantum computer powerful. The answer to this question may not only guide us toward new applications of quantum information science, but may also provide alternative views of the quantum physics underlying these devices.

Quantum entanglement

The implicit parallelism in quantum superpositions is not revolutionary by itself. Indeed, there are many classical wavelike phenomena and analog processing models that involve superposition and interference. The new ingredient offered by quantum superpositions such as eqn (17.2) is that it takes 2^N amplitudes to describe the state of only N qubits. The general state of a quantum computer (eqn (17.2)) exhibits a property not found in classical superpositions: *quantum entanglement*. Entanglement refers to the fact that eqn (17.2) cannot in general be written as a direct product state of the N individual qubits state, which would require only $2N$ amplitudes:

$$\Psi_N^{prod} = (\alpha_1|0\rangle + \beta_1|1\rangle) \otimes (\alpha_2|0\rangle + \beta_2|1\rangle) \otimes \cdots \otimes (\alpha_N|0\rangle + \beta_N|1\rangle). \quad (17.3)$$

The concept of quantum entanglement neatly combines the two properties of quantum mechanics – superposition and measurement – that are by themselves unremarkable, but taken together cause all the usual interpretive conundrums of quantum mechanics. Schrödinger (1935) himself said, "I would not call [entanglement] one but rather *the* characteristic trait in quantum mechanics, the one that enforces an entire departure from all our classical lines of thought." Yet entanglement seems to be one of the most misunderstood concepts in quantum mechanics. There seem to be many levels of definition, with their own supporting assumptions. Below, several possible definitions of entanglement are considered.

The classic case of quantum entanglement is the thought experiment originally proposed by Einstein, Podolsky, and Rosen (Einstein *et al.* 1935). EPR posed a quantum state of two particles expressed in position space as

$$\Psi(x_1, x_2) = \frac{1}{2\pi\hbar} \int e^{i(x_1 - x_2 - s)p/\hbar} dp = \delta(x_1 - x_2 - s), \quad (17.4)$$

where $\delta(x)$ is the Dirac-delta function. The particles are always found to be separated in space by s when their positions are measured, yet they are also found to have precisely opposing momenta (seen by Fourier transforming eqn (17.4)). David Bohm's discrete version of the EPR state (Bohm 1951) is the familiar spin-0 particle decaying into two spin-$\frac{1}{2}$ daughter particles (qubits), represented by the spinor quantum state

$$\Psi(S_1, S_2) = |\uparrow\rangle_1|\downarrow\rangle_2 - |\downarrow\rangle_1|\uparrow\rangle_2, \tag{17.5}$$

where S_1 and S_2 are the spins of the two particles, each taking on one of the two values $\downarrow$ or $\uparrow$. In both cases (eqns (17.4) and (17.5)), the overall quantum state cannot be written as a direct product state of its constituents, and the "essence" of quantum mechanics in these states is the fact that there their correlation is definite, yet the state of the individual particles is not definite. It's tempting to thereby define entanglement as follows:

Definition 1 An entangled state is a quantum state that is not separable.

(For mixed states, this definition can be extended by requiring inseparability of the density matrix.) But this definition is misleading. While the right-hand side of eqn (17.5) certainly cannot be expressed as a direct product state of the spins, the left-hand side of the equation, describing the same state, is obviously not entangled – it is just the simple lone state $\Psi(S_1, S_2)$. For example, in the ground hyperfine states of the hydrogen atom, the entangled singlet state of electron and proton spin is identical to the same state in the usual coupled basis $|J=0, m_J=0\rangle$. Many therefore dismiss the whole notion of entanglement as simply a choice of basis. However, entanglement should not only reflect a nonseparable quantum state, but one in which independent quantum measurements on the individual constituents have taken (or will take) place. This measurement naturally selects the uncoupled basis. It might be unsettling to define a quantity that depends on what the experimenter has done (or will do). But this is exactly how most of us interpret quantum mechanics already.

What makes entanglement interesting is that in almost all cases of quantum states expressed following Definition 1, such as hydrogen ground states in the uncoupled basis, it is virtually impossible to measure particular constituents without directly affecting the others. Unfortunately, it would be quite difficult to prepare a hydrogen atom in the singlet state and subsequently measure the electron spin without affecting the proton spin or vice versa. So we might refine the definition in terms of these measurements:

Definition 2 An entangled state is one that is not separable, where measurements are performed on one constituent without affecting the others.

In order to verify the correlations of the subsystems, there must not be much technical noise associated with the measurement. That is, the detection process itself should not change the quantum state – apart from the usual "wave function collapse" that occurs when a superposition is measured. To be more precise, we require that the probability distribution of measurement results accurately reflect the amplitudes of the original quantum states, and if a subsystem is prepared in a given eigenstate of the measurement operator, our detector should faithfully indicate so. It's reasonable to assume that we cannot tell the difference between a detector that randomly gives incorrect results and a detector that actually influences the quantum state of the system in a random way. Both shortfalls can be lumped into a single parameter known as the detector quantum efficiency, defined as the probability that the detector accurately reflects a measurement of any previously prepared quantum eigenstate.

Definition 3 An entangled state is one that is not separable, where highly quantum-efficient measurements are performed on one constituent without affecting the others.

A quantum computer is nothing more than a device capable of generating an arbitrary entangled state following Definition 3. If the quantum computer consists of N qubits, then the probability that the final measurement accurately reflects the underlying quantum state is η^N, where η is the detector efficiency per qubit. For large numbers of qubits, this requires extremely high detector efficiencies in order to give a reasonable success probability. Even for a 99% efficient detector with each of 1000 qubits, the probability that the complete measurement is not plagued by an error is only 0.000 04.

A more strict definition of entanglement would rule out any possibility of interaction between the constituents during a measurement. This would require that the two subsystems be separated by a spacelike interval (given that we do not abandon relativity). In fact, this condition is the basis for the proof by John Bell that quantum mechanics is an inherently nonlocal theory, and that any extension to quantum mechanics (e.g., involving unobserved "hidden" variables) must itself be nonlocal (Bell 1965). Measurements of Bell's inequality violations are thus very useful measures of entanglement.

Definition 4 An entangled state is one that is not separable, where highly quantum-efficient measurements are performed on one constituent without affecting the others, and where the constituents are spacelike separated during the measurement time.

To date, entangled states following this most strict definition have not yet been created, and no full experimental test of Bell's inequality has been performed (however, see Fry *et al.* (1995)). Entangled states following Definitions 2 and 3 have been created, with a consequent violation of a Bell's inequality under relaxed conditions ("loopholes"). A series of experiments with optical parametric down-conversion have demonstrated spacelike entanglement with poor detectors (Definition 2) (Weihs *et al.* 1998), and an experiment with two trapped atoms has demonstrated entanglement with efficient detectors but without spacelike separations (Definition 3) (Rowe *et al.* 2001).

In general, there is no known measure of *how* much entanglement a given quantum state possesses. An important exception is for the case of pure quantum states that can be represented by a state vector or wave function. Here, the amount of entanglement can be mathematically described as the gain in von Neumann entropy of the state when only a subsystem is considered. This is reasonable, as any pure quantum state has zero entropy, and only when the state is separable does the entropy remain zero when one subsystem is traced over. It is interesting to apply this quantification of entanglement to simple quantum states such as the two entangled states below:

$$\Psi_{\mathrm{A}} = \frac{\downarrow\downarrow\downarrow\downarrow + \uparrow\uparrow\uparrow\uparrow}{\sqrt{2}} \tag{17.6}$$

$$\Psi_{\mathrm{B}} = \frac{\downarrow\downarrow\uparrow\uparrow + \downarrow\uparrow\downarrow\uparrow + \downarrow\uparrow\uparrow\downarrow + \uparrow\downarrow\downarrow\uparrow + \uparrow\downarrow\uparrow\downarrow + \uparrow\uparrow\downarrow\downarrow}{\sqrt{6}} \tag{17.7}$$

Even though state Ψ_{A} appears to have a stronger correlation between the four spins, when a trace is performed over any two spins, state B has slightly more entropy than state A, so Ψ_{B} is *more* entangled than Ψ_{A}. This definition of entanglement for pure quantum states highlights a peculiar feature of quantum mechanics: the entropy of a quantum subsystem can be *more* than the entropy of the complete quantum system. This is in stark contrast to classical systems, where entropy of the whole can only be greater than or equal to the sum of the entropies of the individual parts.

Quantum computer hardware

The more strict definitions of entanglement (3 and 4 above) required for a large-scale quantum computer rule out most physical systems. This can be seen by considering the chief hardware requirements for a quantum information processor (DiVincenzo 2000):

(i) arbitrary unitary operators must be available and controlled to launch an initial state to an arbitrary entangled state (eqn (17.2)), and
(ii) measurements of the qubits must be performed with high quantum efficiency.

From (i), the qubits must be well isolated from the environment to ensure pure initial quantum states and preserve their superposition character, but they must also interact strongly between one another in order to become entangled. On the other hand, (ii) calls for the strongest possible interaction with the environment to be switched on at will. The most attractive physical candidates for quantum information processors are thus fairly exotic physical systems offering a high degree of quantum control.

A collection of laser-cooled and trapped atomic ions represents one of the few developed techniques to store qubits and prepare entangled states of many qubits (Cirac and Zoller 1995; Monroe *et al.* 1995; Wineland *et al.* 1998). Here, electromagnetic fields confine individual atoms in free space in a vacuum chamber, and when multiple ions are confined and laser-cooled, they form simple stationary crystal structures given by the balance of the external confining force of the trap with the mutual repulsion of the atoms (Fig. 17.2). Qubits are effectively stored in internal electronic states of the atoms, typically the same long-lived hyperfine states that are used in atomic clocks. When appropriate laser radiation is directed to the atomic ions, qubit states can be coherently mapped onto the quantum state of collective motion of the atoms and subsequently mapped to other atoms. A single normal mode of collective crystal motion thus behaves as a "quantum data-bus," allowing quantum information to be shared and entangled between remote atomic qubits in the crystal. Finally, the internal states of individual trapped ions can be measured with nearly 100% quantum efficiency (Blatt and Zoller 1988) by applying appropriate laser radiation and collecting fluorescence, as in Fig. 17.2. In certain species atoms, a "cycling" transition allows a large amount of fluorescence to result from one qubit state, while the other remains dark.

Quantum logic gates have been demonstrated with up to four trapped atomic ion qubits, resulting in the generation of particular four-qubit entangled states such as eqn (17.6) (Sackett *et al.* 2000). While this scheme is scalable to arbitrarily large numbers of qubits in principle, the main problems deal with control of the collective motion of the atoms. As more qubits are added to the collection, the density of motional states balloons, and isolation of a single mode of motion (e.g., the center-of-mass) becomes even more slow and difficult (Wineland *et al.* 1998). Moreover, external noisy electric fields tend to compromise the motional coherence of large numbers of trapped atomic ions (Turchette *et al.* 2000a). A promising approach that attacks both problems is the *quantum CCD*, where individual atomic ions are entangled as above, but only among a small collection (under 10) of atomic ions in an "accumulator" (Kielpinski *et al.* 2002). To scale to larger numbers, individual atoms are physically shuttled between the accumulator and a "memory" reservoir of trapped atom qubits. This can be done quickly with externally applied electric fields in elaborate ion trap electrode geometries. The central features of the quantum CCD are that trapped ion shuttling can be done *without perturbing the internal qubits*, and the motional quantum state of the ions factors from the internal qubit states following quantum gate operation. In order to quench this extra motional energy for subsequent logic gates, ancillary ions in the accumulator can be

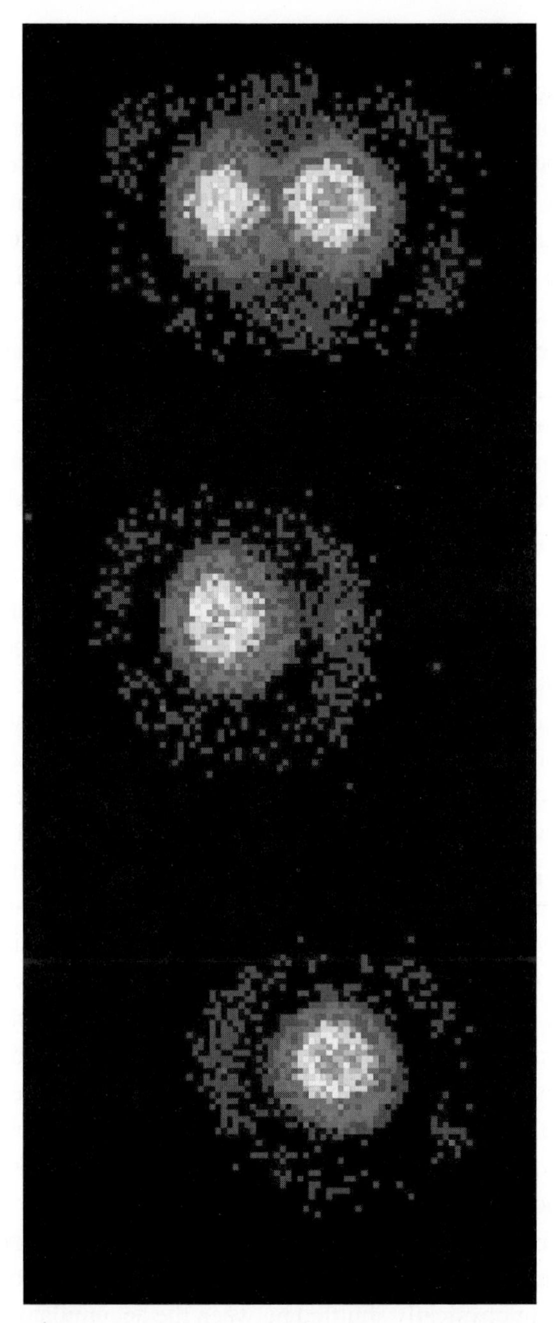

Figure 17.2. Spatial image of two trapped cadmium atomic ions from the University of Michigan Ion Trap Group (Blinov *et al.* 2002). Resonant laser radiation near 215 nm illuminates the atoms, and an imager collects the ultraviolet fluorescence. This image was integrated for about 1 s. The atoms are separated by approximately 2 μm, a balance between the external confining force and the Coulomb repulsion. The breadth of each atom is consistent with diffraction from the imaging optics, and the Airy rings are visible around each atom. The confinement electrodes (not shown at this scale) have a characteristic dimension of 200 μm.

laser-cooled in between gate operations; thus the qubit ions are sympathetically cooled through their strong Coulomb interaction with these extra refrigerator ions (Larson *et al.* 1986; Blinov *et al.* 2002).

Other potential quantum information processor candidates (Monroe 2002) include trapped atoms in optical lattices, trapped photons (cavity QED), and nuclear magnetic resonance (NMR) techniques applied to low temperature samples – nearly identical to the ion trap concept described above. Much less is clear in the domain of many solid-state systems, where quantum mechanics plays only a small role. However, there is exciting current research in exotic condensed-matter systems such as semiconductor quantum dots (Stievater *et al.* 2001) and superconducting current loops (van der Waal *et al.* 2000; Friedman *et al.* 2002) and charge pumps (Nakamura 1999; Vion 2002), which may some day allow the scale-up to a large-scale quantum information processor.

Outlook

There is a proliferation of quantum mechanical interpretations all attempting to address the conceptual problems unifying quantum mechanics with quantum measurement – the so-called measurement problem that plagues the conventional Copenhagen interpretation used by the vast majority of physicists. While current experiments are very far from demonstrating useful quantum information processing, some systems may ultimately put us in a position of questioning (or more likely ruling out) these alternatives to quantum mechanics.

The most popular alternative quantum views include Bohmian mechanics – a nonlocal hidden-variables theory that at least removes indeterminism from quantum mechanics (Albert 1994); the many-worlds interpretation proposing that quantum measurements cause the universe to bifurcate (Everett 1957); the consistent or decoherent histories approach (Griffiths 2001), and the transactional interpretation of quantum mechanics (Cramer 1988). Perhaps the most popular melding of quantum mechanics and quantum measurement is the theory of decoherence (Zurek 1982, 1991). Decoherence theory applies the usual quantum mechanics to a closed system, but when the uncountable degrees of freedom of the environment are inevitably coupled into that system, via noise or a measurement, entanglements form between the system and environment. Now when we perform a trace over the environmental degrees of freedom, we find that the coherence in quantum mechanics decays, or pure states of a closed quantum system continuously evolve into mixtures. Decoherence formalism is a useful method of calculating the dissipation expected in quantum systems when environmental couplings are known, but it certainly does not address the quantum measurement problem. This would be akin to claiming that Newton's law of gravitational attraction $F = Gm_1m_2/r^2$ explains the origin of gravity. In fact, nearly all of the alternative interpretations of quantum mechanics predict the same answers for any conceivable experiment. While some versions have a satisfactory feel to them – perhaps by removing the observer from the theory (Goldstein 1998) – these differing frameworks might seem unremarkable to the experimentalist.

There is at least one alternative to quantum mechanics that *is* testable. It posits that quantum mechanics and classical mechanics are just two limits of the same underlying theory. Small systems such as isolated atoms and electrons are well approximated by quantum mechanics, while large systems like cats are well approximated by classical mechanics. Such a theory predicts a frontier between these two limits where new physics may arise. One example is a class of "spontaneous wave function collapse" theories; the most popular having been put forth by Ghirardi, Rimini, and Weber (GRW) in the last decades (Ghirardi *et al.* 1986; Bell 1987; Pearle 1993). The GRW theory attempts to meld quantum and classical mechanics by adding a nonlinear stochastic driving field to quantum mechanics that randomly localizes or collapses wave functions. This localization acts with an effective spatial dimension a, and the frequency of the collapses is proportional to a rate λ times the number of degrees of freedom N in the system. The fundamental constants a and λ are chosen such that the average time of collapse of simple systems like a single atom or electron is very long, while the average time of collapse of a macroscopic superposition of a body with 10^{20} degrees of freedom is unobservably short (favored values of a and λ are approximately 10^{-5} cm and 10^{-16} Hz, respectively). Admittedly, such a phenomenological theory is not very plausible, but the ad hoc details of GRW's proposal are not the main point. What makes their theory remarkable is that it is testable. Stochastic collapses predicted by GRW indeed imply an upper limit on the size of a quantum computer.

Experiments that may test spontaneous wave function collapse theories are naturally the same systems that are considered as viable future quantum computers. A review on the state-of-the-art in "large superpositions" is considered by Leggett (2002), including an exhaustive definition of what constitutes a "degree of freedom" so critical to the GRW theory. The more notable systems include quantum optics systems of trapped ions (Monroe *et al.* 1996; Myatt *et al.* 2000; Turchette *et al.* 2000b) and the related system of cavity QED (Brune 1996); and superconducting systems of quantum dots (Nakamura *et al.* 1999; Vion *et al.* 2002) and SQUIDs (van der Waal *et al.* 2000; Friedman *et al.* 2002). The quantum optics systems are complementary to the condensed matter systems in the context of attacking the GRW wave function collapse frontier. The superconducting systems deal with superpositions of supercurrents or numbers of Cooper-paired electrons, boasting a very large number N of degrees of freedom. However, access to these individual degrees of freedom through highly efficient measurements has not been demonstrated. This masks the underlying entanglement in the system (see Definition 3 above), admitting a more classical-like description of the observed phenomena. Quantum optics systems, on the other hand, offer highly efficient measurements, but only with a small value for N. All the above systems are prime candidates for quantum computing hardware, and as more qubits are entangled in these (or any quantum hardware), so too will the frontiers of GRW collapse be pushed back.

Of the three possible results in the quest to build a quantum computer, two are tantalizing: either a fully blown large-scale quantum computer will be built, or the theory of quantum mechanics will be found to be incomplete. The third possibility, that the technology will never reach the complexity level required for either of the first possibilities due to economic constraints, has nothing to do with physics, but is probably favored by the majority of physicists. Indeed, it's amusing to see physicists bristle when confronted with the notion of a macroscopic quantum state – a "Schrödinger cat." In *A Brief History of Time*, Stephen Hawking quips that "Whenever I hear a mention of *that cat*, I reach for my gun." Even Schrödinger himself labeled his famous cat as ridiculous, and was so disturbed at this logical path of quantum mechanics, that he switched fields altogether. It is this steadfast parochial view that suggests that we should continue to probe foundational aspects of quantum mechanics, even if the result is only a full-scale quantum information processor.

Acknowledgments

The author acknowledges support from the US National Security Agency, Army Research Office, and National Science Foundation.

References

Albert, D Z (1994) Bohm's alternative to quantum mechanics. *Scient. Am.* **270**, 32.
Bell, J S (1965) On the Einstein–Podolsky–Rosen paradox. *Physics* **1**, 195.
(1987) *Speakable and Unspeakable in Quantum Mechanics*. Cambridge: Cambridge University Press.
Benioff, P (1980) The computer as a physical system: a microscopic quantum mechanical model of computers as represented by Turing machines. *J. Stat. Phys.* **22**, 563.
(1982) Quantum mechanical Hamiltonian models of Turing machines that dissipate no energy. *Phys. Rev. Lett.* **48**, 1581.
Blatt, R and Zoller, P (1988) Quantum jumps in atomic systems. *Eur. J. Phys.* **9**, 250.
Blinov, B, *et al.* (2002) Sympathetic cooling of trapped Cd^+ isotopes. *Phys. Rev.* **A65**, 040304.
Bohm, D (1951) *Quantum Theory*. New York: Dover.
Brune, M, *et al.* (1996) Observing the progressive decoherence of the 'meter' in a quantum measurement. *Phys. Rev. Lett.* **77**, 4887.
Cirac, I and Zoller, P (1995) Quantum computations with cold trapped ions. *Phys. Rev. Lett.* **74,** 4091.

Cramer, J G (1988) An overview of the transactional interpretation of quantum mechanics. *Int. J. Theor. Phys.* **27**, 227.
Deutsch, D (1985) Quantum theory, the Church–Turing principle and the universal quantum computer. *Proc. Roy. Soc. Lond.* **A400**, 97.
DiVincenzo, D (2000) The physical implementation of quantum computation. *Fortschr. Phys.* **48**, 771.
Einstein, A, *et al.* (1935) Can quantum-mechanical description of reality be considered complete? *Phys. Rev.* **47**, 777.
Everett, H (1957) Relative state formulation of quantum mechanics. *Rev. Mod. Phys.* **29**, 454.
(1960) http://www.zyvex.com/nanotech/feynman.html.
Feynman, R P (1982) Simulating physics with computers. *Int. J. Theor. Phys.* **21**, 467.
Friedman, J R, *et al.* (2002) *Nature* **406**, 43.
Fry, E S, *et al.* (1995) Proposal for loophole-free test of the Bell inequalities. *Phys. Rev. Lett.* **52**, 4381.
Ghirardi, G C, *et al.* (1986) Unified dynamics for microscopic and macroscopic systems. *Phys. Rev.* **D34**, 470.
Goldstein, S (1998) Quantum theory without observers. *Phys. Today* March, 42; April, 38.
Griffiths, R B (2001) *Consistent Quantum Theory*. Cambridge: Cambridge University Press.
Grover, L (1997) Quantum mechanics helps in searching for a needle in a haystack. *Phys. Rev. Lett.* **79**, 325.
Kielpinski, D, *et al.* (2002) Architecture for a large scale ion-trap quantum computer. *Nature* **417**, 709.
Larson, D J, *et al.* (1986) Sympathetic cooling of trapped ions: a laser-cooled two-species nonneutral ion plasma. *Phys. Rev. Lett.* **57**, 70.
Leggett, A J (2002) Testing the limits of quantum mechanics: motivation, state of play, prospects. *J. Phys. Condens. Matter* **14**, R415.
Monroe, C (2002) Quantum information processing with atoms and photons. *Nature* **416**, 238.
Monroe, C, *et al.* (1995) Demonstration of a universal quantum logic gate. *Phys. Rev. Lett.* **75**, 4714.
(1996) A Schrödinger cat superposition state of an atom. *Science* **272**, 1131.
Myatt, C, *et al.* (2000) Decoherence of quantum superpositions coupled to engineered reservoirs. *Nature* **403**, 269.
Nakamura, Y, *et al.* (1999) Coherent control of macroscopic quantum states in a single Cooper-pair box. *Nature* **398**, 786.
Nielsen, M A and Chuang, I L (2000) *Quantum Computation and Quantum Information*. Cambridge: Cambridge University Press.
Pearle, P (1993) Ways to describe dynamical state-vector reduction. *Phys. Rev.* **A48**, 913.
Rivest, R, *et al.* (1978) A method for obtaining digital signatures and public-key cryptosystems, *Comm. Ass. Comptg Machinery* **21**, 120.
Rowe, M A, *et al.* (2001) Experimental violation of a Bell's inequality with efficient detectors. *Nature* **409**, 791.
Sackett, C, *et al.* (2000) Experimental entanglement of four particles. *Nature* **404**, 256.
Schrödinger, E (1935) Discussion of probability relations between separated systems. *Proc. Camb. Philos. Soc.* **31**, 555.
Shannon, C E (1948) A mathematical theory of communication. *Bell System Tech. J.* **27**, 379; 623.

Shor, P (1997) Polynomial-time algorithms for prime factorization and discrete logarithms on a quantum computer. *SIAM J. Comp.* **26**, 1484.
Stievater, T, *et al.* (2001) Rabi oscillations of excitons in single quantum dots. *Phys. Rev. Lett.* **87**, 133603.
Turchette, Q, *et al.* (2000a) Quantum heating of trapped ions. *Phys. Rev.* **A61**, 063418.
(2000b) Decoherence of quantum superpositions coupled to engineered reservoirs. *Phys. Rev.* **A62**, 053807.
van der Waal, C H, *et al.* (2000) Quantum superposition of macroscopic persistent-current states. *Science* **290**, 773.
Van Dyck, R S, *et al.* (1987) New high-precision comparison of electron and positron **g** factors. *Phys. Rev. Lett.* **59**, 26.
Vion, D, *et al.* (2002) Manipulating the quantum state of an electrical circuit. *Science* **296**, 886.
Weihs, G, *et al.* (1998) Violation of Bell's inequality under strict Einstein locality conditions. *Phys. Rev. Lett.* **81**, 5039.
Wineland, D, *et al.* (1998) Experimental issues in coherent quantum manipulation of trapped atomic ions. *Nat. Inst. Standards and Tech. J. Res.* **103**, 259.
Zurek, W H (1982) Environment-induced superselection rules. *Phys. Rev.* **D26**, 1862.
(1991) Decoherence and the transition from quantum to classical. *Phys. Today* **44**(10), 36.

Part V

Big questions in cosmology

18

Cosmic inflation and the arrow of time

Andreas Albrecht
University of California

Introduction

One of the most obvious and compelling aspects of the physical world is that it has an "arrow of time." Certain processes (such as breaking a glass or burning fuel) appear all the time in our everyday experience, but the time reverse of these processes is never seen. In the modern understanding, special nongeneric initial conditions of the universe are used to explain the time-directed nature of the dynamics we see around us.

On the other hand, modern cosmologists believe it is possible to explain the initial conditions of the universe. The theory of cosmic inflation (and a number of competitors) claims to use physical processes to *set up* the initial conditions of the standard Big Bang. So in one case initial conditions are being used to explain dynamics, and in the other, dynamics are being used to explain initial conditions. In this chapter I explore the relationship between two apparently different perspectives on initial conditions and dynamics.

My goal in pursuing this question is to gain a deeper insight into what we are actually able to accomplish with theories of cosmic initial conditions. Can these two perspectives coexist, perhaps even allowing one to conclude that cosmic inflation explains the arrow of time? Or do these two different ideas about relating dynamics and initial conditions point to some deep contradiction, leading us to conclude that a fundamental explanation of both the arrow of time and the initial conditions of the universe is impossible? Thinking through these issues also leads to interesting comparisons of different theories of initial conditions (e.g., inflation vs. cyclic models).

Throughout this article, by the "arrow of time" I mean the *thermodynamic* arrow of time. As discussed below, I regard this to be equivalent to the radiation,

Science and Ultimate Reality, eds. J. D. Barrow, P. C. W. Davies and C. L. Harper, Jr. Published by Cambridge University Press.

psychological, and quantum mechanical arrows of time. The cosmic expansion (or the "cosmological arrow of time") may or may not be correlated with the thermodynamic arrow of time, depending on the specific model of the universe in question (see for example Hawking (1994)).

I also should be clear about how I use the phrase "initial conditions." The classical standard Big Bang cosmology has a genuine set of (singular) "initial conditions" in the sense that the model cannot be extended arbitrarily far back in time. Much of my discussion in this chapter involves various ways one can work in a larger context where time, or at least some physical framework, is eternal. In that context, the problem of initial conditions that concerns us here is how some region entered into a state that reflects the "initial" conditions we use for the part of the universe we observe. This state might not be *initial* at all in a global sense, but it still seems like an initial state from the point of view of our observable universe. I will often use the term "initial conditions" to refer to the state at the *end* of inflation which forms the initial conditions for the standard Big Bang phase that follows. I hope in what follows my meaning will be clear from the context.

I hope this chapter will be stimulating and perhaps even provocative for experts on inflation and alternative theories of the initial conditions of the universe. But, in the spirit of the Wheeler volume, I've also tried to make this chapter for the most part accessible to a wider audience of physical scientists who may be experts in other areas but who might find the subject interesting.

This article is organized as follows: I will first set the stage by contrasting a cosmologist's view of initial conditions with that of "everyone else." Next I present the standard modern view of the origin of the arrow of time. I discuss the case where gravity is irrelevant (which covers most everyday intuition), then the case where gravity is dominant, which allows the discussion of the arrow of time to be extended to the entire cosmos. Next I present the inflationary perspective on initial conditions, and contrast them with the perspective taken when discussing the arrow of time. I proceed to show how these two perspectives can coexist (still with some tension) in an overarching "big picture" that allows both an explanation of initial conditions and the arrow of time. Then I will discuss and contrast a variety of different ideas about cosmic initial conditions in light of the insights from earlier sections, plus additional issues, including "causal patch" physics and problems with measures. Finally I spell out some big open questions for the future, and collect my conclusions.

The everyday perspective on initial conditions

Scientists other than cosmologists almost never consider the type of question addressed by cosmic inflation. Cosmic inflation tries to *explain* the initial conditions of the standard big bang phase of the universe. Where else does one try to explain the initial conditions of anything?

The typical perspective on initial conditions is very different. Consider the process of testing a scientific theory in the laboratory. A particular experiment is performed in the laboratory, theoretical equations are solved, and the results of theory and experiment are compared. To solve the equations, the theoretician has to make a choice of initial conditions. The principle guiding this choice is trivial: reproduce the initial conditions of the corresponding laboratory set-up as accurately as necessary. The theoretician might wonder why the experimenter chose a particular set-up and may well have influenced this choice, but the origin of initial conditions does not usually count as a fundamental question that needs to be addressed by major scientific advances. Instead, the initial conditions play a subsidiary role. The comparison of theory and experiment is really seen as a test of the equations of motion. The initial conditions facilitate this test, but are not typically themselves the focus of any fundamental tests.

The choice of vacuum in quantum field theories is an illustration of this point. One can propose a particular field theory as a theory of nature, but the proposal is not complete until one specifies which state is the physical vacuum. That choice determines how to construct excited states which contain particles, thus allowing one to define initial states that correspond to a given experiment.

The conceptual framework from quantum field theory has been carried over, at least loosely, into quantum cosmology where declaring "the state of the universe" based on some symmetry principle or technical definition appears to give initial conditions for the universe (Hartle and Hawking 1983; Linde 1984; Vilenkin 1984). However, as I will emphasize below, such a declaration is nothing like the sort of dynamical explanation of the initial state of the universe that is attempted by inflation, and these two approaches should not be confused with one another.

Another example where the initial conditions play a subsidiary role is in constructing states of matter whose evolution exhibits a thermodynamic arrow of time. In the next section I will discuss this example in detail before turning in the following section to the very different perspective on initial conditions taken in inflation theory.

Arrow of time basics

Overview

In this section I discuss how to construct states of matter that exhibit a thermodynamic arrow of time. The case of systems whose self-gravity can be neglected is simplest and most well understood, so we will consider that case first. All other

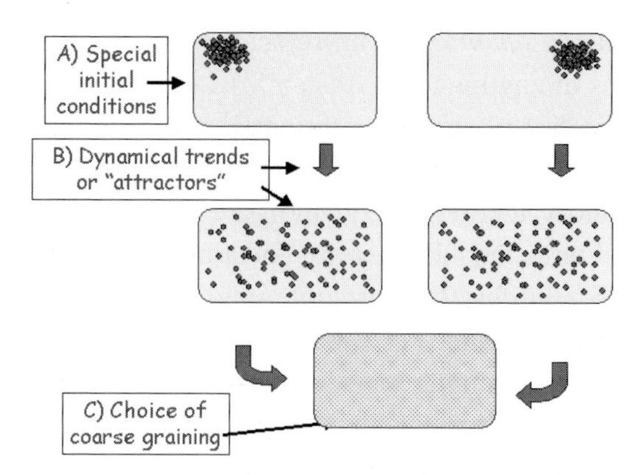

Figure 18.1. A box of gas illustrates the three basic ingredients which allow the arrow of time to emerge from a fundamentally reversible microscopic world. Special out-of-equilibrium initial conditions are required, as are trends (or attractors) in the underlying dynamics drawing the system toward equilibrium. Finally, a choice of coarse graining is essential. Without it, different initial states always evolve into different final states, and attractor-like behavior is impossible to identify.

factors being equal, the importance of gravitational forces between elements of a system is related to the overall size of the system, and the critical size is characterized by a length scale called the "Jeans length" (l_J). For example, for a box of gas with size $l \ll l_J$ self-gravitation is unimportant and the gas pressure can easily counteract any tendency to undergo gravitational collapse. For a larger body of gas with $l \gg l_J$ (but otherwise the same temperature, density, and other local properties) the self-gravity of the larger overall mass will overwhelm the pressure and allow gravitational collapse to proceed (see for example Longair (1998)).

The material in all of this section is pretty standard, and I give only a brief review. A much more thorough treatment of all these issues can be found in a number of excellent books on the subject (Davies 1977; Zeh 1992) which also contain references to the original literature.

Without gravity ($l \ll l_J$)

There are three critical ingredients that go into the arrow of time: special initial conditions, dynamical trends or "attractors" that are intrinsic to the equations of motion, and a choice of coarse graining. I will illustrate how this works in the canonical example of a box of gas.

Figure 18.1 illustrates a box of gas that starts in the two pictured initial states, with all the gas stuck in a corner. In each case the gas spreads out into states that look the same regardless of which corner was the starting point. This system has an arrow of time: a movie shown backwards would show a process that would never spontaneously occur in our everyday experience. Furthermore, once the gas becomes spread out, we can count on it not to spontaneously evolve back into the corner again.

Of course, according to the microscopic theory of the gas the two different initial conditions evolve into different states, and even though both look to us simply as a gas in equilibrium, the microscopic differences are retained forever, albeit in very subtle correlations among the positions and velocities of the gas particles (as well as their internal degrees of freedom). This is where coarse graining is critical[1]. The fact that we ignore subtle differences, such as the ones differentiating the "equilibrium" states corresponding to the two different initial conditions, is the only reason we can conceive of a single stable "equilibrium state." Without coarse graining there would be no such thing as equilibrium, just ever-changing microscopic states. Coarse graining is also essential to identifying the approach to equilibrium. Without coarse graining one could only identify the microscopic evolutions of individual states, not dynamical trends, and there would be no notion of the arrow of time.

The roles of initial conditions, dynamics, and coarse graining are closely interconnected. If the dynamics of the molecules depicted in Fig. 18.1 was different so that, for example, the molecules were constrained to remain in the corner of the box where they started, then a typical initial state like the one depicted would *already* be in equilibrium, and such a system would not exhibit an arrow of time.

Similarly, in principle it is possible to construct formal coarse grainings where one ignores different aspects of the microscopic state and which give arbitrarily different results. One could formally go about this, for example, by choosing some random microscopic state normally associated with equilibrium and declaring microscopic states that were dynamically nearby to that state to be in the same coarse-grained "bin," and by similarly making other coarse-grained bins from other more dynamically distant states. From that particular coarse graining, the box of gas illustrated in Fig. 18.1 would not exhibit an arrow of time.

The fact that a system may or may not exhibit an arrow of time depending on the particular choice of coarse graining creates no problems for people (like me) who are happy to see coarse graining as a natural consequence of what kind of

[1] Coarse graining is basically the act of ignoring certain aspects of microscopic states, so that many different microscopic states are identified with a single coarse-grained state (or are put in the same "coarse-grained bin"). A simple example: take precisely defined values for position and momentum and round those values to a reduced number of digits. That gives coarse-grained coordinates in discrete phase space.

measurements we can actually make (something ultimately related to the nature of the fundamental Hamiltonian). However, those who wish to see the arrow of time defined in more absolute terms are concerned by the fact that in the modern understanding the arrow of time of a given system only exists relative to a particular choice of coarse graining and is likely to only be a temporary phenomenon (Prigogine 1962; Price 1989).

The above construction only buys you a "temporary" arrow of time because according to the microscopic theory, it *is* possible for gas in equilibrium to evolve spontaneously into one corner of its container. It just takes, on average, an incredibly long time before that happens (much longer than the age of the universe). The stability of the equilibrium coarse-grained state is deeply linked with the large number of microscopic states that are associated with the equilibrium state. From the microscopic point of view, one state is constantly evolving into another. The huge degeneracy of microscopic states associated with equilibrium means that there is lots of room to evolve from one state to another without leaving the equilibrium coarse-grained bin.

These features are closely linked with the definition of the statistical mechanical entropy of a coarse-grained state as $\ln(N)$ (where N is the number of microscopic states corresponding to the particular coarse-grained state), and with the fact that the equilibrium state is the coarse-grained state with maximum entropy. The large microscopic degeneracy of the equilibrium state is also closely related to the fact that many different initial states will all approach equilibrium.

The fact that such a large portion of all possible states for the system are associated with equilibrium means that the special out-of-equilibrium initial states required for the arrow of time are very rare indeed. If one watched a random box of gas it would be in equilibrium almost all the time, a state with no arrow of time. At extraordinarily rare moments, there would be large fluctuations out of equilibrium and the transient associated with the return to equilibrium would exhibit an arrow of time.

In fact, due to the long periods of equilibrium the system itself has no overall time direction. The rare fluctuations out of equilibrium actually represent two back-to-back periods, each with an arrow of time pointing in the opposite direction, with each arrow originating at the point of maximum disequilibrium. Such a rare fluctuation is depicted in Fig. 18.2.

Interestingly, actually achieving a state of equilibrium is not absolutely necessary in order to have an arrow of time. For example, consider the generalization of the above discussion to a gas that starts out in the corner of an infinitely large box. This case is depicted in Fig. 18.3. To achieve an arrow of time one must follow a clear dynamical trend, but a final equilibrium end point is not essential. This fact is especially relevant to the self-gravitating case discussed below.

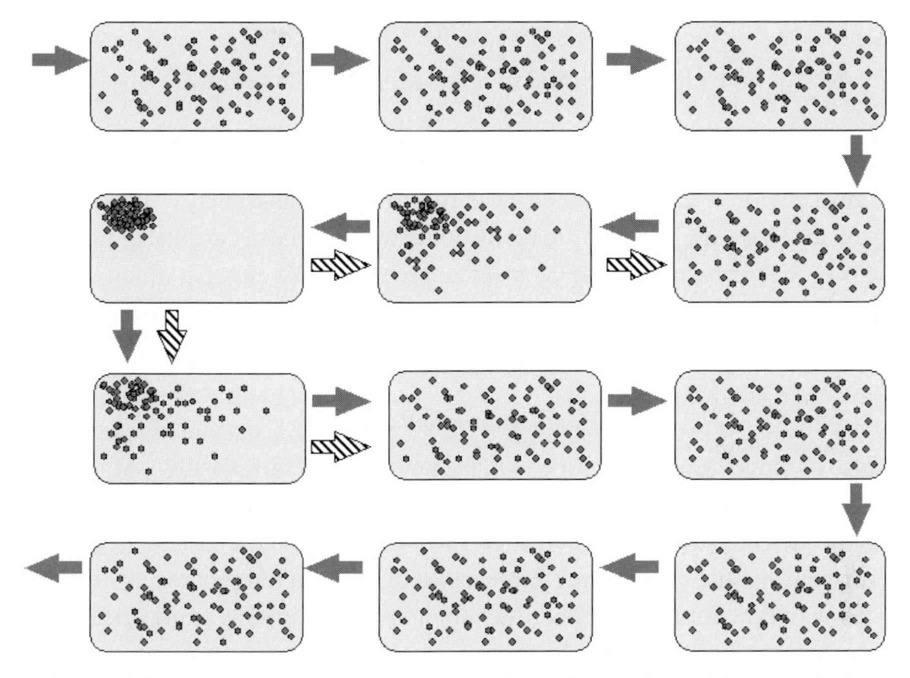

Figure 18.2. A very rare large fluctuation in a box of gas. The solid arrows depict the time series (with a randomly chosen overall direction) and the hashed arrows depict the thermodynamic arrow of time.

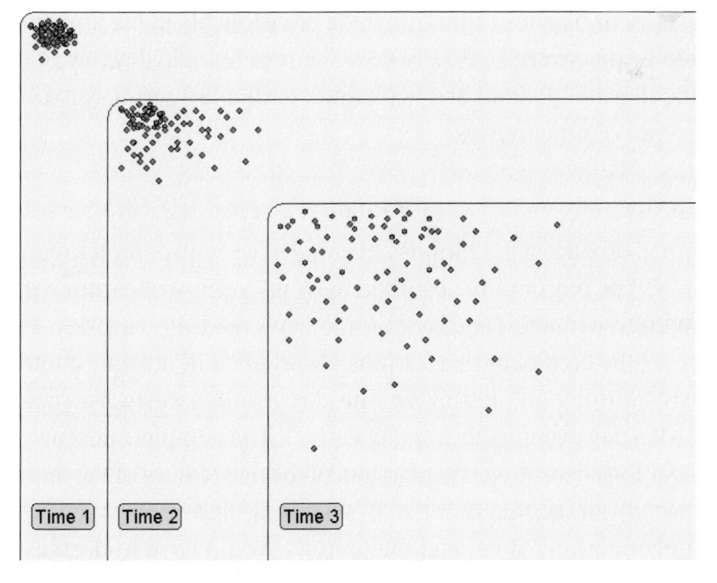

Figure 18.3. A modification of the system depicted in Fig. 18.1 to the case with an infinite box leads to a system that has a definite arrow of time, but which never achieves equilibrium.

The key roles of the arrow of time

The arrow of time plays a key role in many aspects of our world.

Burning fuel

The most obvious example is when we burn fuel to "produce energy" which we then harness in some way. What we are really doing when we burn gasoline or metabolize food is producing *entropy*. The critical resource is not the energy (which after all is conserved), but the reliability of the arrow of time. The presence of fuel and food in our world is part of the special initial conditions that give us the arrow of time.

Computation and thought

We also harness the arrow of time to make key processes irreversible. Make a mark on a page or a blackboard and you can be sure the time-reverse process (the mark popping back up into the pencil or chalk) will never happen. This allows us to make "permanent" records which are a critical part of information processing. The use of the arrow of time for making records is ubiquitous in everyday experience, and this use of the arrow of time has also been formalized in the case of computations in work on "the thermodynamic cost of computation" (Bennett and Landauer 1985).

Given our lack of a fundamental understanding of the process of human thought, there are many different views about the psychological arrow of time. I personally do not expect advances in understanding human thought to bring any new insights into microscopic laws of physics (although there are probably some amazing collective phenomena to be discovered). So I believe the psychological arrow of time is none other than the thermodynamic arrow of time, particularly as it is expressed in the making of records (or memories).

Radiation

A TV station can broadcast the Monday evening news with complete confidence that the radiation will be thoroughly absorbed by whatever it strikes and will not be still around to interfere with the Tuesday evening news the following day. Furthermore, broadcasters can be confident that various absorbers will not cause interference by spontaneously emitting an alternative Tuesday evening newscast (not to mention emitting the Tuesday evening news a day early!). The complete absence of the time-reverse of radiation absorption is understood to be one feature of the thermodynamic arrow of time in our world. A hillside absorbing an evening news broadcast is entering a higher entropy state, and the entropy would have to decrease for any of the troublesome time-reversed cases to take place.

Of course much of a radio signal propagates off into empty space. In that case, the emptiness of the space appears to play a similar role to the infinite box in

Fig. 18.3. Time-reversed solutions, with the evening news broadcasts propagating from outer space back into the "transmitting" antenna *are* legitimate solutions to the equations of motion. But a complete solution could not have such radiation really propagating in from infinity. Instead, it would have to be emitted from some astrophysical object or barring that, from the "surface of last scattering" (the most recent point in the history of the universe when the universe was sufficiently dense to be opaque). Any of these astrophysical or cosmological sources would have to be in a much lower entropy state than we expect if they are to produce time reversed "evening news" radiation. So in the end, the radiation arrow of time is none other than the thermodynamic arrow of time, which is the topic of this chapter. I should note that much of our understanding of the radiation arrow of time was developed by John Wheeler (Wheeler and Feynman 1945, 1949) whom we honor with this volume.

Quantum measurement

An arrow of time is critical to quantum mechanics as we experience it. Once a quantum measurement is made there is no undoing it, and one says the wave function has "collapsed." There are different attitudes about this collapse. One approach is to see this collapse as a consequence of establishing stable correlations: a double-slit electron striking a photographic plate is only a good quantum measurement to the extent that the photographic plate is well constructed, and has a very low probability of re-emitting the electron in the coherent "double-slit" state. Good photographic plates are possible because of the thermodynamic arrow of time: the electron striking the plate puts the internal degrees of freedom of the plate into a higher entropy state, which is essentially impossible to reverse. Furthermore, different electron positions on the plate become entangled with different states of the internal degrees of freedom, so there is essentially no interference between positions of the electron. From this point of view (which I prefer) the quantum mechanical arrow of time is none other than the thermodynamic arrow of time[2]. Others want to establish a quantum arrow of time that is separate from the thermodynamic arrow, but no well-established theory of this type exists so far.

With gravity ($l \gg l_{\mathrm{J}}$)

When the self-gravity of a system is significant, the dynamical trends are very different. While the gas in the box discussed above tended to spread out into a uniform equilibrium state, for gravitating systems the trend is toward gravitational collapse

[2] Significant contributions to this perspective come from John Wheeler and his students (Everett 1957; Wheeler and Zurek 1983; Zurek 1991). See also Albrecht (1992, 1994).

into a state with less homogeneity. Interestingly, when gravitational collapse runs its course, matter also approaches a kind of equilibrium state: the black hole. As is fitting for equilibrium states, one can even define the entropy of a black hole, namely the Bekenstein–Hawking entropy given by

$$S_{\text{bh}} = 4\pi M^2 \tag{18.1}$$

for a black hole of mass M. Although black-hole entropy is not as well understood as the entropy of a box of gas, it certainly fits with the general picture quite well.[3]

As with any other system, gravitating systems will exhibit a thermodynamic arrow of time if they have special "low entropy" initial conditions. The observed universe is an excellent example. The observed universe certainly has a sufficiently strong self-gravity so as to be subject to gravitational collapse (namely $l \gg l_{\text{J}}$). But this trend is in its very early stages, and is very far from having run its course. That is, the observed universe is very far from forming one giant black hole. Penrose (1979) quantified this fact by comparing the entropy of the very early universe (as measured by the ordinary entropy of the cosmic radiation fluid) with the entropy of a black hole with mass equal to the mass of the observed universe. The result is that the entropy of the early universe is 35 orders of magnitude smaller than the maximal entropy black-hole state:

$$S_{\text{Univ}} \approx 10^{-35} S_{\text{bh-Max}} = 10^{-35} 4\pi M_{\text{Univ}}^2. \tag{18.2}$$

As Penrose originally argued, the low entropy of the early universe is the ultimate origin of the arrow of time we experience. Just as the box of gas depicted in Fig. 18.1 evolves reliably toward a more homogeneous state, giving the system an arrow of time, so too does the universe follow its own dynamical trends from a state of homogeneity toward a state of gravitational collapse. In the case of the universe, it is not clear that a final equilibrium black-hole state will be achieved, so it may turn out that the better analogy is the infinite box of gas depicted in Fig. 18.3. The key point is that the universe starts out in a very special state which is far from where the dynamical evolution wants to take it. The realization of this evolution results in an arrow of time.

I conclude this section with an illustration of the relationship between the arrow of time of the universe as a whole, as expressed by a trend from homogeneity toward gravitational collapse, and the simple everyday examples of the arrow of time as discussed above. Figure 18.4 illustrates a process by which we might construct a

[3] This discussion is classical and does not include the effects of Hawking radiation. Including Hawking radiation might make it more difficult to formulate an "ultimate equilibrium" state for general gravitating systems, but that does not matter for the discussion here. Hawking radiation is irrelevant on the temporal and spatial scales over which gravitational collapse defines the arrow of time in the observed universe.

Figure 18.4. One can pump on a box of gas to move all the gas into one corner. When the pumping stops the gas spreads out, exhibiting an arrow of time as depicted in Fig. 18.1. In this example the pump uses electricity generated by fossil fuels, produced from organic matter which originally harnessed solar energy to be created. The hot Sun radiating into cold space is our local manifestation of the ongoing process of gravitational collapse throughout the universe. This example illustrates the links between everyday examples of the arrow of time and the overall arrow of time of the universe, as expressed through gravitational collapse.

box of gas with an arrow of time such as that depicted in Fig. 18.1. The gas is pumped into the corner by an electric pump, with electricity generated by fossil fuels. The organic matter which formed the crude oil that was refined into the fuel was created by photosynthesis which harnessed the Sun's radiation. The hot Sun radiating into cold space is our local manifestation of the ongoing process of gravitational collapse throughout the universe.

As discussed above, what we traditionally call sources of power or energy are really sources of *entropy*, which allow us to harness the arrow of time. Most of our power sources can be traced to radiation from the Sun, as in Fig. 18.4. The exceptions are geothermal power (which harnesses the gravitational collapse that produced the Earth itself) and nuclear fission power (which uses unstable elements produced in the collapse of stars other than the Sun). Fusion energy exploits another sense in which the universe is out of equilibrium: the homogeneous cosmic expansion proceeds too quickly for the nuclei to equilibrate into the most stable element, and instead produces nuclei that are out of chemical equilibrium (i.e., not in the most tightly bound nuclei). This leaves an opportunity to release entropy by igniting fusion processes that bring nuclear matter closer to chemical equilibrium.[4] This issue (and its links to the initial state of the universe) will be discussed below.

Cosmic inflation: preliminaries

The inflationary perspective on initial conditions

In the previous section we discussed how the thermodynamic arrow of time must necessarily be traced to special initial conditions. In particular, we discussed how the overall arrow of time in the universe is linked to special initial conditions for the universe that are far away from the dynamical trend toward gravitational collapse. With this understanding, one can accept these special initial conditions in the usual subsidiary role: experimental data tell us that the universe has an arrow of time, so to model the universe we obviously must choose initial conditions appropriately. To this end, the homogeneous and isotropic expanding initial conditions of the standard Big Bang are a great choice, and they do indeed (when combined with a suitable initial spectrum of small primordial perturbations) give an excellent match to all the observations.

But enthusiasts of cosmic inflation (Guth 1981; Albrecht and Steinhardt 1982; Linde 1982) take a very different view. Typical presentations of cosmic inflation start by presenting a series of cosmological "problems" that appear to be present in the standard Big Bang (see for example Guth (1981) or Albrecht (1999)). Many cosmologists were concerned about these problems even before the discovery of inflation. The first two of these problems (the "flatness" and "homogeneity" problems) basically state that the initial conditions are far removed from the direction indicated by the dynamical trends.

The flatness problem is the observation that the dynamical trend of the universe is away from spatial flatness, yet to match today's observations, the universe must have been spatially flat to extraordinarily high precision.

The homogeneity problem is exactly a restatement of the main point of the preceding section: the universe is in a state far removed from where gravitational collapse would like to take it. The discussion above emphasized that a property of this sort is absolutely required in order to achieve an arrow of time. From that point of view, the special initial conditions of the universe are not a puzzle, but the answer to the question "Where did the arrow of time come from?".

However, most cosmologists would instinctively take a different perspective. They would try and look further into the past and ask how could such strange

[4] The Sun and other stars in fact produce an interesting combination of "gravitational collapse power" enhanced by nuclear fusion power.

"initial" conditions possibly have been set up by whatever dynamical process went before. Since the initial conditions are counter to the dynamical trends, it seems on the face of it that the creation of these initial conditions by dynamics is a fundamental impossibility. In fact, the "horizon problem" adds to this dilemma by observing that there was insufficient time in the early universe for causal processes to determine the initial conditions of what we see, even if somehow there was a way to fight the dynamical trends.

So we have two different points of view. An inflationist wants the initial conditions of the universe to be more natural, but the intellectual descendants of Boltzmann would say they had better *not* appear natural: the unnaturalness of initial conditions is precisely what is necessary in order to have an arrow of time, so it appears the price of "natural" initial conditions is the absence of an arrow of time. In addition, considering the general comments above, the inflationist would seem to be in a weaker position. Certainly the strongest tradition in physics is for initial conditions to play a subsidiary role, unquestioningly assigned whatever form is required for the situation at hand.

The goal of this chapter is to reconcile these points of view. To get started, I will give two illustrations of familiar situations where the initial conditions do play a more critical role, and for which dynamics actually creates special initial conditions. With the lessons learned from these illustrations, we will be ready to scrutinize cosmic inflation.

Illustration 1: Big Bang nucleosynthesis

One of the classic results from cosmology is the synthesis of nuclei in the early universe. Using cross-sections determined in laboratories on earth, one can calculate the cosmic abundances of different nuclei at different times. As the universe cools and becomes more dilute, a point is reached where the mass fraction stops changing. These "frozen-out" values are the predictions of "primordial nucleosynthesis" (Fig. 18.5).

One might very well wonder whether it is really possible to make such predictions. Surely the state at late times depends on what you chose for initial conditions. Would it not be possible to get any prediction you want by choosing suitable initial conditions? In fact, it turns out that the predictions *are* almost entirely independent of the choice of initial conditions: any initial conditions for the nuclear abundances are erased by the drive toward nuclear statistical equilibrium, which *sets up* the "initial conditions" for the subsequent evolution. Figure 18.6 illustrates this effect with different initial conditions.

As discussed above, different initial conditions always do evolve into different states, when viewed at the microscopic level. The two cases in Fig. 18.6 approach the

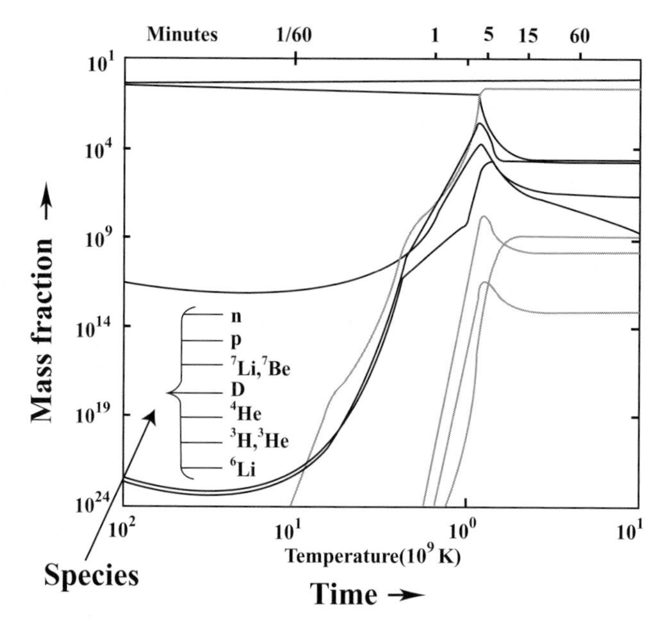

Figure 18.5. The evolution of nuclear species in the early universe: the mass fractions freeze out at specific values, leading to predictions of nuclear abundances from early universe cosmology. (Adapted from Burles *et al.* (2001).)

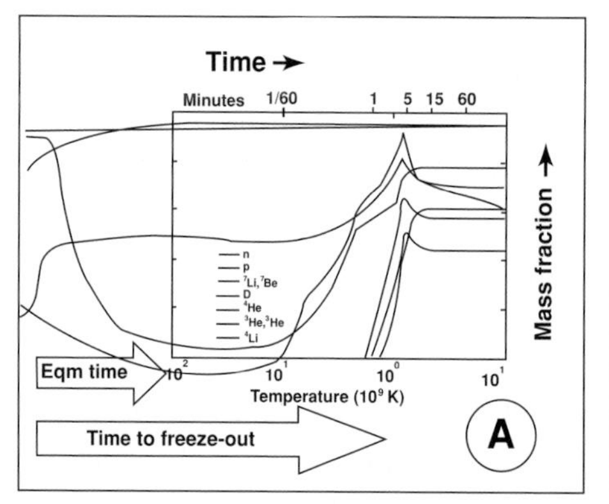

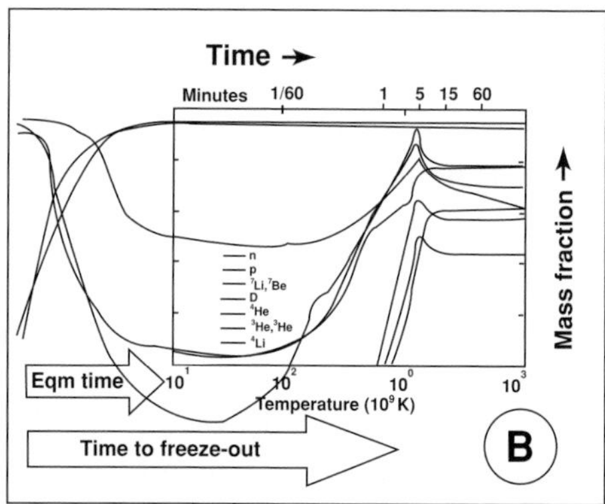

Figure 18.6. Different initial mass fractions of nuclei would all rapidly approach nuclear statistical equilibrium, resulting in the same "initial" conditions for the subsequent process of nucleosynthesis. Thus for nucleosynthesis, the initial conditions are determined by the dynamics of a previous epoch. (Curves outside the box of the original plot are sketched in, and do not represent actual calculations.) (Adapted from Burles *et al.* (2001).)

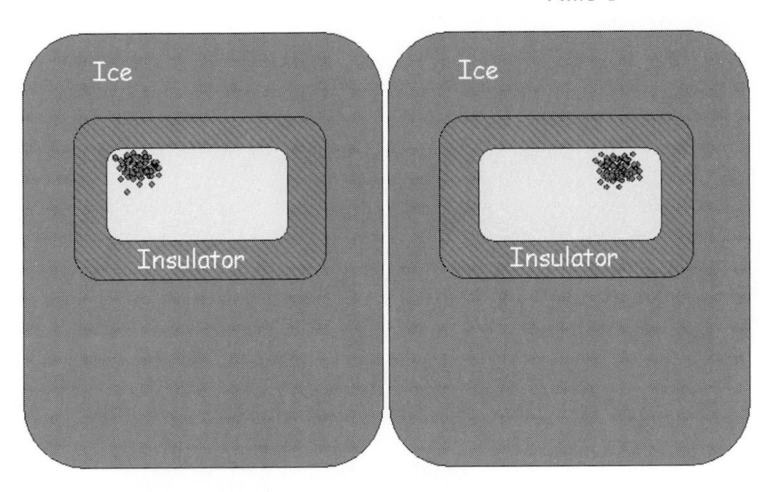

Figure 18.7. Boxes of gas encased in ice.

same equilibrium state only because one is coarse-graining out subtle correlations among particles that carry information about the initial conditions. This coarse graining is implemented by focusing just on the mass fractions, which represent just a small amount of information about a microscopic state.

So Big Bang nucleosynthesis is an example of a situation where "initial" conditions definitely do not play a subsidiary role, but are critical to any claim that one is actually making predictions. In this case, the dynamics of an earlier epoch (namely the approach to equilibrium) step in to set up the subsequent initial conditions, just as one hopes cosmic inflation can set up initial conditions for the Big Bang cosmology.

Illustration 2: Gas in a block of ice

I now discuss an even simpler example, which will turn out to be conceptually very similar to the nucleosynthesis case. Consider two boxes of gas similar to those depicted in Fig. 18.1, but now supposed they are encased in blocks of ice, as shown in Fig. 18.7.

The insulator between the ice and the gas is not perfect, but serves to slow down the equilibration time between the two. Figure 18.8 illustrates the subsequent evolution. The gas has plenty of time to equilibrate, and the equilibration sets up "initial" conditions for the process of condensing and freezing. As a result, the uniform state of the frozen gas at "Time 3" can be *predicted*.

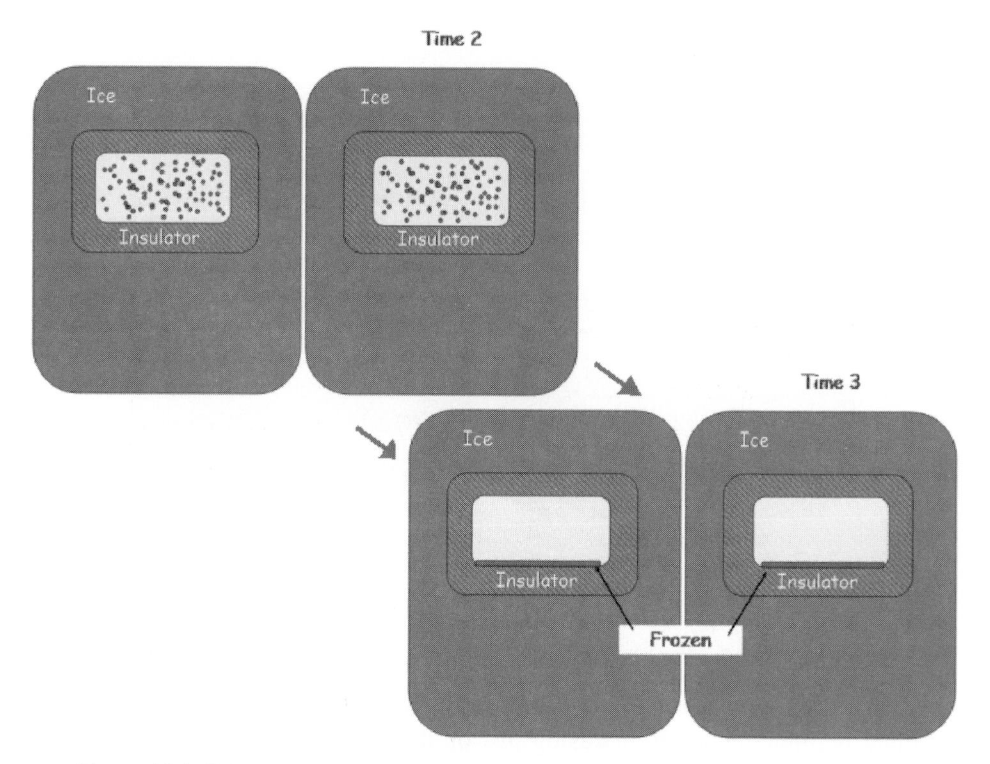

Figure 18.8. Subsequent evolution of the system depicted at "Time 1" in Fig. 18.7. The gas inside the box equilibrates first, setting up the initial conditions for the subsequent condensation and freezing.

Equilibrium and de Sitter space

Both of the above illustrations used an early period of equilibration to set up initial conditions for subsequent evolution. To understand how this concept carries over to the case of inflation, we first have to expand on the discussion above, where equilibrium was discussed for gravitating systems.

Einstein first proposed the "cosmological constant" (known as Λ) early in the days of general relativity. Later, it was realized that certain scalar fields can at least temporarily enter a "potential dominated state" which closely mimics the behavior of a cosmological constant. A cosmological constant, roughly speaking, acts like a *repulsive* gravitational force, and Einstein first proposed it to balance the normal attractive force of gravity in order to model a static universe. However, that idea did not work because such a balance is not stable. The natural evolution of matter in the presence of a non-zero cosmological constant eventually becomes dominated by the repulsive force and is driven apart exponentially fast by the resulting expansion.

The expansion rate is

$$H = \sqrt{\frac{8\pi G}{3}\Lambda}. \tag{18.3}$$

After waiting out a suitable "equilibration time" the exponential expansion will empty out any given region of the universe, leaving nothing but the cosmological constant (or potential dominated matter) which does not dilute with the expansion. This exponentially expanding empty (but for the cosmological constant) spacetime is known as de Sitter space. The approach to de Sitter space is essentially the opposite of the gravitational collapse discussed above: instead of approaching an equilibrium state of total gravitational collapse (a black hole), with a non-zero cosmological constant the universe asymptotically approaches de Sitter space, a state of essentially total "un-collapse."

It seems natural to associate the notion of equilibrium with end-point states toward which many states are dynamically attracted. This perspective makes it natural to think of a black hole as an equilibrium state, and thus it seems natural to define black-hole entropy. Perhaps not surprisingly, similar arguments to the black-hole case produce a definition of the entropy of de Sitter space (Gibbons and Hawking 1977):

$$S_{\mathrm{dS}} = \frac{3\pi}{\Lambda}. \tag{18.4}$$

The statistical foundations of de Sitter space entropy are probably even more poorly understood than the black-hole entropy, but it certainly fits in nicely with the heuristic notion of entropy and equilibrium considered here. Also, the part of de Sitter space toward which a cosmological-constant-dominated universe evolves is homogeneous and flat: two features of the Big Bang cosmology that inflation seeks to explain.

The potential-dominated state

Models of cosmic inflation use a scalar field $\varphi(x)$ (the inflaton) to mimic the behavior of a cosmological constant for a certain period of time. This cosmological-constant-like behavior is achieved when the inflaton is in a potential-dominated state. Specifically, it is the inflaton stress energy, given by an expression like

$$T_{\mu\nu} = \partial_\mu \partial_\nu \varphi(x) + g_{\mu\nu}[g_{\alpha\beta}\partial_\alpha \partial_\beta \varphi(x) + V(\varphi)] \xrightarrow{g\partial\partial\varphi \ll V} g_{\mu\nu} V(\varphi) \tag{18.5}$$

that must be dominated by the potential term $\propto V(\varphi)$ for the inflaton to look like a cosmological constant.[5] So ultimately constraints like

$$V(\varphi) \gg g_{\alpha\beta}\partial_\alpha\partial_\beta\varphi(x) \tag{18.6}$$

must hold. As has been known since the early days of inflation, and has been emphasized over the years (Penrose 1989; Unruh 1997; Trodden and Vachaspati 1999; Hollands and Wald 2002), the potential-dominated state for the inflaton is a very special state. The field $\varphi(x)$ has a huge number of degrees of freedom, and many possible states of excitation. Only a tiny fraction of these will obey the constraints in eqn (18.6) sufficiently strongly to allow the onset of inflation. This fact will be important in what follows.

Cosmic inflation

Basic inflation

The basic idea of inflation is that the universe entered a potential-dominated state at early times. If the potential-dominated phase was sufficiently long, the spacetime would have had a chance to equilibrate toward de Sitter space.[6] The de Sitter space has the flatness and homogeneity properties required for the early stages of the Big Bang, so via the approach to de Sitter space these features are acquired dynamically.[7]

But of course in the early stages of the Big Bang the universe is full of ordinary matter, not potential-dominated matter. A critical part of cosmic inflation is *reheating*: after a sufficient period of inflation the potential-dominated state decays (or "reheats") into ordinary matter in a hot thermal state.

In typical modern inflation models the instability is a classical one of the "slow roll" type illustrated in Fig. 18.9. The critical degree of freedom driving inflation is the homogeneous piece (or average value) of the inflaton field φ, depicted in the figure. This degree of freedom can be thought of as "rolling" in its potential $V(\varphi)$. At the onset of inflation φ starts out in a relatively flat part of $V(\varphi)$ so the small values of the time derivative $\partial_0\varphi$ (required for eqn (18.6) to hold) can be maintained. The field is rolling slowly here, and the potential domination causes exponential expansion to set in. However, φ is never completely stationary, and it eventually reaches a part of the potential that is steeper. At that point φ speeds up,

[5] $\partial_\mu\varphi(x)$ denotes the space and time derivatives of $\varphi(x)$. For further background see for example Kolb and Turner (1990).

[6] The fact that inflation is never in perfect equilibrium has been analyzed by Albrecht *et al.* (2002).

[7] The standard Big Bang models the universe all the way back to an initial singularity of infinite density and temperature. Inflation provides an alternate account of the very early universe, and matches on to the standard Big Bang at a finite time after the initial Big Bang singularity. The question of whether other singularities necessarily precede inflation is under active investigation; see section "Eternal inflation" (pp. 388–9).

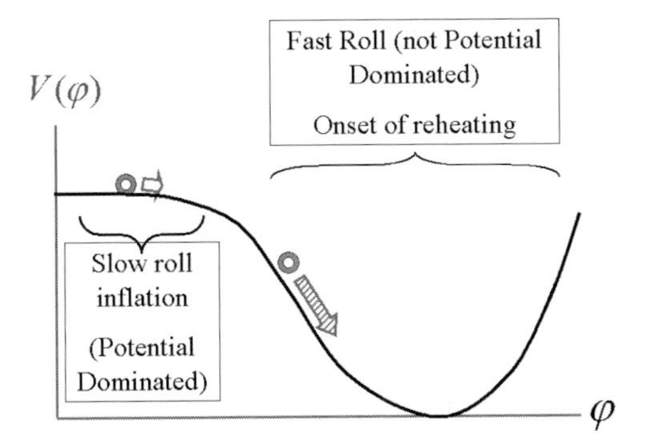

Figure 18.9. The homogeneous piece of the inflaton field (depicted here) controls inflation by first rolling slowly in a flat part of its potential $V(\varphi)$ (allowing potential domination and exponential expansion) and then entering a steeper part of the potential that ends the slow roll and allows reheating to occur.

eqn (18.6) no longer holds, and the exponential expansion is over. If φ is suitably coupled to ordinary matter, energy can couple out of φ and into ordinary matter in the non-slow-roll regime, creating the right conditions for the beginning of the standard Big Bang.

Quantum corrections to the above discussion allow one to predict deviations (or perturbations) from perfect homogeneity produced during inflation, which evolve into galaxies and other structure in the universe. These are discussed further below.

At this stage we do not have a strongly favored "standard model" for the inflaton. There are a huge number of workable proposals for the origin of the scalar field, $V(\varphi)$, etc., but no clear favorite and none that is deeply rooted in well-established theories of fundamental physics. This fact must be regarded as a weakness in the inflationary picture. But to be fair, that situation might be more a reflection of our generally primitive understanding of fundamental physics at the relevant high energy scales rather than anything intrinsically suspect about inflation.

Inflation and the arrow of time

We now have seen three examples where a process of equilibration generates dynamically predicted initial conditions for the next stage of evolution. (The examples are Big Bang nucleosynthesis, the gas in ice, and cosmic inflation.) But how do these examples address the key question of this chapter, namely how can one harness equilibration to make a special initial condition "generic" and still have an arrow of time (that is, nongeneric initial conditions)?

In each of these examples the question is resolved in the same way. The equilibration during the first "initial condition creating" stage is not equilibration of the entire system, but just of a *subsystem*. In each case there are additional degrees of freedom which are never in equilibrium that drive the system and carry information about the arrow of time.

In the case of Big Bang nucleosynthesis, it is the spacetime (or gravitational) degrees of freedom that are out of equilibrium. The universe is not one giant black hole (equilibrium for a normal gravitating system) but rather a homogeneous and isotropic expanding Big Bang state (which, as Penrose taught us, is far out of equilibrium). Against this background, the nuclear reactions are able to maintain chemical equilibrium among nuclear species at early times, when the densities and temperatures are high. As the out-of-equilibrium degrees of freedom (namely the cosmic expansion) cool the universe, the lower energies and densities put matter in a state that can no longer maintain nuclear statistical equilibrium. The expanding spacetime background is the out-of-equilibrium subsystem that drives the change, first allowing the "nuclear species subsystem" to enter chemical equilibrium and then (having thus set up the "initial conditions") driving the nuclear matter toward the out-of-equilibrium conditions that produce the predicted mass fractions from primordial nucleosynthesis.

For the ice and gas (Fig. 18.8), they start far from equilibrium but with a slow equilibration time due to the presence of the insulator. The initial equilibration (of gas within the box) is just the equilibration of the gas subsystem. Viewed as a whole, the ice and gas are still out of equilibrium, even as the gas subsystem spreads out into an equilibrium state within its box (which of course defines the "initial" conditions for what comes next).

For inflation, the *inflaton field* is the out-of-equilibrium degree of freedom that drives other subsystems. The inflaton starts in a fairly homogeneous potential-dominated state which is certainly not a high-entropy state for that field (Trodden and Vachaspati 1999). In a well-designed inflation model the special potential-dominated inflaton state "turns on" an effective cosmological constant and leaves it on for an extended time period, allowing plenty of time for the matter to equilibrate toward de Sitter space. But the slow-roll inflaton instability eventually "turns off" the cosmological constant, and the continued out-of-equilibrium evolution of the inflaton leads to a period of reheating followed by conditions appropriate for the early stages of the standard Big Bang (at which point the spacetime is the out-of-equilibrium degree of freedom driving the subsequent arrow of time).

So while inflation does dynamically "predict" the special initial state of the Big Bang phase, it does not *predict* the arrow of time. Inflation "passes the arrow of time buck" to the special initial conditions of the inflaton field. An arrow of time, by its fundamental nature, requires nongeneric initial conditions. For a Big Bang universe created by inflation, the nongeneric quality of the initial conditions that give us an arrow of time can be traced right back to the special inflaton initial state.

To better understand the role of inflation and its relationship to the arrow of time, it is necessary to put the above discussion in a larger context. That exercise (which is the subject of the following section) will help us understand how inflation has a crucial role, despite the fact that it does not *predict* every aspect of the universe we observe. (There is an early discussion of some of the key issues from this section and the next in a series of papers by Davies (1983, 1984) and Page (1983).)

Initial conditions: the big picture

Data, theory, and the "A" word

To what extent should we use observational data to confront theoretical predictions, and to what extent should those data instead be used as input to theoretical models, in order to constrain free parameters? The debate about this issue can get extremely passionate, and often involves using "the A word" or the "anthropic principle."

I believe that the reality behind the passions is pretty straightforward, and offers clear guidance about how to proceed: every theory known so far requires some observational data to be used as input, to constrain charges, masses, and other parameters. On the other hand, pretty much everyone would agree that if a new theory required fewer data as input (i.e., had fewer free parameters to set) and could in turn predict some of the data that the old theory used as input, then the new theory would simply be better than the old theory, and would supersede it.

A consequence of this line of reasoning is that data should be treated as a precious resource: using up data to set parameters should be avoided if at all possible. It is much better to save up the data to use to test the predictions of your theory after a minimal number of parameters are set. If you are sloppy about this issue, your most serious penalty is not really the harsh criticism you might experience at the hands of physicists with other passionately held views. The real threat is that another approach that is more efficient with the data could simply leave your line of thinking behind in the dust.

So while "pro-anthropic" scientists tend to alarm their colleagues by apparently freely using up precious data to constrain models[8] the anthropicists might be equally indignant that many of their opponents seem unwilling to acknowledge that some data really do need to be used up as input.

A more fruitful approach lies between these two extremes: admit that some of our precious data need to be used up, but work as hard as possible to use up as few data as possible in this manner.

[8] Statements that life could not exist without some detailed property of the known physical world come across as gratuitous to many physicists (including me), since we really do not have a clue what great varieties of "life" might be possible. Without some more concrete expression of this idea, one appears to be simply using the physical property as input, and giving up on actually predicting it.

Using the arrow of time as an input

The arrow of time, as it is currently understood, simply has to be used as an "input" to any theory of the universe. At its most fundamental level, the arrow of time emerges from evolution from a special initial state toward more generic subsequent states (where "generic" and "nongeneric" are defined relative to the natural evolution under the equations of motion and also relative to a particular coarse graining). To have an arrow of time, there must be something nongeneric about the initial state. That property of the initial state must be chosen not because it is a typical property but because that (necessarily atypical) property is required in order to have an arrow of time.

An attractive way of incorporating this line of reasoning into a "big picture" follows up on the discussion of Fig. 18.2. The figure shows a random large fluctuation in an "equilibrium" box of gas creating conditions where there temporarily is an arrow of time. If one thinks of a box of gas sitting there for all eternity, such events, although rare, will occur infinitely many times. In this kind of picture, the special initial conditions that produce an arrow of time are not imposed on the whole system at some arbitrary absolute origin of time. Instead the special "initial" conditions are found by simply waiting patiently until they occur randomly. Boltzmann (1897, 1910) already was thinking along these lines a hundred years ago, but found some aspects of this argument deeply troubling. We will discuss Boltzmann's problem below, and see what inflation has to say about it.

Most modern thinking about inflation borrows at least some aspects from Fig. 18.2. One typically imagines some sort of chaotic primordial state, where the inflaton field is more or less randomly tossed about, until by sheer chance it winds up in a very rare fluctuation that produces a potential-dominated state (Linde 1983). One important difference between the box of gas and the "pre-inflation" state is that it is much easer to calculate things for the box of gas. Although very interesting pioneering work has been done (see for example Linde (1996)), we still do not appear very close to a concrete systematic treatment of a chaotic pre-inflation state.

Of course once it is possible to create a period of inflation, one may not need to know too much about the pre-inflation state. In many models, inflation creates such a large volume of the universe in the inflated domain that the predictions appear to be insensitive to many details of the pre-inflation state. Still, one certainly needs to know enough about the pre-inflation state to establish at least roughly such insensitivity.

But there is an even bigger question lurking behind this issue. If one is willing to concede that even with inflation, special initial conditions must either be stumbled upon accidentally or imposed arbitrarily, what role is left for inflation? Why not simply wait around for the Big Bang itself to emerge directly out of chaos, or impose Big Bang initial conditions directly on the universe, without bothering with an initial period of inflation (see for example Barrow (1995))?[9] The answer is that even though inflation is not all-powerful, and cannot create the Big Bang from absolutely anything, inflation still has a great deal of predictive power which allows one to make more economical use of the data than one could in the absence of inflation.

Predictions from cosmic inflation

Once the special inflaton initial conditions get inflation started, a whole package of predictions is made. The universe is predicted to be homogeneous, with a density equal to the critical density (to better than 0.01% accuracy). A spectrum of perturbations away from perfect homogeneity is also predicted, with a specific "nearly scale-invariant" form. Perhaps most important for this discussion, the volume of a typical region that has these properties is huge, exponentially larger than the entire observed universe. These predictions go well beyond the basic notion of what the standard Big Bang cosmology describes. Taking the standard Big Bang model on its own, there is no particular reason to expect the density to be nearly critical, or to expect a particular form for the spectrum of perturbations. Currently, a large body of data supports the inflationary predictions (see for example Fig. 18.10 as well as Albrecht (2000) for examples and more information).

No inflation model predicts that the entire universe is converted to a Big Bang-like state. In many models, quantum fluctuations take the inflaton back "up the hill" sufficiently frequently that at any time after inflation starts, regions that are still inflating actually dominate the volume of the universe (Linde 1986). But if you do find a region with ordinary matter (as opposed to the potential-dominated inflating state) that region will be exponentially large, and have the properties described in the previous paragraph. (Note however that here I am using additional data as input.)

Inflation is best thought of as the "dominant channel" from random chaos into a Big Bang-like state. The exponentially large volume of the Big Bang-like regions produced via inflation appear to completely swamp any other regions that might have fluctuated into a Big Bang-like state via some other route. So if you went looking around in the universe for a region like the one we see, it would be exponentially more likely to have arrived at that state via inflation, than some other way, and is thus strongly predicted to have the whole package of inflationary predictions.

[9] This question is raised directly by Barrow (1995) and is also closely related to other concerns (Penrose 1989; Unruh 1997; Linde *et al.* 1994, 1996; Vanchurin *et al.* 2000; Hollands and Wald 2002).

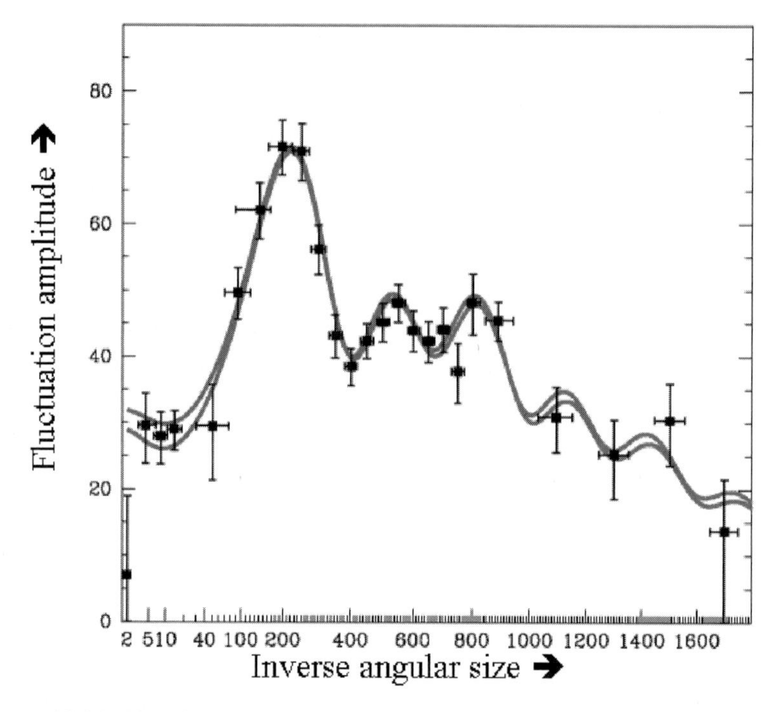

Figure 18.10. This figure shows a compilation of measurements of anisotropies in the cosmic microwave background (points) along with curves from inflation models as a function of inverse angular scale on the sky. The left–right location of the peak structure is very sensitive to the overall density of the universe. Current best estimates show the density is consistent with the critical value predicted by inflation to within error bars around 10% (Wang *et al.* 2001). The oscillatory behavior and the lack of an overall sharp rise or drop across the plot also support the predictions of inflation (Wang *et al.* 2001; Albrecht 2000). (Reproduced with permission from Tegmark and Zaldarriaga (2002).)

Boltzmann's "efficient fluctuation" problem

It is pretty exciting to have a theory of initial conditions with plenty of specific predictions to test. The fact that inflation offers such a theory has had a huge impact on the field of cosmology, and has motivated high ambitions on both the theoretical and observational sides of the field. But inflation also addresses another issue that had Boltzmann worried a century ago.

Boltzmann also was trying to think of the "dominant channel" into a universe like ours, but without the benefit of inflationary cosmology. Boltzmann realized that the only way one could expect Nature to produce the unusual "initial" conditions that lead to an arrow of time is to wait for a rare fluctuation of the sort depicted in Fig. 18.2. But that "dominant channel" also comes with its package of somewhat disturbing predictions. In particular, rare fluctuations in ordinary matter seem to be very stingy about producing regions with an arrow of time. If you use as input the data that you are sitting in a room, like whatever room you are sitting in, and that it has existed for at least an hour, then by far the most likely fluctuation to fit the data is a room that fluctuates alone in the midst of chaos, and is immediately destroyed by the surrounding chaos as soon as the hour is up. If you want to look for a larger piece of your world (the whole building that contains your room, the whole city, the whole planet, etc.) you would have to wait around for an even more rare fluctuation, and by far the most likely fluctuation would just barely fit the input data, and exhibit utter chaos everywhere else.

So Boltzmann (and many others since) worried that if our world really emerged from a random fluctuation, then a strong prediction is made that we exist in the midst of utter chaos. The fact that instead we live in a universe billions of light years in size which is extremely quiet and unchaotic, and that seems to have room for not just our cozy planet, but many more like it seems to be in blatant contradiction to these predictions. To get a rare fluctuation to produce all that, you would have to use up all those features of the universe as input data. None of those features would be predicted. (For a nice account of this issue, see section 3.8 of Barrow and Tipler (1986).)

Cosmic inflation gives a very attractive resolution to this problem. The big picture is similar, in that one has to wait for a rare fluctuation to create the universe we observe. But inflation says the most likely rare fluctuation to produce the world we see is not the random assembly of atoms, molecules, and larger structures directly out of a chaos of ordinary matter. Inflation offers a completely different set of dynamics, where a small fluctuation in the inflaton field gives rise to regions that look like our universe, but which actually generically extend exponentially further beyond what we see. Inflation transforms the large-scale nature of our universe from a mystery into a prediction.

Comparing different theories of initial conditions

My discussion has emphasized four key aspects of the inflationary picture:

1. *Attractor*. Inflation exhibits "attractor" behavior (or equilibration toward de Sitter space) which causes many different states to evolve into states that resemble the early stages of the Big Bang.
2. *Volume factors*. Inflation generates exponentially large volumes, which fact gives extra weight to the inflationary channel into these early Big Bang states.
3. *Arrow of time*. Despite points 1 and 2 the initial conditions for inflation need to be nongeneric to some degree. This is required in order to have an arrow of time.
4. *Predictions*. Still, when points 1–3 are taken together, inflation produces an impressive package of predictions that overall allow one to use up fewer data as input than one would have to do in the standard Big Bang model taken alone.

Today there are a variety of different ideas about initial conditions in play, and it is interesting to consider how different ideas compare with inflation on these four key points.

Chaotic inflation

The discussion in this chapter embraces the ideas put forth by Linde on chaotic inflation (for an overview, with references to the original literature, see Linde (1997)). This chapter should be seen as a further extension of these ideas.

Eternal inflation

There has been a lot of discussion recently about whether it is possible to describe the universe as an eternal inflating state with (exponentially large) islands of reheated matter. This description would allow one to forget about trying to understand the "pre-inflation" state altogether. There simply would be no pre-inflation. Different viewpoints have emerged on this subject. One view states that such an eternally inflating state is impossible to create because singularities necessarily arise. These singularities can take a variety of forms, but in each case the upshot is that additional initial data are required, implying some notion of "pre-inflation" (Borde *et al.* 2001).

Another view is that the very statement that one is looking for an eternally inflating state contains enough information to resolve such singularities. Aguirre and Gratton (2002) claim that when one uses this information to good effect, there is one obvious choice for the "pre-inflation" state. If that choice is made, Aguirre and Gratton argue that a global state is constructed which can, in the end, be thought of as defining an eternally inflating state. The eternally inflating state that emerges from that approach has specific global properties that reflect an arrow of time. In particular, an array of regions of reheating (or decay of the potential-dominated state) must be organized coherently to be pointing in a commonly agreed "forward direction." In fact, there are actually two different "back-to-back" coherent domains in this picture, with arrows of time pointing in opposite directions. The coherence must extend over infinitely many reheated regions, distributed throughout an infinitely large spacetime volume.

Several technical issues remain unanswered (for example whether the construction of Aguirre and Gratton can be implemented at the level of full fundamental equations), but here I simply comment on how these two perspectives relate to the four key points mentioned above. I start with the Aguirre–Gratton picture: (1) The eternal inflation picture specifically avoids needing attractors. By fiat the state of the universe is specified completely, and there is no need to draw other states toward it. (2) In the Aguirre–Gratton picture there is only one way to create Big Bang-like regions, so although the exponentially large volume factors certainly are present, they do not seem to have as crucial a role as they have in a more standard inflationary picture. (3) In the Aguirre–Gratton construction, the arrow of time is put in by hand. One simply declares "the universe is in this state," and it happens to have an arrow of time. The only conceptual difference between the Aguirre–Gratton idea and simply declaring "the universe is in a standard Big Bang state" (in other words, forgetting about inflation altogether) is the claim (still debated) that the eternal model does not have singularities. The Aguirre–Gratton idea specifically tries to eliminate the role of a rare random fluctuation of the inflaton that one sees in the standard discussions of chaotic inflation (and replaces it with a special choice of state for *all* time).

On the other hand, Borde *et al.* (2001) say that singularities exist that make it impossible to extend the inflationary state eternally back in time. This perspective fits perfectly with the picture developed above, where the singularity is resolved by extending back in time not with more inflation, but into some more chaotic state of spacetime and matter (probably with its own naturally occurring singularities that need to be resolved by a more fundamental theory).

The ekpyrotic universe

This idea basically suggests a way of extending the story of the universe backward past the Big Bang phase into an epoch where the universe can be described (presumably at a more fundamental level) by colliding "branes" in a higher-dimensional space (Khoury *et al.* 2001a). (1) The proposed dynamics do *not* contain any attractor behavior. (2) Nor do they have any exponential volume creation. (3) The arrow of time and many other features of the Big Bang cosmology are a direct consequence of very special properties of the initial brane configuration which are put in by hand (or by "principles"). This picture also involves a singularity (when the branes collide; also meant to be the starting point of the standard Big Bang) and considerable controversy surrounds the questions of how this singularity might be resolved (see for example Kallosh *et al.* (2001), Khoury *et al.* (2001b), and Gordon and Turok (2002)). (4) In terms of predictions, much depends on how the singularity is resolved. Certainly the homogeneity and flatness of the universe (predictions of inflation) are put by hand into the initial conditions of this model. Some argue that predictions for cosmic perturbations in these models have already ruled them out (Tsujikawa *et al.* 2002), but others argue that the predictions are consistent with what we know so far, but offer novel differences from inflationary predictions that could be observed in the future (Khoury *et al.* 2002).

Because of the differences on points 1, 2 and 4, the ekpyrotic universe does not represent an alternative mechanism that can replace inflation by doing what inflation does in a different way. As far as initial conditions are concerned it is, much like eternal inflation, a retreat back to the conceptual framework of a stand-alone standard Big Bang, where most of the specifics of the state of the universe are put into the initial conditions by hand. However, as with eternal inflation, if the vision of the original authors pans out this idea will offer a resolution of the big bang singularity. In addition, the ekpyrotic idea suggests intriguing testable predictions for cosmic perturbations.

The cyclic universe

If the singularity of the ekpyrotic universe *can* be resolved in the manner originally proposed, very similar dynamics could also be used to construct a cyclic model of the universe (Steinhardt and Turok 2002a). Although some notion of a cyclic universe has been around for a long time (Tolman 1934), suitable dynamics to turn a contracting universe into an expanding one were always lacking. If the brane-collision picture can be shown to work, it will offer a nice way to construct a universe that bounces from contraction back into expansion. Using this innovation, Steinhardt and Turok (2002a) constructed a cyclic model of the universe which includes a period of inflation late in the cycle. Rather efficiently, this proposal uses today's cosmic acceleration (see Albrecht (2002) for review) as the inflation period for the next cycle. With a period of inflation built into the scenario one might be tempted to view the cyclic universe as a variation on the inflation theme, and indeed, modulo clarifying what happens at the singularity, I regard this as a pretty interesting variation.

However, Turok and Steinhardt originally state that a key feature of the new cyclic scenario was to offer completely eternal cyclic evolution (see for example Steinhardt and Turok (2002b)). In this picture, like the eternal and ekpyrotic scenarios discussed above, there is no pre-inflation state to contend with. One simply declares "This is the state of the universe."

My discussion in this chapter leads to a number of concerns about this claim of eternality. First of all, the claim of eternality is a very extreme one. If there is any non-zero probability, no matter how small, of the model fluctuating (unstably) off its cycle, that fluctuation has all of eternity to get around to happening, and thus it is 100% certain to happen at some time. Any such event will completely destroy any claims to eternality.

The arrow of time is a nice illustration of just this sort of effect. While to some approximation the arrow of time can be regarded as an absolute property of our physical world, a deeper analysis reveals the arrow of time indeed to be only an approximation. Just as it is possible for air to spontaneously rush into one corner of a room, and just as in fact such a rare event is absolutely certain to happen if you wait long enough, there are probably many different ways some "conspiracy" of microscopic degrees of freedom could conspire to divert the oscillating universe from its cycle. To make a compelling case for eternality, one would have to argue that all possible rare events had been completely accounted for. Such a case has certainly not been made so far, and it is very hard to see how such a case ever could be made.

This issue must in some sense be a weakness of the eternal and ekpyrotic scenarios as well, but in those models one controls more aspects of the state of the universe simply by declaration (namely making eternality part of the definition of the state). The new cyclic model is presented in a way that leaves more details in the hands of dynamical evolution (possible because of the attractor behavior during the regular periods of inflation). I feel this greater focus on dynamics is a strength of the cyclic model, but it also makes it easier to formulate the concern that a very rare event could prevent eternality.

To be more specific, one can study the origin of the arrow of time in the cyclic model. A crucial role is played by the assumption that heat (and entropy) is reliably produced upon brane collision but the time reverse (cooling) *never* occurs. In the current literature this feature is put in completely "by hand" and only at the "thermodynamic" level. Namely, the current treatment uses what is effectively a friction term to impose an arrow of time on the cyclic model. Just as a deeper understanding of everyday friction allows for the ridiculously small but non-zero probability that a coherent fluctuation could appear and produce a push in the opposite direction to normal friction, one would expect that whatever microscopic mechanism underlies the friction term in the cyclic universe would be able to do the same. Because eternality is such an extreme claim, one such fluctuation could be enough to destroy eternality. In any case, it would certainly be interesting to learn what microscopic picture the advocates of the eternally cycling model have in mind.

I should reiterate that my criticism of claims of eternality does not detract from the appeal the cyclic model could have as a new mechanism within the normal conceptual framework of inflation, with the arrow of time originating as a rare fluctuation.[10]

Varying speed of light

Another approach to initial conditions is based on the idea that the speed of light could have been faster in the past (Moffat 1993a, b; Albrecht and Magueijo 1999). These models are still in their early stages and a clear picture of the fundamental origin of the varying speed of light (VSL), as well as the origin of perturbations,

[10] I recently learned that authors of the new cyclic model no longer see eternality as an important goal of the model and agree that the cyclic universe is unlikely to be truly eternal (P. Steinhardt and N. Turok, pers. comm.).

has yet to emerge. Still the VSL concept attempts to duplicate the approach of inflation on all four points discussed in this section. In particular one would expect any fundamental theory that allowed the speed of light to vary would make it just as likely to be slower or faster in the past. In that sense the speed of light would have to play a similar role to the inflaton, linking the arrow of time to a rare fluctuation in $c(t)$.

Holographic cosmology

Another intriguing proposal uses the idea of holographic bounds on the entropy of gravitating systems to describe a maximal entropy "black-hole gas" state from which our Big Bang universe emerged (Banks and Fischler 2001). In this work, Banks and Fischler take the view that the causal structure around an observer is absolutely fundamental, and build a physical picture on top of that. The arrow of time appears in their picture as a fundamental feature (not an emergent or approximate one) linked to the causal structure. Thanks to space of states of matter that actually grows with time, the dynamics in this picture are not even reversible at the microscopic level. In this way the holographic cosmology picture is completely different from the standard inflationary picture discussed here.

Still, there are some interesting parallels. In particular, the global properties of the black-hole gas state are very different from the universe we observe, so Banks and Fischler propose a dominant channel (quite different from inflation) whereby rare regions in the black-hole gas evolve into something like the standard Big Bang. Much still needs to be developed in this picture, particularly the origin of cosmic structure on large scale, but already it offers the most dramatic and stimulating departure yet from standard ideas about cosmic initial conditions.

Wave function of the universe

Many have been tempted to think that some argument or principle could define the "wave function of the universe" (Hartle and Hawking 1983; Linde 1984, 1998; Vilenkin 1984; Hawking and Turok 1998; Hawking and Hertog 2002).[11] So far such attempts have yielded different wave functions in the hands of different authors (Vilenkin 1998). On the face of it this approach is fundamentally different from the inflation-based picture discussed in this article. Simply declaring the wave function of the universe is not about dynamical mechanisms that give a preferred channel from chaos into the standard Big Bang. It is about principles simply choosing the state of the universe by one method or another, much as we see in the eternal and ekpyrotic cases.

[11] Of course the starting point is usually the Wheeler–DeWitt equation (DeWitt 1967; Wheeler 1968).

Interestingly, many discussions include both the wave function of the universe and inflation, and use the wave function of the universe to determine the most probable way that inflation gets started. In that work, the wave function of the universe is basically an approach to describing the pre-inflation state. Most proposed wave functions of the universe are not so very sharply peaked, and their breadth might be interpreted as an expression of the "chaos" often assumed for the pre-inflation state, perhaps tempered slightly by some general principles. If things develop in this way, the wave function of the universe idea might turn out to be more closely connected to standard ideas about inflation than superficially appears to be the case.

Another interesting idea advocated in some discussions of the wave function of the universe is that classical spacetime must be automatically correlated with an arrow of time. In these discussions the wave function of the universe is used to provide constraints on classical spacetime as it emerges from a quantum regime, and it is argued that the classical spacetimes that emerge naturally come with low- and high-entropy "ends" and thus an arrow of time (further discussion of this point of view with additional references can be found in Zeh (1992), Gell-Mann and Hartle (1994), and Hawking (1994)). So far these arguments are made in the context of "mini-superspaces" based on a Friedmann–Roberson–Walker (FRW) background, which of course presupposes the homogeneity that accounts for the actual arrow of time in the universe. If these results persist in a more complete theory (with a more complete superspace) this line of reasoning could give key insights into origin of the arrow of time. In such a picture one could predict the arrow of time by simply using the classicality of spacetime as an observational input.

Of course, it is not at all clear that stating the wave function of the universe from first principles will ever take hold as a theory of initial conditions. The physical world clearly has a huge phase space which it tends to explore in a thorough way, and I am far from convinced that principles concocted by humans could really convince the universe to avoid large parts of that phase space. My skepticism is only enhanced by the fact that we do not have one theory of the wave function of the universe, but many, and the community as a whole has not found compelling reasons to choose one over the others.

Chaotic mixing

Cornish *et al.* (1996) argue that "chaotic mixing" which can occur in topologically complex spaces could dynamically "explain" the initial conditions for a special type of inflation. This line of reasoning seems to be in conflict with the idea that there has to be something nongeneric or rare about the initial conditions for inflation in order to have an arrow of time. As far as I can tell, this idea has not been well enough developed to determine where the arrow of time fits in. There seems to be an asymmetric treatment of the "instability" associated with inflation; it also appears that the usual gravitational instability is not accounted for at all in their discussion.

Another somewhat related idea proposes that the homogeneity of the early universe arises from a kind of statistical averaging over higher dimensions (Starkman *et al.* 2001). In that model, the arrow of time is put in by hand via the assumptions about the initial state. They assume FRW topology, as well as a statistical ensemble of states with an average curvature of zero (far from gravitational collapse). Both these assumptions produce an initial state that is effectively "low entropy," and thus generates an arrow of time. The authors argue that this is an explanation of the homogeneity of the universe, but it is hard to imagine this dynamical mechanism competing with inflation. The statistical ensemble of states they require, with no mean curvature over a huge volume, seems to be much lower entropy (and thus much more rare) than the small inflaton fluctuation required for inflation. Also, unlike inflation, Starkman *et al.*'s mechanism does not generate any large volume factors which could leverage their mechanism.

Brane gas cosmology

This idea proposes that the homogeneity of our universe emerges from some kind of equilibration process of branes in higher dimensions (Watson and Brandenberger 2002). As in the cases of chaotic mixing and holographic cosmology, it is not clear that care has been given to arrow-of-time issues in this model. Something has to fluctuate or "be declared" out of equilibrium in order to have the arrow of time we observe. What degree of freedom takes on that role in the brane gas model?

Cardassian expansion

The "cardassian expansion" model (Freese 2002; Freese and Lewis 2002) comes from modified Friedmann equations which have homogeneity built in, and as such does not address the origin of homogeneity in the universe. Unless this model develops into one that does address the homogeneity of the universe (the origin of the arrow of time) it is not possible to analyze the relationship of the cardassian model to the arrow of time and other topics discussed here.

Further discussion

Emergent time and quantum gravity

Microscopic time has long been regarded as a problematic notion for a full theory of quantum gravity. One attractive resolution of the problem of time in quantum gravity is to view microscopic time as an emergent quantity that does not have to be well defined for all states of the universe. Operationally speaking, microscopic time is just a statement about correlations between physical systems designated as clocks, and other physical systems of interest. Perhaps we should understand quantum theory most fundamentally as a theory of correlations. These correlations can only be organized according to a microscopic time parameter under conditions where physical subsystems exist that actually behave like good clocks. Depending on what the complete space of states looks like for a full theory of quantum gravity, states with "good clock" subsystems and thus well-defined microscopic time might only be a small subset of all possible states. (For reviews and further references on the problem of time in quantum gravity including the notion of emergent time, see Kuchar (1992), Isham (1993), and Zeh (1992); see also Albrecht (1995).)

If microscopic time *is* emergent, how could that affect the discussion in this chapter, which for the most part is basically classical? Perhaps not at all: to discuss time, of course one has to restrict oneself to physical states where microscopic time is well defined. But having done so in the context of a full theory of quantum gravity, one may well be faced with the exact situation discussed classically in this chapter, namely one in which almost all possible states do not have a thermodynamic arrow of time, and one has to make an additional selection to identify those special states that do.

Another possibility is that microscopic time emerges prepackaged with a thermodynamic arrow of time, so that "good clock" subsystems naturally come correlated with matter states that are very low entropy at one end of the timeline and high entropy in the other direction. This line of reasoning appears in many discussions of the wave function of the universe in the context of quantum cosmology (see the discussion in section "Wave function of the universe" above).

Of course another alternative is that once we have a full understanding of quantum gravity we will learn that all states have a well-defined microscopic time, which comes automatically correlated with a thermodynamic arrow of time (see for example the discussion in section "Holographic cosmology" above).

Causal patch physics

One popular explanation of today's observed cosmic acceleration posits a non-zero value of the fundamental cosmological constant which is today just starting to dominate over the energy density in ordinary matter. The acceleration might be thought of as a second period of inflation, but if it is driven by a fundamental cosmological constant then unlike inflation the acceleration today will not come to an end.

A non-zero cosmological constant could revolutionize fundamental physics (Banks 2000, 2002; Fischler 2000; Witten 2000). In particular, Banks and Fischler argue that a cosmological constant would place an absolute upper bound on the entropy of the universe, which in turn would imply a finite-dimensional Hilbert space for any fundamental theory. This is related to the fact that one only assigns physical meaning to events to which you are causally connected, that is, your "causal patch."

Dyson *et al.* (2002) have explored the implications of this idea for inflation. The good news is that with a fundamental cosmological constant, some things become simpler. For example, the highest entropy state in such a universe is pure de Sitter space, and so it must be de Sitter space that describes the "pure chaos" that preceded inflation. This situation appears to be theoretically much more tractable than trying to conceive of the perfectly chaotic state in the absence of a cosmological constant.

But challenges arise if one embraces the finite Hilbert space idea. The first, of course, is that no one knows a compelling fundamental theory that fits this constraint. But Dyson *et al.* (2002) argue that even without those details, the causal patch constraint deprives inflation of its huge volume factors. In this picture, the entire volume of the universe is not much larger than what we see, and without the usual exponentially large volumes, Dyson *et al.* argue that inflation is *not* the dominant channel into the standard Big Bang.

So one must abandon either inflation, the non-zero cosmological constant, or the causal patch constraint (at least in the heuristic form used by Dyson *et al.*).

Turok (2002) has also argued the case for excluding the large volume factors using an argument based on causality, without specific reference to a cosmological constant, and similar concerns are raised by Hawking and Hertog (2002).

Measures and other issues

I should acknowledge that much of the discussion of what inflation has to offer (for example that the large volume factors make inflation the dominant channel from chaos into the Big Bang) rests on very heuristic arguments. The program of putting this sort of argument on firmer foundations is in its infancy. Also, there are many poorly developed technical matters related to the origin of perturbations in inflation (see for example Brandenberger and Martin (2002) and Kaloper *et al.* (2002)). Progress on these issues certainly has the potential to overturn many of the beliefs about inflation expressed in this chapter.

Some "Wheeler class" questions

John Wheeler has never shied away from the really tough "big questions." In his honor, I take this section to touch on some deeper questions raised by my discussion above. At this point, I cannot offer answers to any of these questions.

The arrow of time, classicality and microscopic time

I have mentioned the above key role played by the arrow of time in quantum measurement. If one models the universe in a way where the arrow of time is only a transient phenomenon, what then do quantum probabilities mean in the absence of the arrow of time? We seem pretty comfortable working with such probabilities, but perhaps we should be more careful here. (For a recent discussion of some of these issues, see Banks *et al.* (2002).)

Also, we are used to thinking of the time that appears in the microscopic equations (and which is differentiated from space thanks to its different role from space in Lorentz transformations) as being quite different from the arrow of time under discussion here. For example, it is the microscopic arrow of time that allows us to construct a time sequence for a box of gas in equilibrium (such as depicted in Fig. 18.2) even when a thermodynamic arrow of time does not exist.

If one thinks carefully about how one operationally defines this microscopic time, the thermodynamic arrow of time is always required indirectly. One might say something like: "Measure a system at time 1, and the microscopic evolution equations will tell you what the system will look like at time 2." But actually to check that you have to make a good record of the state at time 1, a record that will still be intact at time 2. The thermodynamic arrow of time is essential to making stable records.

So perhaps one cannot really have microscopic time, or even quantum probabilities, without a thermodynamic arrow of time. This idea might connect with other speculation that microscopic time, like the thermodynamic time, could be an emergent feature of the physical world (Gross 2002). On the other hand, the ideas from section "Holographic cosmology" offer a very different angle which also connects microscopic time with the arrow of time in a fundamental way.

The arrow of time in the approach to de Sitter space

To us, the familiar arrow of time is driven by gravitational collapse as it destroys a homogeneous state. In many models of inflation there are huge regions of the universe that are undergoing extremely long epochs of exponential expansion. As time progresses and various imperfections get diluted by the expansion, these regions will asymptotically approach the high-entropy de Sitter state, and by doing so will exhibit increasing entropy. That is another manifestation of an arrow of time, which is quite different from the one we are used to, which is based on gravitational collapse, not dilution. Is it possible for other types of creatures to exist that harness the "dilution" arrow of time as effectively as we harness ours? If the answer is yes, perhaps these creatures will start to evolve as today's cosmic acceleration takes over.

Conclusions

Perhaps the key point of this chapter is that having an arrow of time in the universe places demands on the initial conditions that apparently conflict with the goals of inflationary cosmology. Inflation wants to use dynamics to argue that the initial conditions of the Big Bang are generic, but the arrow of time requires that the initial conditions not be generic in precisely the same sense, namely that the conditions are far from the asymptotic behavior produced by the dynamics.

This conflict is resolved by recognizing that the goal of inflation can never be to make the initial conditions of our observed universe be *completely* generic. To do so would remove the arrow of time. However, inflation teaches us that it *is* possible to make the initial conditions of our universe "more generic": the inflationary dynamics shows that the special initial conditions required for an arrow of time need only appear as special initial conditions for the inflaton field in a small region, not the entire state of matter in the universe. Inflationary dynamics then leverages these special inflaton initial conditions into exponentially large numbers of exponentially large regions that exhibit the properties of the familiar Big Bang.

So one important conclusion is that inflation, or any other attempt to explain dynamically the initial conditions of the observed universe, will necessarily require some special initial conditions itself, in order to have an arrow of time. These special initial conditions are the vestiges of Boltzmann's original "rare fluctuations" which can never be completely excised from this sort of dynamical approach.

This conclusion is particularly directed at those who hold up the special initial conditions of inflation as a serious flaw of the idea. However, I know of no fundamental law that prevents one from hoping that some improved dynamical process could produce the universe we observe using initial fluctuations that are even less rare than those which initiate inflation, so perhaps it is just as well that the critics keep the pressure on.

But the discussion in this chapter also relates to another debate about initial conditions. There are those who find the dynamical approach inherently flawed. Instead, they wish to uncover broad principles or fundamental laws that will uniquely specify the state of the universe (see for example Hollands and Wald (2002)). As I discussed above, several current ideas (such as the ekpyrotic model and eternal inflation) fall under this category. The field seems to be divided among people who strongly favor a dynamical approach, and those who strongly favor defining a unique state of the universe based on principles. I am definitely in the dynamical camp.

My most concrete criticism of the "unique state" approach is that all the dynamics we know, especially when quantum effects are included, tends to spread states out in phase space in a very broad manner. I challenge proponents of the unique-state approach to articulate their ideas in a fully quantum treatment. I suspect that any such attempt will yield a probability distribution for the state of the universe that is broad in many directions, effectively describing something not so different from the pre-inflation chaos discussed above.

Under those conditions, there is no other way of finding our place in the universe besides identifying the dynamical mechanisms that are most likely to produce the Big Bang universe we observe. As I have argued here, the fact that we experience an arrow of time requires that these dynamical mechanisms also need some kind of rare fluctuation to function, a feature that is deeply connected with Boltzmann's original insights into the arrow of time more than a century ago.

Acknowledgments

I would like to thank A. Aguirre, T. Banks, S. Carroll, W. Fischler, N. Kaloper, M. Kaplinghat, A. Linde, A. Olinto, L. Susskind, N. Turok, and W. Zurek for very helpful conversations. I also thank the Aspen Center for Physics, where much of this article was written, and B. Gold for comments and corrections to the manuscript.

References

Aguirre, A and Gratton, S (2002) *Phys. Rev.* **D65**, 083507.
Albrecht, A (1992) *Phys. Rev.* **D46**, 5504.
(1994) In *Physical Origins of Time Asymmetry*, ed. J. Halliwell, J. Perez-Mercader, and W. Zurek. Cambridge: Cambridge University Press. http://www.arxiv.org/abs/gr-qc/9408023.
(1995) gr-qc/9408023.
(1999) In *Structure Formation in the Universe*, ed. R. Crittenden and N. Turok, p. 17, Amsterdam: Kluwer.
(2000) In *Proc. 35th Rencontres de Moriond "Energy Densities in the Universe"*. http://moriond.in2p3.fr/J00/ProcMJ2000/.
(2002) Plenary talk at American Physical Society Division of Particles and Fields (DPF) 2002. http://dpf2002.velopers.net/.
Albrecht, A and Magueijo, J (1999) *Phys. Rev.* **D59**, 043516.
Albrecht, A and Steinhardt, P (1982) *Phys. Rev. Lett.* **48**, 1220.
Albrecht, A, Kaloper, N, and Song, Y-S (2002) hep-th/0211221.
Banks, T (2000) Talks at the *Lennyfest* Spring 2000, and *Michigan Strings Conference* in Summer 2000. http://www.stanford.edu.dept/physics/events/lennyfest.html and http://www.feynman.physics.lsa.umich.edu/strings2000/.
(2002) hep-ph/0203066.
Banks, T and Fischler, W (2001) hep-th/0111142.
Banks, T, Fischler, W, and Paban, S (2002) hep-th/0210160.
Barrow, J (1995) *Phys. Rev.* **D51**, 3113.
Barrow, J and Tipler, F (1986) *The Anthropic Cosmological Principle*. Oxford: Oxford University Press.

Bennett, C H and Landauer, R (1985) *Sci. Am.* **253**, 38.
Boltzmann, L (1897) *Ann. Physik* **60**; English translation by SG Brush (1996) *Kinetic Theory*, Vol. 2, *Irreversible Processes*. Oxford: Pergamon Press.
(1910) *Vorlesungen über Gastheorie*. Oxford: Barth.
Borde, A, Guth, A, and Vilenkin, A (2001) gr-qc/0110012.
Brandenberger, R and Martin, J (2002) *Int. J. Mod. Phys.* **A17**, 3663.
Burles, S, Nollett, K M, and Turner, M S (2001) *Ap. J.* **552**, L1.
Cornish, N, Spergel, D, and Starkman, G (1996) *Phys. Rev. Lett.* **77**, 215.
Davies, P C W (1977) *The Physics of Time Asymmetry*. Berkeley, CA: University of California Press.
(1983) *Nature* **301**, 398.
(1984) *Nature* **312**, 524.
DeWitt, B S (1967) *Phys. Rev.* **160**, 1113.
Dyson, L, Kleban, M, and Susskind, L (2002) hep-th/0208013.
Everett, H (1957) *Rev. Mod. Phys.* **29**, 454.
Fischler, W (2000) Talk at the symposium in honor of Geoffrey West's 60th birthday, Santa Fe.
Freese, K (2002) hep-ph/0208264.
Freese, K and Lewis, M (2002) *Phys. Lett.* **B540**, 1.
Gell-Mann, M and Hartle, J B (1994) In *Physical Origins of Time Asymmetry*, ed. J. Halliwell, J. Perez-Mercader, and W. Zurek, p. 311. Cambridge: Cambridge University Press.
Gibbons, G W and Hawking, S W (1977) *Phys. Rev.* **D15**, 2738.
Guth, A (1981) *Phys. Rev.* **D23**, 347.
Gordon, C and Turok, N (2002) hep-th/0206138.
Gross, D (2002) Closing remarks at Cosmo-02. http://pancake.uchicago.edu/~cosmo02/
Hartle, J B and Hawking, S W (1983) *Phys. Rev.* **D28**, 2960.
Hawking, S W (1994) In *Physical Origins of Time Asymmetry*, ed. J. Halliwell, J. Perez-Mercader, and W. Zurek, p. 346. Cambridge: Cambridge University Press.
Hawking, S W and Hertog, T (2002) hep-th/0204212.
Hawking, S W and Turok, N (1998) *Phys. Lett.* **B425**, 25.
Hollands, S and Wald, R (2002) hep-th/0210001.
Isham, C (1993) In *Integrable Systems, Quantum Groups, and Quantum Field Theories*, ed. L. A. Ibort and M. A. Rodriguez, p. 157. Amsterdam: Kluwer.
Kallosh, R, Kofman, L, and Linde, A (2001) *Phys. Rev.* **D64**, 123523.
Kaloper, N, Kleban, M, Lawrence, A, *et al.* (2002) hep-th/0209231.
Khoury, J, Ovrut, B, Steinhardt, P, *et al.* (2001a) *Phys. Rev.* **D64**, 123522.
(2001b) hep-th/0105212.
(2002) *Phys. Rev.* **D66**, 046005.
Kolb, E and Turner, M (1990) *The Early Universe*. Reading, MA: Addison-Wesley.
Kuchar, K (1992) In *Proc. 4th Canadian Conf. General Relativity and Relativistic Astrophysics*, p. 211.
Linde, A D (1982) *Phys. Lett.* **108B**, 389.
(1983) *Phys. Lett.* **129B**, 177.
(1984) *Lett. Nuovo Cimento* **39**, 401.
(1986) *Phys. Lett.* **B175**, 395.
(1996) *Phys. Rev.* **D54**, 2504.
(1997) In *Critical Dialogues in Cosmology*, ed. N. Turok. Singapore: World Scientific Press.
(1998) *Phys. Rev.* **D58**, 083514.

Linde, A, Linde, D, and Mezhlumian, A (1994) *Phys. Rev.* **D49**, 1783.
(1996) *Phys. Rev.* **D54**, 2504.
Longair, S L (1998) *Galaxy Formation*. New York: Springer-Verlag.
Moffat, J (1993a) *Int. J. Mod. Phys.* **D2**, 351.
(1993b) *Found. Phys.* **23**, 411.
Page, D N (1983) *Nature* **304**, 39.
Penrose, R (1979) In *General Relativity, and Einstein Centenary Survey*, ed. S. Hawking and W. Israel, p. 581. Cambridge: Cambridge University Press.
(1989) In *Proc. 14th Texas Symp. Relativistic Astrophysics*, ed. E. J. Fergus, p. 249. New York: New York Academy of Sciences.
Price, H (1989) *Nature* **340**, 181.
Prigogine, I (1962) *Non-Equilibrium Statistical Mechanics*. New York: John Wiley.
Seiberg, N (2002) Talk presented at Cosmo-02. http://pancake.uchicago.edu/~cosmo02/
Starkman, G, Stojkovic, D, and Trodden, M (2001) *Phys. Rev. Lett.* **87**, 231303.
Steinhardt, P and Turok, N (2002a) *Phys. Rev.* **D65**, 126003.
(2002b) astro-ph/0204479.
Tegmark, M and Zaldarriaga, M (2002) astro-ph/0207047.
Tolman, R (1934) *Relativity, Thermodynamics and Cosmology*. Oxford: Clarendon Press.
Trodden, M and Vachaspati, T (1999) *Mod. Phys. Lett.* **A14**, 1661.
Tsujikawa, S, Brandenberger, R, and Finelli, F (2002) hep-th/0207228.
Turok, N (2002) *Class. Quant. Grav.* **19**, 3449.
Unruh, W (1997) In *Critical Dialogues in Cosmology*, ed. N. Turok, p. 249. Singapore: World Scientific Press.
Vanchurin, V, Vilenkin, A, and Winitzki, S (2000) *Phys. Rev.* **D61**, 083507.
Vilenkin, A (1984) *Phys. Rev.* **D30**, 509.
(1998) *Phys. Rev.* **D58**, 067301.
Wang, X, Tegmark, M, and Zaldarriaga, M (2001) *Phys. Rev.* **D65**, 123001.
Watson, S and Brandenberger, R (2002) hep-th/0207168.
Wheeler, J A (1968) In *Battell Recontres*, ed. C. DeWitt and J. Wheeler, p. 242. New York: Benjamin.
Wheeler, J A and Feynman, R P (1945) *Rev. Mod. Phys.* **17**, 157.
(1949) *Rev. Mod. Phys.* **21**, 425.
Wheeler, J A and Zurek, W H (1983) *Quantum Theory and Measurement*. Princeton, NJ: Princeton University Press.
Witten, E (2000) hep-ph/0002297.
Zeh, H (1992) *The Physical Basis of The Direction of Time*, 2nd ed. New York: Springer-Verlag.
Zurek, W H (1991) *Phys. Today* **44**, 36.

19

Cosmology and immutability

John D. Barrow
University of Cambridge

> If people do not believe that mathematics is simple, it is only because they do not believe how complicated life is.
>
> *John von Neumann*

The ups and downs of oscillating universes

John Wheeler was one of the first to stress the physical significance of the fundamental Planck scales of mass, length, and time. He recognized their quantum gravitational significance and speculated upon the strange things that might happen when the universe crossed that mysterious threshold where general relativity and quantum theory meet to consummate their arranged marriage. For Wheeler, Einstein's conception of cosmology always implied a universe that was finite in size and total lifetime, a "closed" universe evolving from a Big Bang in the past to a Big Crunch in the future.[1] We still do not know whether these two singular points of the evolution signal merely a breakdown of the nonquantum theory of gravity that we are using or whether they have special significance and will remain even in a future quantum theory of cosmology.

If our expanding universe of stars and galaxies did not appear spontaneously out of nothing at all, then from what might it have arisen? One option that has an ancient pedigree is to sidestep the question and propose that it had no beginning. It always existed. A persistently compelling picture of this sort is one in which the universe undergoes a cyclic history, periodically disappearing in a great conflagration before reappearing phoenix-like from the ashes (Eliade 1934; Barrow and

[1] In Barrow and Tipler (1986), we proposed that such "one-shot" universes be called Wheeler universes.

Science and Ultimate Reality, eds. J. D. Barrow, P. C. W. Davies and C. L. Harper, Jr. Published by Cambridge University Press.

Tipler 1986). This stoic scenario has a counterpart in modern cosmological models of the expanding universe.

If we consider closed universes which have an expansion history that expands to a maximum and then contracts back to zero, then there is the tantalizing possibility that this episode of cosmic history might continue to repeat itself into the future. Suppose the universe re-expands and repeats this behavior over and over again. If this can happen then there is no reason why we should be in the first cycle. We could imagine an infinite number of past oscillations and a similar number to come in the future. We are simply ignoring the fact that a singularity arises at the start and the end of each cycle. It could be that repulsive gravity stops the universe just short of the point of infinite density, causing it to bounce back into expansion, or some more exotic passage occurs "through" the singularity, but this is pure speculation at present.

This speculation is not entirely unrestrained, though. Let us assume that one of the central principles governing Nature, the second law of thermodynamics, which tells us that the total entropy (or disorder) of a closed system can never decrease, governs the evolution from cycle to cycle. Gradually, ordered forms of matter will be transformed into disordered radiation and the entropy of the radiation will steadily increase. The result is to increase the total pressure exerted by the matter and radiation in the universe and so increase the size of the universe at each successive maximum point of expansion.[2] As the cycles unfold they get bigger and bigger. The entropy is given by the square of maximum size of the universe in Planck units:

$$S = k_{\rm B}(R_{\rm max}/R_{\rm pl})^2 \tag{19.1}$$

where $R_{\rm pl} = (G\hbar/c^3)^{1/2}$ and $k_{\rm B}$ is Boltzmann's constant.[3] So, if we assume $R_{\rm pl}$ is constant then the increase of S from cycle to cycle means that $R_{\rm max}$ increases from cycle to cycle. Intriguingly, the universe expands closer and closer to the critical state of flatness that also arises as a consequence of inflation. If we follow it backwards in time through smaller and smaller cycles it need never have a beginning at any finite past time although life can only exist after the cycles get big enough and old enough for atoms and biological elements to form (see Fig. 19.1).

For a long time this sequence of events used to be taken as evidence that the universe had not undergone an infinite sequence of past oscillations because the build-up of entropy would eventually make the existence of stars and life impossible

[2] This was first pointed out by Tolman (1931a, b). Recently, a detailed reanalysis and extension was given by Barrow and Dąbrowski (1995).

[3] The key assumption here is that this formula, proved for the event horizons of stationary black holes, can be applied to the particle horizon of an expanding universe. This is a big assumption. Something like this may well be true because as the particle horizon expands more information is contained within it.

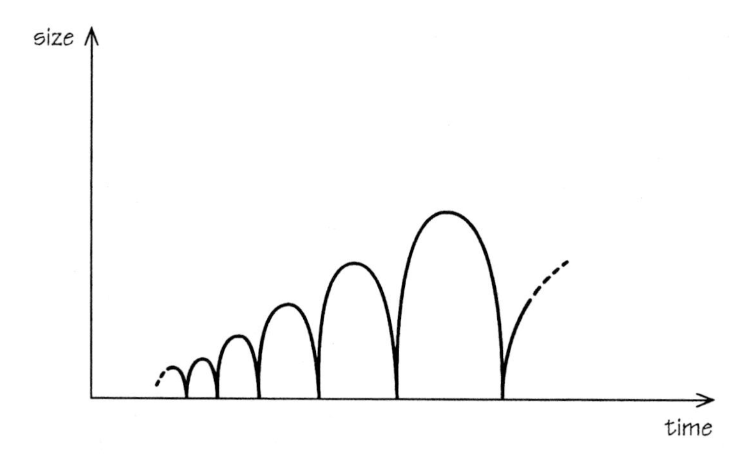

Figure 19.1. The cycles of an oscillating universe increase in size when the entropy of the universe continuously increases in accord with the second law of thermodynamics.

(see, for example, Harrison (1981)) and the number of photons that we measure on average in the universe for every proton (about one billion) gives a measure of how much entropy production there could have been. However, we now know that this measure does not need to keep on increasing from cycle to cycle. It is not a gauge of the increasing entropy. Everything goes into the mixer when the universe bounces and then the number of protons that there are compared with photons gets set by processes that occur early on. One problem of this sort might be the accumulation of black holes. Once large black holes form, like those observed at the centers of many galaxies, including the Milky Way, they will tend to accumulate in the universe from cycle to cycle, getting ever more massive until they engulf the universe, unless they can be destroyed at each bounce or become separate "universes" which we can neither see nor feel gravitationally. Smolin (1984) (see also Barrow 1999) has proposed an adventurous scheme in which black-hole collapses bounce back to produce new expanding universes in which the values of the physical constants are slightly shifted. In the long run this could lead to the population of universes being dominated by those which maximize the production of black holes because small shifts in the values of the constants should reduce the production of black holes in our universe. However, it is possible that universes which maximize black-hole production do not permit observers and the real prediction of the scenario is that we should be in a universe which maximizes the production of black holes given that observers can exist. Some variations in constants might be completely neutral with respect to black-hole production, and so remain untuned by the successive shifts. Worse still, some changes in the values of constants might curtail the formation of black holes. Actual predictions from this type of scenario are difficult to make because they require a complete understanding of the restrictions placed on the possibility of conscious observers by changes in constants.

In the same spirit Edward Harrison (1995) has suggested that it might be possible for intelligent observers to tune the values of the constants. We know that it might well be possible to "create" universes in the laboratory in the sense that very rapid inflationary expansion might be initiated in a part of the universe in a way that determined the values of some of the defining constants in that region. Of course, we don't know how to do this in practice but it is not impossible that more advanced civilizations do. Harrison suggests that if they could do this then they would act to tune the values of the constants of Nature so as to make the evolution of observers like themselves more likely in the future. If these observers in their turn do the same in the far future then eventually their descendants should find themselves inhabiting a universe that possesses many very finely tuned life-supporting coincidences between the values of the constants of Nature. A bit like our observable universe in fact! This type of fine-tuning seems to me to be a better paradigm for a "participatory universe" than Wheeler's original conception of a universe which is brought into being in some sense by observation taking place in a quantum mechanical sense.

It is interesting to compare the Smolin and Harrison scenarios. Both seek an explanation for the apparent fine-tuning of our constants around life-supporting values. Smolin offers a mechanism for converging on particular values but there is no reason why they should be life-supporting (they are black-hole supporting). Harrison offers an explanation for why they are life-supporting but cannot explain why they came to support life in the first place (before they became tunable by intelligent observers).

A curious postscript to the story of cyclic universes was recently discovered by Barrow and Dąbrowski (1995). We showed that if Einstein's cosmological constant exists then, no matter how small a positive value it takes, its repulsive gravitational effect will eventually cause the oscillations of a cyclic universe to cease (see Fig. 19.2). The oscillations get bigger and bigger until eventually the universe becomes large enough for the cosmological constant to dominate over the gravity of matter. When it does so it launches the universe off into a phase of accelerating expansion from which it can never escape unless the vacuum energy creating the cosmological constant stress were to decay mysteriously away in the far future. Thus the bouncing universe can eventually escape from its infinite oscillatory future in the presence of a positive cosmological constant. If there has been a past eternity of oscillations we might expect to find ourselves in the last ever-expanding cycle so long as it is one that permits life to evolve and persist. In fact, our universe does seem to be dominated by a cosmological constant.

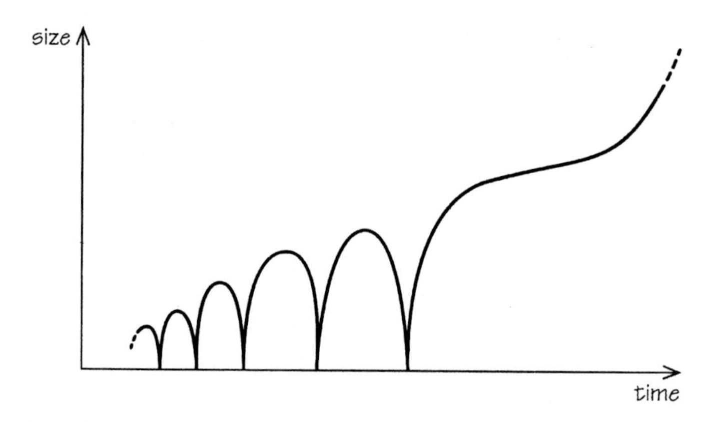

Figure 19.2. If a positive cosmological constant exists then, no matter how small its value, it will eventually bring to an end the cycles of oscillation of a bouncing universe.

Another means by which the universe can avoid having a beginning is to undergo the exotic sequence of evolutionary steps created by the eternal inflationary history (Linde 1994). There seems to be no reason why the sequence of inflations that arise from within already inflating domains should ever have had an overall beginning. It is possible for any particular domain to have a history that has a definite beginning in an inflationary quantum event, but the process as a whole could just go on in a steady fashion for all eternity, past and present. One of the major uncertainties of this scenario is knowing what can vary from inflation to inflation: is it just the size and entropy of the universe or do the variations encompass the numbers of dimensions of space, the values of the constants of Nature, and the laws of physics themselves?

John Wheeler's speculation about oscillating universes was that the constants of Nature themselves might be changed each time the universe bounces from a state of contraction into expansion (Misner *et al.* 1973). This "reprocessing" of the universe and its defining constants and conserved quantities leads inevitably to a "biological selection of physical constants" in the sense that living observers will only inhabit those cycles of the universe in which the constants have fallen out "right" for living complexity to evolve and persist. In order for the process to continue it is also necessary for the constants to fall out in a way that allows collapse of the universe to occur again in the future. As we have seen, this might not happen if, for example, the reprocessing gave rise to a positive cosmological constant.

A key issue in this discussion is clearly the nature of the constants of nature themselves: what are they? Are they truly constant? Do they arise at random as Wheeler implies or are they programmed in by some inflexible self-consistency principle of nature? Let us consider some of the features that are distinctive about the quantities we treat as constants of nature.

Distinctive features of the constants of nature

Our constants of nature help organize our understanding of the world. They are like beacons from which we can take our bearings. Real advances in our understanding of the physical world always seem to involve one of the following:

(i) *Revelation:* the discovery of a new fundamental constant of nature.
(ii) *Elevation*: the enhancement of the status of a known constant.
(iii) *Reduction:* the discovery that the value of one constant of nature is determined by the numerical values of others.
(iv) *Elucidation:* the discovery that an observed phenomenon is governed by a new combination of constants.
(v) *Variation:* the discovery that a quantity believed to be a constant of nature is not truly constant.
(vi) *Enumeration*: the calculation of the value of a constant of Nature from first principles, showing that its value is explained.
(vii) *Transmogrification*: the discovery that our supposed constants are a small part of a vastly more exotic structure.

As an example of *revelation*, we recall how the introduction of the quantum theory by Planck, Einstein, Bohr, Heisenberg, and others introduced us to the new fundamental constant, h, that bears Planck's name. It gave a finite numerical value to something that was previously assumed to be zero: the minimum energy change that is possible in nature.

Another more recent example is suggested by the development of a candidate for the title "theory of everything," called superstring theory, in which the fundamental ingredients of the world are not point particles of mass but loops, or strings, of energy which possess a tension, rather like elastic bands. This string tension is the basic defining constant of the theory. Almost all other properties of the world follow from it (although they are yet to be worked out in most cases). This string tension may prove to be as fundamental as the Planck units of mass and energy. A third example is provided by Einstein's discovery of the theoretical possibility of a cosmological constant, Λ, in the law of gravitation. Only in recent years has the first convincing astronomical evidence been found for its existence from observations of distant supernovae.

As an example of *elevation*, we see how Einstein's development of the theory of special relativity gave a new universal status to the velocity of light in vacuum, c. Einstein revealed its far-reaching significance. He showed that it provides the link between the concepts of mass (m) and energy (E) through his famous formula $E = mc^2$. Einstein did not discover that light moved with a finite speed. That had been observed long before and precise measurements of the speed of light had been made in the nineteenth century. But Einstein's new theory of motion changed the status of the speed of light in vacuum forever. It became the ultimate speed limit. No information can spread faster. More fundamental still, it was the one velocity that all observers, no matter what their own motion, should always measure to be the same. It was unique amongst all velocities. In the future, other known constants might take on a similarly more elevated status. For example, many elementary particles have masses which are believed to be universal. The smallest of these particle masses would be a very special one because the lightest particle would be unable to decay away into anything else – there would be nowhere for it to go. It would inevitably come to dominate the universe.

The discovery of a *reduction* is something that usually comes later in the game than either *revelation* or *elevation*. To carry out a reduction we already need to know some probable constants; then we need to develop a broader explanation that links their domains of application. Often, the constants defining each of the areas that are made to overlap will be found to be linked. This is typically what happens whenever physicists manage to create a theory that "unifies" two, previously distinct, forces of Nature. In the late nineteenth century it was found that the product of the permeability and permittivity of free space equaled the square of speed of light. Clearly there was a hidden link between electromagnetism and light. Likewise, in 1967, a theory was proposed by Glashow, Weinberg, and Salam that linked electromagnetism and the weak force of radioactivity. This theory was successfully tested by observation for the first time in 1983 and it joins together the constants of Nature that label the strengths of the forces of electromagnetism and radioactivity. This fusion reduces the number of independent constants of nature that are believed to exist.

The discovery of an *elucidation* is slightly different to that of a *reduction*, but equally revealing. It occurs when a theory predicts that some observed quantity – a temperature or a mass for example – is given by a new combination of constants. The combination tells us something about the interrelatedness of different parts of science.

A good example is provided by Stephen Hawking's famous prediction, in 1974, that black holes are not entirely black. Thermodynamically, they are black bodies: perfect radiators of heat radiation. Prior to then it was believed that black holes were just cosmic cookie-monsters, swallowing everything that came within their gravitational clutches. Once you fell inside a surface known as the event horizon, there was no return to the outside world.

Hawking succeeded in predicting what would happen if quantum processes were included in the story. Remarkably, black holes then turned out to be not quite black. The strong change in gravity near the event horizon could turn the gravitational energy of the black hole into particles which could be radiated away from the black hole, gradually sapping the mass of the hole, until it disappeared in a final explosion.[4] What is unusual about this evaporation process is that it is predicted to be governed by the simple everyday laws of thermodynamics that apply to all known hot bodies in equilibrium. Thus black holes turn out to be objects that are at once gravitational, relativistic, quantum mechanical, and thermodynamical. The formula which gives the temperature of the radiation that a black hole of mass M radiates away into space by means of Hawking's evaporation process involves the fundamental constants G, h, and c. But it also includes the thermodynamic constant of Boltzmann, k_B, which links energy to temperature. The temperature of a Schwarzschild black hole is

$$T_{bh} = hc^3/16\pi^2 GMk_B. \qquad (19.2)$$

This formula is a spectacular elucidation of the interlinked structure of superficial disparate pieces of nature and a hint about the thermodynamic significance of quantum gravity.

The discovery of a *variation* is quite different to the previous four developments. It means that a quantity that we believed to be constant is discovered to be an imposter, masquerading as a true constant. It turns out to vary slightly in space or in time. Generally, such a step will require the variation to be very small, or the quantity would not have been believed to have been constant in the first place. None of the fundamental constants of nature has so far indubitably suffered this downgrading of its cosmic status. However, as we shall see later on, some are under suspicion as they have had their constancy probed to greater and greater levels of precision.

The prime suspect for tiny variations has always been gravitational constant, G. Gravity is far and away the weakest force of nature and the least closely probed by experiment. If you look up the known values of the major constants in the back of a physics textbook you will discover that G is specified to far fewer decimal places than c, h, or e. In the early 1960s it was thought for a time that Einstein's general theory of relativity disagreed with observations of the rate of precession of perihelion of Mercury's orbit around the Sun. The first thing that was done to reconcile the two was to extend Einstein's theory by allowing G to change with time. Ultimately, the problem was traced to incorrect observations of the shape of the Sun (difficult to make accurately because of surface activity on the Sun) but, like a genie, once the varying G theory was released it couldn't be shut up again.

Although G has withstood assaults on its constancy for longest, the most recent and detailed attacks have been launched against the constancy of α, the fine structure constant. The fine structure constant is a linkage of the speed of light, Planck's

[4] It is not possible so far to predict what should remain after the final explosion. Many different suggestions have been made, ranging from nothing at all, to a hole in space and time, a wormhole into a new universe, or just a stable mass.

constant, and the electron charge. If it varies then we may choose to which of these dimensional constants we attribute the time variation. This is entirely a matter of convention or convenience: only the variation of dimensionless constants has an invariant operational meaning.

All of these five touchstones of progress revolve around constants of Nature and they show the central role that constants play in our appraisal of progress. There is a sixth development on our list. We called it *enumeration*. This is the Holy Grail of fundamental physics and it means the numerical calculation of one of the constants of nature. This has never been done. So far, the only way we can know their values is by measuring them.[5] This seems unsatisfactory. It allows the constants that appear in our theories to have a wide range of different possible values without overthrowing the theory. This is not the situation that Einstein imagined when he embarked on his quest for a "unified field theory." He thought that the true theory should only permit one choice for the constants that define it – the values we observe. Some people share his view today, but it has become increasingly apparent that not all the constants that define the world need be uniquely straitjacketed in this way. It is likely that some are determined in a more liberal fashion by quantum randomness.

Many people hope that a complete theory would allow us to calculate the numerical values of some constants, like c, h, and G, as accurately as we liked. This would also be a wonderful way of testing such a "complete" theory. So far, this is just a dream. None of the constants that we believe to be truly fundamental has been calculated in this way from one of the theories in which it appears. Yet, such a calculation may not be too far away. Just a few years ago physicists were at an impasse with several possible string theories on offer, all seeming to be equally viable "theories of everything." This was odd. Why did our universe use just one of them? Then Edward Witten of Princeton University made a major discovery. He showed that all these superficially different string theories were not different at all. They were just different limiting situations of a single, bigger, deeper theory which we have yet to find. It as if we are illuminating a strange object from many different angles, casting different shadows on a wall. From enough of these shadows it should be possible to reconstruct the illuminated object. This deep theory has become known as M theory (M for mystery or M for matrix depending on your taste). Hidden within its mathematical defenses is an explanation for the numerical values of the constants of nature. So far, no one has been able to penetrate them and extract the information. We know a little about the structure of the M theory but the mathematics needed to elucidate it is formidable. Physicists are used to being able to take mathematics that mathematicians have already developed and use it like a tool to fashion physical theories. For the first time since Newton patterns

[5] We don't know, for example, if the fine structure is a rational or an irrational number.

have been encountered in nature that require the development of new mathematics in order to further our understanding of them. Witten believes we have been lucky to stumble upon M theory about 50 years too early. Others might point to the warning that the most dangerous thing in science is the idea that arrives before its time.

Despite the lack of a fundamental theory with which to pursue a calculation of constants there has been no lack of numerological efforts to explain them. This is an activity that has a history, anthropology, and sociology all of its own. Its fruits are rather unusual, and occasionally fantastic, as we are about to see. But before we do, we should mention the last development on our list, that of *transmogrification*. For it might turn out that our quest for a "theory of everything" shows that our whole conception of constants of nature was extremely limited. Indeed, M theory points us in just such a direction. These theories only seem to exist in finite form if there are many more dimensions of space than the three we inhabit. The true constants of Nature exist only in these higher-dimensional spaces. The quantities that we see in three dimensions are merely pale shadows of the true constants and need not even be constant.

Are our constants constant?

There are many ways in which the quantities we call the constants of nature might come to vary in space or time. With regard to variations in space we might consider that:

- Their origins might be intrinsically quantum mechanical and so they would possess intrinsic randomness which permitted them to be defined only probabilistically. Their distribution might be strongly peaked around the values we observe.
- Some constants might be fixed completely by the self-consistency of the laws of nature whilst others are composed of the sum of two pieces: one fixed, the other random. The random component might arise through a spontaneously broken symmetry during the early history of the universe.
- Some constants may not be specified at all by the self-consistency of a "theory of everything" and so they may be permitted to take on any (or a wide range of) values. This is equivalent to the vacuum state of the underlying "theory of everything" not being uniquely prescribed. Any random process in the early universe might exploit this freedom to make "constants" vary from place to place or from "universe" to "universe" in the eternal or chaotic inflationary scenarios.

With regard to variation in time we have some further possibilities:

- If there are extra dimensions of space (more than three), as the most favored candidates for a "theory of everything" predict, and the true constants of nature are defined in the total number of dimensions, there is no reason why their three-dimensional shadows we see in our laboratories need to be constant. Moreover, in the simplest scenarios, if the extra dimensions change in time then our three-dimensional constants should be observed to change at approximately the same rate.

- Some of the quantities we believe to be constant may simply be variables which asymptote to constant values after long periods of time. The universe is more than 13 billion years (or 10^{60} Planck times) old – plenty of time for variations to have faded away to give the illusion of constancy.
- Our observational limits on the constancy of constants are, for many of them, very weak. Gravitational forces are very weak and our ability to measure the value of Newton's constant of gravitation, G, is very limited. We simply can't turn gravity off and on or reduce its effects to zero as we can with electricity or magnetism because mass (unlike electric charge or magnetic polarity) comes with only one (positive) sign.

These rationalizations are all very well but the truth is we have no understanding of why *any* of the constants of nature we infer take the numerical values that they do – even whether we expect them to be rational or irrational numbers. Our understanding of their origins and interrelationship is therefore very limited. Our best chance of discovering whether they are truly the constants we believe them to be is simply to go out and look with the most sensitive instruments that we have.

It has long been known that the best way to probe the constancy of the constants is to reach for the sky. Watching atoms in a laboratory allows you to test that things don't change over days, or weeks, possibly even months. But observations of the nature of physics in distant astronomical objects enables us to take a snapshot of how physics was more than 10 billion years ago. Back in 1967, Bahcall and Schmidt (1967) observed a pair of oxygen emission lines that appear in the spectra of five galaxies which emit radio waves, located at an average redshift of 0.2 (thus emitting their light about 2 billion years ago) and produced a result consistent with no change in the fine structure constant, α, between a redshift of $z = 0.2$ and now (a redshift of $z = 0$):

$$\alpha(z = 0.2)/\alpha(z = 0) = 1.001 \pm 0.002. \tag{19.3}$$

These ideas set the scene for astronomers to improve our knowledge of the constancy of particular constants of nature by improving the sensitivity of telescopes and electronic detectors which allow observations of faint objects to be made at higher and higher redshifts, so reaching further and further back in time. The general strategy is to compare two atomic transitions in an astronomical site and here and now in the laboratory. For example, if they are doublets of elements like carbon, silicon, or magnesium, which are commonly seen in gas clouds at high redshifts, then the wavelengths of two spectral lines, λ_1 and λ_2 say, will separated by a distance that is proportional to α^2. The relative line shift is given by a formula

$$(\lambda_1 - \lambda_2)/(\lambda_1 + \lambda_2) \propto \alpha^2. \tag{19.4}$$

Now we need to measure the wavelengths λ_1 and λ_2 very accurately in the laboratory here and now and by astronomical observations. By calculating the left-hand side of our formula to high accuracy in both cases we can divide our results to find that

$$[(\lambda_1 - \lambda_2)/(\lambda_1 + \lambda_2)]_{\text{lab}}/[(\lambda_1 - \lambda_2)/(\lambda_1 + \lambda_2)]_{\text{ast}} = \alpha^2_{\text{lab}}/\alpha^2_{\text{ast}}. \tag{19.5}$$

We aim to discover if there is any significant deviation from 1 when we calculate the ratio on the left-hand side. If there is, it tells us that the fine structure constant has changed between the time the light left and the present. In order to be sure that there really is a significant deviation from 1, several things must be under very precise control. We need to be able to measure the wavelengths λ_1 and λ_2 to high accuracy in the laboratory. We also need to be sure that the observations are not being affected by extraneous noise, or biased by some subtle propensity of our instruments to gather certain sorts of evidence more readily than others.

Another approach is to observe (Drinkwater *et al.* 1998) the redshifts of light emitted by molecules like carbon monoxide with that from atoms of hydrogen in the same cloud. In effect, one is measuring the redshift of the same cloud by two means and comparing them. This uses radio astronomy and allows us to compare the value of α^2 here and now[6] with its value at the astronomical sources at redshift 0.25 and 0.68. This leads to a limit on a possible shift in α of

$$\Delta\alpha/\alpha = -(1.0 \pm 1.7) \times 10^{-6}. \tag{19.6}$$

One of the challenges of this method is to make sure that the atomic and molecular observations are looking at atoms and molecules that are moving in the same way at their location.

A third method is to compare the redshift found from 21 cm radio observations of emissions from atoms with optical atomic transitions in the same cloud. The ratio of the frequencies of these signals enables us to compare the constancy of another combination of constants[7]

$$A \equiv \alpha^2 m_e/m_{pr} \tag{19.7}$$

where m_e is the electron mass and m_{pr} is the proton mass. Observation of a gas cloud at a redshift of $z = 1.8$ led to a limit (Cowie and Songalia 1995) on any change in

[6] Actually it measures the constancy of the product $g_p\alpha^2$ where g_p is the proton "g factor." We assume here that g_p is not changing.

[7] Again we assume that g_p is constant.

the combination A of[8]

$$\Delta A/A = [A(z) - A(now)]/A(\text{now}) = (0.7 \pm 1.1) \times 10^{-5}. \qquad (19.8)$$

The important thing to notice about these two results is that the measurement uncertainties are large enough to include the case of *no* variation:

$$\Delta\alpha/\alpha = 0 \quad \text{and} \quad \Delta\text{A}/\text{A} = 0. \qquad (19.9)$$

It is important to stress that over the whole period from 1967 to 1999 when these observations were being made with ever-increasing precision there was never any expectation that a non-zero variation of a traditional constant like α would be found. The observations were pursued as means of improving the limits on what the smallest allowed variations could be. A novelty was that they were so much more restrictive than any limits that could be obtained in the laboratory by direct experimental attack. Just watching the energy of an atom for a few years to see if it drifted just cannot compete with the billions of years of history that astronomical observations can routinely mine.

The fourth and newest method is the most powerful. Again, it looks for small changes in how atoms absorb light from distant quasars. Instead of looking at pairs of spectral lines in doublets of the same element, like silicon, it looks at the separation between lines caused by the absorption of quasar light by *different* chemical elements in clouds of dust in between the quasar and us.

There are a number of big advantages with this new method. It is possible to look at the separations between many absorption lines and build up a much more significant data set. Better still, it is possible to pick the pairs of lines whose separations are being measured so as to maximize the sensitivity of the separations to little shifts in the value of α over time. But there is an unusual extra advantage of this method. The wavelength separations that need to be extracted from the astronomical data and measured in the laboratory depend on α in an unusual way. We can use large computer simulations[9] to discover what would happen to the positions of the lines if a tiny shift was made in the value of α. The shifts are very different for different pairs of lines. Increase α by one part in a million and some separations increase, some decrease, while some are almost unaffected. The whole collection of shifts defines a distinctive fingerprint of a shift in the value of α. Any spurious influence on the data, or messy turbulence at the site where the absorption is occurring out in the universe, seeking to fool us into thinking that α is changing when it isn't has got to mimic the entire fingerprint left on the wavelength separations by α variation.

[8] This limit excludes inclusion of uncertainties associated with possible variations of local velocities of the line sources.

[9] These simulations have been developed to predict the locations of the spectral lines and energy levels of the atoms in the laboratory already and are carried out by Victor Flambaum, Vladimir Dzuba, and their colleagues at the University of New South Wales, Sydney.

This method, which we have called the many-multiplet (MM) method is far more sensitive than the other astronomical methods and uses much more of the information in the astronomical data to be used.[10] It has been applied to observations of 128 quasars, looking at separations between magnesium, iron, nickel, chromium, zinc, and aluminum. When we first developed the MM method we expected that it would lead simply to a further major improvement of the limits on any allowed change in the fine structure constant. But the results gathered and analyzed over two years by our team of astronomers and atomic physicists proved to be unexpected and potentially far-reaching. We find a persistent and significant difference in the separation of spectral lines at high redshift compared to their separation when measured in the laboratory.[11] The complicated fingerprint of shifts matches that predicted to occur if the value of the fine structure constant was *smaller* at the time when the absorption lines were formed by about seven parts in a million. If we combine all the results then the overall pattern of variation that results (Webb *et al.* 2001) is shown in Fig. 19.3.

The first studies using the MM method reported evidence for a variation in the value of the fine structure constant in the past in 1999. Since then the data have steadily increased and better analysis techniques have been employed. Remarkably, the same pattern of results are found from the whole collection of observations of 128 quasars, only now with smaller measurement uncertainties. This is the largest direct observational assault on the question of whether the constants are constant up to 13 billion years ago.

The first striking feature is that if we use them to calculate what the fine structure was in the past we find a period in cosmic history where it appears to be slightly smaller than it is today. The magnitude of the dip in its value is very small, about seven parts in a million, and too small to have been found in any earlier investigations by observers using other methods, or detected in any laboratory experiment. It points to the fine structure constant being slightly smaller in the past. If we take the observations of sources lying between redshifts of 0.5 and 3.5 as a whole, the observed shift is provisionally computed to be[12]

$$\Delta\alpha/\alpha = [\alpha(z) - \alpha(\text{now})]/\alpha(\text{now}) = (-0.57 \pm 0.10) \times 10^{-5}. \quad (19.10)$$

[10] This improved sensitivity arises because the sensitivity to α with respect to relativistic aspects of atomic structure enters as $(\alpha Z)^2$ where Z is the atomic number (number of protons in the nucleus) of the atom. Thus by comparing lines of different atomic species with large and small values of Z a significant gain in sensitivity is obtained over methods that observe doublets of a species with the same Z.

[11] The measurement of the required spectral lines in the laboratory at the required level of accuracy (for which there appears to have been no need before) is very challenging and with more laboratory observations the MM method could extract even more information from the available data.

[12] This can be compared with the results obtained with the first round of observations in 1999:

$$\Delta\alpha/\alpha = [\alpha(z) - \alpha(\text{now})]/\alpha(\text{now}) = -1.09 \pm 0.36 \times 10^{-5}$$

published in Webb *et al.* (1999).

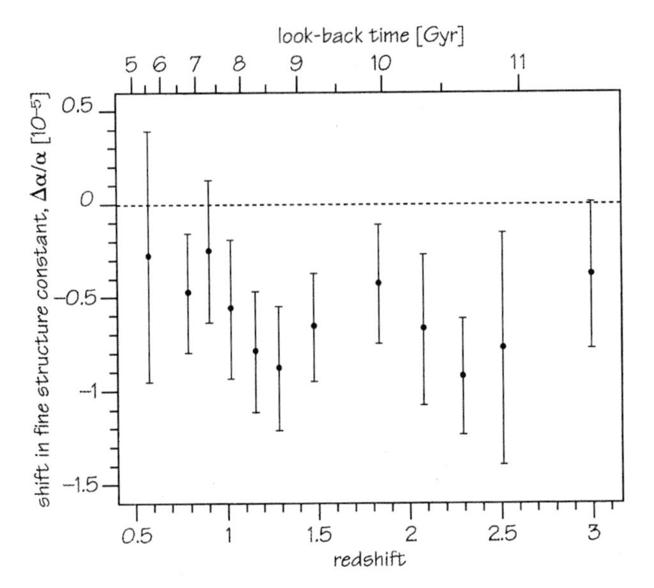

Figure 19.3. The relative shift in the value of the fine-structure constant inferred from 147 observations of quasar absorption spectra referred to in the text. The data points shown here are each formed from ten observations. Negative $\Delta\alpha/\alpha$ indicates that α was smaller in the past (at higher redshift).

If one converts this into a rate of change of α with time it amounts to

$$\{\text{rate of change of } \alpha \text{ in time}\}/\{\text{current value of } \alpha\} \approx 10^{-16} \text{ per year.} \quad (19.11)$$

This rate of change is about one million times slower than the universe is expanding and is far too small to affect any known laboratory measurement.

Although this small variation would have no perceptible effect in laboratory experiments yet, it would have indirect consequences that we have to worry about. We know that geological conditions 2 billion years ago conspired to produce a sequence of spontaneous nuclear chain reactions below the Earth's surface in Oklo, West Africa. The process hinges upon neutron capture by samarium nuclei and requires that a crucial nuclear energy level must have been located very close to where it is today. Its location is determined by a combination of the constants of Nature, including α, as Alex Shlyakhter (1976) first realized (see also Damour and Dyson (1996) and Fujii *et al.* (2002) for more detailed studies). The requirement that this resonance be in place 2 billion years ago with the required precision sets a limit of about

$$\Delta\alpha/\alpha = [\alpha(\text{Oklo}) - \alpha(\text{now})]/\alpha(\text{now}) < 10^{-7}. \quad (19.12)$$

But on reflection they are not in direct conflict. Leaving aside all the uncertainties that go into finding the exact dependence of neutron capture rates in the Oklo reactor on the fine structure constant, the Oklo observations probe the fine structure constant's value only about 2 billion years ago (a redshift of less than about 0.1) whereas the quasar observations span the range from about 3 to 11 billion years ago. The two observations are not necessarily in conflict unless you assume that the fine structure constant always changes at the same rate. But, as we shall see, fortunately we don't have to assume it does or it doesn't. We can predict what will happen.

What do we make of that?

The evidence that the fine structure constant may have been different in the past is impressive but it is statistical in character. It is based upon the totality of astronomical observations of light absorption by many different chemical elements in nearly 128 different dust clouds. In the future more data will be added to the total and the question will be probed by better and better observations. Ideally, other astronomers should repeat our observations and use different instruments and different data analysis techniques to see if they get the same results.

Yet, desirable as they are, more observations and greater accuracy are not panaceas. In observational science one must be aware of different types of uncertainty and "error." First, there is uncertainty introduced by the limiting accuracy of the measurement process This type of uncertainty is usually well understood and can gradually be reduced by improving technology (use a more finely graduated ruler). Second, there is a subtler form of uncertainty, usually called "systematic error" or "bias," which skews the data-gathering process so that you unwittingly gather some sorts of evidence more easily than others. More serious still, it may ensure that you are not observing what you thought you were observing.[13]

All forms of experimental science are challenged by these subtle biases. In down-to-earth laboratory measurements it is usual to repeat experiments in several ways, changing certain aspects of the experimental set-up each time, so as to exclude many types of bias. But in astronomy there is a bit of a problem. There is only one universe. We are able to observe it but we can't experiment with it. In place of experiment we look for correlations between different properties of objects: do all the clouds with particular redshifts have smaller spectral shifts between certain absorption lines, for instance, or are they all located in a particular sector of the sky? One might be aware of a bias and be unable to correct completely for its influence, as in the case of creating a big catalog of galaxies where one is aware of the simple

[13] There are other forms of error that are introduced deliberately, especially by politicians, when treating voting data. For example, a party with a ten-point manifesto assumes without question that if they win the election by an overall majority they have a mandate for all their manifesto policies whereas they might in reality only have a majority vote for a modest fraction of them.

fact that brighter galaxies are easier to see than faint ones. But the real problem is the bias that you *don't* know about. The data used to study the possible variation in the fine structure constant have been subjected to a vast amount of test and scrutiny to evaluate the effects of every imaginable bias. So far, only one significant influence has been found and accounting for it actually makes the deduced variations *bigger*[14] but the search by the observers for subtle sources of bias continues.

The reaction of most physicists or chemists to the idea that the fine-structure constant might be changing by a tiny amount over billions of years is generally one of horror and outright disbelief. The whole of chemistry is founded on the belief in theories which assume that it is absolutely constant. However, a change of a few parts in a million over 10 billion years would have no discernible effect upon any terrestrial physics or chemistry experiment. To see this more clearly it is time to ask what exactly are the best direct experiment limits that we have on the change in the fine structure constant.

Most direct tests of the constancy of the fine structure constant take an atom and monitor it for a given length of time as accurately as the measuring set-up will allow, typically to a few parts in a billion. This amounts to comparing different atomic clocks. This monitoring cannot be carried out for very long because of the need to keep other things constant, and the best results have come from a run of 140 days (Prestage *et al.* 1995). Assuming that the ratio of the electron and proton masses does not change, experimenters find that the stability of the value of an energy transition between hydrogen and mercury means that if the fine structure constant is changing then its rate of change must be less than 10^{-14} per year. This result sounds very strong. It allows the constant to change by only about one part in 10 000 over the whole age of the universe but the astronomical observations are consistent with a variation that is about 100 times smaller still. This gap between laboratory and outer space also illustrates the huge gain in sensitivity that the astronomical observations offer over the direct laboratory experiments. They may not be making measurements of the fine structure constant at the technological limit of sensitivity but they are looking so far back into the past – 13 billion years instead of 140 days – that they provide far more sensitive tests.[15] The universe has to be billions of years old in order that stars have enough time to create the biological elements

[14] This is the effect of refraction of the incoming light which depends upon the depth of atmosphere it has to traverse which depends on the geographical latitude of the telescope. It is a very small effect, usually ignorable in astronomy, but it enters at the same level as the apparent fine-structure variations. If corrected for it makes the value of the fine-structure constant slightly smaller still in the past when compared with its value today.

[15] In the future new atomic interferometers may offer an improvement on the Prestage limit. The current experimental resolution of this technology is sensitive to shifts in α of about 10^{-8} over 1–2 hours, which corresponds to time rate of change levels of about 10^{-14} per year. In the future it may be adapted to test the constancy of α. There is no immediate prospect of it approaching astronomical levels of accuracy though. Motivated by new atomic physics calculation by Dzuba and Flambaum (2000), Torgerson (2000) has discussed the potential of optical cavities to provide improved measurements of α stability with time. He expects laboratory experiments soon to be sensitive to time variations of order 10^{-15} per year.

needed for living complexity to exist within it. If those complicated pieces of chemistry happen to be astrophysicists then it is a nice by-product of the universe's great age that such sensitive probes of nature's constancy will be available to them.

So it seems that we cannot use terrestrial experiments to double-check the apparent changeability of the fine structure constant – we just don't have instruments sensitive enough to pick up a variation at the level seen in the astronomical data. At the moment the best chance of an independent confirmation from a completely different direction would seem to lie with some other astronomical probe.

Our place in history

If the constants of nature are slowly changing then we could be on a one-way slide to extinction. We have learnt that our existence exploits many peculiar coincidences between the values of different constants of nature and the observed values of the constants fall within some very narrow windows of opportunity for the existence of life. If the values of these constants are actually shifting, what might happen? Might they not slip out of the range that allows life to exist? Are there just particular epochs in cosmic history when the constants are right for life?

There are two situations where it is possible to examine the changes in traditional constants in some detail. For only when the fine structure "constant," α, or Newton's gravitational "constant," G, are changing do we have a full theory which accommodates this. These theories[16] are generalizations of the famous general theory of relativity created by Einstein in 1915. They allow us to extend our picture of how an expanding universe will behave to include variations of these constants. If we know something about the magnitude of a variation at one epoch we can use the theory to calculate what should be seen at other times. In this way the hypothesis that the constants are varying becomes much more vulnerable to observational attack.

If constants like G and α do *not* vary in time then the standard history of our universe has a simple broad-brush appearance. During the first 300 000 years the dominant energy in the universe is radiation and the temperature is greater than 3000 degrees and too hot for any atoms or molecules to exist. The universe is a huge soup of electrons, photons of light, and nuclei. We call this the "radiation era" of the universe. After about 300 000 years there is a big change. The energy of matter catches up and overtakes that of radiation. The expansion rate of the universe is now primarily dictated by the density of atomic nuclei of hydrogen and helium.

[16] The theory including varying G is the Brans–Dicke theory of gravity, found by Carl Brans and Robert Dicke (1961). The cosmological theory including varying α was found by Håvard Sandvik, João Magueijo, and Barrow (Sandvik *et al.* 2002) extending developments by Jacob Bekenstein (1982).

Soon the temperature falls off enough for the first simple atoms and molecules to form. Over the next 13 billion years a succession of more complicated structures are formed: galaxies, stars, planets, and, eventually, people. This is called the "matter era" of the universe's history. But the matter era might not continue right up to the present day. If the universe is expanding fast enough then, eventually, the matter will not matter, and the expansion just runs away from the decelerating clutches of gravity, like a rocket launched at more than the escape speed from Earth. When this happens we say the universe is "curvature dominated" because the rapid expansion creates a negative curvature to astronomical space, just like that near the curved neck of a vase.

There are three overall trajectories for an expanding universe to follow. The "closed" universe expands too slowly to overcome the decelerating effects of gravity and eventually it collapses back to high density. The "open" universe has lots more expansion energy than gravitational deceleration and the expansion runs away forever. The in-between world, that is often called the "flat" or "critical" universe, has a perfect balance between expansion energy and gravity and keeps on expanding for ever. Remarkably, our universe is tantalizingly close to this critical or "flat" state today.

Another possibility is that the vacuum energy of the universe can eventually come to dominate the effects of the ordinary matter and cause the expansion of the universe to begin accelerating. Remarkably, recent astronomical observations show that our universe may have begun to accelerate quite recently, when the universe was about three-quarters of its current size. Moreover, these observations imply that the expansion of our universe has not become curvature dominated. The overall pattern of the expansion history since it was about one second old is shown in Fig. 19.4. The observations are telling us that about 70% of the energy in the universe is in the vacuum form which acts to accelerate the expansion whilst almost all the rest is in the form of matter.

What happens to this story if the fine structure constant changes? The expansion is virtually unaffected by the variations in the fine structure constant if they are as small as observations suggest – a million times slower than the universe is expanding – but the expansion dramatically affects how the fine structure "constant" changes. Håvard Sandvik, João Magueijo, and I investigated what would happen to the fine structure constant over billions of years of cosmic history. The conclusions were strikingly simple. During the radiation era there is no significant change at all in a universe like ours. But once the matter era begins, when the universe is about 300 000 years old, the fine structure "constant" starts to *increase* in value very slowly,[17] varying as the logarithm of time. When the curvature era begins, or the vacuum energy begins to accelerate the universe, this increase

[17] It increases in proportion to the logarithm of the age of the universe; for full details see Barrow *et al.* (2002a).

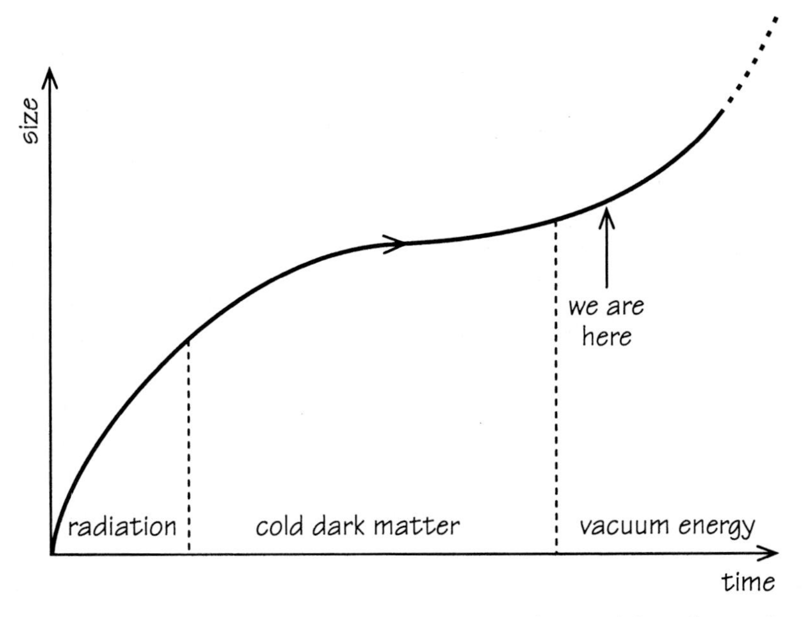

Figure 19.4. The broad-brush picture of the thermal history of the universe showing the evolution from radiation domination to cold-dark-matter domination and the transition to accelerated expansion dominated by a cosmological constant or vacuum energy.

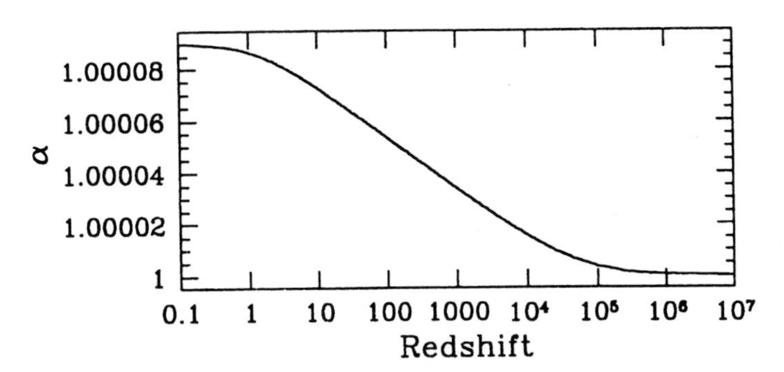

Figure 19.5. The evolution of the fine structure "constant" with redshift in the standard cosmological model of our universe. The fine structure constant stays constant during the radiation era, grows logarithmically in time and decreasing redshift during the dust era, and then becomes constant when the vacuum energy or curvature (shown) dominates the expansion. The evolution is chosen so that α has the observed numerical value today.

stops. This characteristic history is shown in Fig. 19.5 for a universe with matter, radiation, and vacuum energy values equal to those we observe in our universe today.

This is intriguing. It paints a picture that fits all the evidence rather well. Our universe began accelerating at a redshift of about 0.5–0.7 and so there will be no significant variation of the fine structure constant at the time of the Oklo reactor. Over the interval of redshifts corresponding to the quasar observations the variations can be of the form that is seen and α is predicted to be smaller in the past: just what we see. If we keep going back to the redshift around 1100 where the microwave radiation starts flying freely towards us we predict that variation in α should be much smaller than the sensitivity of the present observations.

If α is varying at a rate sufficient to explain the observations of quasar absorption spectra then there is a tantalizingly possible further observational test of the variation. Theories with varying α give rise to violations of the weak equivalence principle. Different materials will fall under gravity in a vacuum with different accelerations. This is because they carry different numbers of charged nucleons in their nuclei and these charged particles couple to the field that carries the variations in α. We predict that if the magnitude of α variation is just that required to match the quasar observations then the relative difference in the free-fall accelerations, a_1 and a_2, of two test materials will be (Magueijo *et al.* 2002):

$$|a_1 - a_2|/|a_1 - a_2| \approx 10^{-13}. \tag{19.13}$$

The current experimental limits are that this quantity be less than 10^{-12}. Future space missions like STEP will have the capability to probe its value down to an accuracy of 10^{-18} and thereby provide an independent experimental test of the direct consequences of varying α. Recent general discussions of varying constants and the constants of Nature in general can be found in the review article by Uzan (2002) and the book by Barrow (2002).

If these variations really are taking place as the universe expands then they have huge consequences for the evolution of life. We know that if the fine structure constant becomes too large then atoms and molecules will be unable to exist and no stars will be able to form because their centers will be too cool to initiate self-sustaining nuclear reactions. Such a universe would be atomless and lifeless.

It is therefore crucial that the dust era of cosmic history during which the fine structure constant increases does not last too long. Without the vacuum energy or the curvature to stop the steady increase in the fine structure constant's value there would come a time when no life is possible. The universe would cease to be habitable by atom-based forms of life who relied upon stars for energy.

Something similar happens if there can be variations in the strength of gravity, represented by the Newtonian "constant" G. During the radiation era it tends to stay constant but when the matter era begins it starts to fall in value until the curvature era begins. If the universe never experiences a curvature era then gravity just keeps getting weaker and weaker and it becomes harder and harder for planets and stars to exist. This behavior is shown for G and α in Fig. 19.6.

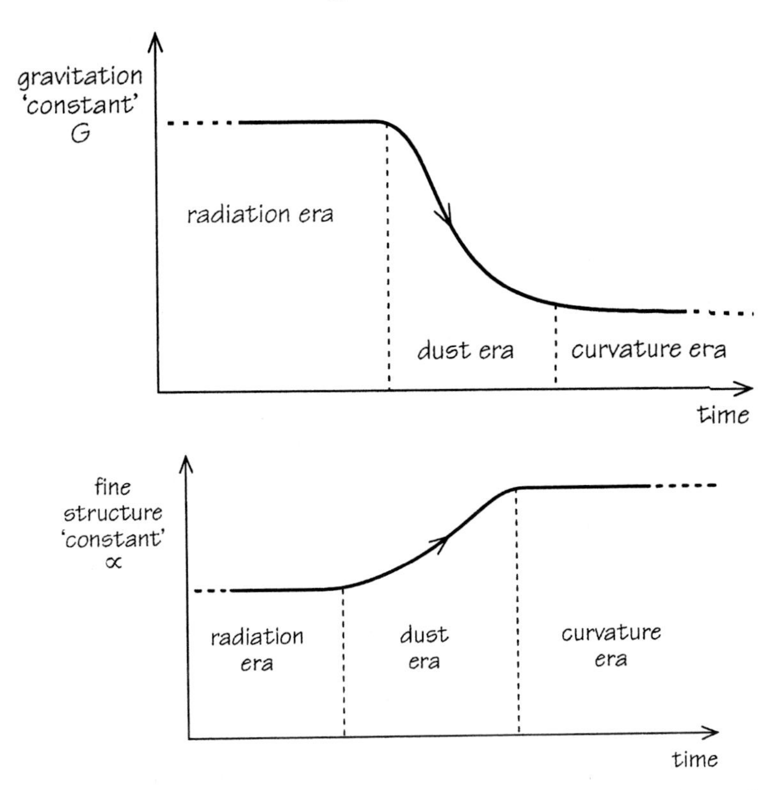

Figure 19.6. The evolution of G (top) and α (bottom) versus time in a universe containing matter and radiation whose expansion becomes dominated by its negative spatial curvature at late times. The constants change slowly during the dust era but become constant when the curvature begins to control the expansion.

This overall behavior is very intriguing. It shows that even when the constants G and α are allowed to vary they are only able to exploit that freedom to vary when the universe is in the matter era.

It has always been something of a mystery why our universe is so close to the critical state of expansion today and why the vacuum energy is so fantastically small. We know that if we were too far from the critical expansion then life would have been far less likely to have evolved on Earth, and would probably be impossible anywhere else in the universe as well. If universes are too curvature dominated then the expansion goes so fast that islands of material cannot overcome the effect of the expansion and contract to form galaxies and stars. On the other had, if the universe expands too slowly it soon collapses back to a Big Crunch. Dense islands of material form too quickly and fall into large black holes before stars and biochemistry have a chance to form; see Fig. 19.6.

The inflationary universe hypothesis provides a good explanation for our close proximity to flatness.

Likewise, with the vacuum energy. If it were ten times bigger it would have started accelerating the universe so early on in its history that galaxies and stars would not have been able to separate out from the overall expansion. Unfortunately, the inflationary universe hypothesis is unable to explain the small value of the cosmological vacuum energy today.

Both these arguments show us that we should not be surprised to find that the deviations from the critical expansion rate or from zero vacuum energy in the universe are *small*. We would not be here if they were not. But the possibility of varying constants provides us with a possible reason why we could not observe the universe to be both exactly critical and to have zero vacuum energy (Barrow *et al.* 2002b). The vacuum energy and the curvature are the brake-pads of the universe that turn off variations in the constants of Nature. They stop the constants changing. If they are not stopped then they will ultimately reach values that prevent the existence of atoms, nuclei, planets, and stars. The inclusion of varying constants in the roster of things that can change over cosmic history opens up new interconnections between properties of the universe that otherwise seem unconnected and arbitrary. In the years to come these problems will be attacked on many fronts: new high-precision observations to test the alleged constancy of the "constants" of nature, new explorations of the status of constants and extra dimensions in string theories, new observational tests of inflationary universe predictions, and, last but not least, daring new ideas of the sort that John Wheeler has been providing us with for more than 50 years.

Acknowledgments

I would like to thank my collaborators H. Sandvik, J. Magueijo, M. J. Drinkwater, J. K. Webb, V. V. Flambaum, C. W. Churchill, M. Dąbrowski, V. Dzuba, and D. Mota for their contributions to all the work described here and to thank John Wheeler for his encouragement and help that began long ago in Austin, Oxford, and Berkeley.

References

Bahcall, J and Schmidt, M (1967) *Phys. Rev. Lett.* **19**, 1294.
Barrow, J D (1999) *Phys. World* 5 **Dec**, 31.
(2002) *The Constants of Nature: From Alpha to Omega*. London: Jonathan Cape.
Barrow, J D and Dąbrowski, M (1995) *Mon. Not. Roy. Astron. Soc.* **275**, 850.
Barrow, J D and Tipler, F J (1986) *The Anthropic Cosmological Principle*. Oxford: Oxford University Press.
Barrow, J D, Sandvik, H, and Magueijo, J (2002*a*) *Phys. Rev.* **D65**, 063504.
(2002b) *Phys. Rev.* **D65**, 123501.
Bekenstein, J (1982) *Phys. Rev.* **25**, 1527.

Brans, C and Dicke, R (1961) *Phys. Rev.* **124**, 924.
Cowie, L L and Songalia, A (1995) *Astrophys. J.* **453**, 596.
Damour, T and Dyson, F (1996) *Nucl. Phys.* **B480**, 37.
Drinkwater, M J, Webb, J K, Barrow, J D, *et al.* (1998) *Mon. Not. Roy. Astron. Soc.* **295**, 457.
Dzuba, V and Flambaum, V (2000) *Phys. Rev.* **A61**, 1.
Eliade, M (1934) *The Myth of the Eternal Return*. New York: Pantheon.
Fujii, Y, Iwanmoto, A, Fukahori, T, *et al.* (2002) *Nucl. Phys* **B573**, 38.
Harrison, E R (1981) *Cosmology*. Cambridge: Cambridge University Press.
(1995) *Q. J. Roy. Astron. Soc.* **36**, 193.
Linde, A (1994) *Sci. American* **5** (May), 32.
Magueijo, J, Barrow, J D, and Sandvik, H (2002) *Phys. Lett.* **B549**, 284. astro-ph/0202374.
Misner, C, Thorne, K, and Wheeler, J A, (1973) *Gravitation*. San Francisco, CA: W. H. Freeman.
Prestage, J D, Tjoelker, R L, and Maleki, L (1995) *Phys. Rev. Lett.* **74**, 18.
Sandvik, H, Barrow, J D, and Magueijo, J (2002) *Phys. Rev. Lett.* **88**, 031302.
Shlyakhter, A (1976) *Nature* **264**, 340.
Smolin, L (1984) *Class. Quant. Gravity* **9**, 173.
Tolman, R C (1931a) *Phys. Rev.* **37**, 1639.
(1931b) *Phys. Rev.* **38**, 1758.
Torgerson, J R (2000) physics/0012054.
Uzan, J-P (2003) *Rev. Mod. Phys.* **75**, 403. hep-ph/0205340.
Webb, J K, Flambaum, V V, Churchill, C W, *et al.* (1999) *Phys. Rev. Lett.* **82**, 884.
Webb, J K, Murphy, M, Flambaum, V, *et al.* (2001) *Phys. Rev. Lett.* **87**, 091301.

20

Inflation, quantum cosmology, and the anthropic principle

Andrei Linde

Stanford University

Introduction

One of the main desires of physicists is to construct a theory that unambiguously predicts the observed values for all parameters of all elementary particles. It is very tempting to believe that the correct theory describing our world should be both beautiful and unique.

However, most of the parameters of elementary particles look more like a collection of random numbers than a unique manifestation of some hidden harmony of Nature. For example, the mass of the electron is 3 orders of magnitude smaller than the mass of the proton, which is 2 orders of magnitude smaller than the mass of the W-boson, which is 17 orders of magnitude smaller than the Planck mass M_p. Meanwhile, it was pointed out long ago that a minor change (by a factor of two or three) in the mass of the electron, the fine structure constant α_e, the strong interaction constant α_s, or the gravitational constant $G = M_p^{-2}$ would lead to a universe in which life as we know it could never have arisen. Adding or subtracting even a single spatial dimension of the same type as the usual three dimensions would make planetary systems impossible. Indeed, in spacetime with dimensionality $d > 4$, gravitational forces between distant bodies fall off faster than r^{-2}, and in spacetime with $d < 4$, the general theory of relativity tells us that such forces are absent altogether. This rules out the existence of stable planetary systems for $d \neq 4$. Furthermore, in order for life as we know it to exist, it is necessary that the universe be sufficiently large, flat, homogeneous, and isotropic. These facts, as well as a number of other observations, lie at the foundation of the so-called anthropic principle (Barrow and Tipler 1986; Rozental 1988; Rees 2000). According to this principle, we observe the universe to be as it is because only in such a universe could observers like ourselves exist.

Science and Ultimate Reality, eds. J. D. Barrow, P. C. W. Davies and C. L. Harper, Jr. Published by Cambridge University Press.

Until very recently, many scientists were ashamed of using the anthropic principle in their research. A typical attitude was expressed in the book *The Early Universe* by Kolb and Turner (1990): "It is unclear to one of the authors how a concept as lame as the 'anthropic idea' was ever elevated to the status of a principle."

This critical attitude is quite healthy. It is much better to find a simple physical resolution of the problem rather that speculate that we can live only in the universes where the problem does not exist. There is always a risk that the anthropic principle does not cure the problem, but acts like a painkiller.

On the other hand, this principle can help us to understand that some of the most complicated and fundamental problems may become nearly trivial if one looks at them from a different perspective. Instead of denying the anthropic principle or uncritically embracing it, one should take a more patient approach and check whether it is really helpful or not in each particular case.

There are two main versions of this principle: the weak anthropic principle and the strong one. The weak anthropic principle simply says that if the universe consists of different parts with different properties, we will live only in those parts where our life is possible. This could seem rather trivial, but one may wonder whether these different parts of the universe are really available. If it is not so, any discussion of altering the mass of the electron, the fine structure constant, and so forth is perfectly meaningless.

The strong anthropic principle says that the universe must be created in such a way as to make our existence possible. At first glance, this principle must be faulty, because mankind, having appeared 10^{10} years after the basic features of our universe were laid down, could in no way influence either the structure of the universe or the properties of the elementary particles within it.

Scientists often associated the anthropic principle with the idea that the universe was created many times until the final success. It was not clear who did it and why it was necessary to make the universe suitable for our existence. Moreover, it would be much simpler to create proper conditions for our existence in a small vicinity of a solar system rather than in the whole universe. Why would one need to work so hard?

Fortunately, most of the problems associated with the anthropic principle were resolved (Linde 1983a, 1984b, 1986a) soon after the invention of inflationary cosmology. Therefore we will remember here the basic principles of inflationary theory.

Chaotic inflation

Inflationary theory was formulated in many different ways, starting with the models based on quantum gravity (Starobinsky 1980) and on the theory of high-temperature phase transitions with supercooling and exponential expansion in the false vacuum

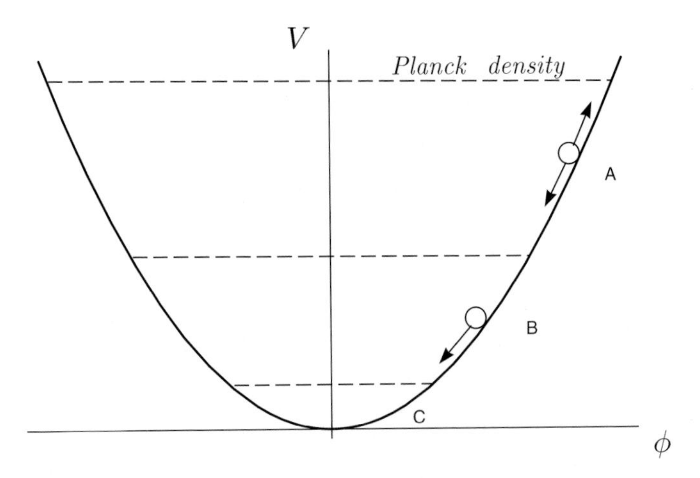

Figure 20.1. Motion of the scalar field in the theory with $V(\phi) = (m^2/2)\phi^2$. Several different regimes are possible, depending on the value of the field ϕ. If the potential energy density of the field is greater than the Planck density $\rho \sim M_p^4 \sim 10^{94}$ g cm^{-3}, quantum fluctuations of spacetime are so strong that one cannot describe it in usual terms. Such a state is called spacetime foam. At a somewhat smaller energy density (region A: $mM_p^3 < V(\phi) < M_p^4$) quantum fluctuations of spacetime are small, but quantum fluctuations of the scalar field ϕ may be large. Jumps of the scalar field due to quantum fluctuations lead to a process of eternal self-reproduction of an inflationary universe which we are going to discuss later. At even smaller values of $V(\phi)$ (region B: $m^2M_p^2 < V(\phi) < mM_p^{-3}$) fluctuations of the field ϕ are small; it slowly moves down as a ball in a viscous liquid. Inflation occurs both in the region A and region B. Finally, near the minimum of $V(\phi)$ (region C) the scalar field rapidly oscillates, creates pairs of elementary particles, and the universe becomes hot.

state (Guth 1981; Albrecht and Steinhardt 1982; Linde 1982a). However, with the introduction of the chaotic inflation scenario (Linde 1983b) it was realized that the basic principles of inflation actually are very simple, and no thermal equilibrium, supercooling, and expansion in the false vacuum is required.

To explain the main idea of chaotic inflation, let us consider the simplest model of a scalar field ϕ with a mass m and with the potential energy density $V(\phi) = (m^2/2)\phi^2$ (see Fig. 20.1). Since this function has a minimum at $\phi = 0$, one may expect that the scalar field ϕ should oscillate near this minimum. This is indeed the case if the universe does not expand. However, one can show that in a rapidly expanding universe the scalar field moves down very slowly, as a ball in a viscous liquid, viscosity being proportional to the speed of expansion.

There are two equations which describe evolution of a homogeneous scalar field in our model, the field equation

$$\ddot{\phi} + 3H\dot{\phi} = -m^2\phi, \tag{20.1}$$

and the Einstein equation

$$H^2 + \frac{k}{a^2} = \frac{8\pi}{3M_p^2}\left(\frac{1}{2}\dot{\phi}^2 + V(\phi)\right). \qquad (20.2)$$

Here $H = \dot{a}/a$ is the Hubble parameter in the universe with a scale factor $a(t)$ (the size of the universe), $k = -1, 0, 1$ for an open, flat or closed universe respectively, M_p is the Planck mass, $M_p^{-2} = G$, where G is the gravitational constant. The first equation becomes similar to the equation of motion for a harmonic oscillator, where instead of $x(t)$ we have $\phi(t)$. The term $3H\dot{\phi}$ is similar to the term describing friction in the equation for a harmonic oscillator.

If the scalar field ϕ initially was large, the Hubble parameter H was large too, according to the second equation. This means that the friction term was very large, and therefore the scalar field was moving very slowly, as a ball in a viscous liquid. Therefore at this stage the energy density of the scalar field, unlike the density of ordinary matter, remained almost constant, and expansion of the universe continued with a much greater speed than in the old cosmological theory. Due to the rapid growth of the scale of the universe and a slow motion of the field ϕ, soon after the beginning of this regime one has $\ddot{\phi} \ll 3H\dot{\phi}$, $H^2 \gg k/a^2$, $\dot{\phi}^2 \ll m^2\phi^2$, so the system of equations can be simplified:

$$3\frac{\dot{a}}{a}\dot{\phi} = -m^2\phi, \qquad (20.3)$$

$$H = \frac{\dot{a}}{a} = \frac{2m\phi}{M_p}\sqrt{\frac{\pi}{3}}. \qquad (20.4)$$

The last equation shows that the size of the universe $a(t)$ in this regime grows approximately as e^{Ht}, where $H = \frac{2m\phi}{M_p}\sqrt{\frac{\pi}{3}}$.

This stage of exponentially rapid expansion of the universe is called inflation. In realistic versions of inflationary theory its duration could be as short as 10^{-35} seconds. When the field ϕ becomes sufficiently small, viscosity becomes small, inflation ends, and the scalar field ϕ begins to oscillate near the minimum of $V(\phi)$. As any rapidly oscillating classical field, it loses its energy by creating pairs of elementary particles. These particles interact with each other and come to a state of thermal equilibrium with some temperature T. From this time on, the corresponding part of the universe can be described by the standard hot universe theory.

The main difference between inflationary theory and the old cosmology becomes clear when one calculates the size of a typical inflationary domain at the end of inflation. Investigation of this issue shows that even if the initial size of inflationary universe was as small as the Plank size $l_P \sim 10^{-33}$ cm, after 10^{-35} seconds of inflation the universe acquires a huge size of $l \sim 10^{10^{12}}$ cm. This makes our universe almost exactly flat and homogeneous on large scale because all inhomogeneities were stretched by a factor of $10^{10^{12}}$.

This number is model-dependent, but in all realistic models the size of the universe after inflation appears to be many orders of magnitude greater than the size of the part of the universe which we can see now, $l \sim 10^{28}$ cm. This immediately solves most of the problems of the old cosmological theory (Linde 1990a).

Consider a universe which initially consisted of many domains with chaotically distributed scalar field ϕ (or if one considers different universes with different values of the field). Those domains where the scalar field was too small never inflated, so they do not contribute much to the total volume of the universe. The main contribution to the total volume of the universe will be given by those domains which originally contained large scalar field ϕ. Inflation of such domains creates huge homogeneous islands out of the initial chaos, each homogeneous domain being much greater than the size of the observable part of the universe. That is why I called this scenario "chaotic inflation."

There is a big difference between this scenario and the old idea that the whole universe was created at the same moment of time (Big Bang), in a nearly uniform state with indefinitely large temperature. In the new theory, the condition of uniformity and thermal equilibrium is no longer required. Each part of the universe could have a singular beginning (see Borde *et al.* (2001) for a recent discussion of this issue). However, in the context of chaotic inflation, this does not mean that the universe as a whole had a single beginning. Different parts of the universe could come into existence at different moments of time, and then grow up to the size much greater than the total size of the universe. The existence of initial singularity (or singularities) does not imply that the whole universe was created simultaneously in a single Big Bang explosion. In other words, we cannot tell any more that the whole universe was born at some time $t = 0$ before which it did not exist. This conclusion is valid for all versions of chaotic inflation, even if one does not take into account the process of self-reproduction of the universe discussed below.

The possibility that our homogeneous part of the universe emerged from the chaotic initial state has important implications for the anthropic principle. Until now we have considered the simplest inflationary model with only one scalar field. Realistic models of elementary particles involve many other scalar fields. For example, according to the standard theory of electroweak interactions, masses of all elementary particles depend on the value of the Higgs scalar field φ in our universe. This value is determined by the position of the minimum of the effective potential $V(\varphi)$ for the field φ. In the simplest models, the potential $V(\varphi)$ has only one minimum. However, in general, the potential $V(\varphi)$ may have many different minima. For example, in the simplest supersymmetric theory unifying weak, strong, and electromagnetic interactions, the effective potential has several different minima of equal depth with respect to the two scalar fields, Φ and φ. If the scalar fields Φ and φ fall to different minima in different parts of the universe (the process called spontaneous symmetry breaking), the masses of elementary particles and the laws describing their interactions will be different in these parts. Each of these parts may become exponentially large because of inflation. In some of these parts, there will be no difference between weak, strong, and electromagnetic interactions, and life of our type will be impossible there. Some other parts will be similar to the one where we live now (Linde 1983c).

This means that even if we are able to find the final theory of everything, we will be unable to uniquely determine properties of elementary particles in our universe; the universe may consist of different exponentially large domains where the properties of elementary particles may be different. This is an important step towards the justification of the anthropic principle. A further step can be made if one takes into account quantum fluctuations produced during inflation.

Inflationary quantum fluctuations

According to quantum field theory, empty space is not entirely empty. It is filled with quantum fluctuations of all types of physical fields. The wavelengths of all quantum fluctuations of the scalar field ϕ grow exponentially during inflation. When the wavelength of any particular fluctuation becomes greater than H^{-1}, this fluctuation stops oscillating, and its amplitude freezes at some non-zero value $\delta\phi(x)$ because of the large friction term $3H\dot{\phi}$ in the equation of motion of the field ϕ. The amplitude of this fluctuation then remains almost unchanged for a very long time, whereas its wavelength grows exponentially. Therefore, the appearance of such a frozen fluctuation is equivalent to the appearance of a classical field $\delta\phi(x)$ produced from quantum fluctuations.

Because the vacuum contains fluctuations of all wavelengths, inflation leads to the continuous creation of new perturbations of the classical field with wavelengths greater than H^{-1}. An average amplitude of perturbations generated during a time interval H^{-1} (in which the universe expands by a factor of e) is given by $|\delta\phi(x)| \approx H/2\pi$ (Linde, 1982c; Vilenkin and Ford 1982).

These quantum fluctuations are responsible for galaxy formation (Mukhanov and Chibisov 1981; Hawking 1982; Guth and Pi 1982; Starobinsky 1982; Bardeen *et al.* 1983; Mukhanov 1985). But if the Hubble constant during inflation is sufficiently large, quantum fluctuations of the scalar fields may lead not only to formation of galaxies, but also to the division of the universe into exponentially large domains with different properties.

As an example, consider again the simplest supersymmetric theory unifying weak, strong, and electromagnetic interactions. Different minima of the effective potential in this model are separated from each other by the distance $\sim 10^{-3} M_p$. The amplitude of quantum fluctuations of the fields ϕ, Φ and φ in the beginning of chaotic inflation can be as large as $10^{-1} M_p$. This means that at the early stages of inflation the fields Φ and φ could easily jump from one minimum of the potential to another. Therefore even if initially these fields occupied the same minimum all over the universe, after the stage of chaotic inflation the universe becomes divided into many exponentially large domains corresponding to all possible minima of the effective potential (Linde 1983c, 1984b).

Eternal chaotic inflation

The process of the division of the universe into different parts becomes even easier if one takes into account the process of self-reproduction of inflationary domains. The basic mechanism can be understood as follows. If quantum fluctuations are sufficiently large, they may locally increase the value of the potential energy of the scalar field in some parts of the universe. The probability of quantum jumps leading to a local increase of the energy density can be very small, but the regions where it happens start expanding much faster than their parent domains, and quantum fluctuations inside them lead to production of new inflationary domains which expand even faster. This surprising behavior leads to the process of self-reproduction of the universe.

This process is possible in the new inflation scenario (Linde 1982b; Steinhardt 1983; Vilenkin 1983). However, even though the possibility to use this result for the justification of the anthropic principle was mentioned in Linde (1982a), this observation did not attract much attention because the amplitude of the fluctuations in new inflation typically is smaller than $10^{-6} M_p$. This is too small to probe most of the vacuum states available in the theory. As a result, the existence of the self-reproduction regime in the new inflation scenario was basically forgotten; for many years this effect was not studied or used in any way even by those who had found it.

The situation changed dramatically when it was found that the self-reproduction of the universe occurs not only in new inflation but also in the chaotic inflation scenario (Linde 1986a). In order to understand this effect, let us consider an inflationary domain of initial radius H^{-1} containing a sufficiently homogeneous field with initial value $\phi \gg M_p$. Equations (20.3) and (20.4) tell us that during a typical time interval $\Delta t = H^{-1}$ the field inside this domain will be reduced by $\Delta\phi = M_p^2/4\pi\phi$. Comparing this expression with the amplitude of quantum fluctuations

$\delta\phi \sim H/2\pi = m\phi/\sqrt{3\pi}\,M_\mathrm{p}$, one can easily see that for $\phi \gg \phi^* \sim (M_\mathrm{p}/2)\sqrt{M_\mathrm{p}/m}$, one has $|\delta\phi| \gg |\Delta\phi|$, i.e., the motion of the field ϕ due to its quantum fluctuations is much more rapid than its classical motion.

During the typical time H^{-1} the size of the domain of initial size H^{-1} containing the field $\phi \gg \phi^*$ grows e times, its volume increases $e^3 \sim 20$ times, and almost in half of this new volume the field ϕ jumps up instead of falling down. Thus the total volume of inflationary domains containing the field $\phi \gg \phi^*$ grows approximately 10 times. During the next time interval H^{-1} this process continues; the universe enters an eternal process of self-reproduction. I called this process "eternal inflation."

In this scenario the scalar field may wander for an indefinitely long time at the density approaching the Planck density. This induces quantum fluctuations of all other scalar field, which may jump from one minimum of the potential energy to another for an unlimited time. The amplitude of these quantum fluctuations can be extremely large, $\delta\varphi \sim \delta\Phi \sim 10^{-1}M_\mathrm{p}$. As a result, quantum fluctuations generated during eternal chaotic inflation can penetrate through any barriers, even if they have Planckian height, and the universe after inflation becomes divided into an indefinitely large number of exponentially large domains containing matter in all possible states corresponding to all possible mechanisms of spontaneous symmetry breaking, i.e., to the different laws of the low-energy physics (Linde 1986a; Linde *et al.* 1994).

A rich spectrum of possibilities may appear during inflation in Kaluza–Klein and superstring theories, where an exponentially large variety of vacuum states and ways of compactification is available for the original 10- or 11-dimensional space. The type of compactification determines coupling constants, vacuum energy, symmetry breaking, and finally, the effective dimensionality of the space we live in. As was shown in Linde and Zelnikov (1988), chaotic inflation at a nearly Planckian density may lead to a local change of the number of compactified dimensions; the universe becomes divided into exponentially large parts with different dimensionality.

Sometimes one may have a continuous spectrum of various possibilities. For example, in the context of the Brans–Dicke theory, the effective gravitational constant is a function of the Brans–Dicke field, which also experiences fluctuations during inflation. As a result, the universe after inflation becomes divided into exponentially large parts with *all* possible values of the gravitational constant G and the amplitude of density perturbations $\delta\rho/\rho$ (Linde 1990b; Garcia-Bellido *et al.* 1994). Inflation may divide our universe into exponentially large domains with continuously varying baryon to photon ratio n_B/n_γ (Linde 1985) and with galaxies having vastly different properties (Linde 1987b). Inflation may also continuously change the effective value of the vacuum energy (the cosmological constant Λ), which is a prerequisite for many attempts to find an anthropic solution of the cosmological constant problem (Linde 1984b, 1986b; Weinberg 1987; Efstathiou 1995; Vilenkin 1995; Martel *et al.* 1998; Garriga and Vilenkin 2000, 2001b, 2002; Bludman and Roos 2002; Kallosh and Linde 2002). Under these circumstances, the most diverse sets of parameters of particle physics (masses, coupling constants, vacuum energy, etc.) can appear after inflation.

To illustrate the possible consequences of such theories in the context of inflationary cosmology, we present here the results of computer simulations of evolution of a system of two scalar fields during chaotic inflation (Linde *et al.* 1994). The field ϕ is the inflaton field driving inflation; it is shown by the height of the distribution of the field $\phi(x, y)$ in a two-dimensional slice of the universe. The field χ determines the type of spontaneous symmetry breaking which may occur in the theory. We paint the surface black if this field is in a state corresponding to one of the two minima of its effective potential; we paint it white if it is in the second minimum corresponding to a different type of symmetry breaking, and therefore to a different set of laws of low-energy physics.

In the beginning of the process the whole inflationary domain was black, and the distribution of both fields was very homogeneous. Then the domain became exponentially large and it became divided into exponentially large domains with different properties (see Fig. 20.2). Each peak of the 'mountains' corresponds to a nearly Planckian density and can be interpreted as a beginning of a new Big Bang. The laws of physics are rapidly changing there, but they become fixed in the parts of the universe where the field ϕ becomes small. These parts correspond to valleys in Fig. 20.2. Thus quantum fluctuations of the scalar fields divide the universe into exponentially large domains with different laws of low-energy physics, and with different values of energy density.

As a result of quantum jumps of the scalar fields during eternal inflation, the universe becomes divided into infinitely many exponentially large domains with different laws of low-energy physics. Each of these domains is so large that for all practical purposes it can be considered a separate universe: its inhabitants will live exponentially far away from its boundaries, so they will never know anything about the existence of other "universes" with different properties.

If this scenario is correct, then physics alone cannot provide a complete explanation for all properties of our part of the universe. The same physical theory may yield large parts of the universe that have diverse properties. According to this scenario, we find ourselves inside a four-dimensional domain with our kind of physical laws not because domains with different dimensionality and with alternative properties are impossible or improbable, but simply because our kind of life cannot exist in other domains.

Figure 20.2. A typical distribution of scalar fields ϕ and χ during the process of self-reproduction of the universe. The height of the distribution shows the value of the field ϕ which drives inflation. The surface is painted black in those parts of the universe where the scalar field χ is in the first minimum of its effective potential, and white where it is in the second minimum. Laws of low-energy physics are different in the regions of different color. The peaks of the "mountains" correspond to places where quantum fluctuations bring the scalar fields back to the Planck density. Each of such places in a certain sense can be considered as a beginning of a new Big Bang.

This provides a simple justification of the weak anthropic principle and removes the standard objections against it. One does not need any more to assume that some supernatural cause created our universe with the properties specifically fine-tuned to make our existence possible. Inflationary universe itself, without any external intervention, may produce exponentially large domains with all possible laws of low-energy physics. And we should not be surprised that the conditions necessary for our existence appear on a very large scale rather than only in a small vicinity of the solar system. If the proper conditions are established near the solar system, inflation ensures that similar conditions appear everywhere within the observable part of the universe.

The new possibilities that appear due to the self-reproduction of the universe may provide a basis for what I have called "the Darwinian approach to cosmology" (Linde 1987a; Garcia-Bellido and Linde 1995; Vilenkin 1995). Mutations of the laws of physics may lead to formation of the domains with the laws of physics that allow a greater speed of expansion of the universe; these domains will acquire greater volume and may host a greater number of observers.

On the other hand, the total volume of domains of each type grows indefinitely large. This process looks like a peaceful coexistence and competition, and sometimes even like a fruitful collaboration, when the fastest-growing domains produce many slower-growing brothers. In this case a stationary regime is reached, and the speed of growth of the total volume of domains of each type becomes equally large for all of the domains (Linde *et al.* 1994).

Baby universes

As we have seen, inflation allows one to justify the weak anthropic principle by ensuring that all vacuum states and, consequently, all possible laws of elementary particle physics that are allowed by the basic theory are realized in some exponentially large and locally uniform parts of our universe.

Note, however, that here we are talking not about the choice among many different theories, but about the choice among many possible vacuum states, or phases, that are allowed by a given theory. This is similar to the possibility to find water in a gaseous, liquid, or solid state. These states look very different (fish cannot live in ice), but their basic chemical composition is the same. Similarly, despite the fact that some of the theories may have extremely large number of vacuum states, our freedom of choice is still limited by the unique fundamental law that is supposed to remain the same in every corner of our universe.

Now it's time to make the next step and ask whether the basic theory was in fact fixed from the very beginning and could not change? A very interesting set of ideas related to this question was developed in the end of the 1980s. It was called the baby universe theory (Banks 1988; Coleman 1988a, b; Giddings and Strominger 1988, 1989). For a short time, this theory was immensely popular, but then it was almost completely forgotten. In our opinion, both extremes were due to the overreaction with respect to the uncritical use of the Euclidean approach to quantum cosmology. But if one distinguishes between this method and the rest of the theory, one can find something very interesting and instructive.

The main idea of the baby universe theory is that our universe can split into disconnected pieces due to quantum gravity effects. Baby universes created from the parent universe can carry from it an electron–positron pair, or some other combinations of particles and fields, unless it is forbidden by conservation laws. Such a process can occur in any place in our universe. Many ways were suggested to describe such a situation. The simplest one is to say that the existence of baby universes leads to a modification of the effective Hamiltonian density:

$$\mathcal{H}(x) = \mathcal{H}_0(\phi(x)) + \sum \mathcal{H}_i[\phi(x)]A_i. \tag{20.5}$$

The Hamiltonian (20.5) describes the fields $\phi(x)$ on the parent universe at distances much greater than the Planck scale. $\mathcal{H}_0$ is the part of the Hamiltonian which does not involve topological fluctuations. $\mathcal{H}_i(\phi)$ are some local functions of the fields ϕ, and A_i are combinations of creation and annihilation operators for the baby universes. These operators do not depend on x since the baby universes cannot carry away momentum. Coleman (1988a, b) argued that the demand of locality, on the parent universe,

$$[\mathcal{H}(x), \mathcal{H}(y)] = 0 \tag{20.6}$$

for spacelike separated x and y, implies that the operators A_i must all commute. Therefore, they can be simultaneously diagonalized by the "α-states":

$$A_i|\alpha_i\rangle = \alpha_i|\alpha_i\rangle. \tag{20.7}$$

If the state of the baby universe is an eigenstate of the A_i, then the net effect of the baby universes is to introduce infinite number of undetermined parameters (the α_i) into the effective Hamiltonian (20.5): one can just replace the operators A_i by their eigenvalues. If the universe initially is not in the A_i eigenstate, then, nevertheless, after a series of measurements the wave junction soon collapses to one of the A_i eigenstates (Coleman, 1988a, b; Giddings and Strominger 1988, 1989).

This gives rise to an extremely interesting possibility related to the basic principles of physics. We were accustomed to believe that the main purpose of physics is to discover the Lagrangian (or Hamiltonian) of the theory that correctly describes our world. However, the question arises: if our universe did not exist sometimes in a distant past, in which sense could one speak about the existence of the laws of Nature which govern the universe? We know, for example, that the laws of our biological evolution are written in our genetic code. But where were the laws of physics written at the time when there was no universe (if there was such time)? The possible answer now is that the final structure of the (effective) Hamiltonian becomes fixed only after measurements are performed, which determine the values of coupling constants in the state in which we live. Different effective Hamiltonians describe different laws of physics in different (quantum) states of the universe, and by making measurements we reduce the variety of all possible laws of physics to those laws that are valid in the (classical) universe where we live.

We will not discuss this issue here any further, since it would require a thorough discussion of the difference between the orthodox (Copenhagen) and the many-world interpretation of quantum mechanics. We would like to mention only that this theory opens a new interesting possibility to strengthen the anthropic principle by allowing all fundamental constants to take different values in different quantum states of the universe.

But if it is so interesting, why don't we hear about this theory any more? In order to answer this question we must remember why it became so popular in the end of the 1980s. The most interesting application of this theory was the possible explanation of the vanishing of the cosmological constant (Coleman 1988a, b). The main idea is closely related to the previous suggestion by Hawking. According to (Hawking, 1984), the cosmological constant, like other constants, can take different values, and the probability to find ourselves in the universe with the cosmological constant $\Lambda = V(\phi)$ is given by

$$P(\Lambda) \sim \exp(-2S_{\mathrm{E}}(\Lambda)) = \exp\frac{3\pi M_P^4}{\Lambda}, \tag{20.8}$$

where S_{E} is the action in the Euclidean version of de Sitter space. However, Coleman pointed out that one should not only take into account one-universe Euclidean configurations. Rather one should sum over all configurations of babies and parents connected by Euclidean wormholes. This finally gives (Coleman 1988a, b).

$$P(\Lambda) \sim \exp\left(\exp\frac{3\pi M_P^4}{\Lambda}\right). \tag{20.9}$$

Equations (20.8) and (20.9) suggest that it is most probable to live in a quantum state of the universe with $\Lambda = 0$. This would be a wonderful solution of the cosmological constant problem.

Unfortunately, the use of the Euclidean approach in this context was not well justified. The whole trick was based on the fact that Euclidean action S_{E} has a wrong (negative) sign (Hartle and Hawking 1983). Usually Euclidean methods work well for $S_{\mathrm{E}} > 0$ and become very problematic for $S_{\mathrm{E}} < 0$ (Linde 1984a, 1998; Vilenkin 1984). After playing with this method for a while, most of the people became dissatisfied and abandoned it. Sometimes one can obtain sensible results by replacing S_{E} by $|S_{\mathrm{E}}|$ (Linde 1984a; Vilenkin 1984), but this would not yield any interesting results with respect to Λ in the context of the baby universe theory. Moreover, current observations suggest that the cosmological constant Λ may be nonvanishing. As a result, the baby universe theory was nearly forgotten.

From our point of view, however, the basic idea that the universe may exist in different quantum states corresponding to different laws of physics may be very productive. But this idea is still somewhat complicated because it preassumes that one can deal with the issues like that only at the level of the so-called third quantization (Banks 1988; Coleman 1988a, b; Giddings and Strominger 1988, 1989), with quantum field theory applied not only to particles but also to the universes. This is a rather radical assumption. A somewhat different approach to quantum cosmology and variation of fundamental constants was suggested later (Linde 1990a; Garcia-Bellido and Linde 1995; Vilenkin 1995). Still it was usually emphasized that these approaches are based on quantum cosmology, which is a rather complicated and controversial science. Thus, it would be helpful to simplify these ideas a bit, and to present them in an alternative form that may allow further generalizations

From the universe to the multiverse

Usually one describes a physical theory by presenting its action. One may write, for example,

$$S = N \int d^4x \sqrt{g(x)} \left(\frac{R(x)}{16\pi G} + L(\phi(x)) \right), \tag{20.10}$$

where N is a normalization constant, $R(x)/2G$ is the general relativity Lagrangian with $G = M_{\mathrm{p}}^{-2}$, and $L(\phi)$ is a Lagrangian for the usual matter fields. One obtains the Einstein equations by variation of the action S with respect to the metric $g^{\mu\nu}$, and one finds the equations of motion for the matter fields ϕ by variation of the action S with respect to ϕ.

Let us now do something very unusual and add to our original action many other actions describing different fields ϕ_i with different Lagrangians L_i living in k different universes of different dimensions n_i with different metrics $g_i^{\mu\nu}$ and different gravitational constants G_i:

$$\begin{aligned} S = N \int d^4x \sqrt{g(x)} \left(\frac{R(x)}{16\pi G} + L(\phi(x)) \right) \\ + \sum_{i=1}^{k} N_i \int d^{n_i}x_i \sqrt{g_i(x_i)} \left(\frac{R(x_i)}{16\pi G_i} + L_i(\phi_i(x_i)) \right). \end{aligned} \tag{20.11}$$

One may wonder whether this modification will affect our life in the universe described by the original action (20.10)? The answer is that it will have no impact whatsoever on the physical processes in our universe. Indeed, equations of motion for ϕ and $g^{\mu\nu}$ will not change because the added parts do not depend on ϕ and $g^{\mu\nu}$, so their variation with respect to ϕ and $g^{\mu\nu}$ vanishes.

This implies that the extended action (20.11) describes all events in our universe in the same way as the original action (20.10). This is very encouraging. So let us continue our exercise and add to this action an *infinite* sum of *all* possible actions describing all possible versions of quantum field theory and M/string theory. If our original theory successfully described our universe, it will continue doing so even after all of these modifications.

But why would anybody want to add all of these extra terms if they do not affect our universe?

There are two related answers. First of all, one may simply reply: Why not? In some countries, everything that is not explicitly allowed, is forbidden. In some other countries (and in science), everything that is not explicitly forbidden, is allowed. We live in one of such countries, so why don't we use the freedom if it does not make us any harm?

But the second answer is more interesting. Now we know that the theory (20.11) and all of its possible extensions are *exactly equivalent* to the theory (20.10) with respect to the processes in our universe (assuming that it is described by (20.10)). So we can take a step back, look at all the different universes described by eqn (20.11), just as we would look for our car among many different cars in a parking lot, and ask: "As a matter of fact, which one of these universes is ours? Are we sure that it is the first one?"

From a purely theoretical point of view, the first universe described by the theory (20.10) is not any better than any other universe. However, we can live only in those universes that are compatible with the existence of life as we know it. When we will search for our universe, first of all we will look for those Lagrangians $L_i(\phi_i(x_i))$ that can describe elementary particles similar to the ones that we see around. Then we will specify our search even further by finding the Lagrangians describing particles with masses and coupling constants that are consistent with our existence. Since we have *all* universes with *all* possible laws of physics described by our extended action, we will certainly find the universe where we can live. But that is exactly what we need to justify the validity of the strong anthropic principle.

Let us summarize our progress so far. Inflationary theory allows our universe to be divided into different parts with different laws of low-energy physics that are allowed by the unique fundamental theory. Most importantly, it makes each of such domains exponentially large, which is a necessary part of justification of the anthropic principle. The diversity of possible laws of physics can be very high, especially in the models of eternal chaotic inflation where quantum fluctuations can have an extremely large amplitude, which makes the transition between all possible states particularly easy.

In addition to that, one can consider different universes with different laws of physics in each of them. This does not necessarily require introduction of quantum cosmology, many-world interpretation of quantum mechanics, and baby universe theory. It is sufficient to consider an extended action represented by a sum of all possible actions of all possible theories in all possible universes. One may call this structure a "multiverse." This could sound like a very complicated and radical proposal, but in fact it is pretty trivial since each part of the infinite sum does not affect other parts. However, it establishes a firm formal background for the further development of the anthropic principle.

But the main reason why we are introducing this structure is not the anthropic principle. As we already mentioned, we need to know what emerged first at the moment of the universe formation: the universe, or the law describing the universe. It is equally hard to understand how any law could exist prior to the universe formation, or how the universe could exist without a law. One could assume that there is only one possible law, and it exists in some unspecified way even prior to the emergence of the universe. However, this would be similar to having elections with only one name on the ballot. Perhaps a better possibility would be to consider all logically possible combinations of the universes, the laws describing them, and the observers populating these universes. Given the choice among different universes in this multiverse structure, we can proceed by eliminating the universes where our life would be impossible. This simple step is sufficient for understanding of many features of our universe that otherwise would seem miraculous.

There are some additional steps that one may want to make. In our analysis we still assumed that any evolution must be described by some kind of action. Meanwhile there are some theories where equations of motion are known even though the action is unavailable. One may consider other models of evolution, based, e.g., on cellular automata. One can go even further, and consider all possible mathematical structures (Tegmark 1998), or, following Wheeler, consider all logical possibilities and the concept of "it from bit" (see Wheeler (1990) and references therein).

But before doing so we would like to show that the concept of a multiverse may have interesting consequences going beyond the justification of the anthropic principle. In order to do it we must learn whether the different universes may interact with each other.

Double-universe model and the cosmological constant problem

Let us consider the double-universe model (Linde 1988). This model describes two universes, X and Y, with coordinates x_μ and y_α, respectively ($\mu, \alpha = 0, 1, \ldots, 3$) and with metrics $g_{\mu\nu}(x)$ and $\bar{g}_{\alpha\beta}(y)$, containing fields $\phi(x)$ and $\bar{\phi}(y)$ with the action of the following unusual type:

$$S = N \int d^4x d^4y \sqrt{g(x)}\sqrt{\bar{g}(y)} \times \left[\frac{M_{\rm P}^2}{16\pi} R(x) + L(\phi(x)) - \frac{M_{\rm P}^2}{16\pi} R(y) - L(\bar{\phi}(y)) \right]. \qquad (20.12)$$

Here N is some normalization constant. This action is invariant under general coordinate transformations in each of the universes separately. A novel symmetry of the action is the symmetry under the transformation $\phi(x) \to \bar{\phi}(x)$, $g_{\mu\nu}(x) \to \bar{g}_{\alpha\beta}(x)$ and under the subsequent change of the overall sign, $S \to -S$. We call this the antipodal symmetry, since it relates to each other the states with positive and negative energies.

An immediate consequence of this symmetry is the invariance under the change of the values of the effective potentials $V(\phi) \to V(\phi) + c$, $V(\bar{\phi}) = V(\bar{\phi}) + c$, where c is some constant. Consequently, nothing in this theory depends on the value of the effective potentials $V(\phi)$ and $V(\bar{\phi})$ in their absolute minima ϕ_0 and $\bar{\phi}_0$. (Note that $\phi_0 = \bar{\phi}_0$ and $V(\phi_0) = V(\bar{\phi}_0)$ due to the antipodal symmetry.) This is the basic reason why it proves possible to solve the cosmological constant problem in our model.

However, our main reason to invoke this new symmetry was not just to solve the cosmological constant problem. Just as the theory of mirror particles originally was proposed in order to make the theory CP-symmetric while maintaining CP-asymmetry in its observable sector, the theory (20.10) is proposed in order to make the theory symmetric with respect to the choice of the sign of energy. This removes the old prejudice that, even though the overall change of sign of the Lagrangian (i.e., both of its kinetic and potential terms) does not change the solutions of the theory, one *must say* that the energy of all particles is positive. This prejudice was so strong, that many years ago physicists preferred to quantize *particles* with *negative energy* as *antiparticles* with *positive energy*, which caused the appearance of such meaningless concepts as negative probability. We wish to emphasize that there is no problem to perform a consistent quantization of theories which describe particles with negative energy. All difficulties appear only when there exist interacting species with both signs of energy. In our case no such problem exists, just as there is no problem of antipodes falling down from the opposite side of the Earth. The reason is that the fields $\bar{\phi}(y)$ do not interact with the fields $\phi(x)$, and the equations of motion for the fields $\bar{\phi}(y)$ are the same as for the fields $\phi(x)$ (the overall minus sign in front of $L(\bar{\phi}(y))$ does not change the Lagrange equations). Similarly, gravitons from different universes do not interact with each other. However, some interaction between the two universes does exist. Indeed, the Einstein equations in our case are:

$$R_{\mu\nu}(x) - \frac{1}{2}g_{\mu\nu}R(x) = -8\pi G T_{\mu\nu}(x) - g_{\mu\nu}\left\langle \frac{1}{2}R(y) + 8\pi G L(\bar{\phi}(y))\right\rangle, \tag{20.13}$$

$$R_{\alpha\beta}(y) - \frac{1}{2}\bar{g}_{\alpha\beta}R(y) = -8\pi G T_{\alpha\beta}(y) - \bar{g}_{\alpha\beta}\left\langle \frac{1}{2}R(x) + 8\pi G L(\phi(x))\right\rangle. \tag{20.14}$$

Here $T_{\mu\nu}$ is the energy–momentum tensor of the fields $\phi(x)$, $T_{\alpha\beta}$ is the energy–momentum tensor of the fields $\bar{\phi}(y)$, the sign of averaging means

$$\langle R(x)\rangle = \frac{\int d^4x\sqrt{g(x)}R(x)}{\int d^4x\sqrt{g(x)}}, \tag{20.15}$$

$$\langle R(y)\rangle = \frac{\int d^4y\sqrt{\bar{g}(y)}R(y)}{\int d^4y\sqrt{\bar{g}(y)}}, \tag{20.16}$$

and similarly for $\langle L(x)\rangle$ and $\langle L(y)\rangle$. Thus, the novel feature of the theory (20.10) is the existence of a *global* interaction between the universes X and Y: the integral *over the whole history* of the Y-universe changes the vacuum energy density of the X-universe.

In general, the computation of the averages of the type (20.15), (20.16) may be a rather sophisticated problem. Fortunately, however, in the inflationary theory (at least, if the universe is not self-reproducing; see below), this task can be rather trivial. Namely, the universe after inflation becomes almost flat and its lifetime becomes exponentially large. In such a case, the dominant contribution to the average values $\langle R\rangle$ and $\langle L\rangle$ comes from the late stages of the universe evolution at which the fields $\phi(x)$ and $\phi(\bar{a})$ relax near the absolute minima of their effective potentials. As a result, the average value of $-L(\phi(x))$ almost exactly coincides with the value of the effective potential $V(\phi)$ in its absolute minimum at $\phi = \phi_0$, and the averaged value of the curvature scalar $R(x)$ coincides with its value at the late stages of the universe evolution, when the universe transforms to the state corresponding to the absolute minimum of $V(\phi)$. Similar results are valid for the average values of $-L(\bar{\phi}(y))$ and of $R(y)$ as well. In such a case one can easily show (Linde 1988) that at the late stages of the universe evolution, when the fields $\phi(x)$ and $\bar{\phi}(y)$ relax near the absolute minima of their effective potentials, the *effective* cosmological constant automatically vanishes,

$$R(x) = -R(y) = \frac{32}{3}\pi G[V(\phi_0) - V(\bar{\phi}_0)] = 0. \tag{20.17}$$

This model provided the first example of a theory with a nonlocal interaction of universes. It inspired the baby-universe scenario, and it was forgotten when the baby-universe scenario failed. However, this model is based on a completely different principle, so it should be considered quite independently.

There are several problems with this model that should be addressed before taking it too seriously. First of all, in order to solve the cosmological constant problem in our universe we added a new universe with negative energy density. At first glance, this may not seem very economical. However, during the past few years the idea that we may have several different interacting universes has become very popular in the context of the brane world scenario (Antoniadis *et al.* 1998; Arkani-Hamed *et al.* 1998, 2000; Randall and Sundrum 1999). The cancellation of the effective cosmological constant on our brane (our universe) is often achieved by the introduction of the negative tension brane (the universe with a negative energy density) (see, e.g., Randall and Sundrum (1999)). It is not quite clear whether any symmetry can protect this cancellation against radiative corrections in the brane world scenario. Meanwhile in our case the theory is fully symmetric with respect to the choice of the sign of energy, which may protect the cosmological constant against radiative corrections.

The second problem is more complicated. If the universe is self-reproducing, one may encounter difficulties when computing the averages (20.15), (20.16), since they may become dominated by eternally inflating parts of the universe with large $V(\phi)$. One can avoid this complication in inflationary theories where $V(\phi)$ grows rapidly enough at large ϕ, since there will be no universe self-reproduction in such theories.

Finally, the cosmological observations indicate that the universe is accelerating as if it has a minuscule positive vacuum energy $V(\phi) \sim 10^{-123} M_{\mathrm{p}}^4$. Thus we need to make the vacuum energy cancellation nonexact. This is quite possible: as we said, the average value of $-L(\phi(x))$ *almost* exactly coincides with the value of the effective potential $V(\phi)$ in its absolute minimum at $\phi = \phi_0$. Also, if $V(\phi)$ is very flat near its minimum, as in the usual dark-energy models, we may move to the minimum very slowly and at any given moment we will still have a small noncompensated positive vacuum energy.

We do not know whether this simple model is going to survive in the future. But this example shows that the multiverse scenario may provide us with new unexpected possibilities that should be considered very seriously.

Now we will make a step back and discuss the anthropic approach to the cosmological constant problem.

Cosmological constant, dark energy, and the anthropic principle

The first attempt to solve the cosmological constant problem using the anthropic principle in the context of inflationary cosmology was made in Linde (1984b, 1986b). The simplest way to do it is to consider inflation driven by the scalar field ϕ

(the inflaton field) and mimic the cosmological constant by the very flat potential of the second scalar field, Φ. The simplest potential of this type is the linear potential (Linde 1986b):

$$V(\Phi) = \alpha M_p^3 \Phi. \tag{20.18}$$

If α is sufficiently small, $\alpha < 10^{-122}$, the potential $V(\Phi)$ is so flat that the field Φ practically does not change during the last 10^{10} years, and its kinetic energy is very small, so at the present stage of the evolution of the universe its total potential energy $V(\Phi)$ acts exactly as a cosmological constant. This model was one of the first examples of what later became known as quintessence, or dark energy.

Even though the energy density of the field Φ practically does not change at the present time, it changed substantially during inflation. Since Φ is a massless field, it has experienced quantum jumps with the amplitude $H/2\pi$ during each time H^{-1}. These jumps move the field Φ in all possible directions. In the context of the eternal inflation scenario this implies that the field becomes randomized by quantum fluctuations: the universe becomes divided into infinitely large number of exponentially large parts containing all possible values of the field Φ. In other words, the universe becomes divided into infinitely large number of "universes" with all possible values of the effective cosmological constant $\Lambda = V(\Phi) + V(\phi_0)$, where $V(\phi_0)$ is the energy density of the inflaton field ϕ in the minimum of its effective potential. This quantity may change from $-M_p^4$ to $+M_p^4$ in different parts of the universe, but we can live only in the "universes" with $|\Lambda| \lesssim O(10)\rho_0 \sim 10^{-28}$ g cm^{-3}, where ρ_0 is the present energy density in our part of the universe.

Indeed, if $\Lambda < -10^{-28}$ g cm^{-3}, the universe collapses within the time much smaller than the present age of the universe $\sim 10^{10}$ years (Linde 1984b, 1986b; Barrow and Tipler 1986). On the other hand, if $\Lambda \gg 10^{-28}$ g cm^{-3}, the universe at present would expand exponentially fast, energy density of matter would be exponentially small, and life as we know it would be impossible (Linde 1984b, 1986b). This means that we can live only in those parts of the universe where the cosmological constant does not differ too much from its presently observed value $|\Lambda| \sim \rho_0$.

This approach constituted the basis for many subsequent attempts to solve the cosmological constant problem using the anthropic principle in inflationary cosmology (Weinberg 1987; Linde 1990a; Vilenkin 1995; Martel *et al.* 1998; Garriga and Vilenkin 2000, 2001b, 2002).

At first glance, an introduction of the minuscule parameter $\alpha < 10^{-122}$ does not provide a real explanation of the equally minuscule cosmological constant $|\Lambda| \sim \rho_0 \sim 10^{-123} M_p^4$. However, exponentially small parameters like that may easily appear due to nonperturbative effects. One could even think that a similar exponential suppression may be the true reason why $|\Lambda|$ is so small. But there are many large contributions to Λ, due to quantum gravity, due to spontaneous symmetry breaking in grand unified theories and in the electroweak theory, due to supersymmetry breaking, quantum chromodynamics effects, etc. One could appeal to the nonperturbative exponential smallness of Λ only if all large contributions to the vacuum energy miraculously cancel, as in the model considered in the previous section. And even if this cancellation is achieved, we still need to explain why $|\Lambda|$ is suppressed exactly to the level when it becomes of the same order as the present energy density of the universe. This coincidence problem becomes resolved in the theory (20.18) for all sufficiently small α; instead of the fine-tuning of α we simply need it to be sufficiently strongly suppressed.

A possible explanation of the extreme flatness of the potential $V(\Phi)$ can be related to the global shift symmetry of the theory under the transformation $\Phi \to \Phi + C$, where C can be any constant. This symmetry implies that the potential $V(\Phi)$ must be absolutely flat. Nonperturbative quantum gravity effects may violate this symmetry, but under certain conditions this violation can be exponentially small, which is exactly what we need. In terms of the "stringy" Planck mass $M_{\rm p}^2 = (8\pi G)^{-1}$ and the string scale $M_{\rm s}$, the nonperturbative effects can be suppressed by the factor of $\exp(-8\pi^2 M_{\rm p}^2/M_{\rm s}^2)$, which, for $M_{\rm s} \ll M_{\rm p}$, is more than adequate (Kallosh *et al.* 1995). A very clear discussion of the issue of fine-tuning versus exponential suppression can be found in (Garriga and Vilenkin 2000) in application to a similar model with the potential $\rho_\Lambda \pm m^2\Phi^2/2$ with $m^2 \ll 10^{-240} M_{\rm p}^6 |\rho_\Lambda|^{-1}$.

Alternative approaches based on the anthropic principle are described in Bousso and Polchinski (2001), Feng *et al.* (2001), and Banks *et al.* (2001). One can also use a more general approach outlined above and consider a baby-universe scenario, or a multiverse consisting of different inflationary universes with different values of the cosmological constant in each of them (Linde 1989, 1990a, 1991). In this case one does not need to consider extremely flat potentials, but the procedure of comparing probabilities to live in different universes with different Λ becomes more ambiguous (Garcia-Bellido and Linde 1995; Vilenkin 1995). However, if one makes the simplest assumption that the universes with different values of Λ are equally probable, one obtains an anthropic solution of the cosmological constant problem without any need of introducing extremely small parameters $\alpha < 10^{-122}$ or $m^2 \ll 10^{-240} M_{\rm p}^6 |\rho_\Lambda|^{-1}$.

The constraint $\Lambda \gtrsim -10^{-28}$ g cm^{-3} still remains the strongest constraint on the negative cosmological constant; for the recent developments related to this constraint see Kallosh and Linde (2002) and Garriga and Vilenkin (2002). Meanwhile, the constraint on the positive cosmological constant, $\Lambda \gtrsim 10^{-28}$ g cm^{-3}, was made much more precise and accurate in the subsequent works.

In particular, Weinberg pointed out that the process of galaxy formation occurs only up to the moment when the cosmological constant begins to dominate the energy density of the universe and the universe enters the stage of late-time inflation (Weinberg 1987). For example, one may consider galaxies formed at $z \gtrsim 4$, when the energy density of the universe was 2 orders of magnitude greater than it is now. Such galaxies would not form if $\Lambda \gtrsim 10^2 \rho_0 \sim 10^{-27}$ g cm^{-3}.

The next important step was made in a series of works (Efstathiou 1995; Vilenkin 1995; Martel *et al.* 1998; Garriga and Vilenkin 2000, 2001b, 2002; Bludman and Roos 2002). The authors considered not only our own galaxy, but all other galaxies that could harbor life of our type. This would include not only the existing galaxies but also the galaxies that are being formed at the present epoch. Since the energy density at later stages of the evolution of the universe becomes smaller, even a very small cosmological constant may disrupt the late-time galaxy formation, or may prevent the growth of existing galaxies. This allows one to strengthen the constraint on the cosmological constant. According to (Martel *et al.* 1998), the probability that an astronomer in any of the universes would find the presently observed ratio Λ/ρ_0 as small as 0.7 ranges from 5% to 12%, depending on various assumptions. For some models based on extended supergravity, the anthropic constraints can be strengthened even further (Kallosh and Linde 2002).

Problem of calculating the probabilities

As we see, the anthropic principle can be extremely useful in resolving some of the most profound problems of modern physics. However, to make this principle more quantitative, one should find a proper way to calculate the probability to live in a universe of a given type. This step is not quite trivial. One may consider the probability of quantum creation of the universe "from nothing" (Hartle and Hawking 1983; Linde 1984a; Vilenkin 1984), or the results of the baby-universe theory (Coleman 1988a, b), or the results based on the theory of the self-reproduction of the universe and quantum cosmology (Garcia-Bellido *et al.* 1994; Linde *et al.* 1994; Garcia-Bellido and Linde 1995; Vilenkin 1995; Linde and Mezhlumian 1996; Vanchurin *et al.* 2000; Garriga and Vilenkin 2001a). Unfortunately, these methods are based on different assumptions, and the results of some of these works significantly differ from each other. This may be just a temporary setback. For example, in our opinion, an interpretation of Euclidean quantum gravity used in Hartle and Hawking (1983) and Coleman (1988a, b) is not quite convincing. The method proposed in Turok (2002) is basically equivalent to the investigation of the probability distribution in comoving coordinates $P_c(\phi, t)$ (Linde 1990a). This approach ignores information about most of the observers living in our universe, so it can hardly have any relation to the standard anthropic considerations and misses the effect of the self-reproduction of the universe. An investigation of creation of the universe "from nothing" (Linde 1984a; Vilenkin 1984) can be very useful, but I believe that it should be considered only as a part of the more general approach based on the stochastic approach to inflation.

It is more difficult to make a definite choice between the different answers provided by the different methods of interpretation of the results obtained by the stochastic approach to inflation (Starobinsky 1986; Garcia-Bellido *et al.* 1994; Linde *et al.* 1994; Garcia-Bellido and Linde 1995; Vilenkin 1995; Linde and Mezhlumian, 1996; Vanchurin *et al.* 2000; Garriga and Vilenkin 2001a). We believe that all of these different answers in a certain sense are correct; it is the choice of the questions that remains problematic.

To explain our point of view, let us study an example related to demographics. One may want to know what is the average age of a person living now on the Earth. In order to find it, one should take the sum of the ages of all people and divide it by their total number. Naively, one could expect that the result of the calculation should be equal to 1/2 of the life expectancy. However, the actual result will be much smaller. Because of the exponential growth of the population, the main contribution to the average age will be given by very young people. Both answers (the average age of a person, and a half of the life expectancy) are correct despite the fact that they are different. None of these answers is any better; they are different because they address different questions. Economists may want to know the average age in order to make their projections. Meanwhile each of us, as well as the people from the insurance industry, may be more interested in the life expectancy.

Similarly, the calculations performed in Garcia-Bellido *et al.* (1994), Linde *et al.* (1994), Garcia-Bellido and Linde (1995), Vilenkin (1995), Linde and Mezhlumian (1996), Vanchurin *et al.* (2000), and Garriga and Vilenkin (2001a) dissect all possible outcomes of the evolution of the universe (or the multiverse) in many different ways. (Unlike the method suggested in Turok (2002), these methods cover the whole universe rather that its infinitesimally small part.) Each of these ways is quite legitimate and leads to correct results, but some additional input is required in order to understand which of these results, if any, is most closely related to the anthropic principle.

In the meantime one may take a pragmatic point of view and consider this investigation as a kind of "theoretical experiment." We may try to use probabilistic considerations in a trial-and-error approach. If we get unreasonable results, this may serve as an indication that we are using quantum cosmology incorrectly. However, if some particular proposal for the probability measure will allow us to solve certain problems that could not be solved in any other way, then we will have a reason to believe that we are moving in the right direction. But we are not sure that any real progress in this direction can be reached and we will be able to learn how to calculate the probability to live in one of the many universes without having a good idea of what is life and what is consciousness (Linde 1990a; Garcia-Bellido and Linde 1995; Linde and Mezhlumian 1996; Linde *et al.* 1996).

A healthy scientific conservatism usually forces us to disregard all metaphysical subjects that seem unrelated to our research. However, in order to make sure that this conservatism is really healthy, from time to time one should take a risk to abandon some of the standard assumptions. This may allow us either to reaffirm our previous position, or to find some possible limitations of our earlier point of view

Does consciousness matter?

A good starting point for our brief discussion of consciousness is quantum cosmology, the theory that tries to unify cosmology and quantum mechanics.

If quantum mechanics is universally correct, then one may try to apply it to the universe in order to find its wave function. This would allow us to find out which events are probable and which are not. However, it often leads to paradoxes. For example, the essence of the Wheeler–DeWitt equation (DeWitt 1967), which is the Schrödinger equation for the wave function of the universe, is that this wave function *does not depend on time*, since the total Hamiltonian of the universe, including the Hamiltonian of the gravitational field, vanishes identically. This result was obtained in 1967 by Bryce DeWitt. Therefore if one would wish to describe the evolution of the universe with the help of its wave function, one would be in trouble: *The universe as a whole does not change in time.*

The resolution of this paradox suggested by Bryce DeWitt (1967) is rather instructive. The notion of evolution is not applicable to the universe as a whole since there is no external observer with respect to the universe, and there is no external clock that does not belong to the universe. However, we do not actually ask why the universe *as a whole* is evolving. We are just trying to understand our own experimental data. Thus, a more precisely formulated question is *why do we see* the universe evolving in time in a given way. In order to answer this question one should first divide the universe into two main pieces: (i) an observer with his clock and other measuring devices and (ii) the rest of the universe. Then it can be shown that the wave function of the rest of the universe does depend on the state of the clock of the observer, i.e., on his "time." This time dependence in some sense is "objective": the results obtained by different (macroscopic) observers living in the same quantum state of the universe and using sufficiently good (macroscopic) measuring apparatus agree with each other.

Thus we see that without introducing an observer, we have a dead universe, which does not evolve in time. This example demonstrates an unusually important role played by the concept of an observer in quantum cosmology. John Wheeler underscored the complexity of the situation, replacing the word *observer* by the word *participant*, and introducing such terms as a "self-observing universe."

Most of the time, when discussing quantum cosmology, one can remain entirely within the bounds set by purely physical categories, regarding an observer simply as an automaton, and not dealing with questions of whether he/she/it has consciousness or feels anything during the process of observation. This limitation is harmless for many practical purposes. But we cannot rule out the possibility that carefully avoiding the concept of consciousness in quantum cosmology may lead to an artificial narrowing of our outlook.

Let us remember an example from the history of science that may be rather instructive in this respect. Prior to the invention of the general theory of relativity, space, time, and matter seemed to be three fundamentally different entities. Space was thought to be a kind of three-dimensional coordinate grid which, when supplemented by clocks, could be used to describe the motion of matter. Spacetime possessed no intrinsic degrees of freedom; it played a secondary role as a tool for the description of the truly substantial material world.

The general theory of relativity brought with it a decisive change in this point of view. Spacetime and matter were found to be interdependent, and there was no longer any question which one of the two is more fundamental. Spacetime was also found to have its own inherent degrees of freedom, associated with perturbations of the metric–gravitational waves. Thus, space can exist and change with time in the absence of electrons, protons, photons, etc.; in other words, in the absence of anything that had previously (i.e., prior to general relativity) been called matter. Of course, one can simply extend the notion of matter, because, after all, gravitons (the quanta of the gravitational field) are real particles living in our universe. On the other hand, the introduction of the gravitons provides us, at best, with a tool for an approximate (perturbative) description of the fluctuating geometry of spacetime. This is completely opposite to the previous idea that spacetime is only a tool for the description of matter.

A more recent trend, finally, has been toward a unified geometric theory of all fundamental interactions, including gravitation. Prior to the end of the 1970s, such a program seemed unrealizable; rigorous theorems were proven on the impossibility of unifying spatial symmetries with the internal symmetries of elementary particle theory. Fortunately, these theorems were sidestepped after the discovery of supersymmetry and supergravity. In these theories, matter fields and spacetime became unified within the general concept of superspace.

Now let us turn to consciousness. The standard assumption is that consciousness, just like spacetime before the invention of general relativity, plays a secondary, subservient role, being just a function of matter and a tool for the description of the truly existing material world. But let us remember that our knowledge of the world begins not with matter but with perceptions. I know for sure that my pain exists, my "green" exists, and my "sweet" exists. I do not need any proof of their existence, because these events are a part of me; everything else is a theory. Later we find out that our perceptions obey some laws, which can be most conveniently formulated if we assume that there is some underlying reality beyond our perceptions. This model of a material world obeying laws of physics is so successful that soon we forget about our starting point and say that matter is the only reality, and perceptions are nothing but a useful tool for the description of matter. This assumption is almost as natural (and maybe as false) as our previous assumption that space is only a mathematical tool for the description of matter. We are substituting *reality* of our feelings by the successfully working *theory* of an independently existing material world. And the theory is so successful that we almost never think about its possible limitations.

Guided by the analogy with the gradual change of the concept of spacetime, we would like to take a certain risk and formulate several questions to which we do not yet have the answers (Linde 1990a; Page 2002):

Is it possible that consciousness, like spacetime, has its own intrinsic degrees of freedom, and that neglecting these will lead to a description of the universe that is fundamentally incomplete? What if our perceptions are as real as (or maybe, in a certain sense, are even more real than) material objects? What if my red, my blue, my pain, are really existing objects, not merely reflections of the really existing material world? Is it possible to introduce a "space of elements of consciousness," and investigate a possibility that consciousness may exist by itself, even in the absence of matter, just like gravitational waves, excitations of space, may exist in the absence of protons and electrons?

Note, that the gravitational waves usually are so small and interact with matter so weakly that we have not found any of them as yet. However, their existence is absolutely crucial for the consistency of our theory, as well as for our understanding of certain astronomical data. Could it be that consciousness is an equally important part of the consistent picture of our world, despite the fact that so far one could safely ignore it in the description of the well-studied physical processes? Will it not turn out, with the further development of science, that the study of the universe and the study of consciousness are inseparably linked, and that ultimate progress in the one will be impossible without progress in the other?

Instead of discussing these issues here any further, we will return to a more solid ground and concentrate on the consequences of eternal inflation and the multiverse theory that do not depend on the details of their interpretation. As an example, we will discuss here two questions that for a long time were considered too complicated and metaphysical. We will see that the concept of the multiverse will allow us to propose possible answers to these questions.

Why is mathematics so efficient?

There is an old problem that bothered many people thinking about the foundations of mathematics: why is mathematics so efficient in helping us to describe our world and predict its evolution?

This question arises at the moment when one introduces numbers and uses them to count. Then a similar question appears when one introduces calculus and uses it to describe the motion of the planets. Somehow there are some rules that help us to operate with mathematical symbols and relate the results of these operations to the results of our observations. Why does it work so well?

Of course, one could always respond that it is just so. But let us consider several other questions of a similar type. Why is our universe so large? Why do parallel lines not intersect? Why do different parts of the universe look so similar? Thirty years ago such questions would look too metaphysical to be considered seriously. Now we know that inflationary cosmology provides a possible answer to all of these questions. Let us try it again.

Before we do it, we should give at least one example of a universe where mathematics would be inefficient. Here it is. Suppose the universe can be in a stable or metastable vacuum state with a Planckian density $\rho \sim M_{\rm p}^4 \sim 10^{94}$ g cm^{-3}. According to quantum gravity, quantum fluctuations of spacetime curvature in this regime are of the same order as the curvature itself. In simple terms, this means that the rulers are bending, shrinking and extending in a chaotic and unpredictable way due to quantum fluctuations, and this happens faster than one can measure the distance. The clocks are destroyed faster than one can measure the time. All records about the previous events become erased, so one cannot remember anything, record it, make a prediction, and compare the prediction with experimental results.

A similar situation occurs in a typical noninflationary closed universe. There is only one natural parameter of dimension of length in quantum gravity, $l_{\rm p} = M_{\rm p}^{-1}$, and only one natural parameter of dimension of energy density, $\rho_{\rm p} = M_{\rm p}^4$. If one considers a typical closed universe of a typical initial size $l_{\rm p}$ with a typical initial density $\rho_{\rm p}$, one can show that its total lifetime until it collapses is $t \sim t_{\rm p} = M_{\rm p}^{-1} \sim 10^{-43}$ seconds, and throughout all of its short history the energy density remains of the order of $M_{\rm p}^4$ or greater. Such a universe can incorporate just a few elementary particles (Linde 1990a), so one cannot live there, cannot build any measuring devices, record any events, or use mathematics to describe events in such a universe.

In the cases described above, mathematics would be rather inefficient because it would not help anybody to relate different things and processes to each other. More generally, if the laws of physics inside some parts of the universe disallow formation of stable long-living structures, then mathematics will not be very useful there, and there will be no observers (long-living conscious beings capable of remembering and thinking) who would be able to tell us about it.

Fortunately, among all possible domains of the universe (or among all possible universes) there are some domains where inflation is possible. Energy density inside such a domain gradually drops down many orders of magnitude below M_{p}^4. These domains become exponentially large and can live for an exponentially long time. Our life is possible only in those exponentially large domains (or universes) where the laws of physics allow formation of stable long-living structures. The very concept of stability implies existence of mathematical relations that can be used for the long-term predictions. The rapid development of the human race became possible only because we live in the universe where the long-term predictions are so useful and efficient that they allow us to survive in the hostile environment and win in the competition with other species.

To summarize, in the context of the multiverse theory, one can consider all possible universes with all possible laws of physics and mathematics. Among all possible universes, we can live only in those where mathematics is efficient.

Why the quantum?

Now we will discuss the famous Wheeler's question: Why the quantum?

Before doing so, I would like to remember the question often asked by Zeldovich: Do we have any experimental evidence of proton instability and baryon nonconservation?

In accordance with the unified theories of weak, strong, and electromagnetic interactions, protons and other baryons can be unstable. They can decay to leptons. But the decay rate is so small that we still have not found any direct evidence of the proton instability. People were watching protons in thousands of tons of water, and did not find any of them decaying. Thus the simple-minded answer to Zeldovich's question would be "No."

However, the true answer is different. To make it sound a little bit more challenging, I will formulate it in a way slightly different from the formulation used by Zeldovich, but conveying the same basic idea: *The main experimental evidence of the baryon number nonconservation is provided by the fact that parallel lines do not intersect.*

What? Is it a joke? What is the relation?

Well, the fact that the parallel lines do not intersect and remain parallel to each other is a consequence of the spatial flatness of the universe. In a closed universe the parallel lines would intersect, in an open universe they would diverge at infinity. The only known explanation of the flatness of the universe is provided by inflationary cosmology. This theory implies that at the end of the exponential expansion of the universe, the number density of all elementary particles becomes vanishingly small.

All matter surrounding us was produced due to the decay of the scalar field after inflation (Abbott *et al.* 1982; Dolgov and Linde 1982; Kofman *et al.* 1994, 1997; Felder *et al.* 2001). The density of protons in our part of the universe is much greater than the density of antiprotons. This means that at the present time the total baryon number density is not zero. It would be impossible to produce these baryons from the post-inflationary state with the vanishing baryon density if the baryon number were conserved.

Thus, the only available explanation of the observed flatness and homogeneity of the universe requires baryon number nonconservation. In this sense, the fact that the parallel lines do not intersect is an observational evidence of the proton instability.

This is a strange and paradoxical logic, but we must get used to it if we want to understand the properties of our universe.

Now let us return to Wheeler's question. At first glance, this question is so deep and metaphysical that we are not going to know the answer any time soon. However, in my opinion, the answer is pretty simple.

The only known way to explain why our universe is so large, flat, homogeneous, and isotropic requires inflation. As we just said, after inflation the universe becomes empty. All matter in the universe was produced due to *quantum* processes after the end of inflation. All galaxies were produced by *quantum fluctuations* generated at the last stages of inflation. There would be no galaxies and no matter in our universe if not for the *quantum effects*. One can formulate this result in the following way: *Without inflation, our universe would be ugly. Without quantum, our universe would be empty.*

But there is something else here. As we already discussed above (pp. 432–6), *quantum* fluctuations lead to the eternal process of self-reproduction of the inflationary universe: *Quantum effects combined with inflation make the universe infinitely large and immortal.*

This provides a possible answer to Wheeler's question.

Isn't it amazing that different, apparently unrelated things can match together to form a beautiful and self-consistent pattern? Are we uncovering the universal truth or simply allowing this beauty to deceive us? This is one of the questions that will remain with us for some time. We need to move carefully and slowly, constantly keeping in touch with solid and well-established facts, but from time to time allowing ourselves to satisfy our urge to speculate, following the steps of John Wheeler.

References

Abbott, L F, Farhi, E, and Wise, M B (1982) Particle production in the new inflationary cosmology. *Phys. Lett.* **B117**, 29.
Albrecht, A and Steinhardt, P J (1982) Cosmology for grand unified theories with radiatively induced symmetry breaking. *Phys. Rev. Lett.* **48**, 1220.
Antoniadis, I, Arkani-Hamed, N, Dimopoulos, S, *et al.* (1998) New dimensions at a millimeter to a Fermi and superstrings at a TeV. *Phys. Lett.* **B436**, 257. arXiv:hep-ph/9804398.
Arkani-Hamed, N, Dimopoulos, S, and Dvali, G R (1998) The hierarchy problem and new dimensions at a millimeter. *Phys. Lett.* **B429**, 263. arXiv:hep-ph/9803315.
Arkani-Hamed, N, Dimopoulos, S, Dvali, G R *et al.* (2000) Manyfold universe. *J. High Energy Phys.* **0012**, 010. arXiv:hep-ph/9911386.
Banks, T (1988) Prolegomena to a theory of bifurcating universes." *Nucl. Phys.* **B309**, 493.
Banks, T, Dine, M, and Motl, L (2001) On anthropic solutions of the cosmological constant problem. *J. High Energy Phys.* **0101**, 031. arXiv:hep-th/0007206.
Bardeen, J M, Steinhardt, P J, and Turner, M S (1983) Spontaneous creation of almost scale-free density perturbations in an inflationary universe. *Phys. Rev.* **D28**, 679.
Barrow, J D, and Tipler, F J (1986) *The Anthropic Cosmological Principle*. Oxford: Oxford University Press.
Bludman, S A, and Roos, M (2002) Quintessence cosmology and the cosmic coincidence. *Phys. Rev.* **D65**, 043503. arXiv:astro-ph/0109551.
Borde, A, Guth, A H, and Vilenkin, A (2001) Inflation is not past-eternal. arXiv:gr-qc/0110012.
Bousso, R and Polchinski, J (2000) Quantization of four-form fluxes and dynamical neutralization of the cosmological constant. *J. High Energy Phys.* **0006**, 006. arXiv:hep-th/0004134.
Coleman, S R (1988a) Black holes as red herrings: topological fluctuations and the loss of quantum coherence. *Nucl. Phys.* **B307**, 867.
(1988b) Why there is nothing rather than something: a theory of the cosmological constant. *Nucl. Phys.* **B310**, 643.
DeWitt, B S (1967) Quantum theory of gravity. 1. The canonical theory. *Phys. Rev.* **160**, 1113.
Dolgov, A D and Linde, A D (1982) Baryon asymmetry in inflationary universe. *Phys. Lett.* **B116**, 329.
Efstathiou, G (1995) *Mon. Not. Roy. Astron. Soc.* **274**, L73.
Felder, G, Garcia-Bellido, J, Greene, P B, *et al.* (2001) Dynamics of symmetry breaking and tachyonic preheating. *Phys. Rev. Lett.* **87**, 011601. hep-ph/0012142.
Feng, J L, March-Russell, J, Sethi, S, *et al.* (2001) Saltatory relaxation of the cosmological constant. *Nucl. Phys.* **B602**, 307. arXiv:hep-th/0005276.
Garcia-Bellido, J and Linde, A D (1995) Stationarity of inflation and predictions of quantum cosmology. *Phys. Rev.* **D51**, 429. arXiv:hep-th/9408023.
Garcia-Bellido, J, Linde, A D, and Linde, D A (1994) Fluctuations of the gravitational constant in the inflationary Brans–Dicke cosmology. *Phys. Rev.* **D50**, 730. arXiv:astro-ph/9312039.

Garriga, J and Vilenkin, A (2000) On likely values of the cosmological constant. *Phys. Rev.* **D61**, 083502. arXiv:astro-ph/9908115.
(2001a) A prescription for probabilities in eternal inflation. *Phys. Rev.* **D64**, 023507. arXiv:gr-qc/0102090.
(2001b) Solutions to the cosmological constant problems. *Phys. Rev.* **D64**, 023517. arXiv:hep-th/0011262.
(2002) Testable anthropic predictions for dark energy. arXiv:astro-ph/0210358.
Giddings, S B and Strominger, A (1988) Loss of incoherence and determination of coupling constants in quantum gravity. *Nucl. Phys.* **B307**, 854.
(1989) Baby universes, third quantization and the cosmological constant. *Nucl. Phys.* **B321**, 481.
Guth, A H (1981) The inflationary universe: a possible solution to the horizon and flatness problems. *Phys. Rev.* **D23**, 347.
Guth, A H and Pi, S Y (1982) Fluctuations in the new inflationary universe. *Phys. Rev. Lett.* **49**, 1110.
Hartle, J B and Hawking, S W (1983) Wave function of the universe. *Phys. Rev.* **D28**, 2960.
Hawking, S W (1982) The development of irregularities in a single bubble inflationary universe. *Phys. Lett.* **B115**, 295.
(1984) The cosmological constant is probably zero. *Phys. Lett.* **B134**, 403.
Kallosh, R and Linde, A D (2002) M-theory, cosmological constant and anthropic principle. arXiv:hep-th/0208157.
Kallosh, R, Linde, A D, Linde, D A, *et al.* (1995) Gravity and global symmetries. *Phys. Rev.* **D52**, 912. arXiv:hep-th/9502069.
Kofman, L, Linde, A D and Starobinsky, A A (1994) Reheating after inflation. *Phys. Rev. Lett.* **73**, 3195. arXiv:hep-th/9405187.
(1997) Towards the theory of reheating after inflation. *Phys. Rev.* **D56**, 3258. arXiv:hep-ph/9704452.
Kolb, E W and Turner, M S (1990) *The Early Universe*. New York: Addison-Wesley.
Linde, A D (1982a) A new inflationary universe scenario: a possible solution of the horizon, flatness, homogeneity, isotropy and primordial monopole problems. *Phys. Lett.* **B108**, 389.
(1982b) Nonsingular regenerating inflationary universe. Print-82-0554 (Cambridge University).
(1982c) Scalar field fluctuations in expanding universe and the new inflationary universe scenario. *Phys. Lett.* **B116**, 335.
(1983a) The new inflationary universe scenario. In *The Very Early Universe*, ed. G.W. Gibbons, S. W. Hawking, and S. Siklos, pp. 205–49. Cambridge: Cambridge University Press.
(1983b) Chaotic inflation. *Phys. Lett.* **B129**, 177.
(1983c) Inflation can break symmetry in susy. *Phys. Lett.* **B131**, 330.
(1984a) Quantum creation of the inflationary universe. *Lett. Nuovo Cim.* **39**, 401.
(1984b) The inflationary universe. *Rep. Prog. Phys.* **47**, 925.
(1985) The new mechanism of baryogenesis and the inflationary universe. *Phys. Lett.* **B160**, 243.
(1986a) Eternally existing self-reproducing chaotic inflationary universe. *Phys. Lett.* **B175**, 395.
(1986b) Inflation and quantum cosmology. In *Three Hundred Years of Gravitation*, ed. S. W. Hawking and W. Israel, pp. 604–630. Cambridge: Cambridge University Press.
(1987a) Particle physics and inflationary cosmology. *Phys. Today* **40**, 61.

(1987b) Inflation and axion cosmology. *Phys. Lett.* **B201**, 437.
(1988) The universe multiplication and the cosmological constant problem. *Phys. Lett.* **B200**, 272.
(1989) Life after inflation and the cosmological constant problem. *Phys. Lett.* **B227**, 352.
(1990a) *Particle Physics and Inflationary Cosmology*. Chur, Switzerland: Harwood Academic.
(1990b) Extended chaotic inflation and spatial variations of the gravitational constant. *Phys. Lett.* **B238**, 160.
(1991) Inflation and quantum cosmology. *Phys. Scripta* **T36**, 30.
(1998) Quantum creation of an open inflationary universe. *Phys. Rev.* **D58**, 083514. arXiv:gr-qc/9802038.
Linde, A D and Mezhlumian, A (1996) On regularization scheme dependence of predictions in inflationary cosmology. *Phys. Rev.* **D53**, 4267. arXiv:gr-qc/9511058.
Linde, A D and Zelnikov, M I (1988) Inflationary universe with fluctuating dimension. *Phys. Lett.* **B215**, 59.
Linde, A D, Linde, D A and Mezhlumian, A (1994) From the Big Bang theory to the theory of a stationary universe. *Phys. Rev.* **D49**, 1783. arXiv:gr-qc/9306035.
(1996) Nonperturbative amplifications of inhomogeneities in a self-reproducing universe. *Phys. Rev.* **D54**, 2504. arXiv:gr-qc/9601005.
Martel, H, Shapiro, P R, and Weinberg, S (1998) Likely values of the cosmological constant. *Astrophys. J.* **492**, 29. arXiv:astro-ph/9701099.
Mukhanov, V F (1985) Gravitational instability of the universe filled with a scalar field. *J. Exp. Theor. Phys. Lett.* **41**, 493.
Mukhanov, V F and Chibisov, G V (1981) Quantum fluctuation and "Nonsingular" universe. *JETP Lett.* **33**, 532.
Page, D N (2002) Mindless sensationalism: a quantum framework for consciousness. In *Consciousness: New Philosophical Essays*, ed. Q. Smith and A. Jokic, pp. Oxford: Oxford University Press. arXiv:quant-ph/0108039.
Randall, L and Sundrum, R (1999) A large mass hierarchy from a small extra dimension. *Phys. Rev. Lett.* **83**, 3370. arXiv:hep-ph/9905221.
Rees, M (2000) *Just Six Numbers: The Deep Forces that Shape the Universe*. New York: Basic Books.
Rozental, I L (1988) *Big Bang, Big Bounce*. New York: Springer-Verlag.
Starobinsky, A A (1980) A new type of isotropic cosmological models without singularity. *Phys. Lett.* **B91**, 99.
(1982) Dynamics of phase transition in the new inflationary universe scenario and generation of perturbations. *Phys. Lett.* **B117**, 175.
(1986) Stochastic de Sitter (inflationary) stage in the early universe. In *Current Topics in Field Theory, Quantum Gravity and Strings*, ed. H. J. de Vega and N. Sanchez, p. 107. Heidelberg: Springer-Verlag.
Steinhardt, P (1983) Natural inflation. In *The Very Early Universe*, ed. G. W. Gibbons, S. W. Hawking, and S. Siklos, p. 251. Cambridge: Cambridge University Press.
Tegmark, M (1998) Is *the theory of everything* merely the ultimate ensemble theory? *Annals Phys.* **270**, 1. arXiv:gr-qc/9704009.
Turok, N (2002) A Critical review of inflation. *Class. Quant. Grav.* **19**, 3449.
Vanchurin, V, Vilenkin, A, and Winitzki, S (2000) Predictability crisis in inflationary cosmology and its resolution. *Phys. Rev.* **D61**, 083507. arXiv:gr-qc/9905097.
Vilenkin, A (1983) The birth of inflationary universes. *Phys. Rev.* **D27**, 2848.
(1984) Quantum creation of universes. *Phys. Rev.* **D30**, 509.

(1995) Predictions from quantum cosmology. *Phys. Rev. Lett.* **74**, 846. arXiv:gr-qc/9406010.
Vilenkin, A and Ford, L H (1982) Gravitational effects upon cosmological phase transitions. *Phys. Rev.* **D26**, 1231.
Weinberg, S (1987) Anthropic bound on the cosmological constant. *Phys. Rev. Lett.* **59**, 2607.
Wheeler, J A (1990) Information, physics, quantum: the search for links. In *Complexity, Entropy and the Physics of Information*, ed. W. H. Zurek, pp. 3–28. New York: Addison-Wesley.

21

Parallel universes

Max Tegmark
University of Pennsylvania

Is there another copy of you reading this article, deciding to put it aside without finishing this sentence while you are reading on? A person living on a planet called Earth, with misty mountains, fertile fields, and sprawling cities, in a solar system with eight other planets. The life of this person has been identical to yours in every respect – until now, that is, when your decision to read on signals that your two lives are diverging.

You probably find this idea strange and implausible, and I must confess that this is my gut reaction too. Yet it looks like we will just have to live with it, since the simplest and most popular cosmological model today predicts that this person actually exists in a galaxy about $10^{10^{29}}$ meters from here. This does not even assume speculative modern physics, merely that space is infinite and rather uniformly filled with matter as indicated by recent astronomical observations. Your *alter ego* is simply a prediction of the so-called concordance model of cosmology, which agrees with all current observational evidence and is used as the basis for most calculations and simulations presented at cosmology conferences. In contrast, alternatives such as a fractal universe, a closed universe, and a multiply connected universe have been seriously challenged by observations.

The farthest you can observe is the distance that light has been able to travel during the 14 billion years since the Big Bang expansion began. The most distant visible objects are now about 4×10^{26} meters away,[1] and a sphere of this radius defines our observable universe, also called our *Hubble volume*, our *horizon volume*, or simply our universe. Likewise, the universe of your above-mentioned twin is a sphere of the same size centered over there, none of which we can see or have any

[1] After emitting the light that is now reaching us, the most distant things we can see have receded because of the cosmic expansion, and are now about 40 billion light years away.

Science and Ultimate Reality, eds. J. D. Barrow, P. C. W. Davies and C. L. Harper, Jr. Published by Cambridge University Press.

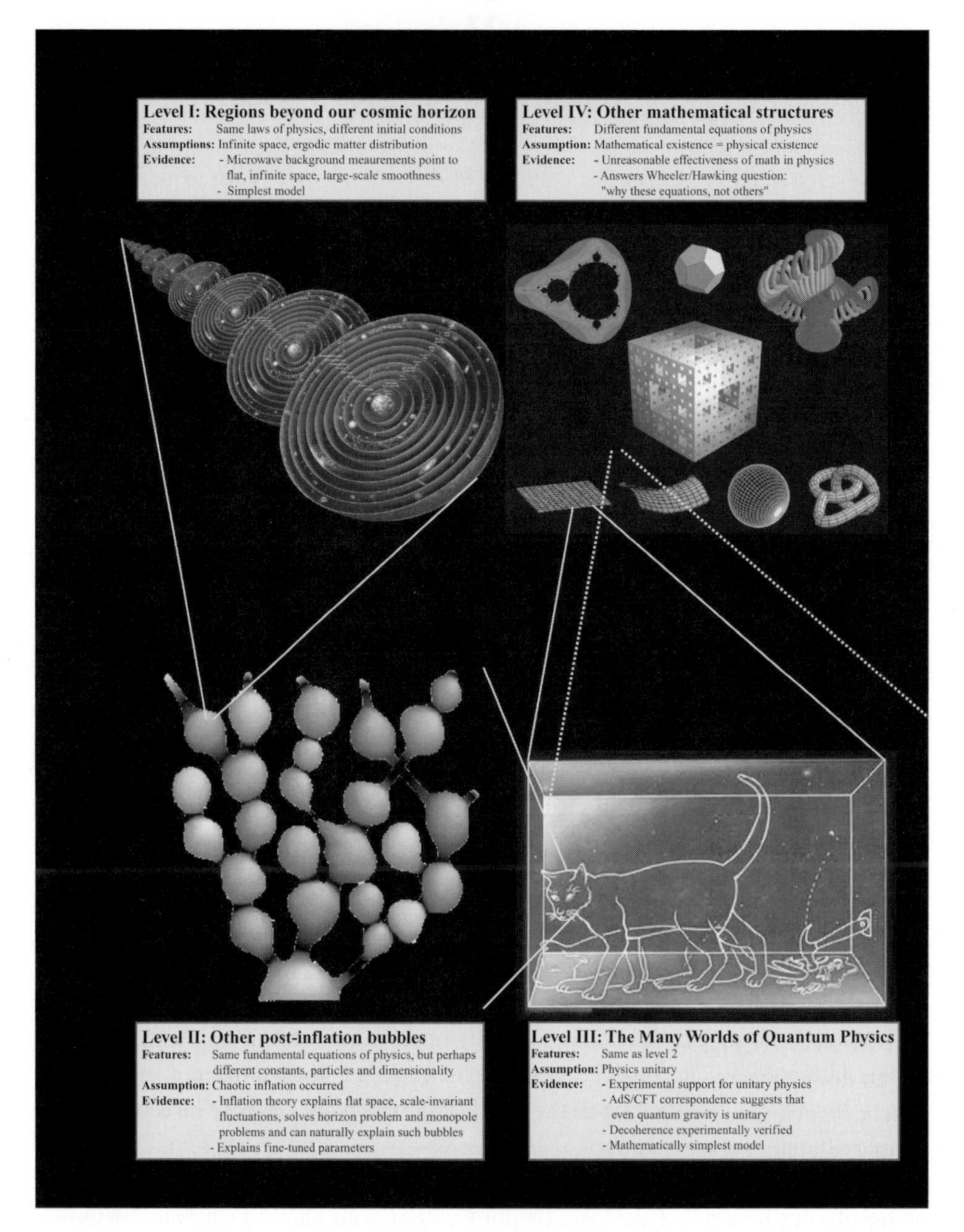

Figure 21.1. Four distinct types of parallel universes.

causal contact with yet. This is the simplest (but far from the only) example of parallel universes.

By this very definition of "universe," one might expect the notion that our observable universe is merely a small part of a larger "multiverse" to be forever in the domain of metaphysics. Yet the epistemological borderline between physics and metaphysics is defined by whether a theory is experimentally testable, not by whether it is weird or involves unobservable entities. Technology-powered experimental breakthroughs have therefore expanded the frontiers of physics to incorporate ever more abstract (and at the time counterintuitive) concepts such as a round rotating Earth, an electromagnetic field, time slowdown at high speeds, quantum superpositions, curved space, and black holes. As reviewed in this chapter, it is becoming increasingly clear that multiverse models grounded in modern physics can in fact be empirically testable, predictive, and falsifiable. Indeed, as many as four distinct types of parallel universes (Fig. 21.1) have been discussed in the recent scientific literature, so that the key question is not whether there is a multiverse (since Level I is rather uncontroversial), but rather how many levels it has.

Level I: regions beyond our cosmic horizon

Let us return to your distant twin. If space is infinite and the distribution of matter is sufficiently uniform on large scales, then even the most unlikely events must take place somewhere. In particular, there are infinitely many other inhabited planets, including not just one but infinitely many with people with the same appearance, name, and memories as you. Indeed, there are infinitely many other regions the size of our observable universe, where every possible cosmic history is played out. This is the Level I multiverse.

Evidence for Level I parallel universes

Although the implications may seem crazy and counterintuitive, this spatially infinite cosmological model is in fact the simplest and most popular one on the market today. It is part of the cosmological concordance model, which agrees with all current observational evidence and is used as the basis for most calculations and simulations presented at cosmology conferences. In contrast, alternatives such as a fractal universe, a closed universe, and a multiply connected universe have been seriously challenged by observations. Yet the Level I multiverse idea has been controversial (indeed, an assertion along these lines was one of the heresies for which the Vatican had Giordano Bruno burned at the stake in 1600[2]), so let us review the status of the two assumptions (infinite space and "sufficiently uniform" distribution).

How large is space? Observationally, the lower bound has grown dramatically (Fig. 21.2) with no indication of an upper bound. We all accept the existence of

[2] Bruno's ideas have since been elaborated by, e.g., Brundrit (1979), Garriga and Vilenkin (2001b), and Davies (1996), all of whom have thus far avoided the stake.

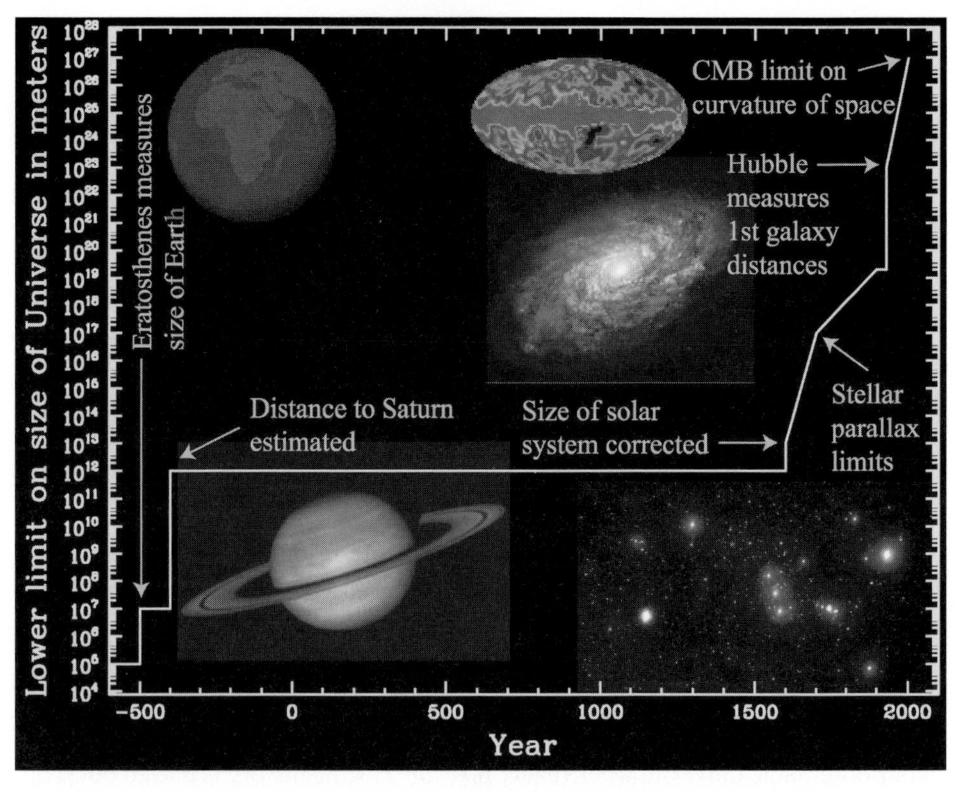

Figure 21.2. Although an infinite universe has always been a possibility, the lower limit on the size of our universe has kept growing.

things that we cannot see but could see if we moved or waited, like ships beyond the horizon. Objects beyond cosmic horizon have similar status, since the observable universe grows by a light year every year as light from further away has time to reach us.[3] Since we are all taught about simple Euclidean space in school, it can therefore be difficult to imagine how space could *not* be infinite – for what would lie beyond the sign saying "*SPACE ENDS HERE – MIND THE GAP*"? Yet Einstein's theory of gravity allows space to be finite by being differently connected than Euclidean space, say with the topology of a four-dimensional sphere or a doughnut so that traveling far in one direction could bring you back from the opposite direction. The cosmic microwave background allows sensitive tests of such finite models, but has so far produced no support for them – flat infinite models fit the data fine and strong limits have been placed on both spatial curvature and multiply connected topologies. In addition, a spatially infinite universe is a

[3] If the cosmic expansion continues to accelerate (currently an open question), the observable universe will eventually stop growing.

generic prediction of the cosmological theory of inflation (Garriga and Vilenkin 2001b). The striking successes of inflation listed below therefore lend further support to the idea that space is after all simple and infinite just as we learned in school.

How uniform is the matter distribution on large scales? In an "island universe" model where space is infinite but all the matter is confined to a finite region, almost all members of the Level I multiverse would be dead, consisting of nothing but empty space. Such models have been popular historically, originally with the island being Earth and the celestial objects visible to the naked eye, and in the early twentieth century with the island being the known part of the Milky Way galaxy. Another nonuniform alternative is a fractal universe, where the matter distribution is self-similar and all coherent structures in the cosmic galaxy distribution are merely a small part of even larger coherent structures. The island and fractal universe models have both been demolished by recent observations as reviewed in Tegmark (2002). Maps of the three-dimensional galaxy distribution have shown that the spectacular large-scale structure observed (galaxy groups, clusters, superclusters, etc.) gives way to dull uniformity on large scales, with no coherent structures larger than about 10^{24} m. More quantitatively, imagine placing a sphere of radius R at various random locations, measuring how much mass M is enclosed each time, and computing the variation between the measurements as quantified by their standard deviation ΔM. The relative fluctuations $\Delta M/M$ have been measured to be of order unity on the scale $R \sim 3 \times 10^{23}$ m, and dropping on larger scales. The Sloan Digital Sky Survey has found $\Delta M/M$ as small as 1% on the scale $R \sim 10^{25}$ m and cosmic microwave background measurements have established that the trend towards uniformity continues all the way out to the edge of our observable universe ($R \sim 10^{27}$ m), where $\Delta M/M \sim 10^{-5}$. Barring conspiracy theories where the universe is designed to fool us, the observations thus speak loud and clear: space as we know it continues far beyond the edge of our observable universe, teeming with galaxies, stars, and planets.

What are Level I parallel universes like?

The physics description of the world is traditionally split into two parts: initial conditions and laws of physics specifying how the initial conditions evolve. Observers living in parallel universes at Level I observe the exact same laws of physics as we do, but with different initial conditions than those in our Hubble volume. The currently favored theory is that the initial conditions (the densities and motions of different types of matter early on) were created by quantum fluctuations during the inflation epoch (see below). This quantum mechanism generates initial conditions that are for all practical purposes random, producing density fluctuations described by what mathematicians call an ergodic random field.[4] *Ergodic* means that if you imagine generating an ensemble of universes, each with its own random initial conditions, then the probability distribution of outcomes in a given volume is identical to the distribution that you get by sampling different volumes in a single universe. In other words, it means that everything that could in principle have happened here did in fact happen somewhere else.

Inflation in fact generates all possible initial conditions with non-zero probability, the most likely ones being almost uniform with fluctuations at the 10^{-5} level that are amplified by gravitational clustering to form galaxies, stars, planets, and other structures. This means both that pretty much all imaginable matter configurations occur in some Hubble volume far away, and also that we should expect our own Hubble volume to be a fairly typical one – at least typical among those that contain observers. A crude estimate suggests that the closest identical copy of you is about $\sim 10^{10^{29}}$ m away. About $\sim 10^{10^{91}}$ m away, there should be a sphere of radius 100 light years identical to the one centered here, so all perceptions that we have during the next century will be identical to those of our counterparts over there. About $\sim 10^{10^{115}}$ m away, there should be an entire Hubble volume identical to ours.[5]

This raises an interesting philosophical point that will come back and haunt us later: if there are indeed many copies of "you" with identical past lives and memories, you would not be able to compute your own future even if you had complete knowledge of the entire state of the cosmos! The reason is that there is no way for you to determine which of these copies is "you" (they all feel that they are). Yet their lives will typically begin to differ eventually, so the best you can do is predict probabilities for what you will experience from now on. This kills the traditional notion of determinism.

How a multiverse theory can be tested and falsified

Is a multiverse theory one of metaphysics rather than physics? As emphasized by Karl Popper, the distinction between the two is whether the theory is empirically

[4] Strictly speaking, the random field is ergodic if (1) space is infinite, (2) the mass fluctuations $\Delta M/M$ approach zero on large scales (as measurements suggest), and (3) the densities at any set of points have a multivariate Gaussian probability distribution (as predicted by the most popular inflation models, which can be traced back to the fact that the harmonic oscillator equation governing the inflaton field fluctuations gives a Gaussian wave function for the ground state). For the technical reader, conditions 2 and 3 can be replaced by the weaker requirement that correlation functions of all orders vanish in the limit of infinite spatial separation.

[5] This is an extremely conservative estimate, simply counting all possible quantum states that a Hubble volume can have that are no hotter than 10^8 K. 10^{115} is roughly the number of protons that the Pauli exclusion principle would allow you to pack into a Hubble volume at this temperature (our own Hubble volume contains only about 10^{80} protons). Each of these 10^{115} slots can be either occupied or unoccupied, giving $N = 2^{10^{115}} \sim 10^{10^{115}}$ possibilities, so the expected distance to the nearest identical Hubble volume is $N^{1/3} \sim 10^{10^{115}}$ Hubble radii $\sim 10^{10^{115}}$ m. Your nearest copy is likely to be much closer than $10^{10^{29}}$ m, since the planet formation and evolutionary processes that have tipped the odds in your favor are at work everywhere. There are probably at least 10^{20} habitable planets in our own Hubble volume alone.

testable and falsifiable. Containing unobservable entities does clearly *not* per se make a theory nontestable. For instance, a theory stating that there are 666 parallel universes, all of which are devoid of oxygen, makes the testable prediction that we should observe no oxygen here, and is therefore ruled out by observation.

As a more serious example, the Level I multiverse framework is routinely used to rule out theories in modern cosmology, although this is rarely spelled out explicitly. For instance, cosmic microwave background (CMB) observations have recently shown that space has almost no curvature. Hot and cold spots in CMB maps have a characteristic size that depends on the curvature of space, and the observed spots appear too large to be consistent with the previously popular "open universe" model. However, the average spot size randomly varies slightly from one Hubble volume to another, so it is important to be statistically rigorous. When cosmologists say that the open universe model is ruled out at 99.9% confidence, they really mean that if the open universe model were true, then fewer than one out of every thousand Hubble volumes would show CMB spots as large as those we observe – therefore the entire model with all its infinitely many Hubble volumes is ruled out, even though we have of course only mapped the CMB in our own particular Hubble volume.

The lesson to learn from this example is that multiverse theories *can* be tested and falsified, but only if they predict what the ensemble of parallel universes is and specify a probability distribution (or more generally what mathematicians call a *measure*) over it. As we will see below, this measure problem can be quite serious and is still unsolved for some multiverse theories.

Level II: other post-inflation bubbles

If you felt that the Level I multiverse was large and hard to stomach, try imagining an infinite set of distinct ones (each symbolized by a bubble in Fig. 21.1), some perhaps with different dimensionality and different physical constants. This is what is predicted by the currently popular chaotic theory of inflation, and we will refer to it as the Level II multiverse. These other domains are more than infinitely far away in the sense that you would never get there even if you traveled at the speed of light forever. The reason is that the space between our Level I multiverse and its neighbors is still undergoing inflation, which keeps stretching it out and creating more volume faster than you can travel through it. In contrast, you could travel to an arbitrarily distant Level I universe if you were patient and the cosmic expansion decelerates.[6]

[6] Astronomical evidence suggests that the cosmic expansion is currently accelerating. If this acceleration continues, then even the Level I parallel universes will remain forever separate, with the intervening space stretching faster than light can travel through it. The jury is still out, however, with popular models predicting that the universe will eventually stop accelerating and perhaps even recollapse.

Evidence for Level II parallel universes

By the 1970s, the Big Bang model had proved a highly successful explanation of most of the history of our universe. It had explained how a primordial fireball expanded and cooled, synthesized helium and other light elements during the first few minutes, became transparent after 400 000 years releasing the cosmic microwave background radiation, and gradually got clumpier due to gravitational clustering, producing galaxies, stars, and planets. Yet disturbing questions remained about what happened in the very beginning. Did something appear from nothing? Where are all the superheavy particles known as magnetic monopoles that particle physics predicts should be created early on (the "monopole problem")? Why is space so big, so old, and so flat, when generic initial conditions predict curvature to grow over time and the density to approach either zero or infinity after of order 10^{-42} seconds (the "flatness problem")? What conspiracy caused the CMB temperature to be nearly identical in regions of space that have never been in causal contact (the "horizon problem")? What mechanism generated the 10^{-5} level seed fluctuations out of which all structure grew?

A process known as *inflation* can solve all these problems in one fell swoop (see reviews by Guth and Steinhardt (1984) and Linde (1994)), and has therefore emerged as the most popular theory of what happened very early on. Inflation is a rapid stretching of space, diluting away monopoles and other debris, making space flat and uniform like the surface of an expanding balloon, and stretching quantum vacuum fluctuations into macroscopically large density fluctuations that can seed galaxy formation. Since its inception, inflation has passed additional tests: CMB observations have found space to be extremely flat and have measured the seed fluctuations to have an approximately scale-invariant spectrum without a substantial gravity wave component, all in perfect agreement with inflationary predictions.

Inflation is a general phenomenon that occurs in a wide class of theories of elementary particles. In the popular model known as *chaotic inflation*, inflation ends in some regions of space allowing life as we know it, whereas quantum fluctuations cause other regions of space to inflate even faster. In essence, one inflating bubble sprouts other inflationary bubbles, which in turn produce others in a never-ending chain reaction (Fig. 21.1, lower left, with time increasing upwards). The bubbles where inflation has ended are the elements of the Level II multiverse. Each such bubble is infinite in size,[7] yet there are infinitely many bubbles since the chain reaction never ends. Indeed, if this exponential growth of the number of bubbles has been going on forever, there will be an uncountable infinity of such parallel universes (the same infinity as that assigned to the set of real numbers, say, which

[7] Surprisingly, it has been shown that inflation can produce an infinite Level I multiverse even in a bubble of finite spatial volume, thanks to an effect whereby the spatial directions of spacetime curve towards the (infinite) time direction (Bucher and Spergel 1999).

is larger than that of the (countably infinite) set of integers). In this case, there is also no beginning of time and no absolute Big Bang: there is, was, and always will be an infinite number of inflating bubbles and post-inflationary regions like the one we inhabit, forming a fractal pattern.

What are Level II parallel universes like?

The prevailing view is that the physics we observe today is merely a low-energy limit of a much more symmetric theory that manifests itself at extremely high temperatures. This underlying fundamental theory may be 11-dimensional, supersymmetric, and involving a grand unification of the four fundamental forces of nature. A common feature in such theories is that the potential energy of the field(s) driving inflation has several different minima (sometimes called "vacuum states"), corresponding to different ways of breaking this symmetry and, as a result, to different low-energy physics. For instance, all but three spatial dimensions could be curled up ("compactified"), resulting in an effectively three-dimensional space like ours, or fewer could curl up leaving a seven-dimensional space. The quantum fluctuations driving chaotic inflation could cause different symmetry breaking in different bubbles, resulting in different members of the Level II multiverse having different dimensionality. Many symmetries observed in particle physics also result from the specific way in which symmetry is broken, so there could be Level II parallel universes where there are, say, two rather than three generations of quarks.

In addition to such discrete properties as dimensionality and fundamental particles, our universe is characterized by a set of dimensionless numbers known as *physical constants*. Examples include the electron/proton mass ratio $m_{\mathrm{p}}/m_{\mathrm{e}} \approx 1836$ and the cosmological constant, which appears to be about 10^{-123} in so-called Planck units. There are models where also such continuous parameters can vary from one post-inflationary bubble to another.[8]

The Level II multiverse is therefore likely to be more diverse than the Level I multiverse, containing domains where not only the initial conditions differ, but perhaps the dimensionality, the elementary particles and the physical constants differ as well.

Before moving on, let us briefly comment on a few closely related multiverse notions. First of all, if one Level II multiverse can exist, eternally self-reproducing

[8] Although the fundamental equations of physics are the same throughout the Level II multiverse, the approximate effective equations governing the low-energy world that we observe will differ. For instance, moving from a three-dimensional to a four-dimensional (noncompactified) space changes the observed gravitational force equation from an inverse square law to an inverse cube law. Likewise, breaking the underlying symmetries of particle physics differently will change the line-up of elementary particles and the effective equations that describe them. However, we will reserve the terms "different equations" and "different laws of physics" for the Level IV multiverse, where it is the fundamental rather than effective equations that change.

in a fractal pattern, then there may well be infinitely many other Level II multiverses that are completely disconnected. However, this variant appears to be untestable, since it would neither add any qualitatively different worlds nor alter the probability distribution for their properties. All possible initial conditions and symmetry breakings are already realized within each one.

An idea proposed by Tolman and Wheeler and recently elaborated by Steinhardt and Turok (2002) is that the (Level I) multiverse is cyclic, going through an infinite series of Big Bangs. If it exists, the ensemble of such incarnations would also form a multiverse, arguably with a diversity similar to that of Level II.

An idea proposed by Smolin (1997) involves an ensemble similar in diversity to that of Level II, but mutating and sprouting new universes through black holes rather than during inflation. This predicts a form of a natural selection favoring universes with maximal black-hole production.

In brane-world scenarios, another three-dimensional world could be quite literally parallel to ours, merely offset in a higher dimension. However, it is unclear whether such a world ("brane") deserves to be called a parallel universe separate from our own, since we may be able to interact with it gravitationally much as we do with dark matter.

Fine-tuning and selection effects

Physicists dislike unexplained coincidences. Indeed, they interpret them as evidence that models are ruled out (Fig. 21.3). We saw earlier how the open universe model was ruled out at 99.9% confidence because it implies that the observed pattern of CMB fluctuations is extremely unlikely, a one-in-a thousand coincidence occurring in only 0.1% of all Hubble volumes.

Suppose you check into a hotel, are assigned room 1967 and, surprised, note that this is the year you were born. After a moment of reflection, you conclude that this is not all that surprising after all, given that the hotel has many rooms and that you would not be having these thoughts in the first place if you'd been assigned another one. You then realize that even if you knew nothing about hotels, you could have inferred the existence of other hotel rooms, because if there were only one room number in the entire universe, you would be left with an unexplained coincidence.

As a more pertinent example, consider M, the mass of the Sun. M affects the luminosity of the Sun, and using basic physics, one can compute that life as we know it on Earth is only possible if M is in the narrow range 1.6×10^{30} kg – 2.4×10^{30} kg – otherwise Earth's climate would be colder than on Mars or hotter than on Venus. The measured value is $M \sim 2.0 \times 10^{30}$ kg. This apparent coincidence of the habitable and observed M-values may appear disturbing given that calculations

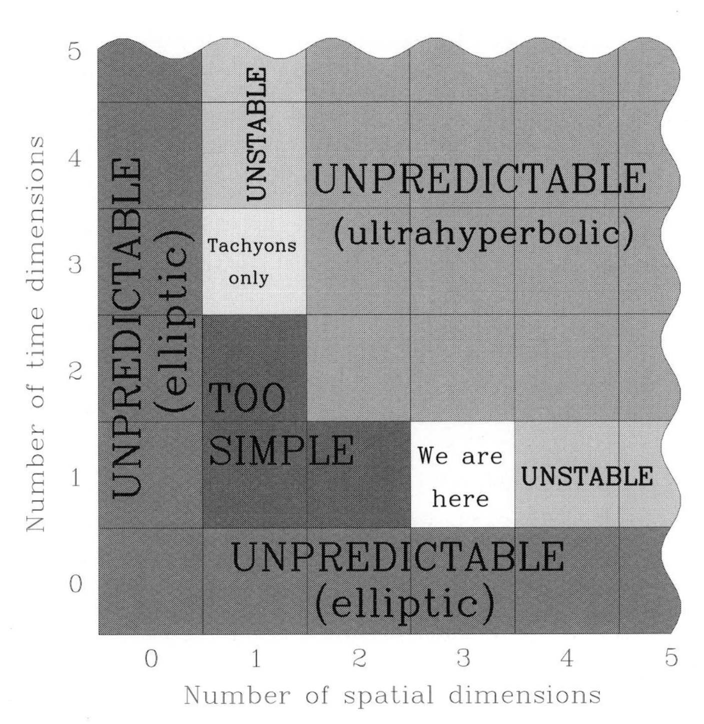

Figure 21.3. Why we should not be surprised to find ourselves living in 3+1-dimensional spacetime. When the partial differential equations of nature are elliptic or ultrahyperbolic, physics has no predictive power for an observer. In the remaining (hyperbolic) cases, $n > 3$ admits no stable atoms and $n < 3$ may lack sufficient complexity for observers (no gravitational attraction, topological problems). (From Tegmark (1997).)

show that stars in the much broader mass range $M \sim 10^{29}$ kg – 10^{32} kg can exist. However, just as in the hotel example, we can explain this apparent coincidence if there is an ensemble and a selection effect: if there are in fact many solar systems with a range of sizes of the central star and the planetary orbits, then we obviously expect to find ourselves living in one of the inhabitable ones.

More generally, the apparent coincidence of the habitable and observed values of some physical parameter can be taken as evidence for the existence of a larger ensemble, of which what we observe is merely one member among many (Carter 1974). Although the existence of other hotel rooms and solar systems is uncontroversial and observationally confirmed, that of parallel universes is not, since they cannot be observed. Yet if fine-tuning is observed, one can argue for their existence using the exact same logic as above (Fig. 21.4). Indeed, there are numerous examples of fine-tuning suggesting parallel universes with other physical constants (see, e.g., Fig. 21.3 and Fig. 21.4), although the degree of fine-tuning is still under active debate and should be clarified by additional calculations – see Rees (2002)

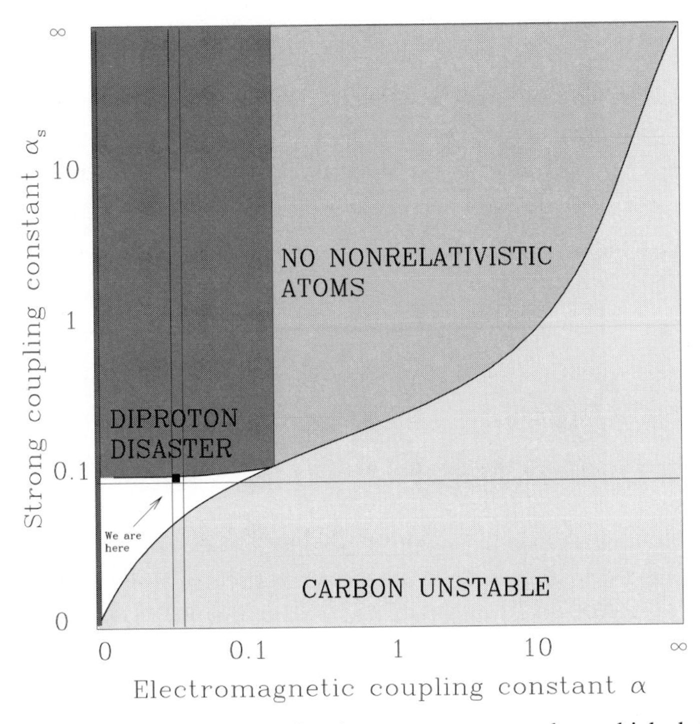

Figure 21.4. Hints of fine-tuning for the parameters α and α_s which determine the strengths of the electromagnetic force and the strong nuclear force, respectively. The observed values $(\alpha, \alpha_s) \approx (1/137, 0.1)$ are indicated with a filled square. Grand unified theories rule out everything except the narrow strip between the two vertical lines, and deuterium becomes unstable below the horizontal line. In the narrow shaded region to the very left, electromagnetism is weaker than gravity and therefore irrelevant. (From Tegmark (1997).)

and Davies (1982) for popular accounts and Barrow and Tipler (1986) for technical details.

For instance, if the electromagnetic force were weakened by a mere 4%, then the Sun would immediately explode (the diproton would have a bound state, which would increase the solar luminosity by a factor 10^{18}). If it were stronger, there would be fewer stable atoms. Indeed, most if not all the parameters affecting low-energy physics appear fine-tuned at some level, in the sense that changing them by modest amounts results in a qualitatively different universe.

If the weak interaction were substantially weaker, there would be no hydrogen around, since it would have been converted to helium shortly after the Big Bang. If it were either much stronger or much weaker, the neutrinos from a supernova explosion would fail to blow away the outer parts of the star, and it is doubtful whether life-supporting heavy elements would ever be able to leave the stars where they were produced. If the protons were 0.2% heavier, they would decay into neutrons unable to hold onto electrons, so there would be no stable atoms around. If the proton-to-electron mass ratio were much smaller, there could be no stable stars, and if it were much larger, there could be no ordered structures like crystals and DNA molecules.

Fine-tuning discussions often turn heated when somebody mentions the "A-word," *anthropic*. The author feels that discussions of the so-called anthropic principle have generated more heat than light, with many different definitions and interpretations of what it means. The author is not aware of anybody disagreeing with what might be termed MAP, the minimalistic anthropic principle:

MAP: When testing fundamental theories with observational data, ignoring selection effects can give incorrect conclusions.

This is obvious from our examples above: if we neglected selection effects, we would be surprised to orbit a star as heavy as the Sun, since lighter and dimmer ones are much more abundant. Likewise, MAP says that the chaotic inflation model is *not* ruled out by the fact that we find ourselves living in the minuscule fraction of space where inflation has ended, since the inflating part is uninhabitable to us. Fortunately, selection effects cannot rescue all models, as pointed out a century ago by Boltzmann. If the universe were in classical thermal equilibrium (heat death), thermal fluctuations could still make atoms assemble at random to briefly create a self-aware observer like you once in a blue moon, so the fact that you exist right now does not rule out the heat death cosmological model. However, you should statistically expect to find the rest of the world in a high-entropy mess rather than in the ordered low-entropy state you observe, which rules out this model.

The standard model of particle physics has 28 free parameters, and cosmology may introduce additional independent ones. If we really do live in a Level II multiverse, then for those parameters that vary between the parallel universes, we will never be able to predict our measured values from first principles. We can merely compute probability distributions for what we should expect to find, taking selection effects into account. We should expect to find everything that can vary across the ensemble to be as generic as is consistent with our existence. As detailed below, this issue of what is "generic" and, more specifically, how to compute probabilities in physics, is emerging as an embarrassingly thorny problem.

Level III: the many worlds of quantum physics

There may be a third type of parallel worlds that are not far away but in a sense right here. If the fundamental equations of physics are what mathematicians call *unitary*, as they so far appear to be, then the universe keeps branching into parallel

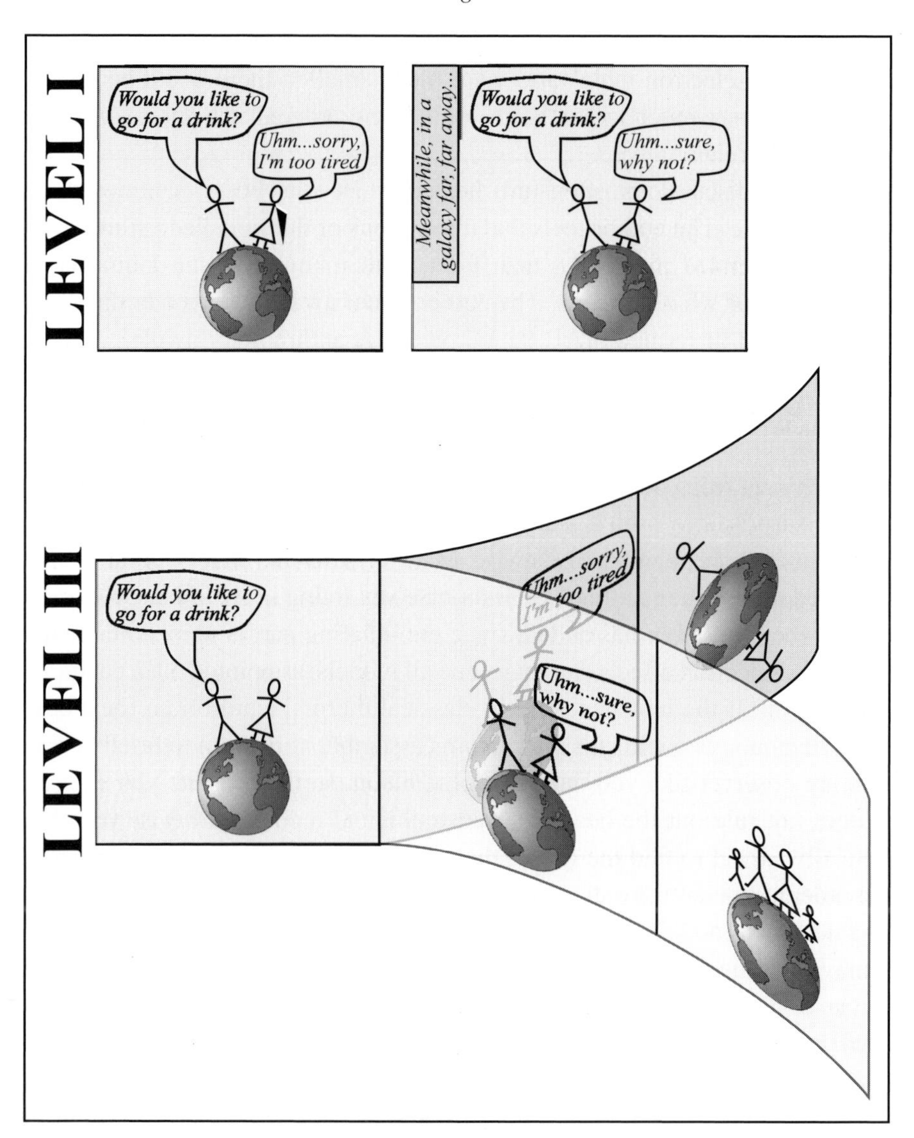

Figure 21.5. Difference between Level I and Level III. Whereas Level I parallel universes are far away in space, those of Level III are even right here, with quantum events causing classical reality to split and diverge into parallel storylines. Yet Level III adds no new storylines beyond Levels I or II.

universes as in the cartoon (Fig. 21.5, bottom): whenever a quantum event appears to have a random outcome, all outcomes in fact occur, one in each branch. This is the Level III multiverse. Although more debated and controversial than Level I and Level II, we will see that, surprisingly, this level adds no new types of universes.

Evidence for Level III parallel universes

In the early twentieth century, the theory of quantum mechanics revolutionized physics by explaining the atomic realm, with applications ranging from chemistry to nuclear reactions, lasers, and semiconductors. Despite the obvious successes in its application, a heated debate ensued about its interpretation – a debate that still rages on. In quantum theory, the state of the universe is not given in classical terms such as the positions and velocities of all particles, but by a mathematical object called a wave function. According to the so-called Schrödinger equation, this state evolves deterministically over time in a fashion termed *unitary*, corresponding to a rotation in Hilbert space, the abstract infinite-dimensional space where the wave function lives. The sticky part is that there are perfectly legitimate wave functions corresponding to classically counterintuitive situations such as you being in two different places at once. Worse, the Schrödinger equation can evolve innocent classical states into such schizophrenic ones. As a baroque example, Schrödinger described the famous thought-experiment where a nasty contraption kills a cat if a radioactive atom decays. Since the radioactive atom eventually enters a superposition of decayed and not decayed, it produces a cat which is both dead and alive in superposition.

In the 1920s, this weirdness was explained away by postulating that the wave function "collapsed" into some definite classical outcome whenever an observation was made, with probabilities given by the wave function. Einstein was unhappy about such intrinsic randomness in nature, which violated unitarity, insisting that "God doesn't play dice," and others complained that there was no equation specifying when this collapse occurred. In his 1957 Ph.D. thesis, Princeton student Hugh Everett III showed that this controversial collapse postulate was unnecessary. Quantum theory predicted that one classical reality would gradually split into superpositions of many (Fig. 21.5). He showed that observers would subjectively experience this splitting merely as a slight randomness, and indeed with probabilities in exact agreement with those from the old collapse postulate (see DeWitt's chapter in this volume). This superposition of classical worlds is the Level III multiverse.

Everett's work had left two crucial questions unanswered: first of all, if the world actually contains bizarre macrosuperpositions, then why don't we perceive them? The answer came in 1970, when Dieter Zeh showed that the Schrödinger equation itself gives rise to a type of censorship effect (Zeh 1970). This effect became known as *decoherence*, and was worked out in great detail by Wojciech Zurek, Zeh, and others over the following decades. Coherent quantum superpositions were found to persist only as long as they were kept secret from the rest of the world. A single collision with a snooping photon or air molecule is sufficient to ensure that our friends in Fig. 21.5 can never be aware of their counterparts in the parallel storyline. A second unanswered question in the Everett picture was more subtle but equally important: what physical mechanism picks out approximately classical states (with each object in only one place, etc.) as special in the bewilderingly large Hilbert space? Decoherence answered this question as well, showing that classical states are simply those that are most robust against decoherence. In summary, decoherence both identifies the Level III parallel universes in Hilbert space and delimits them from one another. Decoherence is now quite uncontroversial and has been experimentally measured in a wide range of circumstances. Since decoherence for all practical purposes mimics wave function collapse, it has eliminated much of the original motivation for nonunitary quantum mechanics and made the Everett's so-called many worlds interpretation increasingly popular. For details about these quantum issues, see Tegmark and Wheeler (2001) for a popular account and Giulini *et al.* (1996) for a technical review.

If the time-evolution of the wave function is unitary, then the Level III multiverse exists, so physicists have worked hard on testing this crucial assumption. So far, no departures from unitarity have been found. In the last few decades, remarkable experiments have confirmed unitarity for ever larger systems, including the hefty carbon-60 Buckyball atom and kilometer-size optical fiber systems. On the theoretical side, a leading argument against unitarity has involved possible destruction of information during the evaporation of black holes, suggesting that quantum-gravitational effects are nonunitary and collapse the wave function. However, a recent string theory breakthrough known as AdS/CFT correspondence has suggested that even quantum gravity is unitary, being mathematically equivalent to a lower-dimensional quantum field theory without gravity (see Maldacena's chapter in this volume).

What are Level III parallel universes like?

When discussing parallel universes, we need to distinguish between two different ways of viewing a physical theory: the outside view or *bird perspective* of a mathematician studying its mathematical fundamental equations and the inside view or *frog perspective* of an observer living in the world described by the equations.[9]

[9] Indeed, the standard mental picture of what the physical world is corresponds to a third intermediate viewpoint that could be termed the *consensus view*. From your subjectively perceived frog perspective, the world turns upside down when you stand on your head and disappears when you close your eyes, yet you subconsciously interpret your sensory inputs as though there is an external reality that is independent of your orientation, your location, and your state of mind. It is striking that although this third view involves both censorship (like rejecting dreams), interpolation (as between eye-blinks), and extrapolation (say attributing existence to unseen cities) of your inside view, independent observers nonetheless appear to share this consensus view. Although the inside view looks black-and-white to a cat, iridescent to a bird seeing four primary colors, and still more different to a bee seeing polarized light, a bat using sonar, a blind person with keener touch and hearing, or the latest

From the bird perspective, the Level III multiverse is simple: there is only one wave function, and it evolves smoothly and deterministically over time without any sort of splitting or parallelism. The abstract quantum world described by this evolving wave function contains within it a vast number of parallel classical storylines (see Fig. 21.5), continuously splitting and merging, as well as a number of quantum phenomena that lack a classical description. From her frog perspective, however, each observer perceives only a tiny fraction of this full reality: she can only see her own Hubble volume (Level I) and decoherence prevents her from perceiving Level III parallel copies of herself. When she is asked a question, makes a snap decision, and answers (Fig. 21.5), quantum effects at the neuron level in her brain lead to multiple outcomes, and from the bird perspective, her single past branches into multiple futures. From their frog perspectives, however, each copy of her is unaware of the other copies, and she perceives this quantum branching as merely a slight randomness. Afterwards, there are for all practical purposes multiple copies of her that have the exact same memories up until the point when she answers the question.

How many different parallel universes are there?

As strange as this may sound, Fig. 21.5 illustrates that this exact same situation occurs even in the Level I multiverse, the only difference being where her copies reside (elsewhere in good old three-dimensional space as opposed to elsewhere in infinite-dimensional Hilbert space, in other quantum branches). In this sense, Level III is no stranger than Level I. Indeed, if physics is unitary, then the quantum fluctuations during inflation did not generate unique initial conditions through a random process, but rather generated a quantum superposition of all possible initial conditions simultaneously, after which decoherence caused these fluctuations to behave essentially classically in separate quantum branches. The ergodic nature of these quantum fluctuations therefore implies that the distribution of outcomes in a given Hubble volume at Level III (between different quantum branches as in Fig. 21.3) is identical to the distribution that you get by sampling different Hubble volumes within a single quantum branch (Level I). If physical constants, spacetime dimensionality, etc. can vary as in Level II, then they too will vary between parallel quantum branches at Level III. The reason for this is that if physics is unitary, then the process of spontaneous symmetry breaking will not produce a unique (albeit random) outcome, but rather a superposition of all outcomes that rapidly decoheres

overpriced robotic vacuum cleaner, all agree on whether the door is open. The key current challenge in physics is deriving this semi-classical consensus view from the fundamental equations specifying the bird perspective. In my opinion, this means that although understanding the detailed nature of human consciousness is an important challenge in its own right, it is *not* necessary for a fundamental theory of physics.

into for all practical purposes separate Level III branches. In short, the Level III multiverse, if it exists, adds nothing new beyond Level I and Level II – just more indistinguishable copies of the same universes, the same old storylines playing out again and again in other quantum branches. Postulating a yet unseen nonunitary effect to get rid of the Level III multiverse, with Occam's razor in mind, therefore would not make Occam any happier.

The passionate debate about Everett's parallel universes that has raged on for decades therefore seems to be ending in a grand anticlimax, with the discovery of a less controversial multiverse that is just as large. This is reminiscent of the famous Shapley–Curtis debate of the 1920s about whether there were really a multitude of galaxies (parallel universes by the standards of the time) or just one, a storm in a teacup now that research has moved on to other galaxy clusters, superclusters, and even Hubble volumes. In hindsight, both the Shapley–Curtis and Everett controversies seem positively quaint, reflecting our instinctive reluctance to expand our horizons.

A common objection is that repeated branching would exponentially increase the number of universes over time. However, the number of universes N may well stay constant. By the number of "universes" N, we mean the number that are indistinguishable from the frog perspective (from the bird perspective, there is of course just one) at a given instant, i.e., the number of macroscopically different Hubble volumes. Although there is obviously a vast number of them (imagine moving planets to random new locations, imagine having married someone else, etc.), the number N is clearly finite – even if we pedantically distinguish Hubble volumes at the quantum level to be overly conservative, there are "only" about $10^{10^{115}}$ with temperature below 10^8 K as detailed above.[10] The smooth unitary evolution of the wave function in the bird perspective corresponds to a never-ending sliding between these N classical universe snapshots from the frog perspective of an observer. Now you're in universe A, the one where you're reading this sentence.

[10] For the technical reader, could the grand superposition of the universal wave functional involve other interesting states besides the semi-classical ones? Specifically, the semi-classical states (corresponding to what we termed the consensus view) are those that are maximally robust towards decoherence (see Zurek's chapter in this volume), so if we project out the component of the wave functional that is spanned by these states, what remains? We can make a hand-waving argument that all that remains is a rather uninteresting high-energy mess which will be devoid of observers and rapidly expand or collapse. Let us consider the special case of the electromagnetic field. In many circumstances (Anglin and Zurek 1996), its semi-classical states can be shown to be generalized coherent states, which have infinite-dimensional Gaussian Wigner functions with characteristic widths no narrower than those corresponding to the local temperature. Such functions form a well-conditioned basis for all states whose wave function is correspondingly smooth, i.e., lacking violent high-energy fluctuations. This is illustrated in Fig. 21.6 for the simple case of a one-dimensional quantum particle: the wave function $\psi(x)$ can be written as a superposition of a low-energy (low-pass filtered) and a high-energy (high-pass filtered) part, and the former can be decomposed as the convolution of a smooth function with a Gaussian, i.e., as a superposition of coherent states with Gaussian wave packets. Decoherence rapidly makes the macroscopically distinct semi-classical states of the electromagnetic field for all practical purposes separate both from each other and from the high-energy mess. The high-energy component may well be typical of the early universe that we evolved from.

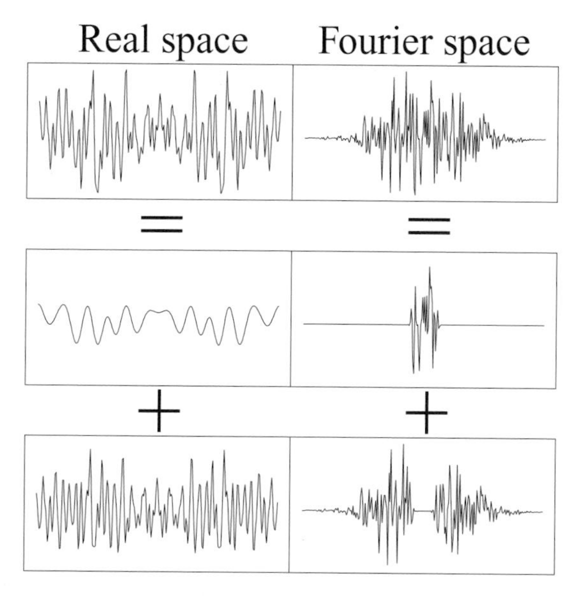

Figure 21.6. Schematic illustration (see footnote 10) of how a wave functional of the Level III multiverse (top row for simple one-dimensional Hilbert space) can be decomposed as a superposition of semi-classical worlds (generalized coherent states; middle row) and a high-energy mess (bottom row).

Now you're in universe B, the one where you're reading this other sentence. Put differently, universe B has an observer identical to one in universe A, except with an extra instant of memories. In Fig. 21.5, our observer first finds herself in the universe described by the left panel, but now there are two different universes smoothly connecting to it like B did to A, and in both of these, she will be unaware of the other one. Imagine drawing a separate dot corresponding to each possible universe and drawing arrows indicating which ones connect to which in the frog perspective. A dot could lead uniquely to one other dot or to several, as above. Likewise, several dots could lead to one and the same dot, since there could be many different ways in which certain situations could have come about. The Level III multiverse thus involves not only splitting branches but merging branches as well (Fig. 21.6).

Ergodicity implies that the quantum state of the Level III multiverse is invariant under spatial translations, which is a unitary operation just as time translation. If it is invariant under time translation as well (this can be arranged by constructing a superposition of an infinite set of quantum states that are all different time translations of one and the same state, so that a Big Bang happens at different times in different quantum branches), then the number of universes would automatically stay exactly constant. All possible universe snapshots would exist at every instant, and the passage of time would just be in the eye of the beholder – an idea explored in the science fiction novel *Permutation City* (Egan 1995) and developed by Deutsch (1997), Barbour (2001), and others.

Two world views

The debate over how classical mechanics emerges from quantum mechanics continues, and the decoherence discovery has shown that there is a lot more to it than just letting Planck's constant $\hbar$ shrink to zero. Yet as Fig. 21.7 illustrates, this is just a small piece of a larger puzzle. Indeed, the endless debate over the interpretation of quantum mechanics – and even the broader issue of parallel universes – is in a sense the tip of an iceberg.

In the science fiction spoof *Hitchhiker's Guide to the Galaxy*, the answer is discovered to be "42," and the hard part is finding the real question. Questions about parallel universes may seem to be just about as deep as queries about reality can get. Yet there is a still deeper underlying question: there are two tenable but diametrically opposed paradigms regarding physical reality and the status of mathematics, a dichotomy that arguably goes as far back as Plato and Aristotle, and the question is which one is correct.

Aristotelian paradigm The subjectively perceived frog perspective is physically real, and the bird perspective and all its mathematical language is merely a useful approximation.

Platonic paradigm The bird perspective (the mathematical structure) is physically real, and the frog perspective and all the human language we use to describe it is merely a useful approximation for describing our subjective perceptions.

What is more basic – the frog perspective or the bird perspective? What is more basic – human language or mathematical language? Your answer will determine how you feel about parallel universes. If you prefer the Platonic paradigm, you should find multiverses natural, since our feeling that say the Level III multiverse is "weird" merely reflects that the frog and bird perspectives are extremely different. We break the symmetry by calling the latter weird because we were all indoctrinated with the Aristotelian paradigm as children, long before we have even heard of mathematics – the Platonic view is an acquired taste!

In the second (Platonic) case, all of physics is ultimately a mathematics problem, since an infinitely intelligent mathematician given the fundamental equations of the cosmos could in principle *compute* the frog perspective, i.e., compute what self-aware observers the universe would contain, what they would perceive, and what language they would invent to describe their perceptions to one another. In other words, there is a "theory of everything" (TOE) at the top of the tree in Fig. 21.7 whose axioms are purely mathematical, since postulates in English regarding

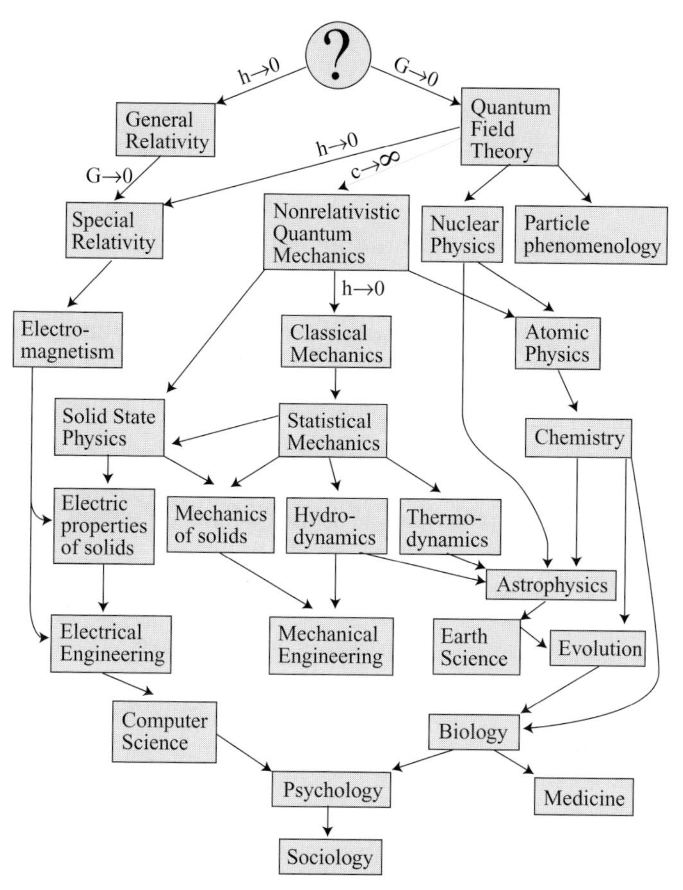

Figure 21.7. Theories can be crudely organized into a family tree where each might, at least in principle, be derivable from more fundamental ones above it. For example, classical mechanics can be obtained from special relativity in the approximation that the speed of light c is infinite. Most of the arrows are less well understood. All these theories have two components: mathematical equations and words that explain how they are connected to what we observe. At each level in the hierarchy of theories, new words (e.g., protons, atoms, cells, organisms, cultures) are introduced because they are convenient, capturing the essence of what is going on without recourse to the more fundamental theory above it. It is important to remember, however, that it is we humans who introduce these concepts and the words for them: in principle, everything could have been derived from the fundamental theory at the top of the tree, although such an extreme reductionist approach would of course be useless in practice. Crudely speaking, the ratio of equations to words decreases as we move down the tree, dropping to near zero for highly applied fields such as medicine and sociology. In contrast, theories near the top are highly mathematical, and physicists are still struggling to understand the concepts, if any, in terms of which we can understand them. The Holy Grail of physics is to find what is jocularly referred to as a "theory of everything," or TOE, from which all else can be derived. If such a theory exists at all, it should replace the big question mark at the top of the theory tree. Everybody knows that something is missing here, since we lack a consistent theory unifying gravity with quantum mechanics.

interpretation would be derivable and thus redundant. In the Aristotelian paradigm, on the other hand, there can never be a TOE, since one is ultimately just explaining certain verbal statements by other verbal statements – this is known as the infinite regress problem (Nozick 1981).

Level IV: other mathematical structures

Suppose you buy the Platonist paradigm and believe that there really is a TOE at the top of Fig. 21.7 – and that we simply have not found the correct equations yet. Then an embarrassing question remains, as emphasized by John Archibald Wheeler: Why these particular equations, not others? Let us now explore the idea of mathematical democracy, whereby universes governed by other equations are equally real. This is the Level IV multiverse. First we need to digest two other ideas, however: the concept of a mathematical structure, and the notion that the physical world may be one.

What is a mathematical structure?

Many of us think of mathematics as a bag of tricks that we learned in school for manipulating numbers. Yet most mathematicians have a very different view of their field. They study more abstract objects such as functions, sets, spaces, and operators and try to prove theorems about the relations between them. Indeed, some modern mathematics papers are so abstract that the only numbers you will find in them are the page numbers! What does a dodecahedron have in common with a set of complex numbers? Despite the plethora of mathematical structures with intimidating names like orbifolds and Killing fields, a striking underlying unity has emerged in the last century: *all* mathematical structures are just special cases of one and the same thing – so-called formal systems. A formal system consists of abstract symbols and rules for manipulating them, specifying how new strings of symbols referred to as theorems can be derived from given ones referred to as axioms. This historical development represented a form of deconstructionism, since it stripped away all meaning and interpretation that had traditionally been given to mathematical structures and distilled out only the abstract relations capturing their very essence. As a result, computers can now prove theorems about geometry without having any physical intuition whatsoever about what space is like.

Figure 21.8 shows some of the most basic mathematical structures and their interrelations. Although this family tree probably extends indefinitely, it illustrates that there is nothing fuzzy about mathematical structures. They are "out there" in the sense that mathematicians discover them rather than create them, and that

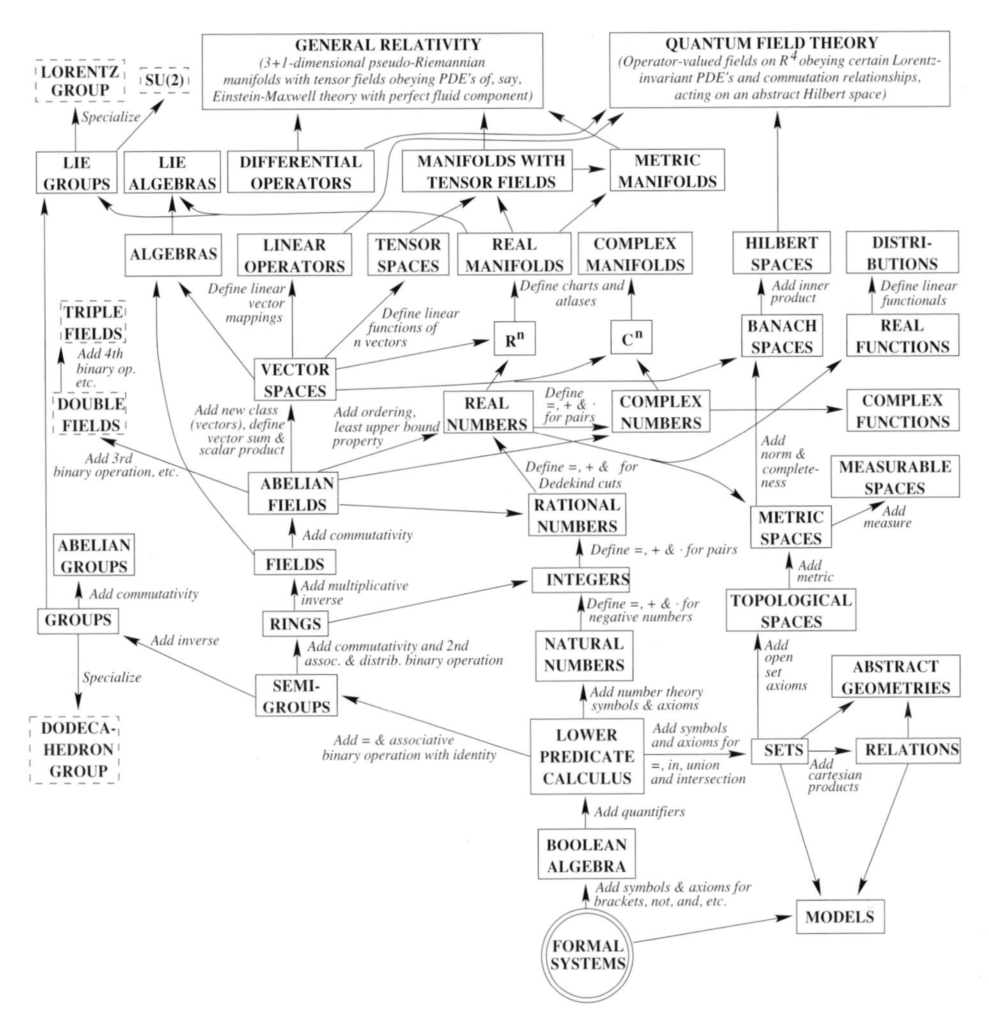

Figure 21.8. Relationships between various basic mathematical structures (Tegmark 1998). The arrows generally indicate addition of new symbols and/or axioms. Arrows that meet indicate the combination of structures; for instance, an algebra is a vector space that is also a ring, and a Lie group is a group that is also a manifold. The full tree is probably infinite in extent – the figure shows merely a small sample near the bottom.

contemplative alien civilizations would find the same structures (a theorem is true regardless of whether it is proven by a human, a computer, or an alien).

The possibility that the physical world is a mathematical structure

Let us now digest the idea that the physical world (specifically, the Level III multiverse) *is* a mathematical structure. Although traditionally taken for granted by many theoretical physicists, this is a deep and far-reaching notion. It means that mathematical equations describe not merely some limited aspects of the physical world, but *all* aspects of it. It means that there is some mathematical structure that is what mathematicians call *isomorphic* (and hence equivalent) to our physical world, with each physical entity having a unique counterpart in the mathematical structure and vice versa. Let us consider some examples.

A century ago, when classical physics still reigned supreme, many scientists believed that physical space was isomorphic to the mathematical structure known as $\mathbf{R}^3$: three-dimensional Euclidean space. Moreover, some thought that all forms of matter in the universe corresponded to various classical *fields*: the electric field, the magnetic field, and perhaps a few undiscovered ones, mathematically corresponding to functions on $\mathbf{R}^3$ (a handful of numbers at each point in space). In this view (later proven incorrect), dense clumps of matter like atoms were simply regions in space where some fields were strong (where some numbers were large). These fields evolved deterministically over time according to some partial differential equations, and observers perceived this as things moving around and events taking place. Could, then, fields in three-dimensional space be the mathematical structure corresponding to the universe? No, since a mathematical structure cannot change – it is an abstract, immutable entity existing outside of space and time. Our familiar frog perspective of a three-dimensional space where events unfold is equivalent, from the bird perspective, to a four-dimensional spacetime where all of history is contained, so the mathematical structure would be fields in four-dimensional space. In other words, if history were a movie, the mathematical structure would not correspond to a single frame of it, but to the entire videotape.

Given a mathematical structure, we will say that it has *physical existence* if any self-aware substructure (SAS) within it subjectively, from its frog perspective, perceives itself as living in a physically real world. What would, mathematically, such an SAS be like? In the classical physics example above, an SAS such as you would be a tube through spacetime, a thick version of what Einstein referred to as a world line. The location of the tube would specify your position in space at different times. Within the tube, the fields would exhibit certain complex behavior, corresponding to storing and processing information about the field values in the surroundings, and at each position along the tube, these processes would give rise to the familiar but mysterious sensation of self-awareness. From its frog perspective, the SAS would perceive this one-dimensional string of perceptions along the tube as passage of time.

Although our example illustrates the idea of how our physical world can *be* a mathematical structure, this particular mathematical structure (fields in four-dimensional space) is now known to be the wrong one. After realizing that spacetime could be curved, Einstein doggedly searched for a so-called unified field theory where the universe was what mathematicians call a $3+1$-dimensional pseudo-Riemannian manifold with tensor fields, but this failed to account for the observed behavior of atoms. According to quantum field theory, the modern synthesis of special relativity theory and quantum theory, the universe (in this case the Level III multiverse) is a mathematical structure known as an algebra of operator-valued fields. Here the question of what constitutes an SAS is more subtle (Tegmark 2000). However, this fails to describe black-hole evaporation, the first instance of the Big Bang, and other quantum gravity phenomena, so the true mathematical structure isomorphic to our universe, if it exists, has not yet been found.

Mathematical democracy

Now suppose that our physical world really is a mathematical structure, and that you are an SAS within it. This means that in the Mathematics tree of Fig. 21.8, one of the boxes is our universe. (The full tree is probably infinite in extent, so our particular box is not one of the few boxes from the bottom of the tree that are shown.) In other words, this particular mathematical structure enjoys not only mathematical existence, but physical existence as well. What about all the other boxes in the tree? Do they too enjoy physical existence? If not, there would be a fundamental, unexplained ontological asymmetry built into the very heart of reality, splitting mathematical structures into two classes: those with and without physical existence. As a way out of this philosophical conundrum, I have suggested (Tegmark 1998) that complete mathematical democracy holds: that mathematical existence and physical existence are equivalent, so that *all* mathematical structures exist physically as well. This is the Level IV multiverse. It can be viewed as a form of radical Platonism, asserting that the mathematical structures in Plato's *realm of ideas*, the *mindscape* of Rucker (1982), exist "out there" in a physical sense (Davies 1993), casting the so-called modal realism theory of David Lewis (1986) in mathematical terms akin to what Barrow (1991, 1992) refers to as "π in the sky." If this theory is correct, then since it has no free parameters, all properties of all parallel universes (including the subjective perceptions of SASs in them) could in principle be derived by an infinitely intelligent mathematician.

Evidence for a Level IV multiverse

We have described the four levels of parallel universes in order of increasing speculativeness, so why should we believe in Level IV? Logically, it rests on two separate assumptions:

Assumption 1 That the physical world (specifically our level III multiverse) is a mathematical structure.

Assumption 2 Mathematical democracy: that all mathematical structures exist "out there" in the same sense.

In a famous essay, Wigner (1967) argued that "the enormous usefulness of mathematics in the natural sciences is something bordering on the mysterious," and that "there is no rational explanation for it." This argument can be taken as support for assumption 1: here the utility of mathematics for describing the physical world is a natural consequence of the fact that the latter *is* a mathematical structure, and we are simply uncovering this bit by bit. The various approximations that constitute our current physics theories are successful because simple mathematical structures can provide good approximations of how an SAS will perceive more complex mathematical structures. In other words, our successful theories are not mathematics approximating physics, but mathematics approximating mathematics. Wigner's observation is unlikely to be based on fluke coincidences, since far more mathematical regularity in nature has been discovered in the decades since he made it, including the standard model of particle physics.

A second argument supporting assumption 1 is that abstract mathematics is so general that *any* TOE that is definable in purely formal terms (independent of vague human terminology) is also a mathematical structure. For instance, a TOE involving a set of different types of entities (denoted by words, say) and relations between them (denoted by additional words) is nothing but what mathematicians call a set-theoretical model, and one can generally find a formal system of which it is a model.

This argument also makes assumption 2 more appealing, since it implies that *any* conceivable parallel universe theory can be described at Level IV. The Level IV multiverse, termed the "ultimate Ensemble theory" in Tegmark (1997) since it subsumes all other ensembles, therefore brings closure to the hierarchy of multiverses, and there cannot be say a Level V. Considering an ensemble of mathematical structures does not add anything new, since this is still just another mathematical structure. What about the frequently discussed notion that the universe is a computer simulation? This idea occurs frequently in science fiction and has been substantially elaborated (e.g., Schmidthuber 1997; Wolfram 2002). The information content (memory state) of a digital computer is a string of bits, say "1001011100111001..." of great but finite length, equivalent to some large but finite integer n written in binary. The information processing of a computer is a deterministic rule for changing each memory state into another (applied over and over again), so mathematically, it is simply a function f mapping the integers onto themselves that gets iterated: $n \mapsto f(n) \mapsto f(f(n)) \mapsto \ldots$. In other words, even the most sophisticated computer simulation is just yet another special case of a mathematical structure, and is already included in the Level IV multiverse. (Incidentally, iterating continuous functions rather than integer-valued ones can give rise to fractals.)

Another appealing feature of assumption 2 is that it provides the only answer so far to Wheeler's question: Why these particular equations, not others? Having universes dance to the tune of all possible equations also resolves the fine-tuning problem once and for all, even at the fundamental equation level: although many if not most mathematical structures are likely to be dead and devoid of SASs, failing to provide the complexity, stability, and predictability that SASs require, we of course expect to find with 100% probability that we inhabit a mathematical structure capable of supporting life. Because of this selection effect, the answer to the question "what is it that breathes fire into the equations and makes a universe for them to describe?" (Hawking 1993) would then be "You, the SAS."

What are Level IV parallel universes like?

The way we use, test, and potentially rule out any theory is to compute probability distributions for our future perceptions given our past perceptions and to compare these predictions with our observed outcome. In a multiverse theory, there is typically more than one SAS that has experienced a past life identical to yours, so there is no way to determine which one is you. To make predictions, you therefore have to compute what fractions of them will perceive what in the future, which leads to the following predictions:

Prediction 1 The mathematical structure describing our world is the most generic one that is consistent with our observations.

Prediction 2 Our future observations are the most generic ones that are consistent with our past observations.

Prediction 3 Our past observations are the most generic ones that are consistent with our existence.

We will return to the problem of what "generic" means below (see section "The measure problem"). However, one striking feature of mathematical structures, discussed in detail in Tegmark (1997), is that the sort of symmetry and invariance properties that are responsible for the simplicity and orderliness of our universe tend to be generic, more the rule than the exception – mathematical structures tend to have them by default, and complicated additional axioms etc. must be added to make them go away. In other words, because of both this and selection effects, we should not necessarily expect life in the Level IV multiverse to be a disordered mess.

Discussion

We have surveyed scientific theories of parallel universes, and found that they naturally form a four-level hierarchy of multiverses (Fig. 21.1) allowing progressively greater differences from our own universe:

- Level I: Other Hubble volumes have different initial conditions
- Level II: Other post-inflation bubbles may have different effective laws of physics (constants, dimensionality, particle content)
- Level III: Other branches of the quantum wave function add nothing qualitatively new
- Level IV: Other mathematical structures have different fundamental equations of physics.

Whereas the Level I universes join seamlessly, there are clear demarcations between those within Levels II and III caused by inflating space and decoherence, respectively. The Level IV universes are completely separate and need to be considered together only for predicting your future, since "you" may exist in more than one of them.

Although it was Level I that got Giordano Bruno in trouble with the Inquisition, few astronomers today would suggest that space ends abruptly at the edge of the observable universe. It is ironic and perhaps due to historic coincidence that Level III is the one that has drawn the most fire in the past decades, since it is the only one that adds no qualitatively new types of universes.

Future prospects

There are ample future prospects for testing and perhaps ruling out these multiverse theories. In the coming decade, dramatically improved cosmological measurements of the cosmic microwave background radiation, the large-scale matter distribution, etc., will test Level I by further constraining the curvature and topology of space and will test Level II by providing stringent tests of inflation. Progress in both astrophysics and high-energy physics should also clarify the extent to which various physical constants are fine-tuned, thereby weakening or strengthening the case for Level II. If the current worldwide effort to build quantum computers succeeds, it will provide further evidence for Level III, since they would, in essence, be exploiting the parallelism of the Level III multiverse for parallel computation (Deutsch 1997). Conversely, experimental evidence of unitarity violation would rule out Level III. Finally, success or failure in the grand challenge of modern physics, unifying general relativity and quantum field theory, will shed more light on Level IV. Either we will eventually find a mathematical structure matching our universe, or we will bump up against a limit to the unreasonable effectiveness of mathematics and have to abandon Level IV.

The measure problem

There are also interesting theoretical issues to resolve within the multiverse theories, first and foremost the *measure problem*. As multiverse theories gain credence, the sticky issue of how to compute probabilities in physics is growing from a minor nuisance into a major embarrassment. The reason why probabilities become so important is that if there are indeed many copies of "you" with identical past lives and memories, you could not compute your own future even if you had complete knowledge of the entire state of the multiverse. This is because there is no way for you to determine which of these copies is "you" (they all feel that they are). All you can predict is therefore probabilities for what you will observe, corresponding to the fractions of these observers that experience different things. Unfortunately, computing what fraction of the infinitely many observers perceive what is very subtle, since the answer depends on the order in which you count them! The fraction of the integers that are even is 50% if you order them 1, 2, 3, 4, . . . , but approaches 100% if you order them alphabetically the way your word processor would (1, 10, 100, 1000, . . .). When observers reside in disconnected universes, there is no obviously natural way in which to order them, and one must sample from the different universes with some statistical weights referred to by mathematicians as a "measure." This problem crops up in a mild and treatable manner in Level I, becomes severe at Level II, has caused much debate within the context of extracting quantum probabilities in Level III (see DeWitt's chapter in this volume), and is horrendous at Level IV. At Level II, for instance, Vilenkin and others have published predictions for the probability distributions of various cosmological parameters by arguing that different parallel universes that have inflated by different amounts should be given statistical weights proportional to their volume (e.g., Garriga and Vilenkin 2001a). On the other hand, any mathematician will tell you that $2 \times \infty = \infty$, so that there is no objective sense in which an infinite universe that that has expanded by a factor of two has gotten larger. Indeed, an exponentially inflating universe has what mathematicians call a timelike Killing vector, which means that it is time-translationally invariant and hence unchanging from a mathematical point of view. Moreover, a flat universe with finite volume and the topology of a torus is equivalent to a perfectly periodic universe with infinite volume, both from the mathematical bird perspective and from the frog perspective of an observer within it, so why should its infinitely smaller volume give it zero statistical weight? Since Hubble volumes start repeating even in the Level I multiverse (albeit in a random order, not periodically) after about $10^{10^{115}}$ m, should infinite space really be given more statistical weight than a finite region of that size? This problem must be solved to observationally test models of stochastic inflation. If you thought that was bad, consider the problem of assigning statistical weights to different mathematical structures at Level IV.

The fact that our universe seems relatively simple has led many people to suggest that the correct measure somehow involves complexity. For instance, one could reward simplicity by weighting each mathematical structure by 2^{-n}, where n is its algorithmic information content measured in bits, defined as the length of the shortest bit string (computer program, say) that would specify it (Chaitin 1987). This would correspond to equal weights for all infinite bit strings (each representable as a real number like .101011101 . . .), not for all mathematical structures. If there is such an exponential penalty for high complexity, we should probably expect to find ourselves inhabiting one of the simplest mathematical structures complex enough to contain observers. However, the algorithmic complexity depends on how structures are mapped to bit strings (Chaitin 1987; Deutsch, this volume), and it far from obvious whether there exists a most natural definition that reality might subscribe to.

The pros and cons of parallel universes

So should you believe in parallel universes? Let us conclude with a brief discussion of arguments pro and con. First of all, we have seen that this is not a yes/no question – rather, the most interesting issue is whether there are 0, 1, 2, 3, or 4 levels of multiverses. Figure 21.1 summarizes evidence for the different levels. Cosmology observations support Level I by pointing to a flat infinite space with ergodic matter distribution, and Level I plus inflation elegantly eliminates the initial condition problem. Level II is supported by the success of inflation theory in explaining cosmological observations, and it can explain the apparent fine-tuning of physical parameters. Level III is supported by both experimental and theoretical evidence for unitarity, and explains the apparent quantum randomness that bothered Einstein so much without abandoning causality from the bird perspective. Level IV explains Wigner's unreasonable effectiveness of mathematics for describing physics and answers the question "Why these equations, not others?".

The principal arguments against parallel universes are that they are wasteful and weird, so let us consider these two objections in turn. The first argument is that multiverse theories are vulnerable to Occam's razor, since they postulate the existence of other worlds that we can never observe. Why should Nature be so ontologically wasteful and indulge in such opulence as to contain an infinity of different worlds? Intriguingly, this argument can be turned around to argue *for* a multiverse. When we feel that Nature is wasteful, what precisely are we disturbed about her wasting? Certainly not "space," since the standard flat universe model with its infinite volume draws no such objections. Certainly not "mass" or "atoms" either, for the same reason – once you have wasted an infinite amount of something, who cares if you waste some more? Rather, it is probably the apparent reduction in simplicity that appears disturbing, the quantity of information necessary to specify all these unseen worlds. However, as is discussed in more detail in Tegmark (1996), an entire ensemble is often much simpler than one of its members. For instance, the algorithmic information content of a generic integer n is of order $\log_2 n$ (Chaitin 1987), the number of bits required to write it out in binary. Nonetheless, the set of all integers 1, 2, 3, . . . can be generated by quite a trivial computer program, so the algorithmic complexity of the whole set is smaller than that of a generic member. Similarly, the set of all perfect fluid solutions to the Einstein field equations has a smaller algorithmic complexity than a generic particular solution, since the former is specified simply by giving a few equations and the latter requires the specification of vast amounts of initial data on some hypersurface. Loosely speaking, the apparent information content rises when we restrict our attention to one particular element in an ensemble, thus losing the symmetry and simplicity that was inherent in the totality of all elements taken together. In this sense, the higher-level multiverses have less algorithmic complexity. Going from our universe to the Level I multiverse eliminates the need to specify initial conditions, upgrading to Level II eliminates the need to specify physical constants, and the Level IV multiverse of all mathematical structures has essentially no algorithmic complexity at all. Since it is merely in the frog perspective, in the subjective perceptions of observers, that this opulence of information and complexity is really there, a multiverse theory is arguably more economical than one endowing only a single ensemble element with physical existence (Tegmark 1996).

The second common complaint about multiverses is that they are weird. This objection is aesthetic rather than scientific, and as mentioned above, really only makes sense in the Aristotelian world view. In the Platonic paradigm, one might expect observers to complain that the correct TOE was weird if the bird perspective was sufficiently different from the frog perspective, and there is every indication that this is the case for us. The perceived weirdness is hardly surprising, since evolution provided us with intuition only for the everyday physics that had survival value for our distant ancestors. Thanks to clever inventions, we have glimpsed slightly more than the frog perspective of our normal inside view, and sure enough, we have encountered bizarre phenomena whenever departing from human scales in any way: at high speeds (time slows down), on small scales (quantum particles can be at several places at once), on large scales (black holes), at low temperatures (liquid helium can flow upward), at high temperatures (colliding particles can change identity), etc. As a result, physicists have by and large already accepted that the frog and bird perspectives are very different. A prevalent modern view of quantum field theory is that the standard model is merely an effective theory, a low-energy limit of a yet to be discovered theory that is even more removed from our cozy classical concepts (involving strings in ten dimensions, say). Many experimentalists are becoming blasé about producing so many "weird" (but perfectly repeatable) experimental results, and simply accept that the world is a weirder place than we thought it was and get on with their calculations.

We have seen that a common feature of all four multiverse levels is that the simplest and arguably most elegant theory involves parallel universes by default, and that one needs to complicate the theory by adding experimentally unsupported processes and ad hoc postulates (finite space, wave function collapse, ontological asymmetry, etc.) to explain away the parallel universes. Our aesthetic judgment therefore comes down to what we find more wasteful and inelegant: many worlds or many words. Perhaps we will gradually get more used to the weird ways of our cosmos, and even find its strangeness to be part of its charm.

Acknowledgments

The author wishes to thank Anthony Aguirre, Aaron Classens, George Musser, David Raub, Martin Rees, and Harold Shapiro for stimulating discussions. This work was supported by National Science Foundation grants AST-0071213 and AST-0134999, NASA grants NAG5-9194 and NAG5-11099, a fellowship from the David and Lucile Packard Foundation, and a Cottrell Scholarship from Research Corporation.

References

Anglin, J R and Zurek, W H (1996) *Phys. Rev.* **D53**, 7327.
Barbour, J B (2001) *The End of Time*. Oxford: Oxford University Press.
Barrow, J D (1991) *Theories of Everything*. New York: Ballantine.
(1992) *Pi in the Sky*. Oxford: Oxford University Press.
Barrow, J D and Tipler, F J (1986) *The Anthropic Cosmological Principle*. Oxford: Oxford University Press.
Brundrit, G B (1979) *Q. J. Royal Astr. Soc.* **20**, 37.
Bucher, M A and Spergel, D N (1999) Inflation in a low-density universe. *Sci. Am.* **280**(1), 62.
Carter, B (1974) In *International Astronomical Union (IAU) Symposium 63*, ed. S. Longair, pp. 291–8. Dordrecht: P. Reidel.
Chaitin, G J (1987) *Algorithmic Information Theory*. Cambridge: Cambridge University Press.
Davies, P C W (1982) *The Accidental Universe*. Cambridge: Cambridge University Press.
(1993) *The Mind of God*. New York: Touchstone.
(1996) *Are We Alone?* New York: Basic Books.
Deutsch, D (1997) *The Fabric of Reality*. New York: Allen Lane.
Egan, G (1995) *Permutation City*. New York: Harper.
Everett, H (1957) Ph.D. thesis, Princeton University.
Garriga, J and Vilenkin, A (2001a) *Phys. Rev.* **D64**, 023507.
(2001b) *Phys. Rev.* **D64**, 043511.

Giulini, D, Joos, E, Kiefer, C, *et al.* (1996) *Decoherence and the Appearance of a Classical World in Quantum Theory*. Berlin: Springer-Verlag.
Guth, A and Steinhardt, P J (1984) *Sci. Am.* **250**, 116.
Hawking, S (1993) *A Brief History of Time*. New York: Touchstone.
Lewis, D (1986) *On the Plurality of Worlds*. Oxford: Blackwell.
Linde, A (1994) *Sci. Am.* **271**, 32.
Nozick, R (1981) *Philosophical Explanations*. Cambridge, MA: Harvard University Press.
Rees, M J (2002) *Our Cosmic Habitat*. Princeton, NJ: Princeton University Press.
Rucker, R (1982) *Infinity and the Mind*. Boston, MA: Birkhauser.
Schmidthuber, J (1997) In *Foundations of Computer Science: Potential-Theory-Cognition, Lecture Notes in Computer Science*, ed. C. Freksa, pp. 201–8. Berlin: Springer-Verlag.
Smolin, L (1997) *The Life of the Cosmos*. Oxford: Oxford University Press.
Steinhardt, P J and Turok, N (2002) *Science* **296**, 1436.
Tegmark, M (1996) *Found. Phys. Lett.* **9**, 25.
(1997) *Class. Quant. Grav.* **14**, L69
(1998) *Ann. Phys.* **270**, 1.
(2000) *Phys. Rev.* **E61**, 4194.
(2002) *Science* **296**, 1427.
Tegmark, M and Wheeler, J A (2001) *Sci. Am.* **2/2001**, 68–75.
Wigner, E P (1967) *Symmetries and Reflections*. Cambridge, MA: MIT Press.
Wolfram, S (2002) *A New Kind of Science*. New York: Wolfram Media.
Zeh, H D (1970) *Found. Phys.* **1**, 69.

22

Quantum theories of gravity: results and prospects

Lee Smolin

Perimeter Institute for Theoretical Physics,
Waterloo, Canada

Introduction

Once, while visiting the University of Texas in 1981, I joined John Wheeler and a group of students and postdocs for lunch. As he often did, John posed a provocative question for discussion. This time he asked something like the following, "Perhaps when we die, Saint Peter gives us a physics test to determine if our time spent on earth searching for knowledge for our fellow human beings has been well spent. Because the experience can be traumatic, and we are likely to forget details, we are allowed to bring along a crib sheet, to jog our memories. But as the point of having laws of physics is that they must be simple and general, the crib sheet is only allowed to be a 3 by 5 inch file card. What would you write down on your card?"

Of course, beyond the theological issues, John was making a simple and fundamental pedagogical point. If we believe that the laws of nature are simple, a measure of our understanding of them is the compactness with which they can be expressed. As individuals and as a community, the better we understand the laws of physics, the less the space that will be required to write them.

We then had a lively discussion of what is the simplest way to write the Einstein equations. We also argued about which formulation of quantum theory is more fundamental. Of course, in retrospect, our answers were pretty silly, for one thing we can be sure of is that neither the Einstein equations nor quantum theory by themselves, qualify as fundamental laws of nature. Each is known to be incomplete, because of the existence of the other. General relativity does not appear to incorporate quantum phenomena and quantum theory has difficulty incorporating relativity's notions of space and time. What is required is a new kind of theory, which unifies quantum theory and Einstein's general theory of relativity into a

Science and Ultimate Reality, eds. J. D. Barrow, P. C. W. Davies and C. L. Harper, Jr. Published by Cambridge University Press.

single theory. Only such a *quantum theory of gravity*, as such a theory is called, has a chance to be a fundamental theory.

So to pass the test, we will have to talk about the quantum theory of gravity. Indeed, John was one of the early pioneers of this subject, with Bryce DeWitt contributing the fundamental equation of a quantum theory of gravity, which we call, with affection, the Wheeler–DeWitt equation.

Since then I, along with many hundreds of other physicists and mathematicians, have followed Bryce and John to devote our efforts to searching for the quantum theory of gravity. So I think it would be more than fair for John to ask us now whether we have learned enough about the quantum theory of gravity that we are able to write its principles on a 3 by 5 inch file card.

I believe the answer is yes, and in the remainder of this essay I would like to explain what I believe we have learned in the 20 years since John posed the file card test to us. Then, at the end I would also like to come back and summarize where we are and what are the key problems that we still have to understand.

Organization of this essay

Quantum gravity is a complicated subject, not only scientifically, but also historically and sociologically. Philosophers and sociologists of science tell us that the histories of science we learn in textbooks are oversimplified to make it seem like physicists are able to move directly from questions to answers, without the politics, mistakes, false turns, and general confusion that accompany progress in other areas such as social theory or art. Certainly they are right, and the story of quantum gravity, so far, serves as a good example of how every sort of human complexity can enter into the story of a search for truth about Nature.

In the case of quantum gravity, there have been many approaches tried over the past half century. Most were abandoned, perhaps half a dozen are being actively pursued. Of these, two stand out as having been the result of sustained study over now more than 18 years, by a large number of physicists and mathematicians. These are string theory and loop quantum gravity. Although these are not the only ideas being studied now, it is fair to say that the majority of results we have concerning quantum gravity are either in one of these two areas, or in an area closely related to one or the other.[1] Because of lack of space, I will discuss only these approaches here.

[1] Among the exceptions to this are the classic results of Bekenstein (1973), Hawking (1975), Unruh (1976), Davies, and others, gotten at the semi-classical level, and the results from the dynamical triangulations program (Agishtein and Migdal 1992; Ambjorn *et al.* 1992; Ambjorn 1995) and causal dynamical triangulations program (Ambjorn and Loll 1998; Ambjorn *et al.* 2000, 2001a, b, c, 2002; Loll 2001, Dittrich and Loll 2002). Three other approaches that have attracted a lot of interest, and which many believe likely contain some of the truth are twistor theory (Penrose and MacCallum 1972; Penrose 1975), noncommutative geometry (Connes 1994) and causal sets (Bombelli *et al.* 1987; Martin *et al.* 2001; Rideout and Sorkin 2001).

I have described these two theories in some detail in my book *Three Roads to Quantum Gravity* (Smolin 2001a), and aspects of each are also described in this volume by Fotini Markopoulou and Juan Maldecena, so I will not take the space here to introduce them.[2] Instead, I will assume that the reader has a rough acquaintance with the basic ideas of each approach, and instead aim to say something about how well each has done, so far, as a candidate for the real theory of quantum gravity. The main question I want to ask is how far each theory is now towards being a complete theory of quantum space and time. By complete I meant that it is precisely formulated and well understood mathematically and conceptually, that there are methods to carry out calculations leading to predictions for real experiments, and that at least a few experiments have been done that either support or falsify the predictions of the theory.

I will thus proceed as follows. In the next section I will list the main questions that a quantum theory of gravity is expected to answer, and explain why each is important. The section after that will describe how a quantum theory of gravity is expected to be tested experimentally. As science is based on experiment, we cannot give a convincing answer to what we have learned, if we are not able to explain how our ideas and calculations will be tested.

Then I will present the basics of loop quantum gravity. I will explain what answers it gives to the major questions, and how it is likely to be tested experimentally. I will also indicate what the key open questions are, which this approach must still answer.

In the following section I will discuss string theory. I will explain the answers it offers to the main questions as well as describe its main open problems.

Finally, I will conclude, with some comments about how, after 20 years of working in this field, its prospects appear.

This essay is one of a pair of papers in which I aim to summarize the state of our knowledge in the field of quantum gravity. The other paper is longer, more technical and detailed, and has many more references to the literature than this one (Smolin 2003). Most of the statements made here are justified in detail there.

What questions should a quantum theory of gravity answer?

In order to know how close the theories are to completion, we should specify what questions the theory is expected to answer. By this time it is possible to make a

[2] Information about string theory and loop quantum gravity is also available on some websites (see http://superstringtheory.com/, www.qgravity.org) and from books (Green *et al.* 1987; Ashtekar 1988; Gambini and Pullin 1996; Polchinski 1998; Greene 1999) and review articles (Smolin 1992, 1997; Rovelli 1998). String theory is not an unemotional subject, and has engendered some controversy; for critical reviews, the reader may want to look at Woit (2001) and Friedan (2002).

list of the questions about Nature that are presently unanswered, which we expect a quantum theory of gravity to resolve.[3]

This is not a short list, but neither is it infinite. We can divide the list into four parts. We begin with those questions about quantum gravity itself.

Questions concerning quantum gravity

The correct quantum theory of gravity must:

1. Tell us whether the principles of general relativity and quantum mechanics are true as they stand, or are in need of modification.
2. Give a precise description of Nature at all scales, including the Planck scale.
3. Tell us what time and space are, in language fully compatible with both quantum theory and the fact that the geometry of spacetime is dynamical. Tell us how light cones, causal structure, the metric, etc are to be described quantum mechanically, and at the Planck scale.
4. Give a derivation of the black-hole entropy and temperature. Explain how the black-hole entropy can be understood as a statistical entropy, gotten by coarse graining the quantum description.
5. Be compatible with the apparently observed positive, but small, value of the cosmological constant. Explain the entropy of the cosmological horizon.
6. Explain what happens at singularities of classical general relativity.
7. Be fully background independent. This means that no classical fields, or solutions to the classical field equations, appear in the theory in any capacity, except as approximations to quantum states and histories.
8. Explain how classical general relativity emerges in an appropriate low-energy limit from the physics of the Planck scale.
9. Predict new physical phenomena, at least some of which are testable in current or near-future experiments.
10. Predict whether the observed global Lorentz invariance of flat spacetime is realized exactly in Nature, up to infinite boost parameter, or whether there are modifications of the realization of Lorentz invariance for Planck scale energy and momenta.
11. Provide precise predictions for the scattering of gravitons, with each other and with other quanta, beyond the classical approximation.

This is a lot of questions, but it is hard to imagine believing in a quantum theory of space and time that did not answer each one. However, there is one that cannot be overemphasized, which is the requirement of background independence. There are two reasons for making this requirement. The first is a matter of principle. Over the whole history of physics, from the Greeks onwards, there have been two

[3] I exclude from this list questions about the theories themselves, that do not directly address questions about the natural world. While important for the internal development of each theory, they should be ignored at this stage. While it is true that a research program may make a lot of progress on such internal questions without actually leading to any new insights about the natural world, this is not necessarily a good thing.

competing views about the nature of space and time. The first is that they are not part of the dynamical system, but are instead eternally fixed, nondynamical aspects of the background, against which the laws of Nature are defined. This was the point of view of Newton and it is generally called the *absolute* point of view. The second view holds that the geometry of space and time are aspects of the dynamical system that makes up the universe. They are then not fixed, but evolve as does everything else, according to law. Further, according to this view, space and time are *relational*. This means there is no absolute meaning to where or when an event occurs, except as so far as can be determined by observable correlations or relations with other events. This was the point of view of Leibniz, Mach, and Einstein and is called the *relational* point of view.

Einstein's theory of general relativity is an instantiation of the relational point of view. The observations that show that gravitational radiation carries energy away from binary pulsars in two degrees of freedom of radiation, exactly as predicted by Einstein's theory, may be considered the experimental death blow to the *absolute* point of view. This means that the metric is a completely dynamical entity, and no component of the metric is fixed and nondynamical. This is expressed also by saying that the physics of the gravitational field is completely background independent. This means that no aspect of the geometry is fixed, independent of the history of the universe. If you take away that part of the geometry of space and time which is the result of dynamical evolution, you are not left with some background geometry, you are left with nothing at all.

This background independence is expressed in general relativity by a certain principle, which is called *diffeomorphism invariance*. It means that there is no fixed, background, structure to space. There are no points, sitting by themselves, with labels on them, at which physical particles and fields come and go. The only things that are well defined are *relationships between dynamical fields*. The only way to speak of a particular point, a particular event, or a particular moment of time, is if it can be distinguished by the values that some dynamically evolving fields happen to take there.

As argued by Einstein and many others since, the diffeomorphism invariance is tied directly to the background independence of the theory. This is shown by the hole argument (see for example Norton (1987), Earman (1989) and Smolin (2001b), and by Dirac's (1964) analysis of the meaning of gauge symmetry. There are good discussions of this by Stachel (1989), Barbour (2000), Rovelli (1991), and others (Smolin, 1997b, 2001a).

Thus, classical general relativity is background independent. The arena for its dynamics is not spacetime, instead the arena is the configuration space of all the degrees of freedom of the gravitational field, which is the metric modulo diffeomorphisms.

Now we can ask, must the quantum theory of gravity also be background independent? To have it otherwise would be as if some particular classical Yang–Mills field was required to *define* the quantum dynamics of quantum chromodynamics (QCD), while no such fixed, nondynamical field need be specified to define the classical theory. Still a number of people have expressed the view that perhaps the quantum theory of gravity requires a fixed nondynamical spacetime background for its very definition. This seems almost absurd, for it would mean taking some particular solution (out of infinitely many) to the classical theory, and making it play a preferred role in the quantum theory. Moreover, there must be no experimental way to discover which classical background was taken to play this preferred role, for if any effect which depended on the fixed background survived in the low-energy limit, it would break diffeomorphism invariance, which is a fundamental gauge symmetry of general relativity. But this would in turn mean that diffeomorphism invariance was not an exact gauge symmetry in the low-energy limit, and this would imply that more than two degrees of freedom of the metric would be excited when matter accelerated. But this would contradict the extreme sensitivity of the agreement between general relativity and the rate of decay of binary pulsar orbits.

Thus, arguments from both principle and from experiment reinforce the conclusion that nature is constructed in such a way that, even in the quantum domain, all the degrees of freedom of the spacetime geometry are dynamical. But if this is the case no fixed classical metric can play any role in the formulation of the quantum theory of gravity.

Questions concerning cosmology

Next we mention cosmological puzzles that are so far unsolved and that are widely believed to require Planck scale physics for their resolution.

1. Explain why our universe apparently began with extremely improbable initial conditions.
2. In particular, explain why the universe had at grand unified times initial conditions suitable for inflation to occur or, alternatively, give an alternate mechanism for inflation or a mechanism by which the successes of inflationary cosmology are duplicated.
3. Explain whether the Big Bang was the first moment of time, or whether there was something before that.
4. Explain what the dark matter is. Explain what the dark energy is. Explain why at present the dark matter is six times as dense as ordinary hadronic matter, while the dark energy is in turn twice as dense as the dark matter.
5. Provide predictions that go beyond those of the currently standard model of cosmology, such as corrections to the cosmic microwave background (CMB) spectra predicted by inflationary models.

Questions concerning unification of the forces

Next, we mention problems in elementary particle physics that must be resolved by any unified theory of all the interactions. As string theory must, if true, be such a theory, it must be evaluated against progress in answering these questions. It is also possible, but not as necessary, that loop quantum gravity offer answers to some of these questions.

1. Discover whether there is a further unification among the forces, including gravity or not.
2. Explain the general features of the standard model of elementary particle physics, i.e., explain why the forces are described by a spontaneously broken gauge theory with group $SU(3) \times SU(2) \times U(1)$, with fermions in the particular chiral representations observed.
3. Explain why there is a large hierarchy in the ratio of masses observed, from the Planck mass, down to the neutrino masses, and finally down to the cosmological constant. Discover the mechanism by which the hierarchy was created, whether by spontaneous breaking of a more unified theory or by other means. Explain why the cosmological constant is so small in Planck units.
4. Explain the actual values of the parameters of the standard model: masses, coupling constants, mixing angles, etc.
5. Tell us whether there is a unique consistent theory of nature that implies unique predictions for all experiments or whether, as has been sometimes proposed, some or all of the questions left open by the standard model of particle physics are to be answered in terms of choices among possible consistent phenomenologies allowed by the fundamental theory.
6. Make some experimental predictions for phenomena that are unique to that theory and which are testable in present or near future experiments.

Foundational questions

Finally, there are the questions in the foundations of quantum theory, which many people believe are closely related to the problem of quantum gravity.

1. Resolve the problem of time in quantum cosmology
2. Explain how quantum mechanics is to be modified to apply to a closed system such as the universe that contains its own observers.

These are lots of questions. Shortly we will see how the different quantum theories of gravity do in giving answers to them.

How are quantum theories of gravity to be tested experimentally?

Until very recently, it was almost universally believed that there was no realistic chance of an experimental test of any quantum theory of gravity in the foreseeable future. This was because there was a simple argument why physical phenomena related to quantum gravity would take place at a scale remote from that which could be probed with our current technology.

One expects that quantum gravity effects will involve phenomena in which all three of the following constants come into physical expressions: Planck's constant $\hbar$ for quantum theory, Newton's constant G for gravity, and the speed of light c for relativity. These together set the scales of distance, time, and energy that we must probe to discover quantum gravitational phenomena. The problem is that the Planck length,

$$l_{\mathrm{Pl}} = \sqrt{\hbar G/c^3} \approx 10^{-33}\,\mathrm{cm} \tag{22.1}$$

is 20 orders of magnitude smaller than an atomic nucleus, while the Planck energy

$$E_{\mathrm{Pl}} = \sqrt{\hbar c^5/G} \approx 10^{19}\mathrm{GeV} \tag{22.2}$$

is more than 15 orders of magnitude larger than the energy that can be created in the largest planned elementary particle accelerator. As a result, it has been almost universally assumed that we would not soon be able to probe these scales, and that we would not be able to test quantum theories of gravity experimentally.

This situation led to a crisis in elementary particle physics, which persisted from the mid 1970s to the present. After the first successful confirmations of the predictions of the standard model of elementary particle physics in the 1970s there has been no instance in which an important new theoretical idea about the fundamental forces and particles has been confirmed experimentally. As a result, while this period has been characterized by an exuberant development of theoretical ideas concerning the unification of the fundamental forces and the quantum theory of gravity, none of these ideas has been confirmed experimentally.

There was however a loophole in this argument, which we were all blind to. This was noticed by several people, including Luis Gonzalez-Mestres (1997a, b), Coleman and Glashow (1997, 1998), and Subir Sarkar, Giovanni Amelino-Camelia and collaborators (Amelino-Camelia *et al.* 1997, 1998; Ellis *et al.* 2000, 2001).[4] These people realized that even if a single quantum gravity event is unobservable, there can be circumstances where there is amplification of effects coming from many quantum gravity effects (Amelino-Camelia *et al.* 1997, 1998; Coleman and Glashow 1997, 1998; Gonzalez-Mestres 1997a, b; Ellis *et al.* 2000, 2001; Amelino-Camelia and Piran 2001).

One such amplification device is a proton decay detector. Proton decay is, if it happens, a consequence of the unification of the four basic interactions. As such it is predicted by some unified theories to take place at a scale about 1000 times less in energy than the Planck scale. This is lower in energy than a quantum gravity effect,

[4] The history of this development is rather intricate; for a review, with many references, see Sarkar (2002).

but still many orders of magnitude away from what could be produced in a particle accelerator. As a result, proton decay events are extremely rare. However, a cubic meter of water contains a huge number of protons. Thus, if you can instrument a large swimming pool in such a way as to pick up the large amount of energy created by a single proton decaying, one has a good chance in a year's observation of seeing such a rare event.

Once one has on board this idea of amplification, there is a clear strategy, which was brought very forcefully to the community of people working in quantum gravity by Amelino-Camelia: write down the laws of physics, as we might have for John Wheeler's file card, then add all possible terms proportional to l_{Pl} consistent with dimensional analysis. Those terms will be tiny for all observable energies involving elementary particles, but one can search for physical situations in which the effects of the small terms is amplified to the point where a deviation from what is normally expected could be observed with current technology.

Here is a simple example: take the energy momentum relationship which, according to the special theory of relativity, holds universally between the energy, E, mass m, and momentum p of any particle:

$$E^2 = p^2c^2 + m^2c^4. \tag{22.3}$$

Add a term proportional to $l_{\mathrm{Pl}} = 1/E_{\mathrm{Pl}}$, which gives

$$E^2 = p^2c^2 + m^2c^4 + \alpha l_{\mathrm{Pl}} E^3 \tag{22.4}$$

where α is a dimensionless constant of order 1.

Now, the physicists among my readers may object because this violates some basic principles. For example, it violates invariance under the Lorentz transformations which tell us how to transform between energy and momentum as measured by different observers. This may seem a fatal objection, but there is a way out, which is that one can also add new terms in l_{Pl} to the transformation laws so that the new expression (22.4) is invariant. This is described in some detail in João Magueijo's chapter in this volume.[5]

Now, for photons, when $m = 0$, expression (22.4) immediately implies that the speed of light depends on energy. We find

$$v = \frac{\partial E}{\partial p} = c(1 + 2\alpha l_{\mathrm{Pl}} E). \tag{22.5}$$

Again, one might object that this violates the theory of relativity, but no, again it turns out that, as explained in Magueijo's chapter, this is compatible with a modification of the principle of relativity.

[5] Exactly how to do this is described also in Bruno *et al.* (2001), Ahluwalia and Kirchbach (2002), Amelino-Camelia (2002), Judes (2002), Judes and Visser (2002), Kowalski-Glikman and Nowak (2002), and Visser (2002).

Now the correction to the energy is a tiny effect for all photons we can observe. For example, for gamma-ray photons it is an effect of the order of one part in 10^{22}. So it might seem unobservable, and certainly it is in the absence of an amplifier.

However, in this case an amplifier is readily available, it is the universe itself. The fact that the speed of light is, according to eqn (22.5) slightly larger or smaller for more or less energetic photons means that if we observe light from a very distant burst of energy, the more energetic photons will arrive slightly before or after the less energetic ones. The effect is tiny, but it amplifies with distance. The question is, then, are there any sources of photons that are far enough away, sharp enough in time, and high enough in energy that the time delay between the arrival of higher and lower energy photons is observable with current technology?

The answer is yes! Gamma-ray bursts are detected coming from cosmological distances about once a day. With present data one can put a limit on α of less than about 1000. The next gamma-ray observatory, scheduled for launch in 2006, will have electronics fast enough to detect the effect, if it exists, and α is of order 1.

There are several other possible observations of this kind, involving photons or cosmic rays coming from distant sources. For more information about them the reader may consult Sarkar (2002). And this is not all. Another amplifier of Planck scale effects is the cosmic inflation which is hypothesized to have occurred very early in the universe. Planck scale effects such as (22.4) do lead to predictions for observations of the cosmic microwave background that may be observable in the next decade. The bottom line is that over the next few years a few quantum gravity effects will, if they exist, become observable. The 25-year period in which fundamental physical theories were developed without check from experiment is about to come to an end!

A quantum theory of gravity: loop quantum gravity

We now come to the quantum theory of gravity itself. In this section I will describe what I think is the best-developed approach to quantum gravity, which is loop quantum gravity.

Loop quantum gravity arose from a merger of two developments, one in general relativity and one in elementary particle physics. From the side of general relativity, Abhay Ashtekar developed in 1985 a reformulation of Einstein's general theory of relativity, which greatly simplified it (Ashtekar 1986). His reformulation was based on a discovery of Amitaba Sen a few years earlier, that certain of the Einstein equations looked very much simpler when written in terms of a special set of variables (Sen 1981, 1982). These special variables are a gauge or Yang–Mills field. These are extensions of the electromagnetic field which allow for there to be more than one kind of charge. Instead of a single charge, there are in this case three charges, which we can think of as red, white, and blue charges.[6]

These gauge fields occur also in the standard model of elementary particle physics, where they describe the different forces in nature. It was then possible immediately in 1985 to take over from elementary particle physics to quantum gravity some of the ideas and techniques that physicists had developed to study forces within the atomic nuclei. Chief of these ideas was an idea that it is useful to describe the quantum physics of gauge fields in terms of loops of electric flux.

Everyone who has taken a high-school science course is familiar with the lines of force of a magnet. If you put some magnetic filings (basically magnetized dust) on a piece of paper, and put under that paper a bar magnet, you see lines of force running from the south pole to the north pole. Now, the teacher will tell you that the lines of force are continuous, and only seem discrete because the magnetic filings have a certain size. But there are circumstances in which the lines of force really are discrete: this takes place in a superconductor. There is in this case a certain quantum of magnetic field, and any magnetic field in a superconductor will come in integer units of that basic quantum. This is by the way, a quantum mechanical effect; it would not happen in classical physics.

In electromagnetism there is also an electric field, and it also can be represented by lines of force. In this case they run between positive and negative charges. Now, we know of no circumstances in which the electric field's lines of force come in discrete quanta. However, for the gauge fields in the standard model this is not the case. There are gauge fields which hold the quarks together inside protons and neutrons, and these seem to come in discrete lines of force, like those in a superconductor. This is sometimes described by saying that the environment inside a proton or neutron is like a "dual" superconductor, dual here means that electric fields have been traded for magnetic fields. Corresponding to this, in the 1970s several physicists[7] had found ways to reformulate the physics of quantum gauge fields in language in which the degrees of freedom correspond to the motions of these discrete lines of force of the electric fields. These were called loop variables.

So when Sen and Ashtekar found that general relativity could be greatly simplified in terms of gauge fields, it was clear that one could apply to the quantization of gravity the physical picture and methods of loops, developed first for gauge fields. So we made the physical hypothesis that empty space is, for these gravitational gauge fields, like a dual superconductor, so that the electric field lines in this case would be quantized.

Now, the physics of the quantized electric field lines is different in the quantum gravity case than the gauge fields because the metric of spacetime plays a different

[6] For a more complete description of gauge fields, suitable for laypeople, the reader may look at Smolin (1997, 2001a).

[7] Sasha Migdal, Sasha Polyakov, Kenneth Wilson, and Stanley Mandelstam.

role. The metric is the variable that describes the geometry of space and time, as such it tells light and other fields how to travel. Einstein's great discovery was that it also is the gravitational field, because he found that the existence of gravity is a consequence of the fact that the geometry of space and time is dynamical, and can evolve in time and adjust itself to the distribution of matter.

When one studies electromagnetism or elementary particle physics, one wants to ignore gravitational phenomena, and so one freezes the metric. It becomes what we call a *background field*, which means that it is treated as fixed and nondynamical, it becomes part of the definition of the context, within which we study the dynamics of the electromagnetic field.

However, we cannot do this when we study quantum gravity, because the gauge fields themselves contain the geometry of space and time. In fact, part of Ashtekar and Sen's great discovery is that the geometry of space is coded inside the electric fields of the new gauge fields that describe gravity. So when we study the quantum mechanics of these fields, the gravitational fields are not part of the fixed background, they are part of the quantum dynamical system we are studying. This is to say that we are taking an approach to quantum gravity which is truly *background independent.*

When I was listing the questions I emphasized the importance of the principle of background independence.

One consequence of this is that the loops of quantized electric field do not live anywhere in space. Instead, their configuration defines space. A second consequence is that because the geometry of space is now coded into the new electric fields, the quantization of the flux of the electric field has consequences for the geometry of space. Because the electric field flux can only come in certain discrete values, the same is true for the geometry of space. A result is the quantization of area and volume. If one measures the area of any surface, the answer can only come in certain discrete units, called *quanta of area.* The same thing is true for volume. The result is that there is a smallest quantum of area and a smallest quantum of volume that the theory predicts exist in nature.

This is a brief sketch of the basic ideas of loop quantum gravity. A more detailed description is given by Fotini Markopoulou's article in this volume, and more can be found in Smolin (2001a).[8] But what we have said here is enough to see how loop quantum gravity answers the questions. Nevertheless, before describing the results I should mention that loop quantum gravity has a rigorous side, and all the key results have been rederived as theorems in a mathematically precise formulation of diffeomorphism invariant quantum field theory. Theimann (2001) and Perez (2003) review the rigorous results on the canonical and path integral side, respectively.

[8] Some of the basic papers in loop quantum gravity are Gambini and Trias (1981, 1983, 1984, 1986), Jacobson and Smolin (1988, 1990), Rovelli and Smolin (1988, 1995a, b), Gambini *et al.* (1989), Smolin (1994).

Loop quantum gravity answers the quantum gravity questions

1. Tell us whether the principles of general relativity and quantum mechanics are true as they stand, or are in need of modification.

Loop quantum gravity answers that the principles can be unified as they stand. Here it must be emphasized that the key principles of general relativity that are taken over are: (1) the geometry of spacetime is completely dynamical, (2) there is no fixed background structure, so the physics must be expressed in a background independent language, and (3) whenever a classical field such as a metric appears, it must be in a diffeomorphism-invariant expression.

We do not take over the idea that the metric is a fundamental field, or that the classical equations that Einstein gave to the classical metric field play a fundamental role. We instead want to impose dynamics that are natural for the quantum gravitational field. Forms for the quantum dynamics can be motivated by the classical Einstein equations, and this has been done very successfully. But it is important to emphasize that as the classical metric plays no fundamental role in loop quantum gravity, what we have taken over are the basic principles of Einstein's theory which apply equally in the quantum and classical theories. These are that the geometry of spacetime is completely dynamical and must be described in a background independent language.

2. Give a precise description of nature at all scales, including the Planck scale.

The precise description is given in terms of the loops of quantized electric flux, which are coded in language that is both completely quantum mechanical and compatible with the principles of general relativity just mentioned, the geometry of space. These loops are generally organized into networks, which are called spin networks. These are graphs, whose edges are labeled by integers, that tell us how many elementary quanta of electric flux are running through the edge. This translates into quanta of areas, when the edge pierces a surface.

There is then a mathematical theorem which says that these spin network states give a complete and orthonormal basis to the Hilbert space of quantum gravity.

Furthermore, the Wheeler–deWitt equation, which is the fundamental equation of quantum gravity, may be solved exactly using the spin network states (Jacobson and Smolin 1988; Rovelli and Smolin 1988; Thiemann 2001). In fact, an infinite number of solutions can be found.

This picture can be extended in a precise way to incorporate matter degrees of freedom, such as all the fields involved in the standard model of elementary particle physics. If one wants to incorporate additional symmetries that have been hypothesized, such as supersymmetry, this can be done as well. This leads to a description of a theory of gravity called supergravity. The quantum mechanics of these theories can also can be completed described in terms of loop quantum gravity.

3. Tell us what time and space are, in language fully compatible with both quantum theory and the fact that the geometry of spacetime is dynamical. Tell us how light cones, causal structure, the metric, etc. are to be described quantum mechanically, and at the Planck scale.

To describe a whole spacetime in loop quantum gravity, one considers an object we call a causal spin foam. These are also a sort of graph, but here the vertices correspond to events in which the geometry of a spin network graph changes. They are described in Markopoulou's chapter in this volume and in Smolin (2001a). These incorporate a completely quantum mechanical and background independent description of light cones, causal structure, etc., completely realizing the principles of both quantum theory and general relativity.

4. Give a derivation of the black-hole entropy and temperature. Explain how the black-hole entropy can be understood as a statistical entropy, gotten by coarse graining the quantum description.

This has been done in loop quantum gravity, giving a picture of the horizons of black holes which applies to any black hole, rotating or not, charged or not (Smolin 1995; Ashtekar *et al.* 1998, 2000; Krasnov 1998; Rovelli 1998b). A famous relation of Bekenstein, which says that the entropy of a black hole must, in Planck units, be equal to one quarter of the area of its horizon is reproduced in all cases.

It must be mentioned that in doing so a free parameter of loop quantum gravity is determined. This parameter is set once and for all, for a single black hole,[9] after which the theory predicts the right value for all black holes. This is roughly like the fact that physical parameters in a quantum field theory, such as charge and mass, must be renormalized, except that a key difference is that the ratio involved is in this case finite rather than infinite, in fact it is a number roughly equal to one, $\sqrt{3}/\ln(2)$.

Once this is done one has a precise physical picture of the physical degrees of freedom of the black hole horizon, which is completely quantum mechanical, and agrees with all known predictions of Bekenstein and Hawking concerning black holes.

5. Be compatible with the apparently observed positive, but small, value of the cosmological constant. Explain the entropy of the cosmological horizon.

There is no problem with incorporating a cosmological constant, in fact when there is one an exact expression is found for the ground state of quantum geometry, discovered by the Japanese theorist Hideo Kodama (see for example Smolin (2002) for a review with references). The entropy of the horizon seen by cosmological observers is explained exactly, using the methods just described. The temperature

[9] Recent work by Olaf Dreyer (2002) shows that the free parameter can be fixed by the correspondence principle in terms of the oscillations of a black hole.

associated by earlier calculations by Gibbins, Hawking, and others to cosmological spacetimes with horizons is also derived from first principles.

6. Explain what happens at singularities of classical general relativity.

Recent work by Martin Bojowald, under the name of loop quantum cosmology, has shown that the quantization of geometry in loop quantum gravity implies that the initial and final cosmological singularities are removed. In their place are bounces, which means that just before the universe began expanding it was contracting (Bojowald 2001a, b, 2002a, b).[10]

7. Be fully background independent. This means that no classical fields, or solutions to the classical field equations appear in the theory in any capacity, except as approximations to quantum states and histories.

As we have said, this requirement is completely satisfied in loop quantum gravity.

8. Explain how classical general relativity emerges in an appropriate low-energy limit from the physics of the Planck scale.

This is a hard question, as it is analogous to questions in condensed matter physics. There we start with the physics of atoms, which is well understood, and try to deduce the macroscopic properties of the different physical phases a material may be found in. In loop quantum gravity we start with the laws for the evolution of the spin network states and try to deduce the behavior of space and time at much larger scales. To answer this question in general is the focus of much current research.

Already there are a number of results that show that at least in some cases the theory does give a positive answer to this question (Smolin 2002). Some of these concern the case that the cosmological constant is positive. In this case there is a proposal for the ground state of the quantum geometry, discovered by Kodama, which gives a microscopic and exact description of the quantum geometry at Planck scales. But using standard methods of quantum physics, the behavior of space and time at macroscopic scales can be computed, and it is found to agree at leading order with the predictions of general relativity. In particular, the large-scale description is that the geometry of space and time are well approximated by a certain solution to the Einstein equations, which is de Sitter spacetime. Furthermore, low-energy excitations of this state are found to propagate to a good approximation as classical gravitational waves on the de Sitter background.

Thus, at least in the particular case of a nonvanishing cosmological constant, the theory does answer the question positively.

[10] Recently, Tsujikawa *et al.* (2003) showed that Bojowald's formulation of loop quantum cosmology leads to corrections to the predictions of inflationary theories for CMB spectra that may be observable – and may even have already been observed.

9. Predict new physical phenomena, at least some of which are testable in current or near future experiments.

In fact, there are calculations which show that loop quantum gravity does predict corrections to the energy–momentum relations of the form of expression (22.4). The first of these was done by Gambini and Pullin (1999); other calculations leading to the same conclusion are reported in several papers (Alfaro *et al.* 2002; Smolin 2002).

These different calculations make different assumptions as to the ground state of the quantum geometry, and consequently different predictions for the value of the parameter α appear. A goal of present work is to see if the theory makes in fact a unique prediction for α. It would be very good if a firm prediction for the value of α was made before the GLASS observatory, which has the possibility of measuring α, is launched.

10. Predict whether the observed global Lorentz invariance of flat spacetime is realized exactly in nature, up to infinite boost parameter, or whether there are modifications of the realization of Lorentz invariance for Planck scale energy and momenta.

This question arises in any quantum theory of gravity because Lorentz invariance cannot be a symmetry of the quantum theory of gravity. First of all, it is not a symmetry in classical general relativity or any other relativistic theory of gravity that incorporates the equivalence principle. Lorentz invariance (by which I mean here global Lorentz invariance) is a symmetry of Minkowski spacetime, which is a particular solution of the classical equations of motion of general relativity. But it has no significance beyond this; it does not come into consideration with studying any other solution, nor does it play any role in the formulation or physical interpretation of the equations of general relativity.

Since a quantum theory of gravity must be, as I have argued above, background independent, no property that applies only to a single classical solution can play any role in the quantum theory. Thus Lorentz invariance, as a global symmetry, can play no role in a quantum theory of gravity. As such it can only be recovered for a particular case, and it can only be meaningful in a low-energy approximation in which the quantum geometry of space and time is approximated in the language of a solution to classical general relativity.

It is therefore an open problem whether Lorentz invariance is in fact recovered, and whether it is recovered in the usual linear form, or as a nonlinear realization.

Now, as I mentioned, there are calculations in loop quantum gravity that indicate the presence of the new term in expression (22.4). This implies that the usual linear action of the Lorentz transformations cannot be a symmetry of the theory. There are then two possibilities: the low-energy limit of the theory may break Lorentz invariance, so that there is a preferred frame, or it may instead signal that there are also Planck scale corrections to the actions of the Lorentz transformations, so that the symmetry is realized nonlinearly. Although the issue is so far not settled, I believe that there is a good argument that the latter must be the case in loop quantum gravity. This is because a simple breaking of Lorentz invariance implies the existence of a preferred reference frame with a preferred notion of at rest. However, there is no place in the structure of the theory for such a preferred reference frame to come in.

There is also a nice argument that suggests that Lorentz invariance is realized nonlinearly in the low-energy limit. As I argued above, the Planck energy and Planck length are to be seen as thresholds, beyond which the classical description of spacetime geometry breaks down and is replaced by a quantum description. Now length and energy are quantities that transform under Lorentz transformations. Normally, observers moving relative to each other do not agree on their values.

This raises a difficult question: if I find new quantum effects when I probe above the Planck energy, do other observers, moving very rapidly relative to me, see the new effects when the energy goes above the Planck energy, as measured in my frame of reference, or theirs? Because energy transforms, they are not the same thing.

One way to resolve this confusion would be if the action of the Lorentz transformations could be modified so that all observers agree on what the Planck energy is. Then this one energy could be a universal, like the speed of light.[11] It turns out that one can require that the relativity principle is modified so that there is one preferred energy scale, which all observers agree on, while preserving both the invariance of the velocity of light (in the limit of low energies) and the equivalence of inertial observers. Such modified forms of special relativity are described in Bruno *et al.* (2001), Ahluwalia and Kirchbach (2002), Amelino-Camelia (2002), Judes (2002), Judes and Visser (2002), Kowalski-Glikman and Nowak (2002), and Visser (2002).

11. Provide precise predictions for the scattering of gravitons, with each other and with other quanta, beyond the classical approximation.

This is the one thing that the theory does not yet do. However, calculations are in progress and I believe that loop quantum gravity will provide finite predictions for all observables including the scattering of gravitons. The reason is that the spin foam formalism is known to give ultraviolet finite results for all physical evolution amplitudes. The issue is then to represent initial and final gravitons in the language of spin networks, so that their transition amplitudes can be computed. We know how to describe linearized gravitons as perturbations of the Kodama state, thus the only

[11] Or, taking into account eqn (22.5), the speed of light in the limit of low energies.

problem that remains is to transform the wave functions for linearized gravitons to spin network amplitudes. This is currently in progress.

The other questions

Loop quantum gravity does appear to address some of the cosmological questions. In recent work, Martin Bojowald (2001a, b, 2002a, b) and collaborators have shown that loop quantum gravity can be applied to study the early universe, and that the result is that quantum effects do eliminate the Big Bang singularity. Although one can no longer describe the geometry of the universe in classical terms, time does continue back through the Big Bang to an earlier collapsing phase. Thus, loop quantum gravity appears to support a picture of the universe in which the Big Bang resulted from gravitational collapse in a previous universe, either the contraction of the whole universe or the collapse of a black hole.

Regarding the issue of unification, loop quantum gravity so far has nothing new to contribute. It is completely compatible with the standard model of elementary particle physics, and all the usual extensions people study, including grand unified theories and supersymmetric theories.

Finally, by providing a completely explicit formulation of a quantum theory of gravity, loop quantum gravity provides a context to study the foundational questions associated with quantum cosmology. A new approach to the foundational problems of quantum mechanics has been proposed by some of the people working in loop quantum gravity, called "relational quantum theory." It is described briefly in Markopoulou's contribution in this volume and in more detail in Smolin (2003).

String theory

We cannot start off the discussion of string theory by giving the postulates of the theory, as we were able to do in the case of loop quantum gravity. The reason is that many string theorists would argue that, to the extent that string theory is the theory of Nature, its postulates have not yet been formulated. Moreover, the conceptual ideas and mathematical language necessary to express string theory in an axiomatic form are widely believed to remain so far undiscovered.

In this way string theory may be compared to previous research programs such as quantum mechanics and general relativity where several years of hard work preceded the formulation of the postulates of the theory. Thus, the *research program* called "string theory" can be taken to be a set of activities in search of the definition of a theory to be called "STRING THEORY." What exists so far is only a collection of results concerning many different "string theories." These are conjectured to be each an approximate descriptions of some sector of the so far undefined STRING

THEORY. Thus, the discovery of the postulates of the theory is likely to occur close to the end of the development of the research program; it may indeed mark its culmination.

There is of course no a priori reason to believe such a research program will not pay off in the end. But this situation can complicate efforts to achieve a consensus or an objective evaluation of the status of the theory. For this reason I propose here carefully to separate results on the table from the exciting conjectures that have been made. Only by doing so can we get a good idea of what needs to be done to prove or disprove the main conjectures and thus see what needs to be done to move closer to the formulation of the theory.

So my goal here will be to evaluate where string theory stands, *now*, with respect to its ability to answer the questions formulated above. To discuss the results on the table, we cannot talk about STRING THEORY, for that does not exist as of this moment. We must instead talk about *string theories*, for these are what the results in hand concern.

These string theories are background-dependent theories. A string theory is very much like a theory of a particle moving in a fixed, spacetime geometry, except that the particle has been replaced by a one-dimensional string. This is to be considered an elementary object, like a point particle, but it has extension in one dimension. The string may be a closed loop, or it may be open, in which case it is a curve with ends, which also move through space.

In the last few years string theory has been extended to include also the propagation of higher-dimensional objects through a background spacetime. These are called branes, and when they are two-dimensional they are called membranes. These branes can play two roles. The ends of open strings can end on these higher-dimensional objects, in which case they are called d-branes. When seen this way the branes are part of the background in which a string moves. But there are also results that indicate that the branes have their own dynamics, and can be considered to be dynamical objects in their own right.

In any case, string theory always starts with the specification of a fixed background geometry for space and time. Thus at least as concerns what is so far understood, string theory is a background-dependent theory. There are conjectures about extensions of string theory that go beyond the background-dependent description, and we will come to these below. But so far as I know, the actual calculations that have so far been done, and all the precisely stated results, concern string theory as a background-dependent theory.

To define a background-dependent string theory, one must first specify a classical background, consisting of a given manifold $\mathcal{M}$, of some dimension d and a metric, g_{ab}. The background fields are often supplemented by certain other fields, which include a scalar field Φ, called the dilaton, and generalizations of electric and magnetic fields, which we will denote generally as A. We then denote a choice of background $\mathcal{B} = \{\mathcal{M}, g_{ab}, \Phi, A\}$.

There are classical theories of the motion of strings, as well as higher-dimensional membranes in such backgrounds. Some of these have been known for centuries. Examples include theories of the stretched strings and membranes used in musical instruments. But what makes string theory challenging is that not all such theories can be cast into the domain of quantum theory. In many cases inconsistencies appear when one attempts to describe a string or membrane stretched in a classical background in the language of quantum mechanics.

But not always. What is remarkable is that there are some string theories that appear to be consistent quantum mechanically. They are what the subject of string theory is all about.

Thus, the important definition to make is that of a *consistent string theory.* This is defined as follows:

- A consistent string theory is a quantum theory of the propagation and interactions of one-dimensional extended objects, closed or open, moving in a classical background, $\mathcal{B}$, which is completely consistent quantum mechanically. In particular it is unitary (which means quantum mechanics preserves the fact that probabilities always add up to 1) and the energy is never negative (Green *et al.* 1987; Polchinski 1998).
- A background $\mathcal{B}$ is called consistent if one may define a consistent perturbative string theory moving on it. Many backgrounds are not consistent. However, there is an infinite list of consistent backgrounds (counting distinct classical backgrounds as distinct), and some backgrounds allow more than one perturbative string theory to be defined on them.
- Consistent string theories are generally characterized by two parameters, which are a length l_{string}, called the string scale, and a dimensionless coupling constant g_{string}, called the string coupling constant. There may also be additional parameters associated with the different backgrounds. These measure aspects of their geometry or the values of the other background fields. In many cases these may be varied continuously without affecting the consistency of the string theory.
- A string theory is called *perturbative* if it describes interactions of strings in terms of a power series in the dimensionless coupling constant g_{string}, such that when $g_{\text{string}} = 0$ there are no interactions.

The basic idea behind the conjecture that string theory is relevant for physics is that a few of the modes of vibrations in a consistent string theory can be interpreted as corresponding to the propagation of a relativistic quantum mechanical particle in the spacetime. Among the modes of strings are particles that appear to correspond to all the particles of the standard model, including fermions, gauge fields, and Higgs particles. It is also the case that the modes of a closed consistent string include massless spin-2 particles. These correspond, under the wave/particle duality, to gravitational waves moving on the fixed spacetime background. Such quantized gravitational waves are called *gravitons*. The basic result that suggests that string theories are relevant for quantum gravity is that they provide in this way a unification of the gravitons with the particles and forces of the standard model of elementary particle physics. The main disadvantage, however, is that so far this unification is understood in a completely background-dependent language.

How string theory answers the quantum gravity questions

1. Tell us whether the principles of general relativity and quantum mechanics are true as they stand, or are in need of modification.

String theory assumes that the principles of quantum theory remain unchanged. But none of the principles of general relativity is assumed to hold exactly in string theory; instead they are generally believed to hold only in the limit of phenomena on scales larger than the string scale. At the same time it is believed that there may be new principles that replace those of general relativity, but no such principles have yet been proposed.

2. Give a precise description of nature at all scales, including the Planck scale.

A consistent string theory, containing gravitons, would appear to answer this question, at least at the background-dependent level at which it is defined.

There are three caveats, however, that must be mentioned.

First, so far string theory is only defined in terms of a power series expansion in the string coupling constant, which is related to the couplings of gauge fields.

The terms in the expansion have only so far been shown to be unambiguously defined – and to give consistent results – to second order in the expansion (D'Hokera and Phong 2002). There have been claims of a proof to all orders since the mid 1980s. But they have not so far been realized.

Here is a summary of the present situation by the authors of the paper (D'Hokera and Phong 2002) that finally presented the proof of consistency to second order:

> Despite great advances in superstring theory, multiloop amplitudes are still unavailable, almost twenty years after the derivation of the one-loop amplitudes by Green and Schwarz for Type II strings and by Gross *et al.* for heterotic strings. The main obstacle is the presence of supermoduli for world-sheets of nontrivial topology. Considerable efforts had been made by many authors in order to overcome this obstacle, and a chaotic situation ensued, with many competing prescriptions proposed in the literature. These prescriptions drew from a variety of fundamental principles such as BRST invariance and the picture-changing formalism, descent equations and Cech cohomology, modular invariance, the light-cone gauge, the global geometry of the Teichmueller curve, the unitary gauge, the operator formalism, group theoretic methods, factorization, and algebraic supergeometry. However, the basic problem was that gauge-fixing required a local gauge slice, and the prescriptions

ended up depending on the choice of such slices, violating gauge invariance. At the most pessimistic end, this raised the undesirable possibility that superstring amplitudes could be ambiguous, and that it may be necessary to consider other options, such as the Fischler–Susskind mechanism.

This situation is a bit disappointing, given that the main claim for string theory as a quantum theory of gravity is that it alone gives a consistent perturbation theory containing gravitons. After all, supergravity theories, which are ordinary field theories that extend general relativity to incorporate supersymmetry, are also consistent in perturbation theory at least to the two-loop level and $N = 8$ supergravity in four dimensions is expected to be consistent at least to five loops (Howe and Stelle 1989; Ber 2002). The difference is that there are reasons to expect that supergravity theories become inconsistent at some point beyond two loops, while there appear to be no strong reasons to believe that the technical difficulties that have blocked a proof of the consistency of perturbative string theory cannot someday be overcome. At the same time, until that is done we cannot be sure that any string theory is well defined, even at the background-independent level.

The second caveat is that it is clear from internal results in string theory that the series expansion that defines string perturbation theory is a divergent series. In cases like this, we know that the series expansion does not define the exact theory. While the first several terms in the expansion may under some circumstances give a good approximation to the exact theory, the series cannot be summed to yield a definition of the exact theory or exact values of any quantity (Gross and Periwal 1988).

It is also the case that any such exact formulation of string theory must be background independent. Thus, until we have a good proposal for such a background-independent formulation of string theory we cannot assert that string theory gives a positive answer to this question.

The third caveat is that so far string theories that are consistent to second order at least do not exist in any background spacetime that has, like our own universe, three dimensions of space and one of time. Instead, it appears to be necessary for consistency that the background spacetime have nine dimensions of space and one of time. There are then six extra dimensions to account for. String theory, it seems, must predict that there are six additional dimensions of space, so far unobserved.

It turns out that there are backgrounds in which these extra dimension are curled up like a higher-dimensional analog of a cylinder, so that the universe is very much smaller in the six extra dimensions than in the three we observe. It can then be argued that we would not so far have observed effects of the extra dimensions.[12]

At the same time, there is no principle or result in string theory that requires the extra dimensions to be curled up or hidden in this way. So string theory, if true, must

[12] It has recently been discovered by Hertog *et al.* (2003) that most known solutions where the extra dimensions are curled up are unstable.

be supplemented by an explanation of why we do not observe all the dimensions that the theory appears to require for its own consistency.

Related to this is the fact that the theory also appears to require for consistency that Nature have a new kind of symmetry, called supersymmetry. This is a symmetry that mixes bosonic and fermionic particles; it also may mix internal symmetries with symmetries of space and time.

So far no evidence for supersymmetry has been observed. This means that if it exists in nature, supersymmetry must be broken in such a way that the symmetry does not hold for low-energy phenomena, but only comes to play above some energy scale which is called the supersymmetry breaking scale.

It is not known whether there are any string theory backgrounds consistent with such broken supersymmetry. None has been so far constructed, but it is widely, although not universally, believed by string theorists that such backgrounds must exist.

Supersymmetries were proposed before string theory, and there are unified theories of the elementary particles that do incorporate supersymmetry. As a result there are experimental searches for evidence of supersymmetry now under way. If supersymmetry is discovered that will be a moral boost to string theory, but it will not be evidence for string theory itself, as there are other supersymmetric theories besides string theory. (In fact loop quantum gravity can be altered to incorporate supersymmetry completely.) Conversely, if the current searches do not find evidence for supersymmetry it only means that the supersymmetry breaking scale, if it exists at all, lies somewhere out of the reach of the experiments. As there are many orders of magnitude for it to hide in between the present experiments and the Planck scale, there is no known test that can disprove the conjecture that Nature is supersymmetric at some scale.

3. Tell us what time and space are, in language fully compatible with both quantum theory and the fact that the geometry of spacetime is dynamical. Tell us how light cones, causal structure, the metric, etc. are to be described quantum mechanically, and at the Planck scale.

String theory does not do this, at least so far, as the description of space and time, light cones, causal structure, etc. is just that of the classical background which is assumed to begin with.

A background-independent formulation of string theory might provide an answer to this question, if one can be constructed in the future.

4. Give a derivation of the black-hole entropy and temperature. Explain how the black-hole entropy can be understood as a statistical entropy, gotten by coarse graining the quantum description.

As it cannot describe spacetime quantum mechanically, string theory does not so far answer this question directly. However, there is a set of results which indicates very impressively that, when string theory is able to describe black holes, it may predict correctly the black-hole entropy (Breckenridge *et al.* 1996; Strominger and Vafa 1996); for a recent review see Das and Mathur (2001).

It does this by the following very clever method. A black hole is characterized by its mass, angular momentum, and charges. There are results that tell us that a black hole cannot have more than a certain amount of charge or angular momentum; roughly speaking, in the right units these quantities cannot exceed the mass of the black hole. Black holes just at the limit are called extremal black holes.

It turns out that systems with these extremal values of charge and angular momentum can be characterized in a very beautiful way in terms of supersymmetry. Such systems can be characterized in terms of the algebra of supersymmetry transformations, in a way that makes it possible to compute some properties of the spectrum of the Hamiltonian with no or little additional information.

What one may then do is to construct a system without gravity, in which Newton's constant is turned off, that has these special values of charges and angular momentum. Such systems may be constructed out of the branes mentioned earlier. Their thermodynamics may be studied, and they turn out to have the same thermodynamic properties as black holes with the same masses, charges, and angular momenta. Moreover, the agreement with the black-hole results persists to a good approximation for systems close to the critical values.

This is very impressive, but also in some sense not surprising, because the algebraic structure of the supersymmetry transformations is enough to determine the thermodynamics. This allows one to conjecture that *if* one could turn on Newton's constant and incorporate relativistic gravitation into these systems, they would become black holes. Further, as this can be done adiabatically, and without violating supersymmetry, one would expect the black holes to have the same thermodynamic properties as the brane systems.

This is good support for string theory, indeed were it not the case there would be a big problem for string theory. At the same time it is more of a consistency check than anything else; it does not comprise a description in string-theory language of black holes. This is borne out by the fact that, despite a lot of effort, it has proved so far not possible to extend the results to black holes that are not at or very close to the extremal cases which supersymmetry characterizes. In particular, these results do not appear to extend to ordinary black holes of the kind that are found in Nature.

5. Be compatible with the apparently observed positive, but small, value of the cosmological constant. Explain the entropy of the cosmological horizon.

String theory, unfortunately, does not appear to allow the cosmological constant to have a positive value. As there is now good observational evidence that the cosmological constant is non-zero and positive, this constitutes a serious difficulty for string theory.

This problem has been commented on and acknowledged by some of the leading string theorists (Pilch *et al.* 1985; Witten 2001). Despite a lot of effort, so far no proposal has appeared that successfully circumvents this problem.[13]

6. Explain what happens at singularities of classical general relativity.

There are a few intriguing calculations in string theory that do show that string theory eliminates some singularities of classical general relativity. However, most of these have to do with singularities in the geometry of the extra six curled-up dimensions. There are a few results that address the problem of the cosmological singularity and the singularities of black holes, but no results that definitely show that string theory resolves these problems.

7. Be fully background independent. This means that no classical fields, or solutions to the classical field equations, appear in the theory in any capacity, except as approximations to quantum states and histories.

String theory so far is completely background dependent. There is a conjecture that the existing string theories are approximations to different phases of a single, unified, background-independent theory. This is sometimes called the $\mathcal{M}$ theory conjecture. There is some evidence for it, coming from symmetries that have been observed to hold between different supersymmetric sectors of different string theories. But so far there is no agreed-upon conjecture that describes $\mathcal{M}$ theory in any detail, or provides its mathematical structure or basic principles.

8. Explain how classical general relativity emerges in an appropriate low-energy limit from the physics of the Planck scale.

Since string theories so far assume the existence of classical spacetime backgrounds they do not address this problem except in one respect, which is that the Einstein equations are, to a certain approximation, a necessary but not sufficient condition for a spacetime to be a consistent background for string theory.

9. Predict new physical phenomena, at least some of which are testable in current or near-future experiments.

String theory predicts lots of new phenomena. It appears to require that space have extra dimensions and that there are symmetries present in Nature not so far

[13] New results of Kachru *et al.* (2003) provide incomplete evidence, at the level of classical supergravity solutions, for string backgrounds with positive vacuum energy. However, it is not known whether actual consistent string theories exist that correspond to these backgrounds because the explicit construction of such string theories appears to be impossible with currently known techniques.

observed. Also, while a few of the modes of strings are identified with currently known particles, all the rest are so far unobserved.

Thus, the first problem faced by string theory is to hide all the new phenomena. This is done first by setting the string scale near the Planck scale, which makes most of the new phenomena unobservable by any currently conceivable experiment. After this one also has to choose the size of the extra dimensions small enough that they also are unobservable, and one also has to choose the supersymmetry breaking scale high enough that supersymmetry would not yet have been observed.

However, string theory itself offers no principle for how the scale of new phenomena should be set. There are consistent string backgrounds with any values of the scales involved, so that no matter how small in distance or high in energy has been probed, it will always be possible to set the scales so that the new phenomena would not have been observed. (Or at least so long as they are below the Planck scale, which is still 15 orders of magnitude away. For this reason it is not currently possible to design an experiment that would falsify string theory.)

10. Predict whether the observed global Lorentz invariance of flat spacetime is realized exactly in Nature, up to infinite boost parameter, or whether there are modifications of the realization of Lorentz invariance for Planck scale energy and momenta.

As a background-dependent theory, string theory usually is studied on backgrounds for which Lorentz invariance is an exact symmetry. Thus, one would not expect to see Lorentz symmetry breaking effects coming from the Planck scale in the current experiments that will search for such effects. This is one way in which string theory appears to lead to predictions that differ from those of other approaches to quantum gravity.

However, having said this, it must be noted that there are papers in the literature that claim that string theory in some cases will lead to such effects. Moreover, it is not clear that given the large number of unobserved fields in the theory, it will not be possible to choose a background to mock up any experimental results seen within string theory.

11. Provide precise predictions for the scattering of gravitons, with each other and with other quanta, beyond the classical approximation.

String theory does this successfully, subject to the caveats described above, i.e., so far only to second order in string perturbation theory, and with the assumption that spacetime is ten-dimensional rather than four-dimensional.

How string theory addresses the other questions

String theory's main strength is that it offers a genuine theory of the unification of the different forces and elementary particles. At the same time, it so far fails to make

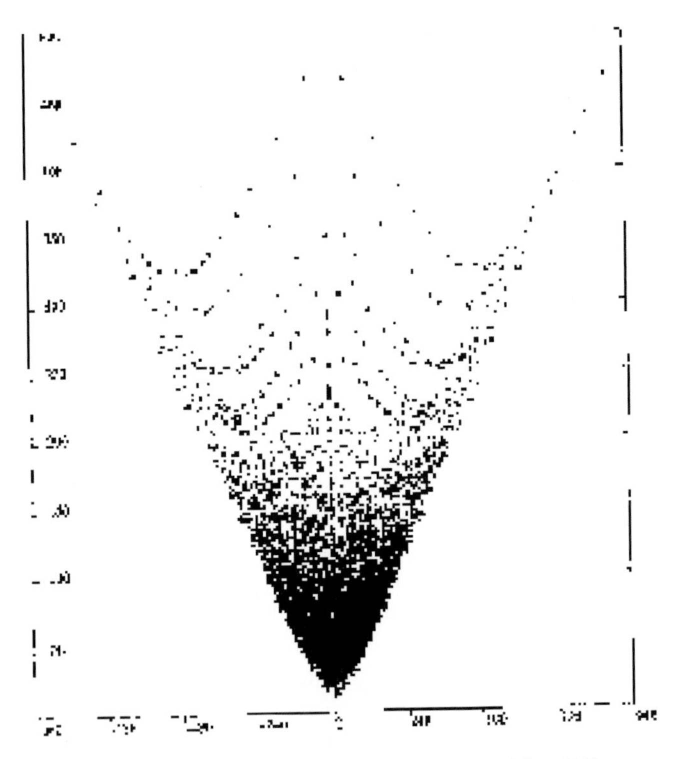

Figure 22.1. A sample of the phenomenology predicted by different consistent string theories. The vertical axis is the number of Higgs fields, up to 480, the horizontal axis is related to the number of left-handed fermion field minus the number of right-handed fermion fields. According to string theory we could equally well live in any of these universes. So far we have observed 48 fermion fields. Experiment gives us indirect evidence for two Higgs fields. (From Klemm and Schimmrigk (1994).)

any falsifiable predictions that would allow us to test the theory against Nature. This deeply frustrating situation is due to the circumstance that string theory is so far only defined at the background-dependent level, in terms of string perturbation theory.

It is unfortunately the case that at this level, string theory is far from a unique construction. There is a huge, possibly even an infinite, number of different string theories which are equally consistent at this level. Given what we know so far, they are all equivalent – any argument for one is an argument for any of them.

These different string theories predict that space has different dimensions, with different spectra of observable elementary particles and forces. An example of the freedom string theory gives the theorist is shown in Fig. 22.1. Here we see a graph of the predictions for the spectra of elementary particles in a particular class of string theories (itself one out of many such classes.) On one axis we have the number of Higgs fields, on the other we see the number of left-handed versus right-handed fermion fields in nature. Nature somehow chose particular values for these quantities; string theory gives us no clue how or why.

Another problem is that each of these string theories has a number of adjustable parameters. While the expectation was that unification would decrease the number of free parameters in physical theory, string theory appears to go in the opposite direction; this is a unified theory that has many more parameters than the total number of parameters in the standard descriptions of the forces it unified.

Finally, all of the many string theories so far constructed disagree with observations in three ways: (1) they have unbroken supersymmetry, whereas supersymmetry is absent or broken in nature, (2) they have many massless scalar fields, whereas none is observed in nature, and (3) they predict that the cosmological constant is zero or negative, whereas it has been measured to be positive.

A lot of effort has gone into constructing string theories without these three problems, but so far no way has been found to do this without giving up the consistency of string perturbation theory.

There is only one way I know to address all these issues, which is by the invention of a background-independent formulation of string theory. The idea is that all the different string theories so far constructed, and certainly many more, would represent approximate descriptions of the physics at different classical solutions of this single, background-independent theory. The classical solutions of this unified theory would then correspond to the different consistent string backgrounds of string perturbation theory. This would be, in essence, a unified theory of string theories.

The hope is that if we had such a theory, we might be able to formulate a principle that would allow us to select which string theory is realized in nature. Such a principle might be mathematical: it might be that in the end there is a unique string theory which minimizes some energy-like variable. We can then hope that this theory would make predictions that would coincide with nature. Or it might be that this principle was contingent and historical, in that different string backgrounds could undergo transitions, analogous to phase transitions, giving rise to others, so that which string background was realized in nature was an historical question, as in multiverse theories or in cosmological natural selection.

The only problem is that so far we do not have any generally agreed-upon proposal for such a unified background-independent string theory, nor do we even have a general method for searching for one. A few people have worked on this problem, I happen to be among them, and I personally think the results so far are encouraging. (I will comment a bit about how we approached this problem in the conclusions, below.) Indeed, there are many relationships that have been discovered between the different perturbative string theories, which suggest the existence of such a unification of string theories, or at least of their supersymmetric sectors.

It may also be hoped that if and when such a unified string theory is invented, we will understand some general principles that lie behind string theory. But so far it is still the case that one cannot speak of a single STRING THEORY, but only of many string theories, and we understand them as the result of an algorithm for constructing approximate theories, and not in terms of any more general principle.

So, while string theory is in principle a unified theory, so far it cannot be said to have answered or provided possible explanations for any of the other questions listed, in phenomenology, cosmology, or the foundations of elementary particle physics.

So what have we learned?

Quantum gravity is still very much an open problem. So, for better or worse, are the big questions of elementary particle physics, cosmology, and the foundations of quantum theory. Theorists have been very busy these last 20 years, but I am not sure we should really be proud of what we have contributed to the store of genuine scientific knowledge. One way to summarize our progress is by listing which of the questions posed in this chapter have been solved. This is done in Table 22.1.

I would like then to close with a few comments, which summarize the situation as I understand it presently.

Regarding quantum gravity, among the many approaches, we have to distinguish approaches that are background independent from approaches that are background dependent. These can be distinguished by the expectations held for the question of whether Lorentz invariance, as formulated originally by Einstein, is a good symmetry of Nature, up to arbitrarily high energies and short distances. If it is then the picture of spacetime as a classical manifold, on which travel excitations of arbitrarily small wavelength, must be assumed to hold, because the difference between large and small energies and large and small distances is not an invariant under the Lorentz transformations. However, if Lorentz invariance breaks down, or is modified at short distances, then a background-dependent theory cannot give very much insight, because there will be no useful notion of a background at scales shorter than the Planck scale, when quantum gravity effects are important.

By far the most hopeful thing I can report is that this question itself is subject to experimental test, and within a few years we will begin to receive relevant data. It is indeed already possible, as Magueijo reports in his contribution to this volume, that such effects have been seen in observations of ultra-high-energy cosmic rays. So it is very encouraging that the key question that distinguishes the different approaches is about to be subject itself to experimental test.

Beyond this, here is the situation.

Table 22.1. *Summary of results: A, solved; B, partial results, or solved in some cases, open in others; C, in progress using known methods; ?, requires the invention of new, so far unknown methods; –, makes no claims to solve*

Question	String theory	Loop quantum gravity
Quantum gravity		
1. General relativity and quantum mechanics true or need modification?	A	A
2. Describes nature at all scales?	B	A
3. Describes quantum spacetime geometry?	B	A
4. Black-hole entropy and temperature explained?	B	A
5. Allows $\Lambda > 0$?	?	A
6. Resolves singularities of general relativity?	B	B
7. Background independent?	?	A
8. New predictions testable now?	?	A
9. General relativity as low-energy limit?	A	B
10. Lorentz invariance kept or broken?	A	A
11. Sensible graviton scattering?	A	C
Cosmology		
1. Explains initial conditions?	?	C
2. Explains inflation?	?	B
3. Does time continue before the Big Bang?	?	A
4. Explains the dark matter and dark energy?	?	?
5. Yields trans-Planckian predictions?	C	C
Unification of forces		
1. Unifies all interactions?	A	–
2. Explains $SU(3) \times SU(2) \times U(1)$ and fermion reps?	?	–
3. Explains hierarchies of scales?	?	–
4. Explains values of standard model parameters?	?	–
5. Unique consistent theory?	?	–
6. Unique predictions for do-able experiments?	?	B
Foundational questions		
1. Resolves problem of time in quantum cosmology?	?	B
2. Resolves puzzles of quantum cosmology?	?	C

String theory is by far the best background-dependent approach as yet invented. But it requires a whole host of extra assumptions, up till now completely unsupported by experiment, stemming from the existence of extra dimensions and extra symmetries, including supersymmetry. Moreover, string theory is a funny kind of unified theory, for to believe in it requires that we believe that the phenomena we observe represent only one set out of a very long list of possible phenomenologies that are logically possible. Our universe's way of existing, and its particular laws, correspond, if string theory is true, to only one of a very large number of solutions of a grander theory.

This situation is radically different from the dream that has motivated most of the people who invented string theory. They were motivated by a traditional reductionist view according to which there would be a single theory that would apply uniquely to our observed universe, and would predict uniquely the values of all the parameters we might measure. There was also a general expectation that as more phenomena were unified, the number of parameters, and choices that the theory allowed, would decrease rather than increase. In string theory the opposite is true. And this is apparently irreversible. Even if a string background that avoids the problems I mentioned and applies to our universe is found, and even if it is found to be the unique solution to some problem of extremization, no one can make all the other string backgrounds go away: those that describe universes with supersymmetry, extra dimensions, massless scalar fields, and a nonpositive cosmological constant. They will remain, all of them, possible universes that, given enough energy, could someday be created, by nature or even by some intelligent beings.

It seems to me that here there is much food for thought and discussion. At least assuming that Lorentz invariance is exact, I find it hard to see how the traditional reductionist idea of a unique theory making unique predictions about a unique universe can be maintained.

At the very least, it is clear that from this point on, string theory is a high-risk enterprise. A lot of energy and time has been devoted to it, over many years, and no solutions to its main problems are in sight. What is certainly clear is that a successful approach to the big open questions of string theory, such as finding a background-independent formulation or a string background, that is consistent, but non-supersymmetric, without massless scalars, but with a small cosmological constant and a large gauge hierarchy, is likely to require completely new methods.

On the other side, once we give up the expectation that Lorentz invariance as given by Lorentz and Einstein is exact at all scales, the situation is much easier. Much of the rationale for believing in extra dimensions or symmetries, possible or actual universes very different from our own, etc. goes away.

Moreover, it turns out that, against many people's expectations, there is at least one way to make a background-independent theory of spacetime geometry. We have an apparently successful answer to the home work question: is there a quantum theory that is, under some definition of quantization, the quantization of general relativity? It is loop quantum gravity, and, so far as we can tell, it describes a perfectly well-defined diffeomorphism-invariant quantum field theory. With the addition of a cosmological constant, it even appears to pass the test of explaining how classical general relativity may emerge in a low-energy limit. There is just one more test it must pass: the prediction of how gravitons scatter, and it is quite possible it will be able to pass this test.

Of course the world is unlikely to be the quantization of general relativity. But this is no argument against loop quantum gravity. It easily incorporates all the standard known and hypothesized ways in which matter might be included in the standard model and supersymmetry. And once we have the mathematical language of spin network states and spin foam histories, it is easy to extend it to a large class of theories. Moreover the study of these theories requires nothing exotic; they are complicated, but they are conventional quantum theories, and they require for their study just the standard methods of quantum field theory and many-body physics.

I think it is fair to say that while string theory is presently limited mainly by the lack of good new ideas for how to address effectively its main open problems, loop quantum gravity is mainly limited by resources. The community of loop theorists is composed of very talented and imaginative people, but they find themselves surrounded by many more good things to do than there is time in which to do them. It will be fascinating to see how this plays out in the next years.

One of the good projects for loop quantum gravity theorists to work on is to make a background-independent form of string theory. So far as I know, all of the good ideas about this problem that have been so far explored are closely connected to loop quantum gravity. This is not surprising, as loop quantum gravity can be understood as the answer to a simple, general question: a background-independent theory of spacetime cannot use the language of manifolds or fields, because any manifold or field represents a particular classical background. So what is left of quantum field theory when manifolds and fields are removed? Whatever is left must be the mathematical language with which to construct a background-independent theory of space, time, and gravity. It, and not some theory that lives in a manifold, must be the language that we use to explain how the physics of space and time is unified with the physics of matter and forces.

It is not hard to answer this question: what is left is only algebra, representation theory of algebras, and combinatorics.

The problem, then, is do we have a general method for constructing quantum theories out of these materials, and studying how they may be derived from classical field theories on manifolds, or else have classical theories as their low-energy limits?

The answer is yes: loop quantum gravity is precisely the method for doing this.

To explain this answer and its power one must have a mathematical language strong enough to explain how structures like manifolds, topology, fields, and quantum amplitudes for them can be associated with algebraic and combinatorial structures. There is in fact such a language: it is the theory of symmetric monoidal categories. I have no space here to introduce it, but the reader may consult Crane (1993a, b, 1995) and Baez (1999). There they will see how the spin foam formalism arises from general considerations about how to make quantum theories with algebra, representation theory, and combinatorics.

So while I am sure that there are still important things about the final answer we may not know, I think there is good reason to expect that the framework of the quantum theory of gravity will have the elements of spin foam models described here.

So what would I put on my 3 by 5 inch file card? Me, I would just put a drawing of a spin foam, like Fig. 24.7, in Markopoulou's chapter in this volume.

Of course, in the end experiment will decide. The most exciting news reported at the conference associated with this book is that intervention from experiment may be coming soon.

Acknowledgments

I would like to thank first of all the many people in the string theory community who over the years have taken the time to explain and teach the ins and outs of a subject they clearly love. These include Tom Banks, John Brodie, Shyamolie Chaudhuri, Michael Douglas, Willy Fischler, Michael Green, Brian Greene, David Gross, Murat Gunyadin, Gary Horowitz, Chris Hull, Clifford Johnson, Renata Kallosh, Juan Maldacena, Djorge Minic, Rob Myers, Herman Nicolai, Amanda Peet, Joe Polchinski, Konstatin Savvidis, Steve Shenker, Kellogg Stelle, Andy Strominger, Arkady Tseytlin, and Edward Witten. If I have still misunderstood something about string theory, it is not their fault, for over the years they have been most generous with time and comments. I am especially grateful to Eric Dowker, Clifford Johnson, Rob Myers, John Schwarz, and Arkady Tseytlin for taking the time to answer questions that came up while writing this review and to Seth Major for very helpful comments on the text. I also must thank my collaborators over the years who have taught me most of what I know about quantum gravity: Stephon Alexander, Matthias Arnsdorf, Abhay Ashtekar, Roumen Borissov, Louis Crane, John Dell, Ted Jacobson, Yi Ling, Seth Major, João Magueijo, Fotini Markopoulou, Carlo Rovelli, and Chopin Soo. Finally I am grateful to the National Science Foundation and the Phillips Foundation for their very generous support which has made my own work possible.

References

Agishtein, M E and Migdal, A A (1992) *Mod. Phys. Lett.* **A7**, 1039.

Ahluwalia, D V and Kirchbach, M (2002) qr-qc/0207004.

Alfaro, J, Morales-Tzcotl, H A, and Urrutia, L F (2002) *Phys. Rev.* **D65**, 103509. hep-th/0108061.

Ambjorn, J (1995) *Class. Quant. Grav.* **12**, 2079.
Ambjorn, J and Loll, R (1998) *Nucl. Phys.* **B536**, 407. hep-th/9805108.
Ambjorn, J, Burda, Z, Jurkiewicz, J, *et al.* (1992) *Acta Phys. Polon.* **B23**, 991.
Ambjorn, J, Dasgupta, A, Jurkiewiczcy, J, *et al.* (2002) hep-th/0201104.
Ambjorn, J, Jurkiewicz, J, and Loll, R (2000) *Phys. Rev. Lett.* **85**, 924. hepth/0002050.
(2001a) *Nucl. Phys.* **B610**, 347. hep-th/0105267.
(2001b) *Phys. Rev.* **D64**, 044011. hep-th/0011276.
(2001c) *J. High Energy Phys.* **9**, 22. hep-th/0106082.
Amelino-Camelia, G (2002) *Nature* **418**, 34.
Amelino-Camelia, G, *et al.* (1997) *Int. J. Mod. Phys.* **A12, 607**.
(1998) *Nature* **393**, 763.
Amelino-Camelia, G and Piran, T (2001) *Phys. Rev.* **D64**, 036005.
Ashtekar, A (1986) *Phys. Rev. Lett.* **57**(18), 2244.
(1988) *New Perspectives in Canonical Gravity*. Naples: Bibliopolis.
(1991) *Lectures on Non-Perturbative Canonical Gravity*, Advanced Series in Astrophysics and Cosmology vol. 6. Singapore: World Scientific Press.
Ashtekar, A, Baez, J, Corichi, A, *et al.* (1998) *Phys. Rev. Lett.* **80**, 904. gr-qc/9710007.
Ashtekar, A, Baez, J, and Krasnov, K (2000) gr-qc/0005126.
Baez, J (1999) gr-qc/9902017.
Barbour, J (1984) In *Proceedings, Quantum Concepts In Space and Time*, Oxford 1984, p. 236.
(2000) *The End of Time*. Oxford: Oxford University Press.
Bekenstein, J D (1973) *Phys. Rev.* **D7**, 2333.
Ber, Z (2002) gr-qc/0206071.
Bojowald, M (2001a) *Class. Quant. Grav.* **18**, L109. gr-qc/0105113.
(2001b) *Phys. Rev. Lett.* **87**, 121301. gr-qc/0104072.
(2002a) *Class. Quant. Grav.* **19**, 2717. gr-qc/0202077.
(2002b) gr-qc/0206054.
Bombelli, L, Lee, J H, Meyer, D, *et al.* (1987) *Phys. Rev. Lett.* **59**, 521.
Breckenridge, J C, Lowe, D A, Myers, R C, *et al.* (1996) *Phys. Lett.* **B381**, 423. hep-th/9603078.
Bruno, N R, Amelino-Camelia, G, and Kowalski-Glikman, J (2001) *Phys. Lett.* **B522**, 133.
Coleman, S and Glashow, S (1997a) hep-ph/9703240.
(1997b) hep-ph/9812418.
Connes, A (1994) *Noncommutative Geometry*. San Diego, CA: Academic Press.
Crane, L (1993a) hep-th/9301061. hep-th/9308126.
(1993b) In *Knot Theory and Quantum Gravity*, ed. J. Baez, p. Oxford: Oxford University Press.
(1995) *J. Math. Phys.* **36**, 6180. gr-qc/9504038.
D'Hokera, E and Phong, D H (2002) *Phys. Lett.* **B529**, 241. hep-th/0110247.
Das, S R and Mathur, S D (2001) hep-th/0105063.
Dirac, P A M (1964) *Lectures on Quantum Mechanics*. Belfer Graduate School of Science Monographs no. 2. New York: Yeshiva University Press.
Dittrich, B and Loll, R (2002) hep-th/0204210.
Dreyer, O (2002) *Phys. Rev. Lett.* **90**, 081301.gr-qc/0211076.
Earman, J (1989) *World Enough and Spacetime: Absolute vs. Relational Theories of Space and Time.* Cambridge, MA: MIT Press.
Ellis, J, *et al.* (2000) *Astrophys. J.* **535**, 139.
(2001) *Phys. Rev.* **D63**, 124025. astro-ph/0108295.
Friedan, D (2002) hep-th/0204131.

Gambini, R and Pullin, J (1996) *Loops, Knots, Gauge Theories and Quantum Gravity*. Cambridge: Cambridge University Press.
(1999) *Phys. Rev.* **D59**, 124021. gr-qc/9809038.
Gambini, R and Trias, A (1981) *Phys. Rev.* **D23**, 553.
(1983) *Lett. Nuovo Cimento* **38**, 497.
(1984) *Phys. Rev. Lett.* **53**, 2359.
(1986) *Nucl. Phys.* **B278**, 436.
Gambini, R, Leal, L, and Trias, A (1989) *Phys. Rev.* **D39**, 3127.
Gonzalez-Mestres, L (1997a) physics/9704017.
(1997b) physics/9705031.
Green, M B, Schwarz, J H, and Witten, E (1987) *Superstring Theory*. Cambridge: Cambridge University Press.
Greene, B R (1999) *The Elegant Universe*. New York: W. W. Norton.
Gross, D J and Periwal, V (1988) *Phys. Rev. Lett.* **60**, 2105.
Hawking, S W (1975) *Commun. Math. Phys.* **43**, 199.
Hertog, J, *et al.* (2003) *J. High Energy Phys.* **305**, 060. hep-th/0304199.
Howe and Stelle (1989) *Int. J. Mod. Phys.* **A4**, 1871.
Jacobson, T and Smolin, L (1988) *Nucl. Phys.* **B299**, 295.
Judes, S (2002) gr-qc/0205067.
Judes, S and Visser, M (2002) gr-qc/0205067.
Kachru, S, *et al.* (2003) hep-th/0301240.
Klemm, A and Schimmrigk, R (1994) *Nucl. Phys.* **B411**, 559.
Kowalski-Glikman, J and Nowak, S (2002) hep-th/0203040.
Krasnov, K (1998) *Gen. Rel. Grav.* **30**, 53. grqc/9605047.
Loll, R (2001) *Nucl. Phys.* **B** (Proc. Suppl.) **94**, 96. hep-th/0011194.
Magueijo, J and Smolin, L (2002a) *Phys. Rev. Lett.* **88**, 190403.
(2002b) gr-qc/0207.
Martin, X, O'Connor, D, Rideout, D P, *et al.* (2001) *Phys. Rev.* **D63**, 084026. arXiv:gr-qc/0009063.
Norton, J D (1987) In *Measurement, Realism and Objectivity*, ed. J. Forge, p. 153. Boston, MA: Reidel.
Penrose, R (1975) In *Proceedings, Oxford Symposium on Quantum Gravity*, p. 268.
Penrose, R and MacCallum, M A (1972) *Phys. Rept.* **6**, 241.
Perez, A (2003) gr-qc/0301113.
Pilch, K, van Nieuwenhuizen, P, and Sohnius, M F (1985) *Commun. Math. Phys.* **98**, 105.
Polchinski, J (1998) *String Theory*, vols. 1 and 2. Cambridge: Cambridge University Press.
Rideout, D P and Sorkin, R D (2001) *Phys. Rev.* **D63**, 104011. arXiv:gr-qc/0003117.
Rovelli, C (1991) *Class. Quant. Grav.* **8**, 297.
(1998) *Living Rev. Rel.* **1**, 1. gr-qc/9710008.
(1998) gr-qc/9603063.
Rovelli, C and Smolin, L (1988) *Phys. Rev. Lett.* **61**, 1155.
(1990) *Nucl. Phys.* **B331**, 80.
(1995a) *Phys. Rev.* **D52**, 5743. gr-qc/9505006.
(1995b) *Nucl. Phys.* **B442**, 593. (Erratum: *Nucl. Phys.* **B456**, 734.
Sarkar, S (2002) *Mod. Phys. Lett.* **A17**, 1025. gr-qc/0204092.
Sen, A (1981) *J. Math. Phys.* **22**, 1781.
(1982) *Phys. Lett.* **B11**, 89.
Smolin, L (1992) In *Quantum Gravity and Cosmology*, ed. J. Perez-Mercader, Singapore: World Scientific Press.
(1994) *Phys. Rev.* **D49**, 4028. arXiv:gr-qc/9302011.

(1995) *J. Math. Phys.* **36**, 6417. gr-qc/9505028.
(1997a) gr-qc/9702030.
(1997b) *Life of the Cosmos*. New York: Oxford University Press.
(2001a) *Three Roads to Quantum Gravity*. New York: Basic Books.
(2001b) arXiv:gr-qc/0104097.
(2002) hep-th/0209079.
(2003a) hep-th/0303/85.
(2003b) *How Close Are We to the Quantum Theory of Gravity?* available online at http://xxx.land.gov.
Stachel, J (1989) In *Einstein Studies*, vol. 1, *Einstein and the History of General Relativity*, ed. D. Howard and J. Stachel, p. Boston, MA: Birkhauser.
Strominger, A and Vafa, C (1996) *Phys. Lett.* **B379**, 99. hep-th/9601029.
Thiemann, T (2001) gr-qc/0110034.
Tsujikawa, K, *et al.* (2003) astr-ph/0311015.
Unruh, W G (1976) *Phys. Rev.* **D14**, 870.
Visser, M (2002) gr-qc/0205093.
Witten, E (2001) hep-th/0106109.
Woit, P (2001) physics/0102051.

23

A genuinely evolving universe

João Magueijo

Imperial College of Science, Technology and Medicine, London

Warm up

A number of surprising observations made at the threshold of the twenty-first century have left cosmologists confused and other physicists in doubt over the reliability of cosmology. For instance the cosmological expansion appears to be accelerating. This is contrary to common sense, as it implies that on large scales gravity is repulsive. Another upheaval resulted from the high redshift mapping of the fine structure constant, when evidence was found for a time dependence of this supposed constant of Nature. Yet another puzzle was the observation of rare *very* high energy cosmic rays. Standard kinematic calculations, based on special relativity, predict a cut-off well below the observed energies, so this may perhaps represent the first experimental mishap of special relativity.

These three surprises are not alone and prompt several questions. Is the universe trying to tell us something radical about the foundations of physics? Or do astronomers merely wish to displease the conservative physicists? It could well be that the strange observations emerging from the new cosmology are correct, and that they provide a unique window into dramatically novel physics. Is the universe trying to give us a physics lesson?

It would be surprising if we already knew everything there is to know about physics. Indeed we expect that currently known physics must break down in the very early universe, or at very high energies. However, no one knows to what extent our current concepts may be inadequate in these extreme situations – the damage caused could be unimaginable. Perhaps Lorentz invariance is broken, energy is not conserved, and the time translational invariance of physics itself lost. The constants of Nature might be lawless dynamical variables, and the observed stability of physics nothing but a sign of old age. Such an extreme picture supports a view

Science and Ultimate Reality, eds. J. D. Barrow, P. C. W. Davies and C. L. Harper, Jr. Published by Cambridge University Press.

of the universe as an evolving being, changing its rules and laws as it goes along. Mutability and evolution may be part of the Big Bang universe at a far more fundamental level – that of physics laws themselves – than ever thought before. It could be that this is precisely the physics lesson to be drawn from the recent cosmological observations.

In this essay I examine this radically new picture of the universe in relation to a class of varying speed of light (VSL) theories, in which mutability is indeed embedded into the formulation of physics. Such theories may explain the recent "unexpected" observations, at least in suitable combinations and with appropriate parameters. In addition, they not only modify the pillars of fundamental physics, but also have profound immediate implications for cosmology and quantum gravity. For example, it could be that a varying c is the ingredient that has been missing in current (mostly failed) attempts to quantize gravity. A number of simple arguments show that special relativity is inconsistent with the existence of an invariant Planck scale, thereby failing to unambiguously separate classical and quantum gravity for all observers. I exhibit a VSL theory capable of resolving this paradox.

But most amazing still are the possible cosmological implications. Time variations in c may cause the energy stored in the cosmological constant to be converted into normal matter. This could account for the creation of the universe, explaining where all the matter in the universe came from – resolving, so to speak, the "Bang" moment of standard cosmology. Or perhaps something even more unpalatable to the unimaginative physicist is behind our existence.

The philosophical background

The issue of mutability in the laws of nature excites considerable passion among scientists and philosophers. Are these laws stable? Or could they change from time to time? Of particular relevance is the fact, discovered last century, that the universe is not static, but is expanding. Nor does the universe seem to be in a state of stationary expansion, that is, it doesn't appear to be expanding at a constant rate. Hence, observations show that *evolution* is a fact of life for the universe, and that the universe looks very different at different times. But could it be that *as the universe undergoes metamorphosis and ages, the laws of nature themselves mutate, in tune with the universe they support?* Such a radical possibility impinges dramatically on a related matter: the nature of physical laws. Could the laws of nature have been different? Could physics have been "otherwise"? Did God have a choice?

In this respect there are two main strands of thought. One is powerfully embodied by John Wheeler's view that in the very early universe, at the dawn of existence, nothing is pre-established. "The only law is that there is no law", or more pictorially "It all comes from higgledy-piggledy" – with the baby universe making up its laws as it goes along. According to this somewhat anarchist conception, the apparent stability of the laws of physics reflects nothing but our old age and cold environment. We live late in the life of a once hot universe, which has cooled down to 3 K from unimaginably higher temperatures. But the hot young universe could on the contrary experience perfect lawlessness.

I am very fond of this view of the nature of physical laws, primarily because I dislike the alternative current of thought: that there is "a law," and that we shall know it; that we are close to the end of theoretical physics; that we may dream of a "final theory." Such mystical views are too "lawyer-minded" for my taste. Why would the universe choose to be so rigid and well behaved, and why should we be the ones privileged to be so close to the Holy Grail? Similar self-serving and self-centered opinions have often been entertained in past history and in different cultures, invariably nurturing diametrically opposed "final answers." Naturally we feel that they were all awfully misguided, whereas of course we aren't. How unbelievably stupid!

I will therefore adopt the first view and examine the implications of anarchy as the only true law of nature, hand in hand with the concept of mutability. This is more easily said than done, because the formalism of physics is made for "setting laws in stone." It is important to recognize that we need to do more than just replace the existing laws by an unchangeable super-law that reigns above all others, telling us how the laws of physics change in time. If we want to have true mutability it is essential that we do something more extreme, and set up existing laws so that their *invariance under time translations is broken*. This is my definition of mutability.

Although implementing this principle within a concrete mathematical scheme is generally hard, an efficient way to do it is to postulate a varying speed of light. Indeed c is so ingrained into the formulation of the laws of physics that if c does vary, say in time, one may expect the collapse of the time translation invariance of physics. VSL will thus provide a framework allowing us to continue our discussion of mutability in the realm of physics rather than metaphysics. We shall start by drawing inspiration from the suggestive observational evidence emerging from modern cosmology.

Some observational puzzles

Over the past few years cosmology has become a field driven by observation, i.e., a proper science. An avalanche of new data from satellite and ground-based observations has shed new light on the high-redshift universe, its large-scale structure, and the history of the cosmological expansion. Some of these results are totally unexpected and defy current theoretical prejudice. Three main observations may be related to the issue of mutability:

- evidence for a changing fine structure constant, α
- ultra high energy cosmic rays
- the accelerating universe

and these shall be the focus of our discussion.

A varying fine structure constant

Webb and collaborators (Webb *et al.* 1999; Murphy *et al.* 2001) have reported evidence for a redshift dependence in the fine structure constant. This evidence was provided by a new observational many-multiplet technique, which exploits the extra sensitivity gained by studying relativistic transitions to different ground states, using absorption lines in quasar spectra at medium redshift. The trend of these results is that the value of α was *lower* in the past, with $\Delta\alpha/\alpha = (-0.72 \pm 0.18) \times 10^{-5}$ for $z \approx 0.5 - 3.5$. In Fig. 23.1 these results are displayed. It is clear that such a result, if true, has tremendous implications. What could be the meaning of a changing α? Looking at the formula $\alpha = e^2/(\hbar c)$, one is immediately faced with a related question – if α is varying, what else must be varying: e, c, $\hbar$, or a combination thereof? Could such a matter be *directly* resolved by experiment?

A moment of thought reveals that this question doesn't make much sense. Whereas α is a dimensionless constant, the three constants that make it up do have dimensions or units. But discussing observational constraints on varying *dimensional* constants is dangerous, because they depend upon the way the units have been defined. Consider for instance the way a meter is currently defined. One takes the period of light from a certain atomic transition as the unit of time, then states that the meter is the distance traveled by light in a certain number of such periods. With this definition it is clear that the speed of light will always be a constant, a statement akin to saying that the speed of light is 1 light year per year. One then does not need to perform any experiment to prove the constancy of the speed of light: it is built into the definition of the units and so has become a tautology.

But if α is seen to vary, the units employed to quote physical measurements may also be expected to vary. A meter stick may elongate or contract and a clock tick faster or slower. Hence under a changing α there is no a priori guarantee that units of length, time, and mass are fixed, and discussing the variability or constancy of a

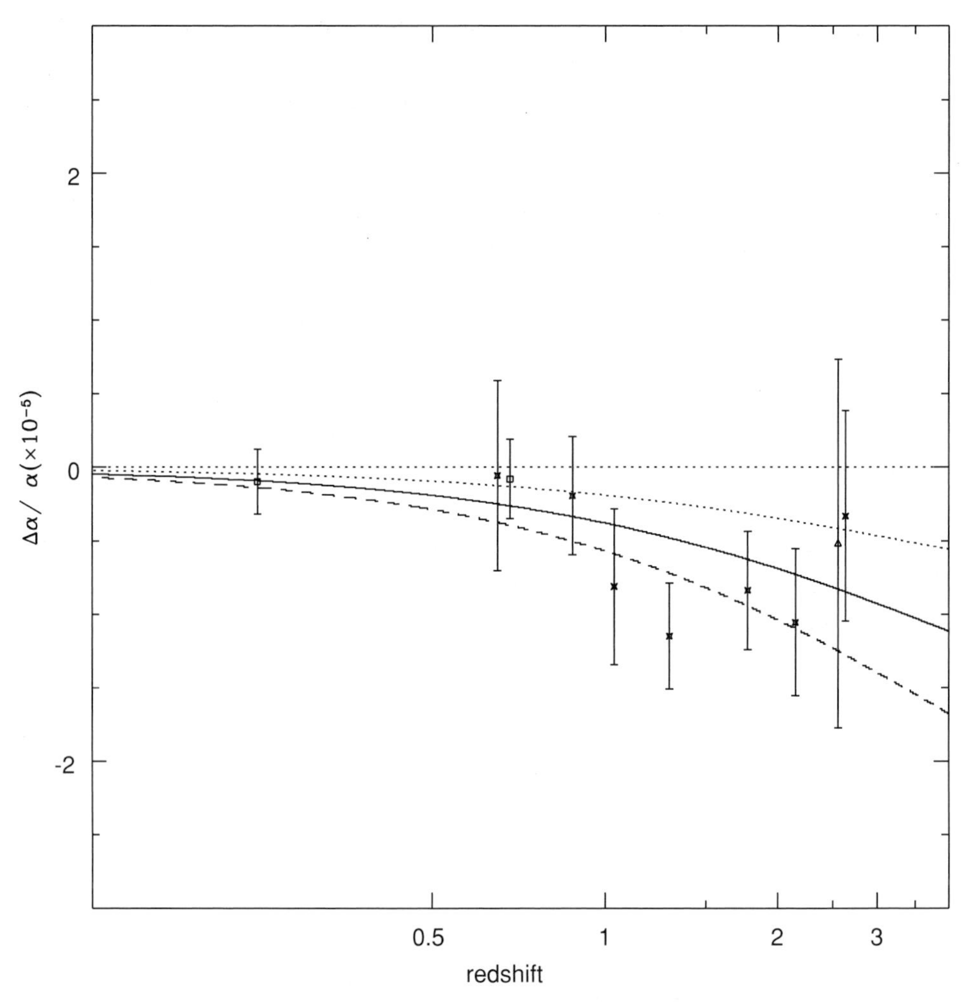

Figure 23.1. The data points are the quasars' results for the changing α. The various lines depict theoretical prediction in several varying-α models.

parameter with dimensions is necessarily circular and depends on the definition of units one has employed. It is precisely to avoid such embarrassing situations that astronomers choose to discuss their observational constraints in terms of parameters, like α, that have no dimensions. They are then testing the true immutability or otherwise of physics, beyond conventions or definitions.

Theorists, however, need dimensional constants in order to set up their theories. Hence the question as to which of e, c, or $\hbar$ is varying is really a question for theorists, at least in the first instance. In order to set up a theory it may be more convenient to make one choice rather than any other. In dilaton theories, or variants thereof (Bekenstein 1982; Barrow *et al.* 2001; Olive and Pospelov 2001;

Sandvik *et al.* 2002), the observed variations in α are attributed to e; VSL theories (Moffat 1993; Barrow and Magueijo 1998, 1999; Albrecht and Magueijo 1999; Barrow 1999; Magueijo 2000) blame c for this variation (and in some cases $\hbar$ too; see Magueijo (2000)). These choices are purely a matter of convenience, and one may change the units so as to convert a VSL theory into a constant c, varying e theory; however such an operation is typically very contrived, with the resulting theory looking extremely complicated. Hence the dynamics associated with each varying α theory "chooses" the units to be used, on the grounds of convenience, and this choice fixes which combination of e, c, and $\hbar$ is *assumed* to vary.

The good news for experimentalists is that once this theoretical choice is made, the different theories typically lead to very different predictions. Dilaton theories, for instance, violate the weak equivalence principle, whereas VSL theories do not (Moffat 2001). VSL theories often entail breaking Lorentz invariance, whereas dilaton theories do not. These differences have clear observational implications, for instance the STEP satellite could soon rule out the dilaton theories capable of explaining the Webb *et al.* (1999) results (STEP 2004). Violations of Lorentz invariance, as we shall see, should also soon be observed – or not.

Hence the question "Is it e or c?", although not directly an observational matter, does return to experiment, and we may hope to get an answer in this respect in the near future.

Threshold anomalies

Another puzzling set of observations are ultra high energy cosmic rays (UHECRs). These are rare events in which one observes showers derived from primary cosmic rays, probably protons, with energies above 10^{11} GeV. At these energies there are no known cosmic ray sources within our own galaxy, so it is expected that in their travels the extragalactic UHECRs should interact significantly with the cosmic microwave background (CMB). These interactions can be shown to impose a hard cut-off above about 10^{11} GeV, the cosmic ray energy at which it becomes kinematically possible to produce a pion. This is the so-called GZK cut-off, and the fascinating result is that UHECRs have been observed beyond this threshold (see Fig. 23.2). Somehow the universe is more transparent to these rays than is predicted.

Several explanations for this result have been advanced, with the most radical noting that the argument leading to the GZK cut-off relies on Lorentz transformations relating the CMB (or cosmological) frame, the proton rest frame, and the center of mass frame. The suggestion is then that the observed threshold anomaly results from quantum gravitational effects breaking Lorentz invariance,

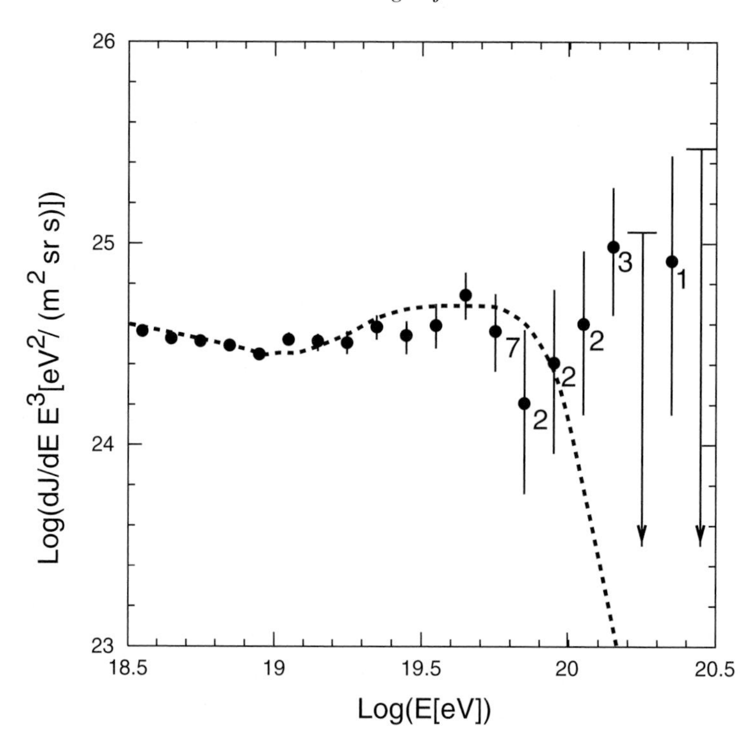

Figure 23.2. The flux of cosmic rays at high energies. The dashed line illustrates the GZK cut-off.

producing corrections to the Lorentz transformations (Amelino-Camelia and Piran 2001).

UHECRs are not lonely freaks: a similar threshold anomaly results from the observation of high-energy gamma rays above 10 TeV. In this case one expects a cut-off due to interactions with the infrared background, with the cut-off energy corresponding to the kinematical condition for production of an electron–positron pair (Finkbeiner *et al.* 2000). For a threshold reaction, the electron–positron pair should have no momentum in the center of mass frame. Hence, the two photons should both have energies of approximately 0.5 MeV in this frame (the electron's rest energy). In order to infer how these energies are perceived in the cosmological frame, one then performs a boost so as to redshift one of the photons to the infrared background energy. The same boost blueshifts the other photon to the expected gamma-ray threshold energy. Could the latter operation differ from the special relativistic prediction?

In both cases it is clear that threshold anomalies may imply corrections to the special relativity formula for boosts at very high energies. Conflict with special relativity leaves three options: the existence of a preferred frame, VSL, or a combination of the two. The first and last possibilities are fascinating: perhaps UHECRs are nothing but our first detection of an "ether wind." But this is not necessary: VSL theories without a preferred frame do exist, and may explain the observed anomalies.

Supernovae results

Recent astronomical observations of distant supernovae light-curves have been realized by the Supernovae Cosmology Project and the High-z Supernova Search (Perlmutter *et al.* 1997, 1998; Garnavich *et al.* 1998; Riess *et al.* 1998; Schmidt 1998). These have extended the reach of the Hubble diagram to high redshifts and provided evidence that the expansion of the universe is accelerating (see Fig. 23.3). This *may* imply that there exists a significant positive cosmological constant, Λ, but any other theory capable of producing a source of repulsive gravity can accommodate these results.

If $\Lambda > 0$, then cosmology faces a very serious fine-tuning problem, and this has motivated extensive theoretical work. There is no theoretical motivation for a value of Λ of currently observable magnitude, a value 10^{120} times smaller than the "natural" Planck scale of density. Such a small non-zero value of Λ is "unnatural" in the sense that making it zero *reduces* the symmetry of spacetime.

One possible explanation is VSL (Barrow and Magueijo 2000). In such theories the energy density in Λ need not remain constant as in the standard theory, and thus does not require fine-tuning. Indeed it is possible to set up theories in which the presence of Λ drives changes in c, which in turn convert the vacuum energy into ordinary matter. In such theories the supernovae results can be explained without any need to fine-tune the initial conditions, and in fact with all parameters of the theory being of order 1 (Barrow and Magueijo 2000).

In addition there is a strange, not often noted coincidence between the redshifts at which the universe starts accelerating and those marking the onset of variations in α. This coincidence can be explained within the framework of these VSL theories: in Barrow and Magueijo (2000) *both* the Webb and supernovae results are fitted using the same set of parameters.

Caveats

All of the above can be seen as evidence for VSL: I have presented three observations and explained how, when put together, they seem to prefer a varying c as a possible explanation. UHECRs support the breaking of Lorentz invariance (implying VSL if we insist on preserving the relativity of motion), which in turn makes a varying c a better theoretical explanation for the changing α results. In addition VSL can

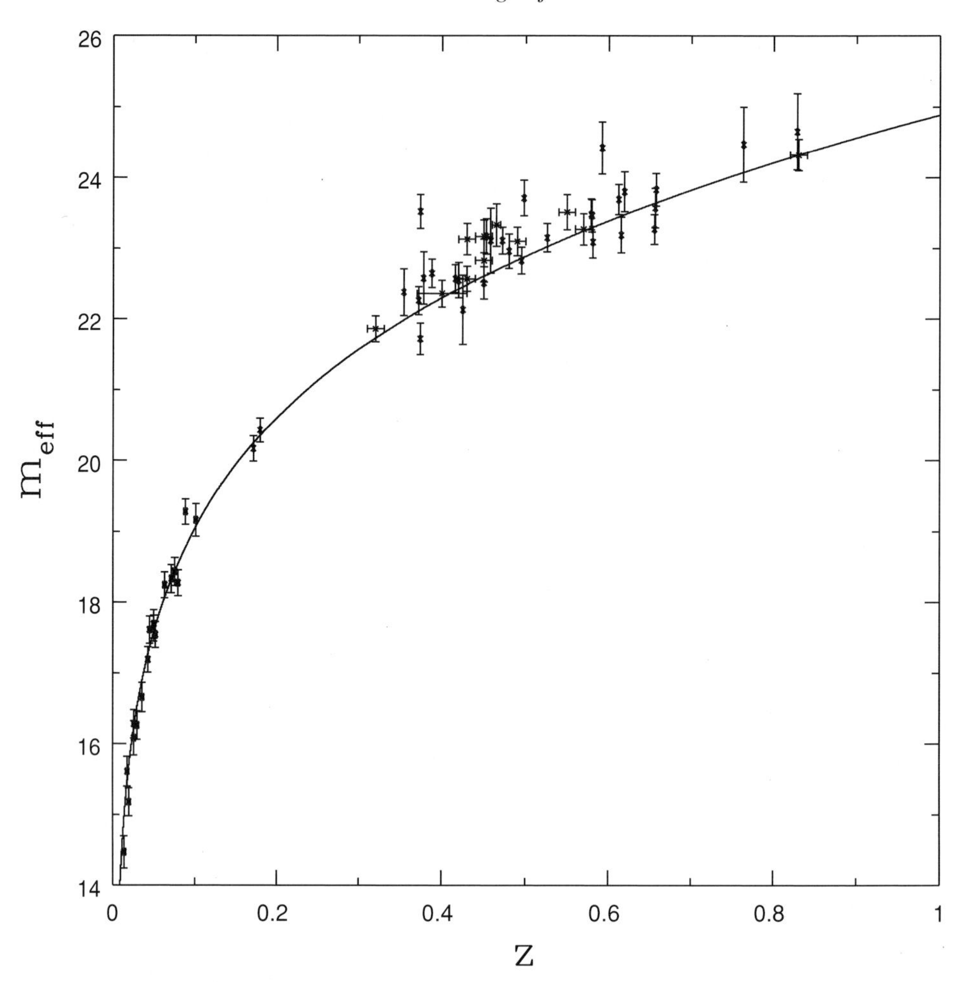

Figure 23.3. The Hubble diagram built from supernovae results (data points) suggests a universe with 30% normal matter and 70% cosmological constant (plotted curve). However, any other form of repulsive gravity could be made to fit the data.

account for the current acceleration of the universe, as well as why this ties in so neatly with the changing α observations.

However, to be honest, there are caveats. Firstly there are more mundane explanations for each of these observations and one may feel that these are still preferable. Changing α results may be explained by dilaton theories, which do not break Lorentz invariance and are very conservative indeed. UHECRs could have an astrophysical explanation; certainly there is no shortage of suggestions, astrophysicists having keen imaginations. Finally the supernovae results can be more modestly explained by quintessence, a replacement for the cosmological constant not dissimilar from the inflaton or dilaton fields (Ratra and Peebles 1988; Wetterich 1988; Zlatev *et al.* 1999).

In addition, and more worryingly, it could also be that all these observations are wrong, due to systematic errors or even more straightforward errors of interpretation. In spite of these caveats, I will proceed with my speculations. For it could also be that these observations are correct and that their more mundane explanations are soon proved wrong. Could new physics be lurking around the corner?

Varying speed of light theories

Now for the theoretical front . . .

I don't think anyone disputes that varying speed of light theories are radical and crazy. Nevertheless there is a growing literature on the subject, and here I categorize the main implementations proposed so far, without trying to be exhaustive. All VSL theories conflict in one way or another with special relativity, but as with deranged people, they are all mad in a very personal way, each VSL theory conflicting with special relativity in its own peculiar fashion. I shall use the type of insult directed at special relativity as my classification criterion for VSL theories.

Recall that special relativity is based upon two *independent* postulates – the relative nature of motion and the constancy of the speed of light. VSL theories do not need to violate the first of these postulates, but in practice one finds it difficult to dispense with the second without destroying the first. This leads to our first criterion for differentiating the various proposals: do they honor or insult the relative nature of motion?

Regarding the second postulate of special relativity, VSL theories behave in a variety of ways, all arising from a careful reading of the small print associated with the constancy of c. Loosely this postulate means that c is a constant, but more precisely it states that the speed of all *massless* particles is the same, regardless of their color (frequency), direction of motion, place, and time, and regardless of the state of motion of observer or emitter. The number of combinations in which this can be violated to accommodate a VSL is large, and explains the large number of theories that have been put forward.

Bearing this in mind we can now distinguish the following main VSL mechanisms:

1. *Hard breaking of Lorentz symmetry*. The most extreme model is that proposed by Albrecht and Magueijo (1999), and studied further by Barrow (1999). In this model both postulates of special relativity are violated: there is a preferred frame in physics, identified with the cosmological frame; the speed of light varies in time, although usually only in the

very early universe; and the invariance in time of physics is broken. In spite of the extreme violence done to relativity, to a large extent this is still the best model as far as cosmological applications are concerned.

2. *Deformed dispersion relations*. This approach was pioneered by Amelino-Camelia and collaborators (Amelino-Camelia *et al.* 1997, 1998; Amelino-Camelia and Majid 2000; Ellis *et al.* 2000, 2001; Amelino-Camelia 2001; Bruno *et al.* 2001). At its most sophisticated it preserves the relative nature of motion, while violating the second postulate of special relativity in the sense that the speed of light is allowed to vary with its color, for frequencies close to the Planck frequency. This is achieved by deforming the photon dispersion relations so that its group velocity acquires an energy dependence. These theories are popular mainly as phenomenological descriptions of quantum gravity. Cosmologies based on them have been constructed (Alexander and Magueijo 2001; Alexander *et al.* 2001) and they could explain the threshold anomalies (Amelino-Camelia and Piran 2001).
3. *Bimetric theories of gravity*. This approach was initially proposed by Clayton and Moffat (1999), and also by Drummond (1999). Again, one does not sacrifice the first principle of special relativity, and special care is taken with the damage caused to the second. In these theories the speeds of the various massless species may be different, but special relativity is still realized within each sector. Typically the speed of the graviton is taken to be different from that of all massless matter particles. This is implemented by introducing two metrics (one for gravity and one for matter), related by the gradients of a scalar field. The greatest achievement of this type of theory is that it is the only VSL theory so far which has led to a model of structure formation (Clayton and Moffat 2002).
4. *"Lorentz invariant" VSL theories*. At the other end of the scale, it is possible to preserve the essence of Lorentz invariance in its totality and still have a varying c. One possibility is that Lorentz invariance is spontaneously broken, as proposed by Moffat in his seminal paper (Moffat 1993). Here the full theory is endowed with exact local Lorentz symmetry; however the vacuum fails to exhibit this symmetry. For example an $O(3, 1)$ scalar field could acquire a timelike vacuum expectation value providing a preferred frame. Another example is the covariant and locally Lorentz invariant theory proposed in Magueijo (2000). Beautiful as these theories may be, their application to cosmology is somewhat cumbersome.
5. *String/M-theory approaches*. In the brane-world scenario we are stuck to a 4-brane which lives in 11 dimensions. Kiritsis (1999) and Alexander (2000) found that if such a brane lives in the vicinity of a black hole it is possible to have perfect 11-dimensional Lorentz invariance (and hence a constant 11-dimensional speed of light), while realizing VSL on the brane. In this approach VSL results from a projection effect, and the Lorentz invariance of the full theory remains unaffected.

For the purpose of our discussion we shall concentrate on the first two approaches. They relate most clearly to the issue of mutability in physics, revealing its close relation to the physics of the early universe and quantum gravity. We examine these two types of implication in turn over the next two sections.

A VSL cosmological model

The VSL cosmological model first proposed in Albrecht and Magueijo (1999) unashamedly makes use of a preferred frame, thereby violating the principle of relativity; in this respect a few remarks are in order. Physicists don't like preferred frames, but they often ignore the very obvious fact that we *have* a great candidate for a preferred frame. Indeed the cosmological frame – the reference frame in which the CMB appears isotropic – is a witness to all our experiments: we have never performed an experiment without the rest of the universe being out there. This has been repeatedly mentioned and then promptly forgotten ever since it was first pointed out by Mach in relation to what he called the "fixed stars."

Of course the problem is that we are generally in motion with respect to this preferred frame, as revealed by the CMB dipole, and this dipole has never been seen to permeate the laws of physics. In addition, every six months our motion around the Sun adds or subtracts a velocity with respect to this frame and we don't see corresponding fluctuations in laboratory physics. The witness is therefore not very talkative, and if the laws of physics are indeed tied to the cosmological frame, its direct influence upon them has to be subtle. But as someone once pointed out, subtle is the Lord.

Having made these remarks we now note that modern physicists invariably *choose* to formulate their laws without reference to our preferred frame. This is more due to mathematical or aesthetic reasons than anything else: covariance and background independence have been regarded as highly cherished mathematical assets since the proposal of general relativity. Naturally, when performing concrete cosmological calculations, everyone loses their prejudices and actually does the sums using the cosmological frame for the sake of mental sanity. But these calculations could in principle have been performed in any other frame, and after much labor they would lead to the same physical result.

The VSL theory proposed in Albrecht and Magueijo (1999) makes a radically different choice in this respect: it ties the formulation of the physical laws to the cosmological frame. The basic postulate is that Einstein's field equations are valid, with minimal coupling (i.e., with c replaced by a field in the relevant equations) in this particular frame. This can only be true in one frame; thus, although the dynamics of Einstein's gravity is preserved to a large extent, it is no longer a geometrical theory. There is still a metric and a curvature, but they are tied to a preferred frame and are no longer geometric objects in the sense of differential geometry. Under such an assumption, if c does not vary then the theory reduces to standard general relativity, and no experiment will reveal the presence of a preferred frame. It is only when c varies significantly that dramatic new effects may be expected.

In the context of the Friedmann metric this means that the dynamical equations are still:

$$\left(\frac{\dot{a}}{a}\right)^2 = \frac{8\pi G}{3}\rho - \frac{Kc^2}{a^2} \tag{23.1}$$

$$\frac{\ddot{a}}{a} = -\frac{4\pi G}{3}\left(\rho + 3\frac{p}{c^2}\right) \tag{23.2}$$

where ρc^2 and p are the energy and pressure densities, $K = 0, \pm 1$, is the spatial "curvature," and a dot denotes a derivative with respect to cosmological time. If the universe is flat ($K = 0$) and radiation dominated ($p = \rho c^2/3$), we have as usual $a \propto t^{1/2}$. At first it might seem that nothing has changed. However the combination of these two equations now leads to:

$$\dot{\rho} + 3\frac{\dot{a}}{a}\left(\rho + \frac{p}{c^2}\right) = \frac{3Kc^2}{4\pi G a^2}\frac{\dot{c}}{c} \tag{23.3}$$

i.e., there is a source term in the energy conservation equation. This turns out to be a general feature of this VSL theory. Energy conservation derives, via Noether's theorem, from the invariance of physics under time translations. The theory badly destroys the latter, so it's not surprising that the former is also not true.

Another way to understand this phenomenon is to note that in general relativity stress-energy conservation results directly from Einstein's equations, as an integrability condition, via Bianchi's identities. By tying our theory to a preferred frame, violations of Bianchi identities must occur, and furthermore the link between them and energy conservation is broken. Hence we may expect violations of energy conservation and these are proportional to gradients of c.

This effect brings out the most physical side of the issue of mutability. If we define mutability as a lack of time translation invariance in physical laws, then what might at first seem to be a metaphysical digression quickly becomes a matter for theoretical physics. We find that lawlessness carries with it shoddy accountancy. Not only do we lose the concept of eternal law, but the book-keeping service provided by energy conservation also goes out the window.

But why would such a feature be desirable? A major source of inspiration in modern cosmology is the flatness problem of Big Bang cosmology, and we find that these violations of energy conservation are coupled to spatial curvature in such a way that they solve this problem. Let ρ_c be the critical density of the universe:

$$\rho_c = \frac{3}{8\pi G}\left(\frac{\dot{a}}{a}\right)^2 \tag{23.4}$$

that is, the mass density corresponding to the flat model ($K = 0$) for a given value of $\dot{a}/a$. Let us quantify deviations from flatness in terms of $\epsilon = \Omega - 1$ with $\Omega = \rho/\rho_c$.

Using eqns (23.1), (23.2), and (23.3) we arrive at:

$$\dot{\epsilon} = (1+\epsilon)\epsilon\frac{\dot{a}}{a}(1+3w) + 2\frac{\dot{c}}{c}\epsilon \tag{23.5}$$

where $w = p/(\rho c^2)$ is the equation of state ($w = 0, 1/3$ for matter/radiation). We conclude that in the standard Big Bang theory ϵ grows like a^2 in the radiation era, and like a in the matter era, leading to a total growth by 32 orders of magnitude since the Planck epoch. The observational fact that ϵ can at most be of order 1 nowadays requires either that $\epsilon = 0$ strictly, or that an amazing fine-tuning must have existed in the initial conditions ($\epsilon < 10^{-32}$ at $t = t_P$). This is the flatness puzzle.

As eqn (23.5) shows, a decreasing speed of light ($\dot{c}/c < 0$) would drive ϵ to 0, achieving the required tuning. If the speed of light changes in a sharp phase transition, with $|\dot{c}/c| \gg \dot{a}/a$, we find that a decrease in c by more than 32 orders or magnitude would suitably flatten the universe. But this should be obvious even before doing any numerics, from inspection of the nonconservation equation eqn (23.3). Indeed if ρ is above its critical value (as is the case for a closed universe with $K = 1$) then eqn (23.3) tells us that energy is destroyed. If $\rho < \rho_c$ (as for an open model, for which $K = -1$) then energy is produced. Either way the energy density is pushed towards the critical value ρ_c. In contrast to the Big Bang model, during a period with $\dot{c}/c < 0$ only the flat, critical universe is stable. This is the VSL solution to the flatness problem.

VSL cosmology has had further success in fighting other problems of Big Bang cosmology that are usually tackled by inflation. It obviously solves the horizon problem (see Figs. 23.4 and 23.5) but it also (nonobviously) solves the homogeneity problem (Albrecht and Magueijo 1999). It solves at least one version of the cosmological constant problem (Albrecht and Magueijo 1999), and in some specific models it can lead to viable structure formation scenarios (Clayton and Moffat 2002).

Perhaps a varying c, along with the mutability in physics it inevitably introduces, is just what the hot phase of the Big Bang Universe is asking for.

The role of VSL in quantum gravity

Most current efforts in quantum gravity and string theory are doomed to fail, because those proposing them have no interest whatsoever in obtaining inspiration from observations. Instead they have arrogantly relied on what might be called "mathematical beauty," usually a byword for theoretical prejudice. Looking first at the world as it is, in particular where it most threatens to destroy our current

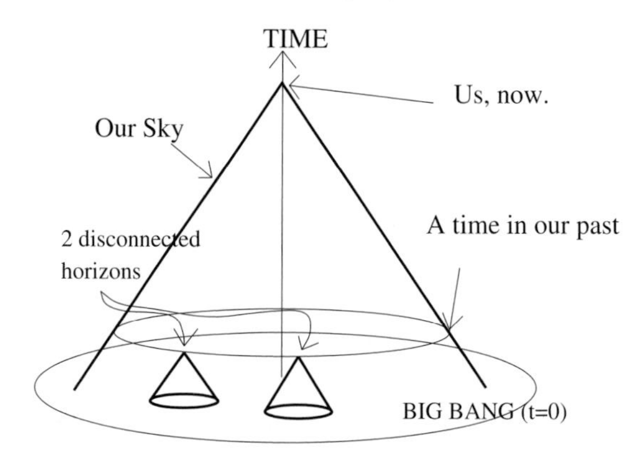

Figure 23.4. Choosing to measure time in years and space in light years, we obtain a diagram in which light travels at 45°. This diagram reveals that the sky is a cone in 4-dimensional spacetime. When we look far away we look into the past; there is an horizon because we can only look as far away as the universe is old. The fact that the horizon is very small in the very early universe, means that we can now see regions in our sky outside each others' horizon. This is the horizon problem of standard Big Bang cosmology.

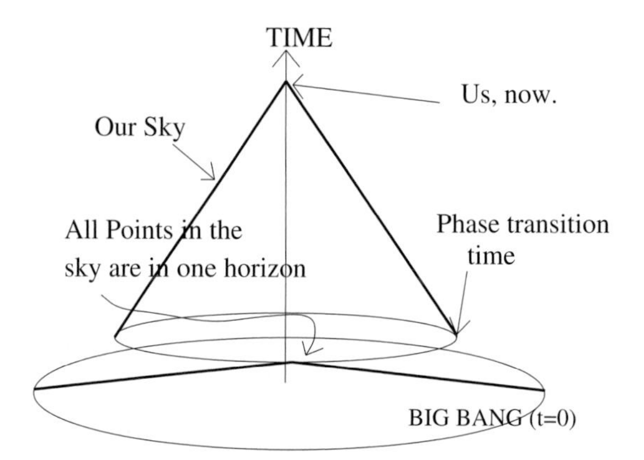

Figure 23.5. Diagram showing the horizon structure in a model in which at time t_c the speed of light changed from c^- to $c^+ \ll c^-$. Light travels at 45° after t_c but it travels at a much smaller angle to the spatial axis before t_c. Hence it is possible for the horizon at t_c to be much larger than the portion of the universe at t_c intersecting our past light cone. All regions in our past have then always been in causal contact. This is the VSL solution of the horizon problem.

perceptions, might be a better strategy. Some puzzling observations have already been mentioned.

This bottom–up approach was pioneered by a number of people, most notably Amelino-Camelia and Kowalski-Glikman. The main idea is disconcertingly simple: the combination of gravity (G), the quantum ($\hbar$), and relativity (c) gives rise to the Planck length, $l_{\rm P} = \sqrt{\hbar G/c^3}$ or its inverse, the Planck energy $E_{\rm P}$. These scales mark thresholds beyond which the classical description of spacetime breaks down and qualitatively new phenomena are expected to appear. No one knows what these new phenomena might be, but both loop quantum gravity (Rovelli 1998; Carlip 2001) and string theory (Polchinski 1996; Forste 2001) are expected to make clear predictions about them once suitably matured.

However, whatever quantum gravity may turn out to be, it is expected to agree with special relativity when the gravitational field is weak or absent, and for all experiments probing the nature of spacetime at energy scales much smaller than E_P. This immediately gives rise to a simple question: *in whose reference frame are* l_P *and* E_P *the thresholds for new phenomena?* For suppose that there is a physical length scale which measures the size of spatial structures in quantum spacetimes, such as the discrete area and volume predicted by loop quantum gravity. Then if this scale is l_P in one inertial reference frame, special relativity suggests it may be different in another observer's frame: a straightforward implication of Lorentz–Fitzgerald contraction.

There are several different answers to this question, the most obvious of which being that Lorentz invariance (both global and local) may only be an approximate symmetry, which is broken at the Planck scale. One may then correct the Lorentz transformations so as to leave the Planck scale invariant, and hope that the modified transformations have something to say about threshold anomalies (Amelino-Camelia 2001; Bruno *et al.* 2001) (but see also Amelino-Camelia *et al.* 1997, 1998; Gambini and Pullin 1999; Adunas *et al.* 2000; Alfaro *et al.* 2000; Ellis *et al.* 2000, 2001; Alexander and Magueijo 2001; Amelino-Camelia and Piran 2001) for other possible experimental implications). Another possibility is that Lorentz invariance gives way to a more subtle symmetry based on a quantum-group extension of the Poincaré or Lorentz group (Amelino-Camelia and Majid 2000; Bruno *et al.* 2001; Lukierski and Nowicki 2002).

But perhaps the most conservative response is the one proposed in Magueijo and Smolin (2002), where it is shown that it is possible to modify the action of the Lorentz group on physical measurements so that a given energy scale, which is taken to be the Planck energy, is left invariant. Hence it is possible to have complete relativity of inertial frames, and have all observers agree that the scale on which a transition from classical to quantum spacetime occurs is the Planck scale, which is the same in every reference frame. At the same time, the familiar and well-tested actions of Lorentz boosts are maintained at large distances and low energy scales.

According to this proposal one simply combines each boost with an energy-dependent dilatation. The boost redshifts the energy; in turn the dilatation blueshifts it, negligibly so for small energies, but just enough to perfectly cancel the standard redshift at the Planck energy, so that this is left invariant. If the ordinary Lorentz generators act as[1]

$$L_{ab} = p_a \frac{\partial}{\partial p^b} - p_b \frac{\partial}{\partial p^a} \tag{23.6}$$

then we consider the modified algebra

$$K^i \equiv L_0{}^i + l_P p^i D \equiv M_0{}^i. \tag{23.7}$$

where the dilatation generator $D = p_a \frac{\partial}{\partial p_a}$. Exponentiation reveals the finite group:

$$p_0' = \frac{\gamma(p_0 - v p_z)}{1 + l_P(\gamma - 1)p_0 - l_P \gamma v p_z} \tag{23.8}$$

$$p_z' = \frac{\gamma(p_z - v p_0)}{1 + l_P(\gamma - 1)p_0 - l_P \gamma v p_z} \tag{23.9}$$

$$p_x' = \frac{p_x}{1 + l_P(\gamma - 1)p_0 - l_P \gamma v p_z} \tag{23.10}$$

$$p_y' = \frac{p_y}{1 + l_P(\gamma - 1)p_0 - l_P \gamma v p_z} \tag{23.11}$$

which reduces to the usual transformations for small $|p_\mu|$. It is not hard to see that the Planck energy is preserved by the modified action of the Lorentz group. In addition, these transformations do not preserve the usual quadratic invariant on momentum space, but preserve instead:

$$||p||^2 \equiv \frac{\eta^{ab} p_a p_b}{(1 - l_P p_0)^2}. \tag{23.12}$$

This invariant is infinite for the new invariant energy scale of the theory $E_P = l_P^{-1}$, and is not quadratic for energies close to or above E_P. This signals the expected collapse in this regime of the usual concept of metric (i.e., a quadratic invariant).

Interestingly, despite the modifications introduced, J^i (the unmodified rotations) and K^i satisfy precisely the ordinary Lorentz algebra:

$$[J^i, K^j] = \epsilon^{ijk} K_k; \; [K^i, K^j] = \epsilon^{ijk} J_k \tag{23.13}$$

(with $[J^i, J^j] = \epsilon^{ijk} J_k$ trivially preserved). However the action on momentum space has become nonlinear due to the term in p^i in eqn (23.7). The new action can be considered to be a nonstandard, nonlinear embedding of the Lorentz group in the conformal group.

[1] Where we assume a metric signature $(+, -, -, -)$ and that all generators are anti-Hermitian; also $a, b, c, = 0, 1, 2, 3$, and $i, j, k = 1, 2, 3$.

This construction can be generalized to incorporate any dispersion relations (which might be measured by experiment) relating E and p by:

$$E^2 f_1^2(E;l_P) - p^2 f_2^2(E;l_P) = m^2 \quad (23.14)$$

where f_1 and f_2 are phenomenological functions. Generally, for massless particles we find that $E/p = f_2/f_1$, so that the group velocity of light $c = dE/dp$ becomes energy dependent at very high energies. Although this is not necessary (and indeed eqn (23.12) is a counterexample), it appears that a varying speed of light may in fact be an essential ingredient in the establishment of an invariant scale (the Planck scale), separating unambiguously the realms of classical and quantum gravity. This unusual approach to quantum gravity, and its implications for threshold anomalies, is currently being actively investigated by a number of groups.

A walk on the wild side

Having shown how VSL may have a say in cosmology and quantum gravity, it is now time to admit that all hell may break loose well beyond these two fields of research. If c and other "constants" of nature do change then the physical support for standard arguments in space science, extraterrestrial biology, information theory, and many others will also be significantly modified. In this section I give two examples, well aware of the wildly speculative nature of these considerations.

I start with the question of life in the universe, and the prospects for establishing communication with alien civilizations. It is not often remarked that these issues depend crucially on our knowledge of physics; yet the most damning constraints sprout directly from the theory of relativity. If, as some have suggested, we are indeed close to the end of physics, then our prospects are dire. If the speed of light really is a speed limit then, combined with the relativistic time-dilation effect, this limits space travel to a range of action of the order of c times the lifetime of the space-traveler (human or alien). Unless there are civilizations out there with lifetimes of millions of years, space travel is essentially impossible. Even in that case, the fact that we do not live millions of years implies that our chances of ever meeting them are very slim indeed.

Of course this dreary landscape changes completely if we are in fact far from the end of physics. This possibility usually leads to wildly unfocused speculations – but I feel it's better form to confine the argument to specific physics models, no matter how speculative. One example of such a concrete model would be wormholes. Here I describe the VSL counterpart: fast-tracks, or VSL cosmic strings.

Cosmic strings are classical solutions to certain particle physics grand unification theories. They are linelike concentrations of energy made up of the grand-unified

Higgs field. Cosmic strings have never been observed, but they are a prediction of the basic mechanism of the standard model: spontaneous symmetry breaking. If one plugs such objects into the equations of some VSL theories, one finds a surprising result (Magueijo 2000). They act as a source for spatial variations in the speed of light, causing in some cases an increase in c along the string, within a radius which could be macroscopic (the string core itself is always microscopic). Hence they create a tunnel within which light travels at much higher speeds, and since in these theories the speed limit is the *local* speed of light, this could make it possible to travel extremely fast throughout the universe without breaking the laws of local causality.

But more importantly in these theories the time dilation effect (and other special relativistic effects) is controlled by v/c where c is the *local* speed of light. Hence such fast space travel would not be afflicted with the annoying side effects of time dilation because we could have v much larger than the asymptotic value of c, but still much smaller than the local value of c. If such cosmic strings existed, the usual constraints affecting space travel would not be valid! This is just one example where the usual framework of current discussions would break down, should the foundations of physics be wildly different than expected, as VSL suggests.

Another matter on which VSL may have some bearing is information theory. Classical and quantum computers have shed new light on the abstract problem of computability, or provability. One usually asks whether or not solving a given problem may be resolved by a code containing a finite number of instructions. An infinite number of operations is always considered to be a physical impossibility.

However suppose that we find a region of spacetime containing a surface where the fine structure parameter α goes like $1/r^{\gamma}$, where r is the distance to the surface and γ an exponent predicted by the theory. One known example is the horizon of a VSL black hole. It was shown in Magueijo (2001) that while this need not be singular, still c must go to *either* zero or infinity (depending upon parameters of the theory) at the horizon.

As we approach such a surface, it is therefore possible to make the rate of electromagnetic interactions diverge, since their timescale is $\tau = \hbar/(\alpha^2 E_e)$, where E_e is the electron's rest energy. Computers (quantum or otherwise) perform operations at a rate directly related to τ, and so by judiciously moving your computer with respect to this surface it would be possible to perform an infinite number of operations. Furthermore such an infinity could be countable or noncountable, and even if it was the latter its divergence could be controlled by a variety of exponents (related to γ).

Could such a far-fetched physical construction destroy our understanding of the limitations of computability? It certainly suggests a breakdown of the equivalence between "impossibility" and "infinity" which is always assumed in discussions of "provability." Yet again shifting the foundations of physics may take a discussion into totally new pastures.

Warm down

In this essay I have examined what might have looked like a metaphysical question – the nature and stability of physical laws – but within the context of a set of physics theories and cosmological observations. I showed that John Wheeler's higgledy-piggledy scenario may be better defined and have more to do with the world of experiment than has been previously thought. VSL theories offer a concrete framework for discussing mutability in the laws of physics. A number of puzzling observations may be about to verify (or rule out!) some of these theories.

I showed how VSL could be useful in the establishment of a healthy Big Bang model, and in the quest for quantum gravity. Curiously no one has attempted to combine VSL theories devised for these two separate purposes. VSL cosmological theories lead to variations in c in space and time, and may explain the Webb and the supernovae results; quantum-gravity driven VSL theories, in contrast, lead to an energy-dependent c and may explain the threshold anomalies. As pointed out in Alexander (2000), it's easy to overlay the two types of theories, but it would be interesting to find a more convincing marriage between the two approaches.

Besides cosmology and quantum gravity, we risked some speculations showing how these theories may also upset the conventional wisdom assumed in discussions held far outside conventional physics. Should VSL be true, quite a lot will have to be rethought.

Nonetheless, at the end of my talk at the symposium several people expressed disappointment – I had promised an anarchist view of physics but had left the basic precepts of logic unshaken. I should stress that my intent was to question the immutability of the laws of physics from the perspective of a physicist. Clearly I could have stripped naked and danced on the table – certainly an act of anarchy, but one which has no bearing upon physics. Likewise one may question the stability of other matters outside physics, such as logic and mathematics, but I am afraid I am not clever enough to do so.

Credits and acknowledgments

I would like to thank the organizers of the symposium *Science and Ultimate Reality* for a unique meeting. As someone pointed out, for once scientists said what they actually thought. Some of the results described here are not my own work, and I apologise if space constraints have limited references. Other results are indeed mine, and I thank my collaborators Andy Albrecht, John Barrow, Håvard Sandvik, Stephon Alexander, and Lee Smolin. I also thank Rachel Bean and Danko Bosanac for discussions relating to the more speculative matters in the section "A walk on the wild side." Finally I would like to thank Kim Baskerville for reading and improving this essay.

References

Adunas, G, *et al.* (2000) *Phys. Lett.* **B485**, 215.
Albrecht, A and Magueijo, J (1999) *Phys. Rev.* **D59**, 043516.
Alexander, S (2000) *J. High Energy Phys.* **0011**, 017. hep-th/9912037.
Alexander, S and Magueijo, J (2001) hep-th/0104093.
Alexander, S, *et al.* (2001) hep-th/0108190.
Alfaro, J, *et al.* (2000) *Phys. Rev. Lett.* **84**, 2318.
Amelino-Camelia, G (2001) *Phys. Lett.* **B510**, 255. gr-qc/0012051.
Amelino-Camelia, G and Majid, S (2000) *Int. J. Mod. Phys.* **A15**, 4301.
Amelino-Camelia, G and Piran, T (2001) *Phys. Rev.* **D64**, 036005.
Amelino-Camelia, G, *et al.* (1997) *Int. J. Mod. Phys.* **A12**, 607.
(1998) *Nature* **393**, 763.
Barrow, J D (1999) *Phys. Rev.* **D59**, 043515.
Barrow, J D and Magueijo, J (1998) *Phys. Lett.* **B443**, 104.
(1999) *Phys. Lett.* **B447**, 246.
(2000) *Astrophys. J. Lett.* **532**, L87-90.
Barrow, J D, *et al.* (2001) astro-ph/0109414.
Bekenstein, J D (1982) *Phys. Rev.* **D25**, 1527.
Bruno, N R, *et al.* (2001) *Phys. Lett.* **B522**, 133.
Carlip, S (2001) *Rept. Prog. Phys.* **64**, 885.
Clayton, M A and Moffat, J W (1999) *Phys. Lett.* **B460**, 263. gr-qc/9910112; gr-qc/0003070.
(2002) astro-ph/0203164.
Drummond, I (1999) gr-qc/9908058.
Ellis, J, *et al.* (2000) *Astrophys. J.* **535**, 139.
(2001) *Phys. Rev.* **D63**, 124025. astro-ph/0108295.
Finkbeiner, D, *et al.* (2000) *Astrophys. J.* **544**, 81.
Forste, S (2001) hep-th/0110055.
Gambini, R and Pullin, J (1999) *Phys. Rev.* **D59**, 124021.
Garnavich, P M, *et al.* (1998) *Astrophys. J. Lett.* **493**, L53.
Kiritsis, E (1999) *J. High Energy Phys.* **9910**, 010. hep-th/9906206.
Lukierski, J and Nowicki, A (2002) hep-th/0203065.
Magueijo, J (2000) *Phys. Rev.* **D62**, 103521.
(2001) *Phys. Rev.* **D63**, 043502.
Magueijo, J and Smolin, L (2002) *Phys. Rev. Lett.* **88**, 190403.
Moffat, J (1993) *Int. J. Mod. Phys.* **D2**, 351.
(2001) astro-ph/0109350.
Murphy, M T, *et al.* (2001) *Mon. Not. Roy. Astron. Soc.* **327**, 1208.
Olive, K and Pospelov, M (2001) hep-ph/0110377.
Perlmutter, S, *et al.* (1997) *Astrophys. J.* **483**, 565.
(1998) *Nature* **391**, 51.
Polchinski, J (1996) hep-th/9611050.
Ratra, B and Peebles, J (1988) *Phys. Rev* **D37**, 321.
Riess, A G, *et al.* (1998) *Astrophys. J.* **116**, 1009.

Rovelli, C (1998) *Living Rev. Rel.* **1**, 1.
Sandvik, H B, *et al.* (2002) *Phys. Rev. Lett.* **88**, 031302.
Schmidt, B P (1998) *Astrophys. J.* **507**, 46.
STEP (2004) http://einstein.stanford.edu/STEP/
Webb, J K, *et al.* (1999) *Phys. Rev. Lett.* **82**, 884.
Wetterich, C (1988) *Nucl. Phys* **B302**, 668.
Zlatev, I, *et al.* (1999) *Phys. Rev. Lett.* **82**, 896.

24

Planck-scale models of the universe

Fotini Markopoulou

Perimeter Institute for Theoretical Physics, Waterloo, Canada

Introduction

Suppose the usual description of spacetime as a $3 + 1$ manifold breaks down at the scale where a quantum theory of gravity is expected to describe the world, generally agreed to be the Planck scale, $l_{\mathrm{p}} = 10^{-35}\ m$. Can we still construct sensible theoretical models of the universe? Are they testable? Do they lead to a consistent quantum cosmology? Is this cosmology different than the standard one? The answer is yes, to all these questions, assuming that quantum theory is still valid. After 80 years work on quantum gravity, we do have the first detailed models for the microscopic structure of spacetime: spin foams.

The first spin foam models (Reisenberger 1994, 1997; Markopoulou and Smolin 1997; Reisenberger and Rovelli 1997; Baez 1998) were based on the predictions of loop quantum gravity, namely the quantization of general relativity, for the quantum geometry at Planck scale. A main result of loop quantum gravity is that the quantum operators for spatial areas and volumes have discrete spectra. (Rovelli and Smolin 1995; for a recent detailed review of loop quantum gravity see Thiemann (2001), and for a nontechnical review of the field see Smolin (2001)). Discreteness is central to spin foams, which are *discrete models* of spacetime at Planck scale. Several more models have been proposed since, based on results from other approaches to quantum gravity, such as Lorentzian path integrals (Ambjørn and Loll 1998; Barrett and Crane 2000; Loll 2001; Perez and Rovelli 2001a; Ambjørn *et al.* 2002.) Euclidean general relativity (Barrett and Crane 1998; Iwasaki 1999; Perez and Rovelli 2001b) string networks (Markopoulou and Smolin 1998), or topological quantum field theory (Baez 2002). For reviews of spin foams see Baez (2002) and Oriti (2001).

Science and Ultimate Reality, eds. J. D. Barrow, P. C. W. Davies and C. L. Harper, Jr. Published by Cambridge University Press.

Spin foam models are background independent, i.e., they do not live in a pre-existing spacetime. Gravity and the familiar $3+1$ manifold spacetime are to be derived as the low-energy continuum approximation of these models. Thus, a spin foam model will be a good candidate for a quantum theory of gravity only if it can be shown to have a good low-energy limit that contains the known theories, namely, general relativity and quantum field theory. One also expects a good model to predict observable departures from these theories.

Spin foams are only a few years old, and progress towards finding their low-energy limit is still in its very early stages. The aim of this chapter is to discuss the basic features of these models, to assess their current status, to point to ways to proceed in future research. The first section contains mostly results on the general formalism of these models. We see that they lead to a novel description of the universe, including a consistent quantum cosmology, in which, in general, there is no wave function of the universe or Wheeler–DeWitt (WDW) equation. In the next section, I note that every spin foam is given by a partition function very similar to that of a spin system or a lattice gauge theory. I argue that this suggests that we should treat this approach to quantum gravity as a problem in statistical physics. However, there is an important difference from systems in statistical physics, the background independence of spin foams. I list features of spin foams relevant to the calculation of their low-energy limit and discuss ways to proceed. Finally, spin foams can address the current challenge that quantum gravity effects, such as breaking of Lorentz invariance, may be observable.[1] Finally I discuss the kind of predictions one could calculate with these models.

No spacetime manifold + quantum theory = spin foams

Several models of the microscopic structure of spacetime have recently been proposed. Different ones were constructed based on different motivations, but they have several features in common which I list here:[2]

A. At energies close to the Planck scale, the description of spacetime as a $3+1$ continuum manifold breaks down. This is the old explanation for the singularities of general relativity and is further supported by the results of loop quantum gravity.

B. At such energies the universe is discrete. This is a simple way to model the idea that in a finite region of the universe there should be only a finite number of fundamental degrees of freedom. This is supported by Bekenstein's arguments, by the black-hole entropy

[1] For possible experiments probing quantum gravity effects see Jacobson *et al.* (2001), Amelino-Camelia (2002), Ellis *et al.* (2002), Kempf (2002), Konopka and Major (2002) and Sarkar (2002).

[2] This is a rather personal interpretation of spin foam models. Several of the models in the spin foam literature are constructed as a path integral formulation of loop quantum gravity, or are modeled on topological quantum field theory, and are not causal, nor is discreteness always considered fundamental. For reviews of spin foams from alternative viewpoints, see Baez (2002) and Oriti (2001).

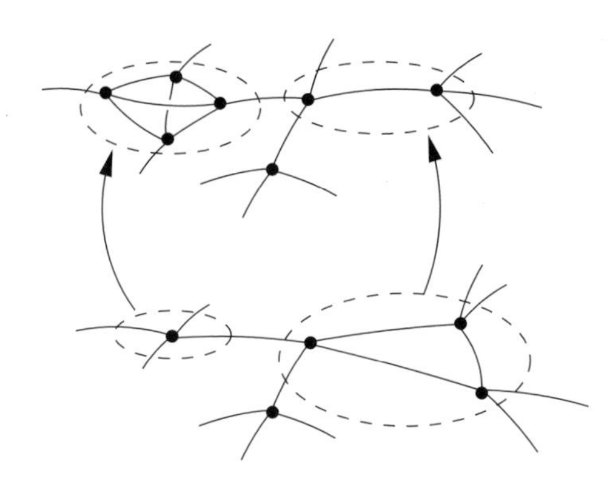

Figure 24.1. One evolution step in a causal spin network model. Only parts of the spin networks are shown.

calculations from both string theory and loop quantum gravity, and by the quantum geometry spectra of loop quantum gravity, and is related to holographic ideas.

C. Causality still persists even when there is no manifold spacetime. How to describe a discrete causal universe has been known for quite some time: it is a *causal set* (Bombelli *et al.* 1987; Sorkin 1990). This is a set of events $p, q, r, \ldots$ ordered by the causal relation $p \leq q$, meaning "p precedes q," which is transitive ($p \leq q$ and $q \leq r$ implies that $p \leq r$), locally finite (for any p and q such that $p \leq q$, the intersection of the past of q and the future of p contains a finite number of events), and has no closed timelike loops (if $p \leq q$ and $q \leq p$, then $p = q$). Two events p and q are unrelated (or spacelike) if neither $p \leq q$ nor $q \leq p$ holds.

Note that the microscopic events do not need to be the same (or a discretization of) the events in the effective continuum theory. Also, the speed of propagation of information in the microscopic theory does not have to be the effective one, the speed of light c.

D. Quantum theory is still valid.

E. Since we are modeling the universe itself, the model should be background independent.

An example of such a model is *causal spin networks* (Markopoulou 1997; Markopoulou and Smolin 1997). Spin networks were originally defined by Penrose as trivalent graphs with edges labeled by representations of SU(2) (Penrose 1971). From such abstract labeled graphs, Penrose was able to recover directions (angles) in three-dimensional Euclidean space in the large spin limit. Later, in loop quantum gravity, spin networks were shown to be the basis states for the spatial geometry states. The quantum area and volume operators, in the spin network basis, have discrete spectra, and their eigenvalues are functions of the labels on the spin network.

Given an initial spin network, to be thought of as modeling a quantum "spatial slice," a causal set is built by repeated application of local moves, local changes of the spin network graph. Each move results in a causal relation in the causal set. An example is shown in Fig. 24.1.

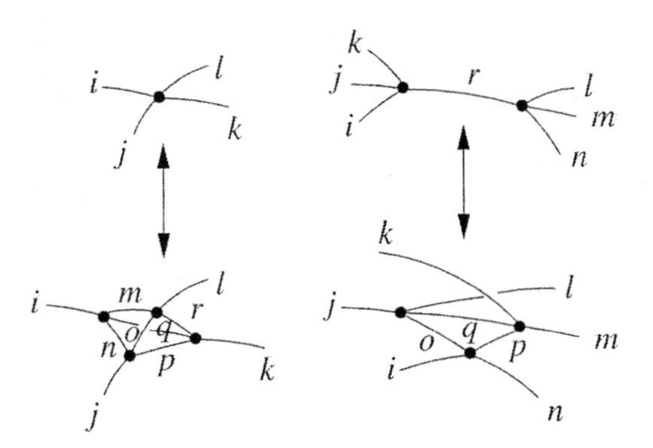

Figure 24.2. The four generating moves for four-valent spin networks.

One can show that a small set of local generating moves can be identified that take us from any given network to any other one. For example, for 4-valent networks, we only need the local moves on pieces of the network that are shown in Fig. 24.2.

One should note that there is no preferred foliation in this model. The allowed moves change the network locally and any foliation consistent with the causal set (i.e., that respects the order in which the moves occurred) is possible. This is a discrete analog of multifingered time evolution. For more details, see Markopoulou (1997).

There is an amplitude A_{move} for each move to occur. The amplitude to go from a given initial spin network S_{i} to a final one S_{f} is the product of the amplitudes for the generating moves in the interpolating history of sequence of such moves, summed over all possible such sequences (or histories):

$$A_{S_{\text{i}}\to S_{\text{f}}} = \sum_{\text{histories } S_{\text{i}}\to S_{\text{f}}} \; \prod_{\substack{\text{moves} \\ \text{in history}}} A_{\text{move}}. \tag{24.1}$$

Explicit expressions for the amplitudes A_{move} have so far been given in Borissov and Gupta (1998), for a simple causal model, in Ambjørn and Loll (1998) (and their higher-dimensional models), with differences in the allowed 2-complexes, and more recently in Livine and Oriti (2002) for the Lorentzian Barrett–Crane model.

The general formalism of the models

With this example in mind, we can write down the formalism of the generic model that has the features **A–E** above. This will let us derive results about the general form of A_{move} and the resulting quantum cosmology.

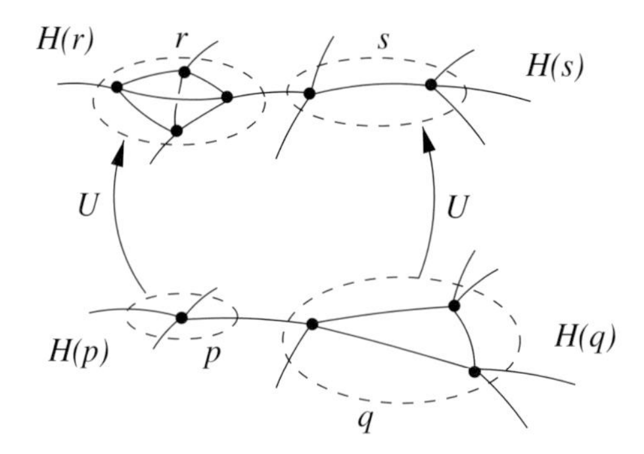

Figure 24.3. A subgraph in the spin network corresponds to a Hilbert space of intertwiners. It is also an event in the causal set. Two spacelike events are two independent subgraphs, and the joint Hilbert space is $H(p \cup q) = H(p) \otimes H(q)$ if they have no common edges, or $H(p \cup q) = \sum_{i_1,\ldots,i_n} H(p) \otimes H(q)$, if they are joined by n edges carrying representations $i_1, \ldots, i_n$. In this example, $H(p \cup q) = \sum_m H(p) \otimes H(q)$.

In the particular example of the causal spin networks, we note that the model really is a causal set "dressed with quantum theory" as follows: a move in the history changes a subgraph of the spin network with free edges. To such a subgraph s is naturally associated a Hilbert space H of so-called intertwiners. These are maps from the tensor product of the representations of SU(2) on the free edges to the identity representation.

The new subgraph, s' has the same boundary as s, the same edges and labels, and therefore corresponds to the same Hilbert space of intertwiners. A move is a unitary operator from a state $|\Psi_s\rangle$ to a new one $|\Psi_{s'}\rangle$ in H (see Fig. 24.3).

Therefore, a causal spin network history is a causal set in which the events are Hilbert spaces and the causal relations are unitary operators.

Is it true for any model with properties **A–E** that it is such an assignment of a quantum theory to a causal set? The answer is yes, although the assignment can be slightly more complicated than for causal spin networks.

We start by interpreting events in C as the smallest Planck scale systems in the quantum spacetime. These, according to **D**, are quantum mechanical. Quantum theory describes the possible states of such a system as states in a Hilbert space, if it is an isolated system, or by a density matrix in the more general case of an open system. It turns out that in our models each event $p \in C$ is best described by a density matrix $\rho(p)$ (for reasons we will explain below).

Going from a causal set C to a quantum spacetime then involves the following steps: (a) To each event $p \in C$ we assign an algebra of operators $A(p)$. Property **B** implies that $A(p)$ is a simple matrix algebra. Any such algebra carries a unique, faithful, normal trace $\tau : A \to \mathbf{C}$ defined by the properties that $\tau(ab) = \tau(ba)$ and $\tau(1) = 1$, and given by the formula $\tau(a) = \mathrm{tr}a/\mathrm{tr}1$. This makes the algebra into a

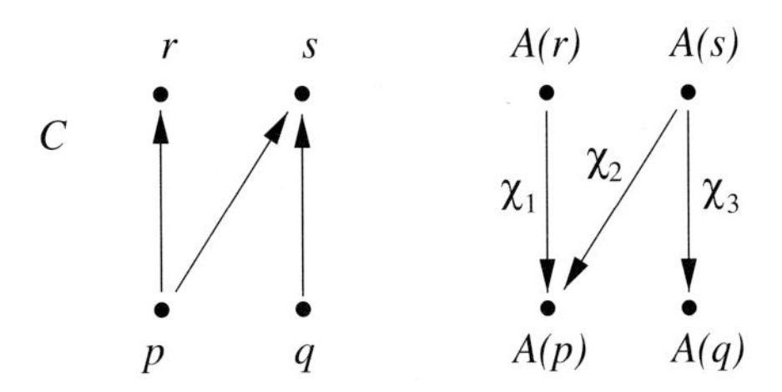

Figure 24.4. On the left we have four events p, q, r, s in a causal set and their causal relations. On the right are the corresponding density matrices and quantum operations χ in the quantum version.

finite-dimensional Hilbert space with inner product $\langle a|\rangle b := \tau(a^b)$. (b) The density matrix $\rho(p)$ representing the state of p is a positive-definite operator in $A(p)$. (c) Two spacelike events p and q are two independent events, and so are in a composite state given by $\rho(p) \otimes \rho(q)$. (d) Every causal relation $p \leq q$ in the causal set corresponds to a *quantum operation* $\chi : A(q) \rightarrow A(p)$ (Fig. 24.4). This is the most general physical transformation that quantum theory allows between two open systems. A quantum operation is a completely positive linear operator, namely: it is linear on the ρ's, it is trace-preserving ($tr(\rho) = tr(\rho') = 1$), positive, and completely positive (if $\chi : \rho(q) \rightarrow \rho(p)$ is positive, then $\chi \otimes \mathbf{1} : \rho(q) \otimes \rho(s) \rightarrow \rho(p) \otimes \rho(s)$ is also positive).

We now have a formalism of models with properties **A–E** as a collection of density matrices connected by quantum operations. When can we have unitary evolution in this quantum spacetime? To answer this, let us first define an *acausal set*. This is a subset $a = \{p, q, r, \ldots\}$ of C with $p, q, r, \ldots$ all spacelike to each other. It is not difficult to check that *unitary evolution is only possible between two acausal sets* a *and* b *that form a complete pair*, namely, every event in b is in the future of some event in a and every event in a is in the past of some event in b. This is because, by construction, information is conserved from a to b. See the example in Fig. 24.5.

The fundamental description of the quantum spacetime as a collection of open systems joined by quantum operations does contain all the relevant physics, including the causal relations and any unitary operations. It is a rather technical construction to discuss here, but one can show that the causal information of C is contained in conditions on the quantum operations, and can prove that given a complete pair a and b, the quantum operations on the causal relations interpolating from a to b compose to give precisely the unitary transformation $U : A(a) \rightarrow A(b)$ (Hawkins *et al.* 2003).

Therefore, our quantum spacetime is a very large set of open systems connected by quantum operations, where unitary evolution arises only as a special case, for a complete pair (the special case of an isolated system). It is interesting to note,

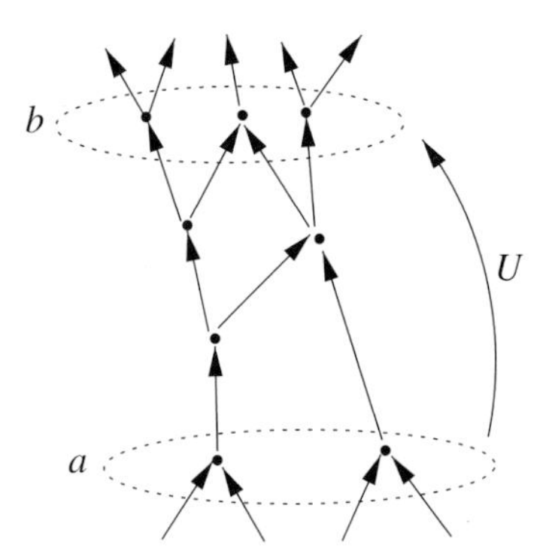

Figure 24.5. The acausal sets a and b form a complete pair.

first, that this description of a quantum spacetime is almost identical to the quantum information theoretic description of noise in a quantum operation (e.g., see Nielsen and Chuang 2000: 353), and, second, that a master equation, already extensively used in quantum cosmology, is a continuum of a quantum operation (Nielsen and Chuang 2000: 386).

These quantum spacetimes lead us to a new quantum cosmology which we describe next.

Quantum cosmology

The standard quantum cosmology is based on the recipe for the canonical, or $3+1$, quantization of gravity. Here one starts with a spacetime with the topology $\Sigma \times R$, where Σ is the three-dimensional spatial manifold. Quantizing the geometry of Σ (identifying variables such as the 3-metric and extrinsic curvature, or Ashtekar's new variables) we obtain the so-called *wave function of the universe* $|\Psi_{\text{univ}}\rangle$. An example of such a state is the Chern–Simons state in loop quantum gravity. This is to "evolve" according to the WDW equation,

$$\hat{H}|\Psi_{\text{univ}}\rangle = 0, \tag{24.2}$$

where $\hat{H}$ is the quantization of the Hamiltonian constraint in the $3+1$ decomposition of the Hilbert–Einstein action of general relativity, a Hermitian operator.

There are several issues with this. First, the simple form of the equation and especially the peculiar right-hand side hides the fact that we need to quantize relativity, a background-independent theory. We only really understand the quantum mechanical evolution of ordinary systems, where an external time is always unambiguously present. Second, eqn (24.2) only works for spacetimes that are globally hyperbolic. Third, one can argue that $|\Psi_{\text{univ}}\rangle$ and eqn (24.2) do not have a satisfactory physical interpretation: $|\Psi_{\text{univ}}\rangle$ is the state of the entire universe and thus only accessible to

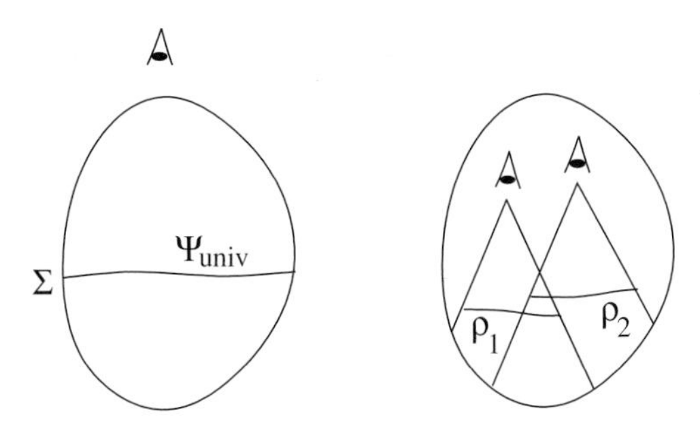

Figure 24.6. A wave function of the universe can only be seen by an observer outside the universe. A quantum theory of cosmology should refer to observables measurable from inside the universe. Inside observers have only partial information of the universe, since only events in their causal past are accessible to them.

an observer *outside* the universe (or specific observers in special universes, such as the final moment of a spacetime with a final crunch, etc.). A satisfactory theory of quantum cosmology has, instead, to refer to physical observations that can be made from inside the universe (Markopoulou 1998) (Fig. 24.6).

In the miscroscopic models we defined, the analog of a spatial slice is an acausal set that is maximal, namely a subset of C such that every other event in C is either in its past or in its future. By tensoring together all the density matrices on each event in this "slice," we could obtain a microscopic $|\Psi_{\text{univ}}\rangle$. However, the causal structure of the generic C is very different than that of a globally hyperbolic spacetime. One can show that, on average, a generic C has very few "slices" (Meyer 1988). And these may cross, i.e., one is partly to the future and partly to the past of the other. All this makes $|\Psi_{\text{univ}}\rangle$ and the WDW equation not very useful for the generic causal set. We cannot restrict to causal sets that admit foliations since these are very special configurations in the partition function of the models.

The interesting fact is that the models provide an alternative quantum cosmology that does not use a wave function of the universe and in fact avoids the issues listed above. The universe is not represented by a global $|\Psi_{\text{univ}}\rangle$, but is instead a collection of ordinary open quantum mechanical systems (all the density matrices on the events of C). Or, at the level of complete pairs, it is a collection of ordinary isolated quantum mechanical systems. There is no WDW equation, but there is *local unitary evolution* and a partition function for the entire system (see Fig. 24.7). These local systems may or may not combine to give an evolving wave function of the universe, depending on the causal structure. As a result, any observables naturally refer to observations made from inside the universe (see Hawkins *et al.* 2003).

A smooth continuous universe with $\Sigma \times R$ topology is what we want to derive in the low-energy limit. Viewed this way, $|\Psi_{\text{univ}}\rangle$ and the WDW equation presuppose the limit we need to derive.

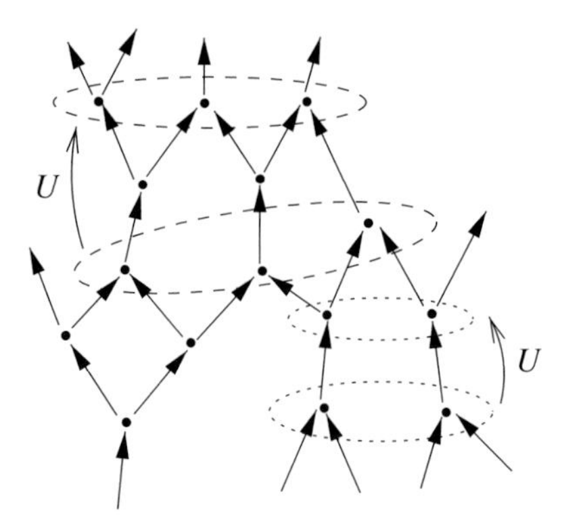

Figure 24.7. Two complete pairs in a quantum spacetime.

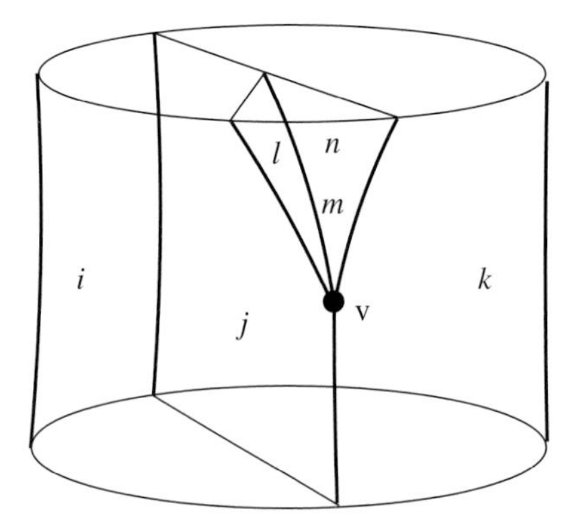

Figure 24.8. A spin foam is a 2-complex whose faces are labeled by group representations as shown. Cuts through a spin foam are spin networks (graphs with edges labeled by the representations on the faces they cut through). Vertices, such as v above, correspond to the moves in the causal spin network example.

Quantum gravity as a problem in statistical physics

We now wish to discuss the problem of calculating the low-energy limit of spin foam models. To do so, we give the general definition of a spin foam, a partition function of which eqn (24.1) is a special case.

A spin foam is a labeled 2-complex whose faces carry representations of some group G, the edges by intertwiners in the group, and the vertices carry the evolution amplitudes. These are functions of the faces and the edges that meet on that vertex and code the evolution dynamics for the model (Fig. 24.8).

A spin foam model is then given by a partition function of the form

$$Z(S_{\mathrm{i}}, S_{\mathrm{f}}) = \sum_{\Gamma} \sum_{\text{labels on } \Gamma} \prod_{f \in \Gamma} \dim j_{\mathrm{f}} \prod_{v \in \Gamma} A_v(j). \tag{24.3}$$

The first sum is over all spin foams Γ interpolating between a given initial spin network $S_{\rm i}$ and a final one $S_{\rm f}$. $\dim j_f$ is the dimension of the G representation labelling the face f of Γ. A_v is the amplitude on the vertex of v of Γ, a given function of the labels on the faces and the edges adjacent to v. A choice of the group G and the functions A_v (and possibly a restriction on the allowed 2-complexes) defines a particular spin foam model. Degrees of freedom such as matter and supersymmetric ones can be introduced by using or adding the appropriate group representations.

Several spin foam models exist in the literature, all candidates for the microscopic structure of spacetime. The very first test such a model has to pass is a tough one: it should have a good low-energy limit, in which it reproduces the known theories, general relativity and quantum field theory. Next, it should predict testable deviations from these theories. The first question is then, what is this limit and how we are going to calculate it.

Note that the models are given by a partition function that is strongly reminiscent of that for a spin system or a lattice gauge theory. This suggests that the problem of their low-energy limit may be best treated as a problem in statistical physics. That is, for spin foams in the correct class of microscopic models, we should find the known macroscopic theory by integrating out microscopic degrees of freedom in Z.

Can we use techniques from statistical physics to test the models? This appears very promising. However, there are important differences between spin foams and the systems studied in condensed matter physics. We can actually list the features of spin foam models, so that we can make a better comparison with the situation in condensed-matter physics. They are:

1. The microscopic degrees of freedom are representations of a group or algebra.
2. The weights in the partition function are amplitudes rather than probabilities.
3. The lattices are the highly irregular spin foam 2-complexes. In general, we cannot simplify the problem by restricting to regular 2-complexes as these are rare configurations in the sum.
4. Spin foams are background independent. This means that we cannot use global external parameters such as time or temperature.
5. There is a minimum length, the Planck length.
6. The partition function contains a *sum* over all 2-complexes with the same given boundary.

Items 1, 2, and 3 above are mainly technical difficulties. Items 4, however, and 6 are novel issues, due to the fact that spin foams are microscopic models of the universe itself. It is possible that one can extend the methods of statistical physics, such as the renormalization group, to deal with background independent systems (Markopoulou 2002).

One thing that is true in statistical physics is that progress is made by analyzing specific models, and the issues mentioned above may or may not turn out to be significant. For example, a very interesting model is the Ambjørn–Loll model of Lorentzian dynamical triangulations (Ambjørn and Loll 1998). In this model, quantum spacetimes of piecewise linear simplicial building-blocks approximate continuum Lorentzian spacetimes. To reflect the causal properties of the continuum spacetimes, the model does not allow any spatial topology change. As a result, it has a foliated structure. It is easy to describe this in $1+1$ dimensions. There, the model is a sum over sequences of discrete one-dimensional spatial slices, namely, closed chains of length L that changes in time.

From the perspective of a relativist, for whom explicit background independence is a necessary condition for a model of the universe to be satisfactory, this model is unpleasant because it appears to have a preferred foliation. The relativist will also question the exclusion of topology change. However, the model is completely well defined, and the suppression of topology change enables us to perform a Wick rotation, solve it analytically, and find that it has a good low-energy limit with very interesting properties. This limit cannot, of course, be classical gravity, since general relativity in $1+1$ is an empty theory. Still, one finds that this model belongs to a different universality class than the well-studied Euclidean (Liouville) two-dimensional quantum gravity, and that it is much better behaved. For example, its Hausdorff dimension[3] is 2, compared to the result for Euclidean two-dimensional histories, which have Hausdorff dimension 4 (reflecting the dominance of fractal geometries). The physically reasonable result of 2 is a direct consequence of the suppression of topology change and the resulting foliated structure.

There are similar results for these models in higher dimensions (see Loll 2001). Certainly, we cannot have a final verdict on this model until we have its solution in four dimensions. However, it raises the possibility that something already familiar from statistical physics, namely, that the properties of the low-energy theory do not have to be present in the microscopic model, may hold even for background independence.

If we regard a spin foam as a statistical physics model, then the phenomenon of universality suggests that it is very likely that models with different microscopic details have the same low-energy limit. This is in contrast with many current arguments for or against specific spin foam models. Most spin foam models are derived from other approaches to quantum gravity (such as path-integral form of loop quantum gravity, deformations of topological quantum field theories, etc.), and so there is attachment to the details of the models. For example, a very popular model is the Lorentzian Barrett–Crane model (Barrett and Crane 2000). It has a partition

[3] The Hausdorff dimension d_H can be measured by finding the scaling behavior of the volumes $V(R)$ of geodesic balls of radius R in the ensemble of Lorentzian geometries: $\langle V(R)\rangle \sim R^{d_H}$.

function of the form (24.3), with representations of the Lorentz group. We know that the Lorentz group is present in the observed low-energy theory (as opposed to Euclidean gravity which is a mathematical construct). This is taken to mean that this model is preferred over the Euclidean Barrett–Crane model (Barrett and Crane 1998). But what is the status of this choice if the Lorentz group appears only in the low-energy theory?

It is my personal opinion that such arguments, for or against particular details of the partition function (24.3), at this stage, miss the point. What is now required is calculations of collective effects in a spin foam. For example, what many spin foam models suppose is that there exist discrete fundamental building-blocks of spacetime. This is more striking than the details of these blocks. Can we demonstrate their existence independently of their detailed structure?

This brings us to the second lesson from statistical physics: experiments are necessary. We currently have several proposals for experiments that will probe the high-energy regime of spacetime. I believe the task at hand is to make contact between the partition function (24.3) and such experiments. Calculation of collective effects in a spin foam can be used to predict, for example, departures from Lorentz invariance. This is not an easy task considering the great gap from the Planck scale to what is currently accessible experimentally, and it is further complicated by questions about what time, temperature, etc. are in these models. But the upside is that, if this works, we will have testable real-physics quantum gravity.

Conclusions

In the last few years, spin foam models have been proposed as the microscopic description of spacetime at Planck scale. I have described their general features and have given the general formalism for such models.

Causal spin foam models provide a new quantum cosmology in which there is no wave function of the universe or WDW equation. Instead the universe is described simply as a collection of ordinary quantum mechanical systems.

Spin foams are best interpreted as *models* of the universe in the statistical physics sense, with gravity and $3 + 1$ spacetime to be derived as the low-energy continuum limit (although this is not the way most were introduced). I believe our best chance to calculate their continuum limit is indeed by importing methods from statistical physics. This immediately implies two things: (a) the microscopic details of the models may not play a role, and (b) further progress should be made by analyzing individual models as well as by experimental input.

It is tempting to compare our current situation to that of the 1900s, shortly before atomism was established. It is hard to believe it now, but at the time, the idea that there could be "*any* hypothesis about the microstructure of matter was opposed on the grounds that (a) such a structure is inherently unobservable; (b) phenomenological theories are quite adequate for the legitimate purposes of science" (quoted from Brush (1986: 92)). Many models of the atomic theory of matter were proposed at the time. "Every Tom, Dick and Harry felt himself called upon to devise his own special combination of atoms and vortices, and fancied in having done so that he had pried out the ultimate secrets of the Creator."[4] It is interesting to see how misguided the physicists were as to the abilities of the experimentalists. They did not think that atoms would be observable in their lifetime. It is also interesting to see that the proof that atoms exist did not involve any particular model. It came with Einstein's theory of Brownian motion, where very basic assumptions on the statistical nature of molecules (that they are identical, interchangeable, etc.), allowed him to calculate a collective effect, that the mean distance traveled by a molecule was proportional to the square root of time, which was *observable* and *different* from the corresponding result for continuous matter.

Again, experience from statistical physics teaches us that experimental input is required to identify the correct models for the systems we are interested in. We are certainly at an early stage but we may well be entering a very exciting period for quantum gravity. There is a real chance that in the next few years experimental data will come in that we will have to explain, with spin foam models, or some other approach. We can start to treat quantum gravity as real physics, where we make contact with experiment and compare predictions with experimental data.

References

Ambjørn, J and Loll, R (1998) Non-perturbative Lorentzian quantum gravity, causality and topology change. *Nucl. Phys.* **B536**, 407. hep-th/9805108.
Ambjørn, J, Jurkiewicz, J, and Loll, R (2002) 3d Lorentzian, dynamically triangulated quantum gravity. *Nucl. Phys. Proc. Suppl.* **106**, 980. hep-lat/0201013.
Amelino-Camelia, G (2002) Quantum-gravity phenomenology: status and prospects. *Mod. Phys. Lett.* **A17**, 899. gr-qc/0204051.
Baez, J (1998) Spin foam models. *Class. Quant. Grav.* **15**, 1827. gr-qc/9709052.
(2002) An introduction to spin foam models of quantum gravity and BF theory. *Lect. Notes Phys.* **543**, 25. gr-qc/9905087.
Barrett, J and Crane, L (1998) Relativistic spin networks and quantum gravity. *J. Math. Phys.* **39**, 3296. gr-qc/9709028.
(2000) A Lorentzian signature model for quantum general relativity. *Class. Quant. Grav.* **17**, 3101. gr-qc/9904025.
Bombelli, L, Lee, J, Meyer, D, *et al.* (1987) Space-time as a causal set. *Phys. Rev. Lett.* **59**, 521.
Borissov, R and Gupta, S (1998) Propagating spin modes in canonical quantum gravity. *Phys. Rev.* **D60**, 024002. gr-qc/9810024.

[4] Boltzmann, L (1899) *Verh. Ges. D. Naturf. Aerzte* **1**, 99. Quoted from Brush (1986: 98).

Brush, S G (1986) *The Kind of Motion We Call Heat*, vol. 1. Amsterdam: North Holland.
Ellis, J, Mavromatos, N E, Nanopoulos, D V, and Sakharov, A S (2002) Quantum-gravity analysis of gamma-ray bursts using wavelets. astro-ph/0210124.
Hawkins E, Markopoulou, F, and Sahlmann, H (2003) Algebraic causal histories. *Class. Quant. Grav.* **20**, 3839.
Iwasaki, J (1999) A surface theoretic model of quantum gravity. gr-qc/9903112.
Jacobson, T, Liberati, S, and Mattingly, D (2001) TeV astrophysics constraints on Planck scale Lorentz violation. hep-ph/0112207.
Kempf, A (2002) On the vacuum energy in expanding space-times. gr-qc/0210077.
Konopka, T J and Major, S A (2002) Observational limits on quantum geometry effects. *New J. Phys.* **4**, 57. hep-ph/0201184.
Livine, E R and Oriti, D (2002) Implementing causality in the spin foam quantum geometry. gr-qc/0210064.
Loll, R (2001) Discrete Lorentzian quantum gravity. *Nucl. Phys. Proc. Suppl.* **94**, 96. hep-th/0011194.
Markopoulou, F (1997) Dual formulation of spin network evolution. gr-qc/9704013.
(1998) The internal logic of causal sets: what the universe looks like from the inside. *Commun. Math. Phys.* **211**, 559. gr-qc/9811053.
(2002) Coarse-graining spin foam models. gr-qc/0203036.
Markopoulou, F and Smolin, L (1997) Causal evolution of spin networks. *Nucl. Phys.* **B508**, 409. gr-qc/9702025.
(1998) Nonperturbative dynamics for abstract (p, q) string networks. *Phys. Rev.* **D58**, 084033. hep-th/9712148.
Meyer, D A (1988) The dimension of causal sets. Ph.D. thesis, Massachussets Institute of Technology.
Nielsen, M A and Chuang, I L (2000) *Quantum Computation and Quantum Information*. Cambridge: Cambridge University Press.
Oriti, D (2001) Spacetime geometry from algebra: spin foam models for non-perturbative quantum gravity. *Rept. Prog. Phys.* **64**, 1489. gr-qc/0106091.
Penrose, R (1971) Theory of quantized directions. In *Quantum Theory and Beyond*, ed. T. Bastin, p. 151. Cambridge: Cambridge University Press.
Perez, A and Rovelli, C (2001a) Spin foam model for Lorentzian general relativity. *Phys. Rev.* **D63**, 041501. gr-qc/0009021.
(2001b) A spin foam model without bubble divergences. *Nucl. Phys.* **B599**, 255. gr-qc/0006107.
Reisenberger, M (1994) Worldsheet formulations of gauge theories and gravity. gr-qc/9412035.
(1997) A lattice worldsheet sum for 4-d Euclidean general relativity. gr-qc/9711052.
Reisenberger, M and Rovelli, C (1997) "Sum over surfaces" form of loop quantum gravity. *Phys. Rev.* **D56**, 3490. gr-qc/9612035.
Rovelli, C and Smolin, L (1995) Discreteness of area and volume in quantum gravity. *Nucl. Phys.* **B442**, 593. Erratum *ibid.* **B456**, 753. gr-qc/9411005.
Sarkar, S (2002) Possible astrophysical probes of quantum gravity. *Mod. Phys. Lett.* **A17**, 1025. gr-qc/0204092.
Smolin, L (2001) *Three Roads to Quantum Gravity*. London: Weidenfeld and Nicholson.
Sorkin, R (1990) Space-time and causal sets. In *Proc. SILARG VII Conf.*, Cocoyoc, Mexico.
Thiemann, T (2001) Introduction to modern canonical quantum general relativity. gr-qc/0110034.

25

Implications of additional spatial dimensions for questions in cosmology

Lisa Randall

Harvard University

Introduction

The topics that have been discussed in this volume are generally very difficult ones. They involve some of the big questions that philosophers have pondered for centuries. The wonderful thing about physics is that sometimes, by pondering "little" tractable problems, you uncover deep truths. Little inconsistencies or new results from old theories can lead to wisdom. These advances are not anticipated, but by having the big questions in mind, one recognizes them when they appear.

In trying to understand deeper truths about cosmology, extra dimensions are a good place to begin. The equations are well grounded in general relativity at scales where quantum gravitational effects should be under control. Nevertheless, by not exclusively focusing on four-dimensional cosmological solutions, one can discover new phenomena. These might even lead to fundamental truths that can impinge on the four-dimensional appearing universe that we observe.

The plan of this chapter is to first go over some of the major questions in cosmology. I will then discuss some new gravitational solutions in more than four dimensions, and what new aspects of gravity they reveal. The other nice thing about these solutions is that they can be used as a testing ground for ideas about gravity that have been developed based on four-dimensional intuition. I will then sketch some of the newer developments in extra dimensions, and how new geometries continue to reveal unanticipated features.

Questions in cosmology

When considering cosmological questions, it is crucial to distinguish the very well-understood late-time cosmology from early cosmology, about which we are much

Science and Ultimate Reality, eds. J. D. Barrow, P. C. W. Davies and C. L. Harper, Jr. Published by Cambridge University Press.

less well informed. Early cosmology deals with physics at the highest energies we can even contemplate. At the highest such energy, as-yet undetermined quantum gravity physics will be relevant, but somewhat below that, general relativity and particle physics should apply. Cosmology will depend on what precisely lies beyond the standard model. There might be a desert, with no new particles or forces appearing until high energy, or there might be more interesting physics possibilities relevant to both particle physics and cosmology.

Late-time cosmology has many notable successes, which include the Hubble expansion of the universe, measurements of the cosmic microwave background radiation (CMBR), and light element abundances. Problems of initial conditions are the horizon, flatness, and homogeneity problems, which are all issues of naturalness. That is, in standard Big Bang cosmology, one would have to take arbitrary and unlikely initial conditions to reproduce what we see. Inflation is from a cosmological vantage point a very nice way to address these problems, and recent CMBR measurements seem to confirm this paradigm. However, it is still an open question what is a natural and believable source of inflation. Any new theory will be constrained to reproduce the successes of late-time cosmology. If it agrees, it is a legitimate candidate for new physics. And we probably need new physics to address some of the remaining big questions.

Many of the big open questions in cosmology involve gravity on long-distance scales, where clearly new matter and energy are needed. What constitutes dark matter? Why isn't there a large cosmological constant? And why does it appear that there is a cosmological term today? Is that really an indication of de Sitter space, or could it be that gravity is somehow modified?

An even more basic question is why the universe appears to be four-dimensional. Although we take this for granted, it is a fact that begs for an explanation.

Black holes also remain to be fully understood. Where and how is the information stored? It appears clear at this point that quantum gravity is relevant to what is happening near the horizon. A perhaps not unrelated question is the potentially holographic nature of the universe.

More generally, of course, we want to probe fully the connections between particle physics and cosmology. We would also like to learn what extra-dimensional physics can tell us. Motivating certain geometries from particle physics can lead to important new insights.

Evolution of extra-dimensional thoughts

There is very strong evidence that makes us think that space is fundamentally three-dimensional. However, earlier in the last century, it was recognized that extra dimensions that are curled up to a very small size (compactified) are completely compatible with what we see for a sufficiently tiny compactification radius. Although below the compactification radius, one sees the higher dimensions, at larger distances, one only sees the extended dimensions. Initially, it was thought these extra dimensions might be very tiny, of order of the Planck length, 10^{-33} cm. It was recognized later on that even scales as big as TeV^{-1} were acceptable, since physics at scales less than 10^{-16} cm, or equivalently, energies greater than TeV, had not been probed (Antoniadis 1990; Lykken 1996). With the advent of "branes," this bound was relaxed significantly further as we will now discuss.

Branes are lower-dimensional objects in a higher-dimensional space. The directions along the brane span only a subset of the full dimensions; the other directions are perpendicular to the brane. A brane is like a membrane, or a curtain. A curtain is a two-dimensional object in three-dimensional space. A brane can be a three-dimensional object in a four-dimensional space. An important property of branes is that particles can be stuck to a brane, much as a bead on a wire is stuck to the one dimension permitted by the set-up. Particles confined to the brane do not probe directly the extra dimensions, in which they are not permitted to propagate.

The particles that can be stuck on a brane include spin-1, spin-$\frac{1}{2}$, and spin-1 particles. This means all particles of the standard model of particle physics conceivably live on a brane. However, the spin-2 graviton is not permitted this option; gravity resides in the entire higher-dimensional space.

There are several important potential properties of branes to know when considering their cosmological implications. First is that rather than curling up extra dimensions, it is an alternative possibility that space is bounded between two branes. That is, appropriate boundary conditions are established so that space ends at branes.

Another significant possibility permitted when there are branes is that extra dimensions can be extremely large, as big as about a tenth of a millimeter (Arkani-Hamed *et al.* 1998; Adelberger 2002). The reason this is permissible is that if only gravity probes the extra dimensions, the constraints on the size of these dimensions is relaxed considerably because gravity is so much weaker, and therefore much less well studied, than the other forces we know.

The third significant implication of branes is that because they carry energy, they can bend space. The brane itself would be bent unless there is also energy in the bulk. It turns out that there is a flat brane solution to Einstein's equations in which there is a negative cosmological constant in the bulk and a positive cosmological term on the brane (Randall and Sundrum 1999a, b). For a suitable tuning between the two energies, one obtains a solution with a flat brane and a horizon an infinite distance from the brane. The metric takes the form $ds^2 = e^{-k|r|}(-dt^2 + d\vec{x}^2) + dr^2$, where r is the distance in the fifth dimension. The coefficient $e^{-k|r|}$ gives the form of the graviton wave function, which in a sense gives the strength with which gravity couples at position r, as illustrated in Fig. 25.1. This says there does exist a graviton

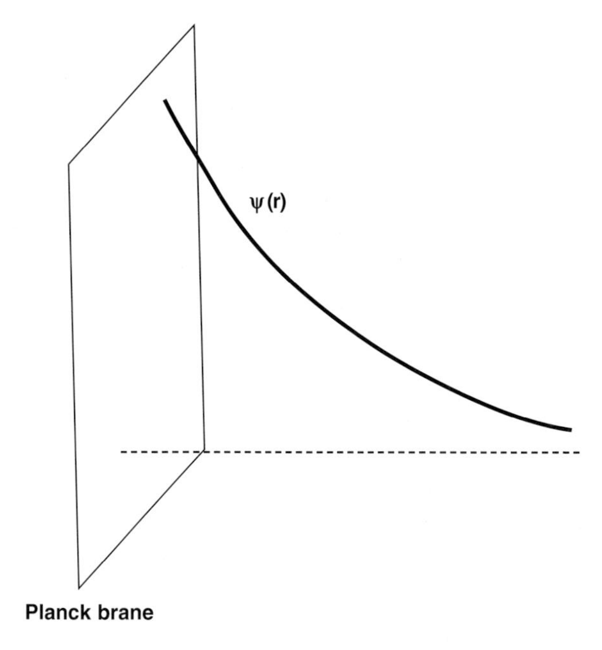

Figure 25.1. $\Psi(r)$ is the graviton wave function.

in the theory, even when the fifth dimension is infinite. It appears that the graviton couples with decreasing strength. This can be reinterpreted as the fact that there is really a single graviton from a four-dimensional perspective with a single M_{Pl}. What the decreasing amplitude tells you is that the maximum energy achieved before quantum gravity effects become important decreases exponentially as you go out into the bulk, away from the brane. That is, the overall energy and length scales are rescaled to maintain the correct graviton coupling.

The fascinating thing about this solution is that although there is an infinite fourth dimension of space, the graviton wave function of the three-dimensional massless graviton mode is nonetheless normalizable, with normalization proportional to M^3/k, where M is the four-dimensional Planck scale and k is the value appearing in the exponential that is determined by the cosmological term in the bulk.

One can calculate the gravitational potential between two masses in this model with a single brane and an infinite fifth dimension. One then finds the usual $1/r^2$ term modified only by a contribution proportional to $1/k^2r^4$ due to the Kaluza–Klein modes, which is very small for k up at the Planck mass.

The Kaluza–Klein spectrum is interesting in and of itself in that there is a continuous, gapless spectrum of modes. Nonetheless, Newton's law holds up to small corrections. The reason for the suppression of the continuum is twofold. First, it is a five-dimensional continuum, which contributes to a more rapid fall-off of gravitational field than the single normalizable zero mode. The fact that there is a further suppression of the Kaluza–Klein modes is due to the fact that unlike the graviton, which is a bound state peaked on the Planck brane, the Kaluza–Klein modes are suppressed on the brane, with a suppression proportional to the square root of the mass of the mode.

The physics revealed by the solution was extremely surprising. The four-dimensional graviton is a bound-state mode. An infinite fifth dimension is completely permissible. It is $1/k$, rather than the size of the space, that sets the normalization of the graviton. The normalizable zero mode means that the gravitational potential throughout the bulk appears to be four-dimensional. However, the gravitational theory becomes strongly coupled at position r at energy of order $M_{Pl}e^{-kr}$. This means the four-dimensional theory can be used only up to lower and lower energy as you move out into the bulk.

So one of the major surprises was that gravity could be localized. Even if gravity is not strictly speaking confined to the brane, its amplitude can be heavily concentrated there. The localized graviton is what permitted an infinite fourth spatial dimension with three-dimensional gravity surviving.

The second major surprise came with the study of a closely related model, that I studied with Andreas Karch (Karch and Randall 2001a). In this example, the relationship between the brane cosmological term and the bulk term is such that the solution to Einstein's equations is an anti-de-Sitter (AdS) brane in a four-dimensional AdS space. This example has the initially confusing property that there is *no* massless normalizable graviton bound state. The form of the solution to Einstein's equations is $ds^2 = \cosh^2((c- |r|)/L)\bar{g}_{ij}dx^i dx^j - dr^2$, where $\bar{g}$ is the three-dimensional AdS metric.

The graviton is again the warp factor, which in this case is $\cosh^2((c - |r|)/L)$. In this case, it is easy to see that this graviton is *not* normalizable, and therefore not in the physical spectrum. This might seem like a paradox, since one can make the four-dimensional AdS curvature arbitrarily small, and still lose the massless four-dimensional graviton. But physics should be continuous, and we should expect to see four-dimensional gravity in this case as well. The resolution to this puzzle is that there does in fact exist a four-dimensional graviton bound state that mediates three-dimensional gravity on and near the brane. The surprising feature of this graviton however is that it is *massive*, with a mass of the order of the three-dimensional AdS curvature. This massive, but very light, mode is what mediates four-dimensional gravity on the brane.

This example is remarkable for several reasons, and teaches us that some of the usual assumptions about gravity are not necessarily correct. The first new thing that we have already mentioned is that the graviton is massive. That the graviton could be massive and still consistently generate four-dimensional gravity was a surprise, and was studied by several authors (Karch and Randall 2001b; Kogan *et al.* 2001).

The second amazing thing in this model is that four-dimensional gravity is a *local* phenomenon. Beyond the turn-around point in the warp factor, there is no evidence of four-dimensional gravity. The theory there just looks like four-dimensional AdS space.

The third interesting thing in this model is that there is truly infinite space, and infinite volume. Many people had tried to argue that the first Minkowski brane example really was equivalent to compactification. The fact that there is infinite volume in this example and that four-dimensional gravity nonetheless locally applies conclusively demonstrates these brane theories with bound-state gravitons are really a new phenomenon, independent of compactification.

The fact that the AdS brane model contains four-dimensional gravity is also important from the vantage point of trying to successfully construct string or supergravity realizations of localized gravity. Previously, it was thought the space had to end at a horizon, imposing overly restrictive boundary conditions. Now we see that the asymptotic behavior of the space is irrelevant; localized gravity is in fact a local phenomenon. This makes a string realization of this model much more viable. An example was conjectured in Karch and Randall (2001a).

One can ask how this example works from a holographic point of view, as I did with Raphael Bousso in (Bousso and Randall 2002). We found the space divides into holographic domains. There is one region of space associated with the brane, while the region of space beyond the turn-around point of the warp factor is associated with the half-boundary of AdS that constitutes the remaining boundary of the full space. The normalization of the four-dimensional graviton is determined by a calculation performed in the first domain, which is why the graviton is normalizable, even in the infinite volume space.

This theory leads to fascinating speculations. It means that the four-dimensions we see might exist on a four-dimensional "island," embedded in a higher-dimensional space. Different locations could see different dimensions. This might even be true for our universe.

What have we learned?

These were two simple solutions to Einstein's equations, but they resulted in some unexpected insights into gravity. Let's review some of the lessons to be taken away.

1. The dimensionality we observe does not necessarily reflect the full dimensionality of space. This is very important since it means there is an alternative to compactification. It could be that additional dimensions are sufficiently curved, rather than curled up. This might help explain why we don't see all the dimensions promised by string theory. It is a new mechanism for rendering extra dimensions invisible. The four-dimensional graviton could be a bound state.

2. The graviton responsible for the four-dimensional gravity we observe could be massive. This was formerly thought to be impossible. Porrati (2002) has shed insight into this result by showing that this is the result of nonstandard boundary conditions and the existence of massless states (in the conformal field theory description). When these states are integrated out, they produce a graviton mass, akin to the Landau mass of the photon.
3. Physical space can be the union of different domains of different apparent dimensionality. This is a fascinating possibility, which is psychologically rather appealing. We only test that space appears four-dimensional locally. How do we know this is true everywhere?
4. There are other things we learn, by turning things around and using the new geometries to test conjectures about gravity. Bousso's prescription for how to associate a light-like region on the boundary with a spatial volume works beautifully on the AdS brane geometry, but not the spatial boundary version of the holographic conjecture. This is the first static example where his prescription was necessary.
5. The explicit AdS examples with branes are also a good place to test the AdS/CFT (conformal field theory) conjecture of Maldacena (Maldacena *et al.* 1997) that relates a theory that looks like $AdS_5 \times S_5$ to a four-dimensional very supersymmetric ($N = 4$) SU(N) gauge theory. More precisely, it allows a test not yet of this specific example, but of the more general conjecture relating AdS_5 to a four-dimensional boundary theory. The branes, which act as spatial cut-offs from the five-dimensional perspective, act as ultraviolet cut-offs from the holographic boundary theory perspective. A second brane, placed where the warp factor is smaller, could be incorporated into either of the geometries I discussed as well. The holographic interpretation of this second brane would be an infrared cut-off on the four-dimensional theory. One can compare gravity from a four-dimensional and five-dimensional perspective when there is only one brane, and one can compare scattering amplitudes and other physical quantities when two branes are present (Arkani-Hamed *et al.* 2001; Rattazzi and Zaffaroni 2001).
6. Although we have not discussed this explicitly in this chapter, from a holographic perspective you might suggest that bulk gauge boson couplings run logarithmically, and can have unified couplings at high energy. With Matthew Schwartz (Randall and Schwartz 2002), I showed this explicitly by a calculation in the five-dimensional bulk in a model with a second brane that also addresses the hierarchy problem of particle physics (why the Higgs is so much lighter than the Planck mass).
7. In the two-brane model that addresses the hierarchy problem, late-time cosmology works as usual when there is a stabilization mechanism for the distance between the two branes. That means cosmology satisfies the most essential requirement.
8. Because the space probes all energies up to the Planck energy, high-scale inflation is straightforward to achieve. However, the cosmology involves a horizon that moves into the infrared as the universe cools.

Multiverses

Of course, extra dimensions introduce the exciting possibility of multiverses. As we have already hinted, there are many more possibilities involving more than one brane. Physics on these branes can be very different. Different forces can live on the different branes. As the locally localized gravity example also demonstrates when a second brane beyond the turn-around point is included, there can be different values of Newton's constant on the two different four-dimensional branes! This fact, that the apparent gravitational coupling can involve different masses, gives a new way of viewing the different masses that appear in our four-dimensional world.

No one has yet explored the full generality of this result. It does however seem clear that if the graviton is a bound state, and is a distinct mode associated with a brane, that space can divide up into spaces with apparently separate gravity. These spaces might see the strength of gravity as different. An even more interesting possibility is that different spaces see gravity with apparently different dimensions. This is true when there is a single brane in the locally localized gravity example. Gravity near the brane appears four-dimensional, while the rest of the space sees five-dimensional AdS gravity. Different spaces, seeing distinct gravitational forces, is a definite possibility.

Of course, the fact that matter and other forces on the brane can be totally different from ours is a fascinating possibility. Entire other worlds can exist without our knowledge. In the minimal scenario, the only thing that connects us is higher-dimensional gravity, and that signal can be extremely weak.

Implications for big questions?

Work on extra dimensions is still in an early stage. Most of the research so far has involved static brane solutions. Some work has been done on time-dependent solutions, but much remains to be studied. With time dependence, it is clear that even more new features will be revealed. The possibilities for cosmology are certainly intriguing.

I will mention what we now know for the implications for cosmology's bigger questions, and speculate about what we might hope to learn.

Dark matter The introduction of extra dimensions brings with it an enormous number of possible candidates for dark matter. Both the bulk and other branes can house matter that we never directly interact with, but for which we experience only the gravitational force. In fact, it is likely that most additional structure will be dark, and not interact with the photons by which we see.

Of course, it is important to have the correct density of dark matter. One needs sufficient dark matter but not so much that you overclose the universe. Fortunately, this is what is likely to happen if the TeV scale is involved. TeV mass stable particles that interact with TeV scale strength are natural dark matter candidates.

Kaluza–Klein modes of this sort or any other stable TeV particle might successfully account for dark matter.

Cosmological constant One of the very interesting things about brane worlds is that the cosmological constant problem is revamped into a new and different problem. Whereas in four dimensions, the question is why is the cosmological constant so extremely tiny relative to what is naturally expected from field theory, with branes the question becomes why are the brane and bulk energies so incredibly closely aligned. With the brane in five-dimensional AdS space, there is a very delicate relation between the brane and bulk cosmological terms that keeps the brane flat. This means that neither energy need be incredibly small, though they do need to be related. However, this different formulation of the problem might possibly be more amenable to solution, something that has evaded the four-dimensional statement of the problem for many years.

Why does the universe appear to be four-dimensional? The examples I presented show that the universe can appear to be four-dimensional, even if it is not. There can be at least one additional infinite dimension, and there are generalizations that permit more. The key is that the state playing the role of the four-dimensional graviton is a bound state. It is not sensitive to the entire geometry of the space. The bound-state graviton is sufficient for the appearance of four-dimensional gravity.

Potentially holographic nature of the universe Having the concrete holographic example allows us to extrapolate to other possible holographic conjectures. By understanding how the degree of freedom counting in the five-dimensional theory turns into a four-dimensional-appearing set-up, we can conjecture possible ideas of how to implement this more generally. The procedure we suggested works when applied to black holes for example (Randall *et al.* 2002).

Relationship between particle physics and cosmology Virtually all new particle physics models in extra dimensions will have implications for cosmology. How could they not? But also, the tools we are learning for particle physics model building might also have applications to cosmology. This might be relevant to cosmological models with many fields for example, where they can be safely separated. In general, it will be of interest to learn whether there are alternative solutions to old problems in the context of extra dimensions, and whether the evolution of the universe can give any signs that that was the case.

More? I have mostly restricted my comments to problems we know about and insights we have had about extra dimensions. But extra dimensions constitutes an entirely different domain in physics, where it is very likely new and unanticipated phenomena will be found. For example, truly new dynamical solutions remain to be understood. They can involve evolving dimensions or moving branes, or effects that appear to be long-distance from a four-dimensional perspective. I look forward to new discoveries in the future.

Conclusions

Clearly, cosmology has had many stunning achievements. But there are many open questions remaining to be resolved. We won't really trust our understanding until all the pieces can be put together. Extra dimensions provide new directions in which to try to construct the successful theory that can explain what we have seen. Clearly, there is much more to come.

References

Adelberger, E G (2002) http://xxx.lanl.gov/hep-ex 0202008
Antoniadis, I (1990) *Phys. Lett.* **B246**, 377.
Arkani-Hamed, N, Dimopoulos, S, and Dvali, G (1998) *Phys. Lett.* **B429**, 263.
Arkani-Hamed, N, Porrati, M, and Randall, L (2001) *J. High Energy Phys.* **0108**, 017.
Bousso, R and Randall, L (2002) *J. High Energy Phys.* **0204**, 057.
Karch, A and Randall, L (2001a) *Phys. Rev. Lett.* **87**, 061601.
(2001b) *J. High Energy Phys.* **0105**, 008.
Kogan, I, Mouslopoulos, S, and Papazoglou, A (2001) *Phys. Lett.* **B501**, 140.
Lykken, J D (1996) *Phys. Rev.* **D54**, 3693.
Maldacena, J M, Strominger, A, and Witten, E (1997) *J. High Energy Phys.* **9712**, 002.
Porrati, M (2002) *J. High Energy Phys.* **0204**, 058.
Randall, L, Sanz, V, and Schwartz, M (2002) *J. High Energy Phys.* **0206**, 008.
Randall, L and Schwartz, M (2002) *Phys. Rev. Lett.* **88**, 081801.
Randall, L and Sundrum, R (1999a) *Phys. Rev. Lett.* **83**, 4690.
(1999b) *Phys. Rev. Lett.* **83**, 3370.
Rattazzi, R and Zaffaroni, A (2001) *J. High Energy Phys.* **0104**, 021.

Part VI

Emergence, life, and related topics

26

Emergence: us from it

Philip D. Clayton
Claremont Graduate University

Emergence, some say, is merely a philosophical concept, unfit for scientific consumption. Or, others predict, when subjected to empirical testing it will turn out to be nothing more than shorthand for a whole batch of discrete phenomena involving novelty, which is, if you will, nothing novel. Perhaps science can study emergence*s*, the critics continue, but not emergence as such.

It's too soon to tell. But certainly there is a place for those, such as the scientist to whom this volume is dedicated, who attempt to look ahead, trying to gauge what are Nature's broadest patterns and hence where present scientific resources can best be invested. John Archibald Wheeler formulated an important motif of emergence in 1989:

> Directly opposite to the concept of universe as machine built on law is the vision of *a world self-synthesized*. On this view, the notes struck out on a piano by the observer–participants of all places and all times, bits though they are, in and by themselves constitute the great wide world of space and time and things.
>
> (Wheeler 1999: 314.)

Wheeler summarized his idea – the observer–participant who is both the result of an evolutionary process and, in some sense, the cause of his own emergence – in two ways: in the famous sketch given in Fig. 26.1 and in the maxim "It from bit." In the attempt to summarize this chapter's thesis with an equal economy of words I offer the corresponding maxim, "Us from it." The maxim expresses the bold question that gives rise to the emergentist research program: *Does nature, in its matter and its laws, manifest an inbuilt tendency to bring about increasing complexity? Is there an apparently inevitable process of complexification that runs from the periodic table of the elements through the explosive variations of evolutionary history to the unpredictable progress of human cultural history, and perhaps even beyond?*

Science and Ultimate Reality, eds. J. D. Barrow, P. C. W. Davies and C. L. Harper, Jr. Published by Cambridge University Press.

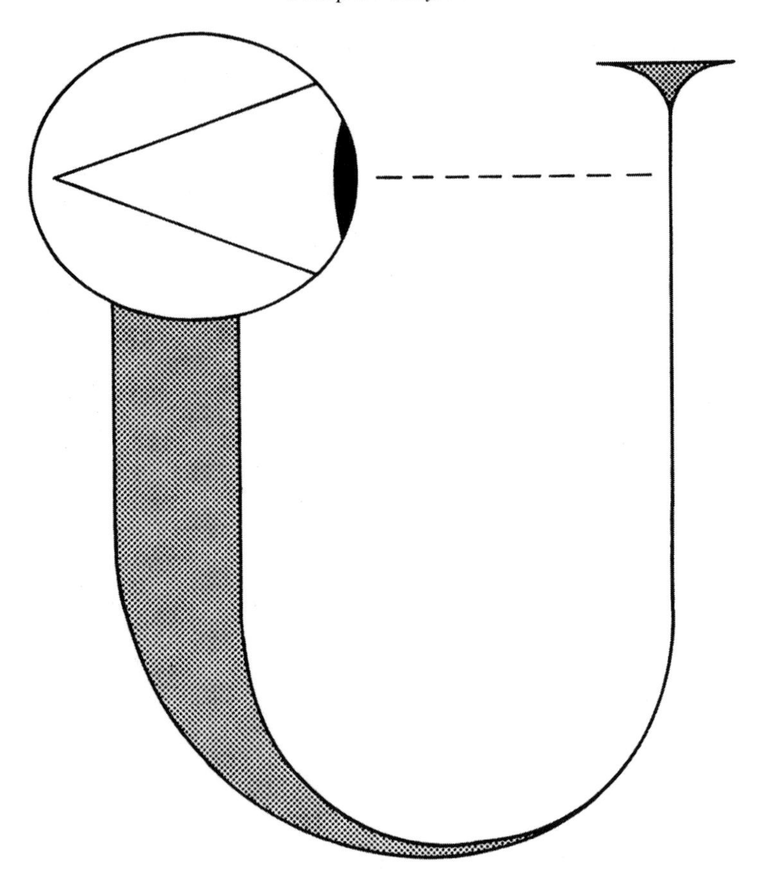

Figure 26.1. "The Wheeler U." (Reproduced with permission from Wheeler, JA (1996) *At Home in the Universe*. © 1996 Springer-Verlag.)

The emergence hypothesis requires that we proceed through at least four stages. The first stage involves rather straightforward physics – say, the emergence of classical phenomena from the quantum world (Zurek 1991, 2002) or the emergence of chemical properties through molecular structure (Earley 1981). In a second stage we move from the obvious cases of emergence in evolutionary history toward what may be the biology of the future: a new, law-based "general biology" (Kauffman 2000) that will uncover the laws of emergence underlying natural history. Stage three of the research program involves the study of "products of the brain" (perception, cognition, awareness), which the program attempts to understand not as unfathomable mysteries but as emergent phenomena that arise as natural products of the complex interconnections of brain and central nervous system. Some add a fourth stage to the program, one that is more metaphysical in nature: the suggestion that the ultimate results, or the original causes, of natural emergence transcend or lie beyond Nature as a whole. Those who view stage-four theories with suspicion should note that the present chapter does not appeal to or rely on metaphysical speculations of this sort in making its case.

Defining terms and assumptions

The basic concept of emergence is not complicated, even if the empirical details of emergent processes are. We turn to Wheeler, again, for an opening formulation:

> When you put enough elementary units together, you get something that is more than the sum of these units. A substance made of a great number of molecules, for instance, has properties such as pressure and temperature that no one molecule possesses. It may be a solid or a liquid or a gas, although no single molecule is solid or liquid or gas.
>
> (Wheeler 1998: 341.)

Or, in the words of biochemist Arthur Peacocke, emergence takes place when "new forms of matter, and a hierarchy of organization of these forms . . . appear in the course of time" and "these new forms have new properties, behaviors, and networks of relations" that must be used to describe them (Peacocke 1993: 62).

Clearly, no one-size-fits-all theory of emergence will be adequate to the wide variety of emergent phenomena in the world. Consider the complex empirical differences that are reflected in these diverse senses of emergence:

- temporal or spatial emergence
- emergence in the progression from simple to complex
- emergence in increasingly complex levels of information processing
- the emergence of new properties (e.g., physical, biological, psychological)
- the emergence of new causal entities (atoms, molecules, cells, central nervous system)
- the emergence of new organizing principles or degrees of inner organization (feedback loops, autocatalysis, "autopoiesis")
- emergence in the development of "subjectivity" (*if* one can draw a ladder from perception, through awareness, self-awareness, and self-consciousness, to rational intuition).

Despite the diversity, certain parameters do constrain the scientific study of emergence:

1. Emergence studies will be scientific only if emergence can be explicated in terms that the relevant sciences can study, check, and incorporate into actual theories.
2. Explanations concerning such phenomena must thus be given in terms of the structures and functions of stuff in the world. As Christopher Southgate writes, "An emergent property is one describing a higher level of organization of matter, where the description is not epistemologically reducible to lower-level concepts" (Southgate *et al.* 1999: 158).
3. It also follows that all forms of dualism are disfavored. For example, only those research programs count as emergentist which refuse to accept an absolute break between neurophysiological properties and mental properties. "Substance dualisms," such as the

Cartesian delineation of reality into "matter" and "mind," are generally avoided. Instead, research programs in emergence tend to combine sustained research into (in this case) the connections between brain[1] and "mind," on the one hand, with the expectation that emergent mental phenomena will not be fully explainable in terms of underlying causes on the other.

4. By definition, emergence transcends any single scientific discipline. At a recent international consultation on emergence theory, each scientist was asked to define emergence, and each offered a definition of the term in his or her own specific field of inquiry: physicists made emergence a product of time-invariant natural laws; biologists presented emergence as a consequence of natural history; neuroscientists spoke primarily of "things that emerge from brains"; and engineers construed emergence in terms of new things that we can build or create. Each of these definitions contributes to, but none can be the sole source for, a genuinely comprehensive theory of emergence.

Physics to chemistry

The following pages focus primarily on examples drawn from artificial systems, biochemistry, biology, and neuroscience. Still, the term "emergence" is not utterly foreign to physics either. Opponents have tried to argue that physical emergence is trivial; they cannot however argue that it is absent from or irrelevant to these scientific domains.

Things emerge in the development of complex physical systems that are understood by observation and cannot be derived from first principles, even given a complete knowledge of the antecedent states. One would not know about conductivity, for example, from a study of individual electrons alone; conductivity is a property that emerges only in complex solid state systems with huge numbers of electrons. Likewise, the fluid dynamics of chaotic hydrodynamic flows with vortices (say, the formation of eddies at the bottom of a waterfall) cannot be predicted from knowledge of the motions of individual particles. The quantum Hall effect and the phenomena of superconductivity are often cited as further examples of emergence.

Such examples are convincing: physicists are familiar with a myriad of cases in which physical wholes cannot be predicted based on knowledge of their parts. Intuitions differ, though, on the significance of this unpredictability. Let's call the two options strong and weak unpredictability (anticipating the heated debate on strong and weak emergence to which we return below). Cases will be unpredictable in a weaker sense if it turns out that one could in principle predict aggregate states given suitably comprehensive information about the parts – even if predictions of

[1] All uses of "brain" in this chapter are meant to include the brain in interaction with the rest of the central nervous system.

system dynamics lie beyond present, or even future, limits on computability. But they will be unpredictable in a much stronger sense – that is, unpredictable even in principle – if the system-as-a-whole is really more than the sum of its parts. In the following pages we consider the evidence for emergence as an attribute of nature in this stronger sense.

Examples like conductivity and fluid dynamics are already familiar to and could be multiplied at will by most readers. Recently, however, more radical claims have been raised about physical emergence. On the standard picture, for example, all that exists emerges from quantum mechanical potentialities, beginning with spacetime itself. For example, Juan Maldacena argued during the conference that inspired this book: "Spacetime appears dynamically, due to the interactions in the quantum field theory at the boundary. It is an 'emergent' property, appearing due to the interactions." General relativity requires that spacetime be treated like a four-dimensional fluid and not as a nonphysical structuring separate from what exists within it (such as mass). Whether spacetime emerges from quantum interactions, as Maldacena claims, is of course a more speculative matter.

In either case, the newer theories certainly require that the classical world be understood to emerge from the quantum world. Andreas Albrecht spoke at the conference of the emergence of classicality in thermodynamics, and Wojciech Zurek, also at the conference, argued that "the path from the microscopic to macroscopic is emergent." Zurek's work since (1991) has demonstrated "the status of decoherence as . . . a key ingredient of the explanation of the emergent classicality" (Zurek 2002: 14). It's thus appropriate, for example, to speak of "the emergence of preferred pointer states" (Zurek 2002: 17): even that paradigmatic touchstone of classical physics, the measure of a macrophysical state by the position of a pointer on a dial, must now be understood as an emergent phenomenon resulting from the decoherence of a quantum superposition.

But even simple examples push toward the same conclusion. Consider, for instance, the Pauli exclusion principle (PEP). PEP is a law of Nature which stipulates that no two electrons of an atom can have the same set of four quantum numbers. Thus a maximum of two electrons can occupy an atomic orbital. This requirement on the way electrons fill up orbitals is basic for understanding modern chemistry. It turns out, based on PEP, that each of the types of sublevel (s, p, d, f) must have its own particular electron capacity. As the orbitals are filled according to this simple rule, beginning with the lowest energy orbitals, the well-known chemical characteristics reflected in the periodic table begin to emerge. A rather simple principle thus has as its outcome the complex chemical distribution of the elements. These emergent qualities are both diverse and unpredictable. (This example again raises the critical question of strong *versus* weak unpredictability discussed above.)

Artificial systems

As one moves toward biological systems, the emerging structures, which are extremely large from a physics point of view, play a larger and larger causal and explanatory role. We consider three examples from work on artificial systems: the emergence of "gliders" in simulated evolutionary systems, the emergence of neural networks, and the emergence of system-level attributes in ant colonies.

Simulated evolutionary systems

Computer simulations study the processes whereby very simple rules give rise to complex emergent properties. John Conway's program "Life," which simulates cellular automata, is already widely known. The program's algorithm contains simple rules that determine whether a particular square on a grid "lights up" based on the state of neighboring squares. When applied, the rules produce complex structures that evidence interesting and unpredictable behaviors. One of these, the "glider," is a five-square structure that moves diagonally across the grid, one step for every four cycles of the program (Fig. 26.2) (Holland 1998: 138).

As in natural systems, further emergent complexity is added by the fact that the program "tiles." This term denotes the phenomenon in which composite structures are formed out of groups of simpler structures and evidence coherent behavior over iterations of the program. What is true of a single square, for example, can also be true of a 3-by-3 array of squares. Now we are dealing with a much more complex system: the resulting tile has 512 states and each of its 8 inputs can take any of 512 values (Holland 1998: 194).

Occurrences of the tiling phenomenon in the natural world, which George Ellis calls "encapsulation" (see his chapter in this volume), reveal why emergent structures in the natural world play such a crucial role in scientific explanations. Composite structures are made up of simpler structures, and the rules governing the behavior of the simple parts continue to hold throughout the evolution of the system. Yet even in as simple a system as Conway's "Life," predicting the movement of larger structures in terms of the simple parts alone turns out to be extremely complex. Thus in the messy real world of biology, behaviors of complex systems quickly become noncomputable in practice. (Whether they are unpredictable in principle, and if so why, remains a central question for emergence theory.) As a result – and, it now appears, necessarily – scientists rely on explanations given in terms of the emerging structures and their causal powers. Dreams of a final reduction "downwards" are fundamentally impossible. Recycled lower-level descriptions cannot do justice to the actual emergent complexity of the natural world as it has evolved.

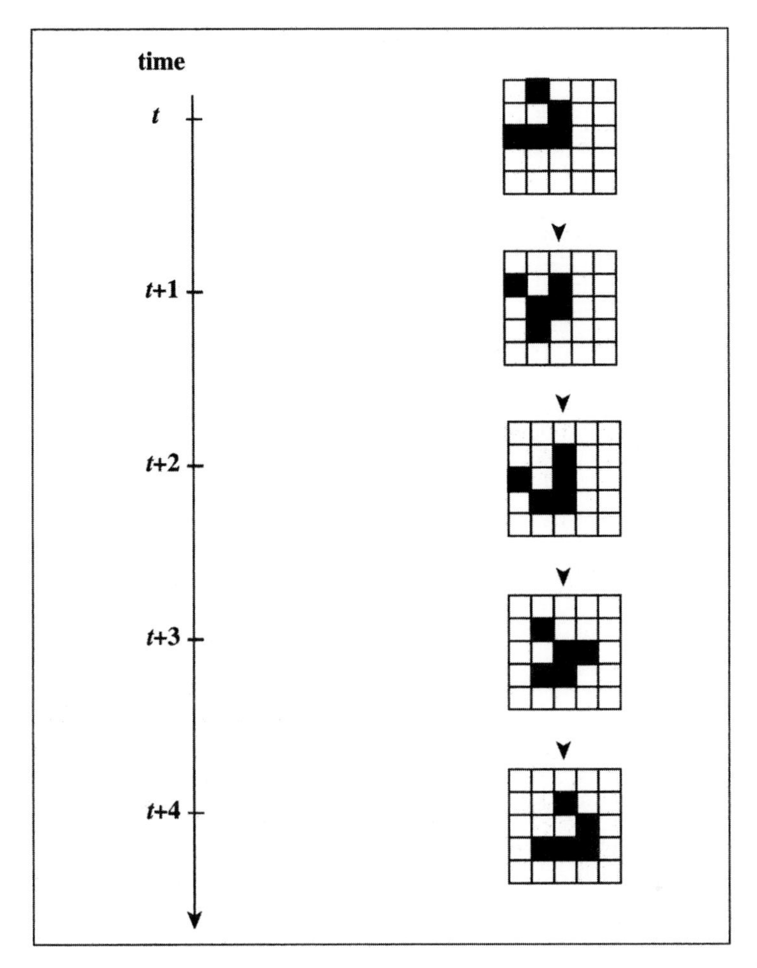

Figure 26.2. "Gliders" in "Life." (Reproduced by permission from Holland, JH (1998) *Emergence: From Chaos to Order*, Perseus Books Publishers, a member of Perseus Books, LLC. © 1998 John H. Holland.)

Stephen Wolfram recently attempted to formulate the core principles of rule-based emergence:

> Even programs with some of the very simplest possible rules yield highly complex behavior, while programs with fairly complicated rules often yield only rather simple behavior. . . . If one just looks at a rule in its raw form, it is usually almost impossible to tell much about the overall behavior it will produce.
>
> (Wolfram 2002: 352.)

As an example of very similar rules producing widely discrepant outputs, Wolfram offers the sequence shown in Fig. 26.3 of elementary cellular automata "whose rules differ from one to the next only at one position" in a Gray code sequence (Wolfram 2002: 352).

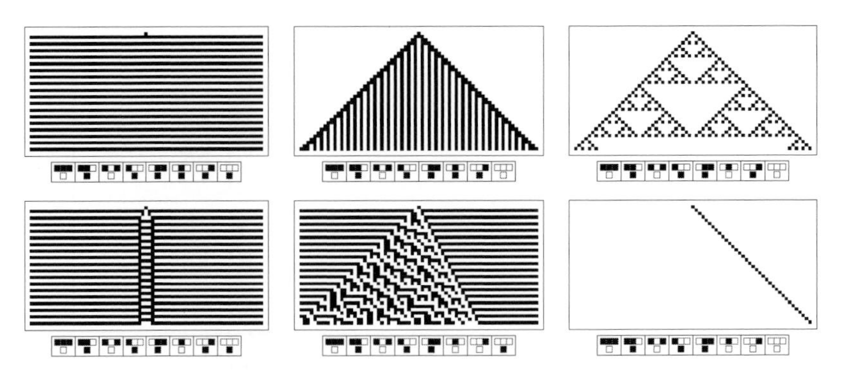

Figure 26.3. Wolfram's cellular automata. (Reproduced with permission from Wolfram, S (2002) *A New Kind of Science*. © 2002 Stephen Wolfram LLC.)

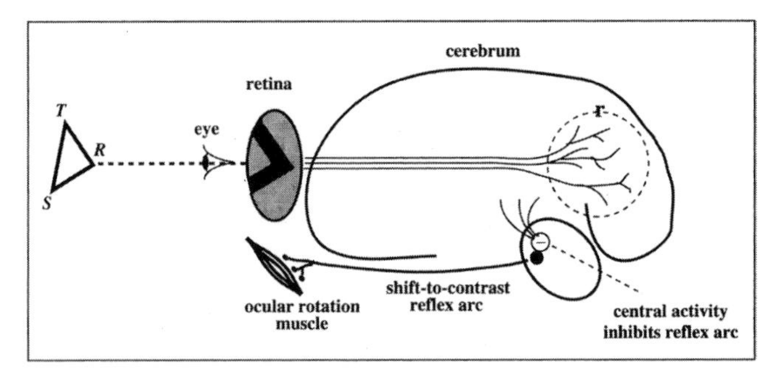

Figure 26.4. Emulating neural fatigue in a neural network. (Reproduced with permission from Holland, JH (1998) *Emergence: From Chaos to Order*, Perseus Books Publishers, a member of Perseus Books, LLC. © 1998 John H. Holland.)

Neural networks

Neural networks research comes to similar results from a very different starting point. Consider John Holland's work on developing visual processing systems. He begins with the simple representation of a mammalian visual system shown in Fig. 26.4 (Holland 1998: 102).

In neural networks research, rather than establishing laws in advance, one constructs a set of random interconnections between a large number of "nodes" to form a network. The researcher then imposes relatively simple processing rules that emulate mammal perception. Crucially, the rules pertain to the synaptic junctions rather than to the overall architecture of the neural network. Thus they might include rules to govern the circulation of pulses based on variable threshold firing, "fatigue" rules to simulate the inhibition of firing after a period of activity, and so forth. Programmers also program a "shift to contrast" reflex, so that the "eye" shifts successively to new points of contrast in the presented image, such as to

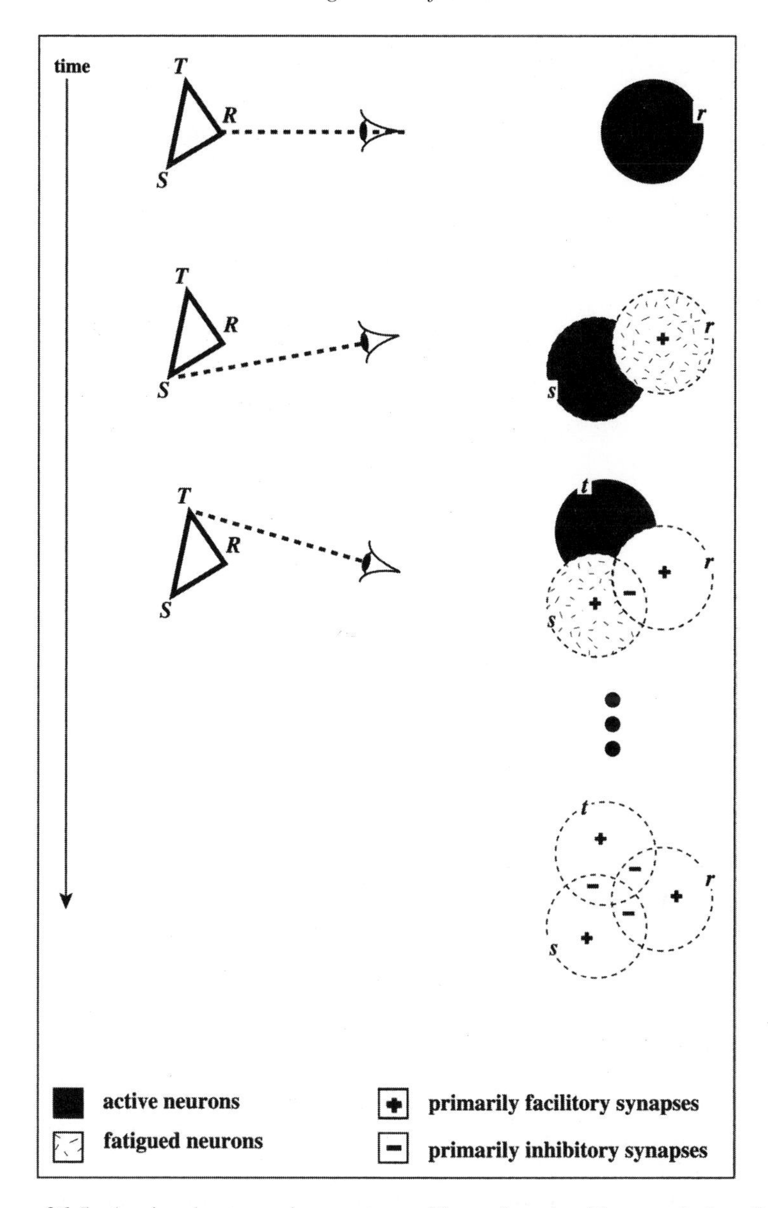

Figure 26.5. A visual processing system. (Reproduced with permission from Holland, JH (1998) *Emergence: From Chaos to Order*, Perseus Books Publishers, a member of Perseus Books, LLC. © 1998 John H. Holland.)

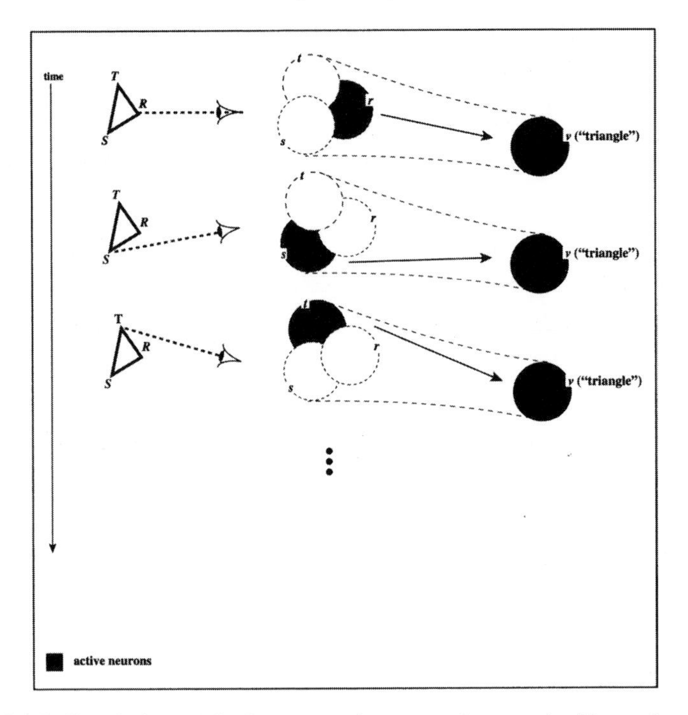

Figure 26.6. Simulating optical memory in a neural network. (Reproduced with permission from Holland, JH (1998) *Emergence: From Chaos to Order*, Perseus Books Publishers, a member of Perseus Books, LLC. © 1998 John H. Holland.)

another vertex in a figure. One then runs multiple trials and measures learning via the system's output.

The idea is to see whether these simple systems can model visual memory in mammals. It seems that they can. Holland's systems, for example, exhibit the features of synchrony, or reverberation, as well as anticipation: groups of "neurons" "prepare" to respond to an expected future stimulus (Fig. 26.5) (Holland 1998: 104). Groups of neurons light up in response to each of the vertices of the triangle, while "fatigued" neurons do not. Particularly fascinating is the phenomenon of hierarchy: new groups of neurons form in response to groups that have already formed (Fig. 26.6) (Holland 1998: 108). Thus a lighting of any of the three original groupings causes a fourth area to light up, which represents the memory for "triangle."

Ant colony behavior

Neural network models of emergent phenomena can model not only visual memory but also phenomena as complex as the emergence of ant colony behavior from the simple behavioral "rules" that are genetically programmed into individual ants. As John Holland's work has again shown, one can program the individual nodes in the simulation with the simple approach/avoidance principles that determine ant behavior (cf. Holland 1998: 228):

- *Flee when you detect a moving object*; but
- *If the object is moving and small and exudes the "friend" pheromone, then approach it and touch antennae.*

The work of ant researchers such as Deborah Gordon (2000) confirms that the resulting program simulates actual ant behaviors to a significant degree. Her work with ant colonies in turn adds to the general understanding of complex systems:

> The dynamics of ant colony life has some features in common with many other complex systems: Fairly simple units generate complicated global behavior. If we knew how an ant colony works, we might understand more about how all such systems work, from brains to ecosystems.
>
> (Gordon 2000: 141.)

Even if the behavior of an ant colony were nothing more than an aggregate of the behaviors of the individual ants, whose behavior follows very simple rules,[2] the result would be remarkable, for the behavior of the ant colony as a whole is extremely complex and highly adaptive to complex changes in its ecosystem. The complex adaptive potentials of the ant colony as a whole are emergent features of the aggregated system. The scientific task is to correctly describe and comprehend such emergent phenomena where the whole is more than the sum of the parts.

Biochemistry

So far we have considered models of how nature *could* build highly complex and adaptive behaviors from relatively simple processing rules. Now we must consider actual cases in which significant order emerges out of (relative) chaos. The big question is how nature obtains order "out of nothing," that is, when the order is not present in the initial conditions but is produced in the course of a system's evolution.[3] What are some of the mechanisms that nature in fact uses? We consider four examples.

Fluid convection

The Bénard instability is often cited as an example of a system far from thermodynamic equilibrium, where a stationary state becomes unstable and then

[2] Gordon (2000: 168) disputes this claim: "One lesson from the ants is that to understand a system like theirs, it is not sufficient to take the system apart. The behavior of each unit is not encapsulated inside that unit but comes from its connections with the rest of the system." I likewise break strongly with the aggregate model of emergence.

[3] Generally this seems to be a question that makes physicists uncomfortable ("Why, that's impossible, of course!"), whereas biologists tend to recognize in it one of the core mysteries in the evolution of living systems.

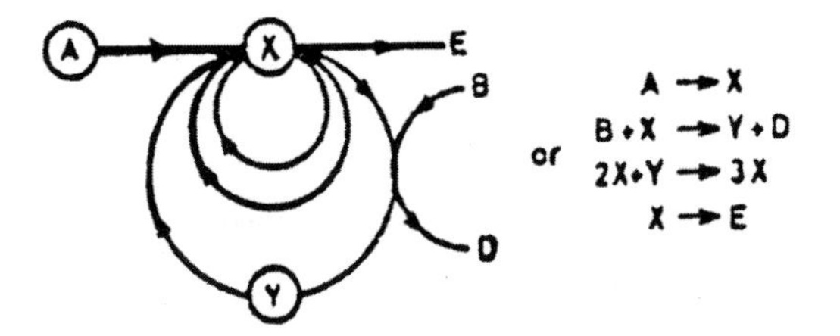

Figure 26.7. A sample autocatalytic process. (Reproduced with permission from Prigogine, I (1998) *Order out of Chaos*. © 1984 Ilya Prigogine.)

manifests spontaneous organization (Peacocke 1994: 153). In the Bénard case, the lower surface of a horizontal layer of liquid is heated. This produces a heat flux from the bottom to the top of the liquid. When the temperature gradient reaches a certain threshold value, conduction no longer suffices to convey the heat upward. At that point convection cells form at right angles to the vertical heat flow. The liquid spontaneously organizes itself into these hexagonal structures or cells.

Differential equations describing the heat flow exhibit a bifurcation of the solutions. This bifurcation represents the spontaneous self-organization of large numbers of molecules, formerly in random motion, into convection cells. This represents a particularly clear case of the spontaneous appearance of order in a system. According to the emergence hypothesis, many cases of emergent order in biology are analogous.

Autocatalysis in biochemical metabolism

Autocatalytic processes play a role in some of the most fundamental examples of emergence in the biosphere. These are relatively simple chemical processes with catalytic steps, yet they well express the thermodynamics of the far-from-equilibrium chemical processes that lie at the base of biology. Much of biochemistry is characterized by a type of catalysis in which "the presence of a product is required for its own synthesis" (Prigogine 1984: 134). Take a basic reaction chain where $A \rightarrow X$, and $X \rightarrow E$, but where X is involved in an autocatalytic process (Fig. 26.7) (Prigogine 1984: 135). For example, molecule X might activate an enzyme, which "stabilizes" the configuration that allows the reaction. Similarly frequent are cases of *crosscatalysis*, namely cases where X is produced from Y and Y from X. In Fig. 26.7 crosscatalysis is represented by the equation $B + X \rightarrow Y + D$, that is, X in the presence of B produces Y and a by-product. The presence of Y in turn produces a higher quantity of X (here, $2X + Y \rightarrow 3X$). The entire reaction loop is autocatalytic in producing E. Such loops play an important role in metabolic functions.

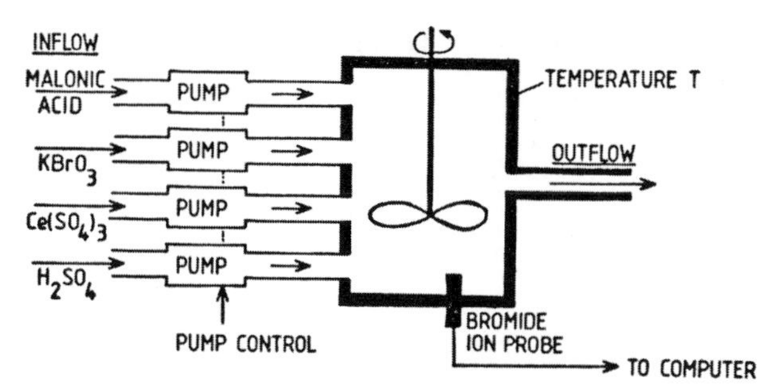

Figure 26.8. The Belousov–Zhabotinsky reaction. (Reproduced with permission from Prigogine, I (1998) *Order out of Chaos*. © 1984 Ilya Prigogine.)

Belousov–Zhabotinsky reactions

The role of emergence becomes clearer as one considers more complex examples. Consider the famous Belousov–Zhabotinsky reaction (Fig. 26.8) (Prigogine 1984: 152). This reaction consists of the oxidation of an organic acid (malonic acid) by potassium bromate in the presence of a catalyst such as cerium, manganese, or ferroin. From the four inputs into the chemical reactor more than 30 products and intermediates are produced. The Belousov–Zhabotinsky reaction provides an example of a biochemical process where a high level of disorder settles into a patterned state.

In more complex chemical systems, multiple states can be achieved far from equilibrium. That is, a given set of boundary conditions can produce one of a variety of stationary outcome states. The chemical composition of these outcome states serves as a "control mechanism" in biological systems. It would be fruitful to explore the similarities between these multiple stationary outcomes and the "attractors" or "strange attractors" that mathematicians have explored in other contexts.

One then wants to ask: what is the general feature of these dissipative structures far from thermodynamic equilibrium? We follow Prigogine's (1984: 171) conclusion:

> One of the most interesting aspects of dissipative structures is their coherence. The system behaves as a whole, as if it were the site of long-range forces. . . . In spite of the fact that interactions among molecules do not exceed a range of some 10^{-8} cm, the system is structured as though each molecule were "informed" about the overall state of the system.

Put in philosophical terms, the data suggest that emergence is not merely epistemological but can also be ontological in nature. That is, it's not just that *we* can't predict emergent behaviors in these systems from a complete knowledge of the structures and energies of the parts. Instead, studying the systems suggests that structural features of the system – which are emergent features of the system as such and *not* properties pertaining to any of the parts – determine the overall state

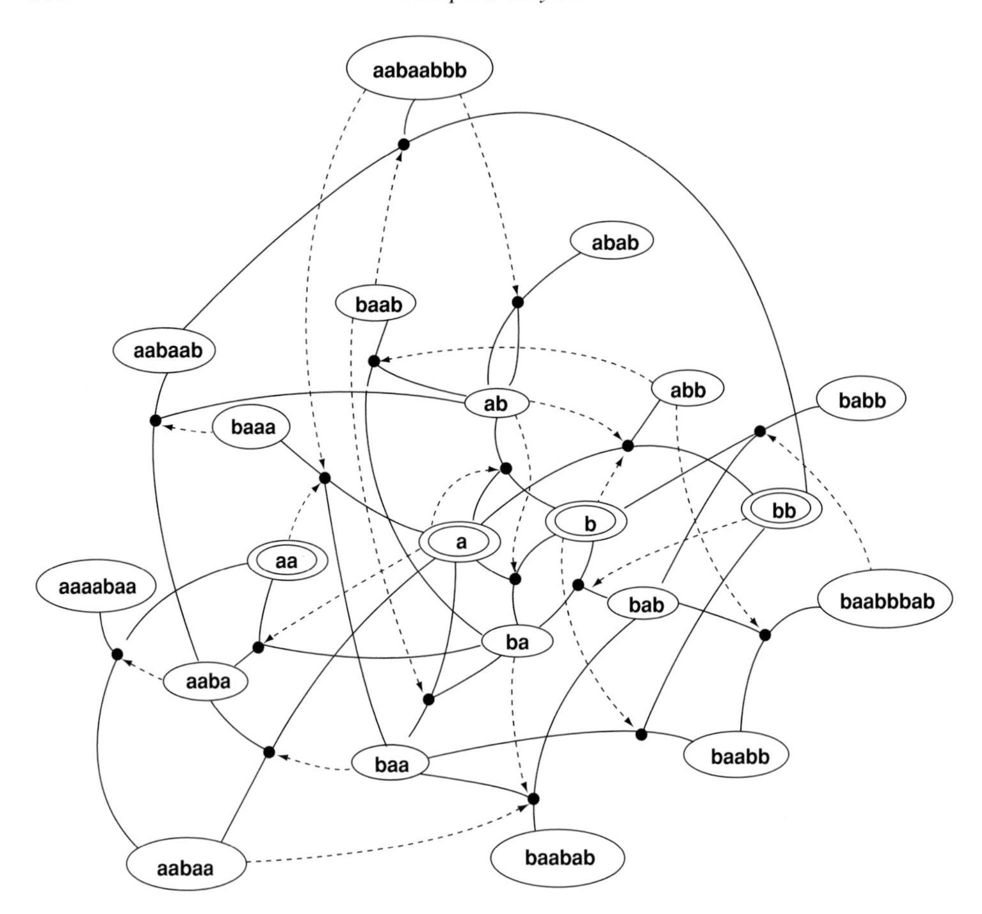

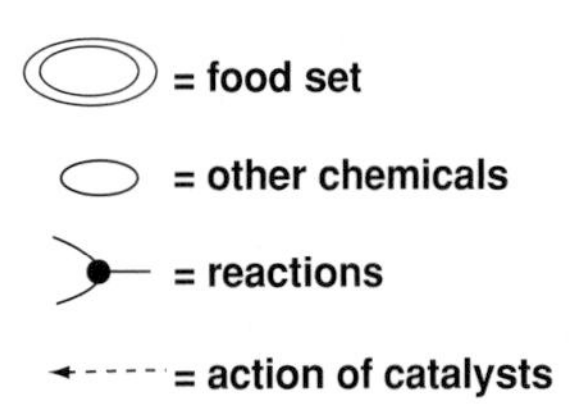

Figure 26.9. Autocatalytic systems in nature. (Adapted with permission from Cowan, G and Pines, D, *Complexity: Metaphors, Models, and Reality*. © Westview Press, a member of Perseus Books, LLC.)

of the system, and hence as a result the behavior of individual particles within the system.

The role of emergent features of systems is increasingly evident as one moves from the very simple systems so far considered to the sorts of systems one actually encounters in the biosphere. Stuart Kauffman (1994: 90) sketches a simple

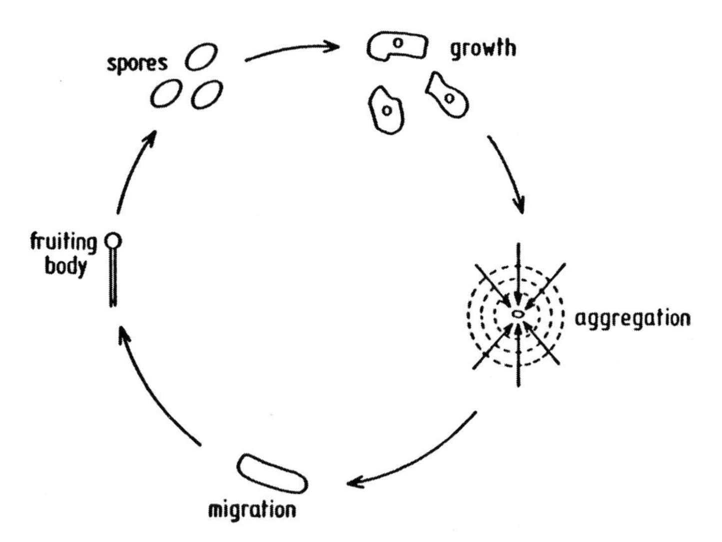

Figure 26.10. The slime mold cycle. (Reproduced with permission from Prigogine, I (1998) *Order out of Chaos*. © 1984 Ilya Prigogine.)

autocatalytic set of the sort that occurs in Nature: see Fig. 26.9. This sketch shows the reactions and the actions of catalysts in a set that involves only four food sets and 17 other chemicals.

The biochemistry of cell aggregation and differentiation

We move finally to processes where a random behavior or fluctuation gives rise to organized behavior between cells based on self-organization mechanisms. Consider the process of cell aggregation and differentiation in cellular slime molds (specifically, in *Dictyostelium discoideum*) (Fig. 26.10). The slime mold cycle begins when the environment becomes poor in nutrients and a population of isolated cells joins into a single mass on the order of 10^4 cells (Prigogine 1984: 156). The aggregate migrates until it finds a higher nutrient source. Differentiation then occurs: a stalk or "foot" forms out of about one-third of the cells and is soon covered with spores. The spores detach and spread, growing when they encounter suitable nutrients and eventually forming a new colony of amoebas.

Note that this aggregation process is randomly initiated. Autocatalysis begins in a random cell within the colony, which then becomes the attractor center. It begins to produce cyclic adenosine monophosphate(AMP). As AMP is released in greater quantities into the extracellular medium, it catalyzes the same reaction in the other cells, amplifying the fluctuation and total output. Cells then move up the gradient to the source cell, and other cells in turn follow their cAMP trail toward the attractor center.

A similar randomly initiated process that produces highly adaptive behavior is found in Coleoptera (termite) larvae (Fig. 26.11) (Prigogine 1984: 182). Here

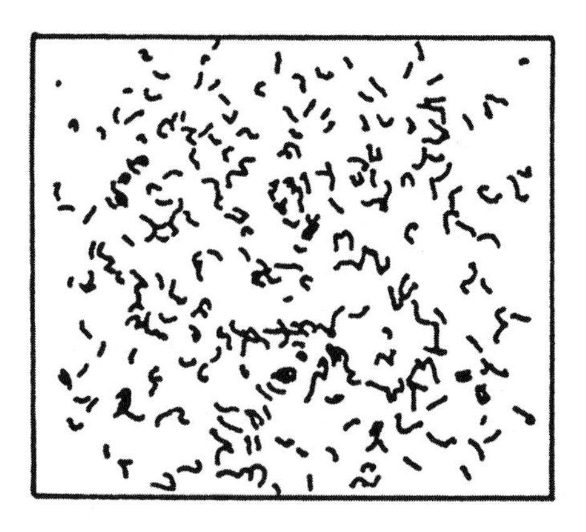

Figure 26.11. Emergent behaviors in Coleoptera larvae. (Reproduced with permission from Prigogine, I (1998) *Order out of Chaos*. © 1984 Ilya Prigogine.)

the aggregation process is induced through the release of a pheromone by the Coleoptera. The higher their nutrition state, the higher the rate of release. Other larvae then move up the concentration gradient. The process is autocatalytic: the more larvae that move into a region, the more the attractiveness of that region is enhanced, until the nutrient source is finally depleted. It is also dependent on random moves of the larvae, since they will not cluster if they are too dispersed.

Biology

Ilya Prigogine did not follow the notion of "order out of chaos" up through the entire ladder of biological evolution. Stuart Kauffman (1995, 2000) and others (Gell-Mann 1994; Goodwin 2001; see also Cowan *et al.* 1994 and other works in the same series) have however recently traced the role of the same principles in living systems. Biological processes in general are the result of systems that create and maintain order (stasis) through massive energy input from their environment. In principle these types of processes could be the object of what Kauffman envisions as "a new general biology," based on sets of still-to-be-determined laws of emergent ordering or self-complexification. Like the biosphere itself, these laws (if they indeed exist) are emergent: they depend on the underlying physical and chemical regularities but are not reducible to them. Kauffman (2000: 35) writes:

> I wish to say that life is an expected, emergent property of complex chemical reaction networks. Under rather general conditions, as the diversity of molecular species in a reaction system increases, a phase transition is crossed beyond which the formation of collectively autocatalytic sets of molecules suddenly becomes almost inevitable.

Until a science has been developed that formulates and tests physics-like laws at the level of biology, the "new general biology" remains an as-yet-unverified, though intriguing, hypothesis. Nevertheless recent biology, driven by the genetic revolution on the one side and by the growth of the environmental sciences on the other, has made explosive advances in understanding the role of self-organizing complexity in the biosphere. Four factors in particular play a central role in biological emergence.

The role of scaling

As one moves up the ladder of complexity, macrostructures and macromechanisms emerge. In the formation of new structures, scale matters – or, better put, changes in scale matter. Nature continually evolves new structures and mechanisms as life forms move up the scale from molecules (*c.* 1 Ångstrom) to neurons (*c.* 100 micrometers) to the human central nervous system (*c.* 1 meter). As new structures are developed, new whole–part relations emerge.

John Holland argues that different sciences in the hierarchy of emergent complexity occur at jumps of roughly three orders of magnitude in scale. By that point systems have become too complex for predictions to be calculated, and one is forced to "move the description 'up a level'" (Holland 1998: 201). The "microlaws" still constrain outcomes, of course, but additional basic descriptive units must also be added. This pattern of introducing new explanatory levels iterates in a periodic fashion as one moves up the ladder of increasing complexity. To recognize the patterns is to make emergence an explicit feature of biological research. As of now, however, science possesses only a preliminary understanding of the principles underlying this periodicity.

The role of feedback loops

The role of feedback loops, examined above for biochemical processes, becomes increasingly important from the cellular level upwards. In plant–environment

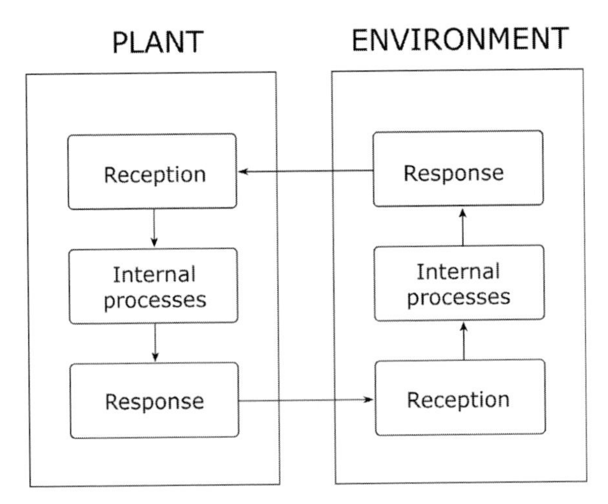

Figure 26.12. Schematic summary of the plant–environment cycle. (© Philip Clayton; redrawn by Ben Klocek.)

interactions, for example, one can trace the interaction of interacting mechanisms, each of which is the complex result of its own internal autocatalytic processes. Plants receive nutrients, process them, and provide new materials to the environment (e.g., oxygen, pollen). The environment in turn takes up these materials and processes them, so that new resources become available to the plant. This is shown schematically in Fig. 26.12. This sort of feedback dynamic is the basis for ecosystems theory: the particular behaviors of organisms bring about changes in their environment, which affect the organisms with which they interact; in turn, these organisms' complex responses, products of their own internal changes, further alter the shared environment, and hence its impact on each individual organism.

The role of local–global interactions

In complex dynamical systems the interlocked feedback loops can produce an emergent global structure. Roger Lewin (1999: 13) offers a schematic representation (Fig. 26.13) derived from the work of Chris Langton. In these cases, "the global property – [the] emergent behavior – feeds back to influence the behavior of the individuals . . . that produced it" (Lewin 1999). The global structure may have properties the local particles do not have. An ecosystem, for example, will usually evidence a kind of emergent stability that the organisms of which it is constituted lack. Nevertheless, it is impossible to predict the global effects because of the sensitive dependence on initial conditions (among other factors): minute fluctuations near the bifurcation are amplified by subsequent states of the system. George Ellis

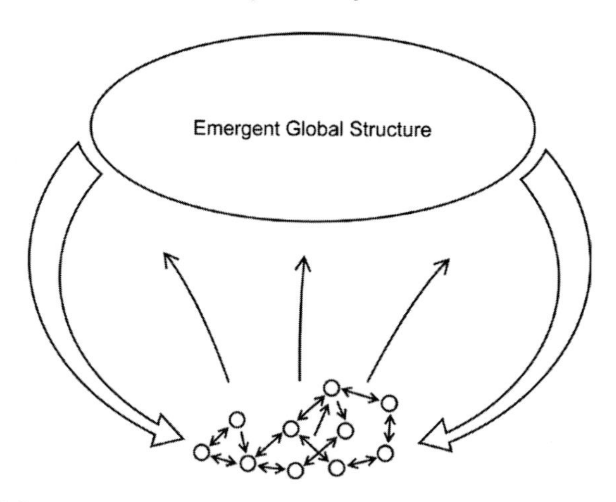

Figure 26.13. Local–global interactions. (Reproduced with permission from Chris Langton who frequently uses this drawing; redrawn by Ben Klocek.)

(this volume) correctly describes this form of "downward" feedback process as representing a case of *downward causation.*

Figure 26.13 schematizes the idea of a global structure. In contrast to Langton, Kauffman insists that an ecosystem is in one sense "merely" a complex web of interactions. Yet consider a typical ecosystem of organisms of the sort that Kauffman (2000: 191) analyzes, as shown in Fig. 26.14. Depending on one's research interests, one can focus attention either on holistic features of such systems or on the interactions of the components within them. Thus Langton's term "global" draws attention to system-level features and properties, whereas Kauffman's "merely" emphasizes that no mysterious outside forces need to be introduced (such as, e.g., Rupert Sheldrake's (1995) "morphic resonance"). Since the two dimensions are complementary, neither alone is scientifically adequate; the explosive complexity manifested in the evolutionary process involves the interplay of *both* systemic features and component interactions.

The role of nested hierarchies

A final layer of complexity is added in cases where the local–global structure forms a nested hierarchy. Such hierarchies are often represented using nested circles: see Fig. 26.15. Nesting is one of the basic forms of combinatorial explosion. Such forms appear extensively in natural biological systems (Wolfram 2002: 357ff.; see his index for dozens of further examples of nesting). Organisms achieve greater structural complexity, and hence increased chances of survival, as they incorporate discrete subsystems. Similarly, ecosystems complex enough to contain a number

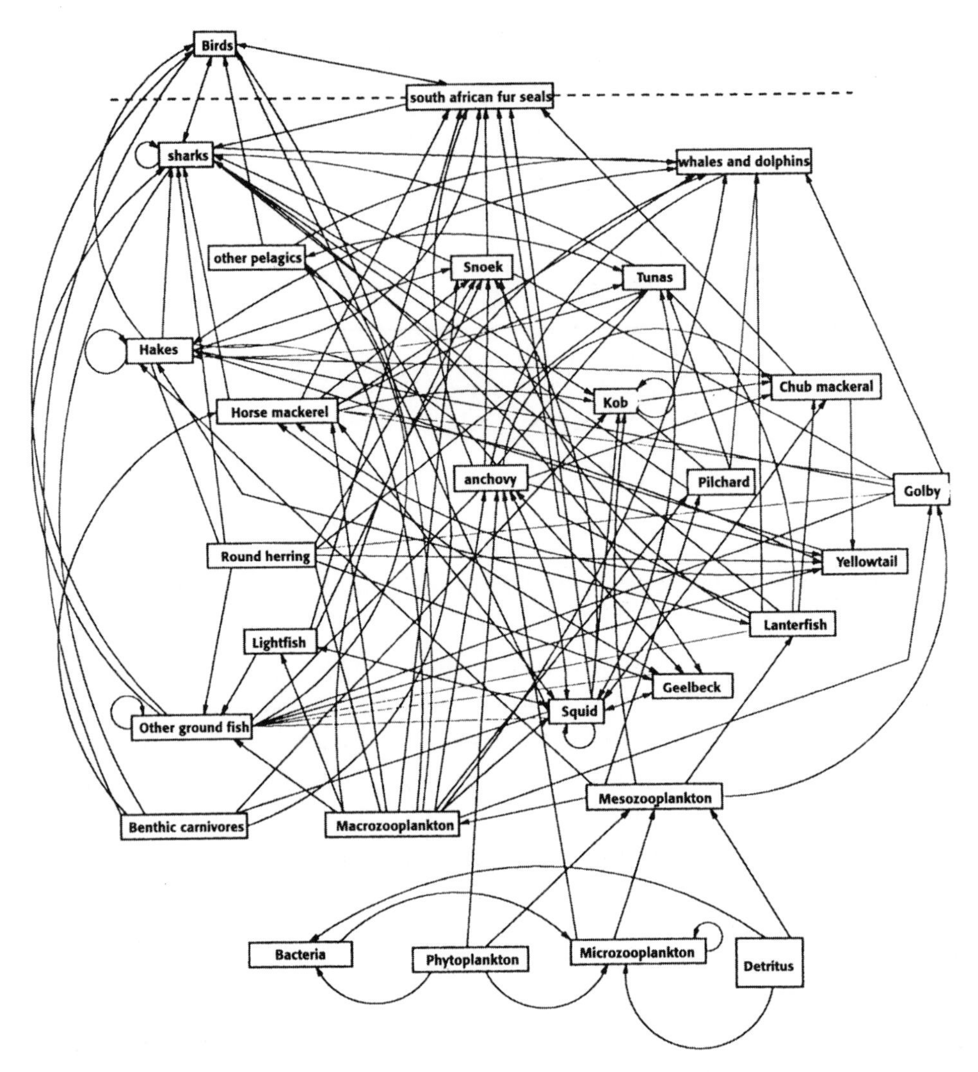

Figure 26.14. Interactions in a typical complex ecosystem. (Reproduced with permission from Kauffman, S (2000) *Investigations*. © 2000 Oxford University Press, Inc.)

of discrete subsystems evidence greater plasticity in responding to destabilizing factors.

"Strong" versus "weak" emergence

The resulting interactions between parts and wholes mirror yet exceed the features of emergence that we observed in chemical processes. To the extent that the evolution of organisms and ecosystems evidences a "combinatorial explosion"

Figure 26.15. Nested hierarchies in biological systems. (Frequently used image of embedding, redrawn by Ben Klocek.)

(Morowitz 2002) based on factors such as the four just summarized, the hope of explaining entire living systems in terms of simple laws appears quixotic. Instead, natural systems made up of interacting complex systems form a multileveled network of *inter*dependency (cf. Gregersen 2003), and each level contributes distinct elements to the overall explanation.

Systems biology, the Siamese twin of genetics, has established many of the features of life's "complexity pyramid" (Oltvai and Barabási 2002; cf. Barabási 2002). Construing cells as networks of genes and proteins, systems biologists distinguish four distinct levels: (1) the base functional organization (genome, transcriptome, proteome, and metabolome); (2) the metabolic pathways built up out of these components; (3) larger functional modules responsible for major cell functions; and (4) the large-scale organization that arises from the nesting of the functional modules. Oltvai and Barabási (2002) conclude that "[the] integration of different organizational levels increasingly forces us to view cellular functions as distributed among groups of heterogeneous components that all interact within large networks." Milo *et al.* (2002) have recently shown that a common set of "network motifs" occurs in complex networks in fields as diverse as biochemistry, neurobiology, and ecology. As they note, "similar motifs were found in networks that perform information processing, even though they describe elements as different as biomolecules within a cell and synaptic connections between neurons in *Caenorhabditis elegans.*"

Such compounding of complexity – the system-level features of networks, the nodes of which are themselves complex systems – is sometimes said to represent only a *quantitative* increase in complexity, in which nothing "really new" emerges. This view I have elsewhere labeled "weak emergence." It is the view held by (among others) John Holland (1998) and Stephen Wolfram (2002). But, as Leon Kass (1999: 62) notes in the context of evolutionary biology, "it never occurred to Darwin that certain differences of degree – produced naturally, accumulated gradually (even incrementally), and inherited in an unbroken line of descent – might lead to a difference in kind . . ." Here Kass nicely formulates the principle involved. As long as nature's process of compounding complex systems leads to irreducibly complex systems with structures and causal mechanisms of their own, then the natural world evidences not just weak emergence but also a more substantive change that we might label *strong emergence*. Cases of strong emergence are cases where the "downward causation" emphasized by George Ellis (this volume) is most in evidence. By contrast, in the relatively rare cases where rules relate the emergent system to its subvening system (in simulated systems, via algorithms; in natural systems, via "bridge laws") a weak emergence interpretation suffices. In the majority of cases, however, such rules are not available; in these cases, especially where we have reason to think that such lower-level rules are impossible in principle, the strong emergence interpretation is suggested.

Neuroscience, qualia, and consciousness

Consciousness, many feel, is the most important instance of a clearly strong form of emergence. Here if anywhere, it seems, nature has produced something irreducible – no matter how strong the biological dependence of mental qualia (i.e., subjective experiences) on antecedent states of the central nervous system may be. To know everything there is to know about the progression of brain states is not to know what it's like to be you, to experience your joy, your pain, or your insights. No human researcher can know, as Thomas Nagel (1980) so famously argued, "what it's like to be a bat."

Unfortunately consciousness, however intimately familiar we may be with it on a personal level, remains an almost total mystery from a scientific perspective. Indeed, as Jerry Fodor (1992) noted, "Nobody has the slightest idea how anything material could be conscious. Nobody even knows what it would be like to have the slightest idea about how anything material could be conscious. So much for the philosophy of consciousness."

Given our lack of comprehension of the transition from brain states to consciousness, there is virtually no way to talk about the "C" word without sliding into the domain of philosophy. The slide begins if the emergence of consciousness is qualitatively different from other emergences; in fact, it begins even if consciousness is different from the neural correlates of consciousness. Much suggests that both

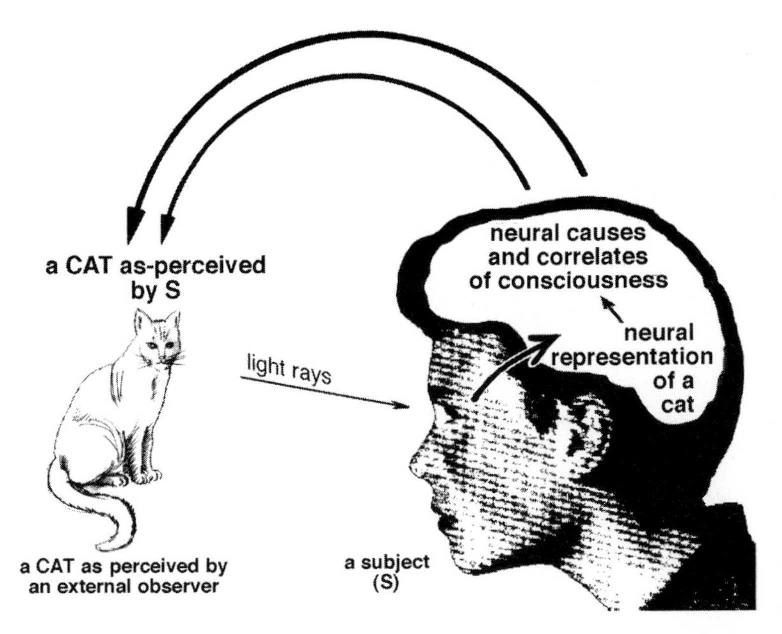

Figure 26.16. Neural representations of objects in the world. (Reproduced with permission from Velmens, M (2000) *Understanding Consciousness*, Routledge. © Thompson Publishing Services.)

differences obtain. How far can neuroscience go, even in principle, in explaining consciousness?

Science's most powerful ally, I suggest, is emergence. As we've seen, emergence allows one to acknowledge the undeniable differences between mental properties and physical properties, while still insisting on the dependence of the entire mental life on the brain states that produce it. Consciousness, the thing to be explained, is different because it represents a new level of emergence; but brain states – understood both globally (as the state of the brain as a whole) and in terms of their microcomponents – are consciousness's sine qua non. The emergentist framework allows science to identify the strongest possible analogies with complex systems elsewhere in the biosphere. So, for example, other complex adaptive systems also "learn," as long as one defines learning as "a combination of exploration of the environment and improvement of performance through adaptive change" (Schuster 1994). Obviously, systems from primitive organisms to primate brains record information from their environment and use it to adjust future responses to that environment.

Even the representation of visual images in the brain, a classically mental phenomenon, can be parsed in this way. Consider Max Velmans's (2000) schema shown in Fig. 26.16. Here the cat-in-the-world and the neural representation of the cat are both parts of a natural system; no nonscientific mental "things" like ideas or forms

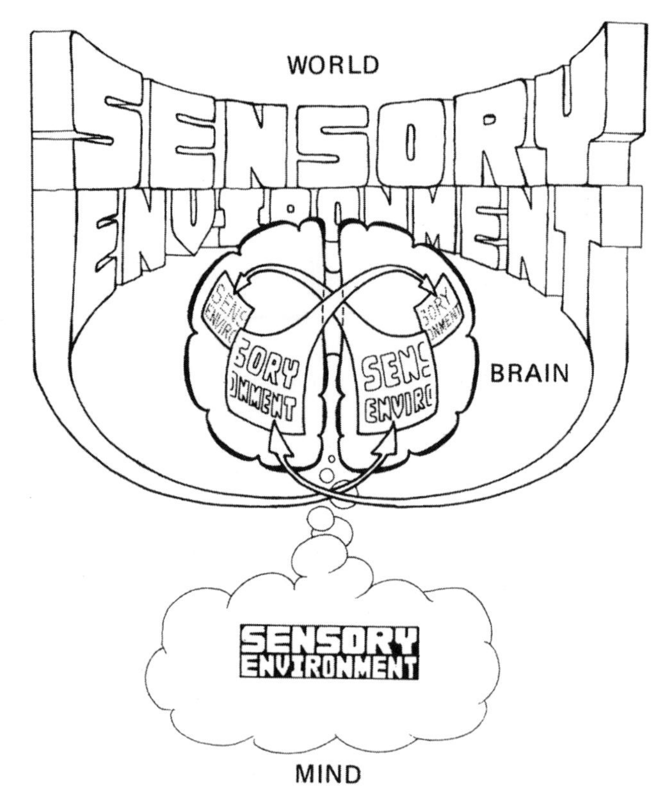

Figure 26.17. "Mind" mirroring the sensory environment. (Reproduced with permission from LeDoux *et al.* (1978) *The Integrated Mind*, Plenum Press. Used by permission of the author.)

are introduced. In principle, then, representation might be construed as merely a more complicated version of the feedback loop between a plant and its environment considered above. Such is the "natural account of phenomenal consciousness" defended by (e.g.) Le Doux (1978). In a physicalist account of mind, no mental causes are introduced. Without emergence, the story of consciousness must be retold such that thoughts and intentions play no causal role. The diagram (Fig. 26.17) nicely expresses the challenge: if one limits the causal interactions to world and brains, mind must appear as a sort of thought-bubble outside the system. Yet it is counter to our empirical experience in the world, to say the least, to leave no causal role to thoughts and intentions. For example, it certainly *seems* that your intention to read this chapter is causally related to the physical fact of your presently holding this book in your hands.

Arguments such as this force one to acknowledge the *disanalogies* between the emergence of consciousness and previous examples of emergence in complex systems. Consciousness confronts us with a "hard problem" different from those already considered (Chalmers 1995: 201):

> The really hard problem of consciousness is the problem of *experience*. When we think and perceive, there is a whir of information-processing, but there is also a subjective aspect. As Nagel has put it, there is *something it is like* to be a conscious organism. This subjective aspect is experience. When we see, for example, we *experience* visual sensations: the felt quality of redness, the experience of dark and light, the quality of depth in a visual field. Other experiences go along with perception in different modalities: the sound of a clarinet, the smell of mothballs. Then there are bodily sensations, from pains to orgasms; mental images that are conjured up internally; the felt quality of emotion, and the experience of a stream of conscious thought. What unites all of these states is that there is something it is like to be in them. All of them are states of experience.

The distinct features of human cognition, it seems, depend on a quantitative increase in brain complexity vis-à-vis other higher primates. Yet, if Chalmers is right (as I fear he is), this particular quantitative increase gives rise to a qualitative change. Even if the development of conscious awareness occurs gradually over the course of primate evolution, the (present) end of that process confronts the scientist with conscious, symbol-using beings clearly distinct from those who preceded them (Deacon 1997). Understanding consciousness even as an emergent phenomenon in the natural world – that is, naturalistically, nondualistically – requires a theory of "felt qualities," "subjective intentions," and "states of experience." Intention-based actions, structures built up out of ideas, and mental causes require new types of explanations and, it appears, a new set of sciences: the social or human sciences. By this point emergence has driven us to a level beyond the natural-science-based framework of the present book. New concepts, new testing mechanisms, and perhaps even new standards for knowledge are now required. From the perspective of physics the trail disappears into the clouds; we can follow it no further.

The five emergences

In the broader discussion the term "emergence" is used in multiple and incompatible senses, some of which are incompatible with the scientific project. Clarity is required to avoid equivocation between five distinct levels on which the term may be applied:

- Let *emergence*$_1$ refer to occurrences of the term within the context of a specific scientific theory. Here it describes features of a specified physical or biological system of which we have some scientific understanding. Scientists who employ these theories claim that the term (in a theory-specific sense) is currently useful for describing features of the natural world. The preceding pages include various examples of theories in which this term occurs. At the level of emergence$_1$ alone there is no way to establish whether the term is used analogously across theories, or whether it really means something utterly distinct in each theory in which it appears.

- *Emergence*$_2$ draws attention to features of the world that may eventually become part of a unified scientific theory. Emergence in this sense expresses postulated connections or laws that may in the future become the basis for one or more branches of science. One thinks, for example, of the role of emergence in Stuart Kauffman's notion of a new "general biology," or in certain proposed theories of complexity or complexification.
- *Emergence*$_3$ is a meta-scientific term that points out a broad pattern across scientific theories. Used in this sense, the term is not drawn from a particular scientific theory; it is an observation about a significant pattern that connects a range of scientific theories. In the preceding pages I have often employed the term in this fashion. My purpose has been to draw attention to common features of the physical systems under discussion, as in (e.g.) the phenomena of autocatalysis, complexity, and self-organization. Each is scientifically understood, and each shares common features that are significant. Emergence draws attention to these features, whether or not the individual theories actually use the same label for the phenomena they describe.

  Emergence$_3$ thus serves a heuristic function. It assists in the recognition of common features between theories. Recognizing such patterns can help to extend existing theories, to formulate insightful new hypotheses, or to launch new interdisciplinary research programs.[4]
- *Emergence*$_4$ expresses a feature in the movement between scientific disciplines, including some of the most controversial transition points. Current scientific work is being done, for example, to understand how chemical structures are formed, to reconstruct the biochemical dynamics underlying the origins of life, and to conceive how complicated neural processes produce cognitive phenomena such as memory, language, rationality, and creativity. Each involves efforts to understand diverse phenomena involving levels of self-organization within the natural world. Emergence$_4$ attempts to express what might be shared in common by these (and other) transition points.

  Here, however, a clear limitation arises. A scientific theory that explains how chemical structures are formed is perhaps unlikely to explain the origins of life. Neither theory will explain how self-organizing neural nets encode memories. Thus emergence$_4$ stands closer to the philosophy of science than it does to actual scientific theory. Nonetheless, it is the sort of philosophy of science that should be helpful to scientists.[5]

[4] For this reason, emergence$_3$ stands closer to the philosophy of science than do the previous two senses. Yet it is a kind of philosophy of science that stands rather close to actual science and that seeks to be helpful to it. By way of analogy one thinks of the work of philosophers of quantum physics such as Jeremy Butterfield or James Cushing, whose work can be and has actually been helpful to bench physicists. One thinks as well of the analogous work of certain philosophers in astrophysics (John Barrow) or in evolutionary biology (David Hull, Michael Ruse).

[5] This as opposed, for example, to the kind of philosophy of science currently popular in English departments and in journals like *Critical Inquiry* – the kind of philosophy of science that asserts that science is a text that needs to be deconstructed, or that science and literature are equally subjective, or that the worldview of Native Americans should be taught in science classes.

- *Emergence*$_5$ is a metaphysical theory. It represents the view that the nature of the natural world is such that it produces continually more complex realities in a process of ongoing creativity. The present chapter does not comment on such metaphysical claims about emergence.[6]

As one moves along the continuum from emergence$_1$ to emergence$_5$, one should acknowledge a transition from specific-domain science to increasingly integrative, and hence increasingly philosophical, concepts.

Conclusion

What do we conclude? Since emergence is used as an integrative ordering concept across scientific fields, its significance is not exhausted by its role within specific scientific theories. (In the terms of the preceding section, emergence may include emergence$_1$ but cannot be limited to it.) It remains, at least in part, a *meta*-scientific term.

Does the idea of distinct levels then conflict with "standard reductionist science"?[7] No, one can believe that there are levels in Nature and corresponding levels of explanation while at the same time working to explain any given set of higher-order phenomena in terms of underlying laws and systems. In fact, isn't the first task of science to whittle away at every apparent "break" in Nature, to make it smaller, to eliminate it if possible? Thus, for example, to study the visual perceptual system scientifically is to attempt to explain it fully in terms of the neural structures and electrochemical processes that produce it. The degree to which downward explanation is possible will be determined by long-term empirical research. At present we can only wager on the one outcome or the other based on the evidence before us.

As the discussion has shown, emergence reflects a pattern shared between a number of specific areas of research. It is most powerful in describing relationships between the domains of nature studied by two neighboring scientific disciplines, for example quantum physics and classical physics, particle physics and chemistry, biochemistry and cell biology, or neurophysiology and cognitive psychology. Research into emergence must thus be satisfied with examining family resemblances between similar but nonidentical sets of relations. As we saw, emergence theory predicts that each relation between neighboring disciplines will exhibit certain general features recognizable across disciplines.

[6] I note only that such extrapolations are neither excluded by good science nor damaging to it – as long as one avoids confusion on which of the five emergences one intends to refer to. Indeed, good reasons are sometimes given to engage in metaphysical speculation based on scientific results, and it is possible that emergence will turn out to be one of these cases.

[7] Bill Newsome (neurobiology, Stanford University), personal correspondence.

In the end, if research on emergence is to interest practicing scientists, it must be useful for understanding specific phenomena in the natural world. One can identify at least three levels of potential usefulness. At the least, emergence should be a useful heuristic.[8] By drawing attention to patterns across disciplines, it can suggest new approaches to solving specific empirical problems. Second, emergence may also contribute to the process of theory selection. This will occur when emergence theory allows knowledge gained about one set of relations between levels of natural phenomena to be applied to analogous sets of relations. Finally, the more that specific and well-defined science-based examples of emergent phenomena are described rigorously, the more the concept may serve as a conceptual basis for framing large breakthrough theories in science.

Ultimately, emergence involves the prediction that increases in complexity will correlate with specific transition points where new types of structural organization or behavior will appear. If this is right, Nature is both continuous and discontinuous. Continuous, because the laws of physics continue to determine the possibility space for everything that emerges in natural history, from cells to birds to brains. Yet discontinuous, if in fact increases in complexity tend to produce distinct forms of organization and behaviors that are most fruitfully studied by distinct sets of scientific disciplines. Each level in Nature requires a corresponding level of scientific explanation. Even future scientific progress will not eliminate the transitions and distinctions between these levels. This, at any rate, is the empirical prediction behind the emergentist research program.

Acknowledgments

I am grateful to Pranab Das, Terry Deacon, Charles Harper, Steven Knapp, Robert Laughlin, and Bill Newsome for extensive criticisms and suggestions during the revisions phase of this research project.

References

Barabási, Al-L (2002) *Linked: The New Science of Networks*. Cambridge, MA: Perseus Books.

Chalmers, D (1995) Facing up to the problem of Consciousness. *Journal of Consciousness Studies* **2**, 200.

Cowan, G, *et al.* (eds.) (1994) *Complexity: Metaphors, Models, and Reality*. Santa Fe Institute Studies in the Sciences of Complexity, Proceedings vol. 19. Reading, MA: Addison-Wesley.

[8] A heuristic is a guide to long-term research choices and agendas. The emergence heuristic focuses study on how complex phenomena grow out of lower-level phenomena; it predicts that at least some of the resulting levels will require explanation in terms of causal forces operating at that particular level.

Earley, J (1981) Self-organization and agency: in chemistry and in process philosophy. *Process Studies* **11**, 242.
Deacon, T (1997) *The Symbolic Species: The Co-Evolution of Language and the Brain*. New York: W. W. Norton.
Fodor, J (1992) The big idea: Can there be a science of mind?, *The Times Literary Supplement*, July 3.
Gell-Mann, M (1994) *The Quark and the Jaguar: Adventures in the Simple and the Complex*. New York: W. H. Freeman.
Goodwin, B (2001) *How the Leopard Changed its Spots: The Evolution of Complexity*. Princeton, NJ: Princeton University Press.
Gordon, D M (2000) *Ants at Work: How an Insect Society Is Organized*. New York: W. W. Norton.
Gregersen, N H (ed.) (2003) From anthropic design to self-organized complexity. In *Complexity to Life: On the Emergence of Life and Meaning*, ed. N. H. Gregersen, p. 206. Oxford: Oxford University Press.
Holland, J (1998) *Emergence: From Chaos to Order*. Cambridge, MA: Perseus Books.
Kass, L (1999) *The Hungry Soul: Eating and the Perfecting of Our Nature*. Chicago, IL: University of Chicago Press.
Kauffman, S (1990) Whispers from Carnot: the origins of order and principles of adaptation in complex nonequilibrium systems. In *Complexity: Metaphors, Models, and Reality*, ed. G. Cowan *et al.*, p. 83. Reading, MA: Addison-Wesley.
(1995) *At Home in the Universe: The Search for Laws of Self-Organization and Complexity*. New York: Oxford University Press.
(2000) *Investigations*. New York: Oxford University Press.
LeDoux, J, *et al.* (1978) *The Integrated Mind*. New York: Plenum Press.
Lewin, R (1999) *Complexity: Life at the Edge of Chaos*, 2nd edn. Chicago, IL: University of Chicago Press.
Milo, R, *et al.* (2002) Network motifs: simple building blocks of complex networks. *Science* **298**, 824.
Morowitz, H (2002) *The Emergence of Everything: How the World Became Complex*. New York: Oxford University Press.
Nagel, T (1980) What is it like to be a bat? In *Readings in Philosophy of Psychology*, vol. 1, ed. Ned Block, pp. 159–68. Cambridge, MA: Harvard University Press.
Oltvai, Z and Barabási, Al-L (2002) Life's complexity pyramid. *Science* **298**, 763.
Peacocke, A (1993) *Theology for a Scientific Age: Being and Becoming Natural, Divine, and Human*, enlarged edn. Minneapolis, MN: Fortress Press.
(1994) *God and the New Biology*. Gloucester, MA: Peter Smith.
Prigogine, I (1984) *Order out of Chaos: Man's New Dialogue with Nature*. New York: Bantam Books.
Schuster, P (1994) How do RNA molecules and viruses explore their worlds? In *Complexity: Metaphors, Models, and Reality*, ed. G. Cowan *et al.*, p. 383. Reading, MA: Addison-Wesley.
Sheldrake, R (1995) *A New Science of Life: The Hypothesis of Morphic Resonance*. Rochester, VT: Park Street Press.
Southgate, C, *et al.* (1999) *God, Humanity, and the Cosmos: A Textbook in Science and Religion*. Harrisburg, PA: Trinity Press.
Velmans, M (2000) *Understanding Consciousness*. London: Routledge.
Wheeler, J A (1996) *At Home in the Universe*. New York: Springer-Verlag.
(1998) (with K. Ford) *Geons, Black Holes, and Quantum Foam: A Life in Physics*. New York: W. W. Norton.

(1999) Information, physics, quantum: the search for links. In *Feynman and Computation: Exploring the Limits of Computers*, ed. A. J. G. Hey, p. 309. Cambridge, MA: Perseus Books.

Wolfram, S (2002) *A New Kind of Science*. Champaign, IL: Wolfram Media.

Zurek, W (1991) Decoherence and the transition from quantum to classical. *Phys. Today* **44**, 36.

(2002) Decoherence and the transition from quantum to classical Ḅ revisited. *Los Alamos Science* **27**, 2.

27

True complexity and its associated ontology

George F. R. Ellis
University of Cape Town

True complexity and the natures of existence

My concern in this chapter is true complexity and its relation to physics. This is to be distinguished from what is covered by statistical physics, catastrophe theory, study of sand piles, the reaction diffusion equation, cellular automata such as "The Game of Life," and chaos theory. Examples of truly complex systems are molecular biology, animal and human brains, language and symbolic systems, individual human behavior, social and economic systems, digital computer systems, and the biosphere. This complexity is made possible by the existence of molecular structures that allow complex biomolecules such as RNA, DNA, and proteins with their folding properties and lock-and-key recognition mechanisms, in turn underlying membranes, cells (including neurons), and indeed the entire bodily fabric and nervous system.

True complexity involves vast quantities of stored information and hierarchically organized structures that process information in a purposeful manner, particularly through implementation of goal-seeking feedback loops. Through this structure they appear purposeful in their behavior ("teleonomic"). This is what we must look at when we start to extend physical thought to the boundaries, and particularly when we try to draw philosophical conclusions – for example, as regards the nature of existence – from our understanding of the way physics underlies reality. Given this complex structuring, one can ask, "What is real?", that is, "What actually exists?", and "What kinds of causality can occur in these structures?"

This chapter aims to look at these issues. It contains further sections covering the following topics: the nature of true complexity; the natures of existence; the nature of causality; and the relation to fundamental physics, including some comments on the relation to ultimate reality.

Science and Ultimate Reality, eds. J. D. Barrow, P. C. W. Davies and C. L. Harper, Jr. Published by Cambridge University Press.

Sociology/Economics/Politics
Psychology
Physiology
Cell biology
Biochemistry
Chemistry
Physics
Particle physics

Figure 27.1. A hierarchy of structure and causation. Each lower level underlies what happens at each higher level, in terms of physical causation. For a much more detailed exploration of such hierarchies, see Ellis (2002).

The nature of true complexity

In broad outline, emergence of complex systems has the following features:

1. Emergence of complexity occurs in terms of (a) function (simple structures causally underlying functioning of more complex structures), (b) development (a single initial cell growing to a complex interlocking set of 10^{13} cells), and (c) evolution (a universe region containing no complex systems evolving to one containing billions of them), each occurring with very different timescales.

Natural selection (see, e.g., Campbell 1991) is seen as the mechanism that allows this all to come into being through an evolutionary process.

2. Complex systems are characterized by (a) hierarchical structures delineating both complexity and causality with (b) different levels of order and descriptive languages plus (c) a relational hierarchy at each level of the structural hierarchy.

This is summarized in the hierarchy of structure/causation relating to human beings (Fig. 27.1).

As expressed by Campbell (1991: 2–3):

> With each upward step in the hierarchy of biological order, novel properties emerge that were not present at the simpler levels of organization. These emergent properties arise from interactions between the components. . . . Unique properties of organized matter arise from how the parts are arranged and interact . . . [consequently] we cannot fully explain a higher level of organization by breaking it down to its parts.

One can't even describe the higher levels in terms of lower-level language.

Furthermore, one can't comprehend the vast variety of objects at each higher level without using a characterization structured in hierarchical terms, e.g.

animal – mammal – domestic animal – dog – guard dog – Doberman – Fred
machine – transport vehicle – automobile – sedan – Toyota – CA 687-455.

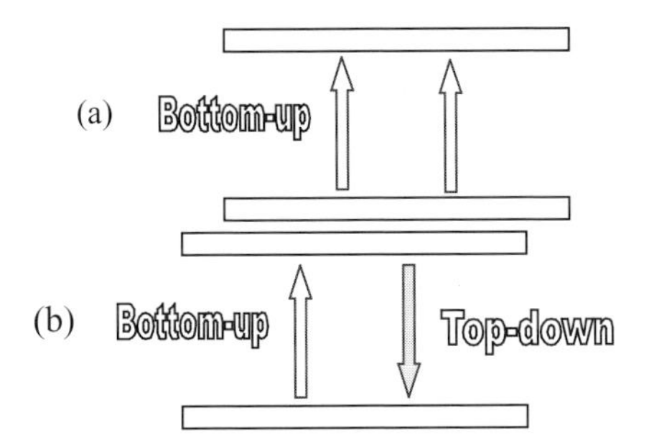

Figure 27.2. Bottom–up and top–down action. The fundamental importance of top–down action is that it changes the causal relation between upper and lower levels in the hierarchy of structure and organization; cf. Fig. 27.2a and Fig. 27.2b.

The hierarchical characterization used may be based on (i) appearance, (ii) structure, (iii) function, (iv) geographic location and/or history (e.g., evolutionary history), or (v) an assigned labeling (e.g., alphabetic or numeric, each themselves hierarchical in nature). Note that in the end these categorizations go from the generic to the individual/specific.

3. These hierarchical structures are modular – made up by structural combinations of simpler (lower-level) components with their own state variables, incorporating encapsulation and inheritance, enabling reuse and modification.

In general many lower-level states correspond to a single higher-level state, because a higher-level state description is arrived at by averaging over lower-level states and throwing away a vast amount of lower-level information ("coarse graining"). The number of lower-level states corresponding to a single higher-level state determines the entropy of that state. This is lower-level information hidden in the higher-level view.

4. Complex emergence is enabled by (a) bottom–up and (b) top–down action, the latter occurring by coordinating lower-level actions according to the system structure and boundary conditions.

Higher-level structures and boundary conditions can structure lower-level interactions in a coordinated way; this is top–down action in the hierarchy. This affects the nature of causality significantly, particularly because inter-level feedback loops become possible (Fig. 27.2). Reliable higher level behavioral laws occur if the variety of lower-level states corresponding to a particular higher-level initial state all lead to the same higher-level final state, thus enabling same-level action.

Causality in coherent complex systems has all these dimensions (bottom–up, same level, top–down): there are explanations of each kind that are simultaneously applicable.

5. Living systems involve purposeful use of information to control physical functions in accord with higher-level goals. They are structured as (a) feedback control systems that (b) can learn by (c) capturing, storing, recalling, and analyzing information which is used to set the system goals; this involves (d) pattern recognition and (e) utilization of simplified predictive models.

It is these capacities that make the difference between complicated and complex systems. They enable strongly emergent phenomena such as the existence of the rules of chess (as well as the resulting strategies of chess players) and of money and exchange rates. There is no implication here as to how the information is stored (it might be encoded in particular system energy levels, sequences of building-block molecules, or synaptic connection patterns, for example). Also there is no implication here that the amount of useful information is described by the Shannon formulae.

This adds up to "***organized complexity***" (Sellars 1932; Simon 1962, 1982; Churchman 1968; Flood and Carson 1990; Bar-Yam 2000). Here we see nonmaterial features such as information and goals having causal effects in the material world of forces and particles, which means they have an ontological reality. We now examine aspects of these various features in more detail.

Complexity and structure

The essential point of systems theory is that the value added by the system must come from the relationships between the parts, not from the parts per se (Emery 1972; von Bertalanffy 1973). True complexity, with the emergence of higher levels of order and meaning, occurs in *modular hierarchical structures*, because these form the only viable ways of building up and utilizing real complexity. The principles of hierarchy and modularity have been investigated usefully in the context of computing, and particularly in the discussion of *object-oriented programming* (see Booch 1994), and it is helpful to see how these principles are embodied in physical and biological structures.

Modularity (Booch 1994: 12–13, 54–59) The technique of mastering complexity in computer systems and in life is divide and rule – decompose the problem into smaller and smaller parts, each of which we may then refine independently (Booch 1994: 16). By organizing the problem into smaller parts, we break the informational bottleneck on the amount of information that has to be received, processed, and remembered at each step; and this also allows specialization of operation. This implies creation of a set of specialized modules to handle the smaller problems that together comprise the whole. In building complex systems from simple ones, or improving an already complex system, you can reuse the same modular components in new combinations, or substitute new more efficient components, with the same functionality, for old ones. Thus we can benefit from a library of tried and trusted components. However the issue then arises as to how we structure the relationships between the modules, and what functional capacity we give them. Complex structures are made of modular units with *abstraction*, *encapsulation*, and *inheritance*; this enables the modification of modules and reuse for other purposes.

Abstraction and labeling (Booch 1994: 20, 41–48) Unable to master the entirety of a complex object, we choose to ignore its inessential details, dealing instead with a generalized idealized model of the object. *An abstraction* denotes the essential characteristics of an object that distinguishes it from all other kinds of objects. An abstraction focuses on the outside view of the object, and so serves to separate its essential behavior from its implementation; it emphasizes some of the system's details or properties, while suppressing others. A key feature is that *compound objects can be named and treated as units.* This leads to the power of abstract symbolism and symbolic computation.

Encapsulation and information hiding (Booch 1994: 49–54) Consumers of services only specify what is to be done, leaving it to the object to decide how to do it: "No part of any complex system should depend on the internal details of any other part." *Encapsulation* means that the internal workings are hidden from the outside, so its procedures can be treated as black-box abstractions. To embody this, each class of object must have two parts: an *interface* (its outside view, encompassing an abstraction of the common behavior of all instances of the class of objects) and an *implementation* (the internal representations and mechanisms that achieve the desired behavior). Efficiency and usability introduce the aim of reducing the number of variables and names that are visible at the interface. This involves *information hiding*, corresponding to coarse graining in physics; the accompanying loss of detailed information is the essential source of entropy.

Inheritance (Booch 1994: 59–62) In heritance is the most important feature of a classification hierarchy: it allows an object class – such as a set of modules – to inherit all the properties of its superclass, and to add further properties to them (it is a "is a" hierarchy). This allows similarities to be described in one central place and then applied to all the objects in the class and in subclasses. It makes explicit the nature of the hierarchy of objects and classes in a system, and implements generalization/ specialization of features (the superclass represents generalized abstractions, and subclasses represent specializations in which variables and behaviors are added, modified, or even hidden). It enables us to understand something as a modification of something already familiar, and saves unnecessary repetition of descriptions or properties.

Hierarchy (Booch 1994: 59–65) Hierarchical structuring is a particularly helpful way of organising the relationship between the parts, because it enables building up of higher-level abstractions and permits relating them by inheritance. A hierarchy represents a decomposition of the problem into constituent parts and processes to handle those constituent parts, each requiring less data and processing, and more restricted operations than the problem as a whole. The success of hierarchical structuring depends on (a) implementation of modules which handle these lower-level processes, and (b) integration of these modules into a higher-level structure.

The basic features of how hierarchies handle complexity are as follows:

> Frequently, complexity takes the form of *a hierarchy*, whereby a complex system is composed of inter-related subsystems that have in turn their own subsystems, and so on, until some lowest level of component is reached.
>
> (Courtois 1985.)

> The fact that many complex systems have a nearly decomposable hierarchic structure is a major facilitating factor enabling us to understand, describe, and even "see" such systems and their parts.
>
> (Simon 1982.)

Not only are complex systems hierarchic, but the levels of this hierarchy represent *different levels of abstraction*, each built upon the other, and each understandable by itself (and each characterized by a different phenomenology). This is the phenomenon of *emergent order*. All parts at the same level of abstraction interact in a well-defined way (which is why they have a reality at their own level, each represented in a *different language* describing and characterizing the causal patterns at work at that level).

> We find separate parts that act as independent agents, each of which exhibit some fairly complex behaviour, and each of which contributes to many higher level functions. Only through the mutual co-operation of meaningful collections of these agents do we see the higher-level functionality of an organism. This is *emergent behaviour* – the behaviour of the whole is greater than the sum of its parts, and cannot even be described in terms of the language that applies to the parts. Intra-component linkages are generally stronger than inter-component linkages. This fact has the effect of separating the *high-frequency dynamics* of the components – involving their internal structure – from the low-frequency dynamics – involving interactions amongst components.
>
> (Simon 1982.)

(This is why we can sensibly identify the components.) In a hierarchy, through encapsulation, objects at one level of abstraction are shielded from implementation details of lower levels of abstraction.

Bottom–up and top–down action

Bottom–up action

A fundamental feature of the structural hierarchy in the physical world is *bottom–up action*: what happens at each higher level is based on causal functioning at the level below, hence what happens at the highest level is based on what happens at the bottom-most level. This is the profound basis for reductionist world-views. The successive levels of order entail chemistry being based on physics, material science on both physics and chemistry, geology on material science, and so on.

Top–down action

The complementary feature to bottom–up action is *top–down action,* which occurs when the higher levels of the hierarchy direct what happens at the lower levels in a coordinated way (Campbell 1974; Peacocke 1993). For example depressing a light switch leads to numerous electrons systematically flowing in specific wires leading from a power source to a light bulb, and consequent illumination of a room.

In general many lower-level states correspond to a single higher-level state, because a higher-level state description is arrived at by averaging over lower-level states and throwing away a vast amount of lower-level information (coarse graining). Hence, specification of a higher-level state determines a family of lower-level states, any one of which may be implemented to obtain the higher-level state (the light switch being on, for example, corresponds to many billions of alternative detailed electron configurations). The specification of structure may be loose (attainable in a very large number of ways, e.g., the state of a gas) or tight (defining a very precise structure, e.g., particular electrons flows in the wiring of a VLSI chip in a computer). In the latter case both description and implementation require far more information than in the former.

The dynamics acts on a lower-level state L_i to produce a new lower-level state L_i' in a way that depends on the boundary conditions and structure of the system. Thus specifying the upper state H_1 (for example by pressing a computer key) results in some lower-level state L_i that realizes H_1 and then consequent lower-level dynamic change. The lower-level action would be different if the higher-level state were different. It is both convenient and causally illuminating to call this *top–down action*, and to explicitly represent it as an aspect of physical causation; this emphasizes how the lower-level changes are constrained and guided by structures that are only meaningful in terms of a higher-level description.

The question now is whether all the states L_i corresponding to a single initial upper state H_1 produce new lower-level states that correspond to the same higher level state H_1'. Two major cases arise. Different lower-level realizations of the same higher-level initial state may result, through microphysical action, in different

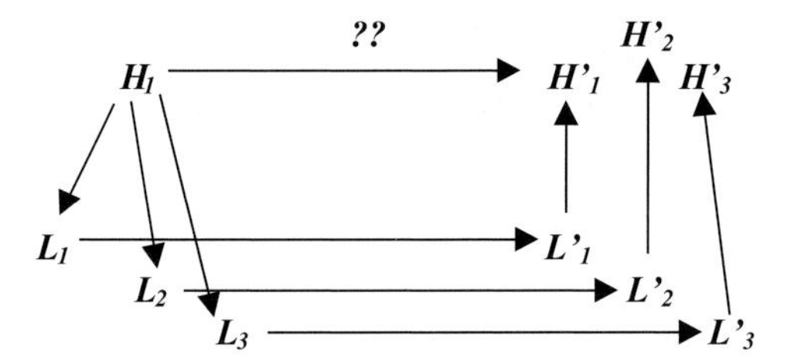

Figure 27.3. Low-level action that does not result in coordinated high-level action.

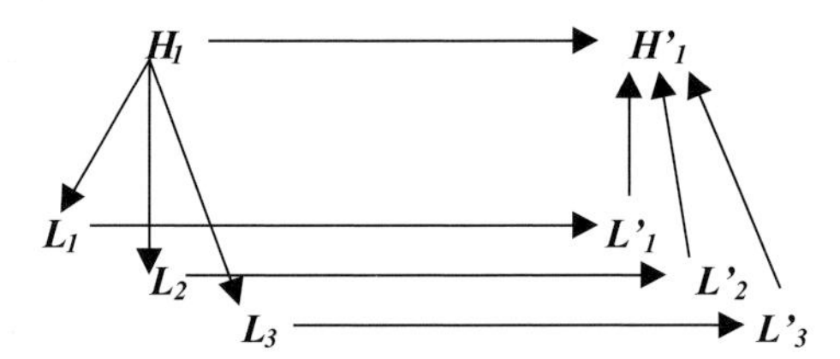

Figure 27.4. Same-level action results from coordinated low-level action that results in coordinated high-level action.

higher-level final states (Fig. 27.3). Here there is no coherent higher-level action generated by the lower-level actions. On the other hand, top–down action generating same-level action occurs when each lower-level state corresponding to a specific higher-level state results in the same higher-level state, so that every lower-level implementation of the initial higher-level state gives the same higher-level outcome (Fig. 27.4). In this case, consistent behavior occurs at the higher level, regarded as a causal system in its own right – there is now effective higher-level autonomy of *same-level action*, which we can consider in its own right independent of the lower levels:

$$H_1 \longrightarrow H_1'$$

This higher-level action is effective by coordinating actions at the lower levels, resulting in coherent higher-level action. Whether this happens or not may depend on the particular coarse graining (i.e., higher-level description) chosen.

Multiple top–down action enables various higher levels to coordinate action at lower levels in a complex system in a coherent way, and so gives them their causal effectiveness. It is prevalent in the real physical world and in biology, because no real physical or biological system is isolated. Boundary conditions as well as structural relations effect top–down action. I will illustrate this with a series of examples.

1. Gas in a cylinder with piston The cylinder walls together with the piston position determine the gas pressure and temperature. Both are macro concepts that make no sense at the micro level.

2. Interaction potentials Potentials in the Schrödinger equation, or in the action for the system, represent the summed effects of other particles and forces, and hence are the way that the nature of both simple and complex structures can be implemented (from a particle in a box to a computer or a set of brain connections). These potentials describe the summed interactions between microstates, enabling internal top–down effects. Additionally one may have external potentials imposed in the chosen representation, representing top–down effects from the environment on the system.

3. Nucleosynthesis in the early universe The rates of nuclear interactions depend on the density and temperature of the interaction medium. The microscopic reactions that take place in the early universe, and hence the elements produced, thus depend on the rate of expansion of the universe (determined by the Friedmann equation).

4. Quantum measurement Top–down action occurs in the quantum measurement process – the collapse of the wave function to an eigenstate of a chosen measurement system (Penrose 1989; Isham 1997). The experimenter chooses the details of the measurement apparatus – e.g., aligning the axes of polarization measurement equipment – and that decides what set of microstates can result from a measurement process, and so crucially influences the possible microstate outcomes of the interactions that happen. Thus the quantum measurement process is partially a top–down action controlled by the observer, determining what set of eigenstates are available to the system during the measurement process. The choice of Hilbert space and the associated operators and functions is made to reflect the experimenter's choice of measurement process and apparatus, thus reflecting this top–down action.

5. The arrow of time Top–down action occurs in the determination of the arrow of time. One cannot tell how a macrosystem will behave in the future on the basis of the laws of physics and the properties of the particles that make up the system alone, because time-reversible microphysics allows two solutions – one the time reverse of the other – but the macrosystem allows only one of those solutions (Davies 1974; Zeh 1992). The prohibition of one of the allowed microsolutions is mathematically put in by hand to correspond to the real physical situation, where only entropy-increasing solutions in one direction of time occur at the macrolevel; this does not follow from the microphysics. For example, Boltzmann brilliantly proved the H-theorem (increase of macroscopic entropy in a gas on the basis of microscopic molecular collisions); but Loschmidt pointed out that this theorem works equally well with both directions of the arrow of time. Physically, the only known solution to this arrow of time problem seems to be that there is top–down action by the universe as a whole, perhaps expressed as boundary conditions at the beginning of spacetime (Penrose 1989), that allows the one solution and disallows the other. This is related to the quantum measurement issue raised above: collapse of the wave function takes place with a preferred direction of time, which is not determined if we look at the microlevel of the system alone.

6. Evolution Top–down action is central to two main themes of molecular biology: First, the development of DNA codings (the particular sequence of bases in the DNA) occurs through an evolutionary process that results in adaptation of an organism to its ecological niche (Campbell 1991). This is a classical case of top–down action from the environment to detailed biological microstructure – through the process of adaptation, the environment (along with other causal factors) fixes the specific DNA coding. There is no way in which one could ever predict this coding on the basis of biochemistry or microphysics alone (Campbell 1974).

7. Biological development Second, the reading of DNA codings. A second central theme of molecular biology is the reading of DNA by an organism in the developmental process (Gilbert 1991; Wolpert, 1998). This is not a mechanistic process, but is context dependent all the way down, with what happens before having everything to do with what happens next. The central process of developmental biology, whereby positional information determines which genes get switched on and which do not in each cell, so determining their developmental fate, is a top–down process from the developing organism to the cell, based on the existence of gradients of positional indicators in the body. Without this feature organism development in a structured way would not be possible. Thus the functioning of the crucial cellular mechanism determining the type of each cell is controlled in an explicitly top–down way. At a more macro level, recent research on genes and various hereditary diseases shows that existence of the gene for such diseases in the organism is not a sufficient cause for the disease to occur: outcomes depend on the nature of the gene and on the rest of the genome and on the environment. The macro situations determine what happens, not specific micro features by themselves, which do work mechanistically but in a context of larger meaning that largely determines the outcome. The macro environment includes the result of conscious decisions (the patient will or will not seek medical treatment for a hereditary condition, for example), so these too are a significant causal factor.

8. Mind on body Top–down action occurs from the mind to the body and thence into the physical world. The brain controls the functioning of the parts of the body through a hierarchically structured feedback control system, which incorporates the idea of decentralized control to spread the computational and communication load and increase local response capacity (Beer 1972). It is a highly specific system in that dedicated communication links convey information from specific areas of the brain to specific areas of the body, enabling brain impulses to activate specific muscles (by coordinated control of electrons in myosin filaments in the bundles of myofibrils that constitute skeletal muscles), in order to carry out consciously formulated intentions. Through this process there is top–down action by the mind on the body, and indeed on the mind itself, both in the short term (immediate causation through the structural relations embodied in the brain and body) and in the long term (structural determination through imposition of repetitive patterns). An example of the latter is how repeated stimulation of the same muscles or neurons encourages growth of those muscles and neurons. This is the underlying basis of both athletic training and of learning by rote. Additionally, an area of importance that is only now beginning to be investigated by Western medicine is the effect of the mind on health (Moyers 1993), for example through interaction with the immune system (Sternberg 2000).

9. Mind on the world When a human being has a plan in mind (say a proposal for a bridge to be built) and this is implemented, then enormous numbers of microparticles (comprising the protons, neutrons, and electrons in the sand, concrete, bricks, etc. that become the bridge) are moved around as a consequence of this plan and in conformity with it. Thus in the real world, the detailed microconfigurations of many objects (which electrons and protons go where) are to a major degree determined by the macroplans that humans have for what will happen, and the way they implement them. Some specific examples of top–down action involving goal choice are:

(i) *Chess rules*. These are socially embodied and are causally effective. Imagine a computer or alien analyzing a large set of chess games and deducing the rules of chess (i.e., what physical moves of the pieces are allowed and what are not). It would know these are inviolable rules governing these moves but have no concept of their origins, i.e., whether they were implied by modification of Newton's laws, some potential fields that constrain the motion of the chess pieces, or a social agreement that restricts their movement and can be embodied in computer algorithms. Note that the chess rules exist independent of any particular mind, and indeed may survive the demise of the society that developed them (they can for example be written in a book that becomes available to other societies).

Cosmology	***Ethics***
Astronomy	***Sociology***
Earth Science	***Psychology***
Geology	***Physiology***
Materials	***Biochemistry***

Chemistry
Physics

Figure 27.5. Branching hierarchy of causal relations. The hierarchy of physical relations (Fig. 27.1) extended to a branching hierarchy of causal relations. The left-hand side involves only (unconscious) natural systems; the right-hand side involves conscious choices, which are causally effective. In particular, the highest level of intention (ethics) is causally effective.

(ii) *The Internet*. This embodies local action in response to information requests, causing electrons to flow in meaningful patterns in a computer's silicon chips and memory, mirroring patterns thousands of miles away, when one reads web pages. This is a structured influence at a distance due to channeled causal propagation and resulting local physical action.
(iii) *Money and associated exchange rates*. The money is a physical embodiment of the economic order in a society, while the exchange rates are socially embodied but are also embodied for example in ink on newspaper pages, and in computer programs utilized by banks.
(iv) *Global warming*. The effect of human actions on the earth's atmosphere, through the combined effect of human causation moving very large numbers of microparticles (specifically, chlorofluorocarbons) around, thereby affecting the global climate. The macroprocesses at the planetary level cannot be understood without explicitly accounting for human activity (Schellnhuber 1999).
(v) *Hiroshima*. The dropping of the nuclear bomb at Hiroshima was a dramatic macro-event realized through numerous micro-events (fissions of uranium nuclei) occurring through a human-based process of planning and implementation of those plans.

Because of this, the structural hierarchy, now interpreted as a causal hierarchy, bifurcates (see Fig. 27.5). The left-hand side, representing causation in the natural world, does not involve goal choices: all proceeds mechanically. The right-hand side, representing causation involving humans, is to do with choice of goals that lead to actions. Ethics is the high-level subject dealing with the choice of appropriate goals. Because this determines the lower-level goals chosen, and thence the resulting actions, ethics is causally effective in the real physical world. This is of course obvious as it follows from the causation chain, but to make the point absolutely clear: a prison may have present in its premises the physical apparatus of an execution chamber, or maybe not; whether this is so or not depends on the ethics of the country in which the prison is situated.

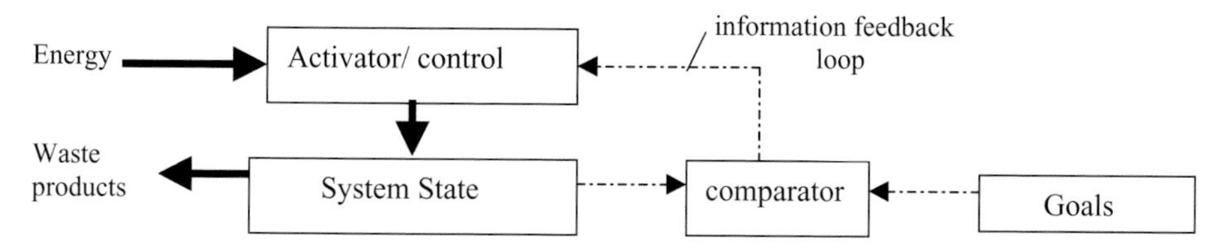

Figure 27.6. The basic feedback control process. The second law of thermodynamics requires energy input and heat output in active processes, which must occur then in an open system.

Information and goal seeking

A key feature is the use of current and past information to set goals that are then implemented in feedback control systems. This is the context in which stored information is the effective core of complex history-dependent behavior.

Feedback control

The central feature of organized action is *feedback control*, whereby setting of goals results in specific actions taking place that aim to achieve those goals (Ashby 1958; Beer 1966, 1972). A comparator compares the system state with the goals, and sends an error message to the system controller if needed to correct the state by making it a better approximation to the goals (Fig. 27.6). The linkages to the comparator and thence to the controller are *information linkages* rather than power and/or material linkages like that from the activator to the system (the information flow will use a little power but only that required to get the message to where it is needed). Examples are controlling the heat of a shower, the direction of an automobile, or the speed of an engine. Thus it is here that the *key role of information* is seen: *it is the basis of goal choice in living systems* (and artifacts that embody feedback control). The crucial issue now is what determines the goals: where do they come from? Two major cases need to be distinguished.

Homeostasis: in-built goals There are numerous systems in all living cells, plants, and animals that automatically (i.e., without conscious guidance) maintain homeostasis – they keep the structures in equilibrium through multiple feedback systems that fight intruders (the immune system), and control energy and material flows, breathing and the function of the heart, body temperature, and blood pressure, etc. They are effected through numerous enzymes, antibodies, regulatory circuits of all kinds (for example those that maintain body temperature and blood pressure) (Milsum 1966). The inbuilt goals that guide these activities are implicit rather than explicit, for example the temperature of the human body is maintained at 37.0 °C with great accuracy without that figure being explicitly preset in some control apparatus, but certainly these goals are identifiable and very efficiently attained. They have developed in the course of time through the processes of evolution, and so are historically determined in particular environmental context, and are unaffected by individual history. In manufactured artifacts, the goal may be explicitly stated and controllable (e.g., the temperature setting of a thermostat). Not only are the feedback control systems themselves emergent systems, but also the implied goals are emergent properties that guide numerous physical, chemical, and biochemical interactions in a teleological way. They represent distilled information about the behavior of the environment in relation to the needs of life, and so they represent implicit information processing by the organism. At the higher levels they include the instinctive behavior of animals. These feedback control loops are hierarchically structured with maximum decentralization of control from the higher to the lower levels, as is required both in order to handle requisite variety (Ashby 1958) and the associated information loads (Beer 1972), and for maximal local efficiency (ability to respond to local conditions).

Goal seeking: chosen goals However, at higher levels in animals, important new features come into play: there are now explicit behavioral goals, that are either learnt or are consciously chosen. It is in the choice of these goals that explicit information processing comes into play. Information arrives from the senses and is sorted and either discarded or stored in long-term and short-term memory. Conscious and unconscious processing of this information sets up the goal hierarchy (with ethics the topmost level) which then controls purposeful action.

Information origin and use

Responsive behavior depends on purposeful use of information: capture, storage, transmission, recall, and assessment to control physical functions in accord with higher-level goals. The computations are based on stored variables and structured information flows, so hidden internal variables affect external behavior. Current information is filtered against a relevance pattern, the irrelevant being discarded, the moderately significant being averaged over and stored in compressed form, the important being selectively amplified and used in association with current expectations to assess and revise immediate goals. The relevance pattern is determined by basic emotional responses such as those delineated by Panksepp (1998), which provide the evaluation function used in a process of neural Darwinism (Edelmann 1990) that determines the specifics of neural connections in the brain. In this way emotional responses underlie the development of rationality. Expectations are based on causal models based on past experience ("frames," e.g., how to behave in a restaurant), constantly revised on the basis of newer experience and information.

Thus feedback control systems based on sophisticated interpretation of present and past data enable purposeful (teleological) behavior. Memory allows both the long-term past and the immediate environmental context to be taken into account in choosing goals, providing historical information used to shape these goals in conjunction with present data. Long-term memory allows a nonlocal (in time) kind of causation that enables present and future behavior to be based on interpretations of long past events (e.g., remembering that an individual let one down in important ways years ago). Learning allows particular responses to develop into an automatic skill, in particular allowing some responses to become inbuilt and so able to be rapidly deployed (e.g., driving a car, many sports moves, and so on).

Symbolic representation

At the highest level, the process of analysis and understanding is driven by the power of symbolic abstraction (Deacon 1997), codified into language embodying both syntax and semantics. This underpins other social creations such as specialized roles in society and the monetary system on the one hand, and higher-level abstractions such as mathematics, physical models, and philosophy on the other – all encoded in symbolic systems.

Information guides all this, and is manifestly real in that it has a commercial value that underlies development of a major part of the international economy (the information technology sector). The meta-question of how context influences behavior is guided and constrained by a system of ethics based on an overall world-view associated with meaning. This will also be encoded in language and symbols.

These are all strongly emergent phenomena that are causally effective. They exist as nonmaterial effective entities, created and maintained through social interaction and teaching, and are codified in books and sometimes in legislation. Thus while they may be represented and understood in individual brains, their existence is not contained in any individual brain and they certainly are not equivalent to brain states. Rather the latter serve as just one of many possible forms of embodiment of these features.

Mathematical and physical description

How to model all this? There are two approaches to quantitative modeling of hierachical systems and emergent properties that may be useful. On the one hand, network mathematics and related network thermodynamics tackle the problem directly, see, e.g., Peacocke (1989) and Holland (1998); and on the other some of these issues are tackled in the studies of neural networks and genetic algorithms (Carpenter and Grossberg 1991; Bishop 2000) and of control systems (e.g., Milsum 1966). What are needed are computer hierarchical models plus heuristic understanding

- ***World 1:*** **Matter and forces**
- ***World 2:*** **Consciousness**
- ***World 3:*** **Physical and biological possibilities**
- ***World 4:*** **Mathematical reality**

Figure 27.7. The different kinds of reality implied by causal relationships can be characterized in terms of four worlds, each representing a different kind of existence.

of interplay of components, together with mathematical models of specific subsystems and networks and physical models of molecular structure and interactions (needing mathematical models of three-dimensional structure) that allow this to come into existence in complex systems.

The natures of existence

In this section I propose a holistic view of ontology, building on the previous proposals by Popper and Eccles (1977) and Penrose (1997). I clearly distinguish between ontology (existence) and epistemology (what we can know about what exists). They should not be confused: whatever exists may or may not interact with our senses and measuring instruments in such a way as to demonstrate clearly its existence to us.

A holistic view of ontology

I take as given the reality of the everyday world – tables and chairs, and the people who perceive them – and then assign a reality additionally to each kind of entity that can have a demonstrable causal effect on that everyday reality. The problem then is to characterize the various kinds of independent reality that may exist in this sense. Taking into account the causal efficacy of all the entities discussed above, I suggest as a possible completion of the proposals by Popper and Eccles and Penrose that the four worlds indicated in Fig. 27.7 are ontologically real. These are not different causal levels within the same kind of existence, rather they are quite different kinds of existence, but related to each other through causal links. The challenge is to show firstly that each is indeed ontologically real, and secondly that each is sufficiently and clearly different from the others that it should be considered as separate from them. I now discuss them in turn.

Matter and forces

World 1: *The physical world of energy and matter*

This is hierarchically structured to form lower and higher causal levels whose entities are all ontologically real.

This is the basic world of matter and interactions between matter, based at the micro level on elementary particles and fundamental forces, and providing the ground of physical existence. It comprises three major parts:

World 1a: Inanimate objects, both naturally occurring and manufactured.
World 1b: Living things, apart from humans (amoeba, plants, insects, animals, etc.).
World 1c: Human beings, with the unique property of being self-conscious.

All these objects are made of the same physical stuff, but the structure and behavior of inanimate and living things (described respectively by physics and inorganic chemistry, and by biochemistry and biology) are so different that they require separate recognition, particularly when self-consciousness and purposive activity (described by psychology and sociology) occurs. The hierarchical structure in matter is a real physical structuration, and is additional to the physical constituents that make up the system themselves. It provides the basis for higher levels of order and phenomenology, and hence of ontology.

There is ontological reality at each level of the hierarchy. Thus we explicitly recognize as being real, quarks, electrons, neutrinos, rocks, tables, chairs, apples, humans, the world, stars, galaxies, and so on. The fact that each is composed of lower-level entities does not undermine its status as existing in its own right (Sellars 1932). We can attain and confirm high representational accuracy and predictive ability for quantities and relations at higher levels, independent of our level of knowledge of interactions at lower levels, giving well-validated and reliable descriptions at higher levels accurately describing the various levels of emergent nonreducible properties and meanings. An example is digital computers, with their hierarchical logical structure expressed in a hierarchy of computer languages that underlie the top-level user programs. The computer has a reality of existence at each level that enables one to meaningfully deal with it as an entity at that level (Tannenbaum 1990). The user does not need to know machine code, and indeed the top-level behavior is independent of which particular hardware and software underlie it at the machine level. Another example is that a motor mechanic does not have to study particle physics in order to ply her trade.

Consciousness

World 2: *The world of individual and communal consciousness*

This consists of ideas, emotions, and social constructions. It again is ontologically real (it is clear that these all exist), and causally effective.

This world of human consciousness can be regarded as comprising three major parts:

World 2a: Human information, thoughts, theories, and ideas.
World 2b: Human goals, intentions, sensations, and emotions.
World 2c: Explicit social constructions.

These worlds are different from the world of material things, and are realized through the human mind and society. They are not brain states, although they can be represented as such, for they do not reside exclusively in any particular individual mind. They are not identical to each other: World 2a is the world of rationality, World 2b is the world of intention and emotion, and so comprehends nonpropositional knowing, while World 2c is the world of consciously constructed social legislation and convention. Although each individually and socially constructed in a complex interaction between culture and learning, these are indeed each capable of causally changing what happens in the physical world, and each has an effect on the others. Each is described in more detail below.

World 2a: The world of human information, thoughts, theories, and ideas This world of rationality is hierarchically structured, with many different components. It includes words, sentences, paragraphs, analogies, metaphors, hypotheses, theories, and indeed the entire bodies of science and literature, and refers both to abstract entities and to specific objects and events. It is necessarily socially constructed on the basis of varying degrees of experimental and observational interaction with World 1, which it then represents with varying degrees of success. World 2a is represented by symbols, particularly language and mathematics, which are arbitrarily assigned and which can themselves be represented in various ways (sound, on paper, on computer screens, in digital coding, etc.).

Thus each concept can be expressed in many different ways, and is an entity in its own right independent of which particular way it is coded or expressed. These concepts sometimes give a good correspondence to entities in the other worlds, but the claim of ontological reality of entities existing in World 2a makes no claim that the objects or concepts they refer to are real. Thus this world equally contains concepts of rabbits and fairies, galaxies and UFOs, science and magic, electrons and aether, unicorns and apples; the point being that all of these certainly exist *as concepts*. That statement is neutral about whether these concepts correspond to objects or entities that exist in the real universe (specifically, whether there is or is not some corresponding entity in World 1) or whether the theories in this world are correct (that is, whether they give a good representation of World 1 or not).

All the ideas and theories in this world are ontologically real in that they are able to cause events and patterns of structures in the physical world. First, they may all occur as descriptive entries in an encyclopaedia or dictionary. Thus each idea has causal efficacy as shown by existence of the resulting specific patterns of marks on paper (these constellations of microparticles would not be there if the idea did not exist, as an idea). Second, in many cases they have further causal power as shown by the examples of the construction of the Jumbo Jet and the destruction of Dresden. Each required both an initial idea, and resulting detailed plan and an intention to carry it out. Hence such ideas are indubitably real in the sense that they must be included in any complete causal scheme for the real world. You can if you want to deny the reality of this feature – and you will end up with a causal scheme lacking many causal features of the real world (you will have to say that the Jumbo Jet came into existence without a cause, for example!).

World 2b: The world of human goals, intentions, sensations, and emotions This world of motivation and senses is also ontologically real, for it is clear that they do indeed exist in themselves, for example they may all be described in novels, magazines, books, etc., thus being causally effective in terms of being physically represented in such writings. Additionally many of them cause events to happen in the physical world – for example the emotion of hate can cause major destruction both of property and lives, as in Northern Ireland and Israel and many other places. In World 2b, we find the goals and intentions that cause the intellectual ideas of World 2a to have physical effect in the real world.

World 2c: The world of explicit social constructions This is the world of language, customs, roles, laws, etc., which shapes and enables human social interaction. It is developed by society historically and through conscious legislative and governmental processes. It gives the background for ordinary life, enabling Worlds 2a and 2b to function, particularly by determining the means of social communication (language is explicitly a social construction). It is also directly causally effective, for example speed laws and exhaust emission laws influence the design both of automobiles and road signs, and so get embodied in the physical shapes of designed structures in World 1; the rules of chess determine the space of possibilities for movements of chess pieces on a chess board. It is socially realized and embodied in legislation, roles, customs, etc.

Physical and biological possibilities

World 3: *The world of Aristotelian possibilities*

This characterizes the set of all physical possibilities, from which the specific instances of what actually happens in World 1 are drawn.

This world of possibilities is ontologically real because of its rigorous prescription of the boundaries of what is possible – it provides the framework within which

World 1 exists and operates, and in that sense is causally effective. It can be considered to comprise two major parts:

World 3a: The world of physical possibilities, delineating possible physical behavior.
World 3b: The world of biological possibilities, delineating possible biological organization.

These worlds are different from the world of material things, for they provide the background within which that world exists. In a sense they are more real than that world, because of the rigidity of the structure they impose on World 1. There is no element of chance or contingency in them, and they certainly are not socially constructed (although our understanding of them is so constructed). They rigidly constrain what can happen in the physical world, and are different from each other because of the great difference between what is possible for life and for inanimate objects. Each is described in more detail below.

World 3a: The world of physical possibilities This delineates possible physical behavior (it is a description of all possible motions and physical histories of objects). Thus it describes what can actually occur in a way compatible with the nature of matter and its interactions; only some of these configurations are realized through the historical evolutionary process in the expanding universe. We do not know whether laws of behavior of matter as understood by physics are prescriptive or descriptive, but we do know that they rigorously describe the constraints on what is possible (you cannot move in a way that violates energy conservation; you cannot create machines that violate causality restrictions; you cannot avoid the second law of thermodynamics; and so on). This world delineates all physically possible actions (different ways particles, planets, footballs, automobiles, aircraft can move, for example); from these possibilities, what actually happens is determined by initial conditions in the universe, in the case of interactions between inanimate objects, and by the conscious choices made, when living beings exercise volition.

If one believes that physical laws are prescriptive rather than descriptive, one can view this world of all physical possibilities as being equivalent to a complete description of the set of physical laws (for these determine the set of all possible physical behaviors, through the complete set of their solutions). The formulation given here is preferable, in that it avoids making debatable assumptions about the nature of physical laws, but still incorporates their essential effect on the physical world. Whatever their ontology, what is possible is described by physical laws such as the second law of thermodynamics:

$$dS > 0,$$

Maxwell's laws of electromagnetism:

$$F_{[ab;c]} = 0, \qquad F^{ab}{}_{;b} = J^a, \qquad J^a{}_{;a} = 0,$$

and Einstein's law of gravitation:

$$R_{ab} - \frac{1}{2} R\, g_{ab} = k\, T_{ab}, \qquad T^{ab}{}_{;b} = 0.$$

These formulations emphasize the still mysterious extraordinary power of mathematics in terms of describing the way matter can behave, and each partially describes the space of physical possibilities.

World 3b: The world of biological possibilities This delineates all possible living organisms. It defines the set of potentialities in biology, by giving rigid boundaries to what is achievable in biological processes. Thus it constrains the set of possibilities from which the actual evolutionary process can choose – it rigorously delineates the set of organisms that can arise from any evolutionary history whatever. This "possibility landscape" for living beings underlies evolutionary theory, for any mutation that attempts to embody a structure which lies outside its boundaries will necessarily fail. Thus even though it is an abstract space in the sense of not being embodied in specific physical form, it strictly determines the boundaries of all possible evolutionary histories. In this sense it is highly effective causally.

Only some of the organisms that can potentially exist are realized in World 1 through the historical evolutionary process; thus only part of this possibility space is explored by evolution on any particular world. When this happens, the information is coded in the hierarchical structure of matter in World 1, and particularly in the genetic coding embodied in DNA, and so is stored via ordered relationships in matter; it then gets transformed into various other forms until it is realized in the structure of an animal or plant. In doing so it encodes both a historical evolutionary sequence, and structural and functional relationships that emerge in the phenotype and enable its functioning, once the genotype is read. This is the way that directed feedback systems and the idea of purpose can enter the biological world, and so distinguishes the animate from the inanimate world. The structures occurring in the nonbiological world can be complex, but they do not incorporate "purpose" or order in the same sense. Just as World 3a can be thought of as encoded in the laws of physics, World 3b can be thought of as encoded in terms of biological information, a core concept in biology (Kuppers 1990; Pickover 1995; Rashidi and Buehler 2000) distinguishing the world of biology from the inanimate world.

Abstract (Platonic) reality

World 4: *The Platonic world of (abstract) realities*

These are discovered by human investigation but are independent of human existence. They are not embodied in physical form but can have causal effects in the physical world.

World 4a: Mathematical forms The existence of a Platonic world of mathematical objects is strongly argued by Penrose (1989), the point being that major parts of mathematics are discovered rather than invented (rational numbers, zero, irrational numbers, and the Mandelbrot set being classic examples). They are not determined by physical experiment, but are rather arrived at by mathematical investigation. They have an abstract rather than embodied character; the same abstract quantity can be represented and embodied in many symbolic and physical ways. They are independent of the existence and culture of human beings, for the same features will be discovered by intelligent beings in the Andromeda galaxy as here, once their mathematical understanding is advanced enough (which is why they are advocated as the basis for interstellar communication). This world is to some degree discovered by humans, and represented by our mathematical theories in World 2; that representation is a cultural construct, but the underlying mathematical features they represent are not – indeed like physical laws, they are often unwillingly discovered, for example irrational numbers and the number zero (Seife 2000). This world is causally efficacious in terms of the process of discovery and description (one can for example print out the values of irrational numbers or graphic versions of the Mandelbrot set in a book, resulting in a physical embodiment in the ink printed on the page).

A key question is what if any part of logic, probability theory, and physics should be included here. In some as yet unexplained sense, the world of mathematics underlies the world of physics. Many physicists at least implicitly assume the existence of the following.

World 4b: Physical laws These underlie the nature of physical possibilities (World 3a). Quantum field theory applied to the standard model of particle physics is immensely complex (Peskin and Schroeder 1995). It conceptually involves, *inter alia*,

- Hilbert spaces operators, commutators, symmetry groups, higher dimensional spaces
- particles/waves/wave packets, spinors, quantum states/wave functions
- parallel transport/connections/metrics
- the Dirac equation and interaction potentials, Lagrangians, and Hamiltonians
- variational principles that seem to be logically and/or causally prior to all the rest.

Derived (effective) theories, including classical (nonquantum) theories of physics, equally have complex abstract structures underlying their use: force laws, interaction potentials, metrics, and so on.

There is an underlying issue of significance: *What is the ontology/nature of existence of all this quantum apparatus, and of higher-level (effective) descriptions?* We seem to have two options:

(A) These are simply our own mathematical and physical constructs that happen to characterize reasonably accurately the physical nature of physical quantities.
(B) They represent a more fundamental reality as Platonic quantities that have the power to control the behavior of physical quantities (and can be represented accurately by our descriptions of them).

On the first supposition, the "unreasonable power of mathematics" to describe the nature of the particles is a major problem – if matter is endowed with its properties in some way we are unable to specify, but not determined specifically in mathematical terms, and its behavior happens to be accurately described by equations of the kind encountered in present-day mathematical physics, then that is truly weird! Why should it then be possible that *any* mathematical construct whatever gives an accurate description of this reality, let alone ones of such complexity as in the standard theory of particle physics? Additionally, it is not clear on this basis why all matter has the same properties – why are electrons here identical to those at the other side of the universe? On the second supposition, this is no longer a mystery – the world is indeed constructed on a mathematical basis, and all matter everywhere is identical in its properties. But then the question is how did that come about? How are these mathematical laws imposed on physical matter? And which of the various alternative forms (Schrödinger, Heisenberg, Feynman, Hamiltonian, Lagrangian) is the "ultimate" one? What is the reason for variational principles of any kind?

World 4c: Platonic aesthetic forms These provide a foundation for our sense of beauty.

In this chapter, those further possibilities will not be pursued. It is sufficient for my purpose to note that the existence of a World 4a of mathematical forms, which I strongly support, establishes that this category of world indeed exists and has causal influence.

Existence and epistemology

The overall family of worlds

The major proposal of this section is that all these worlds exist – Worlds 1 to 4 are ontologically real and are distinct from each other, as argued above.

These claims are justified in terms of the effectiveness of each kind of reality in influencing the physical world. What then of epistemology? Given the existence of the various worlds mentioned above, the proposal here is that as follows.

Epistemology

Epistemology is the study of the relation between World 2 and Worlds 1, 3, and 4. It attempts to obtain accurate correspondences to quantities in all the worlds by means of entities in World 2a.

This exercise implicitly or explicitly divides World 2a theories and statements into (i) true/accurate representations, (ii) partially true/misleading representations, (iii) false/poor/misleading representations, and (iv) ones where we don't know the situation. These assessments range from statements such as "It is true her hair is red" or "There is no cow in the room" to "Electrons exist," "Newtonian theory is a very good description of medium-scale physical systems at low speeds and in weak gravitational fields," and "The evidence for UFOs is poor." This raises interesting issues about the relation between reality and appearance, e.g., everyday life gives a quite different appearance to reality than microscopic physics – as Eddington (1928) pointed out a table is actually mostly empty space between the atoms that make up its material substance, but in our experience is a real hard object. As long as one is aware of this, it can be adequately handled.

There is a widespread tendency to equate epistemology and ontology. This is an error, and a variety of examples can be given where it seriously misleads. This is related to a confusion between World 2 and the other Worlds discussed here which seems to underlie much of what has happened in the so-called "Science Wars" and the Sokal affair (Sokal 2000a, b). The proposal here strongly asserts the existence of independent domains of reality (Worlds 1, 3, and 4) that are not socially constructed, and implies that we do not know all about them and indeed cannot expect to ever understand them fully. That ignorance does not undermine their claim to exist, quite independently of human understanding. The explicit or implicit claim that they depend on human knowledge means we are equating epistemology and ontology – just another example of human hubris.

The nature of causality

The key point about causality in this context is that simultaneous multiple causality (inter-level, as well as within each level) is always in operation in complex systems. Any attempt to characterize any partial cause as the whole (as characterized by the phrase "nothing but") is a fundamentally misleading position (indeed this is the essence of fundamentalism). This is important in regard to claims that any of physics, evolutionary biology, sociology, psychology, or whatever are able to give *total* explanations of any specific properties of the mind. Rather they each provide partial and incomplete explanations. There are always multiple levels of explanation that all hold at the same time: no single explanation is complete, so one can have a top–down system explanation as well as a bottom–up explanation, both being simultaneously applicable.

Analysis "explains" the properties of the machine by analyzing its behavior in terms of the functioning of its component parts (the lower levels of structure). Systems thinking (Churchman 1968) "explains" the behavior or properties of an entity by determining its role or function within the higher levels of structure (Ackoff 1999). For example, the question: "Why does an aircraft fly?" can be answered:

- in terms of bottom–up explanation: it flies because air molecules impinge against the wing with slower-moving molecules below creating a higher pressure as against that due to faster-moving molecules above, leading to a pressure difference described by Bernoulli's law, etc.
- in terms of same-level explanation: it flies because the pilot is flying it, and she is doing so because the airline's timetable dictates that there will be a flight today at 16h35 from London to Berlin
- in terms of top–down explanation: it flies because it is designed to fly! This was done by a team of engineers working in a historical context of the development of metallurgy, combustion, lubrication, aeronautics, machine tools, computer aided design, etc., all needed to make this possible, and in an economic context of a society with a transportation need and complex industrial organizations able to mobilize all the necessary resources for design and manufacture.

These are all simultaneously true explanations. The higher-level explanations rely on the existence of the lower-level explanations in order that they can succeed, but are clearly of a quite different nature than the lower-level ones, and certainly not reducible to them nor dependent on their specific nature. They are also in a sense deeper explanations than the lower-level ones.

The point is fundamental. The analytic approach ignores the environment and takes the existence of the machine for granted; from that standpoint, it enquires as to how the machine functions. This enables one to understand its reliable replicable behavior. But it completely fails to answer why an entity exists with that specific behavior. Systems analysis in terms of purpose within the higher-level structure, where it is one of many interacting components, provides that answer – giving another equally valid, and in some ways more profound, explanation of why it has the properties it has. This approach determines the rationale, the *raison d'être* of the entity; given that purpose, it can usually be fulfilled in a variety of different ways in terms of structure at the micro level.

Finally, it is not just matter or information that has physical effect. It is also thoughts and emotions, and so intentions. Although physicists don't usually recognize all of these realities, their causal models of the real world will be incomplete unless they include them. Human thoughts can cause real physical effects; this is a top–down action from the mind to the physical world. At present there is no way to express this interaction in the language of physics, even though our causal schemes are manifestly incomplete if this is not taken into account. The minimum requirement to do so is to include the relevant variables in the space of variables considered. That then allows them and their effects to become a part of physical theories – perhaps even of fundamental physics

The relation to fundamental physics

Fundamental physics underlies this complexity by determining the nature and interactions of matter. The basic question for physicists is what are the aspects of fundamental physics that allow and enable this extraordinarily complex modular hierarchical structure to exist, where the higher levels are quite different from the lower levels and have their own ontology; and what are the features that allow it to come into being (i.e., that allow its historical development through a process of spontaneous structure formation)?

A "theory of everything"

The physical reasons allowing this independence of higher-level properties from the nature of lower-level constituents have been discussed by Anderson (1972), Schweber (1993), and Kadanoff (1993), focusing particularly on the renormalization group; however more than that is needed to create fundamentally different higher-order structures than occur at lower levels. This is a fundamentally important property of physics, underlying our everyday lives and their reality. Its source is the nature of quantum field theory applied to the microproperties of matter as summarized in the standard model of particle physics. It would be helpful to have more detailed studies of which features of quantum field theory on the one hand, and of the standard particle/field model with all its particular symmetry groups, families of particles, and interaction potentials on the other, are the keys to this fundamental feature.

What is "fundamental physics" in the sense: what feature of physics is the key to existence of truly complex structures? What for example allows modular separation of subnuclear, nuclear, atomic, and molecular properties from each other in such a way as to allow the development and functioning of DNA, RNA, proteins, and living cells? Whatever it is, this must claim to be the "truly fundamental" feature of physics – it is the foundation of the complexity we see. Is the key to such a "theory of everything":

- the general nature of *quantum theory* (e.g., superposition, entanglement, decoherence) and its classical limit?
- the specific nature of *quantum field theory* and *quantum statistics* (certainly required for the stability of matter) and/or of Yang–Mills *gauge theory*?
- the specific *potentials* and *interactions* of the standard particle physics model and its associated *symmetry groups*?
- the *basic particle properties* (existence of three families of quarks, leptons, and neutrinos, for example)?
- the *basic properties of fundamental forces* (effective existence of four fundamental forces; their unification properties)?
- the specific *masses* and *force strengths* involved?
- the value of specific constants such as the fine structure constant?
- or is it not any one of these, but rather the combination of all of them?

The latter seems most plausible: but if so, then why do they all work together so cunningly as to allow the high-level emergence of structure discussed in this chapter?

Whatever the conclusion here, we ultimately face the fundamental metaphysical issue: what chooses this set of laws/behaviors, and holds it in existence/in operation? And on either view the anthropic issue remains: Why does this specific chosen set of laws that has come into being allow intelligent life to exist by allowing this marvellous hierarchical structuring? These issues arise specifically in terms of current scientific speculations on the origin of the universe by numerous workers (Tryon, Hartle, Hawking, Gott, Linde, Turok, Gasperini, and others; see Ellis (2000) for a brief summary). Most of these proposals either envisage a creation process starting from some very different previous state, which then itself needs explanation, and is based in the validity of the present laws of physics (which are therefore invariant to some major change in the status of the universe, such as a change of spacetime signature from Euclidean to hyperbolic); or else represent what is called "creation from nothing," but in fact envisages some kind of process based on all the apparatus of quantum field theory mentioned above – which is far from nothing! In both cases the laws of physics in some sense pre-exist the origin of the present expansion phase of the universe. In the case of ever-existing universes, the same essential issue arises: some process has chosen amongst alternative possibilities. Whatever these processes are, they do not obviously by themselves imply the possibility of the structuring discussed here.

A further interesting question is how views on all of this would change if indeed a successful physical "theory of everything" as usually envisaged by particle physicists – perhaps based on M theory or superstring theory – were to be developed. Physically speaking, this would have a logically and causally superior status to the rest of physics in the sense of underlying all of physics at a fundamental level. The

Metaphysics ***Ethics***
Cosmology ***Sociology***
Astronomy ***Psychology***
Geology ***Physiology***
Materials ***Biochemistry***
Chemistry
Physics
Particle Physics
"Theory of Everything"
Metaphysics

Figure 27.8. The hierarchy of causal relations (Fig. 27.5) extended to show the metaphysical issues that arise both at the foundations of the physics and in terms of causation in cosmology. In both cases issues arise that are beyond investigation by the scientific process of experimental testing; hence they are metaphysical rather than physical.

puzzle regarding complexity would not be solved by existence of such a theory, it would be reinforced, for such a theory would in essence have the image of humanity built into it – and why that should be so is far from obvious, indeed it would border on the miraculous if a logically unique theory of fundamental physics were also a "theory of everything" in the sense envisaged above. That would be a coincidence of the most extraordinary sort.

The relation to ultimate reality

If we ask the question: "What is 'ultimate reality'?", we find a delightful ambiguity:

- Is it the fundamental physics that allows all this to happen? – its physical causal foundations?
- Is it the highest level of structure and complexity it achieves? – which is the ultimate in emergent structure and behavior?
- Is it the ethical basis that ultimately determines the outcome of human actions and hence of social life, and whatever may underlie this basis?
- Is it the metaphysical underpinning of the fundamental physics, on the one hand, and of cosmology on the other – whatever it is that "makes these physical laws fly" (as John Wheeler put it), rather than any others?

From a physical viewpoint, it can be suggested it is the latter: the causal hierarchy rests in metaphysical ultimate reality as indicated in Fig. 27.8. Here the unknown metaphysical issues that underlie both the choice of specific laws of physics on the one hand, and specific initial conditions for cosmology on the other, are explicitly recognized. It is possible that information is the key in both cases. Others at this meeting discuss the "it from bit" idea in the context of physics; in this chapter I have emphasized it in the context of complex systems, where information plays a central role in their emergence. The loop would close if it figured in a fundamental way in both arenas.

It is a pleasure to dedicate this chapter to John Wheeler, who has done so much to explore the nature of fundamental issues underlying physics.

References

Ackoff, R (1999) *Ackoff's Best: His Classic Writings in Management*. New York: John Wiley.

Anderson, P W (1972) More is different. *Science* **177**, 377. (Reprinted in Anderson, P. W. *A Career in Theoretical Physics*. Singapore: World Scientific Press.)

Ashby, R (1958) *An Introduction to Cybernetics*. London: Chapman and Hall.

Bar-Yam, Y (ed.) (2000) *Unifying Themes in Complex Systems*. Cambridge, MA: Perseus Press.

Beer, S (1966) *Decision and Control*. New York: John Wiley.

(1972) *Brain of the Firm*. New York: John Wiley.

Bishop, C (2000) *Neural Networks for Pattern Recognition*. Oxford: Oxford University Press.

Booch, G (1994) *Object Oriented Analysis and Design with Applications*. Menlo Park, CA: Addison Wesley.

Campbell, D T (1974) Downward causation. In *Studies in the Philosophy of Biology: Reduction and Related Problems*, ed. F. J. Ayala and T. Dobhzansky, p. 181. Berkeley, CA: University of California Press.

Campbell, N A (1991) *Biology*. New York: Benjamin Cummings.

Carpenter, G A and Grossberg, S (1991) *Pattern Recognition by Self-Organizing Neural Networks*. Cambridge, MA: MIT Press.

Churchman, C W (1968) *The Systems Approach*. New York: Delacorte Press.

Courtois, P (1985) On time and space decomposition of complex structures. *Commun. ACM* **28**(6), 596.

Davies, P C W (1974) *The Physics of Time Asymmetry*. Berkeley, CA: University of California Press.

Deacon, T (1997) *The Symbolic Species: The Co-Evolution of Language and the Human Brain*. London: Penguin Books.

Eddington, A S (1928) *The Nature of the Physical World*. Cambridge: Cambridge University Press.

Edelmann, G (1990) *Neural Darwinism*. Oxford: Oxford University Press.

Ellis, G F R (2000) Before the beginning: emerging questions and uncertainties. In *Toward a New Millenium in Galaxy Morphology*, ed. D. Block, I. Puerari, A. Stockton and D. Ferreira, p. 693. Dordrecht: Kluwer.

(2002): The universe around us: an integrative view of science and cosmology. http://www.mth.uct.ac.za/~ webpages/ellis/cos0.html.

Emery, F (ed.) (1972) *Systems Thinking*. London: Penguin Books.

Flood, R L and Carson, E R (1990) *Dealing with Complexity: An Introduction to the Theory and Application of Systems Science*. New York: Plenum Press.

Gilbert, S F (1991) *Developmental Biology*. Sunderland, MA: Sinauer.

Holland, J H (1998) *Emergence: from Chaos to Order*. Cambridge, MA: Perseus Press.

Isham, C J (1997) *Lectures on Quantum Theory*. London: Imperial College Press.

Kadanoff, L (1993) *From Order to Chaos: Essays Critical, Chaotic, and Otherwise*. Singapore: World Scientific Press.

Kuppers, B O (1990) *Information and the Origin of Life*. Cambridge, MA: MIT Press.
Milsum, J H (1966) *Biological Control Systems Analysis*. New York: McGraw Hill.
Moyers, B (1993) *Healing and the Mind*. New York: Doubleday.
Panksepp, J (1998) *Affective Neuroscience: The Foundation of Human and Animal Emotions*. Oxford: Oxford University Press.
Peacocke, A R (1989) *An Introduction to the Physical Chemistry of Biological Organization*. Oxford: Oxford University Press.
(1993) *Theology for a Scientific Age: Being and Becoming*. Minneapolis, MN: Fortress Press.
Penrose, R (1989) *The Emperor's New Mind*. Oxford: Oxford University Press.
(1997) *The Large, The Small, and the Human Mind*. Cambridge: Cambridge University Press.
Peskin, M E and Schroeder, D V (1995) *An Introduction to Quantum Field Theory*. Cambridge, MA: Perseus Press.
Pickover, C A (1995) *Visualizing Biological Information*. Singapore: World Scientific Press.
Popper, K and Eccles, J (1977) *The Self and its Brain: An Argument for Interactionism*. Berlin: Springer-Verlag.
Rashidi, H H and Buehler, L K (2000) *Bioinformatics Basics: Applications in Biological Science and Medicine*. Boca Raton, FL: CRC Press.
Schellnhuber, H J (1999) Earth system analysis and the second Copernican revolution. *Nature* **402** (Suppl.), C19.
Schweber, S (1993) Physics, community, and the crisis in physical theory. *Physics Today*, **Nov**, 34.
Seife, J (2000) *Zero: The Biography of a Dangerous Idea*. London: Penguin Books.
Sellars, R W (1932) *The Philosophy of Physical Realism*. New York: Russell and Russell.
Simon, H A (1962) The architecture of complexity. *Proc. Am. Phil. Soc.* **106**, 467.
(1982) *The Sciences of the Artificial*. Cambridge, MA: MIT Press.
Sokal, A (2000a) http://www.physics.nyu.edu/faculty/sokal/index.html
(2000b) http://zakuro.math.tohoku.ac.jp/~ kuroki/Sokal/
Sternberg, E (2000) *The Balance Within: The Science Connecting Health and Emotions*. New York: W. H. Freeman.
Tannenbaum, A S (1990) *Structured Computer Organization*. Englewood Cliffs, NJ: Prentice Hall.
von Bertalanffy, L (1973) *General Systems Theory*. London: Penguin Book.
Wolpert, L (1998) *Principles of Development*. Oxford: Oxford University Press.
Zeh, H D (1992) *The Physical Basis of the Direction of Time*. Berlin: Springer-Verlag.

28

The three origins: cosmos, life, and mind

Marcelo Gleiser

Dartmouth College

The three origins: how come us?

We cannot but wonder about our origin. Cultures throughout history have created mythical narratives that attempt to answer this most vexing of questions, "Why is there something rather than nothing?" Hard questions inspire further thought, and the harder they are, the more inspiring they can be. The rich variety of creation myths is testimony to this. Most of the myths bypass the issue of "something from nothing" by eliminating the nothing: an absolute reality, in the form of a deity or deities, exists outside space and time and originates the cosmos, the order of material things. Creation involves the transition from the absolute to the relative, from a spaceless and timeless reality to a reality within space and time. Myths that don't invoke a deity presume some form of absolute reality which encompasses all opposites, order and chaos, light and darkness. The cosmos emerges spontaneously out of the tension between the opposites, and differentiation follows. A curious exception comes from the Maori people of New Zealand, who describe the origin of all things as coming from nothing, without the action of a god: the cosmos simply comes into being out of a universal urge to exist, a sort of irresistible impulse of creation (Gleiser 1997).

A more detailed study of creation myths shows that they describe the origin of the cosmos through two distinct uses of the concept of emergence: driven emergence and spontaneous emergence. Driven emergence refers to having a being responsible for the appearance of the cosmos, of order, while spontaneous emergence refers to having order appear on its own, without an external cause or agent. Clearly, as the "driver" in the driven emergence has an implicit supernatural origin, any scientific model of cosmogenesis should fall within the "spontaneous emergence" category. Indeed, it is fair to say that a scientific account of creation attempts precisely to do

Science and Ultimate Reality, eds. J. D. Barrow, P. C. W. Davies and C. L. Harper, Jr. Published by Cambridge University Press.

away with the assumption of a driver who is somehow "outside" physical reality. For science, the first cause must come from within physical reality: the universe must come to be on its own.

The question of cosmic origins can only fit the scientific discourse after a major change of focus: "why" questions, as in "Why is there something rather than nothing?" are often problematic. We are much better at dealing with "how" questions. Science is a language we invented to describe Nature; it is not divine in origin, it is not derived from supernatural revelation, but from the methodic application of thought and experimentation. In short, science is a very human enterprise and, as such, limited in its scope. Thus, we may not know "why" two masses attract each other, but we know "how" they do it, and with increasing accuracy as we go from Newton to Einstein. Knowing the "how" has served science well, and will continue to do so in the future. In order to set the tone of this chapter, I propose from the outset to leave behind the "Why are we here?" kind of question. Instead, I would like to propose a substitute question, amenable to a scientific approach and faithful to Wheeler's spirit: "How come us?"

As soon as this shift in focus is accepted, we realize that the question of "How come us?" is, in reality, not one but three questions of origins, intertwined and mutually dependent: "How come us?" implies the existence of (i) a universe capable of (ii) harboring life which, furthermore, is (iii) intelligent enough to ask about its origins. Thus, "How come us?" encompasses all three origins, of cosmos, life, and mind; they may be (and often are) treated separately, but they are part of an indivisible whole.

I like to think of science as part of a long tradition of thought which can be traced back to the creation myths of ancient cultures. This statement must be read carefully, as it is not meant to imply that science is in any way related to mysticism or supernatural beliefs or practices. Quite the contrary, as Lucretius proclaimed some 60 years before the dawn of the Christian era, reason should serve as a beacon against obscurantism, as a path towards the autonomy and self-growth of the individual:

> For the mind wants to discover by reasoning what exists in the infinity of space that lies out there, beyond the ramparts of this world – that region into which the intellect longs to peer and into which the free projection of the mind does actually extend its flight.

The connection with ancient myths comes through the shared curiosity about origins, about who we are and where we are in this vast cosmos. The languages and goals are clearly different, as religion – broadly speaking – is based on faith and divine revelation, while science is based on empirical validation. However, at their most fundamental level, science and religion try to answer the same questions that have afflicted and inspired humankind since the dawn of civilization. How come us?

The three origins as emergence

Cosmos

We have learned that the universe originated some 14 billion years ago out of a very dense and hot initial state, and that it has been expanding and cooling ever since (Padmanabhan 1998: Peacock 1999). There are two observational pillars of the Big-Bang model which allow us to reconstruct the physical properties of the primordial cosmos: the cosmic microwave background radiation (Lachièze-Rey and Gunzig 1999), left over photons from the epoch of hydrogen-atom formation some 300 000 years after the "bang"; and the abundance of light nuclei, such as deuterium, helium-3, and helium-4, synthesized during the short period between 0.01 to a few minutes after the "bang" (Kolb and Turner 1990). Of course, we would like to extend our knowledge to the very first instants of creation, even if this means extrapolating our present knowledge of physical processes well beyond what we currently can test and validate in the laboratory.

Modern cosmological modeling describes the history of the universe as a history of increasing complexity of its material forms. In this chapter, I will use the term "complex" to characterize a structure that cannot be reduced to the sum of its parts. Thus, a hydrogen atom is not just an electron and a proton, but a bound state of an electron and a proton interacting electromagnetically, capable of preserving its structure in time unless disturbed by external forces. The key point is that a complex structure has properties that cannot be predicted by knowing the properties of its individual constituents. An obvious and extreme example is the brain: its function cannot be understood by describing it as an assembly of neurons, and much less as an assembly of atoms or of elementary particles.

A possible way of measuring the complexity of a structure is by the number of parameters needed to describe its physical state. Clearly, as the number of constituents grows, so does the complexity of the structure. As with any definition of complexity, this one has its shortcomings. One concept that is missing is that of function: some structures have functional complexity not just morphological complexity. A proper definition of complexity should encompass both. Note that these two measures are not necessarily independent; more often than not, the morphological complexity of a given structure is linked to its functional complexity. The reader can clearly see the complexity of defining complexity.

The two observational pillars of the Big Bang model mentioned above, the formation of light atomic nuclei and of hydrogen atoms, are examples of the increasing material complexity of the cosmos. The earlier you go, the simpler is the organization of matter, in principle all the way down to its most basic constituents at the earliest moments. For example, before 10^{-5} s we expect that protons, neutrons and other nucleons get dissociated into their constituent quark and gluons, while before 10^{-12} s, two of the four fundamental forces, the weak and the electromagnetic, get unified into a single *electroweak* interaction. Very close to the beginning, it is expected that all four interactions are unified into a single interaction, although the inclusion of gravity into the unification scheme is plagued by serious conceptual difficulties, as no fully operational theory of quantum gravity exists at present. (See the chapters by Lee Smolin and Juan Maldacena in this volume for current efforts to bridge these difficulties.)

When discussing cosmic origins, it is important to distinguish between a classical and a quantum universe. By classical universe I mean a universe described by a classical pseudo-Riemannian metric, where the spatial geometry and the flow of time may have the usual plasticity dictated by Einstein's general theory of relativity, but may not fluctuate quantum mechanically (at least not with any appreciable probability). The emergence of our universe is, then, the emergence of a classical universe out of a quantum foam of universes, a quantum tunneling event (Hartle and Hawking 1983; Vilenkin 1988; Linde 1990). Within these scenarios, time, in the familiar classical sense, emerges with the classical universe as part of its Riemannian spacetime geometry. An alternative view, which has been derived from a version of the inflationary cosmological model, is that of a multiverse which encompasses all (infinitely many?) universes, including ours. The emergence of a universe such as ours is related to the existence and values of a primordial scalar field, whose origin is attributed (somewhat loosely) to a unified field theory, possibly derivative of string theory (see Linde, this volume).

The usual picture describing tunneling in quantum field theory relies on the spontaneous appearance of a coherent field configuration, sometimes called an instanton or an Euclidean vacuum bubble (Rajaraman 1987). If we think of the field as a superposition of Fourier modes, each with momentum $\mathbf{k}$, a coherent, "bubble-like" configuration will be characterized by modes in a narrow interval, in the neighborhood of $|\mathbf{k}| \sim 2\pi/\lambda_c$, where λ_c is the approximate length-scale associated with the configuration. For example, the kink solution in a one-dimensional ϕ^4 theory is $\phi(x) \sim \tanh(xm)$, where m is the characteristic mass-scale of the system, so that $|\mathbf{k}| \sim m$.

The extrapolation of tunneling in quantum field theory to tunneling in quantum cosmology relies on a drastic reduction of the effective degrees of freedom of the theory, known as the mini-superspace model. A single degree of freedom satisfies the Wheeler–DeWitt equation, which has tunneling solutions (DeWitt 1967; Wheeler 1968). This degree of freedom represents the dominant mode of the tunneling state, the one with largest Fourier amplitude. Several possibilities have been investigated in the literature, which rely on different choices of boundary conditions (Kolb and Turner 1990; Linde 1990). Thus, the question of the emergence of the universe out of a quantum initial state depends on an appropriate choice of boundary conditions. A crucial point of these scenarios is that the tunneling solutions originate from zero-energy Schrödinger-like equations; the universe tunnels "out of nothing," albeit nothing here means a quantum vacuum of metrics (Tryon, 1973; Vilenkin 1988).

As mentioned above, these tunneling solutions, elegant as they are, still depend on the choice of "appropriate" boundary conditions. In the absence of a guiding physical principle, choices must rely on (highly subjective) good sense and compatibility with the observed classical universe. The choice of appropriate boundary conditions can be thought of as the scientific equivalent of the quest for the ultimate first cause, the *primum mobile* of quantum cosmology, a philosophical problem well known to Aristotle. The question remains open. At a more fundamental level, Paul Davies has argued that even if a reasonable physical model of the quantum origin of the universe could be obtained, based on a proper marriage of quantum mechanics and general relativity, it would still leave unanswered the question "How come this universe?" (Davies 1992). Science is intrinsically unable to describe its own origins, the processes that selected the set of physical laws which seem to rule our universe so effectively. Sometimes this issue is discarded with a simple argument, based on a multiverse explanation: if there are infinitely many universes out there, there will be infinitely many choices of physical law, ours being but one of them. I find this dismissive argument incomplete at best, as the multiverse hypothesis itself is a result of our set of physical laws. Instead, I suggest that the set of physical laws is also an emergent property of the cosmos, together with the cosmos itself. This echoes Saint Augustine, who, when asked what was happening before Creation, wrote that time itself originated with it and hence that it was meaningless to ask about "before." Likewise, it may be meaningless to talk about physical laws before the existence of a physical reality where these laws are enacted: I propose that the laws and the material reality they describe can only exist together. The emergence of cosmic order may be the result of an optimization process rooted on the evolution of complex material structures, resulting from random experimentation between form and functionality. The laws we use to describe nature are a consequence of this optimization process, not its cause. That is, the only truly fundamental principle in nature is that of economy of performance, what could be called a cosmic optimization principle, or COP. Operationally, this principle implies that all material structures always organize and function in the least wasteful manner possible. (The only wasteful activities I am aware of are due to humans.) Thus, the laws ruling the physical nature of this universe are necessarily conducive to the hierarchical complexification of material structures, since these structures represent the most efficient collective organization of energy. This optimization relates not only to form but also to function. Paraphrasing Einstein, one cannot but wonder if there could have been any other choice.

Of bubbles, life, and mind

While physical law has been shown to permeate the universe as a whole – from subatomic particles to distant quasars – biological law has been so far restricted to one example, Earth. Exobiologists confidently claim that the discovery of extraterrestrial life is just a matter of time, that even Mars or Jupiter's moon Europa may harbor or may have harbored life in the past (Darling 2001). However, there is a huge jump from the complexification of subatomic particles into nuclei and atoms to that of atoms into organic macromolecules present in every known living organism. And an even bigger jump from life to intelligent life. Considering the dramatic history of the evolution of life on Earth, and of intelligent life on Earth, a blind optimism that life could easily flourish elsewhere may be somewhat naive (Gleiser 2002). That said, life does exhibit amazing adaptability and resiliency. Even if life is rare in outer space, it would be quite remarkable if Earth were the only planet carrying it and we the only intelligent species. Apart from anything else, it would place humankind in a unique position in the cosmos, contradicting what 400 years of modern science has shown us, that we are not so special. Since Stuart Kauffman's essay in this volume describes his idea that life is an emergent phenomenon controlled by autocatalytic biochemical reactions, I would like to focus on certain aspects of living systems from a nonequilibrium statistical physics perspective.

Perhaps the most intriguing fact about life is that it has still not been properly defined: what defines a living organism? At what level of complexity does a system of inanimate organic matter become a living organism? Could it be when a certain level of morphological complexity is reached that allows for the spontaneous unfolding of functional complexity, function arising from form? Optimization may indeed play a key role. Living organisms must multiply and interact with their environments; they are clearly out of thermodynamic equilibrium, fighting continuously against their entropic decay. In fact, it is their ability to exchange energy with an external environment that maintains them out of equilibrium. Life would be impossible otherwise. Living organisms must also keep their coherent structures, their "identities," as time goes by. In short, they are biomolecular machines capable of preserving spatiotemporal information (their coherence), of passing information (through reproduction), and of interacting with their environments (feeding and refuse). Note, however, that reproduction is essential for the preservation of a species but not for an organism belonging to a given species to be alive. Preservation of biochemical identity and interaction with the environment seem to suffice.

There is one common thread linking complex material structures from atomic nuclei (and even nucleons such as protons and neutrons), atoms, planets, stars, all the way to living organisms: they are all structures of localized complexity, bound states of autonomous but interacting parts. In physical systems, the formation of bound states can be considered an optimization process, where the many parts of the system benefit from being together, due to the existence of a confining interaction: it costs less energy to have a bound state of closely interacting parts than to have a collection of independent parts. In fact, form is clearly a result of this optimization, as seen, e.g., in spherical soap bubbles or stars, and in different solitonic configurations. Furthermore, the formation of such bound states often arises due to the influence of environmental interactions. Thus, we may learn plenty about two key properties of living systems – identity preservation and interactions with an external environment – from the study of much simpler physical systems, those giving rise to bound states due to interactions with an environment. Let me briefly illustrate this point with an example. The next section will be dedicated to a more detailed analysis.

Consider a physical system described by a scalar field $\phi(x, t)$ which has interactions controlled by an asymmetric double-well potential,

$$V(\phi) = \frac{a}{2}\phi^2 - \frac{b}{3}\phi^3 + \frac{c}{4}\phi^4, \tag{28.1}$$

where a, b, and c are constants, and, for simplicity, they are all positive. If $b^2 > 4ac$ this potential has two minima, one at $\phi_0 = 0$ and the other at $\phi_+ = b/2c[1 + \sqrt{1 - 4ac/b^2}]$. One can choose the parameters such that the minimum at ϕ_+ is the global minimum. If the system is initially prepared in the ϕ_0 minimum, it will decay into the global minimum by quantum-mechanical or thermal fluctuations. In both cases the decay is promoted by the appearance of a critical bubble (or nucleus) of the global minimum inside the local one; the critical bubble is a saddle point in the free-energy functional, signaling a state of unstable equilibrium (Langer 1991). It grows, converting the system into the state described by the global minimum. In quantum field theory, each minimum is referred to as a vacuum of the theory, while in statistical mechanics, each minimum is referred to as a phase of the system, such as vapor and water. The phase change has a free energy cost given by the energy of the bubble configuration, E_b. The probability for a fluctuation to reach critical status is controlled by the Boltzmann factor, $P[\phi_b] \sim \exp[-E_b/k_B T]$, where k_B is Boltzmann's constant and T is the temperature. The interesting point is that the appearance of the critical bubble can be thought of as an emergence process, due to the stochastic interactions between the field ϕ and an external environment (Alford and Gleiser 1993). (In practice, this environment may be given by short-wavelength modes of the field ϕ itself, which were integrated out of the potential $V(\phi)$ (a coarse-graining procedure), or it may represent the interaction of ϕ with other fields, or, if one thinks of ϕ as a scalar order parameter, it may represent an external heat bath.) The bubble – a spatially coherent field configuration – is the most energy-efficient way of promoting this phase change, the minimum of all the maxima, that is, the shortest path to the new phase in field configuration space. Any other configuration would be wasteful.

In the example above, the emergence of localized complexity (the bubble) is due to a combination of two key factors: self-interactions of the field dictated by the potential $V(\phi)$, and the interactions of ϕ with an external environment at temperature T. The morphological complexity of the bubble is due to it being a coherent state; it is made out of a superposition of several modes with the same approximate momentum, bound together by the nonlinear interactions of the field. As with most coherent configurations, it is clearly out of equilibrium, as its modes do not satisfy the equipartition condition $E_k = T/2$, where E_k is the kinetic energy of the k-mode. Note also that the critical bubble has functional complexity: it promotes a phase change, controlled essentially by the free-energy barrier separating the two phases (and possibly by interactions with the environment).

In his book *Stairway to the Mind*, Alwyn Scott suggests that "consciousness is an *emergent* phenomenon, one born of many discrete events fusing together as a single experience" (Scott 1995). Of all three origins, the emergence of consciousness is, perhaps, the hardest to conceptualize. First, consciousness is very difficult to define: over 15 000 articles have been written on this very topic, and consensus is far from being reached. From the perspective of modern-day cognitive neuroscience, it is clear that the task of applying a reductionist approach to understanding the workings of the brain is doomed to fail (Gazzaniga *et al.* 1998). Most cognitive neuroscientists agree that mind – in all its manifestations – is a property of the brain, the result of both local and long-range interactions of neuronal clusters. As more advances are made in the field, through experimentation and real-time monitoring of brain activity such as functional magnetic resonance imaging and positron-electron tomography scans, it is believed that the myriad collective manifestations of these neuronal clusters will display some fundamental operational laws. These laws will be irreducible laws, representing new levels of organized complexity. This should not be a surprise, as such stratification of ordered behavior is well known from physical systems: collective phenomena often display behavior not predictable from their basic constituents. Just think of the properties of water and try to relate them to the properties of individual protons, neutrons, and electrons. Or how different substances belonging to the same universality class display identical critical phenomena (Stanley 1971; Goldenfeld 1992).

To connect the problem of consciousness with the previous discussion of nonlinearity and environmental interactions, I propose to model neuronal clusters as coherent states, where activity is related to the mean amplitude of the field representing the coherent state. The reader may also consult the work by Hopfield and Brody (2000) on the synchronization of neurons. Thus, a cluster with many neurons firing together is represented by a coherent superposition of many degrees of freedom with the same average properties and with a mean amplitude above a certain threshold. These neuronal clusters self-organize in the most efficient way possible; the behavior of individual clusters is controlled both by local nonlinear interactions (which determine the typical size of the cluster) and by the "environment," which may summarize the collection of outside impulses being processed at different locations in the brain. The coupling to other clusters is due to long-range cluster–cluster interactions, which may be activated through resonant frequencies. This field-theoretical brain model – at this point – is not being proposed as a quantitative explanation of brain activity. (This author is presently working on its implementation.) However, the nonlinear dynamics of coupled clusters of oscillators, driven by interactions with an external "sensorial" environment, is bound to produce irreducible emergent complex behavior which may, in part, illustrate some of the neuronal behavior observed through imaging techniques of brain activity. Could such models include memory? Possibly, if somehow memory could be related to resonant behavior: certain coherent states, once excited, can only be re-excited by a specific combination of stimuli, somewhat like the combination of a safe, which entails a very specific string of instructions.

A field model exhibiting local and global emergent complex behavior

This chapter would not be complete without an explicit demonstration of how complex behavior may emerge from a simple physical system due a combination of its intrinsic nonlinearities and its coupling to an external environment. The model below may be thought of as having some of the essential characteristics needed to explain the emergence of self-organized spatiotemporal structures, be they physical systems, living organisms, or thinking brains. It is a modest first step that hopefully will inspire further research on this promising topic. More details can be found in Gleiser and Howell (2002).

Consider a two-dimensional real scalar field $\phi(t, x, y)$ with a double-well potential $V(\phi) = \frac{1}{4}(\phi^2 - 1)^2$. (All quantities have been made dimensionless.) It has been shown that for certain initial conditions, this system evolves into time-dependent spatially localized configurations named *oscillons* in two (Gleiser and Sornborger 2000) and in three (Gleiser 1994; Copeland *et al.* 1995) spatial dimensions. These configurations are characterized by their extreme longevity: in three dimensions, oscillons may live 3 to 4 orders of magnitude longer than the typical oscillation timescale in the system (of $\mathcal{O} \sim 1$ for the potential above), while in two dimensions they live at least 7 orders of magnitude longer, without decaying.

Here we will focus on oscillons from scalar fields satisfying the nonlinear Klein–Gordon equation, although their universal features are shared by many nonlinear systems, ranging from vibrating grains (Umbanhowar *et al.* 1996) to acoustic waves in astrophysical plasmas (Umurhan *et al.* 1998). So far, oscillons have been found during the deterministic evolution of the system, starting from ordered initial conditions. The field is prepared as a localized symmetric state, such as a Gaussian profile, $\phi(r) = \phi_0 \exp[-r^2/R^2] - 1$, and quickly evolves into an oscillon configuration, as long as the following conditions are satisfied: (1) the initial configuration's energy must be larger than the energy of the oscillon; (2) the initial radius must be above the unstable bifurcation value R_c (Gleiser and Sornborger 2000). More recently, it has been shown that oscillons can also emerge from elliptically deformed initial configurations, and are stable against small arbitrary radial and angular perturbations, behaving as attractors in field configuration space (Adib *et al.* 2002).

The combination of longevity and spatiotemporal order suggests that oscillons may play an important role in systems exhibiting hierarchical complexity, that is, different layers of self-organization that are not easily predictable from the bottom up (Laughlin *et al.* 2000). We found that this is indeed the case: oscillons not only spontaneously emerge during the evolution of the system (local spatiotemporal order), but they tend to initially emerge nearly synchronized through the lattice (global spatiotemporal order). The field – properly discretized in a squared lattice – was coupled to an external heat bath via a Langevin equation with additive noise and viscosity satisfying the fluctuation–dissipation relation

$$\frac{\partial^2 \phi}{\partial t^2} = \nabla^2 \phi - \eta \frac{\partial \phi}{\partial t} - \frac{\partial V}{\partial \phi} + \xi(\mathbf{x}, t), \tag{28.2}$$

where the viscosity coefficient η is related to the stochastic force of zero mean $\xi(\mathbf{x}, t)$ by the fluctuation–dissipation relation

$$\langle \xi(\mathbf{x}, t)\xi(\mathbf{x}', t') \rangle = 2\eta T \delta(\mathbf{x} - \mathbf{x}')\delta(t - t'). \tag{28.3}$$

Note that the environment above is highly simplistic, white noise with no intrinsic spatial or temporal memory, although it is guaranteed to take the system to its final equilibrium state. A rich variation of behavior can be achieved by modifying the properties of the external bath (colored noise) or its coupling to the field ϕ (multiplicative noise). To date, these variations remain largely unexplored, at least in the context of field theory.

We performed two classes of numerical experiments, modeling open (canonical) and closed (microcanonical) thermodynamic systems. In both classes the field is initially thermalized in a single well potential centered at the left minimum, $\phi = -1$, until the average kinetic energy per degree of freedom satisfies equipartition, $E_k = T/2$. (Within a certain window of accuracy, which was chosen to be 10%. This is not an absolute measure of equipartition, but it is an extremely useful indicator of how energy is being distributed through the system's many degrees of freedom.) In both classes of experiments, after the field is thermalized the potential is switched to the double-well potential. The key difference between the two classes of experiments concerns the coupling to the external environment. While in the first class the coupling is switched off as the potential changes from single to double-well, in the second class it is kept on throughout the simulation. Thus, after the initial thermalization is completed, the first class of experiments corresponds to closed energy-conserving thermodynamic systems, while the second corresponds to open systems.

The switch from single to double-well tosses the field out of thermal equilibrium and the energy exchange between its modes will lead it once again toward equipartition. We introduce a measure of the entropy of the discretized field based on the number of degrees of freedom outside equipartition, $N_{\rm ne}(t)$. Writing the total number of degrees of freedom (lattice points) as $N_T = (L/dx)^2$, where L is the linear size of the square lattice and dx the lattice spacing, we write the normalized entropy of the field as

$$S(t) = \frac{\ln\left[2 - N_{\rm ne}(t)/N_T\right]}{\ln 2}. \tag{28.4}$$

$S(t)$ varies between 0 (no modes are thermalized, $N_{\rm ne} = N_T$) and 1 (all modes are thermalized, $N_{\rm ne} = 0$). In Fig. 28.1 we show the difference between the entropy during the initial thermalization in a single well ($S_{\rm th}(t)$) and during the double-well evolution ($S_{\rm dw}(t)$) for the open and closed systems, $S_{\rm th} - S_{\rm dw}$. There is a marked difference between the time evolution of the two entropies for the closed system (upper curve), while for the open system it evolves practically uniformly for both potentials. For the closed system, the slower evolution toward equipartition with the nonlinear potential is directly related to the spontaneous emergence of oscillon configurations (Fig. 28.2).

In contrast, no oscillons emerged during the evolution of the open system, before or after equipartition. (Due to ergodicity, they would eventually appear with a rate per unit area proportional to the Boltzmann factor, $\exp[-E_{\rm osc}/T]$, where $E_{\rm osc}$ is the energy of the oscillon configuration. But since we work with fairly low temperatures so that $E_{\rm osc}/T \gg 1$, the typical timescale is many orders of magnitude larger than our simulation times.) Contrasting the results between open and closed systems, we conclude that rapid, nonergodic spontaneous self-organization emerges only if the system is cut off from the external environment: it needs the

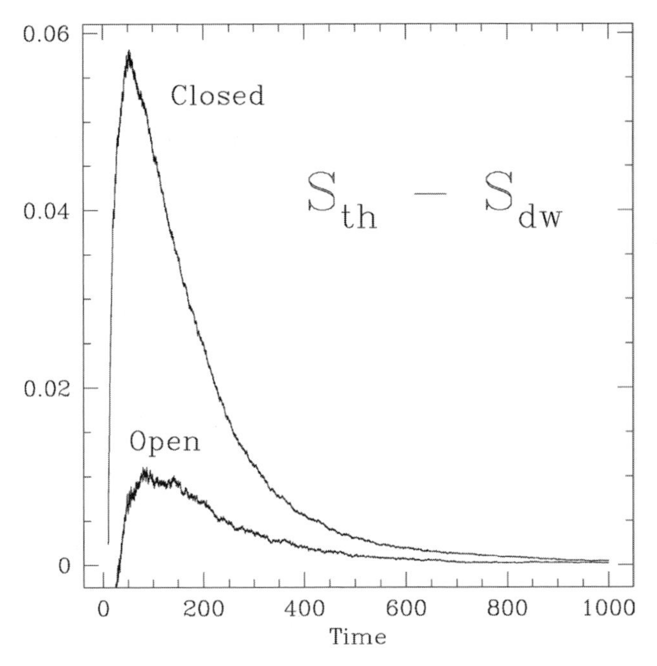

Figure 28.1. Entropy difference for open and closed systems. For each case, the entropy was computed separately during the evolution with a single (S_{th}) and with a double-well (S_{dw}) potential. The time and entropy were reset to zero when the potential was switched to a double-well to allow a direct comparison. Note the marked departure from equipartition for the closed system, peaking at $t \simeq 50$. A lattice point was set to satisfy energy equipartition if its kinetic energy was within 10% of $T/2$. The numerical experiment was performed using a leapfrog routine in a squared lattice of side $L = 100$ and lattice spacing $dx = 0.1$. The time step was $dt = 0.01$.

initial energy influx from this environment, but cannot self-organize in the constant presence of incoherent noise. These structures reappear once the coupling with a heat bath is periodically reactivated creating renewed nonequilibrium conditions, in a suggestive analogy with living systems that must feed periodically to survive.

In order to identify the large-amplitude fluctuations observed in the numerical experiments (Fig. 28.2) with long-lived oscillons we need to estimate their size and their typical oscillation frequency. First we smeared the field over a length-scale based on the average oscillon size obtained in previous work. We then counted the number of lattice points of the smeared field with amplitude above zero. Figure 28.3 shows the time evolution of those large-amplitude fluctuations. There are two main features: first, large-amplitude fluctuations are absent after equipartition (approximately for $t > 600$); second, there is a clear periodicity, related to the synchronized emergent properties of the system, displayed in Fig. 28.2. The periodicity directly

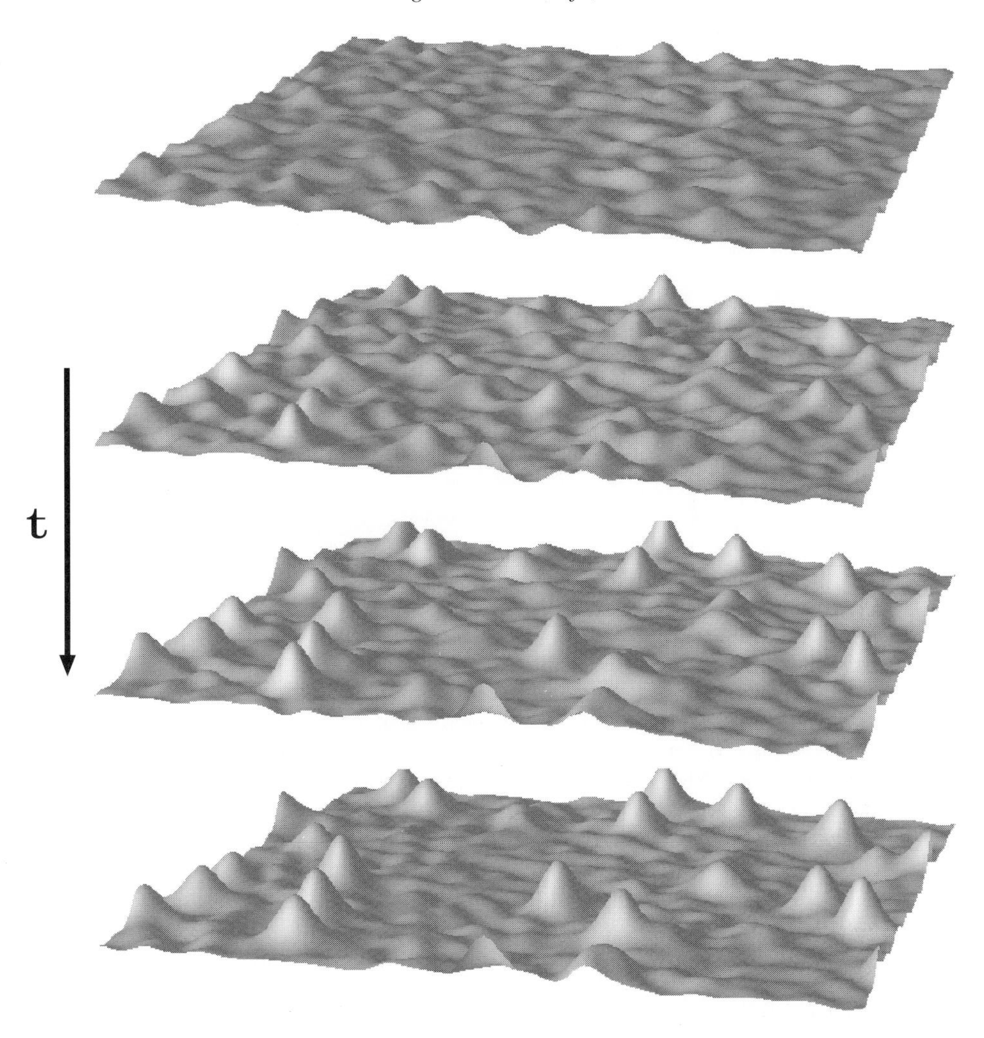

Figure 28.2. Sequence of time snapshots showing the synchronized emergence of oscillons on the two-dimensional lattice. (Animations can be seen at www.dartmouth.edu/~cosmos/oscillons.)

correlates with that of oscillons, as can be seen in Fig. 28.4, which shows both oscillation periods, extracted by a fast Fourier transform method. We conclude that the local large-amplitude fluctuations are oscillons, which, furthermore, tend to initially emerge nearly in phase throughout the lattice. Not surprisingly, the emergent oscillons do not live as long as those obtained under ideal infinite lattices and ordered initial conditions. This initial synchronization can be attributed to the existence of a global nearly harmonic mode which takes the field above the instability point, or the spinodal. More details can be found elswhere (Gleiser and Howell 2002).

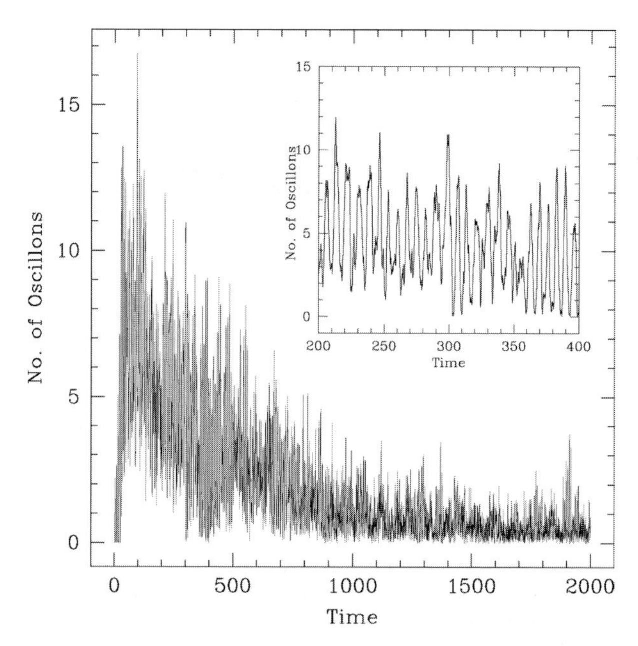

Figure 28.3. Number of oscillons as a function of time. We estimated the number of lattice points in an oscillon configuration compatible with our choice of lattice spacing; only points above $\phi(x, y) \geq 0$ were counted. Note that oscillons only appear when the system is out of equilibrium (compare with Fig. 28.1). Insert covers the period $200 \leq t \leq 400$, where the near-periodicity of the system is still seen.

This investigation demonstrates that local and global ordered spatiotemporal structures may spontaneously emerge during the evolution of a two-dimensional nonlinear scalar field on its way to equipartition. The extension to three dimensions should be straightforward. Since ordered structures with similar spatiotemporal properties have been found in a variety of physical systems, we expect that our results represent a universal feature of self-organizing systems with amplitude-dependent nonlinearities, and not just a consequence of a specific set of initial conditions and nonlinear interactions. This possibility is certainly worth pursuing, as is the potential role of the periodic coupling of the external environment to the reemergence and control of order in the system.

Some closing thoughts

One of the great paradoxes of life is that it is, at its fundamental level, dependent on absolutely dumb atoms and molecules. How can it be that such simple structures can perform collectively to generate a living organism? Or, even more dramatically, a thinking being? The answer is not going to be found in the laws describing the

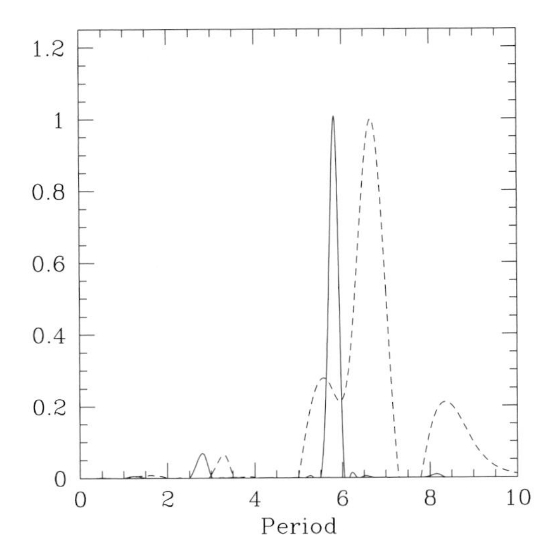

Figure 28.4. Periods of regular long-lived oscillons (continuous lines) and of the emergent oscillons (dashed lines) obtained in the present numerical experiments (Fig. 28.3) extracted by a fast Fourier transform. The amplitudes were normalized to unity. The difference is due to the fact that the oscillons in the present system approach asymptotically $\langle\phi\rangle_T$, that is, the temperature-dependent value of the vacuum state, and not the zero-temperature value $\phi = -1$.

behavior of individual atoms and molecules. New laws are necessary to describe a living and/or thinking assembly of atoms and molecules, all coupled to each other and to an environment. These laws will have to explain how form is maintained, what I call the identity of a structure, and how the emergence of function is a result of the morphological complexity of the structure. Different layers of organization have different sets of laws, which are irreducible; one cannot go continuously from elementary particles to DNA, and then to cells and neurons, to arrive, finally, at life and mind.

To many scientists, it is indeed quite difficult to look beyond reductionism. After all, most of what has been accomplished so far in the natural sciences owes much of its success to the diligent application of reductionism in the description of natural phenomena. There is surely a path to be followed by the application of reductionistic techniques to Nature, which follows on the venerable historical tracks of the pre-Socratic atomists; there is much that we still don't know about the inner structure of matter, from many properties of its most fundamental constituents (including who they are) to their mutual interactions and the parameters describing them. There is, however, a confusion in associating the reductionistic approach to simplicity and everything else to unnecessary complication; it is simply not correct to assume that less is always better. Sometimes, less is just not enough. This is certainly the case with living systems and the mind, and with complex physical systems that cannot be broken down into smaller parts in order to be studied. For these systems, more is better, and their rich behavior(s) is to be extracted from collective phenomena, irreducible to the properties of its smallest constituents. Much of the complexity of Nature springs from dumb atoms and molecules whose behavior is dictated by independent sets of very smart laws, operating at different length scales. It is not a matter of reductionism versus complexity, but of a complementary description of natural phenomena calling for one or the other approach, or, sometimes, for both at once. After all, even within complex phenomena, one is searching for the smallest number of laws, or the most efficient and general one(s). The main split here is not in the need to simplify, which is a given in science, but in the route to understanding. Emergenticists do not believe that all behavior observed in Nature can be traced down to a system's individual components and the laws controlling their behavior. It is true that emergent science is still in its infancy, and that grand claims have been made with not so much substance to support them. However, new challenges call for new approaches, for news ways of thinking. It may very well be that we are witnessing a transitional time in the natural sciences, where some of the traditional ways of analysis need to be complemented by new ones. After all, Nature is always a step ahead of us.

Cosmos, life, and mind are linked through our existence. The answer to "How come us?" must come from this link. Whatever the final answer may be, it will have to incorporate what I believe is the most fundamental principle of Nature, that Nature is the ultimate cheapskate: complexity is the result of a constant struggle to save resources. And the laws that describe the emergence of complexity emerge, themselves, as the result of this principle, which I have called the cosmic optimization principle (COP). We thus move from prediction to description, no less in awe with the amazing creativity of Nature. And with our capacity to understand some of it.

Acknowledgments

I would like to express my gratitude to the organizers of this unique event for the invitation to honor Professor John A. Wheeler's work. I also thank Steve Weinstein for his comments upon reading this manuscript. This work was supported in part by National Science Foundation grants PHY-0070554 and PHY-0099543, and by the Mr. Tompkins Fund for Cosmology and Field Theory at Dartmouth College.

References

Adib, A, Gleiser, M, and Almeida, C A S (2002) Long-lived oscillons from asymmetric bubbles: existence and stability. *Phys. Rev.* **D66**, 085011.

Alford, M and Gleiser, M (1993) *Phys. Rev.* **D48**, 2838.

Copeland, E J, Gleiser, M, and Muller, H-R (1995) *Phys. Rev.* **D52**, 1626.
Darling, D (2001) *Life Everywhere*. New York: Basic Books.
Davies, P C W (1992) *The Mind of God: The Scientific Basis for a Rational World*. New York: Simon and Schuster.
DeWitt, B S (1967) *Phys. Rev.* **160**, 1113.
Gazzaniga, M S, Ivry, R B, and Mangun, G R (1998) *Cognitive Neuroscience: The Biology of the Mind*. New York: W. W. Norton.
Gleiser, M (1994) *Phys. Rev.* **D49**, 2978.
(1997) *The Dancing Universe: From Creation Myths to the Big Bang*. New York: Dutton.
(2002) *The Prophet and the Astronomer: A Scientific Journey to the End of Time*. New York: W. W. Norton.
Gleiser, M and Howell, R (2002) hep-ph/0209176.
Gleiser, M and Sornborger, A (2000) *Phys. Rev.* **E62**, 1368.
Goldenfeld, N (1992) *Lectures on Phase Transitions and the Renormalization Group*. Reading, MA: Addison Wesley.
Hartle, J B and Hawking, S W (1983) *Phys. Rev.* **D28**, 2960.
Hopfield, J J and Brody, C D (2000) *Proc. Nat. Acad. Sci. USA* **97**, 13919.
Kolb, E W and Turner, M S (1990) *The Early Universe*. Reading, MA: Addison Wesley.
Lachièze-Rey, M and Gunzig, E (1999) *The Cosmological Background Radiation: Echo of the Early Universe*. Cambridge: Cambridge University Press.
Laughlin, R B, Pines, D, Schmalian, J, *et al.* (2000) *Proc. Nat. Acad. Sci. USA* **97**, 32.
Langer, J S (1991) In *Solids Far from Equilibrium*, ed. C. Godrèche. Cambridge: Cambridge University Press.
Linde, A (1990) *Inflation and Quantum Cosmology*. Boston, MA: Academic Press.
Padmanabhan, T (1998) *After the First Three Minutes: The Story of the Universe*. Cambridge: Cambridge University Press.
Peacock, J A (1999) *Cosmological Physics*. Cambridge: Cambridge University Press.
Rajaraman, R (1987) *Solitons and Instantons*. Amsterdam: North-Holland.
Scott, A (1995) *Stairway to the Mind*. New York: Copernicus.
Stanley, H E (1971) *Introduction to Phase Transitions and Critical Phenomena*. Oxford: Oxford University Press.
Tryon, E (1973) *Nature* **246**, 396.
Umbanhowar, P, Melo, F, and Swinney, H (1996) *Nature* **382**, 793.
Umurhan, M O, Tao, L, and Spiegel, EA (1998) *Ann. N. Y. Acad. Sci.* **867**, 298.
Vilenkin, A (1988) *Phys. Rev.* **D37**, 888.
Wheeler, J A (1968) In *Batelle Rencontres*, ed. C. DeWitt and J. A. Wheeler. New York: Benjamin.

29

Autonomous agents

Stuart Kauffman
Santa Fe Institute

Writing in Dublin in 1944, Erwin Schrödinger sought the source of order in biological systems. Given the recent radiation mutagenic evidence on the target size of a gene showing that a gene had at most a few thousand atoms, Schrödinger argued that the familiar order due to square root N fluctuations around an equilibrium was insufficient because the fluctuations were too large to account for the hereditary order seen in biology. He argued that quantum mechanics, via stable chemical bonds, was essential for that order. Then he made his brilliant leap. Noting that a periodic crystal could not "say" very much, he opted for genes as aperiodic crystals which, via the aperiodicity, would carry a microcode specifying the ontogeny of the organism. It was a mere two decades until the structure of the aperiodic double helix of DNA and much of the genetic code were known.

But did Schrödinger's book, *What Is Life?* actually answer his core question? I think not, and the aim of this chapter is to propose a different definition, one concerning what I call an "autonomous agent," that may have stumbled on an adequate definition of life. I will not insist that I have succeeded, but at a minimum the definition leads in many useful and unexpected directions with import for physics, chemistry, biology, and beyond.

Our questions drive much of our science. The material in this chapter derives from my third book, *Investigations* (Kauffman 2000). In it, I am driven by an initial question: consider a bacterium swimming upstream in a glucose gradient. We would, and do, all say that the bacterium is going to get food. That is to say, the bacterium is acting on its own behalf in an environment. I will call a system that can act on its own behalf in an environment an "autonomous agent." I do no mean that it is alone in its environment, but that it can act on its own behalf.

Science and Ultimate Reality, eds. J. D. Barrow, P. C. W. Davies and C. L. Harper, Jr. Published by Cambridge University Press.

But the bacterium is "just" a physical system. So my question became: what must a physical system be to constitute an autonomous agent?

I will jump to the definition I found my way to after much consideration: an autonomous agent is a self-replicating system that is able to perform at least one thermodynamic work cycle. Importantly, all free-living entities fit this description. Thus, I may have stumbled upon an adequate definition of life, but, as noted, will not insist upon it.

It is a stunning fact that the biosphere is filled with autonomous agents that continuously reach out and manipulate the universe on their own behalf. Further, it is deeply interesting that with autonomous agents, the concept of "doings" as opposed to mere happenings, takes its place in our conceptual system.

Note two features of my definition. First, it is a definition. Definitions are neither true nor false, but can be useful or useless. I hope mine is useful. Second, the definition subtly leaps Hume's naturalistic fallacy. Hume argued that we cannot deduce "ought" from "is." But my definition jumps this gap definitionally. For once we admit that the autonomous agent is acting on its own behalf, we have a locus of value in a world of fact. Is this legitimate? I suspect that what is going on is, roughly, the following. Wittgenstein taught us about language games that were not reducible one to another. I suspect that my definition gives the minimum physical conditions for a physical system about which the language game of doing, acting, and value becomes natural.

You may cavil, but the language of doing, acting, and value is the language with which we talk about autonomous agents. Thus, since the bacterium is just a physical system, physics talk, which has only happenings, must broaden to include doings.

We have no theory of organization

I wish to make a central point, that we have no theory of organization, by starting with the familiar Carnot cycle, to which I make a single addition: a handle. Figure 29.1 shows the apparatus of the Carnot cycle, a hot reservoir, and cold reservoir, and a cylinder and piston located between the two. I show a handle extending from the cylinder to emphasize the need for organization of the processes that constitute the work cycle. State 1 of the work cycle has the piston high in the cylinder, the working gas compressed and as hot as the hot reservoir. Now I pull on the handle and bring the cylinder into contact with the hot reservoir, then allow the working gas to expand in the isothermal part of the power stroke. As the gas expands and cools, heat flows from the hot reservoir into the working gas and tends to maintain it at the hot temperature of the reservoir. Half way down the power stroke, I push on the handle, and move the cylinder to a position between the two reservoirs, touching neither. The gas continues to expand in the adiabatic part of the power stroke and,

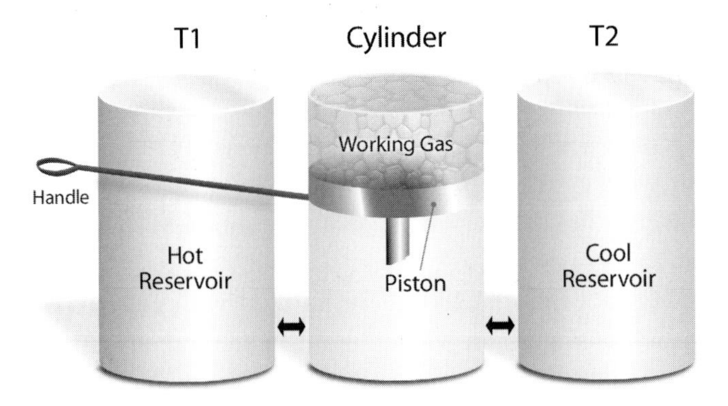

Figure 29.1. The Carnot cycle with a handle. See text for description.

because the cylinder is out of contact with the hot reservoir, the gas cools to the temperature of the cold reservoir at the end of the power stroke. Now I push the handle, pushing the cylinder into contact with the cold reservoir, then I push on the piston upwards, recompressing the working gas. As the gas compresses, it tends to heat up, but thanks to contact with the cold reservoir, the gas is held close to and somewhat above its cold temperature in the isothermal part of the compression stroke. Half way through the compression stroke, I pull on the handle, pulling the cylinder out of contact with the cold reservoir, then again push the piston up toward the cylinder head, thereby further compressing the working gas. Due to the compression, the working gas heats up to the temperature of the hot reservoir, completing the work cycle.

Two points require emphasis for our current purpose. First, the work cycle is a cyclic linkage of a spontaneous process, the power stroke, and a nonspontaneous process, my pushing on the piston to recompress the gas. Second, the handle and my pushing and pulling on it play a critical role, for they organize the flow of the linked processes. Now, in a real engine, gears, escapements, and chains, and so forth replace my pushing and pulling on the handle. Thus, the gears and so forth organize the processes of the work cycle. We understand, more or less, matter, energy, entropy, and information. But none of these is organization of process. As far as I can tell, we have no theory of such organization. Indeed, at present, it is unclear even what mathematical framework might be appropriate to a theory of organization. This will become all the more important in considering autonomous agents next.

A hypothetical autonomous agent

Our next task is to exhibit a conceptual example of a molecular autonomous agent. The example is not meant to be realistic in the detailed chemistry among the

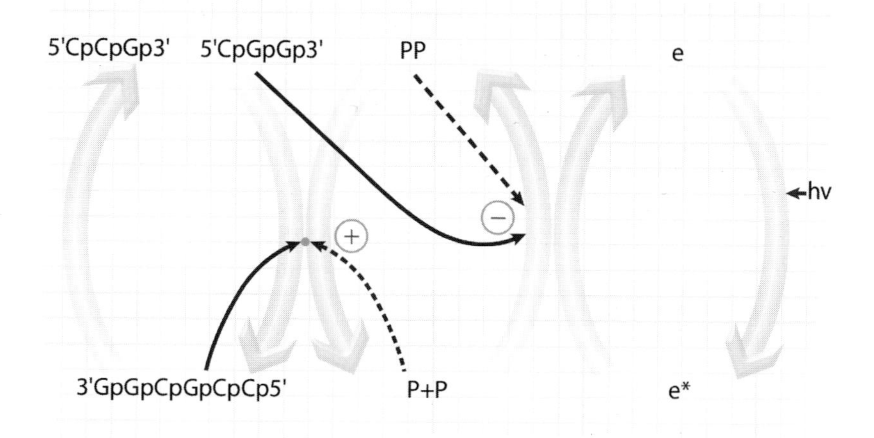

Figure 29.2. An hypothetical molecular autonomous agent. See text for description. For clarity, the reaction of the two DNA trimers to form the hexamer without coupling to pyrophosphate (PP) breakdown is not shown.

molecular components that it proposes. Rather it is meant to make concrete the concepts involved. Figure 29.2 shows a hypothetical molecular autonomous agent. It consists in a self-reproducing molecular system comprised of a single-stranded DNA hexamer and two trimers which are its Watson–Crick complements, coupled to a molecular motor that drives excess replication of the DNA hexamer. The entire system is a new class of open thermodynamic chemical reaction networks. This system takes in matter and energy in the forms of the two DNA trimers and a photon stream. As a preamble, I would note that self-reproducing molecular systems have been achieved experimentally, as have molecular motors. It remains to put the two together in a single system.

First, some familiar points should be made. The system I exhibit is a chemical reaction network some of whose components are enzymes. Enzymes do not change the equilibrium of a chemical reaction, they merely speed the approach to equilibrium. As will be familiar to most readers, a chemical reaction can approach equilibrium from either an initial excess of substrates or products. As a simple example, suppose A converts to B, and B converts to A. If one starts with pure A, A is converted to B. As B builds up, B is converted to A. Equilibrium is achieved when the net rate of conversion of A to B equals the net rate of conversion of B to A. Physical chemists draw this as an energy diagram, with the reaction coordinate on the X-axis and free energy on the Y-axis (Fig. 29.3). A spontaneous chemical reaction always proceeds "exergonically," in the direction of losing free energy until the minimum free energy is reached at equilibrium. If the reaction system is to be driven beyond equilibrium, say in order to synthesize more B that would be achievable by the undriven system, energy must be added to the system to drive it beyond equilibrium in an endergonic process.

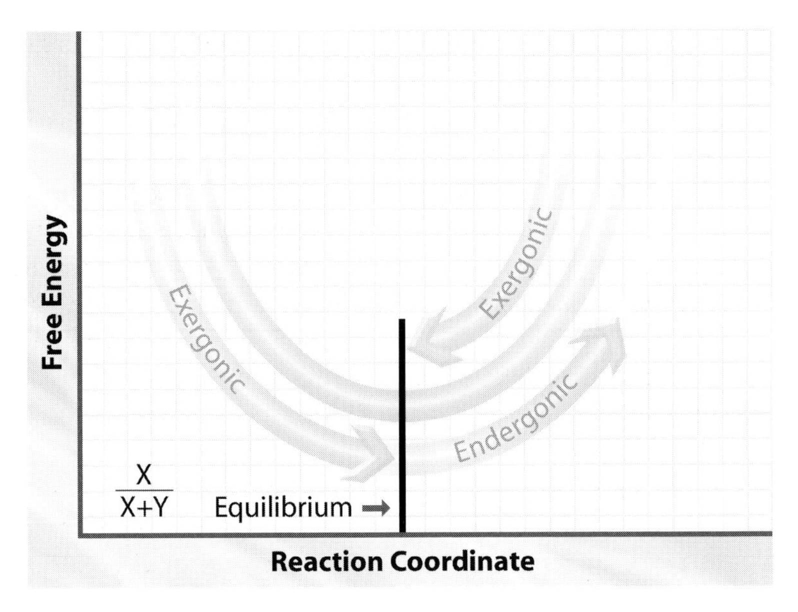

Figure 29.3. Exergonic and endergonic chemical reactions. See text for description.

Returning to the hypothetical autonomous agent, the DNA hexamer is capable of acting as an enzyme linking the two trimers into a hexamer which is a second copy of the initial hexamer, hence autocatalysis, or self-replication, is achieved. (This physical system, without coupling to pyrophosphate, PP, has actually been realized experimentally (von Kiedrowski 1986).) In the example, the synthesis of hexamer by linking the two trimers is coupled to the exergonic breakdown of PP to two monophosphates, P + P. This exergonic breakdown is utilized to drive endergonic excess synthesis of the hexamer compared to the equilibrium ratio of hexamer and trimers that would characterize the equilibrium of that subsystem. So, just as I pushed on the piston, the exergonic breakdown of PP is used to push excess replication of hexamer compared to equilibrium.

Once the pyrophosphate has been used to drive the endergonic synthesis of excess hexamer, it is necessary to restore the pyrophosphate concentration to its former level by driving the endergonic synthesis of PP from the two monophosphates. To supply the energy for this endergonic synthesis, I imagine an electron which absorbs a photon, is driven to an excited state, and, when it spontaneously and thus exergonically falls back to its ground state, uses that loss of free energy in a coupled reaction which drives the endergonic synthesis of pyrophosphate. I invoke one of the trimers as the catalyst that speeds up this coupled reaction.

Just as gears and escapements coordinate the flow of processes in the real Carnot engine, I invoke their analogs in the hypothetical autonomous agent. Specifically,

I want the forward reaction synthesizing excess hexamer to happen rapidly, then I want the reverse reaction resynthesizing pyrophosphate, PP, to occur. Accordingly, I assume that monophosphate, P, feeds back as an allosteric activator of the hexamer enzyme. Thus, the forward reaction proceeds slowly until the concentration of P rises sufficiently, then, thanks to the feedback activation of the hexamer enzyme, the forward reaction "flushes through," yielding excess synthesis of the new copies of the hexamer. Simultaneously, to stop the reaction resynthesizing PP from $P + P$, I invoke PP itself as an allosteric inhibitor of the enzyme catalyzing the resynthesis of PP from monophosphate. Thus, the forward reaction synthesizing hexamer occurs, then, after the concentration of PP falls, the resynthesis of PP occurs. My colleagues and I have written down the appropriate differential equations for this system and it behaves as described (Daley *et al.* 2002).

This system has a self-reproducing subsystem, the trimers and hexamer. And it has a chemical motor, the $PP \longleftrightarrow P + P$ reaction cycle. The engine's running can be seen by the fact that there is a net rotation of P counterclockwise around this reaction cycle.

Thus, this hypothetical system exhibits a molecular autonomous agent. Several points need to be made. First, this is a perfectly legitimate, if unstudied and new, class of open thermodynamic chemical reaction networks. Second, the system does not cheat the second law. The system eats trimers and photons, and, via the work cycle, pumps that energy into the excess synthesis of hexamer. Third, the system only works if displaced from chemical equilibrium in the "right" direction, an excess of trimers and photons. Agency only exists in systems displaced from equilibrium. Fourth, as pointed out to me by Phil Anderson, there is energy stored in the excess concentration of hexamer compared to its equilibrium concentration. That energy could later be used to correct errors, as happens in contemporary cells. Fifth, like the Carnot cycle, the autonomous agent contains a reciprocal and cyclic linking of spontaneous and nonspontaneous processes. Sixth, like the gears of a real Carnot engine, the allosteric couplings achieve the organization of the processes that is integral to the autonomous agent.

I would note that we are likely in the next decades to construct autonomous agents. Such systems actually do work and reproduce. They promise a technological revolution. More, if I have stumbled onto an adequate definition of life, they will constitute novel life forms. Sometime or another we will find or make novel life forms, and the way will be open to the creation of a general biology, freed from the constraints on our imagination of the only biology we know.

Considerations about the concept of work

I find work a puzzling concept. It is defined, of course, as force acting through a distance. But there are several unsettling aspects to the concept. Consider a concrete case of work. I lift a pen, doing work on it lifting it in a gravitational potential. In any specific case of work, there is an organization to the process that is not captured in the definition of work. It is true that physicists invoke initial and boundary conditions in any specific case, but ultimately we want to consider the entire evolution of the universe, and it is precisely the coming into existence of those initial and boundary conditions that is in question in the evolving universe and biosphere.

Consider an isolated thermodynamic system, say gas in a large box. Now, an isolated thermodynamic system can perform no work. But if the box be divided by a membrane into two subcompartments, and the pressure is higher in one compartment, the membrane will push into the other compartment, doing work on it. Thus, it appears that for work to occur, the universe must be divided into two regions. Where did that come from?

The definition of work that I find most congenial is due to Atkins in his book on the Second Law. He points out that work is a "thing"; specifically, it is the constrained release of energy into relatively few degrees of freedom. So consider the cylinder and piston, with the working gas compressed into the head of the cylinder. As the gas expands, the chaotic thermal motion of the gas molecules is released into the translational motion of the piston.

But what are the constraints in the cylinder and piston system? Obviously, the cylinder and piston, and the location of the piston inside the cylinder are among the constraints. And now I ask a new question: where did the constraints come from? Well, it took work to build the cylinder and the piston, and work to place the piston inside the cylinder. So it appears to take work to make constraints and constraints to make work. I do not want to say that it always takes work to make constraints. One might start with a nonequilibrium system at high energy that falls to a lower energy state in which constraints are constructed. Nevertheless, it typically takes work to make constraints and constraints to make work. This is certainly true in real cells as I note below.

I said we have no theory of organization, but I have the deep suspicion that this reciprocal linking of work and constraints on the release of energy that constitutes work is part of that theory. If so, notice that this is not part of physics at present, nor of chemistry, nor of biology.

I want next to show that a cell can and does accomplish a kind of propagating work and constraint construction until the cell, astonishingly, builds a rough copy of itself. Figure 29.4 shows an example. A cell does thermodynamic work to synthesize lipid molecules from their building blocks. The lipid molecules then fall to a low energy state creating a bilipid membrane hollow sphere called a liposome. Inside the aqueous interior are two small organic molecular species, A and B. These can undergo three different reactions. A and B can convert to C and D; A and B can convert to E; A and B can convert to F and G. Each of these three reactions has its

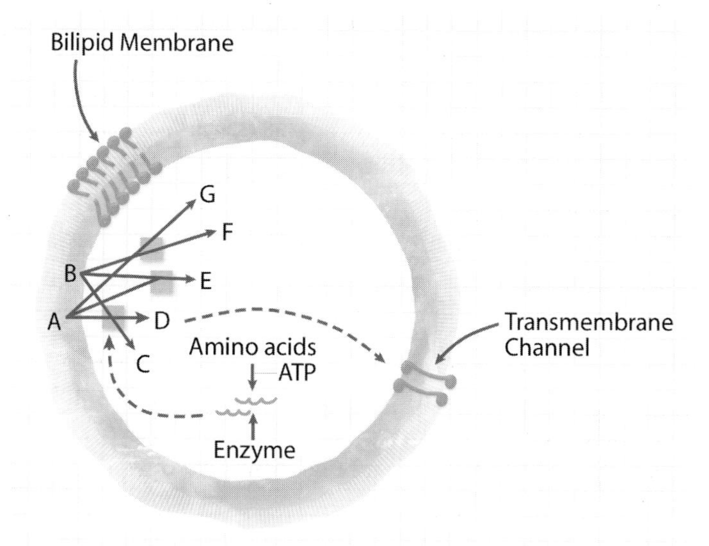

Figure 29.4. Propagating work and constraint construction in cells. See text for description.

own reaction coordinate and free energy diagram, with the substrates and products in potential wells separated by an energy barrier.

Now allow A and B to diffuse into the bilipid membrane. When that happens, the translational, vibrational, and rotational motions of A and B molecules alter. In turn, this modifies the reaction coordinate free energy profiles of the three reactions, perhaps raising some potential barriers and lowering others. But the raising and lowering of such barriers is precisely the manipulation of constraints. So cells do work to build constraints and manipulate them. Here the cell does work to construct a membrane which affords constraints and modifications of constraints. But more: the cell does thermodynamic work to link amino acids into an enzyme that happens to catalyze the conversion of A and B to C and D. Thus, free energy is released in constrained ways – the other two reactions are not catalyzed. But that released energy can propagate and do more work. For example, D may diffuse across the cell and bind to a transmembrane channel, giving up some of its vibrational energy to do work to open the channel to the ingress of an ion species. Hence "propagating" work has been done. In turn the ion species may cause further constraint construction and release of energy or other work. In short, a cell does in fact do work to construct constraints on the release of energy which, when released, does work to construct more constraints on the release of energy, which propagates until the cell completes a set of propagating work tasks and builds a copy of itself.

Note that we have, as yet, no developed language in physics, chemistry, or biology to discuss these matters. Consider also the miracle of the cell building a copy of itself, then the two repeat the process to make four cells, then eight, then a bacterial colony. I can only stumble with ordinary English: the cell achieves a propagating organization of building, work, and constraint construction that completes itself by the formation of a second cell. Is this matter alone, energy alone, entropy alone, or information alone? No. Do we have a formulated concept for what I just described? No.

Yet just such propagating organization occurs. Kant told us the same thing long ago. We have no language, no mathematical framework that I know of, which captures this process. It appears to be a new state of matter – call it the living state.

Maxwell's demon flummoxed

I now make a detour to discuss Maxwell's demon in a nonequililbrium setting. Good work shows that the demon cannot "win" in an equilibrium setting. It now appears that as the demon performs measurements on a system, for every one bit of entropy reduction there is a corresponding one bit increase in the most compressed description of the system that encodes what has been learned by the measurements. Since it ultimately costs energy to erase these bits, it does not pay the demon to measure in an equilibrium setting.

Now the way physicists always seem to phrase the issue about the demon is this: "And then, in principle, the demon can extract work."

I therefore want to consider the demon in a nonequilibrium setting and raise two important new questions. Consider the demon with the familiar box separated into two chambers with a flap valve between them. Let each chamber have the same number of particles, N. But let those in the right chamber be hotter, hence faster, than those in the left box. Thus, the system is nonequilibrium and work could be extracted. For example, let a small windmill be placed near the flap valve. Let the demon open the valve and a transient wind will flow from the right to the left box. The vane on the back of the windmill will "measure" the wind and orient the windmill perpendicular to the wind. So the system has detected and measured a source of energy, the wind. Then the wind will blow on the blades of the windmill and cause them to rotate. Thus the system detects a source of energy, responds to it "appropriately," and actually extracts energy to do work.

Now consider the demon confronting the hot and cold chambers, each with N particles in it. Let him perform an heroic experiment: he measures the instantaneous positions of all the N particles in both the right and left chambers. Note that, from this experiment, he cannot deduce that the particles in the right box are moving faster than those in the left box. Hence, he has carried out an ambitious measurement, but failed to detect a source of energy from which work can be extracted. Had he measured the positions of all the particles at two moments, he could have deduced that the right box was hotter, and would have detected a source of energy. Or he might have measured momentum transfer to the partition, or walls of the chambers and detected a source of energy.

How does the demon know what to measure?

The demon does not know what to measure. And that is the point. Next, how does the demon actually rig up a device to capture the energy and do work? Not so clear. And remember it takes work to rig the device. Does it pay in the sense that more work can be extracted than was used to construct and position the device? Now consider the biosphere. Let a mutation happen in some autonomous agent, say a protist, that allows it to detect, capture, and use a new source of energy to do some work of selective significance to the protist. Then natural selection will amplify this lucky mutation, and a new source of energy and work will become part of the biosphere. The combined system of autonomous agents, mutations, and selection does the job of picking out the useful measurements, detections of sources of energy, and getting work done that is useful. So the biosphere has and continues, presumably, to do just this. Think of the linked spontaneous and nonspontaneous chemical processes in a bit of biosphere on Darwin's tangled bank of an ecosystem. Sunlight falls, redwood trees grow.

I suspect that there is a deeper theory to be had here, but cannot prove it and, for reasons given shortly below, do not know how to construct it. The intuition, at least for a biosphere, and perhaps the universe, is that this process is part of the universe becoming complex and diverse. Consider the demon and the transient wind. Any flake of material, say mica, would flutter in the wind, hence extract work. Not much sophistication is required to detect the source of energy and extract work. Now consider an antiferromagnet not in its ground state. Thus, if the system could be triggered to fall towards its ground state, energy would be available to do work. But, intuitively, it takes a system as subtle in structure and behavior as the antiferromagnet to detect the displacement from equilibrium, trigger the fall toward equilibrium, and extract work. For example, a second antiferromagnet at the ground state might be in the vicinity of the nonequilibrium antiferromagnet, trigger the latter to fall toward the ground state, and use some of that energy to lift the ground state ferromagnet to a non-ground state. A source of free energy would have been detected and work would have been done.

Somehow, we need a theory of how such presumably increasingly subtle entities come into existence, couple spontaneous and nonspontaneous processes, and progressively build the complex structures of the universe and of a biosphere. I know of no such theory.

We cannot finitely prestate the future of the biosphere

I come now to the most troubling of the implications of autonomous agents. I suspect that we cannot prestate the future of the biosphere.

Consider the heart. Darwin would have told us that the function of the heart is to pump blood. That is, this is the causal consequence for which the heart was selected. But the heart also makes heart sounds. Heart sounds are not the function of the heart. Already this has important consequences. The function of a part of an organism is a subset of its causal consequences. To discover the function, we must study the whole organism in its environment. There is an unavoidable holism to biology.

Darwin also considered what he called preadaptations, and Stephen Jay Gould named exaptations. A part of an organism might have a causal consequence that is not of functional significance in the normal environment, but might prove useful in a different environment, hence be selected.

To make myself the butt of my example, I tell the following tale. The heart is a resonant chamber. Suppose I have a mutant heart, due to a single Mendelian dominant mutation, which renders my heart able to detect earthquake pre-tremors. I am in Los Angeles and feel something odd in my chest. "Uh oh," I think, "It must be an earthquake coming." I run into an open field. Millions die, but I am safe. Word leaks out that I detected the pre-tremor and I am invited onto "Good Morning America." I become famous. Women flock to my side. I mate with many. (This is necessary for my story.) Soon there are lots of little boy and girl progeny that have my mutant heart. If earthquakes happened often enough that this was of selective significance, soon the biosphere would sport earthquake detectors in humans.

Now my question is this: do you think that you can say, ahead of time, all the possible Darwinian preadaptations of, say, the 30 000 000 to 100 000 000 extant species? Can you say ahead of time all possible preadaptations for all the species that have lived? More formally, could you finitely prestate all the possible preadaptations of such species?

I think the answer is "No." Indeed, I have not found anyone who thinks the answer is "Yes." At least part of the problem is that I have no idea how to get started trying to state all possible environments for all members of all species. I want to say that we cannot finitely prestate the configuration space of the biosphere. Now you may tell me that, speaking classically, the configuration space is just some vast $6N$-dimensional phase space. Perhaps. But I will then respond by saying that you cannot prestate the relevant macroscopic collective variables of the biosphere that drive its further evolution, such as earthquake detecting hearts.

Next note that most major adaptations are Darwinian preadaptations. Most minor adaptations may also be preadaptations. Thus arose flight, hearing, lungs, etc.

Let us suppose the answer is, in fact "No." I want first to note that I do not know that the answer is really "No." Then I note that I do not understand whether this is a mathematical statement, perhaps akin to the halting problem or Gödel's theorem, or an empirical statement. Nor do I know if it is due to the finite computing power of the entire universe, should we be able to harness that power. It may be related to the fact that, for many algorithms, there is no shorter way to find out what they will do than to run the program. In any case, we seem to be precluded from knowing the future state of the biosphere, not due to quantum uncertainty, nor to chaotic dynamics, but because we do not have the concepts to pick out the relevant collective variables ahead of time.

Consider the frequency interpretation of probability. One begins by stating the space of possibilities. We cannot do this for the biosphere. The economists have a concept called Knightian uncertainty. Roughly it corresponds to the unknown that we do not yet know about. I suspect Knightian uncertainty is linked to what I am here describing. I am not a physicist, but my impression is that in Newtonian physics, quantum mechanics, statistical mechanics, and general relativity, one begins by prestating the relevant configuration space. If so, and if I am right, the biosphere seems to escape how we have been taught to do science.

But our incapacity to prestate the relevant collective variables is not slowing down the evolution of the biosphere. It therefore appears that we cannot prestate the future possibilities of the biosphere.

The Adjacent Possible

Finally, I want to touch on the concept of the Adjacent Possible. Consider a box filled with 1000 species of organic molecules. Call these the Actual. These molecules can undergo reactions creating, probably, molecules that are novel with respect to the Actual. Call the novel molecular species, reachable in a single reaction step from the Actual, the Adjacent Possible. The biosphere, 4.8 billion years ago, had only a few hundred organic molecular species. Now it has trillions. So the biosphere has been advancing into the Adjacent Possible over its history. The next point is that this advance is grossly nonergodic. Consider all possible proteins length 200. There are 20 raised to the 200th power, or 10 to the 260th power such sequences. We can make any single one we wish with today's technology. Now consider chemical reactions on a femtosecond timescale, and the estimated 10 to the 80th particle diversity of the known universe. Suppose the universe were busy building only proteins of length 200. Forget that distant particles cannot interact. The maximum number of pairwise interactions on a femtosecond timescale since the Big Bang is about 10 raised to the 193rd power – a vast number, but tiny compared to 10 raised to the 260th power. In short, it would take at least 10 raised to the 67th power repetitions of the history of the universe to make all possible proteins length 200. In short, we are on a unique trajectory, once one considers entities of complexity significantly greater than atomic species.

Thus, we are, the universe is, advancing into an Adjacent Possible on some unique trajectory that is grossly nonrepeating, hence nonergodic. We are entitled to wonder whether there may be laws that govern how the expansion into the Adjacent Possible happens. No one knows, of course. But I want to close with a candidate law for biospheres: as a secular trend, (there are major extinction events that may interrupt this trend), it may be the case that a biosphere expands into the Adjacent Possible as fast as it can without destroying the order it has already assembled. Of course, this candidate law is still poorly stated, but the rough intuition is that biospheres anywhere in the cosmos, as a secular trend, maximize the diversity of what can happen next.

Conclusion

I have drawn attention to the issue of autonomous agents, systems that can act on their own behalf in an environment. All free-living organisms are autonomous agents. I have offered a definition of an autonomous agent as a system able to reproduce itself and carry out at least one work cycle. The issue is whether the defintion is useful or not. It appears to me that the concept leads us towards puzzles that we have not seen before, despite the fact that those puzzles are right in front of us. Central to this is a missing theory of organization. In discussing the flummoxed Maxwell's demon, I opined that there must be a theory for why the universe gets more complex. My difficulty in constructing such a theory is, at least, that I cannot see how to begin when I may not be able to finitely prestate what the entities are that will come into existence, detect sources of energy, link to them, and build yet new, unforeseen entities. Yet the biosphere, for 4.8 billion years, has been doing just this. But the biosphere is "just" a physical system. So, in the construction of a general biology, we must lift physics and chemistry, let alone biology, to some new level that can deal with biospheres anywhere in the cosmos.

References

Atkins, P W (1994) *The Second Law: Energy, Chaos, and Form*, revd edn. New York: W. H. Freeman.
Daley, A J, Grivin, A, Kauffman, S A, *et al.* (2002) Simulation of a chemical autonomous agent. *Zeits. Phys. Chem.* **216**, 41.
Darwin, C (1859) *On The Origin of Species*. Cambridge, MA: Harvard University Press. (Facsimile edn. London: John Murray).
Gould, S J (1977) *Ontogeny and Phylogeny*. Cambridge, MA: Harvard University Press.
Kauffman, S (2000) *Investigations*. New York: Oxford University Press.
von Kiedrowski,G (1986) A self-replicating hexadeoxynucleotide. *Angew. Chem. Int. Ed. Eng.* **25**, 982.

30

To see a world in a grain of sand

Shou-Cheng Zhang
Stanford University

Introduction

Modern physics is built upon three principal pillars, quantum mechanics, special relativity, and general relativity. Historically, these principles were developed as logically independent extensions of classical Newtonian mechanics. While each theory constitutes a logically self-consistent framework, unification of these fundamental principles encountered unprecedented difficulties. Quantum mechanics and special relativity were unified in the middle of the last century, giving birth to relativistic quantum field theory. While tremendously successful in explaining experimental data, ultraviolet infinities in the calculations hint that the theory can not be in its final form. Unification of quantum mechanics with general relativity proves to be a much more difficult task and is still the greatest unsolved problem in theoretical physics.

In view of the difficulties involved with unifying these principles, we can ask a simple but rather bold question: is it possible that the three principles are not logically independent, but rather that there is an hierarchical order in their logical dependence? In particular, we notice that both relativity principles can be formulated as statements of symmetry. When applying nonrelativistic quantum mechanics to systems with a large number of degrees of freedom, we sometimes find that symmetries can emerge in the low-energy sector, which are not present in the starting Hamiltonian. Therefore, there is a logical possibility that one could start from a single nonrelativistic Schrödinger equation for a quantum many-body problem, and discover relativity principles emerging in the low-energy sector. If this program can indeed be realized, a grand synthesis of fundamental physics can be achieved. Since nonrelativistic quantum mechanics is a finite and logically self-consistent

Science and Ultimate Reality, eds. J. D. Barrow, P. C. W. Davies and C. L. Harper, Jr. Published by Cambridge University Press.

framework, everything derived from it should be finite and logically consistent as well.

The standard model in particle physics is described by a relativistic quantum field theory and is experimentally verified below the energy scale of 10^3 GeV. On the other hand, the Planck energy scale, where quantum gravitational force becomes important, is at 10^{19} GeV. Therefore, we need to extrapolate 16 orders of magnitude to guess the new physics beyond the standard framework of relativistic quantum field theory. It is quite conceivable that Einstein's principle of relativity is not valid at the Planck energy scale; it could emerge at energies much lower compared to the Planck energy scale through the magic of renormalization group flow. This situation is analogous to one in condensed-matter physics, which deals with phenomena at much lower absolute energy scales. The "basic" laws of condensed matter physics are well known at the Coulomb energy scale of $1 \sim 10$ eV; almost all condensed-matter systems can be well described by a nonrelativistic Hamiltonian of the electrons and the nuclei (Laughlin and Pines 2000). However, this model Hamiltonian is rather inadequate to describe the various emergent phenomena, like superconductivity, superfluidity, the quantum Hall effect (QHE), and magnetism, which all occur at much lower energy scales, typically of the order of 1 milli-electron volt (meV). These systems are best described by "effective quantum field theories," not of the original electrons, but of the quasi-particles and collective excitations. In this chapter, I shall give many examples where these "effective quantum field theories" are relativistic quantum field theories or topological quantum field theories, bearing great resemblance to the standard model of elementary particles. The collective behavior of many strongly interacting degrees of freedom is responsible for these striking emergent phenomena. The laws governing the quasi-particles and the collective excitations are very different from the laws governing the original electrons and nuclei (Anderson 1972). This observation inspires us to construct models of elementary particles by conceptually visualizing them as quasi-particles or collective excitations of a quantum many-body system, whose basic constituents are governed by a simple nonrelativistic Hamiltonian. This point of view is best summarized by the following diagram:

Planck energy at 10^{19} GeV	$\Leftrightarrow$	Coulomb energy at 10 eV
$\uparrow ?$		$\downarrow$
Standard model at 10^3 GeV	$\Leftrightarrow$	Superconductivity, QHE, magnetism, etc. at 1 meV
Relativistic quantum field theory of elementary particles		Effective quantum field theory of quasi-particles

The conceptual similarity between particle physics and condensed matter physics has played a very important role in the history of physics. A crucial ingredient of the standard model, the idea of spontaneously broken symmetry and the Higgs mechanism, first originated from the Bardeen–Cooper–Schrieffer (BCS) theory of superconductivity. This example vividly shows that the physical vacuum is not empty, but a condensed state of many interacting degrees of freedom. Another fundamental concept is the idea of renormalization group transformation, which was simultaneously developed in the context of particle physics and in the study of critical phenomena. From the theory of renormalization group, we learned that symmetries can emerge at the low-energy sector, without being postulated at the microscopic level. Today, as physicists face unprecedented challenges of unifying quantum mechanics with relativity, and tackling the problem of quantum gravity, it is useful to look at these historic successes for inspiration. A new era of close interaction between condensed-matter physics and particle physics could shed light on these grand challenges of theoretical physics.

Examples of emergence in condensed matter systems

In this section, we review some well-known examples in condensed-matter physics, where one starts from a quantum many-body system at high energies and arrives at a relativistic or topological field theory of the low-energy quasi-particles and elementary excitations. The high-energy models often look simple and innocuous, yet the emergent low-energy phenomena and their effective field theory description are profound and beautiful.

2 + 1 dimensional QED from superfluid helium films

Let us first start from the physics of a superfluid film. The mean velocity of the helium atoms is significantly lower compared to the speed of light, so relativistic effects of the atoms can be completely neglected. The basic nonrelativistic Hamiltonian for this system of identical bosons can be expressed in the following closed form:

$$H = \frac{1}{2m}\sum_n \vec{p}_n^{\,2} + \sum_{n<n'} V(x_n - x_{n'}) \qquad (30.1)$$

where V is the inter-atomic potential, whose form depends on the details of the system. However, for a large class of generic interaction potentials, the system flows towards a universal low-energy attractive fix point, namely the superfluid ground state. At typical inter-atomic energy scales of a few eV's, helium atoms are the correct dynamic variables, and the Hamiltonian (30.1) is the correct model Hamiltonian. However, at the energy scale characteristic of the superfluid transition,

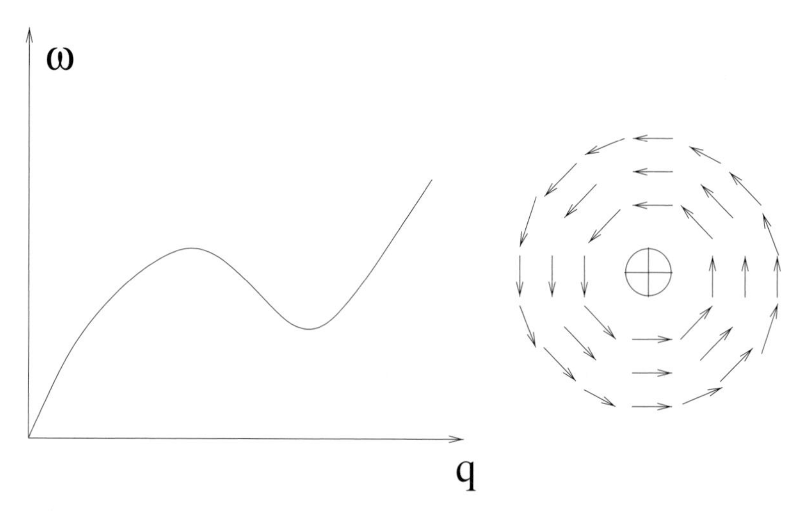

Figure 30.1. Collective excitations of a neutral two-dimensional superfluid film are the sound waves and the vortices. In the long wave length limit, the sound wave maps onto the Maxwell fields, while vortices map onto electric charges.

which is of the order of $1K \sim 10^{-4}$ eV, the correct dynamical variables are sound wave modes and the vortices of the superfluid film (Fig. 30.1).

The remarkable thing is that the effective field theory model for these low-energy degrees of freedom is exactly the relativistic quantum electrodynamics (QED) in $2+1$ dimensions! This connection was established by the work of Ambegaokar, Halperin, Nelson, and Siggia (Ambegaokar *et al.* 1980) and derived from the point of view of vortex duality (Fisher and Lee 1989). To see how this works, let us recall that the basic hydrodynamical variables of the superfluid film are the density $\rho(x)$ and the velocity $v_i(x)$ fields, $(i = 1, 2)$, satisfying the equation of continuity

$$\partial_t \rho + \partial_i (\bar{\rho} v_i) = 0 \tag{30.2}$$

where $\bar{\rho}$ is the average density of the fluid. Now let us recall that in $2+1$ dimensions, the electric field E_i has two components while the magnetic field B has only one component, which can therefore be identified as a scalar. Faraday's law of induction is given by the Maxwell equation:

$$\frac{1}{c}\partial_t B + \epsilon_{ij} \partial_i E_j = 0 \tag{30.3}$$

where ϵ_{ij} is the antisymmetric tensor in two dimensions. Therefore, if we make the following identification,

$$B \Leftrightarrow -c\frac{\rho}{\bar{\rho}} \qquad E_i \Leftrightarrow \epsilon_{ij} v_j \tag{30.4}$$

we see that the equation of continuity of the superfluid film agrees exactly with Faraday's law as expressed in the Maxwell's equation (30.3). Next we examine the fluid velocity in the presence of a vortex, with unit vorticity, located at the position x_n. The superfluid state has a well-defined U(1) order parameter, and the velocity field can be expressed in terms of the phase, ϕ, of the U(1) order parameter:

$$v_i = \frac{\hbar}{m}\partial_i \phi. \tag{30.5}$$

Because of the single valuedness of the quantum mechanical wave function, $e^{i\phi}$ must be single valued. Therefore, the superflow around a vortex is *quantized*:

$$\int \vec{v}\cdot d\vec{l} = 2\pi \frac{\hbar}{m} q \tag{30.6}$$

where q is an integer. For elementary vortices, $q = \pm 1$. The differential form of this integral equation is

$$\epsilon_{ij}\partial_i v_j = 2\pi \rho_v(x) \tag{30.7}$$

where $\rho_v(x) = \sum_n q_n \delta(x - x_n)$ is the density of the vortices and $q_n = \pm 1$ is the vorticity. If we identify the vortex density with the electric charge density in Maxwell's equations, we see that eqn (30.7) is nothing but Gauss's law in $2 + 1$ dimensions:

$$\partial_i E_i = 2\pi \rho_v(x). \tag{30.8}$$

Next let us investigate the dynamics of the superfluid velocity v_i, through the Josephson equations of superfluidity. The first Josephson equation relates the superfluid velocity to the gradient of the superfluid phase, ϕ, as expressed in eqn (30.5). The second Josephson equation relates the time derivative of the phase to the chemical potential $\hbar\partial_t \phi = -\mu$. Combining the two Josephson equations, we obtain

$$\partial_t v_i = \frac{\hbar}{m}\partial_t \partial_i \phi = -\frac{1}{m}\partial_i \mu = -\frac{\kappa}{m\bar{\rho}}\partial_i \rho \tag{30.9}$$

where we use the compressibility $\kappa = \bar{\rho}\frac{\partial \mu}{\partial \rho}$ to express the chemical potential μ in terms of the density ρ. This equation agrees exactly with the source-free Maxwell equation

$$c\epsilon_{ij}\partial_j B = \partial_t E_i \tag{30.10}$$

provided one identifies the speed of light as $c^2 = \kappa/m$. This equation needs to be modified in the presence of the vortex flow J_i^v, which unwinds the U(1) phase by 2π each time a vortex passes. The vortex current satisfies the equation of continuity

$$\partial_t \rho_v + \partial_i J_i^v = 0. \tag{30.11}$$

Therefore, the source-free Maxwell equation (30.10) acquires a additional term, in order to be compatible with both eqns (30.11) and (30.8):

$$c\epsilon_{ij}\partial_j B = \partial_t E_i + 2\pi J_i^v. \tag{30.12}$$

This is nothing but Ampère's law, supplemented by Maxwell's displacement current.

This proves the complete equivalence between the superfluid equations and Maxwell's equations in $2+1$ dimensions. Interestingly enough, we seem to have completed a rather curious loop. Starting from the relativistic standard model of the quarks and leptons, one arrives at an effective nonrelativistic model of the helium atoms (30.1). However, as one reduces the energy scale further, the effective low-energy degrees of freedom become the sound modes and the vortices, which are described by the field theory of $2+1$ dimensional quantum electrodynamics, very similar to the model we started from in the first place! A "civilization" living inside the helium film would first discover the Maxwell's equations, and then, after much harder work, they would establish eqn (30.1) as their "theory of everything."

Superfluid ^{4}He films are relatively simple because the ^{4}He atom is a boson. The superfluidity of ^{3}He is much more complex, with many competing superfluid phases. In fact, Volovik (2001) has pointed out that many phenomena of the superfluid phase of ^{3}He share striking similarities with the standard model of elementary particles. These similarities inspired him to pioneer a program to address cosmological questions by condensed matter analogs.

Dirac fermions of d *wave superconductors*

Having considered the low-energy properties of a superfluid, let us now consider the low-energy excitations of a superconductor, with *d* wave pairing symmetry. In this case, there are low-energy fermionic excitations besides the bosonic excitations. This system is realized in the high T_c superconductors. The microscopic Hamiltonian is the two-dimensional (2D) Hubbard model, or the $t-J$ model, expressible as

$$H = -t\sum_{\langle ij\rangle,\sigma} c_{i\sigma}^\dagger c_{j\sigma} + J\sum_{\langle ij\rangle} \mathbf{S}_i\cdot\mathbf{S}_j \tag{30.13}$$

where $c_{i\sigma}^\dagger$ is the electron creation operator on site i with spin σ, $\mathbf{S}_i$ is the electron spin operator and $\langle ij\rangle$ denotes the nearest neighbor bond on a square lattice. Double occupancy of a single lattice site is forbidden.

This model is valid at the energy scale of 150 meV, which is the typical energy scale of the antiferromagnetic exchange J. When the filling factor x lies between 10% and 20%, the ground state of this model is believed to be a *d* wave superconductor. There is indeed overwhelming experimental evidence that the pairing symmetry of the high T_c superconductor is d wave-like. Remarkably, the elementary excitations in this case can be described by the $2+1$ dimensional Dirac Hamiltonian. In contrast to the $t-J$ model, which is valid at the energy scale of 100 meV, the effective Dirac Hamiltonian for the d wave quasi-particles is valid at much lower energy, typically of the order of 30 meV, which is the maximal gap. While the connection between the $t-J$ model and d wave superconductivity still needs to be firmed established, the connection between the d wave BCS quasi-particle Hamiltonian and the Dirac equation is well known in the condensed-matter community (Volovik 1993; Simon and Lee 1997; Balents *et al.* 1998; Franz *et al.* 2002). Here we follow a pedagogical presentation by Balents, Fisher, and Nayak (Balents *et al.* 1998).

The BCS mean field Hamiltonian for a d wave superconductor is given by

$$H = \sum_{k\alpha} \epsilon_k c^\dagger_{k\alpha} c_{k\alpha} + \sum_k \left[\Delta_k c^\dagger_{k\uparrow} c^\dagger_{-k\downarrow} + \Delta^*_k c_{-k\downarrow} c_{k\uparrow}\right] \tag{30.14}$$

where ϵ_k is the quasi-particle dispersion relation, and Δ_k is the d wave pairing gap, given by

$$\epsilon_k = -2t(\cos k_x + \cos k_y), \qquad \Delta_k = \Delta_0(\cos k_x - \cos k_y). \tag{30.15}$$

One can introduce a four-component spinor

$$\Upsilon_{a\alpha}(\vec{k}) = \begin{bmatrix} \Upsilon_{11} \\ \Upsilon_{21} \\ \Upsilon_{12} \\ \Upsilon_{22} \end{bmatrix} = \begin{bmatrix} c_{k\uparrow} \\ c^\dagger_{-k\downarrow} \\ c_{k\downarrow} \\ -c^\dagger_{-k\uparrow} \end{bmatrix}. \tag{30.16}$$

which doubles the number of degrees of freedom. This can be compensated by summing over only half of the Brillouin zone, say $k_y > 0$. In terms of these variables, the BCS Hamiltonian becomes

$$H = \sum_{k,k_y>0} \Upsilon^\dagger_{a\alpha}(\vec{k})[\tau^z\epsilon_k + \tau^+\Delta_k + \tau^-\Delta^*_k]_{ab}\Upsilon_{b\alpha}(\vec{k}), \tag{30.17}$$

where $\vec{\tau}_{ab}$ are the standard Pauli matrices acting in the particle/hole subspace.

The d wave nodes are approximately located near the special wave vectors $\mathbf{K}_1 = (\pi/2, \pi/2)$, $\mathbf{K}_2 = (-\pi/2, \pi/2)$, $\mathbf{K}_3 = -\mathbf{K}_1$ and $\mathbf{K}_4 = -\mathbf{K}_2$. In order to obtain a long wavelength and low-energy description, we can expand around the nodal points $\mathbf{K}_1$ and $\mathbf{K}_2$, which satisfy the $k_y > 0$ constraint. The nodal points $\mathbf{K}_3$ and $\mathbf{K}_4$ are automatically taken into account in the Υ spinor.

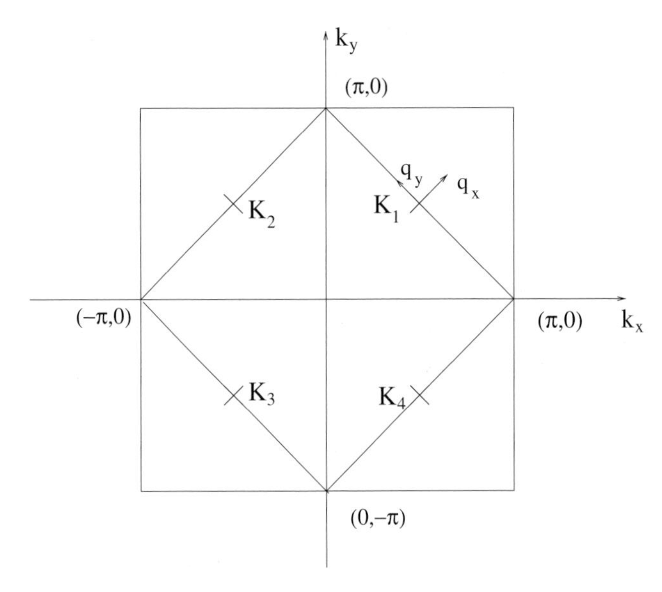

Figure 30.2. A 2D d wave superconductor has four nodes, indicated by K_1, K_2, K_3, and K_4. Around these nodal points, BCS quasi-particles obey the massless Dirac equation.

Introducing the rotated coordinates q_x and q_y, as indicated in Fig. 30.2, and the effective spinors

$$\Psi_{1a\alpha}(\vec{q}) = \Upsilon_{a\alpha}(\vec{K}_1 + \vec{q}), \qquad \Psi_{2a\alpha}(\vec{q}) = \Upsilon_{a\alpha}(\vec{K}_2 + \vec{q}) \tag{30.18}$$

we obtain

$$H = \sum_{q \in K_1} \Psi^{\dagger}_{1a\alpha}(\vec{q})\left[\tau^z \epsilon_{K_1+q} + \tau^+ \Delta_{K_1+q} + \tau^- \Delta^*_{K_1+q}\right]_{ab} \Psi_{1b\alpha}(\vec{q}) + (1 \leftrightarrow 2). \tag{30.19}$$

Here $q \in K_1$ denotes a momentum sum near the vector K_1. Expansion near K_1 gives

$$\epsilon_{K_1+q} \approx v_F q_x, \qquad \Delta_{K_1+q} \approx \Delta q_y. \tag{30.20}$$

A similar expansion applies for K_2. Going to the continuum limit, we obtain the Hamiltonian density

$$\begin{aligned} \mathcal{H} = \Psi^{\dagger}_{1a\alpha}[v_F \tau^z i\partial_x + (\tilde{\Delta}\tau^+ + \tilde{\Delta}^* \tau^-) i\partial_y]_{ab} \Psi_{1b\alpha} \\ + (1 \leftrightarrow 2; x \leftrightarrow y), \end{aligned} \tag{30.21}$$

which is exactly the Dirac Hamiltonian density in $2+1$ dimensions. Once again, we see the emergent relativistic behavior of a quantum many-body system. We start from a nonrelativistic interacting fermion problem at higher energies, but recover a relativistic Dirac equation at low energies.

Emergence of a topological quantum field theory

When Einstein first wrote down his field equation of general relativity, he said that the left-hand side of the equation that had to do specifically with geometry and gravity was beautiful – it was as if made of marble. But the right-hand side of the equation that had to do with matter and how matter produces gravity was ugly – it was as if made of wood. Taking Einstein's aesthetic point of view one step further, one is tempted to construct a fundamental theory by starting with the description of the topology, or a topological field theory without matter and without even geometry from the start. Having demonstrated that the relativistic Maxwell equation and Dirac equation can emerge in the low-energy sector of a quantum many-body problem, I now give an example demonstrating how a topological quantum field theory, namely the Chern–Simons (CS) theory, can emerge from the matter degrees of freedom in the low energy sector of the QHE. The CS topological quantum field theory was derived microscopically by Zhang, Hansson, and Kivelson (Zhang *et al.* 1989), and is reviewed extensively in Zhang (1992).

The basic Hamiltonian of quantum Hall effect (QHE) is simply that of a 2D electron gas in a perpendicular magnetic field

$$H = \frac{1}{2m}\sum_n \left[\vec{p}_n - \frac{e}{c}\mathbf{A}(x_n)\right]^2 + \sum_n eA_0(x_n) + \sum_{n<n'} V(x_n - x_{n'}) \quad (30.22)$$

where $\mathbf{A}$ is the vector potential of the external magnetic field, which in the symmetric gauge can be expressed as

$$A_i = \frac{1}{2}B\epsilon_{ij}x_j. \quad (30.23)$$

A_0 is the scalar potential of the external electric field, $E_i = -\partial_i A_0$, and $V(x)$ is the interaction between the electrons. For high magnetic fields, the electron spins are polarized along the same direction. Since the spin wave function is totally symmetric, the Hamiltonian (30.22) operates on orbital wave functions that are totally antisymmetric. This model is valid at the Coulombic energy scale of several eV's and has no particular symmetry or topological properties. Since the external magnetic field breaks time-reversal symmetry, an invariant tensor ϵ_{ij} can be introduced into the response function, and in particular, one can have a current J_i, which flows transverse to the applied electric field E_j, given by

$$J_i = \rho_{\mathrm{H}}^{-1}\epsilon_{ij}E_j \quad (30.24)$$

where $\rho_{\rm H}$ is defined as the Hall resistance. Since the electric field is perpendicular to the induced current, it does no work on the electrons, and the current flow is dissipationless. The 2D electron density n in a magnetic field B is best measured in terms of a dimensionless quantity called the filling factor $\nu = n/n_B$, where $n_B = B/\phi_0 = eB/hc$ is the magnetic flux density. QHE is the remarkable fact that the coefficient of the Hall response is quantized, given by

$$\rho_{\rm H} = \nu^{-1}\frac{h}{e^2} \tag{30.25}$$

when the filling fraction is near a rational number $\nu = p/q$ with odd denominator q. QHE at fractional values of ν is referred to as the fractional QHE (FQHE).

FQHE is described by Laughlin's celebrated wave function. There is also an alternative way to understand this profound effect by the Chern–Simons–Landau–Ginzburg (CSLG) effective field theory (Zhang 1992). The idea is to perform a singular gauge transformation on eqn (30.22), and turn electrons into bosons. This is only possible in $2+1$ dimensions. Consider another Hamiltonian

$$H' = \frac{1}{2m}\sum_n \left[\mathbf{p}_n - \frac{e}{c}\mathbf{A}(x_n) - \frac{e}{c}\,\mathbf{a}(x_n)\right]^2 + \sum_n eA_0(x_n) + \sum_{n<n'} V(x_n - x_{n'}). \tag{30.26}$$

Every symbol in H' has the same meaning as in H, except the new vector potential $\vec{a}$, which describes a *gauge interaction* among the particles and is given by

$$\mathbf{a}(x_n) = \frac{\phi_0}{2\pi}\frac{\theta}{\pi}\sum_{n'\neq n}\vec{\nabla}\,\alpha_{nn'} \tag{30.27}$$

where $\phi_0 = hc/e$ is the unit of flux quantum and $\alpha_{nn'}$ is the angle sustained by the vector connecting particles n and n' with an arbitrary vector specifying a reference direction, say the $\hat{x}$ axis. The crucial difference here is while H operates on a fully antisymmetric fermionic wave function, H' operates on a fully symmetric bosonic wave function. One can prove an exact theorem which states that these two quantum eigenvalue problems are equivalent to each other when $\theta/\pi = (2k+1)$ is an odd integer. In this case, each electron is attached to an odd number of fictitious quanta of gauge flux (cause by a), so that their exchange statistics in $2+1$ dimensions becomes bosonic (see Fig. 30.3). These bosons, called composite bosons (Girvin and Macdonald 1987; Read 1989; Zhang *et al.* 1989), see two different types of gauge fields: the external magnetic field A, and an internal statistical gauge field a. The average of the internal statistical gauge field depends on the density of the electrons. When the external magnetic field is such that the filling fraction $\nu = n_B/n = 1/(2k+1)$ is the inverse of an odd integer, we can always choose $\theta = (2k+1)\pi$ so that the net field seen by the composite bosons cancels each other on average.

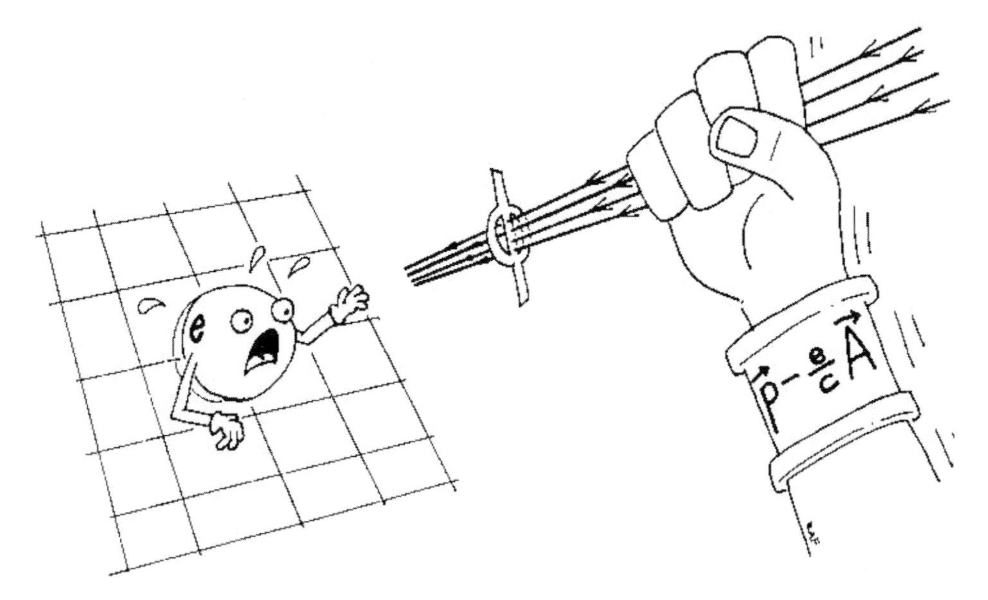

Figure 30.3. An electron just before the flux transmutation operation. (Taken from the Ph.D. thesis of D. Arovas, illustrated by Dr. Roger Freedman).

The statistical transmutation from electrons to composite bosons can be naturally implemented in quantum field theory through the CS term. The CS Lagrangian is given by

$$\mathcal{L} = \frac{1}{2}\left(\frac{\pi}{\theta}\right)\frac{1}{\phi_0}\,\varepsilon^{\mu\upsilon\rho}\,a_\mu\,\partial_\upsilon\,a_\rho - a_\mu\,j^\mu \tag{30.28}$$

where j^μ is the current of the composite boson, and $\mu = 0, 1, 2$ is the spacetime index in $2+1$ dimensions. The equation of motion for the a_0 field is

$$\epsilon^{ij}\,\partial_i\,a_j(x) = \phi_0\,\frac{\theta}{\pi}\,\rho(x) \tag{30.29}$$

whose solution for $\rho(x) = \sum_n \delta(x - x_n)$ exactly gives the statistical gauge field in eqn (30.27).

Now we can present the key argument of the CSLG theory (Zhang 1992) of QHE. Even though of course the statistical transformation can be performed in any $2+1$ dimensional system, this does not mean that the low-energy limit of any $2+1$ dimensional system is given by a CS theory, since the partition function also involves the integration over the matter fields j^μ in the second term of eqn (30.28). The key observation is that at the special filling factors of $\nu = 1/(2k+1)$, the combined external and statistical magnetic field seen by the composite boson vanishes, therefore composite bosons naturally condense into a superfluid state. This is the "magic" of the magic filling factors $\nu = 1/(2k+1)$. We already showed above that the effective field theory of a $2 + 1$ dimensional bosonic condensate is the $2 + 1$ dimensional Maxwell theory. Therefore, the integration over the matter fields in eqn (30.28) gives the Maxwell Lagrangian, $f_{\mu\nu}^2$. In $2 + 1$ dimensions, the CS term contains one fewer derivative compared with the Maxwell term; it therefore dominates in the long wave length and low-energy limit. Therefore, the effective Hamiltonian of FQHE is just the topological CS theory, without the matter current term in eqn (30.28).

Matter degrees of freedom in the starting Hamiltonian (30.22) are magically turned into topological degrees of freedom of the CS field theory. Alchemy works! Wood is turned into marble! Many people argued that a quantum theory of gravity should be formulated independently of the background metric. The emergent CSLG theory starts from matter degrees of freedom in a background setting, but the resulting effective field theory is independent of the background metric. This demonstrates that in principle, a background-independent theory can indeed be constructed from nonrelativistic quantum many-body systems. In fact, the CSLG theory also leads to a beautiful duality symmetry based on the discrete $SL(2, Z)$ group, very similar to the duality symmetry in the 4D Seiberg–Witten theory. This duality symmetry is again emergent, and it predicts the global phase diagram of the QH Hall system. The phase diagram has a beautiful fractal structure, with one phase nested inside each other, iterated *ad infinitum* (Kivelson *et al.* 1992).

The four-dimensional quantum Hall effect

In the previous sections we saw that the collective behavior of quantum many-body systems often gives rise to novel emergent phenomena in the low-energy sector, which are described in terms of relativistic or topological quantum field theories. Therefore, one can't help but wonder if the standard model could also work this way. The problem is that the well-understood examples of emergent relativistic behaviors in quantum many-body systems work only for lower dimensions, and these models do not have sufficient richness yet. In order for the standard model to appear as emergent behavior, we are led to study higher-dimensional quantum many-body systems, specially the higher-dimensional generalizations of QHE.

The model

Of all the novel quantum many-body systems, QHE plays a very special role: it is the only well-understood condensed-matter system whose low-energy limit is a topological quantum field theory. Unlike most other emergent phenomena, like superconductivity and magnetism, QHE works only in two spatial dimensions. There are various ways to see this. First of all, the Hall current is nondissipative.

For the electric field to do no work on the current, the current must flow in a direction perpendicular to the direction of the electric field. In two spatial dimensions, given the direction of the electric field, there is a unique transverse direction for the Hall current, given by eqn (30.24). Since the current and the electric field both carry spatial vector indices, the response must therefore be a rank-two tensor. But there are no natural rank-two antisymmetric tensors in higher dimensions! Second, both the single-particle wave function and Laughlin's many-body wave function make extensive use of complex coordinates of particles, which can only be done in two spatial dimensions. This suggests that the higher-dimensional generalization of QHE would necessarily involve a higher-dimensional generalization of complex numbers and analytic functions. In fact, both of these considerations lead to the same higher-dimensional structure, as we shall explain below.

In higher dimensions, given a direction of the electric field, there is no unique transverse direction for the Hall current to flow. However, this statement holds only if we consider the U(1) charge current. If the underlying particles – and the associated currents – carry a non-Abelian, e.g., SU(2) quantum number, an unique prescription for the current can be given in *four* dimensions. In four dimensions, given a fixed direction of the electric field, say along the x_4 direction, there are three transverse directions. If the current carries a SU(2) isospin label, it also has three internal isospin directions. In this case, the current can flow exactly along the direction in which the isospin is pointing. In this prescription, no preferential direction in space or isospin is picked. The system is invariant under a *combined* rotation of space and isospin. To be more precise, the mathematical generalization of eqn (30.24) in four dimensions is

$$J^i_\mu = \sigma \eta^i_{\mu\nu} E_\nu. \tag{30.30}$$

Here σ is the generalized Hall conductivity, $\eta^i_{\mu\nu}$ is the 't Hooft tensor, explicitly given by $\eta^i_{\mu\nu} = \epsilon_{i\mu\nu 4} + \delta_{i\mu}\delta_{4\nu} - \delta_{i\nu}\delta_{4\mu}$ and J^i_μ is the isospin current and E_ν is the electric field. Here $\mu, \nu = 1, 2, 3, 4$ label the spatial directions and $i = 1, 2, 3$ label the isospin directions. From eqn (30.30), we see easily that if E_ν points along the x_4 direction, the current flows along the $x_{1,2,3}$ directions, explicitly determined by the direction of the isospin. Therefore, the 't Hooft tensor is exactly the rank-two antisymmetric tensor we were looking for! The occurrence of the 't Hooft tensor suggests that this problem must have something to do with the SU(2) instanton (Belavin *et al.* 1975), where the 't Hooft tensor was first introduced. It is not only an *invariant* tensor under combined spatial and isospin rotations, it also satisfies a self-duality condition:

$$\eta^i_{\mu\nu} = \epsilon_{\mu\nu\rho\lambda} \eta^i_{\rho\lambda}. \tag{30.31}$$

Self-duality and anti-self-duality are the hallmarks of the SU(2) Yang–Mills instanton.

Now let us motivate the problem from the point of view of generalizing complex numbers. The natural generalizations of complex numbers are quaternionic numbers, first discovered by Hamilton. A quaternionic number is expressed as $q = q_0 + q_1 i + q_2 j + q_3 k$, where i, j, k are the three imaginary units. This again suggests that the most natural generalization of QHE is from 2D to 4D, where quaternionic numbers can be interpreted as the coordinates of particles in four dimensions. Unlike complex numbers, quaternionic numbers do not commute with each other. In fact, the three imaginary units of quaternionic numbers can be identified with the three generators of the SU(2) group. This suggests that the underlying quantum mechanics problem should involve a non-Abelian SU(2) gauge field.

Our last motivation to generalize QHE comes from its geometric structure. As pointed out by Haldane (1983), a nice way to study QHE is by mapping it to the surface of a 2D sphere S^2, with a Dirac magnetic monopole at its center (Fig. 30.4). The Dirac quantization condition implies that the product of the electric charge, e, and the magnetic charge, g, is quantized, i.e., $eg = S$, where $2S$ is an integer. The number $2S + 1$ is the degeneracy of the lowest Landau level. The reason for the existence of a magnetic monopole over S^2 is a coincidence between algebra and geometry. In order for the monopole potential to be topologically nontrivial, the gauge potentials extended from the north pole and the south pole have to match nontrivially at the equator. Since the equator, S^1, and the gauge group, U(1), are isomorphic to each other, a nontrivial winding number exists. Therefore, one may ask whether there are other higher-dimensional spheres for which a similar monopole structure can be defined. This naturally leads to the requirement that the equator of a higher-dimensional sphere be isomorphic to a mathematical group. This coincidence occurs only for the four sphere, S^4, whose equator, S^3, is isomorphic to the group SU(2). This coincidence between algebra and geometry leads to the first two Hopf maps, $S^3 \to S^2$ and $S^7 \to S^4$.

Therefore, all three considerations – the physical motivation of the transverse current, the mathematical motivation of generalizing complex numbers to quaternionic numbers and the geometric consideration of nontrivial monopole structures – lead to the same conclusion: a nontrivial QHE liquid can be defined in four spatial dimensions (4D) with a SU(2) non-Abelian gauge group. Recently, Hu and I (ZH) indeed succeeded in constructing such a model for the 4D QHE (Zhang and Hu 2001). The microscopic Hamiltonian describes a collection of N fermionic particles moving on S^4, interacting with a SU(2) background isospin gauge potential A_a. It is explicitly defined by

$$H = \frac{\hbar^2}{2MR^2} \sum_{a<b} \Lambda_{ab}^2 \tag{30.32}$$

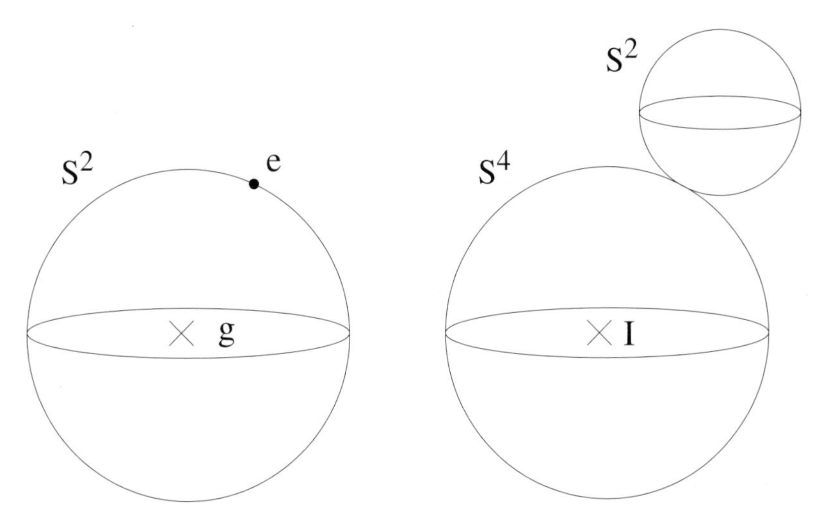

Figure 30.4. The 2D QHE consists of electrons e on the surface of a 2D sphere S^2, with a U(1) magnetic monopole g at its center. Similarly, the 4D QHE can be defined on the surface of a 4D sphere S^4, with a SU(2) monopole I at its center. In the large I limit, the SU(2) isospin degree of freedom is S^2.

where M is the mass of the fermionic particle, R is the radius of S^4, and $\Lambda_{ab} = -i(x_a D_b - x_b D_a)$ is the gauge covariant angular momentum operator. Here x_a is the coordinate of the fermionic particle and $D_a = \partial_a + A_a$ is the gauge invariant momentum operator. The gauge potential A_a $(a = 1, 2, 3, 4, 5)$ is given by

$$A_\mu = \frac{-i}{1 + x_5}\eta^i_{\mu\nu} x_\nu I_i, \qquad A_5 = 0 \tag{30.33}$$

where I_i are the generators of the SU(2) gauge group. An important parameter in this problem is I, the isospin quantum number carried by the fermionic particle. The eigenstates and the eigenvalues of this Hamiltonian can be solved completely, and the spectrum shares many properties with the Landau levels in the 2D QHE problem. In particular, when I becomes large, the ground state of this problem is massively degenerate, with the degeneracy scaling like $D \sim I^3$. In order to keep the energy levels finite in the thermodynamical limit, one is required to take the limit $I \to \infty$ as $R \to \infty$, such that

$$R^2/2I = l^2 \tag{30.34}$$

is finite. l, called the magnetic length, defines the fundamental length scale in this problem. It gives a natural ultraviolet cut-off in this theory, without breaking any rotational symmetries of the underlying Hamiltonian.

While the 4D QH liquid can be elegantly defined on S^4, with the full isometry group as the symmetry of the Hamiltonian, it can also be defined on R^4, with more restricted symmetries. This construction has recently been given by Elvang and Polchinski (2002).

Properties of the model

The 2D QH liquid has many interesting properties including incompressibility of the quantum liquid, fractional charge and statistics of elementary excitations, a topological field theory description of the low-energy physics, a realization of noncommutative geometry, and relativistic chiral excitations at the edge of the QH droplet. Most of these properties also carry over to the QH liquid constructed by ZH. When one completely fills the massively degenerate lowest energy ground states with fermionic particles, with filling factor $\nu \equiv N/D = 1$, one obtains an incompressible quantum liquid, with a finite excitation gap towards all excited states. FQH states can also be constructed for filling fractions $\nu = 1/k^3$, where k is an odd integer. Explicit microscopic wave functions, similar to Laughlin's wave function for the 2D QHE, can be constructed for these incompressible states. The elementary excitations of the FQH states also carry fractional charge $1/k^3$, providing the first direct generalization of fractional charge in a higher-dimensional quantum many-body system.

As discussed above, the low-energy physics of the 2D QHE can be described by a topological quantum field theory, the CSLG theory. A natural question is whether the QH liquid constructed by ZH can be described by a topological quantum field theory as well. This construction has indeed been accomplished recently, by Bernevig, Chern, Hu, Toumbas, and myself (Bernevig *et al.* 2002). As explained earlier, while the underlying orbital space for our QH liquid is four-dimensional, the fermionic particles also carry a large internal isospin degree of freedom I. Since I scales like R^2, the internal space is 2D, which makes the total configuration space a six-dimensional (6D) manifold. Therefore, our QH liquid can either be viewed as a 4D QH liquid with a large internal SU(2) isospin degrees of freedom, or equivalently, as a 6D QH liquid without any internal degree of freedom. The 6D manifold is CP_3, the complex projective space with three complex (and therefore six real) dimensions. This manifold is locally isomorphic to $S^4 \times S^2$. The deep connection between the four-sphere S^4 and the complex manifold CP_3 was first introduced to physics through the twistor program of Penrose (Penrose and MacCallum 1972) and has been exploited extensively in the mathematical literature. Sparling (2002) has recently pointed out the close connection between the twistor theory and the 4D QHE. Our recent work shows that the low-energy effective field theory of our QH liquid is given by an Abelian CS theory in 6 + 1 dimensions

$$S = \nu \int dt d^6 x A \wedge dA \wedge dA \wedge dA \tag{30.35}$$

where A is an Abelian U(1) gauge field over the total configuration space CP_3, and ν is the filling factor. This theory can also be dimensionally reduced to a SU(2) non-Abelian CS theory in 4 + 1 dimensions, given by

$$S = \frac{4\pi\nu}{3}\int dt d^4 x Tr\left(\mathbf{A}\wedge d\mathbf{A}\wedge d\mathbf{A} - \frac{3i}{2}\mathbf{A}\wedge\mathbf{A}\wedge\mathbf{A}\wedge d\mathbf{A} - \frac{3}{5}\mathbf{A}\wedge\mathbf{A}\wedge\mathbf{A}\wedge\mathbf{A}\wedge\mathbf{A}\right) \quad (30.36)$$

where $\mathbf{A}$ is a SU(2) matrix-valued gauge field over the orbital space S^4. The precise equivalence of these two models parallels the two equivalent views of our QH liquid mentioned earlier.

An interesting property which arises from this field theory is the concept of duality. As discussed above, there is a natural particle–flux duality in the 2D QHE problem: an electron can be represented as a boson with an odd number of flux quanta attached to it. In the new QH liquid, there are other extended objects, namely 2-branes and 4-branes besides the basic fermionic particle, which can be viewed as a 0-brane. Each one of these extended objects is dual to a generalized flux, according to the following table:

Particle	$\Longleftrightarrow$	6-flux
Membrane	$\Longleftrightarrow$	4-flux
4-brane	$\Longleftrightarrow$	2-flux

In the 2D QH problem, the Laughlin quasi-particles obey fractional statistics in 2 + 1 dimensions. It is natural to ask how fractional statistics generalize in our QH liquid. It turns out that the concept of fractional statistics of point particles can not be generalized to higher dimensions, but fractional statistics for extended objects exist in higher dimensions (Wu and Zee 1988; Tze and Nam 1989). In our case, 2-branes have nontrivial statistical interactions which generalize statistical interactions of Laughlin quasi-particles.

Extended objects like D-branes have been studied extensively in string theory; however, a full quantum theory describing their interactions still needs to be developed. The advantage of our approach is that the underlying microscopic quantum physics is completely specified. Since the extended topological objects emerge naturally from the underlying microscopic physics, there is hope that a full quantum theory can be developed in this case.

The study of 4D QHE is partially motivated by the possibility of emergent relativistic behavior in 3 + 1 dimensions. There are several ways to see the connection. First of all, the eigenstates and the eigenfunctions of the Hamiltonian (30.32) have a natural interpretation in terms of the 4D Euclidean quantum field theory. If we consider a Euclidean quantum field theory as obtained from a Wick rotation of a 3 + 1 dimensional compactified Minkowskian quantum field theory, one is naturally led to consider the eigenvalues and the eigenfunctions of the Euclidean Dirac, Maxwell, and Einstein operators on S^4. It turns out that the these eigenvalues and eigenfunctions coincide exactly with the eigenvalues and eigenfunctions of the 4D QHE Hamiltonian (30.32), where the spins of the relativistic particles are identified with the isospin quantum number, I. The eigenvalue problems of the Dirac, Maxwell, and Einstein operators can be directly identified with the Hamiltonian eigenvalue problems for $I = 1/2$, 1, and 2. We mentioned earlier that the underlying fermionic particles constituting our QH liquid have high isospin quantum numbers. However, collective excitations of this QH liquid, which are formed as composite particles, can have low isospin quantum numbers. It is therefore tempting to identify the collective excitations of the QH liquid with the relativistic particles we are familiar with. However, this equivalence is only established in Euclidean space. In order to consider the relationship to Minkowski space, we are naturally led to the excitations at the boundary, or the edge of our QH liquid.

Let us first review the collective excitations at the edge of a 2D QH liquid. The 2D QH liquid can be confined by a one-body confining potential V. A density excitation is created by removing a particle from the QH liquid and placing it outside the QH liquid. This way, we have created a particle–hole excitation. If the particle–hole pair moves along a direction parallel to the edge, with a center of mass momentum q_x, the Lorentz force due to the magnetic field acts oppositely on the particle–hole pair, and tries to stretch the pair in the direction perpendicular to the edge. This Lorentz force is balanced by the electrostatic attraction due to the force of the confining potential. Therefore, a unique dipole moment, or a finite separation y of the particle–hole pair, is obtained in terms of q_x:

$$y = l^2 q_x. \tag{30.37}$$

On the other hand, the energy of the dipole pair is simply given by $E = V'y$. Here V' is the derivative of the potential evaluated at the edge. Therefore, we obtain a relativistic dispersion relation for the dipole pair

$$E = V'y = l^2 V' q_x \tag{30.38}$$

with the speed of light given by $c = l^2 V'$. Since the cross product of the gradient of the potential and the magnetic field selects a unique direction along the edge, the excitation is also chiral. In this problem, it can also be shown that not only the dispersion, but also the full interaction is relativistic in the low-energy limit. Therefore, the physics at the edge of a 2D QH liquid provides another example of emergent relativistic behavior (Stone 1990; Wen 1990).

The physics of the edge excitations of a 2D QH liquid *partially* carries over to our 4D QH liquid (Zhang and Hu 2001; Elvang and Polchinski 2002; Hu and Zhang 2002). Here we can also introduce a confining potential, say around the north pole of S^4, and construct a droplet of the QH fluid. Since our QH liquid is incompressible, the only low-energy excitations are the volume preserving shape distortions at the surface. These surface waves can be formed from the particle–hole excitations similar to the ones we described for the 2D QH liquid. A natural speed of light can be introduced, and is given by $c = l^2 V'$. Since our underlying particles carry a large isospin, I, the bosonic composite particle–hole excitations carry all isospins, ranging from 0 to $2I$. The underlying fermionic particles have a strong coupling between their orbital and isospin degrees of freedom. This coupling translates into a relativistic spin–orbital coupling of the bosonic collective excitations. Therefore, excitations with $I = 0, 1, 2$ obey the *free* relativistic Klein–Gordon, Maxwell, and Einstein equations. This is an encouraging sign that one might be able to construct an emergent relativistic quantum field theory from the boundary excitations of our 4D QH liquid.

However, there are also many complications that are not yet fully understood in our approach. The most fundamental problem is that the particles of our 4D QH liquid carry a large internal isospin, which makes the problem effectively a 6D one. This is the basic reason for the proliferation of higher-spin particles in our theory, an "embarrassment of riches." In addition, there is an incoherent fermionic continuum besides the bosonic collective modes. All these problems can only be addressed when one studies the effects of the interaction carefully. In fact, single particle states in the lowest-Landau-level (LLL) have the full symmetry of SU(4), which is the isometry group of the 6D CP_3 manifold. In order to make the problem truly 4D, one needs to introduce interactions which breaks the SU(4) symmetry to a SO(5) symmetry, the isometry group of S^4. This is indeed possible. SO(5) is isomorphic to the group $Sp(4)$. $Sp(4)$ differs from SU(4) by an additional reality condition, implemented through a charge conjugation matrix R. Therefore, any interactions which involve this R matrix would break the symmetry from SU(4) to SO(5), and the geometry of S^4 would emerge naturally. In the strong coupling limit, low-energy excitations are not particles but membranes. This reduces the entropy at the edge from $R^3 \times R^2$ to R^3, and is the first step towards solving the problem of "embarrassment of riches."

Space, time, and the quantum

The 2D QH problem gives a precise mathematical realization of the concept of non-commutative geometry (Douglas and Nekrasov 2001). In the limit of high magnetic field, we can take the limit of $m \to 0$, so that all higher Landau levels are projected out of the spectrum. In this limit, the equation of motion for a charged particle in a uniform magnetic field B and a scalar potential $V(x, y)$ is given by

$$\dot{x} = l^2 \frac{\partial V}{\partial y}, \qquad \dot{y} = -l^2 \frac{\partial V}{\partial x}. \tag{30.39}$$

We notice that the equations for x and y look exactly like the Hamilton equations of motion for p and q. Therefore, this equation of motion can be derived as quantum Heisenberg equations of motion if we postulate a similar commutation relation:

$$[x, y] = il^2. \tag{30.40}$$

Therefore, the 2D QHE provides a physical realization of the mathematical concept of noncommutative geometry, in which different spatial components do not commute. Early in the development of quantum field theory, this feature has been suggested as a way to cut off the ultraviolet divergences of quantum field theory. In quantum mechanics, the noncommutativity of q and p leads to the Heisenberg uncertainty principle and resolves the classical catastrophe of an electron falling towards the atomic nucleus. Similarly, noncommutativity of space and time could cut off the ultraviolet spacetime fluctuations in quantum gravity (Douglas and Nekrasov 2001). However, the problem is that eqn (30.40) can not be easily generalized to higher dimensions, since one needs to pick some fixed pairs of non-commuting coordinates. Our QH liquid provides a physical realization of non-commutative geometry in four dimensions. The generalization of eqn (30.40) becomes

$$[X_\mu, X_\nu] = 4il^2\eta^i_{\mu\nu}n_i \tag{30.41}$$

where X_μ's are the four spatial coordinates and n_i is the isospin coordinate of a particle. This structure of noncommutative geometry is invariant under a combined rotation of space and isospin and treats all these coordinates on equal footing. It is tempting to identify l in eqn (30.41) as the Planck length, which provides the fundamental cutoff of the length scale according to the quantization rule (30.41). In our theory, however, we know what lies beyond the Planck length: the degrees of freedom are those associated with the higher Landau levels of the Hamiltonian (30.32).

At this point, it would be useful to discuss the possible implications of eqn (30.41) on the quantum structure of spacetime. In the 4D QH liquid, there is no concept of time. Since all eigenstates in the LLL are degenerate, there is no energy difference that can be used to measure time according to the quantum relation $\Delta t = \hbar/\Delta E$. However, at the boundary of the 4D QH liquid, an energy difference is introduced through the confining potential. The left-hand side of eqn (30.41) involves four coordinates. Three of them are the spatial coordinates parallel to the boundary. The fourth coordinate, perpendicular to the boundary, measures the energy difference, and therefore measures time. The commutator among these coordinates implies a quantization procedure. The right-hand side of this equation involves the Planck length and the spin. Therefore, this simple equation seems to unify all the fundamental physical concepts: space, time, the quantum, the Planck length, and spin in

Table 30.1. *Connections among algebra, geometry, quantum liquids, and concepts in quantum field theory*

Division algebras	Real numbers	Complex numbers	Quaternions	Octonions
Hopfmaps	$S^1 \to S^1$	$S^3 \to S^2$	$S^7 \to S^4$	$S^{15} \to S^8$
QH liquids	Luttinger liquid?	Laughlin liquid	ZH liquid	?
Random matrix ensembles	Orthogonal	Unitary	Symplectic	?
Fractional statistics:	Kink soliton	particles	membranes	?
Geometric phase:	Z_2	U(1)	SU(2)	?
Non-commutative geometry:	?	$[X_i, X_j] = il^2\epsilon_{ij}$	$[X_\mu, X_\nu] = 4il^2\eta^i_{\mu\nu}n_i$	G2
Twistor transformation:	$SO(2,1) = SL(2,\mathbf{R})$	$SO(3,1) = SL(2,\mathbf{C})$	$SO(5,1) = SL(2,\mathbf{H})$	$SO(9,1) = SL(2,\mathbf{O})$
$N = 1$ SUSY Yang–Mills:	$d = 2+1$	$d = 3+1$	$d = 5+1$	$d = 9+1$
Green–Schwarz superstring	$d = 2+1$	$d = 3+1$	$d = 5+1$	$d = 9+1$

a simple and elegant fashion. It would be nice to use it as a basis to construct a fundamental physical theory.

Magic liquids, magic dimensions, magic convergence

So far our philosophical point of view and our model seem to be drastically different from the approach typical of string theory. However, after the discovery of the new QH liquid, a surprising pattern starts to emerge. Soon after the construction of the new 4D QH liquid, Fabinger (2002) found that it could be implemented as certain solutions in string theory. Moreover, close examination of this pattern reveals remarkable mathematical similarities not only between these two approaches, but also with other fundamental ideas in algebra, geometry, supersymmetry, and the twistor program on quantum spacetime. Table 30.1 summarizes the connections.

The construction of the twistor transformation, the $N = 1$ supersymmetric Yang–Mills theory and the Green–Schwarz superstring rely on certain identities of the Dirac Gamma matrices, which work only in certain magic dimensions. In these dimensions, there is an exact equivalence between the Lorentz group and the special linear tranformations of the real, complex, quaternionic and octonic numbers. Our work shows that QH liquids work only in certain magic dimensions exactly related to the division algebras as well! In fact the *transverse* dimensions $((D + 1) - 2)$ of these relativistic field theories match exactly with the *spatial* dimensions of the quantum liquids. The missing entries in Table 30.1 strongly suggests that an octonionic version of the QH liquid should exist and may be deeply related to the superstring theory in $d = 9 + 1$. QH liquids exist only in magic dimensions, have membranes, and look like a matrix theory. They may be mysteriously related to the M theory after all!

Conclusion

Fundamental physics is faced with historically unprecedented challenges. Ever since the time of Galileo, experiments have been the stepping stones in our intellectual quest for the fundamental laws of nature. With our feet firmly on the ground, there is no summit too high to reach. However, the situation is drastically different in the present day. We are faced with a gap of 16 orders of magnitude between the energy of our experimental capabilities and the summit of Mount Planck. Without experiments, we face the impossible mission of climbing up a waterfall!

But maybe there is an alternate passage to Mount Planck. The logical structure of physics may not be a simple one-dimensional line, but rather has a multiply connected or braided topology, very much like Escher's famous *Waterfall*. Instead of going up in energy, we can move down in energy! Atoms, molecules, and quantum liquids are made of elementary particles at very high energies. But at low energies, they interact strongly with each other to form quasi-particles, which look very much like the elementary particles themselves! Over the past 40 years, we have learned that the strong correlation of these matter degrees of freedom does not lead to ugliness and chaos, but rather to extraordinary beauty and simplicity. The precision of flux quantization, Josephson frequency, and quantized Hall conductance are not properties of the basic constituents of matter, but rather are emergent properties of their collective behavior. Therefore, by exploring the connection between elementary particle and condensed-matter physics, we can use experiments performed at low energies to understand the physics at high energies. By carrying out the profound implications of these experiments to their necessary logical conclusions, we may learn about the ultimate mysteries of our universe.

Throughout John Wheeler's life, he tackled the big questions of the universe with an unorthodox vision and a poetic flair. Lacking John's eloquence, I simply conclude this tribute to him by reciting William Blake's timeless lines:

To see a World in a Grain of Sand
And a Heaven in a Wild Flower,
Hold Infinity in the palm of your hand
And Eternity in an hour.

Acknowledgments

I would like to thank A. Bernevig, C.H. Chern, J.P. Hu, R. B. Laughlin, J. Polchinski, P. SanGiorgio, L. Smolin, L. Susskind, N. Toumbas, and G. Volovik for stimulating discussions. This work is supported by the National Science Foundation under grant number DMR-9814289.

References

Ambegaokar, V, Halperin, B I, Nelson, D R, *et al.* (1980) *Phys. Rev.* **B21**(5), 1806.
Anderson, P W (1972) *Science* **177**(4047), 393.
Balents, L, Fisher, M P A, and Nayak, C (1998) *Int. J. Mod. Phys.* **B12**, 1033.
Belavin, A A, Polyakov, A M, Schwartz, A S, *et al.* (1975) *Phys. Lett.* **B59**, 85.
Bernevig, B A, Chern, C-H, Hu, J-P, *et al.* (2002) *Ann. Phys.* **300**, 185.
Douglas, M R and Nekrasov, N A (2001) *Rev. Mod. Phys.* **73**, 977.
Elvang, H and Polchinski, J (2002) hep-th/0209104.
Fabinger, M (2002) hep-th/0201016.

Fisher, M P A, and Lee, D H (1989) *Phys. Rev.* **B39**, 2756.
Franz, M, Tesanovic, Z, and Vafek, O (2002) cond-mat/0203333.
Girvin, S M and Macdonald, A H (1987) *Phys. Rev. Lett.* **58**, 1252.
Haldane, F D M (1983) *Phys. Rev. Lett.* **51**, 605.
Hu, J P and Zhang, S C (2002) *Phys. Rev.* **B66**, 125301.
Kivelson, S, Lee, D H, and Zhang, S C (1992) *Phys. Rev.* **B46**, 2223.
Laughlin, R B and Pines, D (2000) *Proc. Nat. Acad. Sci. USA* **97**, 28.
Penrose, R and MacCallum, M (1972) *Phys. Rep.* **6**, 241.
Read, N (1989) *Phys. Rev. Lett.* **62**, 86.
Simon, S H and Lee, P A (1997) *Phys. Rev. Lett.* **78**, 1548.
Sparling, G (2002) cond-mat/0211679.
Stone, M (1990) *Phys. Rev.* **B42**, 8399.
Tze, C H and Nam, S (1989) *Ann. Phys.* **193**, 419.
Volovik, G E (1993) *JETP Lett.* **58**, 469.
(2001) *Phys. Reps.* – Review section of *Phys. Lett.* **351**, 195.
Wen, X G (1990) *Phys. Rev. Lett.* **64**, 2206.
Wu, Y and Zee, A (1988) *Phys. Lett.* **B207**, 39.
Zhang, S C (1992) *Int. J. Mod. Phys.* **B6**, 25.
Zhang, S C and Hu, J P (2001) *Science* **294**(5543), 823.
Zhang, S C, Hansson, T H, and Kivelson, S (1989) *Phys. Rev. Lett.* **62**, 82.

Appendix A

Science and Ultimate Reality Program Committees

The *Science and Ultimate Reality* program began with the symposium *Science and Ultimate Reality: Celebrating the Vision of John Archibald Wheeler*, held March 15–18, 2002 in Princeton, New Jersey, USA. The members of the Program Oversight Committee and the four Program Development Committees, many of whom are contributors to this volume, are listed below.

For more information about the symposium, and to order the entire proceedings on DVD, see www.metanexus.net/ultimate_reality/.

Program Oversight Committee

Freeman J. Dyson, Chair: Institute for Advanced Study
Max Tegmark, Deputy Chair: University of Pennsylvania
John D. Barrow: University of Cambridge
George F. R. Ellis: University of Cape Town
Robert B. Laughlin: Stanford University
Charles W. Misner: University of Maryland
William D. Phillips: National Institute of Standards and Technology (USA)
Charles H. Townes: University of California, Berkeley

Program Development Committees

Andreas Albrecht: University of California, Davis
John D. Barrow: University of Cambridge
Raymond Y. Chiao: University of California, Berkeley
Philip D. Clayton: Claremont School of Theology and Claremont Graduate University
Paul C. W. Davies: Macquarie University
Freeman J. Dyson: Institute for Advanced Study
Artur Ekert: University of Oxford
George F. R. Ellis: University of Cape Town
Serge Haroche: College of France
Stuart Kauffman: Santa Fe Institute and Bios Group
Robert B. Laughlin: Stanford University

William D. Phillips: National Institute of Standards and Technology (USA)
Lee Smolin: Perimeter Institute for Theoretical Physics and the University of Waterloo
Max Tegmark: University of Pennsylvania
H. Dieter Zeh: University of Heidelberg
Anton Zeilinger: University of Vienna
Wojciech H. Zurek: Los Alamos National Laboratory

John Archibald Wheeler at age 23 as a National Research Council Fellow at the Institute for Theoretical Physics (later known as the Niels Bohr Institute) in Copenhagen (1934–5). (Photograph by kind permission of AIP Emilio Segrè Visual Archives, Wheeler Collection.)

Appendix B

Young Researchers Competition in honor of John Archibald Wheeler for physics graduate students, postdoctoral fellows, and young faculty

A "Young Researchers Competition" was held in conjunction with the *Science and Ultimate Reality* symposium that took place in Princeton, New Jersey, USA, in March 2002. Like the entire Science and Ultimate Reality program, the competition was focused on the future of innovative research into the nature of "quantum reality" and related challenges inspired by Wheeler's "Really Big Questions" (see the Editors' Preface at the front of this book).

Of the 64 original applicants who submitted abstracts in an open competition worldwide, the applications of the 15 young research scientists born on or after January 1, 1970 that were chosen as finalists demonstrated work that is innovative and substantively engaged with the ideas raised by Wheeler's questions related to quantum reality. They also, therefore, related to one or more of the four main themes on which both this book and the symposium were based. The finalists made their presentations in 12-minute time slots (8 minutes plus 4 minutes for questions and answers) at the symposium on Sunday, March 17, 2002.

After evaluating the finalists' research accomplishments, records of achievement, and symposium presentations, appointed judges selected from among the symposium participants (all of whom contributed chapters to this volume) awarded eight prizes, six of $5000 each and a first-place prize of $7500 shared by the top two presenters, on the last day of the symposium on Monday, March 18. Winners were selected on the basis of outstanding merit. The 15 participants, the eight winners, and the topics of their presentations are listed below, followed by a listing of the competition overseers.

For more information about the competition, which is also included on the symposium DVD, see www.metanexus.net/ultimate_reality/competition.htm.

Young Researchers Competition Finalists

Two first-place winners (tied)

Raphael Bousso – Kavli Institute for Theoretical Physics, University of California, Santa Barbara, USA; born May 31, 1971, Israel (also a citizen of Germany): "The holographic principle."

Fotini Markopoulou – Perimeter Institute for Theoretical Physics and Department of Physics, University of Waterloo, Canada; born April 3, 1971, Greece: "Models of Planck-scale spacetime and quantum cosmology."

Six second-place winners

Nicole Bell – NASA/Fermilab, Theoretical Astrophysics Group, Batavia, Illinois, USA; born December 12, 1975, Australia: "Coherence, decoherence and oscillating neutrinos – from quantum Zeno to getting in sync."

Steven Gubser – Department of Physics, Princeton University, USA; born May 4, 1972, USA: "On the connection between gauge theory and gravity."

Olga Khovanskaya – Sternberg Astronomical Institute, Moscow State University, Russia; born April 23, 1977, Russia: "Dilatonic black holes in string gravity and their relation with parameters of the early universe."

Michael Murphy – School of Physics, University of New South Wales, Sydney, Australia; born March 17, 1977, Australia: "Do the fundamental constants vary in spacetime?"

Jonathan Oppenheim – Racah Institute of Physics, The Hebrew University, Jerusalem, Israel; born November 24, 1970, Canada: "Bit from it."

Mark Topinka – Stanford University, Stanford, California, USA; born April 24, 1970, USA: "Imaging flow through electronic wave functions."

Seven semifinalists

Anita Goel – Department of Physics, Harvard University, Cambridge, Massachusetts, USA; born August 22, 1973, USA: "The physics of life."

Jiangping Hu – Department of Physics, Stanford University, Stanford, California, USA; born January 3, 1972, China: "An essay on space, time and the quantum."

Jeremy O'Brien – Centre for Quantum Computer Technology, Department of Physics, University of Queensland, Australia; born November 7, 1975, Australia: "Exploration of the quantum nature of nature and the fabrication of a quantum computer."

Jianwei Pan – Institut für Experimentalphysik, Universität Wien; born March 11, 1970, China: "Multi-photon interferometry and quantum nonlocality."

Mary Rowe – National Institute of Standards and Technology, Boulder, Colorado, USA; born January 5, 1970, USA: "Experimental violation of Bell's inequalities with efficient detection."

André Stefanov – GAP (Group of Applied Physics) – Optique, Université de Genève, Suisse; born August 11, 1975, Switzerland: "Quantum correlations with spacelike beam-splitters in motion."

Vlatko Vedral – Optics Section, Blackett Laboratory, Imperial College, London, UK; born August 19, 1971, the former Yugoslavia: "Probabilities from amplitudes via information theory and thermodynamics."

Young Researchers Competition Chair and Judges

Christopher Monroe, Chair: University of Michigan, Ann Arbor
Marcelo Gleiser: Dartmouth College
Hideo Mabuchi: California Institute of Technology
João Magueijo: Imperial College, London
Wojciech Zurek: Los Alamos National Laboratory

Young Researchers Pre-Competition Screening Panel

Anthony Aguirre: Institute for Advanced Study
Arthur Kosowsky: Rutgers University
Horatiu Nastase: Institute for Advanced Study
Max Tegmark: University of Pennsylvania

Index

Page numbers in italics refer to illustrations.